MW00592429

STRUCTURAL DAMPING

Applications in Seismic Response Modification

Advances in Earthquake Engineering
Series Editor: Franklin Y. Cheng

Structural Damping: Applications in Seismic Response Modification
Zach Liang, George C. Lee, Gary F. Dargush, and Jianwei Song

Seismic Design Aids for Nonlinear Pushover Analysis of Reinforced Concrete and Steel Bridges
Jeffrey Ger and Franklin Y. Cheng

Seismic Design Aids for Nonlinear Analysis of Reinforced Concrete Structures
Srinivasan Chandrasekaran, Luciano Nunziante, Giorgio Serino, and Federico Carannante

STRUCTURAL DAMPING

Applications in Seismic Response Modification

Zach Liang
George C. Lee
Gary F. Dargush
Jianwei Song

CRC Press
Taylor & Francis Group
Boca Raton London New York

CRC Press is an imprint of the
Taylor & Francis Group, an **informa** business

MATLAB® is a trademark of The MathWorks, Inc. and is used with permission. The MathWorks does not warrant the accuracy of the text or exercises in this book. This book's use or discussion of MATLAB® software or related products does not constitute endorsement or sponsorship by The MathWorks of a particular pedagogical approach or particular use of the MATLAB® software.

CRC Press
Taylor & Francis Group
6000 Broken Sound Parkway NW, Suite 300
Boca Raton, FL 33487-2742

© 2012 by Taylor & Francis Group, LLC
CRC Press is an imprint of Taylor & Francis Group, an Informa business

No claim to original U.S. Government works

Version Date: 20111012

International Standard Book Number: 978-1-4398-1582-3 (Hardback)

This book contains information obtained from authentic and highly regarded sources. Reasonable efforts have been made to publish reliable data and information, but the author and publisher cannot assume responsibility for the validity of all materials or the consequences of their use. The authors and publishers have attempted to trace the copyright holders of all material reproduced in this publication and apologize to copyright holders if permission to publish in this form has not been obtained. If any copyright material has not been acknowledged please write and let us know so we may rectify in any future reprint.

Except as permitted under U.S. Copyright Law, no part of this book may be reprinted, reproduced, transmitted, or utilized in any form by any electronic, mechanical, or other means, now known or hereafter invented, including photocopying, microfilming, and recording, or in any information storage or retrieval system, without written permission from the publishers.

For permission to photocopy or use material electronically from this work, please access www.copyright.com (http://www.copyright.com/) or contact the Copyright Clearance Center, Inc. (CCC), 222 Rosewood Drive, Danvers, MA 01923, 978-750-8400. CCC is a not-for-profit organization that provides licenses and registration for a variety of users. For organizations that have been granted a photocopy license by the CCC, a separate system of payment has been arranged.

Trademark Notice: Product or corporate names may be trademarks or registered trademarks, and are used only for identification and explanation without intent to infringe.

Library of Congress Cataloging-in-Publication Data

Structural damping : applications in seismic response modification / Zach Liang ... [et al.].
p. cm. -- (Advances in earthquake engineering)
"A CRC title."
Includes bibliographical references and index.
ISBN 978-1-4398-1582-3 (hardcover : alk. paper)
1. Earthquake resistant design. 2. Damping (Mechanics) I. Liang, Zach.

TA658.44.S77 2012
624.1'762--dc23 2011036557

Visit the Taylor & Francis Web site at
http://www.taylorandfrancis.com

and the CRC Press Web site at
http://www.crcpress.com

Contents

PART II Principles and Guidelines for Damping Control

PART III Design of Supplemental Damping

Series Preface

The major goal of earthquake engineering is to limit seismic-induced structural damages at reasonable cost. To achieve this goal, one of the most effective approaches to reduce structural response is to utilize supplemental dampers, provided such a damping system is properly designed. In this sense, understanding structural damping is critically important for successful seismic design. This book *Structural Damping: Applications in Seismic Response Modification* is a valuable addition to the Advances in Earthquake Engineering Series for it serves the goal of optimal response control.

The book was written under the direction of Dr. George C. Lee, SUNY distinguished professor, who had served as chair of the Department of Civil Engineering, dean of the School of Engineering and Applied Science, and director of the Multidisciplinary Center for Earthquake Engineering Research. He and his coauthors have integrated the essential materials developed by his research team in the past 20 years. The key features of this book may be briefly summarized as follows:

1. The book provides an integrated systematic presentation of the dynamic response of structures with nonproportional or nonlinear damping, as well as with overdamped vibration modes. This theoretical base is important for understanding the dynamic behavior of structures with large damping, prior to the development of aseismic design using damping devices.
2. This volume is a pioneering work to provide comprehensive design principles of structural damping, including design procedures and guidelines for aseismic design of structures with enhanced damping.
3. Based on a comprehensive formulation, limitations of current design practice for large damping are clearly illustrated, and improvements to handle enhanced damping are given.
4. This volume offers a discussion of the safety issues of structures with enhanced damping, based on theoretical formulation and practical design consideration.

This book is useful not only in the practicing engineering community but also to researchers and educators, because numerous research and development challenges remain to be pursued.

Preface

Today, earthquake engineering research is devoting a major effort toward establishing seismic performance requirements associated with large inelastic deformations of the structure. At the same time, structural response modification systems (certainly, passive energy dissipation and seismic isolation systems) have been widely used. It is reasonable to expect that the next research emphasis in performance-based engineering is to integrate both frontiers for more optimal seismic performance of structures. As far as safety, performance, and cost for both components of inelastic deformation and structural response control are concerned, structural damping is a core knowledge area. In this book, recent advances in structural damping are presented and their applications to the design of passive structural response modification devices are given to complement the current supplemental damping design practice for high-level damping. Integration with seismic performance requirement is not addressed in this book.

Aseismic design using supplemental dampers has steadily gained popularity in the earthquake engineering profession over the past several decades. Many practical applications of various dampers can be found worldwide and, in the United States, damper design has been included in building codes. To date, damping design is primarily based on the concepts of the energy equation, effective proportional damping, and simplified linear single-degree-of-freedom (SDOF) response spectrum. These concepts, along with their associated underlying assumptions, support the idea that installing supplemental dampers in structures will dissipate energy. Nonlinear damping is presented by an effective damping ratio through linearization schemes, and the damping coefficients of structures are assumed to be classic damping matrices in order to establish a procedure for damping design. In addition, it is assumed that statistical procedures that use earthquake records can be carried out through proportional scaling of their amplitudes.

These assumptions have enabled us to develop a design procedure for supplemental dampers. However, it is not well understood that some of these assumptions work well only when the amount of damping in structures is within a specific low level. The first main objective of this book is to provide a theoretical foundation on the role of damping in the dynamic response of structures, especially when the level of damping is high or when nonlinearities become important design issues. The second main objective is to provide response spectra–based design principles and guidelines for practical applications of damping devices to reduce earthquake-induced structural vibration.

Generally speaking, structural responses under seismic excitations are dynamic processes. There are three resisting force components to counter the earthquake load, one of which is the damping force. While damping technology has been developed and advanced in a range of mechanical and aerospace engineering applications over many years, it has become a popular approach in structural engineering only in the later part of the twentieth century.

While the development and application of energy dissipation devices in structural engineering continue to expand, there are a number of fundamental issues related to the dynamic behavior of structures with supplemental damping as a system that require further study. Limitations and impacts of using energy dissipation devices need to be clarified and established. This book intends to fill the knowledge gap by helping earthquake engineers to better understand the dynamic behavior of structures and to more effectively use the design codes for dampers.

The key elements in this book are summarized as follows.

A straightforward concept often advocated in damping design is that "more energy dissipated by the added dampers will result in less vibration energy remaining in the structure, and thus the structural response is reduced." This is not always true. A higher level of damping may not effectively reduce the responses of a given structure. In some cases, high-level damping may even magnify the responses, because the level of structural response depends not only on local energy dissipation, but also on energy input and its redistribution. Thus, minimizing the conservative energy of the vibrating system in structures with supplemental damping is a more appropriate general guiding principle.

Several other basic issues in structural damping are carefully reviewed. These include the maximum energy dissipation principle under preset damping force and allowed structural displacement; the damping adaptability of devices that can operate in a large dynamic range of earthquake loads; the viscoelastic behavior of any damping elements that take the supporting stiffness as well as installation practice into consideration; nonproportional damping that needs to be minimized as much as possible in design practice; the limitation of using damping force that provides a practical engineering limit beyond which adding more damping provides diminishing gains; and the problem of damping and stiffness nonlinearity that cannot be accurately approximated by today's design approaches. In addition, a design principle based on energy distribution is discussed, which may be useful for generally damped multi-degree-of-freedom (MDOF) systems.

The characteristics of nonlinear damping and nonlinear structures are complex issues to address in damping design. In this book, nonlinearity is considered in three cases. The first case involves a linear structure with small nonlinear damping, the second case is for a linear structure with larger nonlinear damping, and the third case applies when both the structure and the damping are nonlinear. In the first case, because the damping force is rather small, almost any type of linearization can be used without causing any significant design discrepancies. In the second case, care must be taken to choose proper linearization; employing nonlinear design spectra can often be a reasonable approach. In the third case, linearization methods, though adopted by many building codes, can provide highly inaccurate results. Methods such as the equal displacement approach (using R-factors), equal energy approach, and pushover approach all have their limitations. Since the nonlinear design spectra approach requires too many response spectra, not only for specific damping and stiffness, but also for specific levels of ground excitations, it is not useful in practice. Thus, nonlinear time history analysis must be used. This latter method, though always workable for the first two cases, can yield an unacceptable computation burden, making it unattractive for use in day-to-day practical damping design. For this reason, time history analysis is not emphasized in all discussions.

Two types of design approaches are provided in this book. The first is the design response spectra approach. Specifically, the design is modified by a simplified factor, the damping ratios of the first several vibration modes of the structure. For readers familiar with the design response spectra method for an SDOF linear system, this approach provides a good design when damping is small and for structural responses in elastic ranges. The modified approach addresses cases where damping is large and nonlinear. Examples of how to modify the response spectra design are also provided.

The second type of damper design is based on time history analysis, for which several issues are important. These issues include how to select and scale earthquake records to be consistent with the response spectra, accuracy of modeling of dampers as well as the structure–damper system both as elastic and inelastic systems, and interpretation of response time histories and peak values.

There are many issues related to the role of large damping in the design of earthquake protective systems. Some involve fundamental theories, while others focus on practical details, such as device installations. In this book, focus is first given to the fundamental issues. Detailed technical descriptions and step-by-step design procedures are developed based on the basic principles. These fundamental issues are limited to areas within the scope of structural dynamics principles, although attention is also given to related topics of damper selection, damper specifications, and damper installation.

The arrangement of this book is as follows. Part I provides a foundation for generally damped MDOF systems by emphasizing damping force, energy dissipation, and structural impedances, which are important in structural dynamics and damping control. In Chapter 1, the necessary background of linear SDOF systems, including the concepts of natural frequency and damping ratio, is introduced. Free and forced harmonic vibrations are discussed, and the concepts of damping force and energy dissipation are systemically explained. The effect of damping on free and harmonic vibration is reviewed. Effective damping, as systemically described by Timoshenko, is the fundamental formula in currently used damper design. This concept is analyzed and an alternative force-based approach for linearization of damping is also provided.

In Chapter 2, arbitrary excitations for SDOF systems, including periodic, transient, and random excitations, are considered and the earthquake responses of structures are examined. Parallel with the description of these three types of arbitrary excitations and the corresponding responses of SDOF systems, mathematical tools are introduced. In particular, Fourier series and Fourier/Laplace transforms, which are basic approaches to represent vibration signals, are reviewed. The integral transforms allow a different idea to be considered; that is, the view in the frequency domain is seen as a modal vibration model, whereas the direct response of the system provides the response model in the time domain. Furthermore, to account for random vibrations, the concepts of correlation analysis and spectral analysis are introduced. These necessary mathematical tools are used in the rest of the book. In the last section of this chapter, earthquake response is discussed with a primary focus on the response spectrum. Again, the effect of damping is emphasized.

In Chapter 3, linear MDOF systems with proportional damping are introduced. First, the undamped system is examined and the fundamental approach of eigen-decomposition is considered. The concept of the Rayleigh quotient is introduced as the foundation of modal analysis. Then, proportional damping is discussed, followed by modal analysis and system decoupling. For practical applications, modal participation, modal truncation, and modal parameter estimation, as well as several forms of proportional damping matrices, are presented.

Nonproportionally damped and overdamped systems are discussed in Chapter 4. Although most structures are more or less generally damped, only the equations that are needed for practical use are presented. Theoretical developments on generally damped MDOF systems are not covered here in significant depth, but are available elsewhere (e.g., Liang et al. 2007). Instead, explanations for engineering application of the theoretical principles and design examples are provided.

Part II introduces some design principles and guidelines for damping control. The focus is on using damping force more accurately and effectively in the design of structures with supplemental damping. In Chapter 5, the basic principles of damper design and damping devices are given. The first group of principles are associated with various dampers. These include generic modeling of damping force for dampers and damping parameters of structures with added dampers; conventionally used Timoshenko effective damping ratio based on energy and an alternative approach for the effective damping ratio based on damping force; maximum energy dissipation per device per cycle, which leads to the rectangular law that provides a method for optimal damping design and the upper limit of damping vibration control; damping adaptability, which provides another basic rule for damper selection; damping ratio affected by the physical parameters of the total system and the effectiveness of the structural parameters in affecting the damping ratio; similarity and difference of response spectrum and dynamic stiffness, which provide an alternative rule other than response spectrum as a design criteria; and the relationship between damping and stiffness, which is an often overlooked issue in practical damper design.

Chapter 6 is a continuation of the discussion on design principles, but the focus is on the nonlinearity and damping irregularity of the total system of structures with supplemental damping. The pros and cons of currently used simplified damper design procedures are presented and discussed. In the simplified design procedure, the designers do not have to obtain the exact mode shapes for modes higher than the fundamental one, nor do they have to calculate the exact first mode shape. However, for systems with larger damping, special considerations must be given. These considerations result in modifications to the currently used simplified approach based on the design spectra approach. These suggested modifications are presented in the NEHRP (2009) provisions and are discussed in Part III.

In general, Part III provides more detailed design procedures based on the classification of specific damping devices. In this book, dampers are classified based on their linearity and rate dependency to facilitate the subsequent development of design guidelines in a logical manner, rather than on their displacement dependency and velocity dependency. Chapter 7 deals with linear damping and linearized nonlinear damping. When the supplemental damping is not sufficiently large, this approach can greatly reduce the computational burden, yet provide reasonable accuracy. The simplified design approach of current codes is introduced. While the design logic is virtually identical among these codes, some improvements are suggested to enhance the design procedure. The first part of Chapter 7 presents an approach for using SDOF systems, which is directly related to the design spectrum. This serves as the basis for the entire simplified design process, as well as provides an initial estimation of whether supplemental damping should be used. The multiple-story-single-period (MSSP) structures and multiple modes are then introduced as the main platform for developing design guidelines. Chapter 7 also presents more general linear damping.

When the added damping is sufficiently large, to avoid the errors introduced by the linearization process, nonlinear damping needs to be considered. This is examined in Chapter 8. One typical nonlinear response is the parallelogram-form force–displacement relationship, referred to as the bilinear damper. To estimate the structural responses, specific bilinear response spectra are used. To further obtain the response vectors of nonlinear MDOF systems, separation of the displacement and acceleration is performed and the combination of the first several effective "modes" is rendered. In Chapter 8, another important type of damper, the sublinear damper, is discussed and expressed by the sublinear response spectra. Unlike bilinear damping, in which all the effective "modes" of interest can be treated as bilinear, sublinear damping rarely contributes accurate information for higher "modes." Therefore, an alternative approach of an equivalent linear MDOF system, which likely will be generally damped, is used. Detailed design steps on mode shape normalization, general damping indices, and response computation, as well as selecting damper specifications, are discussed. For sublinear systems, an iterative design procedure for a nonlinearly damped structure is also suggested. This includes the identification of the model, initial design, and response estimation. Note that the spectra-based estimation proposed in this chapter provides simplified calculations with considerably less computational burden. It should be used together with time history analysis for design safety, efficiency of damper use, and cost-effective optimizations.

The materials covered in Part III are incomplete. Much remain to be fully developed. Because most design professionals are familiar with the current design codes (e.g., NEHRP 2009), an approach that follows the NEHRP provisions with added "notations" and "recommendations" is followed in presenting the materials in Chapters 7 and 8. It is hoped that this information will be useful as a supplement to the existing NEHRP provisions. It is obvious that many research and development challenges remain to be faced by the earthquake engineering research community and codification professional groups. This book will hopefully also help clarify some of these future research needs.

The materials presented in this book were gradually developed by the authors during the past 20 years in conjunction with their research activities sponsored by the National Science Foundation and the Federal Highway Administration through the National Center for Earthquake Engineering Research and subsequently, the Multidisciplinary Center for Earthquake Engineering Research. The authors would like to acknowledge these funding agencies for the opportunity to work on structural damping–related subjects, and the Samuel P. Capen Endowment fund of the University at Buffalo, State University of New York, for partial financial support. They would like to express their appreciation to professors Joseph Penzein of the University at California, Berkeley, and Masanobu Shinozuka of the University of California, Irvine, for helpful technical discussions and to many of their colleagues at the University at Buffalo, in particular, professors Michael Constantinou, Andre Filiatrault, Andrei Reinhorn, T. T. Soong, and Andrew Whittaker, for their inspiring discussions and pioneering research efforts in damping design and related areas that greatly benefited this writing effort. In addition, the authors are indebted to the following individuals for their invaluable assistance in technical editing and formatting of the manuscript: Jane Stoyle Welch, Shuchuan Zhang, Nasi Zhang, Yihui Zhou, Hao Xue, Dezhang Sun, and Chao Huang. Last but not the least, the authors express their sincere appreciation and affection to their wives, Yiwei, Grace, Andrea, and Li, for their patience, encouragement, and love, as the authors devoted countless evenings and weekends while this book was being written.

Zach Liang, George C. Lee, Gary F. Dargush, and Jianwei Song
University at Buffalo
State University of New York

MATLAB® is a registered trademark of The MathWorks, Inc. For product information, please contact:

The MathWorks, Inc.
3 Apple Hill Drive
Natick, MA 01760-2098 USA
Tel: 508 647 7000
Fax: 508-647-7001
E-mail: info@mathworks.com
Web: www.mathworks.com

Series Editor

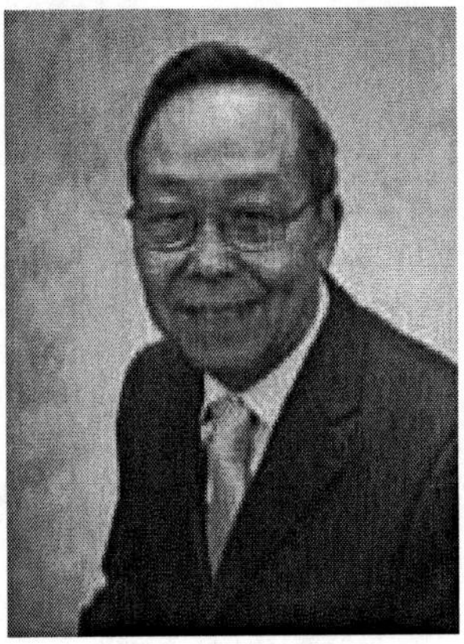

Dr. Franklin Cheng received a BS degree (1960) from the National Cheng-Kung University, Taiwan, and an MS degree (1962) from the University of Illinois at Urbana-Champaign. He gained industrial experience with C. F. Murphy and Sargent & Lundy in Chicago, Illinois. Dr. Cheng then received a PhD degree (1966) in civil engineering from the University of Wisconsin–Madison. Dr. Cheng joined the University of Missouri, Rolla (now named Missouri University of Science and Technology) as assistant professor in 1966 and then associate professor and professor in 1969 and 1974, respectively. In 1987, the Board of Curators of the University appointed him curators' professor, the highest professorial position in the system comprising four campuses. He has been Curators' Professor Emeritus of Civil Engineering since 2000. In 2007, the American Society of Civil Engineers recognized Dr. Cheng's accomplishments by electing him to honorary membership, which is now renamed as distinguished membership. Honorary membership is the highest award the Society may confer, second only to the title of ASCE President. Honorary members on this prestigious and highly selective list are those who have attained acknowledged eminence in a branch of engineering or its related arts and sciences. Until 2007, there have been 565 individuals who were elected to this distinguished grade of membership since 1853. For the year of 2007, only 10 honorary members were selected from more than 14,000 members.

Dr. Cheng was honored for his significant contributions to earthquake structural engineering, optimization, nonlinear analysis, and smart structural control; and for his distinguished leadership and service in the international engineering community; as well as for being a well-respected educator, consultant, author, editor, and member of numerous professional committees and delegations. His cutting edge research helped recognize the vital importance of the possibilities of automatic computing in the future of civil engineering. He was one of the pioneers in allying computing expertise to the design of large and complex structures against dynamic loads. His research expanded over the years to include the important topics of structural optimization and design of smart structures. In fact, he is one of the foremost experts in the world on the application of structural dynamics and optimization to the design of structures. Due to the high caliber and breadth of his research expertise, Dr. Cheng has been regularly invited to serve on the review panels for the National Science Foundation (NSF), hence setting the direction of future structural research. In addition, he has been instrumental in helping the NSF develop collaborative research programs with Europe, China, Taiwan, Japan, and South Korea. Major industrial corporations and government agencies have sought Dr. Cheng's consultancy. He has consulted with Martin Marietta Energy Systems, Inc., Los Alamos National Laboratory, Kjaima Corporation, Martin & Huang International, Inc., and others.

Dr. Cheng received four honorary professorships from China and chaired 7 of his 24 NSF delegations to various countries for research cooperation. He is the author of more than 280 publications, including 5 textbooks: *Matrix Analysis of Structural Dynamics – Applications and Earthquake Engineering, Dynamic Structural Analysis, Smart Structures – Innovative Systems for Seismic Response Control, Structure Optimization – Dynamic and Seismic Applications*, and *Seismic Design Aids for Nonlinear Pushover Analysis of Reinforced Concrete and Steel Bridges*. Dr. Cheng has been recipient of numerous honors and awards including Chi Epsilon, MSM–UMR Alumni

Merit for Outstanding Accomplishments, Faculty Excellence Award, Halliburton Excellence Award, and recognitions in 21 biographical publications such as *Who's Who in Engineering*, and *Who's Who in the World*. He has twice been the recipient of the ASCE State-of-the-Art Award in 1998 and 2004.

Authors

Dr. Zach Liang is a research professor in the Department of Mechanical and Aerospace Engineering, University at Buffalo. He has an MS degree in Mechanical Engineering, Tianjin University, China, 1982 and a PhD in Mechanical and Aerospace Engineering, University at Buffalo, 1987. His research interests include structural dynamics, vibration and control, damping of systems and devices, reliability design and random process, earthquake engineering, testing and measurement, and nondestructive evaluation.

Dr. George C. Lee is a SUNY distinguished professor in the Department of Civil, Structural, and Environmental Engineering, University at Buffalo. Previously, he served as director of the Multidisciplinary Center for Earthquake Engineering Research (MCEER). He received a BS degree from the National Taiwan University, and MS and PhD degrees from Lehigh University. His academic and research interests include steel structures, earthquake engineering, bridge engineering, and multihazard bridge design of principles.

Dr. Gary F. Dargush is a professor and the department chairman of Mechanical and Aerospace Engineering, University at Buffalo. He has a PhD in Civil Engineering, University at Buffalo, 1987. His research interests include computational mechanics, boundary element methods, finite element methods, earthquake engineering, structural dynamics, geomechanics, thermomechanics, and computational fluid dynamics.

Dr. Jianwei Song is a senior research scientist in the Department of Civil and Environmental Engineering, University at Buffalo. He received a PhD in Mechanical Engineering, Tianjin University, China, 1989. His research interests include structural dynamics and control, vibration modal analysis and signal processing, advanced experimental system development, earthquake engineering, and nondestructive test/evaluation.

Part I

Vibration Systems

In Part I, the fundamental concepts for modeling of structures under dynamic excitation are presented. Structural systems represented by linear single- or multiple-degree of freedom models, under both free and forced vibrations, are considered. The corresponding eigen-problem is examined in some detail for both the proportional and nonproportional damping cases.

Part I only presents necessary theories and formulae for damper design. In order to systematically explain the basic concept of structural dynamics, especially for multi-degree-of-freedom system with large damping, the materials discussed are arranged to follow self-complete logics, in both mathematics and vibration theories.

1 Free and Harmonic Vibration of Single-Degree-of-Freedom Systems

The major focus of this book is on the theory, principles, and procedures for the design of earthquake response modification technologies or earthquake protective systems. In particular, special emphasis is placed on the application of various damping devices in the seismic vibration control of civil engineering structures.

In this chapter, several fundamental concepts and governing equations of single-degree-of-freedom (SDOF) systems are reviewed, which serve as basic knowledge for advanced damping concepts and structural dynamics in general. While the presentation is intended to be reasonably self-contained, some detailed derivations and explanations, which can be found in standard structural dynamics reference books, are not included (Inman 2007; Chopra 2006; Clouph and Penzien 1993).

1.1 MODEL OF LINEAR SDOF VIBRATION SYSTEMS

1.1.1 Equation of Motion and Basic Dynamic Parameters

When a structure or a system is subjected to dynamic load, that is, subjected to time-varying load, it will have time-varying responses. The amplitudes of the responses not only will depend on external excitations, but will also be a function of the system itself. This system has three types of internal forces: the inertial force, the damping force, and the restoring force, which may be conceptually explained by a linear single mass-damper-spring (m-c-k) system displayed in Figure 1.1a or a one-story building model shown in Figure 1.1b. Generally, such a system, which contains only one mass and therefore has only one displacement variable, is known as an SDOF system. This simple dynamical system is the focus of this chapter.

1.1.1.1 Equilibrium of Vibration Forces

The linear SDOF m-c-k system can be further modeled as shown in Figure 1.1c. By using the D'Alembert principle, the summation of all the forces acting on the mass must be balanced, where the product of mass m and acceleration $\ddot{x}$ is treated as an inertial force. That is, mathematically, the linear SDOF vibration can be described by the equation of motion obtained through balancing the various forces, that is

$$m\ddot{x}(t) + c\dot{x}(t) + kx(t) = f(t) \tag{1.1}$$

in which, m, c, and k are *mass*, *linear viscous damping coefficient*, and *stiffness*, respectively. It is assumed that all three are constants in Equation 1.1, and that in Figure 1.1b the parameters c and k represent the damping coefficient and stiffness of the structure respectively. Meanwhile, $\ddot{x}(t)$, $\dot{x}(t)$, and x(t) are the *acceleration*, *velocity*, and *displacement*, respectively. The superposed dots, $(\dot{.})$ and $(\ddot{.})$, stand for the first and second derivatives respectively with respect to time t. On the right side, f(t) represents the *external forcing function*. In earthquake engineering, x(t) denotes the relative displacement between the mass and the ground. Since the acceleration, velocity, displacement, and

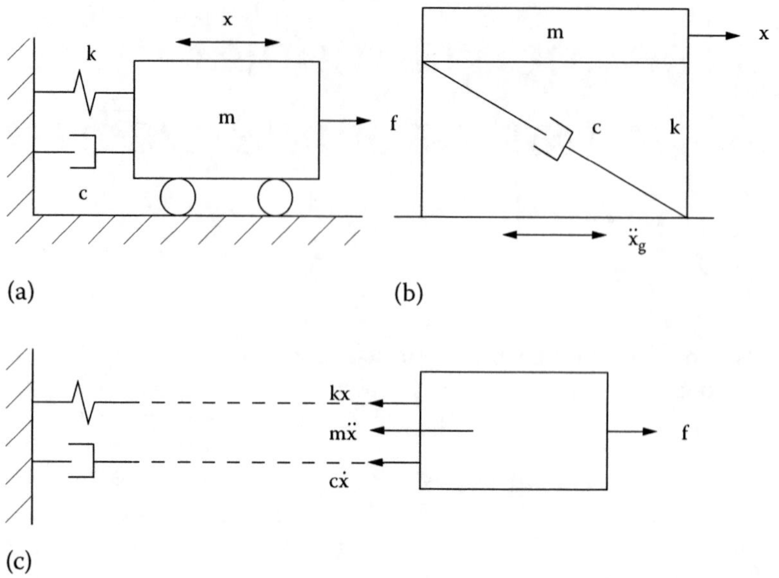

FIGURE 1.1 (a–c) SDOF vibration system.

forcing function are all temporal variables, the symbol (t) is used here. In the following text, for simplicity, the symbol (t) may be omitted at times. Note that Equation 1.2 is a linear, second-order ordinary differential equation with constant coefficients.

In Equation 1.1, the left side consists of three internal forces, which must balance the external force f. The sum can be called the *total internal force*, f_I, that is,

$$f_I = m\ddot{x}(t) + c\dot{x}(t) + kx(t) \tag{1.2}$$

Several other groupings of the vibrational forces are also of interest. For example, the combination of forces other than those associated with the external excitation and the force generated by acceleration can be called the *structural force*, f_S. Thus,

$$f_S = c\dot{x}(t) + kx(t) \tag{1.3}$$

Similarly, the *inertia force* is given by, $(m > 0)$

$$f_i(t) = m\ddot{x}(t) \tag{1.4}$$

In the context of earthquake engineering, $f_i(t)$ is the inertia force generated by the ground excitation. The *damping force* can also be written in the form of *linear viscous damping* as $(c \geq 0)$

$$f_d(t) = c\dot{x}(t) \tag{1.5}$$

Furthermore, the *linear spring* or *restoring force* may be given as $(k > 0)$

$$f_r(t) = kx(t) \tag{1.6}$$

Comparing Equations 1.5 and 1.6 with Equation 1.3, it is seen that the structural force is a combination of the damping and restoring forces. The stiffness and damping of the structural members, along with any added dampers, contribute to this structural force. In a later section, it will be seen that the structural force is used to resist the earthquake load generated by the absolute inertia force, an important concept in the design of dampers.

1.1.1.2 Basic Parameters of the Physical Model

Since $m \neq 0$, the monic and homogeneous form of Equation 1.1 can be written as

$$\ddot{x}(t)+\frac{c}{m}\dot{x}(t)+\frac{k}{m}x(t)=0 \tag{1.7}$$

Since both m and k are positive, the ratio of k to m can be denoted by a positive term ω_n^2,

$$\frac{k}{m}=\omega_n^2 \tag{1.8}$$

The term $2\xi\omega_n$ is used to denote twice the ratio c/m. Thus,

$$\frac{c}{m}=2\xi\omega_n \tag{1.9}$$

or

$$c=2\xi\omega_n m \tag{1.10}$$

along with

$$k=\omega_n^2 m \tag{1.11}$$

Here ξ is the *damping ratio* and ω_n is the *angular natural frequency* of the system described in Equation 1.7. From the relationships above,

$$\omega_n=\sqrt{\frac{k}{m}} \tag{1.12}$$

and

$$\xi=\frac{c}{2\sqrt{mk}}=\frac{c}{2\omega_n m} \tag{1.13}$$

The nature of the m-c-k system can be studied without considering the external force, and the m-c-k model is often referred to as the *physical model.*

In addition to the angular natural frequency ω_n, the *natural frequency* f_n is also often used to indicate how many cycles the system vibrates per each unit of time, usually measured in seconds. Thus,

$$f_n=\frac{1}{2\pi}\omega_n \tag{1.14}$$

In many cases, for convenience, as ω_n and f_n are so closely related, they are only distinguished as described in Equation 1.14. They are both defined as the natural frequency. Note, however, that the unit of ω_n is radian per second (rad/s) and that of f_n is 1/second (1/s) or hertz (Hz). The reciprocal of f_n has unit seconds (s) and is defined as the *natural period*, that is,

$$T_n = 1/f_n \tag{1.15}$$

For SDOF systems, it is seen that the basic parameters are the mass m, damping c, and stiffness k coefficients.

Of course, the natural frequency is a very important parameter in describing an SDOF vibration system. From Equations 1.8 through 1.12, it is further realized that the magnitude of the natural frequency is solely determined by stiffness k, which is related to potential energy, and mass m, which is related to kinetic energy, when the system vibrates. In other words, a system vibrates because the exchange of potential and kinetic energies exists.

Example 1.1

Suppose a building can be modeled as an undamped SDOF system. When equipment weighing 100 (t) is moved into the building, the corresponding natural period is measured to be $T_1 = 1.46$ (s). When an additional 100 (t) of equipment is moved in, the natural period is increased further to $T_2 = 1.50$ (s). Determine the mass and the stiffness of the SDOF building system.

The mass and stiffness of the building are denoted as m and k, respectively. The first 100 (t) mass is denoted as m_1, and the total additional 200 (t) mass as m_2. Also, the natural frequencies of the building alone, the building with the first 100 (t) mass, and the building with the additional 100 (t) mass are denoted as ω_n, ω_1, and ω_2, respectively. Then,

$$\sqrt{\frac{k}{m+m_1}} = \omega_1$$

$$\sqrt{\frac{k}{m+m_2}} = \omega_2$$

Thus,

$$k = (m+m_1)\omega_1^2$$

and

$$k = (m+m_2)\omega_2^2$$

Therefore, from the above two equations,

$$(m+m_1)\omega_1^2 = (m+m_2)\omega_2^2$$

or

$$m = \frac{m_1\omega_1^2 - m_2\omega_2^2}{\omega_2^2 - \omega_1^2} = \frac{m_1T_1^{-2} - m_2T_2^{-2}}{T_2^{-2} - T_1^{-2}} = \frac{m_1T_2^2 - m_2T_1^2}{T_1^2 - T_2^2} = 1{,}700 \quad (t)$$

Furthermore,

$$k = (m+m_1)\omega_1^2 = (m+m_1)\frac{4\pi^2}{T_1^2} = 33{,}343 \quad (kN/m)$$

Therefore, the natural frequency of the building alone is

$$\omega_n = \sqrt{\frac{k}{m}}$$

so that

$$f_n = \frac{\omega_n}{2\pi} = \frac{1}{2\pi}\sqrt{\frac{k}{m}} = 0.7 \quad (\text{Hz})$$

Example 1.2

Suppose a base isolated building can be modeled as an underdamped SDOF system with k = 18,935 (kN/m) and m = 2,000 (t). (In later sections, the concept of an underdamped system and base isolation system will be explained in detail.) With the isolation bearings, the damping ratio is measured as 13%. However, according to the design, a 30% damping ratio is needed. Therefore, the design engineer decides to use a linear viscous damper to increase the damping ratio. Calculate the required damping coefficient.

The original damping coefficient can be calculated as

$$c_0 = 2\times 0.13\sqrt{mk} = 1{,}600 \quad (\text{kN-s/m})$$

The ratio of the original damping ratio ξ_0 to the required damping ratio ξ_{design} is

$$\gamma = \xi_0/\xi_{design} = 0.13/0.3 = 0.43$$

Therefore, the required damping coefficient c can be calculated as

$$c = c_0\left(\frac{1}{\gamma} - 1\right) = 1600\times 1.3077 = 2{,}093.3 \quad (\text{kN-s/m})$$

1.1.1.3 Characteristic Equation and Modal Model

To seek possible solutions of the homogeneous differential equation, assume that

$$x(t) = Ae^{\lambda t} \tag{1.16}$$

where A is a displacement amplitude. Then, substituting Equation 1.16 into Equation 1.7, it can be determined whether a possible solution described by Equation 1.16 exists. The idea of using the assumption such as described in Equation 1.16 is referred to as the *semidefinite method*. Here, λ is a complex number whose physical meaning is discussed later. With the help of Equation 1.16,

$$\dot{x}(t) = \lambda Ae^{\lambda t} \text{ and } \ddot{x}(t) = \lambda^2 Ae^{\lambda t} \tag{1.17}$$

Substituting Equations 1.16 and 1.17, as well as Equations 1.8 and 1.9, into Equation 1.7 yields

$$\left(\lambda^2 + 2\xi\omega_n\lambda + \omega_n^2\right)Ae^{\lambda t} = 0 \tag{1.18}$$

Since $Ae^{\lambda t}$ does not equal zero, each side of Equation 1.18 is divided by this factor and results in

$$\lambda^2 + 2\xi\omega_n\lambda + \omega_n^2 = 0 \tag{1.19}$$

Equation 1.19 is referred to as the *characteristic equation* of the system. Solving Equation 1.19 for λ gives:

$$\lambda = -\xi\omega_n \pm \sqrt{\xi^2 - 1}\,\omega_n \tag{1.20}$$

From Equation 1.20, it is seen that when the damping ratio ξ is smaller than 1, that is,

$$\xi < 1, \tag{1.21}$$

Equation 1.20 can then be rewritten as

$$\lambda_{1,2} = -\xi\omega_n \pm j\sqrt{1 - \xi^2}\,\omega_n \tag{1.22}$$

where

$$j = \sqrt{-1} \tag{1.23}$$

In Equation 1.22, λ_1 and λ_2 are complex conjugates. Both λ_1 and λ_2 belong to two distinct *vibration modes*, which implies the aforementioned energy exchange. In fact, the energy exchange occurs between the potential and the kinetic energies. Once again, a vibration system does not exist without the energy exchange. Furthermore, the energy exchange is described if and only if the inequality shown in Equation 1.21 holds. In this case, the system is *underdamped*. In engineering practice, underdamped systems are more common. In fact, ξ can be much smaller than 1. In that case, no matter what type of damping forces (e.g., viscous, viscoelastic, hysteretic, quadratic) actually exist, the response x will not be greatly influenced by the same "effective" damping ratio (see Section 1.3). The assumption of viscous damping can provide fairly accurate results of responses. The approach of using linear equations should be, from a mathematical standpoint, the most convenient. This will be examined in more detail in Chapter 5, where specificities about the various damping devices are first presented.

From the above discussion, it is seen that the damping ratio and natural frequency are parameters of the system itself, which will not be affected by external conditions. They are referred to as the basic dynamic parameters or *eigen-parameters* of the system; therefore, the term λ is called the *eigenvalue*. Figure 1.2 depicts the eigenvalues for the undamped case in the complex plane. Furthermore, the model described by Equation 1.19 is referred to as the *modal model* because it can be described by the complex conjugate modes determined by parameters ξ and ω_n. Additionally, ξ and ω_n, the damping ratio and the natural frequency, are referred to as the *modal parameters.*

Note that there are three basic parameters of the physical model for the SDOF system, whereas for the modal model there are only two.

When the dynamic behavior of linear multi-degree-of-freedom systems is studied, a complete set of modes or a modal model can be used to represent the entire system. This is discussed in Chapters 2 and 3. Here, for the SDOF system, there is only one mode for consideration in engineering practice. Complex conjugates modes do not need to be distinguished in particular and are usually referred to as a *single identical mode.*

On the other hand, when

$$\xi = 1, \tag{1.24}$$

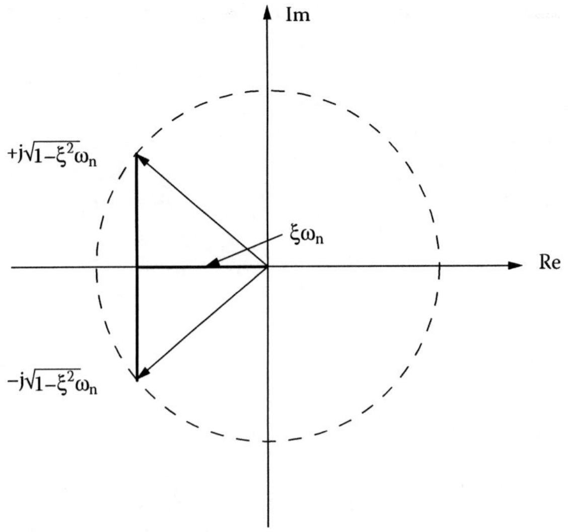

FIGURE 1.2 Eigenvalues of an underdamped system.

Equation 1.20 can be rewritten as

$$\lambda_{1,2} = -\xi\omega_n \tag{1.25}$$

Thus, the characteristic equation has two identical real-valued solutions, and the corresponding system is said to be *critically damped.* From Equation 1.13, it is known that when $\xi = 1$, $c = 2\sqrt{km}$. To denote the critically damped case, this particular damping coefficient is shown with subscript "cr" as

$$c_{cr} = 2\sqrt{km} \tag{1.26}$$

With this notation, the damping ratio for a general system with damping coefficient c can be rewritten as

$$\xi = \frac{c}{2\sqrt{km}} = \frac{c}{c_{cr}} \tag{1.27}$$

Thus, the damping ratio is formally referred to as the *critical damping ratio* in many references. In addition, from Equation 1.27, it is seen that the damping ratio denotes the ratio of the damping coefficient c and twice the geometric mean of the mass m and the stiffness k, namely, $2\sqrt{km}$.

Lastly, when

$$\xi > 1, \tag{1.28}$$

Equation 1.20 can be rewritten as

$$\lambda_{1,2} = -\xi\omega_n \pm \sqrt{\xi^2 - 1}\,\omega_n \tag{1.29}$$

Thus, the characteristic equation has two distinct real-valued solutions, and the corresponding system is said to be *overdamped.*

Notice that in both the critically damped and the overdamped cases, complex-valued eigenvalues no longer exist. The systems are reduced to two first-order real-valued *pseudo modes*, and there will be no energy exchange between the two modes. In other words, both types of systems are no longer vibration systems. Although in these two cases, the mass can still be made to oscillate from external forces, these systems will no longer be able to oscillate in free vibration.

Example 1.3

Suppose a characteristic equation is given by

$$m\lambda^2 + c\lambda + k = 0$$

Find the quantities of $\lambda_1\lambda_2$ and $\lambda_1 + \lambda_2$, where λ_1 and λ_2 are the solutions of the above equation. From the general properties of the quadratic equation, it is known that

$$\lambda_1\lambda_2 = \frac{k}{m}$$

and

$$\lambda_1 + \lambda_2 = -\frac{c}{m}$$

The above expressions are always true regardless of whether the system is underdamped or overdamped. However, if the system is underdamped, then

$$\lambda_{1,2} = -\xi\omega_n \pm j\sqrt{1-\xi^2}\,\omega_n$$

Therefore,

$$\lambda_1\lambda_2 = \lambda_1\lambda_1^* = \omega_n^2 = \frac{k}{m}$$

where superscript * stands for the operation of taking the complex conjugate. In this case, it is easily seen that

$$\frac{k}{m} = \omega_n^2$$

Furthermore,

$$\lambda_1 + \lambda_2 = -\xi\omega_n + (-\xi\omega_n) = -2\xi\omega_n$$

Thus,

$$-2\xi\omega_n = -\frac{c}{m}$$

or

$$\frac{c}{m} = 2\xi\omega_n$$

1.1.2 Homogeneous Solution, Free-Decay Vibration, and the Response Model

The solutions of Equation 1.1 represent the vibration responses, which can be classified as: (1) the transient response, including free-decay vibration and forced vibration excited by the transient forcing function; (2) periodic vibration, the steady-state response due to periodic excitations; and (3) random vibration, due to random excitations.

It is understandable that the system will not vibrate unless a certain external input is applied. The input can be either the initial condition of velocity and/or displacement or the forcing functions. Note that a forcing function can also cause a *forced initial condition*. If the input is the initial conditions only, then free-decay vibrations will occur. That is, Equation 1.1 can have a free-decay solution, if the system, e.g., the cart in Figure 1.1a, is excited by an initial force, or has an initial velocity or displacement; and after the initial excitation, no external force is added to the system. In this case, Equation 1.7 will be used. An SDOF vibration system with m = 2, k = 100, and initial unit velocity is used as an example. Suppose there are two cases of damping magnitude, the first with a 5% damping ratio and the second with a 50% damping ratio. The responses of the two systems are plotted in Figure 1.3. It is seen that the vibration levels continuously decrease in both cases. However, the vibration with a larger damping ratio decays much faster.

Using the above mentioned semi-definite method, the free-decay displacement under certain initial conditions is written as follows:

$$x(t) = Ae^{-\xi\omega_n t}\sin(\omega_d t + \varphi) \tag{1.30}$$

with the initial conditions

$$\begin{cases} x(0) = d_0 \\ \dot{x}(0) = v_0 \end{cases} \tag{1.31}$$

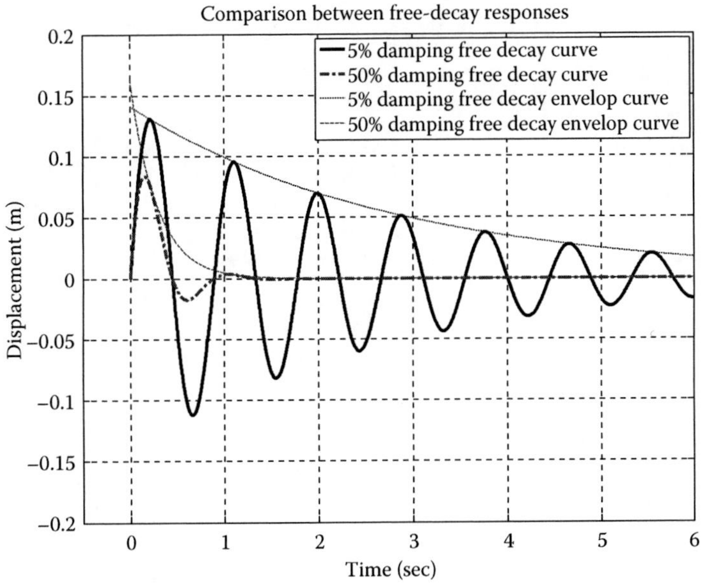

FIGURE 1.3 Free vibration with decay.

Here, d_0 and v_0 are, respectively, the initial displacement and velocity, and

$$\omega_d = \sqrt{1-\xi^2}\,\omega_n \tag{1.32}$$

is called the *damped natural frequency*, which is the imaginary part of the eigenvalue described in Equation 1.22. In Equation 1.30, A and φ are, respectively, the *amplitude* and *phase angle* constants and can be determined by the initial conditions given in Equation 1.31. That is,

$$A = \frac{1}{\omega_d}\sqrt{\left(v_0 + \xi\omega_n d_0\right)^2 + \left(\omega_d d_0\right)^2} \tag{1.33}$$

and

$$\varphi = \tan^{-1}\left(\frac{\omega_d d_0}{v_0 + \xi\omega_n d_0}\right) + h_\varphi \pi \tag{1.34}$$

Note that the period of the tangent function is π, and the arctangent function has multiple values. However, the period of the sine and cosine functions is 2π. Therefore, the Heaviside function, h_φ, cannot be arbitrarily chosen. Based on the fact that most computational programs, such as MATLAB®,* calculate the arctangent by limiting the values from $-\pi/2$ to $+\pi/2$, h_φ is defined as

$$h_\varphi = \begin{cases} 0, & v_0 + \xi\omega_n d_0 \geq 0 \\ 1, & v_0 + \xi\omega_n d_0 < 0 \end{cases} \tag{1.35}$$

As shown in Figure 1.4, the phase angle φ can have four cases, namely, the combination of $\omega_d d_0$ and $v_0 + \xi\omega_n d_0$ to be positive and negative numbers. Despite the values of $\omega_d d_0$, it is seen from Figure 1.4 and Equation 1.35 that the sign of $v_0 + \xi\omega_n d_0$ determines the choice of h_φ.

From Equation 1.30, it is seen that the vibration will have an envelope of $Ae^{-\xi\omega_n t}$. As time goes on, the level of the free vibration will decrease. The rate of the decay per cycle depends on the value

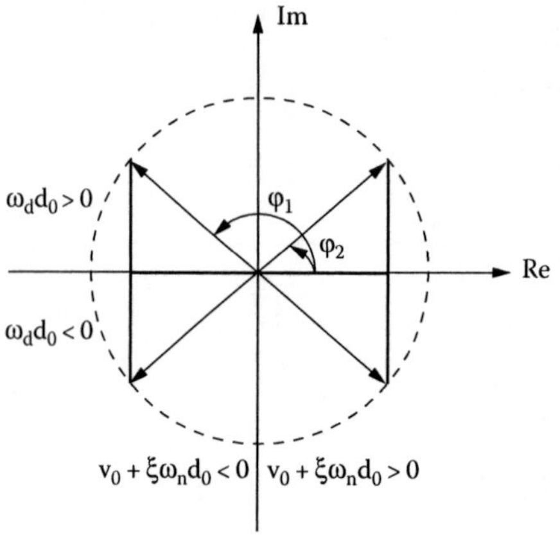

FIGURE 1.4 Phase angle φ.

* The mathematical software package MATLAB is often referenced throughout this book. The reader is encouraged to become familiar with this useful tool, which is used in many of the exercises.

of the damping ratio. In Figure 1.3, it is seen that the responses of the systems with a 5% damping ratio and with a 50% damping have corresponding envelopes.

In addition, when the damping ratio is larger, the first peak value is smaller. Furthermore, the peak value appears earlier and the corresponding damped frequency is smaller.

Taking the time derivative of the displacement in Equation 1.30, the velocity becomes

$$\dot{x}(t) = A\omega_n e^{-\xi\omega_n t}\cos(\omega_n t + \varphi + \theta) \tag{1.36}$$

where

$$\theta = \tan^{-1}\left(\frac{\xi}{\sqrt{1-\xi^2}}\right) \tag{1.37}$$

Comparing Equation 1.30 and Equation 1.36, it is seen that the time variables are sine and cosine functions. The two functions, $\sin(\omega_d t + \varphi)$ and $\cos(\omega_d t + \varphi + \theta)$, are trigonometric functions with a phase difference of $90° + \theta$.

In many civil engineering structures, the damping ratio is a rather small number, therefore,

$$\theta \approx \tan^{-1}(\xi) \approx \xi \tag{1.38}$$

Furthermore, the velocity and displacement will have a phase shift θ close to 90°. That is, since the damping ratio is a small number, in the case of free-decay vibration, the velocity and the displacement will have a *nearly 90°* phase difference, and the smaller the damping is, the closer to 90° the phase difference will be. Note that the above discussion is only for linear systems. This conclusion can be visualized by using the examples of free vibration decay shown in Figure 1.5. In Figure 1.5a, the damping ratio is taken to be 0.01 and in Figure 1.5b, the damping ratio is 0.3. Under initial conditions of $d_0 = 0.1$ (m) and $v_0 = 0$ (m/s), the velocity (shown in broken lines) of the system with a lower damping ratio is close to 90° ahead of the corresponding displacement (shown in solid lines). Meanwhile, for the more heavily damped system, velocity noticeably leads displacement by <90°.

Example 1.4

Suppose the free-decay peak responses of an SDOF system measured at the second cycle and the tenth cycle are, respectively, 20 and 0.1 (mm), whereas the damped natural frequency is 1 (Hz). Find the damping ratio.

Denote the peak response at the second cycle and the tenth cycle to be x_2 and x_{10}, respectively. Also, denote the corresponding time points to be t_2 and t_{10}. From Equation 1.30,

$$x_2 = Ae^{-\xi\omega_n t_2}\sin(\omega_d t_2 + \varphi) = 20 \quad \text{(mm)}$$

and

$$x_{10} = Ae^{-\xi\omega_n t_{10}}\sin(\omega_d t_{10} + \varphi) = 0.1 \quad \text{(mm)}$$

Thus, the ratio of x_2 to x_{10} is further taken to be

$$x_2/x_{10} = e^{\xi\omega_n(t_{10}-t_2)} = 200$$

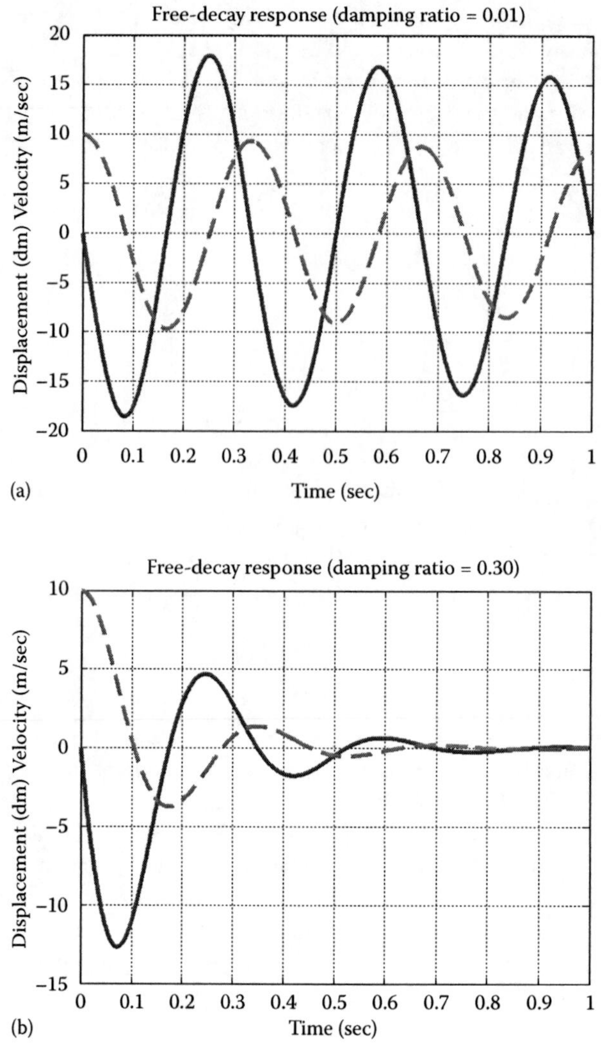

FIGURE 1.5 Free-decay velocities and displacements (a) $\xi=0.01$ and (b) $\xi=0.30$.

Therefore

$$\ln(x_2/x_{10}) = \xi\omega_n(t_{10} - t_2) = \ln(200) = 5.298$$

Hence,

$$\xi = \frac{\ln(x_2/x_{10})}{\omega_n(t_{10} - t_2)} = \frac{5.298}{\omega_n(t_{10} - t_2)}$$

Note that from t_2 to t_{10}, there are eight complete cycles, with each cycle occupying the duration of one period $2\pi/\omega_d$. Therefore,

$$\omega_n(t_{10} - t_2) = \frac{2\pi(10-2)}{\sqrt{1-\xi^2}} = \frac{16\pi}{\sqrt{1-\xi^2}}$$

The damping ratio is calculated to be 10.5%.

Example 1.5

Consider two cases of an SDOF system with different initial conditions. The first case has a zero initial velocity ($v_0 = 0$), while in the second case, zero displacement ($d_0 = 0$) is imposed. Find the relationship between the damping ratio and amplitude A as well as the peak values of the corresponding displacements for these two cases.

The first case when the initial displacement is d_0 is examined. From Equation 1.33,

$$A = \frac{1}{\omega_d}\sqrt{(v_0 + \xi\omega_n d_0)^2 + (\omega_d d_0)^2} = \frac{d_0}{\sqrt{1-\xi^2}} \approx \left(1 + \frac{\xi^2}{2}\right)d_0$$

Therefore, the amplitude of displacement A is approximately proportional to the term $1 + \xi^2/2$. In other words, the larger the damping ratio, the larger the resulting amplitude A is in this case, despite the common sense notion that larger damping will always result in decreased response amplitude.

Note, however, that a larger amplitude A does not mean that the response will have a larger value, since the free-decay response due to initial displacement only will never be larger than the initial displacement. This can be proven by taking the derivative of the displacement described in Equation 1.38 while assigning $v_0 = 0$ and solving for time t. In fact, the left side of Equation 1.36 can be zero to locate the extreme value; that is,

$$A\omega_n e^{-\xi\omega_n t}\cos(\omega_n t + \varphi + \theta) = 0$$

or $\cos(\omega_n t + \varphi + \theta) = 0$, which gives $\omega_n t + \varphi + \theta = \pi/2$. However, with $v_0 = 0$,

$$\varphi + \theta = \tan^{-1}\left(\frac{\sqrt{1-\xi^2}}{\xi}\right) + \tan^{-1}\left(\frac{\xi}{\sqrt{1-\xi^2}}\right) = \pi/2$$

Therefore, t must be zero in order to obtain the extreme value, which is understood to be the peak value of the displacement, denoted by x_{max}. Additionally, when $t = 0$, the displacement is nothing but the initial displacement d_0. That is,

$$x_{max} = d_0$$

In other words, with zero initial velocity, the peak value of the displacement is always the initial displacement, independent of the value of the damping ratio. However, the damping ratio affects the amplitude A, and larger damping yields larger amplitude.

The second case, when the initial velocity v_0 is nonzero, is examined next. From Equation 1.33,

$$A = \frac{1}{\omega_d}\sqrt{(v_0 + \xi\omega_n d_0)^2 + (\omega_d d_0)^2} = \frac{|v_0|}{\omega_d} = \frac{|v_0|}{\sqrt{1-\xi^2}\,\omega_n}$$

and from Equations 1.34 and 1.35 with positive v_0,

$$\varphi = 0$$

Thus, from Equation 1.30, the displacement is given by

$$x(t) = \frac{v_0}{\omega_d}e^{-\xi\omega_n t}\sin(\omega_d t)$$

To find the peak value of the displacement, the derivative of the above equation is taken and

$$\frac{d}{dt}\left[x(t)\right] = \frac{v_0}{\omega_d}\left[(-\xi\,\omega_n)e^{-\xi\omega_n t}\sin(\omega_d t) + e^{-\xi\omega_n t}\omega_d\cos(\omega_d t)\right] = 0$$

From which,

$$\frac{\sin(\omega_d t)}{\cos(\omega_d t)} = \frac{\sqrt{1-\xi^2}}{\xi}$$

This equation yields time t when the displacement reaches peak value, which is

$$t = \frac{1}{\omega_d}\tan^{-1}\frac{\sqrt{1-\xi^2}}{\xi}$$

and with a small damping ratio the corresponding peak value of the displacement, x_{max}, is

$$x_{max} = \frac{v_0}{\omega_d}e^{-\xi\omega_n\left((1/\omega_d)\tan^{-1}\left(\sqrt{1-\xi^2}/\xi\right)\right)}\sin\left(\omega_d\frac{1}{\omega_d}\tan^{-1}\frac{\sqrt{1-\xi^2}}{\xi}\right)$$

$$= \frac{v_0}{\omega_n\sqrt{1-\xi^2}}e^{-\left(\left(\xi/\sqrt{1-\xi^2}\right)\tan^{-1}\left(\sqrt{1-\xi^2}/\xi\right)\right)}\sqrt{1-\xi^2}$$

$$= \frac{v_0}{\omega_n}e^{-\left(\left(\xi/\sqrt{1-\xi^2}\right)\tan^{-1}\left(\sqrt{1-\xi^2}/\xi\right)\right)} \approx \frac{v_0}{\omega_n}e^{-\xi\pi/2}$$

From the above equation, the peak displacement of the SDOF system with initial velocity only is approximately inversely proportional to its natural frequency ω_n and directly proportional to the amplitude of the initial velocity and the term $e^{-\xi\pi/2}$. That is, the larger the damping of the system, the smaller the peak response.

1.1.3 Forced Vibration with Harmonic Excitation

When the external input to the system continues to be applied, the system will vibrate in a forced vibration mode. The case when the linear system defined by Equation 1.1 is excited with harmonic forcing functions is discussed in the following subsections.

1.1.3.1 Steady-State Response

First, the steady-state response is considered, namely, when a harmonic forcing function is applied with sufficiently long duration. Letting $f = f_0\sin(\omega_f t)$, where f_0 and ω_f are, respectively, the amplitude and the driving frequency of the forcing function; the steady-state solution x_{ps} is given by

$$x_{ps} = x_0\sin(\omega_f t + \phi) \tag{1.39}$$

Thus, the velocity is

$$\dot{x}_{ps} = \omega_f x_0\cos(\omega_f t + \phi) \tag{1.40}$$

and the acceleration is

$$\ddot{x}_{ps} = -\omega_f^2 x_0 \sin(\omega_f t + \phi) \tag{1.41}$$

Substituting Equations 1.39 through 1.41 into Equation 1.1 with $f = f_0 \sin(\omega_f t)$ yields

$$-\omega_f^2 m x_0 \sin(\omega_f t + \phi) + \omega_f c x_0 \cos(\omega_f t + \phi) + k x_0 \sin(\omega_f t + \phi) = f_0 \sin(\omega_f t)$$

Dividing both sides of the above equation by m results in

$$-\omega_f^2 x_0 \sin(\omega_f t + \phi) + 2\xi\omega_f\omega_n x_0 \cos(\omega_f t + \phi) + \omega_n^2 x_0 \sin(\omega_f t + \phi) = f_0/m \sin(\omega_f t)$$

Furthermore,

$$\left[\sin(\omega_f t + \phi - \phi)\right] x_0 = \frac{f_0/m \cdot \sin(\omega_f t)}{\sqrt{(\omega_n^2 - \omega_f^2)^2 + (2\xi\omega_n\omega_f)^2}}$$

Since $\sin(\omega_f t)$ cannot always be zero, the amplitude x_0 for Equation 1.39 is derived as

$$x_0 = \frac{f_0/m}{\sqrt{(\omega_n^2 - \omega_f^2)^2 + (2\xi\omega_n\omega_f)^2}} \tag{1.42}$$

Meanwhile, the angle ϕ stands for the phase difference between the excitation force and the response displacement, which can be written as

$$\phi = \begin{cases} \tan^{-1}\left[\dfrac{2\xi\omega_n\omega_f}{\omega_f^2 - \omega_n^2}\right] + h_\phi \pi, & \omega_n \neq \omega_f \\ -\pi/2, & \omega_n = \omega_f \end{cases} \tag{1.43}$$

Mathematically, the expression of angle ϕ, as described in the second line of Equation 1.43, does not have to be written. However, for practical computations, when $\omega_n = \omega_f$, the term in the bracket will have a denominator of zero, which is often not allowed in practical computational programs. In addition, the arctangent function has multiple values, whereas most computational programs only provide the solution in one or two quadrants. (For example, MATLAB provides the solution in the first and the fourth quadrants.) Therefore, a Heaviside function, h_ϕ, is used to handle this situation, which is defined as follows:

$$h_\phi = \begin{cases} 0, & \omega_n < \omega_f \\ -1, & \omega_n > \omega_f \end{cases} \tag{1.44}$$

In the next subsection, the amplitude x_0 and the phase ϕ are studied as functions of the frequency and damping ratio. Here, from the time history of the steady-state solution, the velocity is found to always be 90° ahead of the displacement. Accordingly, in the case of the steady-state harmonic response, the damping force will also be 90° ahead of the restoring force.

1.1.3.2 Method of Complex Response for Steady-State Displacement

The system $m\ddot{x} + c\dot{x} + kx$ may be subjected to the aforementioned excitation $f = f_0 \sin(\omega_f t)$. It may also have another harmonic excitation, $f = f_0 \cos(\omega_f t)$. In the latter case,

$$x_{ps} = x_0 \cos(\omega_f t + \phi)$$

It is easy to see that the amplitude x_0 and the phase ϕ in the above equation can also be expressed by Equations 1.42 and 1.43, respectively, similar to the case of $f = f_0 \sin(\omega_f t)$.

Since the vibration system is linear, when a forcing function is written as a complex combination of

$$f = f_0 \cos(\omega_f t) + j f_0 \sin(\omega_f t) = f_0 e^{j\omega_f t} \tag{1.45}$$

the response can be written as

$$x_{ps} = x_0 \cos(\omega_f t + \phi) + j x_0 \sin(\omega_f t + \phi) = x_0 e^{j(\omega_f t + \phi)} = x_0 e^{j\phi} e^{j\omega_f t} = x_{p0} e^{j\omega_f t} \tag{1.46}$$

Here, x_{p0} is the complex-valued amplitude of x_{ps},

$$x_{p0} = x_0 e^{j\phi}$$

namely,

$$|x_{p0}| = x_0 \tag{1.47}$$

and

$$\angle(x_{p0}) = \phi \tag{1.48}$$

In Equation 1.47, the symbol $|(\cdot)|$ stands for the absolute value of the complex variable $(\cdot)$. In Equation 1.48, the symbol $\angle(\cdot)$ stands for the angle of the complex variable $(\cdot)$. It can be shown that the amplitude x_0 and the phase angle ϕ can also be represented by Equations 1.42 and 1.43, respectively.

Figure 1.6 shows the relationship between the real and the imaginary parts of the forcing functions and the responses. The forcing function, f(t), can be represented by a rotating vector with amplitude f_0 and angular speed ω_f, whose projection on the real axis, marked "Re," is $f_0 \cos \omega_f t$, and on the imaginary axis, marked "Im," is $m f_0 \sin \omega_f t$. At any time t, the angle $\omega_f t$, being the product of the driving frequency ω_f and time t, can be seen as a special angle, as shown in Figure 1.6.

Furthermore, the restoring force kx(t) can also be represented by a vector, with amplitude kx_0 and phase angle ϕ compared to force f(t). The restoring force also has the identical angular speed ω_f. It is seen that it has the projection $kx_0 \cos(\omega_f t + \phi)$ on the Re axis, as shown in Figure 1.6. Its projection on the Im axis must be $kx_0 \sin(\omega_f t + \phi)$, which is not shown.

Similarly, the damping force $c\dot{x}(t)$ can also be represented as the third vector with amplitude $c\omega_f x_0$ and phase angle $\pi/2$ compared to the force kx(t). The damping force also has the identical angular speed ω_f. It is seen that it has the projection $c\omega_f x_0 \cos(\omega_f t + \phi)$ on the Re axis, also shown in Figure 1.6. Its projection on the Im axis is $c\omega_f x_0 \sin(\omega_f t + \phi)$, which is not shown.

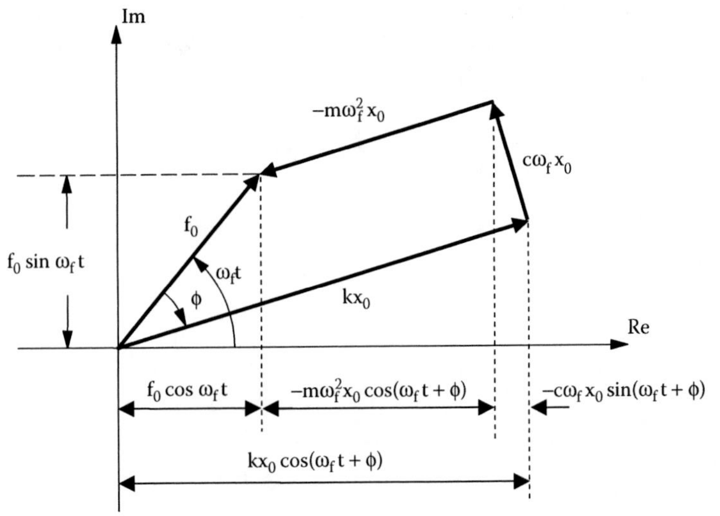

FIGURE 1.6 Vector quadrilaterals at time t.

The inertia force $m\ddot{x}(t)$ can be represented as a fourth vector with amplitude $-m\omega_f^2 x_0$ and phase angle $\pi/2$ compared to damping $c\dot{x}(t)$. The inertia force also has the identical angular speed $\omega_f t$. It is seen that it has the projection $-m\omega_f^2 x_0 \cos(\omega_f t + \phi)$ on the Re axis, again shown in Figure 1.6. Its projection on the Im axis is $-m\omega_f^2 x_0 \sin(\omega_f t + \phi)$, which is not shown.

Figure 1.6 shows that once the position of the vector that represents the forcing function is determined, the vector quadrilateral is determined for a specific m-c-k system. This means that both the amplitudes and the phase differences for the restoring force, the damping force, and the inertia force are determined. Determination of the amplitudes can be obtained through x_0 in Equation 1.42. Determination of the phase angles can be obtained through ϕ in Equation 1.43.

These parameters can also be obtained by plotting the vector quadrilateral, such as that shown in Figure 1.6, which is then referred to as the *geometric method* for the steady-state responses.

When Equations 1.42 and 1.43 are derived, the responses either on the Im axis or on the Re axis are used. Figure 1.6 shows that once the position of the vectors of the forces is determined, so are their projections. Namely, on the Re axis, the cosine functions, and on the Im axis, the sine functions, are established. On the other hand, once the cosine functions on the Re axis are known, together with the angle $\omega_f t$ and the phase difference ϕ, the force vectors and the sine functions can be known on the Im axis and vice versa. These facts imply that the cosine functions or the sine functions or their complex-valued combinations can all be used to represent the steady-state solutions. In other words, once one set of functions (cosine, sine, or complex) is determined, so are the other two sets of functions. Thus, these three sets of representations are uniquely related.

In many cases, response calculations using complex functions can be simpler. Whenever the real-valued formulas are needed, either the real or the imaginary part of the complex functions can be taken without loss of information, since the real and imaginary portions are uniquely related. This is referred to as the *method of complex response*.

1.1.3.3 Response of Harmonic Excitation with Zero Initial Condition

In the literature, when the initial conditions are zero, namely

$$\begin{cases} m\ddot{x} + c\dot{x} + kx = f_0 \cos(\omega_f t) \\ x(0) = 0 \\ \dot{x}(0) = 0 \end{cases} \tag{1.49}$$

the response caused by the forcing function $f_0 \sin \omega_f t$ only is called the *particular solution* of Equation 1.49, which will not only have the steady-state solution described by Equation 1.40a, but will also have a transient solution $x_{pt}(t)$, that is,

$$x(t) = x_{pt}(t) + x_{ps}(t) \tag{1.50}$$

For Equation 1.49, the transient solution $x_{pt}(t)$ can be written as

$$x_{pt}(t) = Ae^{-\xi\omega_n t} \sin(\omega_d t + \Phi) \tag{1.51}$$

where

$$A = \begin{cases} \dfrac{-x_0 \cos\phi}{\sin\Phi} = -\dfrac{x_0\sqrt{(\omega_d \cos\phi)^2 + (\xi\omega_n \cos\phi - \omega_f \sin\phi)^2}}{\omega_d} & \phi \neq -\pi/2 \\ -\dfrac{x_0\omega_f}{\omega_d} & \phi = -\pi/2 \end{cases} \tag{1.52}$$

and the phase Φ is

$$\Phi = \tan^{-1}\left(\frac{\omega_d \cos\phi}{\xi\omega_n \cos\phi - \omega_f \sin\phi}\right) \quad -\frac{\pi}{2} \leq \Phi \leq \frac{\pi}{2} \tag{1.53}$$

In Equation 1.50, the steady-sate solution x_{ps} (t) can be written as

$$x_{ps}(t) = x_0 \cos(\omega_f t + \phi) \tag{1.54}$$

For Equations 1.53 and 1.54, the amplitude x_0 and angle ϕ are defined, respectively, in Equations 1.42 and 1.43. The total particular solution $x_p(t)$ can be written as

$$x_p(t) = x_{pt}(t) + x_{ps}(t) = x_0\left[-\frac{\cos\phi}{\sin\Phi} e^{-\xi\omega_n t} \sin(\omega_d t + \Phi) + \cos(\omega_f t + \phi)\right] \tag{1.55}$$

At the resonance point, when $\omega_f = \omega_n$, Equation 1.55 should be rewritten as

$$x_p(t) = x_{pt}(t) + x_{ps}(t) = x_0\left[-\frac{e^{-\xi\omega_n t}}{\sqrt{1-\xi^2}} \sin(\omega_d t) + \sin(\omega_f t)\right] \tag{1.56}$$

In both cases, since the transient responses of $x_{pt}(t)$ will soon fade out, the steady-state solution will prevail so that the velocity will also lead displacement by a 90° phase difference, as mentioned before.

For the transient process, a special effect of damping is realized by examining the terms Φ, ϕ, and $e^{-\xi\omega_n t}$ in Equation 1.55.

It is seen that the steady-state response $x_{ps}(t)$ is periodic. In fact, it is the simplest of periodic responses. In the following, $x_{ps}(t)$ is studied in more detail.

First, consider the effect on the phase angles. If $\omega_f < \omega_n$, Φ will have a positive sign. As ω_f increases from rather small values and approaches ω_n, the angle Φ gradually decreases from the

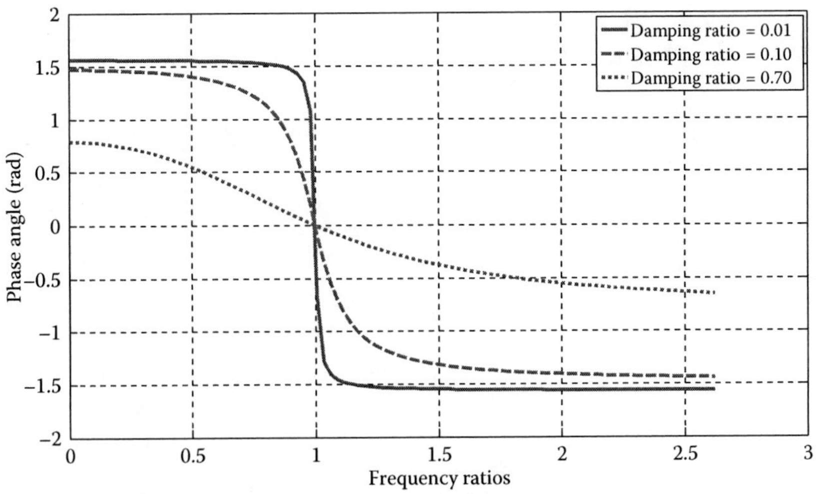

FIGURE 1.7 Phase angle Φ vs. frequency ratio.

value of π/2. The smaller the damping ratio, the closer the original limited value π/2. In addition, the larger the damping ratio, the slower the rate of decrease. On the other hand, with a smaller damping ratio, the rate of decrease becomes faster. These phenomena can be shown by using an example, where the natural frequency is set to be 3 (Hz). In Figure 1.7, the values of the phase angle Φ are plotted vs. the frequency ratio of ω_f/ω_n, where three curves are shown, corresponding to the damping ratios of 0.01, 0.1, and 0.7, respectively.

The value of ϕ will decrease from zero when the frequency ratio is increased. The rate of the increase is also affected by the value of the damping ratio. In Figure 1.8, the values of the phase angle φ are plotted vs. the frequency ratio of ω_f/ω_n. Again, three curves are shown, corresponding to the damping ratios of 0.01, 0.1, and 0.7, respectively.

Next, the effect of the damping ratio on the phase angle is considered. In Figure 1.9a, the values of the phase angle Φ are plotted vs. the damping ratio of ξ. In Figure 1.9b, the values of the phase angle φ are plotted vs. the damping ratio of ξ. In each figure, four curves are shown, corresponding to the frequency ratios $r = \omega_f/\omega_n$ of 1/2, 1/1.5, 1.5, and 2, respectively. From these

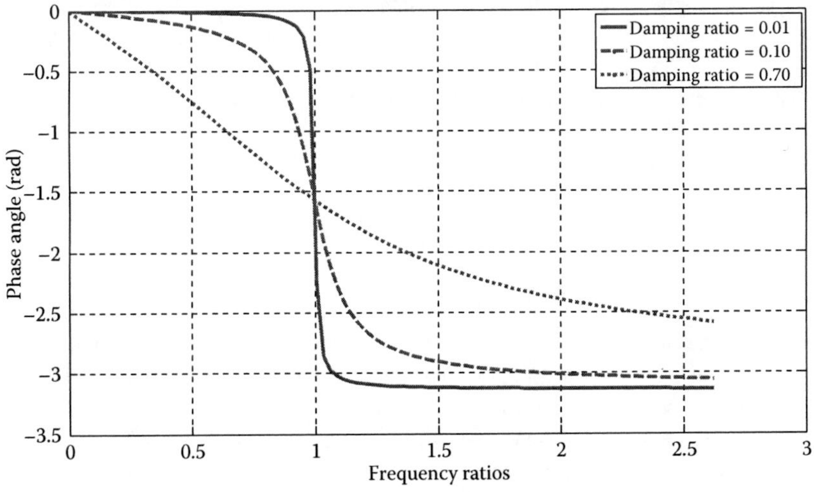

FIGURE 1.8 Phase angle ϕ vs. frequency ratio.

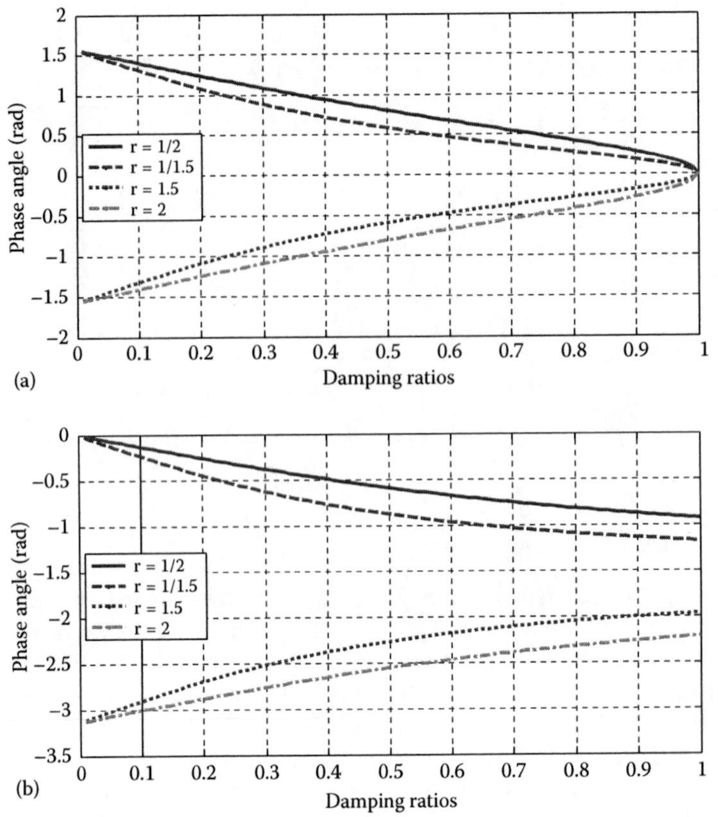

FIGURE 1.9 Phase angle vs. damping ratio (a) Φ and (b) ϕ.

plots, it is seen that when the damping ratio increases, the phase angles will change their values accordingly.

When the damping ratio is small, the phase angle Φ is close to $\pm\pi/2$. When the damping ratio becomes larger to near 1, the phase angle is close to zero. That is, the trends of changing angle Φ can be decreased or increased, depending on whether the frequency ratio is greater or smaller than 1. On the other hand, when the damping ratio is small, the phase angle ϕ is close to 0 or $-\pi$. When the damping ratio becomes larger, the phase angle ϕ approaches zero. That is, the trend of change of the angle ϕ can also be decreased or increased, depending on whether the frequency ratio is smaller or greater than 1. The variation trends of angles Φ and ϕ are similar.

When the damping ratio is sufficiently small, at the resonance point when $\omega_f = \omega_n$, Equation 1.51 can be written as

$$x_p(t) = x_{pt}(t) + x_{ps}(t) = x_0\left(1 - e^{-\xi\omega_n t}\right)\sin(\omega_n t) \tag{1.57}$$

Next, the transient response in the case where the driving frequency is close to the natural frequency is considered. In Equation 1.57, the term $(1 - e^{-\xi\omega_n t})$ is found, which implies an increasing amplitude with respect to time. In Figure 1.9, a system with a natural frequency of 3 (Hz) or 6π (rad/s) with excitation frequency 3.03 (Hz) is used to show the transient responses (assume $m = 1$ and $f_0 = 1$ for simplicity). Comparisons between the vibratory responses and the growing signal $x_0(1 - e^{\xi\omega_n t})$ are made in Figure 1.10. In Figure 1.10a, the damping ratio is taken to be 0.50 and in Figure 1.10b, the damping ratio is 0.05.

In these plots, the broken lines are the increasing amplitudes and the solid lines are the vibrational responses. From Figure 1.10, the increasing signal can be approximately seen as an envelope

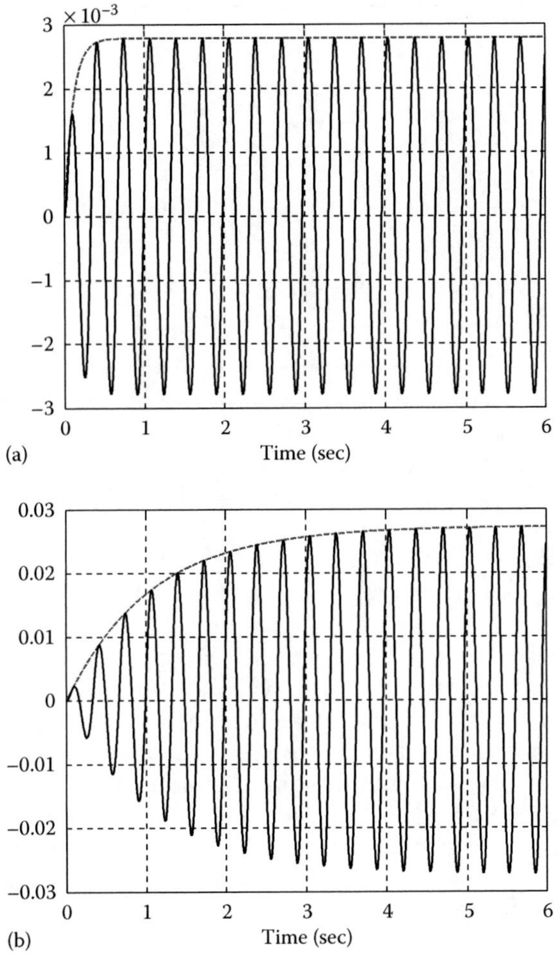

FIGURE 1.10 Transient responses (a) $\xi = 0.50$ and (b) $\xi = 0.05$.

of the vibratory responses. And, when the damping ratio is smaller, the increasing signal is closer to the envelope of the peak vibrations. However, when the damping is larger, at the beginning, there can be a large difference. Next, it is observed that when the damping ratio is large, the vibration reaches its steady state quicker than with smaller damping. A third observation is that large damping ensures a smaller vibration amplitude of the steady-state responses.

1.1.3.4 Responses with Nonzero Initial Conditions

Now, the harmonic excitation together with the nonzero initial conditions x(0) and $\dot{x}(0)$ is further considered. That is,

$$\begin{cases} m\ddot{x} + c\dot{x} + kx = f_0 \cos(\omega_f t) \\ x(0) = d_0 \\ \dot{x}(0) = v_0 \end{cases} \tag{1.58}$$

The solution of Equation 1.58 can be written as

$$x(t) = e^{-\xi\omega_n t}\left[A\sin(\omega_d t) + B\cos(\omega_d t)\right] + x_0 \cos(\omega_f t + \phi) \tag{1.59}$$

Here, amplitude x_0 and angle ϕ are also defined, respectively, in Equations 1.42 and 1.43; additionally, in Equation 1.59,

$$A = \frac{v_0 + \xi\omega_n B + \omega_f x_0 \sin\phi}{\omega_d} \tag{1.60}$$

and

$$B = d_0 - x_0 \cos\phi \tag{1.61}$$

From Equation 1.59, the damping effect on the general harmonic response can be seen. First, the effect on the steady-state portion of the response is the same as discussed before, which is further explored in the next subsection.

Secondly, the transient part has the common term $e^{-\xi\omega_n t}$, which quickly reduces the displacement caused by the initial condition. The larger the damping ratio of the system, the quicker the decay occurs.

At the same time, this term will also affect the growth of the particular response until it reaches steady state. Similarly, the larger the damping ratio of the system, the quicker its growth will be.

Since the transient responses of $e^{-\xi\omega_n t}$ $[A\sin(\omega_d t) + B\cos(\omega_d t)]$ decay quickly, the steady-state solution prevails so that the velocity is 90° ahead of the displacement, as previously mentioned.

From the above discussion, the sine and cosine time variables of the displacement and velocity, respectively, mean that during the vibration, the velocity and the displacement have a 90° phase difference. That is, the velocity is 90° ahead of displacement. For a linear time-invariant system, both the damping coefficient c and spring coefficient k are constants. In Equation 1.5, the product of the damping coefficient and velocity is the damping force. Also, from Equation 1.6, the product of the spring coefficient and displacement is the restoring force. Therefore, the damping force and the restoring force also have a 90° phase difference.

Under earthquake or other random excitations, the response of the system is also random. However, for an SDOF system, the relationship described above can be approximately used. Figure 1.11a shows certain structural steady-state responses for short time durations. This structure has a damping ratio of 5% and is under sinusoidal excitation. Figure 1.11b shows the responses of the same structure under Northridge earthquake excitations. The solid lines are velocities and the broken lines are displacements. From these figures, it is seen that under random excitation, the response is no longer pure sinusoidal signals. Therefore, the concept of phase cannot be used to describe exactly the relationship between the velocity and the displacement. However, maximum displacement always happens when velocity reaches zero; and, in most cases, maximum velocity occurs when the displacement is close to zero. It is still seen that the velocities have a nearly 90° phase ahead of the displacements.

1.1.4 Ground Excitation

1.1.4.1 Governing Equation

Figure 1.1a shows a case that is subjected to an external force, and in the previous sections, the response when the external force is harmonic was discussed. However, in Figure 1.1b, the SDOF system is excited by ground accelerations, as opposed to an external force excitation. Now, only harmonic ground excitation is considered. In fact, this case relates to earthquake excitations.

Denote the ground acceleration as $\ddot{x}_g$, and the absolute acceleration of the mass shown in Figure 1.1b is written as

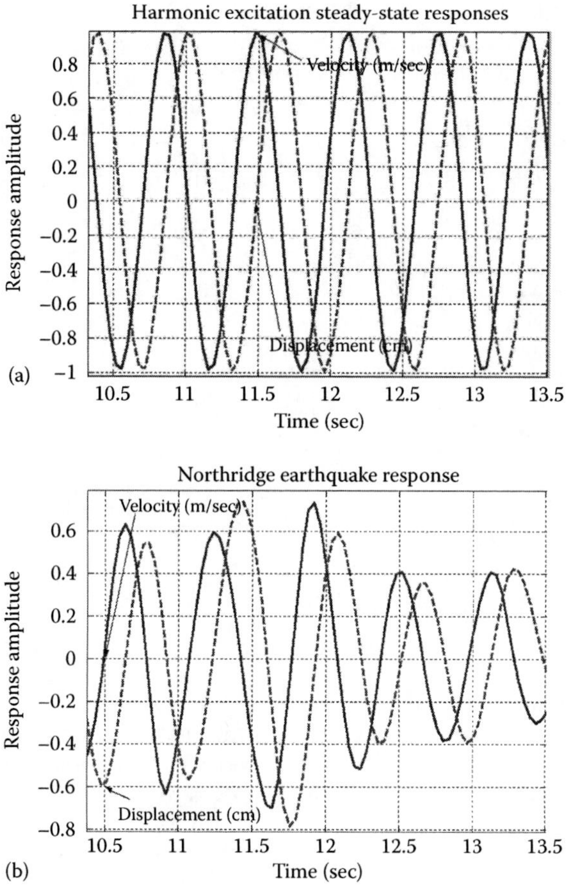

FIGURE 1.11 Structural responses (a) harmonic responses and (b) earthquake responses.

$$\ddot{x}_A(t) = \ddot{x}(t) + \ddot{x}_g(t) \tag{1.62}$$

where $\ddot{x}$ is the acceleration of the mass relative to the ground. Furthermore, let x and $\dot{x}$ denote the displacement and velocity of the mass relative to the ground. Then, the governing equation of motion for the mass can be written as

$$m\ddot{x}_A(t) + c\dot{x}(t) + kx(t) = 0 \tag{1.63}$$

Substituting Equation 1.62 into Equation 1.63 and rearranging the equation, results in

$$m\ddot{x}(t) + c\dot{x}(t) + kx(t) = -m\ddot{x}_g(t) \tag{1.64}$$

Compared to Equation 1.1, the similarity is realized, except in this case, the external force is specified as

$$f(t) = -m\ddot{x}_g(t) \tag{1.65}$$

Namely, the force f(t) is equivalent to the negative value of the product of mass and ground acceleration. In the previous section, when the external force in Equation 1.1 was assumed to be harmonic, the steady-state solution was obtained. Now, if the system is excited by the ground acceleration that is sinusoidal, similar responses can be found. Note that in earthquake engineering, the excitation forcing function is not sinusoidal. Rigorously speaking, the force is not "external" either. However, harmonic excitation can reveal the basic essence of the earthquake response. Consequently, this case is first examined to explore the nature of the seismically induced force by considering sinusoidal ground excitation, followed by using the concept of random vibration.

In Figure 1.1b, the excitation f is due to the ground motion $\ddot{x}_g$. This is caused by the inertial force of the mass. In fact, if the mass is separated from the ground, it is seen that no external force is applied to this system. Instead, there are three forces; the inertia force,

$$f_i = m\ddot{x}_A \tag{1.66}$$

the damping force,

$$f_d = c\dot{x} \tag{1.67}$$

and the spring or restoring force,

$$f_r = kx \tag{1.68}$$

In structural engineering, the spring or restoring force is often used as a design control parameter, which is a general term representing the element forces and stresses, such as axial forces, bending moments, and shears. This method is also adopted by earthquake engineers. In the latter case, the stiffness k is specified as lateral stiffness and the force is referred to as the *equivalent static force*, $f_{ST.}$ That is,

$$f_{ST} = f_r = kx \tag{1.69}$$

Note that as a design control parameter, the force f_{ST} in Equation 1.69 does not include the damping force. This is because for typical civil engineering structures, the damping force is quite small and can be ignored. That is,

$$f_{ST} = kx \approx -m\ddot{x}_A \tag{1.70}$$

In most building codes, f_{ST} is used as the *base shear.* When the damping force is very small, Equation 1.70 provides the formula to compute the base shear through the product $m\ddot{x}_A$.

However, if dampers are added, the damping force may no longer be negligibly small and the use of Equation 1.70 to determine the equivalent static force should be examined. For example, the "base shear" of the SDOF structure defined by Figure 1.1b will be a function of $m\ddot{x}_A$ as modified by the damping force (Mohraz and Sadek 2000). This issue is discussed in detail in the following chapters. In fact, the *total base shear,* denoted by f_{TBS}, by the reaction of the system is the product of the mass and the ground acceleration, as shown in Figure 1.12a for an SDOF structure and Figure 1.12b for a definition of total base shear, where $f_{ST} = (½kx + ½kx) < kx + c\dot{x} = -m\ddot{x}_A = f_{TBS}$.

1.1.4.2 Responses of Harmonic Ground Excitations

If the ground acceleration is sinusoidal, then

$$m\ddot{x}(t)+c\dot{x}(t)+kx(t) = -mx_{g''}\sin(\omega_f t) = m\omega_f^2 x_g \sin(\omega_f t) \tag{1.71}$$

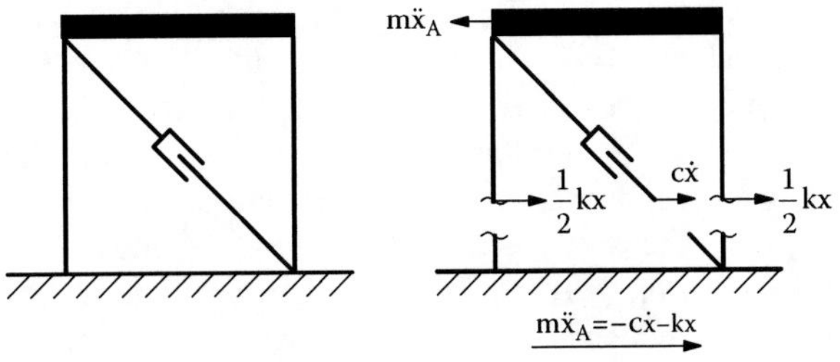

FIGURE 1.12 Total base shear.

where

$$x_{g''} = -\omega_f^2 x_g \tag{1.72}$$

is the amplitude of the ground acceleration and x_g is the amplitude of the ground displacement. Dividing both the left and the right side of Equation 1.72 by m, the monic form is as follows:

$$\ddot{x}(t) + 2\xi\omega_n\dot{x}(t) + \omega_n^2 x(t) = \omega_f^2 x_g \sin(\omega_f t) \tag{1.73}$$

Similarly, the steady-state responses as expressed in Equation 1.40 are

$$x_{ps} = x_0 \sin(\omega_f t + \phi) \tag{1.74}$$

$$\dot{x}_{ps} = \omega_f x_0 \cos(\omega_f t + \phi) \tag{1.75}$$

and

$$\ddot{x}_{ps} = -\omega_f^2 x_0 \sin(\omega_f t + \phi) \tag{1.76}$$

In this case, the amplitude of the relative displacement x(t), also denoted by x_0, has a slightly different expression from Equation 1.42, that is

$$x_0 = \frac{x_g \omega_f^2}{\sqrt{(\omega_n^2 - \omega_f^2)^2 + (2\xi\omega_n\omega_f)^2}} \tag{1.77}$$

On the other hand, the phase angles are identical to Equation 1.43 and repeated as such. In this case, the angle ϕ stands for the phase difference between the excitation ground motion acceleration and the response displacement, written as

$$\phi = \begin{cases} \tan^{-1}\left[\dfrac{2\xi\omega_n\omega}{\omega_f^2 - \omega_n^2}\right] + h_\phi \pi, & \omega_n \neq \omega_f \\ -\dfrac{\pi}{2}, & \omega_n = \omega_f \end{cases} \tag{1.78}$$

As mentioned before, the value of the Heaviside function, h_ϕ (either 1 or zero), depends on the sign of $\omega_n^2 - \omega_f^2$ (e.g., see Equation 1.44).

In this case, the solution of the equation written in pure sinusoidal form is shown in Equation 1.74. However, earthquake ground motions are random rather than sinusoidal. In Chapter 2, arbitrary excitations including the random input will be introduced. Here, the random excitation is briefly discussed by comparing the response with harmonic excitations.

In the case of random excitation, there is no deterministic phase ϕ, which is now rather random. Also, there is no deterministic frequency ω_f. However, when the damping of the structure is small, say, $\xi < 0.1$, the frequency of the response is very close to the natural frequency of the system, which is often referred to as the *narrow band system* (Clough and Penzien 1993).

Figure 1.12a shows the first 10 (s) of an earthquake response history of a narrow band system, whose natural frequency is 5 (Hz) with a damping ratio of 2%. It is seen that the amplitude vs. the time is somewhat random. However, a very clear frequency can also be seen. In Figure 1.12b, the power spectrum of the response is shown. It is seen that in the neighborhood of 5 (Hz), the density of the power spectrum is highly concentrated. Because the band of this concentrated area is rather narrow, the system is thus called a narrow band system.

From Figure 1.13a, the variation of the amplitude indicates the difficulty in finding the peak response. As a matter of fact, for a given vibration system, single or multiple degrees of freedom, linear or nonlinear, together with a given amplitude of the earthquake excitation, it is mathematically impossible to determine exactly the amplitude of the peak response. In other words, under

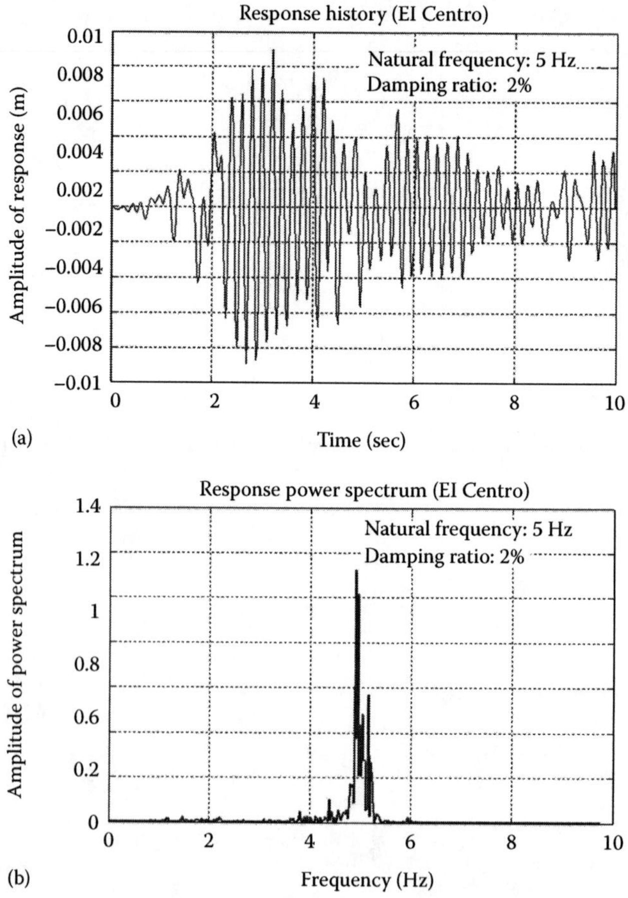

FIGURE 1.13 Response of narrow band system (a) time history and (b) response power spectrum.

random vibration, the fact that *bounded input does not yield a bound on the output* is one of the significant properties of seismic structural responses.

To distinguish the case of the forcing function as described in Equation 1.1 and the case expressed in Equation 1.71, the first case is referred to as a general external excitation and the second case as ground excitation.

Example 1.6

In the above discussion, relative displacement, velocity, and acceleration were used as temporal variables to generate the governing equations. However, the absolute acceleration also needs to be calculated.

According to Equation 1.62, the absolute acceleration is

$$\ddot{x}_A(t) = \ddot{x}(t) + \ddot{x}_g(t)$$

Suppose there is an SDOF system with $m = 2$, $c = 20$, and $k = 200$ as the basic parameters in nondimensional form. With harmonic excitation $\omega_f = 10$,

$$x_g(t) = 5\sin(10t)$$

Note that the undamped natural frequency is $\omega_n = (200/20)^{1/2} = 10$, so that the frequency ratio $r = 1$. Furthermore, the damping ratio is

$$\xi = \frac{c}{2\omega_n m} = \frac{20}{40} = 0.5$$

With zero initial conditions, the amplitude of the steady-state relative displacement is

$$x_0 = \frac{5\omega_f^2}{\sqrt{\left(\omega_n^2 - \omega_f^2\right)^2 + \left(2\xi\omega_n\omega_f\right)^2}} = \frac{5}{2 \times 0.5 \times 100} = 5$$

while the phase angle is

$$\phi = -\pi/2$$

Therefore, the relative acceleration is

$$\ddot{x}(t) = -\omega_f^2(5)\sin(10t - \pi/2) = 500\cos(10t)$$

Thus, the absolute acceleration is

$$\ddot{x}_A(t) = -500\sin(10t) + 500\cos(10t) = 707.1\cos(10t + 0.785)$$

The difference may be observed by comparing the relative and the absolute accelerations.

1.2 DYNAMIC MAGNIFICATION

In the above discussion, the steady-state responses of both general excitation and ground excitation were reviewed. It is helpful to further examine the amplitude of the responses and to compare the results with the static response.

1.2.1 Dynamic Magnification Factor, General Excitation

1.2.1.1 Dynamic Magnification Factor of Displacement

First, the case of general excitation, namely, the amplitude x_0 described in Equation 1.42 is examined. Dividing ω_n^2 simultaneously into both the numerator and denominator on the right side of Equation 1.42 results in

$$x_0 = \frac{f_0/(m\omega_n^2)}{\sqrt{\left(1-\frac{\omega_f^2}{\omega_n^2}\right)^2+\left(2\xi\frac{\omega_f}{\omega_n}\right)^2}} = \frac{f_0/k}{\sqrt{\left(1-\frac{\omega_f^2}{\omega_n^2}\right)^2+\left(2\xi\frac{\omega_f}{\omega_n}\right)^2}} \tag{1.79}$$

Denoting the term multiplying f_0/k on the right side of Equation 1.79 by β_d results in

$$\beta_d = \frac{1}{\sqrt{\left(1-\frac{\omega_f^2}{\omega_n^2}\right)^2+\left(2\xi\frac{\omega_f}{\omega_n}\right)^2}} = \frac{1}{\sqrt{(1-r^2)^2+(2\xi r)^2}} \tag{1.80}$$

Here, r is the frequency ratio with

$$r = \frac{\omega_f}{\omega_n} \tag{1.81}$$

Therefore, the amplitude x_0 can be rewritten as

$$x_0 = \frac{1}{\sqrt{\left(1-\frac{\omega_f^2}{\omega_n^2}\right)^2+\left(2\xi\frac{\omega_f}{\omega_n}\right)^2}}\frac{f_0}{k} = \frac{1}{\sqrt{(1-r^2)^2+(2\xi r)^2}}\frac{f_0}{k} = \beta_d\frac{f_0}{k} \tag{1.82}$$

In Equation 1.82, the term f_0/k can be seen as a static displacement, when a system has stiffness k and is subjected to a force with amplitude f_0. That is, the amplitude of the dynamic displacement x(t), namely, x_0, can be seen as the static displacement times a special factor β_d, which is now referred to as the *dynamic magnification factor for the displacement.*

From Equation 1.82, it is seen that the dynamic magnification factor is a function of: (1) the ratio of driving frequency to natural frequency $r = \omega_f/\omega_n$; and (2) the damping ratio ξ. That is,

$$\beta_d = \beta_d(\xi, r) \tag{1.83}$$

Furthermore, Equation 1.82 can be written as

$$x_0 = \frac{f_0}{k/\beta_d} = \frac{f_0}{k_D} \tag{1.84}$$

where k_D can be referred to as the *apparent stiffness* and

$$k_D = \frac{k}{\beta_d} = \frac{f_0}{x_0} \tag{1.85}$$

In Equation 1.85, it is seen that k_D is the amplitude of the force per unit dynamic response of displacement. It is also the static stiffness modified by dividing the dynamic magnification factor. Generally, it can also be called the *dynamic stiffness.*

This dynamic stiffness is a function of: (1) ratio r, the ratio of driving frequency to natural frequency ω_f/ω_n; (2) the damping ratio ξ; and (3) the static stiffness k. That is,

$$k_D = k_D(\xi, r, k) \tag{1.86}$$

Figure 1.14 shows a set of curves of the dynamic magnification factors with different damping ratios.

Figure 1.14a shows that the dynamic magnification factors β_d can be either larger or smaller than unity. When the frequency ratio is close to 1, the underdamped system will always have the value of β_d greater than unity. This phenomenon is defined as resonance. The smaller the damping, the larger the value of β_d found when resonance occurs. Pure resonant behavior occurs for an ideal system with zero damping; in which case, the response becomes limitless. When the damping ratio becomes larger than 0.707, the value of β_d will never be larger than unity. Furthermore, when the damping ratio is greater than 1, the system is overdamped.

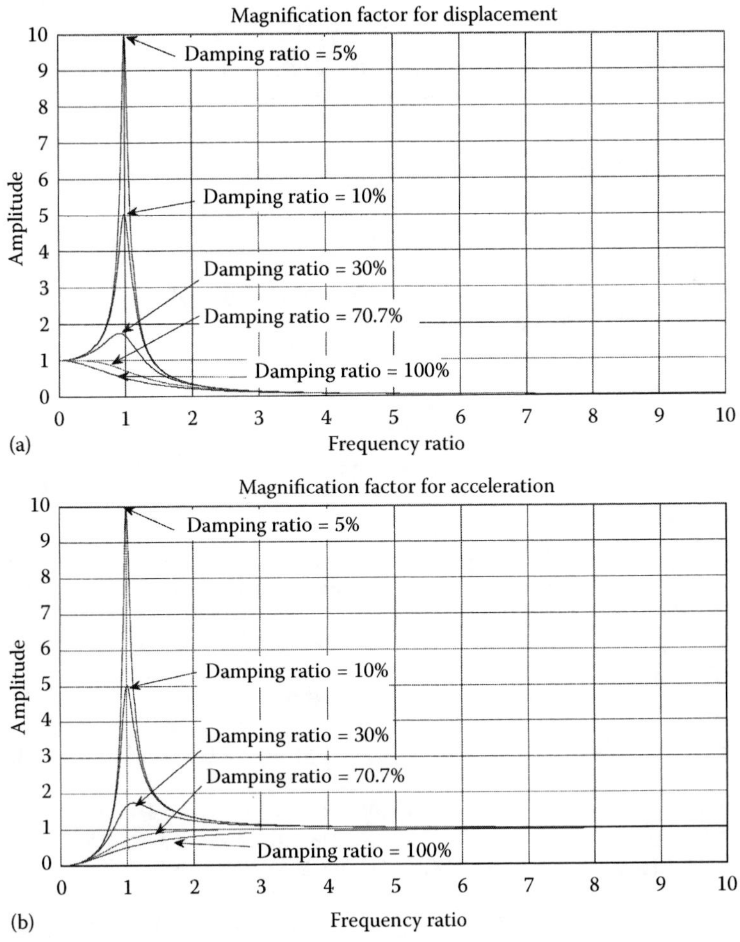

FIGURE 1.14 Dynamic magnification factor (a) displacement and (b) acceleration.

1.2.1.2 Dynamic Magnification Factor of Acceleration

In many cases, the acceleration amplitude, denoted by a_0, needs to be calculated in addition to the amplitude of the displacement. The quantity a_0 is a time-invariant constant. In the case of a sinusoidally excited steady-state response, from Equation 1.40, $a_0 = \omega_f^2 x_0$.

Therefore, in Equation 1.41, the amplitude of the acceleration can be written as

$$a_0 = \frac{f_0}{k}\left[\frac{\omega_f^2}{\sqrt{\left(1-\frac{\omega_f^2}{\omega_n^2}\right)^2+\left(2\xi\frac{\omega_f}{\omega_n}\right)^2}}\right] = \frac{f_0}{m}\left[\frac{\frac{\omega_f^2}{\omega_n^2}}{\sqrt{\left(1-\frac{\omega_f^2}{\omega_n^2}\right)^2+\left(2\xi\frac{\omega_f}{\omega_n}\right)^2}}\right] \tag{1.87}$$

The terms in the bracket on the right side of Equation 1.87 can be denoted by β_a, as

$$\beta_a = \frac{\frac{\omega_f^2}{\omega_n^2}}{\sqrt{\left(1-\frac{\omega_f^2}{\omega_n^2}\right)^2+\left(2\xi\frac{\omega_f}{\omega_n}\right)^2}} = \frac{r^2}{\sqrt{\left(1-r^2\right)^2+\left(2\xi r\right)^2}} \tag{1.88}$$

Here β_a is referred to as the *dynamic magnification factor for the acceleration*, which is shown in Figure 1.14b. By using β_a, the amplitude of the acceleration can be written as

$$a_0 = \beta_a \frac{f_0}{m} = \frac{f_0}{m_D} \tag{1.89}$$

From Equation 1.89, the dynamic response of the acceleration can be seen as the product of the dynamic magnification factor and the term f_0/m. This latter term can be seen as an acceleration of mass m resulting from a force f_0.

Furthermore, in Equation 1.89, the term m_D can be referred to as the *apparent mass*, where

$$m_D = \frac{m}{\beta_a} = \frac{f_0}{a_0} \tag{1.90}$$

From Equation 1.85, m_D is the amplitude of the force per unit dynamic response of acceleration. It is also the mass modified by dividing by the dynamic magnification factor. Generally, it can also be referred to as a special type of dynamic stiffness.

Example 1.7

A machine can be modeled as an SDOF system with m = 500 (kg), c = 707.1 (N-s/m), and k = 2,500,000 (N/m).

Suppose this machine is subjected to a harmonic excitation with an amplitude of 4000 (N) and a driving frequency equal to 10 (Hz). The maximum allowed displacement and acceleration of the machine are, respectively, 1 (cm) and 2 (g) (1g = 9.8 m/s^2). Check if these requirements are satisfied.

It can be calculated that $\beta_d = 4.7353$ and $\beta_a = 3.7388$. Therefore, the displacement is 0.0076 m = 7.6 (mm); and the acceleration is 29.91 (m/s^2) = 3.05 (g). Consequently, the displacement is acceptable, but the acceleration exceeds the allowed value.

1.2.1.3 Peak Values of Dynamic Magnification Factors

The next step is to find the frequency ratio when the maximum value of the dynamic magnification factor of displacement β_d is reached, that is,

$$\frac{1}{\sqrt{\left(1-r^2\right)^2+\left(2\xi r\right)^2}}=\left(\beta_d\right)_{max} \tag{1.91}$$

Let

$$\frac{d}{dr}\left[\frac{1}{\sqrt{\left(1-r^2\right)^2+\left(2\xi r\right)^2}}\right]=0 \tag{1.92}$$

To avoid a complicated mathematical derivation, find the minimum value inside the square root of the denominator; that is, let

$$\left(1-r^2\right)^2+\left(2\xi r\right)^2=\min \tag{1.93}$$

Namely, let the derivative of the term on the left side of Equation 1.93 with respect to variable r equal zero. Thus,

$$\frac{d}{dr}\left[\left(1-r^2\right)^2+\left(2\xi r\right)^2\right]=0 \tag{1.94}$$

Therefore, for the underdamped case,

$$r=\sqrt{1-2\xi^2} \tag{1.95}$$

or

$$\omega_f=\sqrt{1-2\xi^2}\,\omega_n \tag{1.96}$$

the peak value (the resonant point) is reached.

Equations 1.95 and 1.96 imply that the maximum value of the dynamic magnification factor is not at exactly the frequency where $\omega_f = \omega_n$. Usually, the frequency at the peak resonance is used as the natural frequency. However, when the damping ratio becomes large, the resonant frequency should be modified by Equation 1.96. Therefore, a special symbol, ω_D, is used to denote the peak resonance point as

$$\omega_D=\sqrt{1-2\xi^2}\,\omega_n \tag{1.97}$$

where ω_D is the *displacement resonant frequency*. From Equation 1.95, it is seen that only if

$$1-2\xi^2>0$$

or

$$\xi<\sqrt{2}/2=0.707 \tag{1.98}$$

then $r > 0$.

Thus, only if Equation 1.98 holds, can the system have a peak or resonant value. Otherwise, the magnification factor will be always smaller than unity, except at $r = 0$.

To determine the resonant value, substituting Equation 1.95 into Equation 1.80 provides

$$\beta_d(\omega_D) = \frac{1}{2\xi\sqrt{1-\xi^2}} \tag{1.99}$$

That is, when $r = \sqrt{1-2\xi^2}$, from Equation 1.82,

$$x_0 = \frac{f_0}{k}\frac{1}{2\xi\sqrt{1-\xi^2}} \tag{1.100}$$

Also, from Equation 1.82, it is seen that when $r = 1$, the displacement can still be very large, although it has passed the peak value. That is,

$$x_0 = \frac{f_0}{k}\frac{1}{2\xi} \tag{1.101}$$

Note that in the above development, the damping ratio is always smaller than 1. Comparing Equation 1.100 with Equation 1.101, it is seen that the peak value is $1/\sqrt{1-\xi^2}$ times the value when $r = 1$. Therefore, x_0 defined in Equation 1.100 is always larger than the one defined in Equation 1.101.

Similarly, the *acceleration resonant frequency*, ω_A, is defined by making

$$\frac{d}{dr}\left[\frac{r^2}{\sqrt{(1-r^2)^2+(2\xi r)^2}}\right] = 0 \tag{1.102}$$

and for the underdamped case,

$$r = \frac{1}{\sqrt{1-2\xi^2}} \tag{1.103}$$

or

$$\omega_A = \frac{1}{\sqrt{1-2\xi^2}}\omega_n \approx \sqrt{1+2\xi^2}\,\omega_n \tag{1.104}$$

the resonant point is reached.

Substituting Equation 1.103 into Equation 1.88 yields

$$\beta_a(\omega_A) = \frac{1}{2\xi\sqrt{1-\xi^2}} \tag{1.105}$$

That is, when $r = (1/\sqrt{1-2\xi^2})$, from Equations 1.87 and 1.88,

$$a_0 = \frac{f_0}{m}\frac{1}{2\xi\sqrt{1-\xi^2}} \tag{1.106}$$

Also, from Equation 1.88, it is seen that when r = 1, the acceleration becomes very large, although it has not yet reached the peak value. That is,

$$a_0 = \frac{f_0}{m}\frac{1}{2\xi} \tag{1.107}$$

Again, since the damping ratio is always <1, from Equations 1.106 and 1.107, the peak acceleration value is $1/\sqrt{1-\xi^2}$ times the value when r = 1 and the value obtained from Equation 1.106 is always the larger of the two.

Comparing Equation 1.106 with Equation 1.100, and also comparing Equation 1.107 with Equation 1.101, the similarities between the acceleration resonance and the displacement resonance are found.

Similar to the case of displacement magnification, the condition described by Equation 1.98 must hold to have a nonzero real-valued frequency ratio. That is, if the damping ratio is greater than or equal to 0.707, the acceleration resonance will not be reached.

From Equations 1.104 and 1.98, it is seen that when the damping ratio becomes larger, errors will occur if $\omega_f = \omega_n$ is used to estimate the resonant frequency for acceleration and displacement. For example, if the damping ratio is larger than 30%, there will be more than a 10% difference between the ω_A (or ω_D) and ω_n frequencies. For the sake of simplicity, ω_n is often used in this book unless otherwise specified. However, note that under earthquake excitations, even when it is different from sinusoidal forcing functions, the peak values of the acceleration and of the displacement are influenced by the facts described in Equations 1.106 and 1.100.

Note that the velocity resonance frequency ω_V is not affected by damping ratios, that is,

$$\omega_V = \omega_n \tag{1.108}$$

1.2.1.4 Dynamic Magnification Factors Reaching Unity

1.2.1.4.1 Dynamic Magnification Factor of Displacement

When the frequency ratio r = 0, the dynamic magnification factor of displacement, β_d, equals unity. Now, it is time to find a second frequency point at which β_d returns to unity again. Solving

$$\frac{1}{\sqrt{\left(1-r^2\right)^2+\left(2\xi r\right)^2}} = 1 \tag{1.109}$$

results in obtaining the frequency point as

$$r = \sqrt{2}\sqrt{1-2\xi^2} \tag{1.110}$$

Therefore, when

$$\omega_f = \sqrt{2}\sqrt{1-2\xi^2}\,\omega_n \tag{1.111}$$

the dynamic magnification factor of displacement reaches unity. In other words, when the damping ratio is smaller than 0.707, and the driving frequency is smaller than $\sqrt{2}\sqrt{1-2\xi^2}\,\omega_n$, β_d will always be greater than unity, except at $\omega_f = 0$. On the other hand, if the driving frequency is larger than $\sqrt{2}\sqrt{1-2\xi^2}\,\omega_n$, β_d will always be smaller than unity. This behavior can be observed in Figure 1.14a.

1.2.1.4.2 Dynamic Magnification Factor of Acceleration

When the frequency ratio $r = 0$, the dynamic magnification factor of acceleration, β_a, equals zero. When r increases toward infinity, β_a will be unity. To find another frequency point at which β_a approaches one, the following equation may be solved:

$$\frac{r^2}{\sqrt{\left(1-r^2\right)^2+\left(2\xi r\right)^2}} = 1 \tag{1.112}$$

which gives the frequency point

$$r = \frac{1}{\sqrt{2}\sqrt{1-2\xi^2}} \tag{1.113}$$

Therefore, when

$$\omega_f = \frac{1}{\sqrt{2}\sqrt{1-2\xi^2}}\,\omega_n \tag{1.114}$$

the dynamic magnification factor of acceleration reaches unity. That is, when the damping ratio is smaller than 0.707, and if the driving frequency is smaller than $1/(\sqrt{2}\sqrt{1-2\xi^2})\omega_n$, β_a will always be smaller than unity. Alternatively, if the driving frequency is larger than $1/(\sqrt{2}\sqrt{1-2\xi^2})\omega_n$, β_a will always be greater than unity. Several examples of this behavior are illustrated in Figure 1.14b.

1.2.1.5 Half-Power Points and Resonant Region

1.2.1.5.1 Half-Power Points of Dynamic Magnification Factor of Displacement

Of special interest is when the amplitude of the dynamic magnification factor reaches the value of $\sqrt{2}/2$ times its peak value. That is, when

$$\frac{1}{\sqrt{\left(1-r^2\right)^2+\left(2\xi r\right)^2}} = \frac{\sqrt{2}}{2}\frac{1}{2\xi\sqrt{1-\xi^2}} \tag{1.115}$$

Solving Equation 1.115 to find the value of the frequency ratio r,

$$r_{1,2}^2 = 1-2\xi^2 \mp 2\xi\sqrt{1-\xi^2} \tag{1.116}$$

Since the damping ratio is often very small, the approximate result is

$$r_{1,2} \approx 1-\xi^2 \mp \xi\sqrt{1-\xi^2} \tag{1.117}$$

Here,

$$r_1 = \frac{\omega_1}{\omega_n} \tag{1.118}$$

and

$$r_2 = \frac{\omega_2}{\omega_n} \tag{1.119}$$

both denoting the special frequency point when the amplitude of the magnification factor reaches $\sqrt{2}/2$ times its peak value, which is referred to as the *half-power points*. Then, ω_1 and ω_2 are the corresponding driving frequencies defined as the half-power point frequencies.

From Equations 1.118 and 1.119,

$$r_2 - r_1 = 2\xi\sqrt{1-\xi^2} \approx 2\xi \tag{1.120}$$

and

$$r_2 + r_1 = 2\left(1-\xi^2\right) \approx 2 \tag{1.121}$$

Therefore,

$$\xi \approx \frac{r_2 - r_1}{r_2 + r_1} = \frac{\omega_2 - \omega_1}{\omega_2 + \omega_1} \approx \frac{\omega_2 - \omega_1}{2\omega_n} \tag{1.122}$$

Equation 1.122 indicates that, if the half-power frequencies ω_1 and ω_2 and the natural frequency ω_n can be measured, the damping ratio can be approximately determined. In fact, since at the half-power points and at the resonance point the amplitude of the response is large, the measurements of signal-to-noise ratios are rather high. Therefore, when the damping ratio is sufficiently small, say, smaller than 30%, Equation 1.122 can provide a good estimation of the damping ratio.

Note that the approximation for obtaining Equation 1.122 is based on small damping ratios. When the damping ratio is sufficiently large, the first half-power point will no longer exist. In addition, large errors will result when estimating the damping ratio.

From Equation 1.116, when

$$1 - 2\xi^2 - 2\xi\sqrt{1-\xi^2} = 0 \tag{1.123}$$

then

$$\xi = \frac{\sqrt{2-\sqrt{2}}}{2} = 0.3827 \tag{1.124}$$

where the first half-power point vanishes.

Figure 1.15 is a plot of the percentage error of the damping ratio estimation vs. the true damping ratio. It is seen that when the damping ratio is smaller than 18.6%, the estimation error is smaller than 10%.

1.2.1.5.2 Half-Power Points of Dynamic Magnification Factor of Acceleration

Similarly, the half-power points of the dynamic magnification factor of the acceleration can be used to estimate the damping ratio points. Let

$$\frac{r^2}{\sqrt{\left(1-r^2\right)^2 + \left(2\xi r\right)^2}} = \frac{\sqrt{2}}{2}\frac{1}{2\xi\sqrt{1-\xi^2}} \tag{1.125}$$

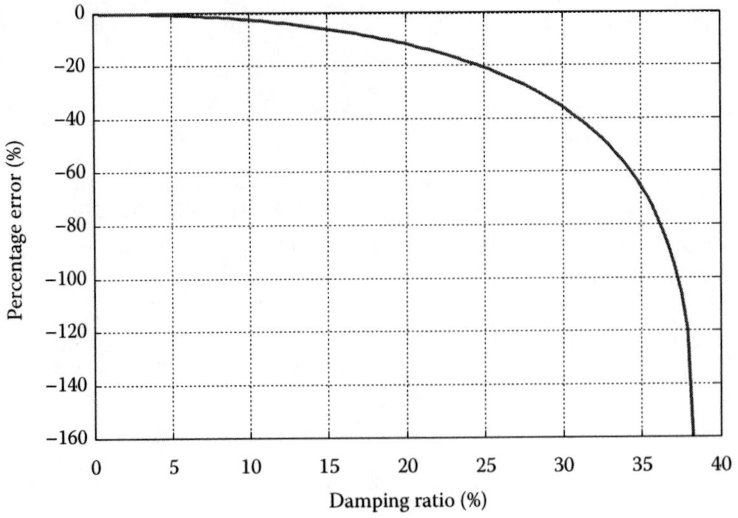

FIGURE 1.15 Percentage error of damping ratio estimation.

Then, the frequency ratio r is given by

$$r_{1,2}^2 = \frac{1-2\xi^2 \mp 2\xi\sqrt{1-\xi^2}}{8\xi^4 - 8\xi^2 + 1} \tag{1.126}$$

Again, since the damping ratio is often very small, it can be approximated as

$$r_{1,2} \approx 1-\xi^2 \mp \xi\sqrt{1-\xi^2} \tag{1.127}$$

Comparing Equation 1.127 with Equation 1.117, it is seen that it is the same formula as described in Equation 1.117 in the case of acceleration magnification. Note that to derive Equation 1.127 from Equation 1.126, it is assumed that

$$8\xi^4 - 8\xi^2 + 1 \approx 1 \tag{1.128}$$

which is not true with larger damping ratios.

1.2.1.6 Response Reduction due to Increase of Damping

By examining both the dynamic magnification factors of the displacement and the acceleration, it is observed that when the damping ratio is increased, the vibration response can be reduced. In the following discussion, the effectiveness of response reduction by increasing the damping ratio in different frequency regions is considered.

First, consider the neighborhood of the resonant frequency. From Equations 1.101 and 1.107, it is seen that when $\omega_f = \omega_n$, the amplitudes of the displacement and the acceleration are inversely proportional to the damping ratio. That is, when the damping is doubled, the amplitude will be one-half of the original value. Such a phenomenon is given in the second column of Table 1.1. Because the relationship between the displacement amplitude and the damping ratio when $r \neq 1$ is not as explicit, as described in Equations 1.101 and 1.107, the amplitude reductions when the damping ratio is doubled at several different frequency ratios are provided.

TABLE 1.1
Percentage Displacement Reduction ($1 - x_0(2\xi_0)/x_0(\xi_0)$) vs. Frequency Ratios

ξ_0 (%)	r=1	$r = \sqrt{1-2\xi^2}$	r_1	r_2	$r_{0.1}$ and r_{10}
1	0.5	0.5000	0.36	0.37	6×10^{-6}
5	0.5	0.4995	0.35	0.38	1.5×10^{-4}
10	0.5	0.4981	0.33	0.39	6×10^{-4}
30	0.5	0.4804	0.26	0.44	0.005

In Table 1.1, the first column is the original damping ratio. When the damping ratio is doubled, it becomes 2%, 10%, 20%, and 60%, respectively. The third column is the percentage displacement reduction ($1 - x_0(2\xi_0)/x_0(\xi_0)$) at the displacement resonance point. It is seen that the reduction ratio is slightly smaller than 0.5, which is not as high as the point when r = 1.

The fourth column is the percentage displacement reduction ($1 - x_0(2\xi_0)/x_0(\xi_0)$) at the first half-power point. The half-power point is a point at a specific frequency ratio, where the amplitude of the displacement is 0.707 times the peak displacement. The fifth column is the percentage displacement reduction ($1 - x_0(2\xi_0)/x_0(\xi_0)$) at the second half-power point.

It can be proven that the highest displacement reduction ratio occurs at the resonance point r = 1. However, within the region between the two half-power points, the displacement reductions are rather large. Generally, this frequency range is referred to as the *resonant region.*

In the last column of Table 1.1, the percentage displacement reduction when r = 0.1 or r = 10 is listed, where these two points are normally outside the range between the two half-power points. It is seen that these reductions are rather minimal.

In Figure 1.16, several curves are plotted to compare the response reduction with the increased damping ratio at different frequency points. These curves, denoted by $R(\xi)$, are percentage reductions, namely, the values of the Y-axis are

$$R(\xi) = \frac{100\left[\beta_d(\xi, r) - \beta_d(0.01, r)\right]}{\beta_d(0.01, r)} \tag{1.129}$$

and the X-axis is 100ξ. In Figure 1.16, the frequency ratios are chosen to be 1.5, 1.2, 1.1, and 1.05. The corresponding curves are plotted and marked with these frequency ratios. It can be proven that the reduction curves of r = 1.5 and r = 1/1.5 are identical, and so on.

In the resonant region, the percentage reductions vs. the damping ratio are close, since the original damping is very same. In addition, outside the resonant region, increasing the damping ratio becomes less effective. Farther away from the resonant region, the effectiveness of increasing damping is further reduced.

From Table 1.1 and Figure 1.16, it is realized that using large damping to reduce the displacement can be effective in the resonant range. However, the effectiveness is reduced outside this resonance frequency range.

Example 1.8

From Example 1.7, it was found that the acceleration level of the SDOF system exceeds the required value. If the mass and the stiffness are constant, the proper damping can be chosen to meet the requirement using the dynamic magnification factors.

Suppose the response level of the newly assigned damping ratio should be R_D times that of the original damping ratio, while the frequency ratio must remain the same.

Denote the original damping ratio and the newly assigned damping ratio as ξ_0 and ξ_{design}, respectively. Then,

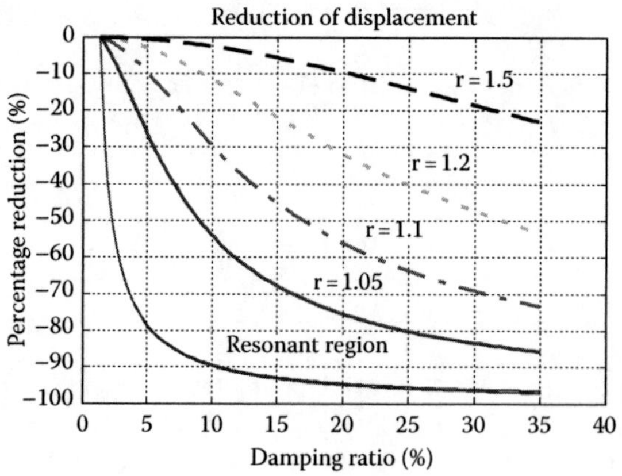

FIGURE 1.16 Reduction of dynamic magnification factor.

$$\frac{\beta_d(\xi_{design}, r)}{\beta_d(\xi_0, r)} = R_D$$

for the displacement reduction and

$$\frac{\beta_a(\xi_{design}, r)}{\beta_a(\xi_0, r)} = R_A$$

for the acceleration reduction.

Notice that both of the above equations yield the same expression

$$\frac{\sqrt{(1-r^2)^2 + (2\xi_0 r)^2}}{\sqrt{(1-r^2)^2 + (2\xi_{design} r)^2}} = R_D = R_A = R$$

Therefore,

$$\xi_{design} = \frac{\sqrt{(1-r^2)^2(1-R^2) + (2\xi_0 r)^2}}{2Rr}$$

In Example 1.7, the result is $r = 0.8886$ and $\xi_0 = 0.01$, with the required ratio $R = 2/3.05$. Thus,

$$\xi_{design} = 0.1372$$

Therefore, if the damping ratio is increased to 13.72%, then the acceleration level will be 2g. In addition, the displacement is further reduced to 4.97 (mm).

1.2.2 Dynamic Magnification Factor, Ground Excitation

In earthquake engineering, the relative displacement vs. the ground displacement is a useful relationship, which can provide information on structural displacement. In addition, it can also be used to study the relative displacement of structure bearings when seismic isolation is used.

Another useful relationship is the ground acceleration vs. the absolute acceleration of the mass. In earthquake engineering, the ground acceleration is often used as the input. The absolute acceleration is used to further obtain the force acting on a given structure. In the case of seismic isolation, the absolute accelerations of the superstructure are studied to try to reduce the forces.

1.2.2.1 Dynamic Magnification Factor of Relative Displacement

The special case of ground harmonic excitation is considered next. Dividing both the numerator and denominator in Equation 1.77 by ω_n^2,

$$x_0 = \frac{x_g \dfrac{\omega_f^2}{\omega_n^2}}{\sqrt{\left(1 - \dfrac{\omega_f^2}{\omega_n^2}\right)^2 + \left(2\xi \dfrac{\omega_f}{\omega_n}\right)^2}} = \frac{r^2}{\sqrt{(1-r^2)^2 + (2\xi r)^2}} x_g = \beta_{dB} x_g \tag{1.130}$$

in which

$$\beta_{dB} = \frac{r^2}{\sqrt{(1-r^2)^2 + (2\xi r)^2}} \tag{1.131}$$

The term β_{dB} can be referred to as the dynamic magnification factor of the relative displacement of the system subjected to ground excitation with amplitude x_g. That is, the relative displacement can be seen as a product of the dynamic magnification factor and the ground amplitude. In Figure 1.17a, the values of the dynamic magnification factor are plotted vs. the frequency ratio for several damping ratios. It is seen that when the frequency ratio is close to zero (i.e., the driving frequency is very small), the dynamic magnification factor is close to zero. However, when the driving frequency is close to the natural frequency, the dynamic magnification factor can be very large, if the damping is small. Similar to the case of general excitation, resonant regions are also present. When the driving frequency becomes very large, the values of the dynamic magnification factor will converge to 1, despite having different damping ratios.

In Figure 1.17b, the phase angles of the displacement are plotted vs. the frequency ratio for several damping ratios. From this figure, it is seen that before the resonant region, the phase difference of the response and the ground displacement is smaller than −90° (−π/2). When damping is small, the phase angle is close to 0° (0). After the resonant region, the phase differences are greater than −90° (−π/2). The smaller the damping is, the closer the phase angle will approach −180° (−π).

Comparison with the case of general excitation shows that the dynamic magnification factor of the relative displacement of the system subjected to ground excitation is, in terms of formulation, equal to the dynamic magnification factor of the absolute acceleration of a system subjected to general excitation. That is,

$$\beta_{dB} = \beta_a \tag{1.132}$$

Therefore, the analysis of β_a is also applicable to β_{dB}. For example, when

$$r = \frac{1}{\sqrt{1 - 2\xi^2}} \tag{1.133}$$

or the acceleration resonant frequency of ground excitation, ω_A, denote that

$$\omega_A = \frac{1}{\sqrt{1 - 2\xi^2}} \omega_n \tag{1.134a}$$

the resonant point is reached.

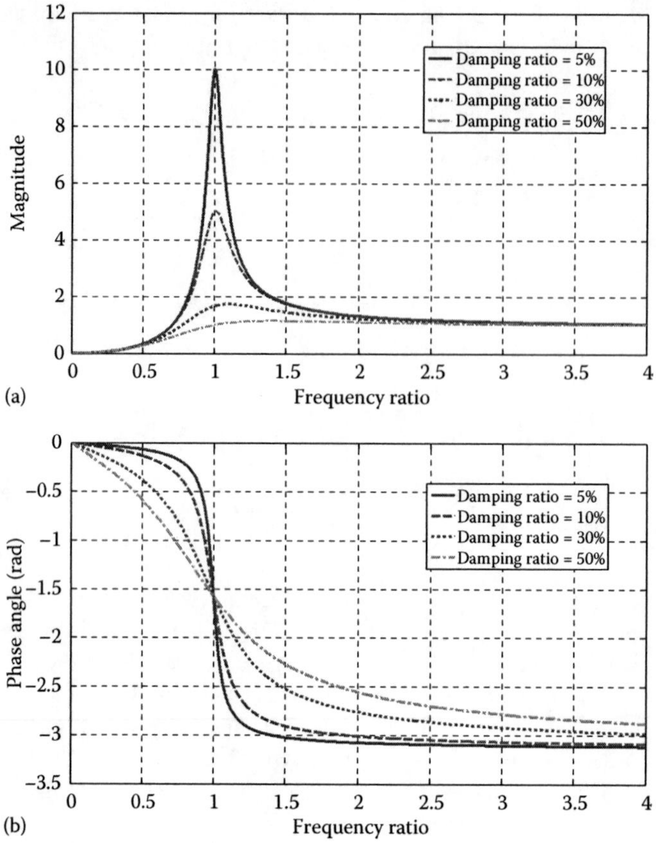

FIGURE 1.17 Frequency responses due to harmonic ground excitations (a) dynamic magnification factor and (b) phase angle.

Note that when the damping ratio is smaller than 40%,

$$\omega_A \approx \sqrt{1+2\xi^2}\,\omega_n \tag{1.134b}$$

with an error <5.55%.

Using Equation 1.34a, the peak value of the magnification factor is calculated as

$$\beta_{dB}(\omega_A) = \frac{1}{2\xi\sqrt{1-\xi^2}} \tag{1.135}$$

That is, when Equation 1.133 holds,

$$x_0 = \frac{1}{2\xi\sqrt{1-\xi^2}}\,x_g \tag{1.136}$$

Also, when

$$r = 1 \tag{1.137}$$

the displacement can still be very large, although it has passed the peak value. That is,

$$x_0 = \frac{1}{2\xi} x_g \tag{1.138}$$

Again, since the damping ratio is normally greater than zero, from Equations 1.136 and 1.138, it is seen that the peak acceleration value is $1/\sqrt{1-\xi^2}$ times the value when r = 1, That is, the value described in Equation 1.136 is always larger than the evaluation from Equation 1.138.

Secondly, when

$$r = \frac{1}{\sqrt{2}\sqrt{1-2\xi^2}} \tag{1.139}$$

or

$$\omega_f = \frac{1}{\sqrt{2}\sqrt{1-2\xi^2}} \omega_n \tag{1.140}$$

the dynamic magnification factor of the relative displacement reaches unity. That is, when the damping ratio is smaller than 0.707, if the driving frequency is smaller than $1/(\sqrt{2}\sqrt{1-2\xi^2})\omega_n$, then β_{dB} will always be smaller than unity. Otherwise, if the driving frequency is larger than $1/(\sqrt{2}\sqrt{1-2\xi^2})\omega_n$, then β_{dB} will always be greater than unity.

Example 1.9

Suppose a set of equipment in a special room of a building requires a very low level of acceleration, so that the entire floor needs to be isolated. The design frequency is 3 (Hz) and the total floor mass is 10,000 (kg). The design damping ratio is 7%. This floor is subjected to a harmonic excitation of 1 (cm) displacement and 10 (Hz) driving frequency. What is the relative displacement of the isolators? If the allowed displacement must be <1.05 (cm), how should the natural frequency of the isolation system be chosen, while keeping the damping ratio unchanged?

According to the given parameters, the frequency ratio is 10/3 and the damping ratio is 0.07. From Equation 1.121, the dynamic magnification factor is

$$\beta_{dB} = \frac{r^2}{\sqrt{(1-r^2)^2 + (2\xi r)^2}} = 1.10$$

Therefore, the relative displacement is 1 × (1.10) = 1.10 (cm), which is larger than the allowed displacement of 1.05 (cm).

Generally speaking, if the displacement needs to be R times the original level, while the damping ratio remains unchanged, the original frequency ratio is denoted as r_0 and the frequency ratio is to be designated as r_{design}. Then,

$$\frac{\beta_{dB}(r_{design})}{\beta_{dB}(r_0)} = R$$

Therefore,

$$\frac{r_{design}^4\left[(1-r_0^2)^2 + (2\xi r_0)^2\right]}{r_0^4\left[(1-r_{design}^2)^2 + (2\xi r_{design})^2\right]} = R^2$$

Solving this equation,

$$r_{design} = \sqrt{\frac{b + \sqrt{b^2 - a}}{a}}$$

where

$$a = 1 - \frac{\left(1 - r_0^2\right)^2 + \left(2\xi r_0\right)^2}{r_0^4 R^2}$$

and

$$b = 1 - 2\xi^2$$

In the above example, $r_0 = 10/3$, $R = 1.05/1.1$, so that $a = 0.089$ and $b = 0.990$. Therefore,

$$r_{design} = 4.656$$

Comparing with 10/3 = 3.333, the newly selected frequency needs to have a comparatively lower natural frequency of 2.15 (Hz). Note that this implies a significant shift in natural frequency to achieve a small reduction in relative displacement. This occurs because the original design was already far from the resonant peak.

1.2.2.2 Dynamic Magnification Factor of Absolute Acceleration

Now, the relationship between the ground excitation and the absolute acceleration of mass m is considered. This relationship is a significant quantity in seismic isolation, and a more detailed derivation of the corresponding dynamic magnification is given.

1.2.2.2.1 Definition of Dynamic Magnification Factor of Absolute Acceleration

Using the method of complex response for the steady-state response (see Equations 1.45 and 1.46), results in

$$m\ddot{x} + c\dot{x} + kx = m\omega_f^2 x_g e^{j\omega_f t} \tag{1.141}$$

Note that in Equation 1.141, x, $\dot{x}$, and $\ddot{x}$ are relative displacement, velocity, and acceleration, respectively. Denote

$$x = x_{p0} e^{j\omega_f t} \tag{1.142}$$

where x_{p0} is the complex-valued amplitude of the relative displacement. Also,

$$\ddot{x} = a_{p0} e^{i\omega_f t} = -x_{p0}\omega_f^2 e^{j\omega_f t} \tag{1.143}$$

Here, a_{p0} is the complex-valued amplitude of the relative acceleration, where

$$a_{p0} = -\omega_f^2 x_{p0} \tag{1.144}$$

Also, the amplitude of the ground acceleration can be written as

$$a_g = \omega_f^2 x_g \tag{1.145}$$

Note that in Equations 1.144 and 1.145, a_{p0} and x_{p0} are complex valued, whereas a_g and x_g are real valued.

From Equation 1.130, the relationship between the real-valued amplitude of the relative displacement and the ground displacement is given by

$$x_0 = \beta_{dB} x_g \tag{1.146}$$

Now, for the complex-valued base excitations,

$$x_{p0} = \beta_{dB}^C x_g \tag{1.147}$$

Here β_{dB}^C is a complex-valued term, instead of the real-valued dynamic magnification factor, β_{dB}, expressed in the previous discussion; superscript C denotes that it is a complex-valued term. The result is

$$\beta_{dB}^C = \frac{x_{p0}}{x_g} = \frac{r^2}{1 - r^2 + 2j\xi r} \tag{1.148}$$

Consider the relationship between the relative acceleration and the ground acceleration using Equations 1.145 and 1.146:

$$\frac{a_{p0}}{x_{g''}} = \frac{\omega_f^2 x_{p0}}{\omega_f^2 x_g} = \beta_{dB}^C \tag{1.149}$$

Therefore, the dynamic magnification factor of the real-valued relative acceleration, denoted by β_{aB}, can be written as

$$\beta_{aB} = \left|\frac{a_{p0}}{x_{g''}}\right| = \frac{|x_{p0}|}{x_g} = \left|\beta_{dB}^C\right| = \frac{r^2}{\sqrt{\left(1 - r^2\right)^2 + \left(2\xi r\right)}} = \beta_{dB} \tag{1.150}$$

Furthermore, the complex-valued amplitude of the absolute acceleration can be written as $a_{p0} + x_{g''}$. Therefore,

$$\frac{a_{p0} + x_{g''}}{x_{g''}} = \frac{a_{p0}}{x_{g''}} + 1 = \frac{r^2}{1 - r^2 + 2j\xi r} + 1 = \frac{1 + 2j\xi r}{1 - r^2 + 2j\xi r} \tag{1.151}$$

The real-valued amplitude of the absolute acceleration of mass m, denoted by $x_{A''}$, can be obtained by taking the absolute value of $(a_{p0} + x_{g''})$, that is,

$$x_{A''} = \left|a_{p0} + x_{g''}\right| \tag{1.152}$$

Therefore, the dynamic magnification factor of the absolute acceleration for the base excitation, denoted by β_{AB}, can be obtained as follows:

$$\beta_{AB} = \frac{|a_{p0} + x_{g''}|}{x_{g''}} = \left|\frac{a_{p0}}{x_{g''}} + 1\right| = \left|\frac{1+2j\xi r}{1-r^2+2j\xi r}\right| = \sqrt{\frac{(1+2j\xi r)(1-2j\xi r)}{(1-r^2+2j\xi r)(1-r^2-2j\xi r)}} \tag{1.153}$$

where subscript A denotes absolute, whereas the lowercase subscripts stand for relative quantities. Therefore,

$$\beta_{AB} = \frac{x_{A''}}{x_{g''}} = \sqrt{\frac{1+(2\xi r)^2}{(1-r^2)^2+(2\xi r)^2}} \tag{1.154}$$

By using the term β_{AB}, the real-valued amplitude of the absolute acceleration can be written as

$$x_{A''} = \beta_{AB} x_{g''} \tag{1.155}$$

The phase angle of term $|x_{''} + x_{g''}|/x_{g''}$ is

$$\angle\left\{\frac{|x_{''} + x_{g''}|}{x_{g''}}\right\} = \begin{cases} \tan^{-1}\left[\dfrac{2\xi r^3}{(1-4\xi^2)r^2-1}\right] & (1-4\xi^2)r^2-1 \le 0 \\ \tan^{-1}\left[\dfrac{2\xi r^3}{(1-4\xi^2)r^2-1}\right]-\pi & (1-4\xi^2)r^2-1 > 0 \end{cases} \tag{1.156}$$

In Figure 1.18a, the dynamic magnification factor β_{AB} is plotted vs. the frequency ratio for several different damping ratios. In Figure 1.18b, the phase angles are plotted vs. the frequency ratio.

From Figure 1.18a, it is seen that despite the damping ratios, the dynamic magnification factors all start from the value of unity when $r = 0$. After that, the magnitudes of the factors become larger and larger until they reach resonance points. Similar to all the various dynamic magnification factors, when the damping ratio is small, the magnitude of the dynamic magnification factor of the absolute acceleration can be very large. As a difference from the dynamic magnification factors β_a and/or β_{dB}, which will not have resonance when the damping ratio is >0.707, when the damping ratio reaches unity, β_{AB} will still be greater than unity.

By continuously increasing the frequency ratio after the resonance point, the magnitude decreases. At a special frequency point, the value returns to unity. Interestingly, this frequency point, which is derived in the next subsection, is independent of the damping ratio. Consequently, all curves in Figure 1.18 pass through this unique point. Beyond that frequency point, the value of β_{AB} will be less than unity. However, in contrast to all the dynamic magnification factors discussed above, which always reduce the magnitudes as the damping increases, the magnitude of β_{AB} will have a different trend after this frequency point. That is, the larger the damping ratio, the smaller the reduction of the dynamic magnification factor.

Note that again, the phase angle should have a minus sign. Thus, by using Figure 1.18b, the plots of the phase angle vs. the frequency ratio show that as the frequency ratio increases, the phase difference between the mass and the ground will decrease from zero to $-\pi$. In this figure, the effect of the damping ratio is also realized. Generally, when the damping ratio is small, the phase will have a sharp turning point close to $r = 1$. However, when the damping ratio is large, the phase value will be increased more gradually with a less sharp turning frequency point. In

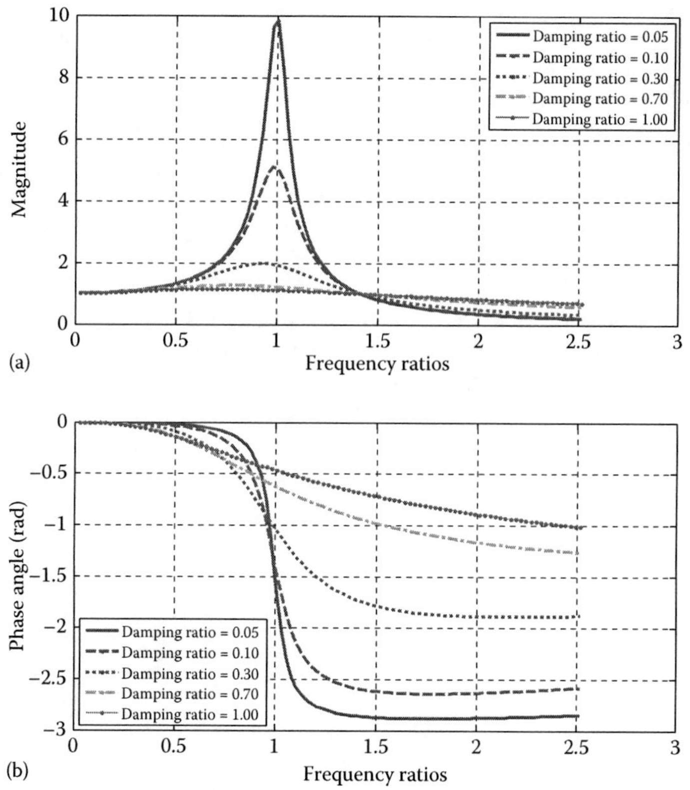

FIGURE 1.18 Magnitude and phase of absolute acceleration due to ground excitation (a) dynamic magnification factor and (b) phase angle.

addition, when the damping is small, the phase difference will quickly reach $-\pi$ or $-180°$ as the frequency ratio becomes larger than unity. However, with a large damping ratio, the phase angle reaches a smaller maximum value in a more gradual manner. That is, the maximum value is also reduced. For example, when $\xi = 1$, the maximum phase difference is no longer $-180°$; instead, it becomes $-90°$.

1.2.2.2.2 Peak Value of the Dynamic Magnification Factor of Absolute Acceleration

In the preceding section, the nature of the dynamic magnification factor of absolute acceleration underground excitation was discussed in general terms. Now, a more detailed analysis is carried out at several frequency points. The first important point is when r = 0. It is easy to see that, in this case,

$$\beta_{AB}\,|_{r=0} = 1 \tag{1.157}$$

Next, the peak value of the dynamic magnification factor of absolute acceleration and the corresponding frequency ratio are determined. The derivative of β_{AB} is taken with respect to r and the result is set equal to zero. Thus,

$$\frac{d}{dr}\beta_{AB} = \frac{d}{dr}\sqrt{\frac{1+(2\xi r)^2}{\left(1-r^2\right)^2+(2\xi r)^2}} = 0 \tag{1.158}$$

Solving Equation 1.158 to find the proper value of r, results in

$$r = \left[\frac{\sqrt{1+8\xi^2}-1}{4\xi^2} \right]^{1/2} \tag{1.159}$$

From Equation 1.159, it is realized that since

$$\xi > 0 \tag{1.160}$$

the frequency ratio is always real valued. This fact implies that Equation 1.158 will always have a meaningful solution, so that regardless of the value of the damping ratio, the dynamic magnification factor of the absolute acceleration will always have a resonant peak value.

Substituting Equation 1.159 into Equation 1.153, the peak value of the dynamic magnification factor is given by

$$\beta_{AB}(\omega_A) = 2\sqrt{2}\xi^2 \left(8\xi^4 - 4\xi^2 - 1 + \sqrt{1+8\xi^2} \right)^{-1/2} \tag{1.161}$$

In Equation 1.161, ω_A is the resonant frequency given by

$$\omega_A = \left[\frac{\sqrt{1+8\xi^2}-1}{4\xi^2} \right]^{1/2} \omega_n \tag{1.162}$$

It can be seen that when the damping ratio is small,

$$\beta_{AB}(\omega_A) \approx \frac{1}{2\xi\sqrt{1-\xi^2}} \tag{1.163}$$

Thus, according to Equation 1.163, $1/2\xi\sqrt{1-\xi^2}$ seems to be a peak value of the terms β_d, β_a, β_{dB} and β_{AB}.

However, it is also observed that no matter how the damping ratio is chosen, the following is always the case:

$$\beta_{AB}(\omega_A) > \frac{1}{2\xi\sqrt{1-\xi^2}} \tag{1.164}$$

Inequality Equation 1.164 indicates that the peak value of $\beta_{AB}(\omega_A)$ is always larger than the other types of dynamic magnification factors. When the damping ratio is sufficiently large, the difference becomes more significant. Furthermore, when r = 1,

$$\beta_{AB}(1) = \frac{\sqrt{1+4\xi^2}}{2\xi} \tag{1.165}$$

which is $\sqrt{1+4\xi^2}$ times the value of the other dynamic magnification factors, such as that of displacement, at r = 1.

1.2.2.2.3 Frequency Point of Dynamic Magnification Factor of Absolute Acceleration being Unity
Consider the case when the value of the dynamic magnification factor of absolute acceleration reaches unity, besides the point when r = 0. Let

$$r = \sqrt{2} \tag{1.166}$$

Then, substituting Equation 1.165 into Equation 1.154 yields

$$\beta_{AB}\left(\sqrt{2}\right) = 1 \tag{1.167}$$

Therefore, despite the value of the damping ratio, when Equation 1.165 is satisfied, the value of the dynamic magnification factor of absolute acceleration reaches unity again. Beyond this point,

$$\beta_{AB}\left(\omega_f > \sqrt{2}\omega_f\right) < 1 \tag{1.168}$$

That is, the value of the dynamic magnification factor of absolute acceleration can be smaller than unity, only if Equation 1.168 is satisfied. Consequently, in base isolation design, the natural period, T_n, needs to be $\sqrt{2}$ times larger than the major driving period. Otherwise, the acceleration of the superstructure will not be reduced, but will instead be magnified.

1.2.2.2.4 Half-Power Points of Dynamic Magnification Factor of Absolute Acceleration
To obtain the approximation of the half-power points, let

$$\sqrt{\frac{1+(2\xi r)^2}{\left(1-r^2\right)^2+(2\xi r)^2}} = \frac{\sqrt{2}}{2}\left[2\sqrt{2}\xi^2\left(8\xi^4-4\xi^2-1+\sqrt{1+8\xi^2}\right)^{-1/2}\right] \tag{1.169}$$

By solving Equation 1.169,

$$r_{1,2} = \left\{\frac{-b \mp \sqrt{b^2-4ac}}{2a}\right\}^{1/2} \tag{1.170}$$

where

$$a = 4\xi^4 \tag{1.171a}$$

$$b = -16\xi^6 + 8\xi^4 + 4\xi^2 - 4\xi^2\sqrt{1+8\xi^2} \tag{1.171b}$$

$$c = -4\xi^4 + 4\xi^2 + 1 - \sqrt{1+8\xi^2} \tag{1.171c}$$

When the damping ratio is sufficiently small, it can be proven that the frequency ratios calculated by Equation 1.170 can be used to determine the damping ratio with the help of Equation 1.122. In fact, when the actual damping ratio is <21%, there can be a <10% error in damping ratio overestimation. When the actual damping ratios are 30%, 40%, and 50%, the errors in damping ratio overestimation will be 18.82%, 31.11%, and 50.33%, respectively.

Example 1.10

A computer is mounted on a floor with total mass 25 (kg), which is subjected to ground harmonic excitation with an amplitude of 1 (g) and a driving frequency of 4 (Hz). The computer only allows 0.3 (g) of acceleration so that it is base isolated. The base isolator provides a stiffness of 3,000 (N/m) and a damping ratio of 10%. If the allowed relative displacement of the isolator is <1.6 (cm), the base isolation must be checked to determine if it can satisfy the required parameters.

With the given parameters, the natural frequency of the isolation system is 1.7 (Hz). The frequency ratio is then calculated to be 2.29, $\beta_{dB} = 1.23$, and $\beta_{AB} = 0.257$. Therefore, the amplitude of acceleration is 0.257 (g) < 0.3 (g), whereas the displacement of isolation is 1.9 > 1.6 (cm).

It is seen that although the acceleration satisfies the required level, the displacement does not. Therefore, a different group of design data must be chosen. This time, k = 500 (N/m) is supposed. The natural frequency of the isolation system is 0.712 (Hz). The frequency ratio is then calculated to be 5.62. When the damping ratio is chosen to be 0.7, $\beta_d = 1.001$, $\beta_{dB} = 1.23$, and $\beta_{AB} = 0.251$. Therefore, the amplitude of acceleration is 0.251 (g) < 0.3 (g), whereas the displacement of isolation is 1.55 < 1.6 (cm).

This example implies that when the level of acceleration is to be reduced, which is often the main goal of base isolation, the relative displacement of the isolator must be checked. In fact, in the design stage, it is best to consider the reduction of the acceleration and the regulation of the displacement simultaneously.

1.3 ENERGY DISSIPATION AND EFFECTIVE DAMPING

In Example 1.10, the vibration of a linear SDOF system under harmonic excitations was discussed. When the responses reach a steady state, the input energy and the energy dissipated by damping of the system are balanced. The steady-state response of a system can be used as the basis to study the effect of damping. In this section, the energy dissipation during a complete vibrating cycle is used as a basis to explore the function of damping in forced vibration.

1.3.1 Energy Dissipated per Cycle

1.3.1.1 Linear Viscous Damping

1.3.1.1.1 Damping in SDOF Systems: Qualitative and Quantitative Definitions

In a vibration system, there are three internal forces to balance the excitation load. Unlike the inertial and restoring forces, the damping force is nonconservative and the corresponding work done is dissipative. The simplest damping of an SDOF system is linear viscous damping. Generally speaking, the damping force is notably smaller than other internal forces. Therefore, using linear viscous damping to approximate other types of damping will not introduce significant errors in response estimations.

In this section, the concepts of damping and vibration in SDOF systems are reviewed by qualitative and quantitative definitions. Many of the definitions and formulas can also be found in textbooks (i.e., Inman 2007; Chopra 2006; Clough and Penzien 1993). However, for convenience in further discussions, some necessary equations are given for later reference. The important concepts of damping are also marked in italics.

Primarily, damping is a *measurement of the capability of a system to dissipate dynamic (vibration) energy*. Therefore, damping can be represented by energy dissipation, or by a force that works to dissipate energy. Generally speaking, for an SDOF system, the larger the damping force, the higher the capability of the system to resist external input energy. Hence, the remaining energy to vibrate the structure will be smaller under a given driving frequency and amplitude of excitation. This is conventionally believed to be the reason why damping can reduce the vibration level or it can control earthquake-induced structural response. Here, the way a system resists external dynamic loading is examined first.

1.3.1.2 Energy Dissipated by Damping Force

The energy dissipation by linear viscous damping is considered. When the harmonic response of a system reaches the steady state, the energy dissipated by damping force during a complete cycle can be calculated. That is,

$$E_d = \oint f_d dx = \int_0^{2\pi/\omega} c\dot{x}\dot{x}dt = cx_0^2 \int_0^{2\pi/\omega_f} \cos^2(\omega_f t + \phi)dt = c\omega_f \pi x_0^2$$

The energy dissipated in the complete cycle under harmonic excitation with the driving frequency equal to the natural frequency, $E_d(\omega_n)$, is called the *damping capacity* of the SDOF system. That is,

$$E_d(\omega_n) = c\omega_n \pi x_0^2 \quad (1.172)$$

Since $c = 2\xi\omega_n m$,

$$E_d = 2\xi\omega_n m\omega_f \pi x_0^2 = 2\xi \frac{\omega_f}{\omega_n} \pi k x_0^2 = 2\xi\pi k r x_0^2 \quad (1.173)$$

Next, the work done by the external harmonic force $f_0 \sin \omega_f t$ is examined. Denoting this work by W_f,

$$W_f = \oint f_0 \sin \omega_f t dx = 4\int_0^{\pi/2\omega} f_0 \sin \omega_f t x_0 \sin(\sin \omega_f t + \phi)dt = f_0 x_0 \pi \sin \phi \quad (1.174)$$

From Equation 1.43, it is seen that

$$\phi = \tan^{-1}\left[\frac{-2\xi\omega_n\omega}{\omega_n^2 - \omega_f^2}\right] = \tan^{-1}\left[\frac{-2\xi r}{1 - r^2}\right] \quad (1.175)$$

Therefore,

$$\sin \phi = \frac{2\xi r}{\sqrt{(1 - r^2)^2 + (2\xi r)^2}} = 2\xi r \beta_d \quad (1.176)$$

Furthermore, from Equation 1.84,

$$f_0 = x_0 k / \beta_d \quad (1.177)$$

Substituting Equations 1.176 and 1.177 into Equation 1.174 yields

$$W_f = \frac{x_0 k}{\beta_d} \pi x_0 \, 2\xi r \beta_d = 2\xi\pi k r x_0^2 \quad (1.178)$$

Comparing Equation 1.178 with Equation 1.173, results in the conclusion that at the steady state,

$$W_f = E_d \quad (1.179)$$

Namely, the work done by the external force or the input energy is totally dissipated by the damping force. On the other hand, the change of the potential energy and that of the kinetic energy, denoted by ΔE_p and ΔE_k, respectively, are both equal to zero. Thus,

$$\Delta E_p = \oint f_S dx = 0 \tag{1.180}$$

and

$$\Delta E_k = \oint f_I dx = 0 \tag{1.181}$$

1.3.1.3 Damping Coefficient and Damping Ratio of Linear Viscous Damping

In the above example, when the damping coefficient is known, the damping capacity can be determined by Equation 1.172. On the other hand, if the damping capacity is known, the damping coefficient can be determined as follows:

$$c = \frac{E_d}{\pi \omega_f x_0^2} \tag{1.182}$$

Note that the damping coefficient c of linear viscous damping is a constant, but in Equation 1.182, it appears to be a function inversely proportional to the driving frequency ω_f. This is because the energy dissipated per cycle is also a function of the driving frequency. Consequently, Equation 1.182 is not in a convenient form to use. In the following discussion, a different approach is given.

The damping ratio of the linear viscous damping system can be written as

$$\xi = \frac{E_d}{\pi \omega_f x_0^2} \frac{1}{2\sqrt{km}} = \frac{E_d}{2\pi\omega_f \frac{k}{k}\sqrt{km}\, x_0^2} = \frac{E_d}{2\pi\omega_f \sqrt{\frac{m}{k}}\, k x_0^2} = \frac{E_d}{2\pi \frac{\omega_f}{\omega_n} 2 \frac{k x_0^2}{2}}$$

Therefore,

$$\xi = \frac{E_d}{4\pi r E_p} \tag{1.183}$$

where E_p is the maximum potential energy with

$$E_p = \frac{k x_0^2}{2} \tag{1.184}$$

Substituting Equations 1.172 and 1.184 into Equation 1.183 yields

$$\xi = \frac{c\pi\omega_f x_0^2}{4\pi \frac{\omega_f}{\omega_n} \frac{k x_0^2}{2}} = \frac{c}{2\sqrt{km}} \tag{1.185}$$

Comparing Equation 1.183 with Equation 1.13 – the definition of the damping ratio – it becomes obvious that Equation 1.183 can be used to represent the damping ratio. In addition, from the definition of Equation 1.13, the damping ratio depends only on the physical parameters of the system, namely, mass m, damping coefficient c, and stiffness k. From Equation 1.185, it is seen that regardless of the value of the driving frequency, the resulting ratio of E_d and $4\pi r E_p$ will be the damping ratio ξ. In other words, the damping ratio ξ of the linear viscously damped system is not a function of external force, such as the driving frequency ω_f. However, in Equation 1.185, the frequency ratio is involved, which seems to mean that the driving frequency must be considered. Such an involvement can introduce unnecessary constraints and difficulties for practical applications, because both harmonic and arbitrary excitations will exist. For example, in earthquake excitations, it is difficult to specify the driving frequency ω_f.

One way to avoid this problem is to use the damping capacity $E_d(\omega_n)$; that is, let the driving frequency equal the natural frequency. This is experimentally sound because when $\omega_f = \omega_n$ (i.e., at the resonance point), a high signal-to-noise ratio can be used for measurement. Furthermore, the damping capacity is also applicable for random excitations, in which case the system is likely to have vibration with frequencies around the natural frequency. Therefore, Equation 1.183 is rewritten as

$$\xi = \frac{E_d}{4\pi E_p} \tag{1.186}$$

For simplicity, unless specifically mentioned, E_d is used instead of $E_d(\omega_n)$ to denote the special energy dissipation or the damping capacity.

1.3.1.4 Linear System

The above discussion is limited to linear systems. That is, both the stiffness and the damping are linear. The following discussion addresses nonlinear damping. Before undertaking a detailed study, a more rigorous explanation of the condition of the linearity function is necessary.

Generally speaking, if a function

$$y = f(x) \tag{1.187}$$

is linear, the relationship between the variable x and function y should satisfy certain conditions. To examine these conditions, denote

$$y_i = f(x_i) \tag{1.188a}$$

$$y_j = f(x_j) \tag{1.188b}$$

Here, x_i and x_j are different variables within the domain of functions f, and y_i and y_j are the corresponding functions. In addition, the variable x_k is introduced in the same domain as x_i and x_j. If the function f is linear, the following conditions must be satisfied:

$$f(x_i + x_j) = f(x_j) + f(x_i) = y_j + y_i \tag{1.189}$$

$$f(x_i + x_j + x_k) = f(x_i) + f(x_j + x_k) \tag{1.190}$$

$$f(\alpha x_i) = \alpha y_i \tag{1.191}$$

$$f((\alpha + \beta) x_i) = \alpha y_i + \beta y_i \tag{1.192}$$

$$f(\alpha(\beta x_i)) = (\alpha\beta) y_i \tag{1.193}$$

$$1 y_i = y_i \tag{1.194}$$

$$0 y_i = 0 \tag{1.195}$$

where 0 is the null function; α and β are scalars and

$$0 + y_i = y_i \tag{1.196}$$

The above equations indicate that all these functions are additive and multiplicative. The results of these summations and products are still in the same range. From the viewpoint of engineering applications, a single equation can be used to summarize the above conditions, that is,

$$f(\alpha x_i + \beta x_j) = \alpha y_i + \beta y_j \tag{1.197}$$

In a previous subsection on the method of complex response for steady-state displacement, Equation 1.197 was used to examine the combined forcing function $f(t) = f_0 \cos(\omega_f) + j f_0 \sin(\omega_f t)$ and the resulting response as $x(t) = x_0 \cos(\omega_f) + j x_0 \sin(\omega_f t)$. That is, the forcing function f can be viewed as an input variable and the response x as its function.

Similarly, the restoring force is a linear function of the displacement, the damping force is a linear function of the velocity, and so on. For example, since

$$f_d(t) = c\dot{x}(t) \tag{1.198}$$

if the velocity is multiplied by a scalar α,

$$c[\alpha\dot{x}(t)] = \alpha c[\dot{x}(t)] \tag{1.199}$$

Specifically, for a forced vibration system, an equivalent condition can exist, which is both necessary and sufficient; that is, a vibration system is linear if and only if the steady-state response contains a single frequency component of ω_f when a harmonic excitation with driving frequency ω_f is applied.

1.3.2 Damping and Seismic Force

1.3.2.1 Parametric Equation

In the above discussion, it is seen that both the damping force and the displacement are functions of time. Therefore, using the time variable as a parameter, the parametric equation of the damping force and the displacement can be obtained, which can be a useful tool to investigate the relationship between the force and the displacement.

By using the parametric equation, the curve of the damping force vs. the displacement can be plotted. In the case of linear viscous damping, during a vibration cycle of steady-state response, the curve must be closed. It will be shown that the closed curve will cover an elliptic area, which is the energy dissipated by the damping force. Generally, the parametric equation of an ellipse can be written as

$$\begin{cases} x = X_m \cos(\tau) \\ f = F_m \sin(\tau + \phi) \end{cases} \tag{1.200}$$

where X_m and F_m are, respectively, the maximum values of displacement and force; τ is the parameter of the parametric equation; x and f are variables; and ϕ is the *damping phase shift*. When $\tau = \pi/2$, $x(\pi/2) = 0$ and $f(\pi/2) = f|_{x=0} = F_m \cos\phi$. Thus, the phase shift is

$$\phi = \cos^{-1}\left(\frac{f|_{x=0}}{F_m}\right) \tag{1.201}$$

Q is denoted as

$$Q = f|_{x=0} \tag{1.202}$$

where Q is called the *characteristic strength*. Therefore,

$$\cos\phi = \frac{Q}{F_m} \tag{1.203}$$

For the physical meaning of the damping phase shift, ϕ, consider an SDOF system with ground motion excitation. Rearranging Equation 1.63 as

$$-m\ddot{x}_A(t) = c\dot{x}(t) + kx(t) \tag{1.204}$$

the combination of the damping force $c\dot{x}(t)$ and the spring force $kx(t)$, previously defined as the structural force, can be seen as the resistance against the inertial force $m\ddot{x}_A(t)$, which is referred to as the *seismic force*. Note that the seismic force and the structural force have identical magnitudes, but opposite signs.

For steady-state responses under harmonic excitation, the displacement and velocity can be represented by

$$x = X_m \cos(\omega_f t) \tag{1.205}$$

and

$$\dot{x} = -X_m \omega_f \sin(\omega_f t) \tag{1.206}$$

The structural force can be written as

$$\begin{aligned} c\dot{x}(t) + kx(t) &= -cX_m\omega_f \sin(\omega_f t) + kX_m \cos(\omega_f t) \\ &= \sqrt{c^2\omega_f^2 + k^2}\, X_m \sin(\omega_f t + \phi) \end{aligned} \tag{1.207}$$

Comparing Equation 1.205 with the first equation in Equation 1.200, and Equation 1.207 with the second equation in Equation 1.200, it can be realized that in Equation 1.207,

$$F_m = \sqrt{c^2\omega_f^2 + k^2}X_m \tag{1.208}$$

and

$$\phi = -\tan^{-1}\frac{k}{c\omega_f} \tag{1.209}$$

The parameter τ is used to represent $\omega_f t$, that is,

$$\tau = \omega_f t \tag{1.210}$$

In Figure 1.19, the idealized ellipse is shown as a solid line. It can be proven that the area where the maximum energy is dissipated by the structural force, or the maximum work done by the seismic force denoted by W_S, is represented by

$$W_S = X_m F_m \pi \cos\phi = QX_m\pi \tag{1.211}$$

In Figure 1.19, the second curve is shown with a broken line. This curve is plotted by letting $\phi = 0$, in which case, it can be realized that $k = 0$. In other words, the broken line stands for the damping force only vs. the displacement. It can be seen that the area is also $QX_m\pi$.

Therefore, from the comparison of the seismic force and the pure viscous damping force, it is seen that both have the same characteristic strength Q and the areas of energy dissipation are identical.

Thus, by using Equation 1.211, the maximum seismic work can be calculated through the relationship between the absolute acceleration and the displacement with specified period and damping. Namely, from the time histories of absolute acceleration and displacement, the maximum displacement X_m and characteristic strength Q can be found. From these two quantities, the maximum seismic work can be defined.

In order to review the effect of the damping phase shift, consider the normalized seismic work, where the seismic force is normalized to unity. The parametric equation is

$$\begin{cases} \bar{x} = x/F_m = a\cos(\tau) \\ \bar{f} = f/F_m = \sin(\tau + \phi) \end{cases} \tag{1.212}$$

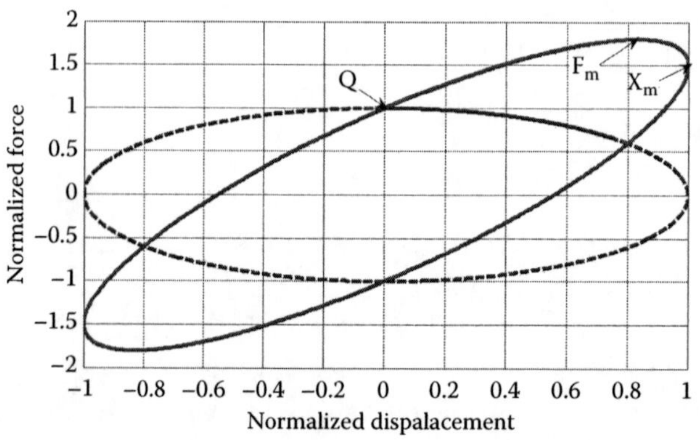

FIGURE 1.19 Idealized elliptic energy dissipation.

where

$$a = X_m / F_m \tag{1.213}$$

In this circumstance, the normalized work is

$$w_S = a\pi \cos\phi = F_m^{-2} W_S \tag{1.214}$$

Figure 1.20 shows the normalized work vs. the phase shift.

Nine curves are shown in Figure 1.20. They represent, respectively, the cases of $a = 0.9, 0.8, 0.7, 0.6, 0.5, 0.4, 0.3, 0.2$, and 0.1. From Figure 1.19, it is seen that when the damping phase shift is zero, the work done, represented by the area of these ellipses, has the maximum value. When the phase shifts are close to $\pi/2$ or 90°, the corresponding work done is close to zero.

In certain cases, using the following formula can yield more numerically accurate results, that is,

$$W_S = \pi\sqrt{\left[X_m^2 \cos^2(\psi) + F_m^2 \sin^2(\psi + \phi)\right]\left[X_m^2 \sin^2(\psi) + F_m^2 \cos^2(\psi + \phi)\right]} \tag{1.215}$$

where

$$\psi = \frac{1}{2}\tan^{-1}\left[\frac{F_m^2 \sin(2\phi)}{X_m^2 - F_m^2 \cos(2\phi)}\right] \tag{1.216}$$

Figure 1.20 shows that the damping phase shift plays an important role that affects the seismic work. Next, the impact of this parameter is discussed in detail by considering the governing equation of an SDOF system as previously described.

Assume that the SDOF system is near resonance at the steady state. In this case, the displacement is close to a sinusoidal function, and ω_f is replaced by ω_n, that is,

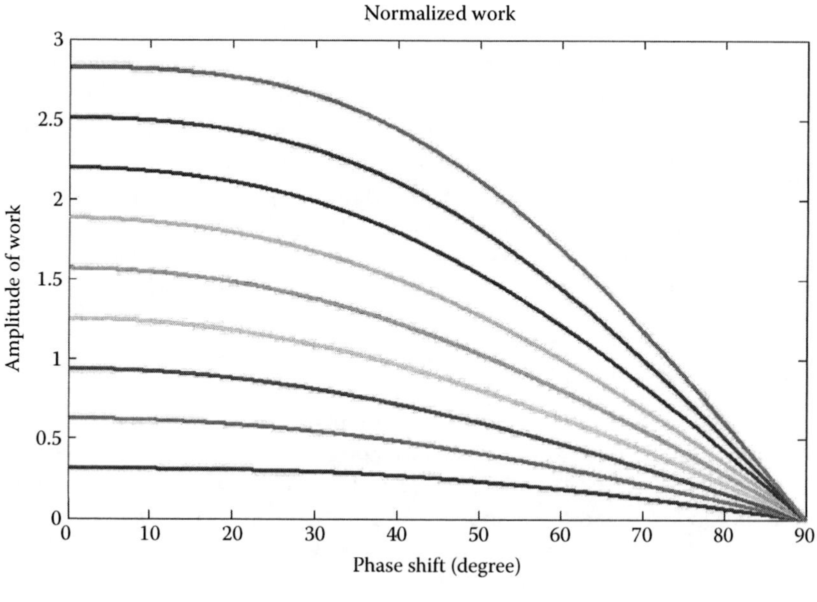

FIGURE 1.20 Effect of damping phase shift.

$$x = X_m \cos(\omega_n t) \tag{1.217}$$

Further, the seismic force is

$$\left[2\xi\omega_n^2 X_m \sin(\omega_n t) + \omega_n^2 X_m \cos(\omega_n t)\right] = F_m \sin(\phi + \omega_n t) \tag{1.218}$$

Here

$$F_m = \sqrt{1+4\xi^2}\,\omega_n^2 X_m \tag{1.219}$$

and

$$\phi = \cos^{-1}\left(\frac{2\xi}{\sqrt{1+4\xi^2}}\right) \tag{1.220}$$

Comparing Equation 1.217 with the first equation in Equation 1.200, it is realized that these equations are essentially the same if $\tau = \omega_n t$. Therefore, by using Equations 1.220 and 1.203, it can be written that

$$\frac{2\xi}{\sqrt{1+4\xi^2}} = \frac{Q}{F_m} \tag{1.221}$$

or

$$\hat{\xi} = \frac{1}{2}\sqrt{\frac{Q^2}{F_m^2 - Q^2}} \tag{1.222}$$

where $\hat{\xi}$ stands for the calculated value.

As seen in Equation 1.222, if the characteristic strength Q and the maximum displacement from the time history plot can be measured, the corresponding damping ratio can be computed, if the SDOF system is near resonance. The condition of "near resonance" is reached by the maximum value of the acceleration response spectrum at a different period T. In Figure 1.21, the calculated damping ratios $\hat{\xi}$ are plotted as a function of the original assigned damping ratio ξ in Equation 1.222. The heavy straight line is the theoretical damping ratio. The group of small circles represent the calculated damping ratios with the periods taken from 0.5 to 3.0 (s). It is seen that when the damping ratios are smaller than 0.7, errors are acceptable.

1.3.3 Effective Damping

The above discussion shows that the damping capacity plays an important role in linearizing most nonlinear damping. This is important because most commonly used damping cannot be expressed as the linear viscous damping described by Equation 1.5. That is, often the damping force cannot simply be written as being proportional to the velocity, with the proportional coefficient being the damping ratio. More generally, the damping force can be written as

$$f_d(t) = f_d\left(x(t), \dot{x}(t)\right) \tag{1.223}$$

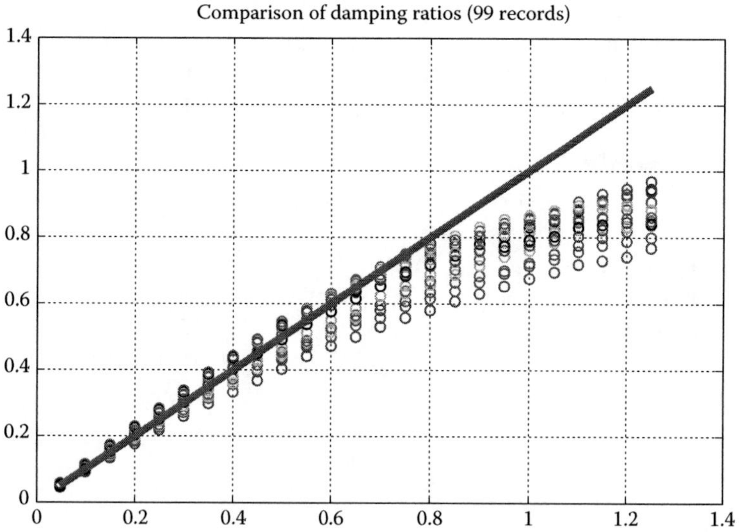

FIGURE 1.21 Calculated damping ratios.

In these cases, the entire system will no longer be linear, so that under harmonic excitation, both the driving frequency and the amplitude should be examined.

1.3.3.1 Effective Damping Coefficient

The first approach to linearized damping for practical applications is the concept of *effective damping coefficient.* It specifies a linear system with precisely the same mass and stiffness and with the damping coefficient, such that the system has the exact energy dissipation per cycle as the nonlinear damping system. Let the energy dissipation per cycle of the system with nonlinear damping be E_d. The damping coefficient of the equivalent linear system, which is now referred to as an *effective* system, is denoted by c_{eff},

$$E_d = c_{eff}\omega_f \pi x_0^2 \tag{1.224}$$

Therefore, the *effective damping coefficient* is given by

$$c_{eff} = \frac{E_d}{\omega_f \pi x_0^2} \tag{1.225}$$

Again, ω_f is the driving frequency. Thus, from this definition, it is seen that the effective damping coefficient is, in general, not the same constant as that of the linear viscous damping system used in Equation 1.5. Instead, it is often a function of the driving frequency ω_f.

In practical applications, pure harmonic excitation with a fixed driving frequency, ω_f, rarely occurs. Therefore, the damping coefficient defined in Equation 1.225 is not very convenient to use. In many cases, the natural frequency, ω_n, is used to replace the driving frequency in Equation 1.225,

$$c_{eff} = \frac{E_d}{\omega_n \pi x_0^2} \tag{1.226}$$

Here, the term E_d is taken to be the energy dissipation with driving frequency ω_n.

Note that the concept of a damping coefficient was originally used to define the property of an individual damper but not a vibration system. However, the concept of an effective damping coefficient here has become a parameter of the entire nonlinear system.

1.3.3.2 Effective Damping Ratio

Equation 1.186 states that for a linear viscously damped system, the damping ratio can be represented by the ratio of damping capacity and 4π times the maximum potential energy. From Equation 1.226, by calculating the energy dissipated per cycle, the concept of an effective damping coefficient can be realized. Thus, these two concepts of calculating the damping ratio and the energy dissipation can be related together and the effective damping ratio can be defined.

That is, if the response reaches the steady state in forced vibration, Timoshenko (1937) suggested the following formula for the damping ratio:

$$\xi_{eff} = \frac{E_d}{4\pi E_p} \tag{1.227}$$

where E_d is the energy dissipated during one cycle and E_p is the total conservative energy. This formula is widely used. The damping ratio defined in Equation 1.227 is called *Timoshenko damping.** For linear viscous damping, Timoshenko damping has exactly the same value as defined in Equation 1.13. In more general cases, other types of damping exhibit a different nature than that of linear viscous damping. Timoshenko defined an equivalent system that is linear and has the damping ratio defined in Equation 1.227, which dissipates energy during one cycle under sinusoidal excitations equal to the energy dissipation of this general system.

Figure 1.22 shows the area of the oval-shaped energy dissipation loop, $E_d = \pi c\omega x_0^2$, for which the amplitude of the damping force is $f_{d0} = c\omega x_0$. The figure also shows the triangular-shaped maximum potential energy, $E_p = ½kx_0^2$, for which the amplitude of restoring force is $f_R = kx_0$.

As shown in Equation 1.227, the energy, E_p, is often evaluated by using the instantaneous potential energy when the displacement reaches the maximum value, which is a state quantity. However, the energy, E_d, is evaluated through the entire vibration cycle, which is a process quantity.

In earthquake engineering, the focus is on randomly forced vibration, which is more or less different from harmonically excited steady-state responses. This issue is discussed in detail in the next chapter.

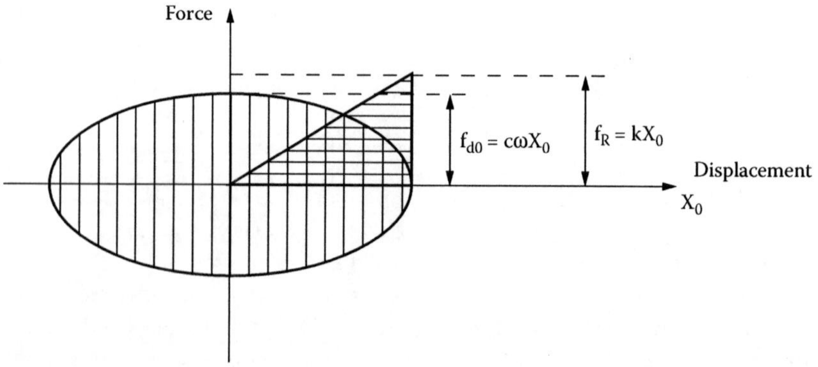

FIGURE 1.22 Dissipated energy and potential energy in steady-state vibration.

* In 1930, Jacobsen (1930) first suggested this method. However, it did not receive significant attention until 7 years later, when Timoshenko (1937) provided a more systematic description in his textbook *Vibration Problems in Engineering*, 2nd ed.

1.3.3.3 General Forced Vibration, Energy Dissipation

In more general cases, when a structural force f does not synchronize with the displacement, there is a chance that some energy will be dissipated, which can be seen conceptually in Figure 1.23a. That is, if a system with damping is forced to have displacement back and forth in a complete cycle, the force vs. the displacement will form a closed loop, called the *hysteresis loop*, or sometimes called the energy dissipation loop.

Figure 1.23a shows that displacement x_1 is associated with two forces, f_1 and f_2; that is, the energy dissipation loop will be associated with loading and unloading forces, respectively. The loading and unloading curves are not identical. The areas under these different curves are the work done by the forces f_1 and f_2 through distance x_1, denoted by E_1 and E_2, respectively. The difference between these areas, $E_1 - E_2$, denotes the energy difference or the energy dissipation in the first quadrant. That is, when a system has a displacement moving away from its equilibrium origin, the corresponding force will define the first curve. When the system has a displacement returning to its equilibrium position, the corresponding force will define the second curve. If the force associated with the first curve is larger than the second one, that is, if $f_1 > f_2$ as shown in Figure 1.23a, then energy dissipation occurs. In this case, there is a *positive damping*. This phenomenon also holds true in the remaining quadrants.

Note that in Figure 1.23, the force f may actually contain two parts. In the case of a linear system with viscous damping, this point is rather clear; see Equation 1.3. In the more general case, this statement is also true. The first part of the force is dissipative, which means that such a force, f_d, dissipates energy. Here, the symbol stands for a more general dissipative force, different from the specific linear damping force defined in Equation 1.5. However, a second type of force, f_c, may exist as a conservative force. That is,

$$f = f_c + f_d \tag{1.228}$$

The system with damping can be forced to vibrate over many cycles. In many cases, it can be assumed that at a given displacement, the amplitude of the damping force remains constant, despite specific cycles. These cases are referred to as the steady state of the vibration systems. The sign of the damping force depends on the direction of movement of the system. For a steady-state system at any point, the amplitude of the conservative force will always be identical.

Suppose that the energy dissipation loop is contributed by both damping and restoring forces. Also, assume that the restoring force is linear with zero mean. At the equilibrium position, there are only damping forces. Assuming that the damping forces are symmetrical with respect to the X-axis, the conservative force can be computed from the following equations:

$$f_c \approx \frac{f_1 - f_2}{2} \tag{1.229a}$$

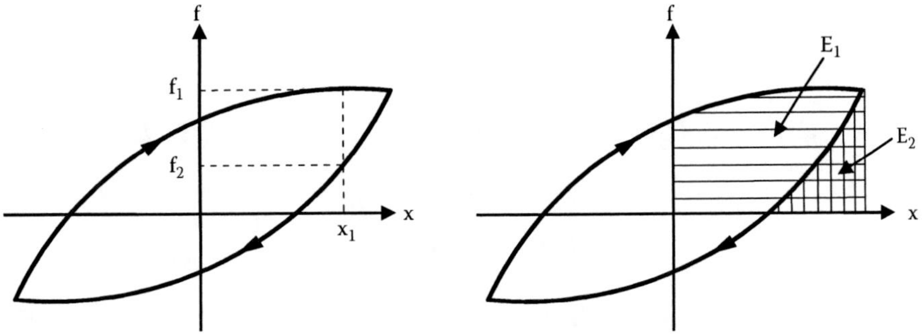

FIGURE 1.23 Energy dissipation loop.

and

$$f_d \approx \frac{f_1 + f_2}{2} \tag{1.229b}$$

To use these equations, the sign of the conservative force needs to be defined as positive if the force exists in the first and fourth quadrants, and vice versa. Note that both f_1 and f_2 also have their own signs, which are defined as follows: when the curve is formed as the system is going forward from its origin, the force has a positive sign. When the curve is formed as the system returns to its origin, the force has a negative sign.

If the energy dissipation loop is known to have contributions from both the aforementioned restoring and damping forces, then Equations 1.229a and b are used. However, in some cases, it is not certain if the loop contains both forces. In these circumstances, Equation 1.228 may not hold. Therefore, Equations 1.229a and b cannot be used. However, Equations 1.228 and 1.229 imply that the structural force f_S may contain forces that are essentially different in nature. Once the dissipative force exists, the total structural force will be distinct from the conservative force; and a certain area will be formed under the loop of the structural force and the displacement. Therefore, it is seen that the existence of the dissipative force creates a phase difference between the structural force and the displacement.

1.3.3.4 Alternative Form of Damping Ratio

In Equation 1.227, the damping ratio is expressed as the ratio of the dissipated and maximum potential energies with a proportionality coefficient of $\pi/4$. This formula is obtained through linear SDOF systems when their steady-state responses are reached under sinusoidal excitation. This equation is very important in providing a measurement of damping ratio for systems that may not be linear, and their responses may not be at a steady state under random excitations.

When the damping ratio of a system is comparatively small, this equation can provide sufficiently accurate results in damper design. However, when the damping ratio becomes larger, this equation may overestimate the damping effect. Therefore, an alternative equation is needed to measure the damping ratio.

Under sinusoidal excitation, when a linear system resonates, that is, when $\omega_f = \omega_n$,

$$x_0 = \frac{f_0}{2\xi k} \tag{1.230}$$

From Equation 1.230, it can be further written that

$$\xi = \frac{f_0}{2kx_0} = \frac{1}{2}\frac{f_0}{f_R} \tag{1.231}$$

Here, f_R is the amplitude of the linear spring force or restoring force $f_r(t)$, which was previously defined, that is,

$$f_r(t) = kx(t)$$

Different from the Timoshenko damping, the alternative formula in Equation 1.231 is based on the ratio of the damping force to the spring force. Therefore, the damping ratio defined in this equation may be referred to as the *force-based effective damping ratio.*

Furthermore, at the resonance point, the amplitude of the external force is equal to the damping force, f_D, that is,

$$f_0 = f_D \tag{1.232}$$

Note that,

$$f_D = c\omega_n x_0 \tag{1.233}$$

and

$$c = 2\xi\omega_n m \tag{1.234}$$

Therefore,

$$\xi = \frac{f_D}{2kx_0} = \frac{1}{2}\frac{f_D}{f_R} \tag{1.235}$$

Furthermore, it can be seen that

$$\xi = \frac{2\xi\omega_n\omega_n m}{2kx_0} = \frac{2\xi}{2}\frac{\omega_n^2}{\omega_n^2} = \xi$$

Therefore, when the steady-state responses of a linear SDOF system are considered under sinusoidal excitation, the damping ratio described in Equation 1.235 and the Timoshenko damping defined in Equation 1.227 are equivalent. Equation 1.235 uses the ratio of the dissipative and conservative forces to represent the damping ratio. Compared to Equation 1.227, Equation 1.235 is an alternative equation that represents the damping ratio, which uses the ratio of the amplitude of the excitation force and the restoring force. Equations 1.227 and 1.235 will be used later to approximate nonlinear systems with the concept of effective damping ratio. In Chapter 2, nonlinear damping forces and use of the Fourier series to represent the damping force are discussed. These parameters will be further used in Chapter 5 to model effective damping ratios for nonlinear damping.

1.4 SUMMARY

In this chapter, a number of fundamental concepts have been presented, along with their corresponding governing equations. These concepts will serve as the basis for the study of earthquake protective systems. The focus in this chapter has been on SDOF systems under free vibration and forced vibration under harmonic loading. The presentation has emphasized the concepts of dynamic amplification and energy dissipation mechanisms. In the next chapter, these same systems are examined under arbitrary excitations.

REFERENCES

Chopra, A.K. 2006. *Dynamics of Structures: Theory and Applications to Earthquake Engineering*. 3rd ed. Upper Saddle River, NJ: Prentice Hall.

Clough, R.W. and Penzien, J. 1993. *Dynamics of Structures*. 2nd ed. NY: McGraw-Hill.

Inman, D.J. 2007. *Engineering Vibration*. 3rd ed. Hoboken, NJ: Pearson Prentice Hall.

Jacobsen, L.S. 1930. Steady forced vibration as influenced by damping. *Transaction ASME* 52:169.

Mohraz, B. and Sadek, F. 2001. Earthquake ground motion and response spectra. Chapter 2 in *The Seismic Design Handbook*, F. Naeim (ed.). 2nd ed. Boston: Kluwer Academic.

Timoshenko, S.P. 1937. *Vibration Problems in Engineering*. 2nd ed. NY: Van Nostrand.

2 Linear Single-Degree-of-Freedom Systems with Arbitrary Excitations

In Chapter 1, the linear single-degree-of-freedom (SDOF) system and its vibration responses due to initial conditions and harmonic excitations were briefly defined and reviewed. These serve as the theoretical basis for complex excitations, such as an earthquake ground motion. In this chapter, the forcing functions and their structural responses are classified into three categories, each requiring special treatment.

The first type of forcing function is time periodic. The basic approach to examining the system response is to use the combinations of sine and cosine functions as a series to represent a periodic function. The corresponding mathematical tool is the Fourier series, which will be further used as a basis for integral transforms, including the Fourier and Laplace transforms.

The second type of forcing function is transient, which does not have a period. In other words, the period of these functions is infinitely long. However, the functions are deterministic. Mathematically, the Fourier integral is used instead of the Fourier series to represent these temporal functions. The integral transform and the convolution of the forcing function and impulse response function, as well as the corresponding transfer functions, are the mathematical tools.

The third type of forcing function, the random function, is directly applicable for earthquake excitations. Such functions often do not have Fourier transforms. In fact, the analysis of these functions is essentially different from that of the above two deterministic signals. Furthermore, the analysis must be based on random processes, so a statistical study must be applied, including ensemble average, correlation functions of the excitations and responses, and their Fourier transforms, such as power spectrum density functions. Finally, using the spectral analysis approach, earthquake response spectra are discussed, which is a special case of ensemble average with respect to periods and is used to account for the excitation of random ground accelerations.

2.1 PERIODIC EXCITATIONS

In Chapter 1, it is shown that the dynamic response of a linear SDOF system can be expressed as a product of the static response and a dynamic magnification factor. The dynamic magnification factor can be determined solely by the physical properties of the system itself. In this chapter, it is shown that periodic forcing functions and their responses can be represented by the summation of a series of harmonic functions. From now on, for convenience, these temporal functions are called "signals."

2.1.1 Periodic Signals

2.1.1.1 Periodic Functions

A *periodic signal* is a function that repeats itself in time. That is, any signal for which a deterministic time T exists, such that

$$f(t) = f(t + nT), \quad n = \pm 1, 2, \ldots \tag{2.1}$$

Here, time T is the minimum possible value and is called the *period* of function, f(t).

For example, a sine function with frequency ω has a period of 2π/ω and a tangent function with frequency ω has a period of π/ω. That is,

$$f(t) = \sin(\omega t) = \sin(\omega t + 2n\pi), \quad n = \pm 1, 2, \ldots \tag{2.2}$$

Therefore,

$$\sin(\omega t + 2\pi) = \sin\left[\omega\left(t + 2\pi/\omega\right)\right] \tag{2.3}$$

Also, since

$$f(t) = \tan(\omega t) = \tan(\omega t + n\pi), \quad n = \pm 1, 2, \ldots \tag{2.4}$$

then

$$\tan(\omega t + \pi) = \tan\left[\omega\left(t + \pi/\omega\right)\right] \tag{2.5}$$

The above equations show that as long as these sine and cosine functions have the identical frequency ω, they will have the identical period 2π/ω.

For example, let

$$f(t) = a\sin(\omega t) + b\sin(\omega t + \theta) \tag{2.6}$$

then,

$$\begin{aligned} f(t) &= a\sin(\omega t) + b\cos(\omega t)\sin(\theta) + b\sin(\omega t)\cos(\theta) \\ &= \left[a + b\cos(\theta)\right]\sin(\omega t) + \left[b\sin(\theta)\right]\cos(\omega t) \\ &= c\sin(\omega t + \psi) = c\sin(\omega t + \psi + 2\pi) \\ &= c\sin\left[\omega\left(t + 2\pi/\omega\right) + \psi\right] \end{aligned} \tag{2.7}$$

where

$$c = \left\{\left[a + b\cos(\theta)\right]^2 + \left[b\sin(\theta)\right]^2\right\}^{1/2} \tag{2.8}$$

and

$$\psi = \tan^{-1}\left[\frac{b\sin(\theta)}{a + b\cos(\theta)}\right] \tag{2.9}$$

Therefore, the phase angle ψ will not affect the period, as long as the sine and cosine functions share the identical frequency. (Here, it is assumed that $a + b\cos(\theta) > 0$.)

Furthermore, consider

$$f(t) = a\sin(\omega_1 t) + b\sin(\omega_2 t) \tag{2.10}$$

It is seen that as long as the ratio of ω_1/ω_2 is a rational number, the signal f(t) can have a period T. Thus, the condition can be regarded as an expression of ω_1/ω_2, which represents the quotient of two integer quantities, n and m, that is,

$$\frac{\omega_1}{\omega_2} = \frac{n}{m} \tag{2.11}$$

Letting n and m be the smallest possible integer, such that n and m do not have a common divisor, the period T can be written as

$$T = 2n\frac{\pi}{\omega_1} = 2m\frac{\pi}{\omega_2} \tag{2.12}$$

Example 2.1

Given

$$f(t) = a\sin(1.5t) + b\sin(2.5t)$$

then

$$\frac{\omega_1}{\omega_2} = \frac{n}{m} = \frac{3}{5}$$

Consequently, the period T is

$$T = 2\times 3\frac{\pi}{1.5} = 4\pi \quad (s) \tag{2.13}$$

When a signal contains more than two sine or cosine functions, the corresponding period can still be found using the same procedure by dealing with the terms one by one. Thus, suppose a function has

$$f(t) = a_1\sin(\omega_1 t) + a_2\sin(\omega_2 t) + a_3\sin(\omega_3 t) + \cdots + a_n\sin(\omega_n t) \tag{2.14}$$

As long as a series of integers n_1, n_2, $n_3,\ldots,n_n$, can be found such that

$$n_1/\omega_1 = n_2/\omega_2 = n_3/\omega_3 \cdots = n_n/\omega_n \tag{2.15}$$

The period T can be defined by

$$T = 2\pi\frac{n_i}{\omega_i} \quad (s) \tag{2.16}$$

In addition, when the constant a_0 is added on the right side of Equation 2.14, the period T will not be affected.

From the above discussion, it is seen that a function consisting of sine and cosine functions with frequencies, ω_1, $\omega_2,\ldots,\omega_n$, as well as a constant, is periodic, as long as the condition in Equation 2.15 is satisfied.

2.1.2 Fourier Series

2.1.2.1 Fourier Coefficients

The next step is to explore whether a periodic function can always be represented by the summation of a constant and a series of sine and cosine functions. Suppose the periodic function f(t) has the following form:

$$f(t) = \frac{a_0}{2} + \sum_{n=1}^{\infty} (a_n \cos n\omega_T t + b_n \sin n\omega_T t) \tag{2.17}$$

Two groups of conditions, referred to as the *Dirichlet conditions*, are sufficient for the assumption described in Equation 2.17: First f(t) must be bounded and have a finite number of extrema and a finite number of discontinuities in any given interval. Secondly, f(t) must be absolutely integrable over a period. Generally speaking, most engineering vibration signals satisfy the first group of conditions but they may not satisfy the second one, which is discussed in Equation 2.139 and further in Subsection 2.2.3.

In Equation 2.17, the period of f(t) is T and the basic frequency ω_T is

$$\omega_T = \frac{2\pi}{T} \tag{2.18}$$

Equation 2.17 is completely defined if the coefficients a_0, a_n, and b_n are determined.

Note that the sine and cosine functions have the following complementary properties:

$$\int_{-T/2}^{T/2} \sin n\omega_T t\, dt = 0, \quad n = 1, 2, \ldots \tag{2.19}$$

$$\int_{-T/2}^{T/2} \cos n\omega_T t\, dt = 0, \quad n = 1, 2, \ldots \tag{2.20}$$

$$\int_{-T/2}^{T/2} \sin n\omega_T t \sin m\omega_T t\, dt = \begin{cases} 0, & n \neq m \\ \frac{T}{2}, & n = m \end{cases} \tag{2.21}$$

$$\int_{-T/2}^{T/2} \cos n\omega_T t \cos m\omega_T t\, dt = \begin{cases} 0, & n \neq m \\ \frac{T}{2}, & n = m \end{cases} \tag{2.22}$$

and

$$\int_{-T/2}^{T/2} \sin n\omega_T t \cos m\omega_T t\, dt = 0 \tag{2.23}$$

In the above equations, conditions in Equations 2.19 and 2.20 define the fundamental nature of the sine and cosine functions, since

$$\sin(-\theta) = -\sin(\theta) \tag{2.24}$$

and

$$\cos(-\theta) = \cos(\theta) \tag{2.25}$$

Thus, the corresponding integral in a complete period must be zero.

Conditions in Equations 2.21 through 2.23 are *orthogonal conditions*. Furthermore, from Equations 2.19 and 2.20, it is realized that the sine and cosine functions are also orthogonal to a constant, so the orthogonality condition must also include Equations 2.19 and 2.20.

Through the conditions in Equations 2.19 and 2.20, it can be realized that

$$\frac{2}{T}\int_{-T/2}^{T/2} f(t)\,dt = a_0 \tag{2.26}$$

This is because in the complete cycle of period T from −T/2 to T/2, the integration of all sine and cosine functions must be zero, except for the constant term, which remains.

Furthermore,

$$\frac{2}{T}\int_{-T/2}^{T/2} f(t)\cos n\omega_T t\,dt = a_n, \quad n = 1, 2, \ldots \tag{2.27}$$

This is possible due to the orthogonality conditions, since in the complete cycle with period T, all the cosine terms with $\cos(n\omega_T T)$ ($m \neq n$) and all the sine terms will vanish.

Similarly,

$$\frac{2}{T}\int_{-T/2}^{T/2} f(t)\sin n\omega_T t\,dt = b_n, \quad n = 1, 2, \ldots \tag{2.28}$$

Equations 2.26 through 2.28 determine the corresponding *Fourier Coefficients* a_0, a_n, and b_n. Therefore, the existence of Equation 2.17 is proven. In other words, as long as a function is periodic and integrable over the period, it can be represented by the series described in Equation 2.17, which is now referred to as the *Fourier series*. All the orthogonal terms, namely, the x_{max_i}, are the base of the Fourier series, which can now be regarded as a set.

From the discussion of the Fourier series, it is further realized that such a set can be fully represented by the linear summation of the base. In this case, the coefficients of these bases are a_0, a_n, and b_n. These bases of the Fourier series set representing a periodic function are periodic themselves. These bases are orthogonal; in other words, they are independent of each other. This independency means that any terms of $\cos(n\omega_T T)$ and/or $\sin(n\omega_T T)$ are isolated functions, which will not contain other bases. Figure 2.1 conceptually shows this independency with a diagram of several amplitudes of cosine functions at corresponding frequencies 0, ω_T, $2\omega_T$, $3\omega_T$, etc. Only the cosine terms are plotted.

From Figure 2.1, it is realized that these coefficients of the cosine functions form a special spectrum. This spectrum has the frequency interval

$$\Delta\omega = n\omega_T - (n-1)\omega_T = \omega_T \tag{2.29}$$

At each frequency point

$$\omega_n = n\omega_T \tag{2.30}$$

is the amplitude a_n. However, in between $n\omega_T$ and $(n-1)\omega_T$, there is nothing. That is, all these spectral lines are isolated and have no relationship with other spectral lines. In other words, all these spectral lines are independent. The same is true for sine functions.

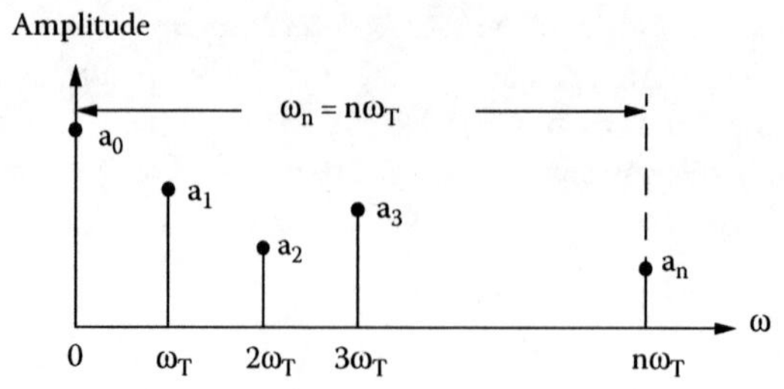

FIGURE 2.1 Cosine spectrum.

By means of Euler's equations:

$$\cos\theta = \frac{e^{j\theta} + e^{-j\theta}}{2} \tag{2.31}$$

and

$$\sin\theta = \frac{e^{j\theta} - e^{-j\theta}}{2j} = -j\frac{e^{j\theta} - e^{-j\theta}}{2} \tag{2.32}$$

Equation 2.17 is rewritten as

$$\begin{aligned} f(t) &= \frac{a_0}{2} + \sum_{n=1}^{\infty}\left(a_n\frac{e^{jn\omega_T t} + e^{-jn\omega_T t}}{2} - jb_n\frac{e^{jn\omega_T t} - e^{-jn\omega_T t}}{2}\right) \\ &= \frac{a_0}{2} + \sum_{n=1}^{\infty}\left(\frac{a_n - jb_n}{2}e^{jn\omega_T t} + \frac{a_n + jb_n}{2}e^{-jn\omega_T t}\right) \end{aligned} \tag{2.33}$$

Now, let

$$c_0 = \frac{a_0}{2} = \frac{1}{T}\int_{-T/2}^{T/2} f(t)dt \tag{2.34}$$

and

$$\begin{aligned} c_n = \frac{a_n - jb_n}{2} &= \frac{1}{T}\left[\int_{-T/2}^{T/2} f(t)\cos n\omega_T t\,dt - j\int_{-T/2}^{T/2} f(t)\sin n\omega_T t\,dt\right] \\ &= \frac{1}{T}\left\{\int_{-T/2}^{T/2} f(t)\left[\cos n\omega_T t - j\sin n\omega_T t\right]dt\right\} \\ &= \frac{1}{T}\left\{\int_{-T/2}^{T/2} f(t)e^{-jn\omega_T t}dt\right\} \quad n = 1,2,\ldots \end{aligned} \tag{2.35}$$

Also

$$c_{-n} = \frac{a_n + jb_n}{2} = \frac{1}{T}\int_{-T/2}^{T/2} f(t)e^{jn\omega_T t}dt \quad n = 1,2,\ldots \tag{2.36}$$

With the help of Equation 2.30, Equations 2.35 and 2.36 can be combined as

$$c_n = \frac{1}{T}\int_{-T/2}^{T/2} f(t)e^{-jn\omega_T t}dt \quad n = 1, 2, \ldots \tag{2.37}$$

Thus, Equation 2.17 can then be rewritten as

$$f(t) = c_0 + \sum_{n=1}^{\infty}\left(c_n e^{jn\omega_T t} + c_{-n}e^{-jn\omega_T t}\right) = \sum_{n=-\infty}^{\infty} c_n e^{jn\omega_T t}$$

That is,

$$f(t) = \sum_{n=-\infty}^{\infty} c_n e^{jn\omega_T t} \tag{2.38}$$

This is the Fourier series in the form of complex exponentials. In this case, the bases are in terms of $e^{jn\omega_T t}$ and the complex-valued constants c_n are also called the *Fourier coefficients*.

Example 2.2

Consider a function written as

$$f(t) = 10 + 5\sin(1.5t) + 2\sin(2.5t + 1) - 0.5\cos(4t - 2)$$

Find the Fourier series that represents the above signal.

First, the period f(t) is found. Since in this case, $\omega_1 = 1.5$ (rad/s), $\omega_2 = 2.5$ (rad/s), and $\omega_3 = 4$ (rad/s), the result is

$$\frac{15}{\omega_1} = \frac{25}{\omega_2} = \frac{40}{\omega_3}$$

The integers 15, 25, and 40 have the maximum common divisor 5. Therefore,

$$\frac{3}{\omega_1} = \frac{5}{\omega_2} = \frac{8}{\omega_3}$$

Therefore, the period T is $2\pi \times (3/1.5) = 4\pi$ (s), so that $\omega_T = (2\pi/T) = 0.5$ (rad/s). Thus, this signal is periodic. Furthermore, for cosine coefficients, $a_0 = 20$, $a_1 = a_2 = a_3 = a_4 = 0$, $a_5 = 2\sin(1)$, $a_6 = a_7 = 0$, and $a_8 = -0.5\cos(2)$. For sine coefficients, $b_1 = b_2 = 0$, $b_3 = 5$, $b_4 = 0$, $b_5 = 2\cos(1)$, $b_6 = b_7 = 0$, and $b_8 = 0.5\sin(2)$. Any term with $n > 3$ is null. In Figure 2.2, these coefficients are plotted vs. the corresponding frequencies. The vertical axis denotes the amplitudes. The a_0 axis is the coefficient of the cosine terms and the b_0 axis is the coefficient of the sine terms. It is seen that the Fourier coefficients and the corresponding frequencies form a 3D spectrum.

In the above examples, the Fourier coefficients are computed directly from the original signals, rather than using the formulas described in Equations 2.26 through 2.28, 2.34, and 2.37. In the following subsection, further discussion about the Fourier series using these formulas is provided.

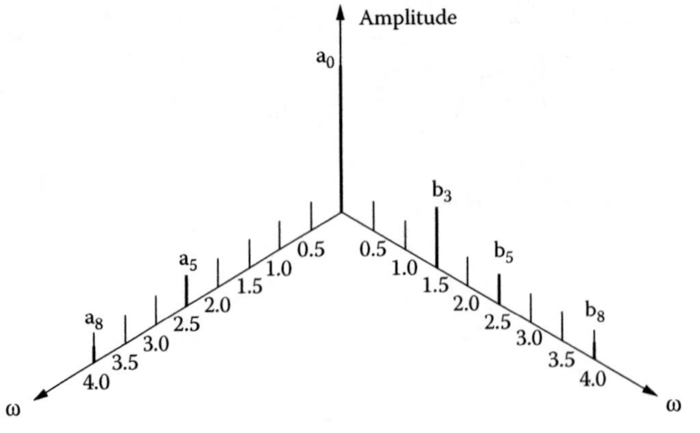

FIGURE 2.2 3D Fourier spectra.

2.1.3 Discrete Fourier Transform

In the above discussion, the Fourier series is used to represent periodic signals. In order to carry out numerical analysis using digital computers, these periodic signals must be converted into computerized signals. That is, the signal in the continuous time domain, which is called an *analog signal*, will be digitized and thus becomes a digital signal in the discrete time domain. This process is also referred to as *sampling*. When an analog signal is sampled into a digital signal, it will usually have a new period. If its original period is infinite, then this new period will be the length of the total samples.

Suppose the total sampling length is T. With such a sampling procedure, the digitized signal will be forced to have a sampling period T, except in the following case. Theoretically, a signal may have T/n as its own period, with integer n. Then, after sampling, the digitized signal will retain the period T/n. This is referred to as *complete period sampling*. Practically, an exact complete period sampling is very rare, except for samples with carefully selected intervals. Thus, most signals that are digitized, despite their own periods, will have periods T, so that they can be represented by the aforementioned Fourier series.

2.1.3.1 Discretization of Signals

In the real world, a dynamic signal x is likely to be a signal in the continuous time domain. Thus, it is denoted as

$$x = x(t) \tag{2.39}$$

Furthermore, let

$$\Delta t \to dt \tag{2.40}$$

except for certain signals that may exhibit discontinuities. Note that, a specific process of sampling must have a fixed time interval, that is,

$$\Delta t = \text{constant} \tag{2.41}$$

In other words, the signal values between Δt will be ignored. In Figure 2.3a, the solid line represents an analog signal, whereas the dots stand for the digitized discrete signals.

From Figure 2.3a, it is realized that once the time interval Δt is determined, the analog time history can be sampled at Δt, 2Δt, 3Δt,... The corresponding x(Δt), x(2Δt), x(3Δt),..., are the samples

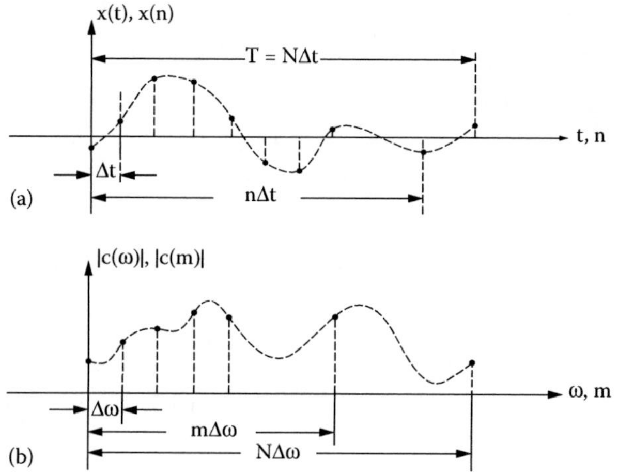

FIGURE 2.3 Discretization of a signal in both time and frequency domains: (a) time domain and (b) frequency domain.

taken from the time history x(t). As a difference from the continuous time history, the collection of x(Δt), x(2Δt), x(3Δt),…, x(nΔt) represents discrete individual values. Thus, the signal x(t) is referred to in the continuous time domain and the series x(nΔt) in the discrete time domain. It is clear that only at the sampling time points,

$$x(t) = x(n\Delta t) \tag{2.42}$$

and

$$t = n\Delta t \tag{2.43}$$

For convenience, the signal is denoted in the discrete time domain as simply x(n).

The inverse of the sampling time interval Δt is referred to as the *sampling frequency*, denoted by f_{SP}, that is,

$$f_{SP} = 1/\Delta t \tag{2.44}$$

From the above discussion, it is seen that the collection of x(nΔt) may not necessarily represent the signal x(t). This is because x(nΔt) is not continuous and certain important time points may be missing. In fact, the condition to recover the analog time history from the digitized series x(nΔt) can be written as follows:

$$f_{SP} \geq 2f_H \tag{2.45}$$

where f_H is the highest frequency contained in the signal x(t). If the condition described by Equation 2.45 is violated, then the signal x(t) will not be recoverable, a condition that is referred to as signal *aliasing*. The corresponding theory is based on the *Nyquist sampling theorem*.

In practical applications, any collection of the values of x(nΔt) has a limited number of samples. Suppose a total of N samples are taken. The time duration will be (see Figure 2.3a)

$$T = N\Delta t \tag{2.46}$$

where T is the aforementioned sampling period. Therefore, the sampled signal is forced to have the period T, unless the case of complete period sampling applies.

2.1.3.2 Discrete Fourier Series

Mathematically, a discrete signal with period T can be expressed using the complex Fourier series described in Equation 2.38. Thus,

$$x(n) = x_n = \sum_{m=0}^{N-1} c_m e^{jm\omega_T n\Delta t} = \sum_{k=0}^{N-1} c_m e^{2\pi jmn/N} \tag{2.47}$$

Here, c_m is a complex-valued Fourier coefficient. The absolute value of $c_m = c(m)$ can be plotted, as in Figure 2.3b. Here, ω_T satisfies Equation 2.18 and $m\omega_T$ is the mth harmonic frequency. The term c_m can be described as

$$c_m = c_m(m) = \frac{1}{T}\sum_{n=0}^{N-1} x_n e^{-jn\omega_T m\Delta t}\Delta t = \frac{1}{N}\sum_{n=0}^{N-1} x_n e^{-2\pi jnm/N} \tag{2.48}$$

Equations 2.47 and 2.48 define the *discrete Fourier transform* (DFT) *pair.* In the next section on transient signals, the Fourier transform in the continuous time and frequency domains is further explored. The relationship between the Fourier pair is discussed in more detail. Here, note that once the period is determined by the actual sampling process, any signal will be practically treated as a periodic signal with a sampling period T, whether it has a true period or not. Figure 2.4 shows the mathematical expansion.

From Figures 2.3a and b, it is visualized that for a limited number of samples, the sampling period T is limited as is the maximum measurable frequency f_H or ω_H. In fact, Equation 2.45 has implied this limitation. That is, generally,

$$f_H = 1/T \tag{2.49}$$

or

$$\omega_H = 2\pi/T \tag{2.50}$$

Taking a digital signal from an analog signal is often referred to as A-to-D (A/D) conversion. In the digitizing process of A/D conversion, the time interval Δt is the *time resolution.* In the spectral analysis, the frequency interval Δf or $\Delta\omega$ is called the *frequency resolution.* These two resolutions are obviously related by the given number of total samples N. That is,

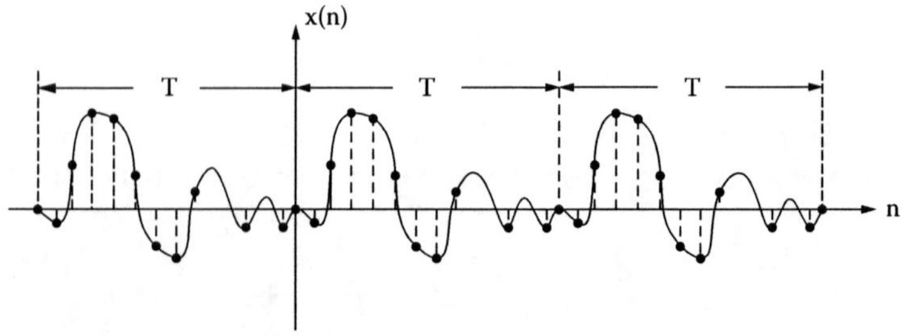

FIGURE 2.4 Periodic expansion of x(n).

$$\Delta f = 1/T \tag{2.51}$$

$$\Delta\omega = 2\pi/T \tag{2.52}$$

$$f_H = (N/2 - 1)\Delta f \tag{2.53}$$

$$\omega_H = (N/2 - 1)\Delta\omega \tag{2.54}$$

Based on the above discussion, it is clear that due to limited sampling time or number of samples, the digitized signal is artificially assigned with the period T, in order to have a Fourier series represent the signals in the discrete time domain.

In vibration measurement and testing, either the forcing function or the response may have a period of its own other than T. In this case, there may be measurement and testing errors. Therefore, for true periodic signals, it is better to adjust the time interval and the total sampling length to be close to the true period.

In Equation 2.47, a total of N terms of c_n are required to represent a single value of x(n). Similarly, from Equation 2.48, it is seen that it takes N terms of x_n to represent a single value of c(n). If there are N terms of x_n, then the N terms of c_n are determined and vice versa. In other words, the N terms of x_n and the N terms of c_n are mutually determined through a full rank linear transfer. Therefore, a full rank n × n matrix **G** can be found such that

$$\begin{Bmatrix} x_0 \\ x_1 \\ \dots \\ x_{n-1} \end{Bmatrix} = \mathbf{G} \begin{Bmatrix} c_0 \\ c_1 \\ \dots \\ c_{n-1} \end{Bmatrix} \tag{2.55}$$

and

$$\begin{Bmatrix} c_0 \\ c_1 \\ \dots \\ c_{n-1} \end{Bmatrix} = \mathbf{F} \begin{Bmatrix} x_0 \\ x_1 \\ \dots \\ x_{n-1} \end{Bmatrix} \tag{2.56}$$

Here

$$\mathbf{F} = \frac{1}{N} \begin{bmatrix} 1 & 1 & \dots & 1 \\ 1 & e^{-j\omega_T \Delta t} & \dots & e^{-j(n-1)\omega_T \Delta t} \\ 1 & e^{-j(n-1)\omega_T \Delta t} & \dots & e^{-j(n-1)^2 \omega_T \Delta t} \end{bmatrix} \tag{2.57a}$$

Note that the Fourier coefficients are complex numbers and Equation 2.56 provides the complex pairs. Their real values are located symmetrically about the point N/2. The imaginary values are also located inverse symmetrically about the point N/2. To plot these coefficients vs. the frequency as a spectrum, the X-axis or the frequency axis needs to be determined. The unit of the frequency axis is also referred to as the frequency resolution or frequency interval, which is given in Equation 2.51. The total length of the frequency range is NΔf.

Often, when only a real or imaginary or absolute value is needed, only half of the spectrum needs to be plotted. In this case, the total length of the frequency range or frequency band is NΔf/2. Thus, the full amplitude of the absolute values of the Fourier coefficients needs to be plotted; that is, instead of using Equation 2.57a,

$$\mathbf{F} = \frac{2}{N}\begin{bmatrix} 1 & 1 & \ldots & 1 \\ 1 & e^{-j\omega_T \Delta t} & \ldots & e^{-j(n-1)\omega_T \Delta t} \\ 1 & e^{-j(n-1)\omega_T \Delta t} & \ldots & e^{-j(n-1)^2 \omega_T \Delta t} \end{bmatrix} \tag{2.57b}$$

Further, from Equations 2.55 and 2.56, it is seen that

$$\mathbf{F} = \mathbf{G}^{-1} \tag{2.58}$$

Example 2.3

Suppose there is a signal

$$x(t) = 5\sin(2\pi t) - 3\cos(2\pi \times 3 \times t + 1) + 2\sin\left(2\pi \times \sqrt{21} \times t - 2\right)$$

Since $\sqrt{21}$ is an irrational number, this signal has no period. However, if 300 samples are taken with a time interval $\Delta t = 10/300$ (s), the sampling period is 10 (s). Using Equations 2.57 and 2.56, the corresponding Fourier coefficients c_n can be calculated. The real part and imaginary part of the Fourier coefficients are plotted, respectively, in Figure 2.5a and b where the x axes denote the sample points.

By using Equation 2.57b, the absolute values of the Fourier coefficients are plotted in Figure 2.6, with a frequency band from 0 to NΔf/2 = 150. Note that since T = 10 (s), according to Equation 2.51, Δf = 0.1 (Hz).

Note that in Figure 2.6, the amplitudes at 1 and 3 (Hz) are exactly 5 and 3. This is because these frequency components are complete or full-period sampled. However, the amplitude of $\sqrt{21}$ (Hz) is 1.9057, instead of 2, because the irrational frequency component cannot be full-period sampled. This phenomenon is called *power leakage*.

Also note that there is no amplitude at the period T = 10 (s) in Figure 2.6. This means that, although the digitized signal x(n) is forced to have a period 10 (s), there is no corresponding amplitude shown in the spectrum of its Fourier series.

2.1.4 General Damping Force

The nonlinear damping force can be represented using the Fourier series. Note that two different situations may exist. The first is when the damping force is generated by an SDOF system, for which, with or without the nonlinear damping, the total system including the damping force is nonlinear. The response of such a system under a sinusoidal excitation contains more frequency components than the driving frequency. In other words, the response displacement or velocity will not be a pure sinusoidal signal.

In the second case, the nonlinear damping force is caused by nonlinear dampers only. In this case, such a nonlinear damper is often tested by a forced sinusoidal movement. In other words, the damper velocity is sinusoidal. In Chapter 1, the concept of effective damping is based on such assumptions. Namely, the relative displacement or velocity of the moving end of the damper (compared with the fixed end of the damper) is forced to be sinusoidal and the damping force is measured accordingly. Since the nature of the damper is nonlinear, the damping force becomes nonlinear.

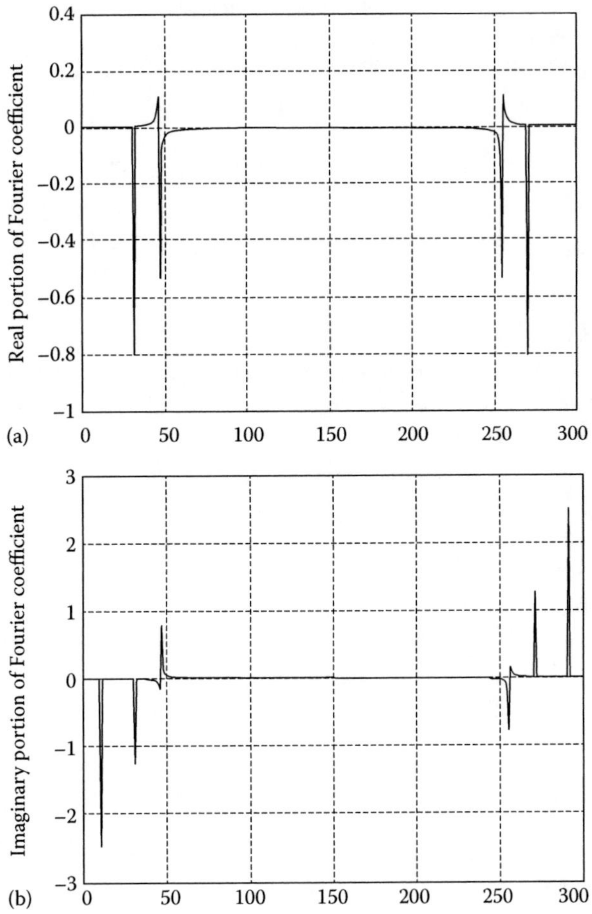

FIGURE 2.5 Plot of complex-valued Fourier coefficients: (a) real part and (b) imaginary part.

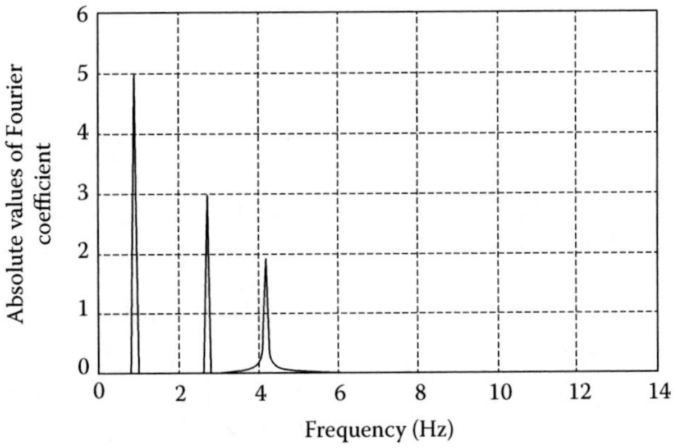

FIGURE 2.6 Fourier spectrum.

In this section, only the second case of a nonlinear damping force is discussed. Suppose the damper under test is forced to have a sinusoidal movement in the form

$$x(t) = x_0 \sin(\omega_f t) \tag{2.59}$$

Thus, the velocity is

$$\dot{x}(t) = x_0 \omega_f \cos(\omega_f t) \tag{2.60}$$

2.1.4.1 Linearity of Fourier Series

The linearity of the transfer matrix **F** expressed in Equations 2.56 and 2.57 actually implies the linear relationship between the temporal variable x(t) and the Fourier coefficient c_n. In the following discussion, only the case of DFT is described. Denoting

$$\mathbf{x} = [x_1, x_2, \ldots, x_n]^T \tag{2.61}$$

and

$$\mathbf{c} = [c_1, c_2, \ldots, c_n]^T \tag{2.62}$$

Equation 2.56 is rewritten as

$$\mathbf{c} = \mathbf{F}\mathbf{x} \tag{2.63a}$$

Equation 2.63a implies that the matrix of the Fourier series **c** can be seen as a result of an operation $\mathcal{F}$ acting on **x**, that is,

$$\mathbf{c} = \mathcal{F}(\mathbf{x}) \tag{2.63b}$$

Here, $\mathcal{F}$ is also used to denote Fourier transformation in the following text. Apparently,

$$\mathcal{F}(a\mathbf{x} + b\mathbf{y}) = a\mathcal{F}(\mathbf{x}) + b\mathcal{F}(\mathbf{y}) \tag{2.64}$$

Here,

$$\mathbf{y} = [y_1, y_2, \ldots, y_n]^T \tag{2.65}$$

where the term $y_{(.)}$ are a different series from a temporal signal.

The above-mentioned linearity of operation $\mathcal{F}$ can be used to deal with the summation of the damping force and the spring forces, namely, the seismic force.

2.1.4.2 Steady-State Structural Force

The steady-state structural force of a linear m-c-k system excited by a harmonic force f_0 in the form of $\sin(\omega_f t)$ can be represented by a Fourier series. Denoting the amplitude of the displacement to be x_0, the displacement $x(t) = x_0 \sin(\omega_f t + \phi)$ and the structural force are given by (see Equation 1.3):

$$f_S(t) = c\dot{x}(t) + kx(t) = c\omega_f x_0 \cos(\omega_f t + \phi) + kx_0 \sin(\omega_f t + \phi)$$

$$= x_0 \frac{1}{\sqrt{c^2\omega_f^2 + (k - m\omega_f^2)^2}} \left\{ \left[c\omega_f (k - m\omega_f^2) - kc\omega_f \right] \cos(\omega_f t) + \left[(c\omega_f)^2 + k(k - m\omega_f^2) \right] \sin(\omega_f t) \right\} \tag{2.66}$$

The period of the structural force is $2\pi/\omega_f$ and the basic frequency is $\omega_T = \omega_f$. Therefore,

$$f_S(t) = a_1 \cos(\omega_f t) + b_1 \sin(\omega_f t) \tag{2.67}$$

Here, a_1 and b_1 are coefficients and

$$a_1 = x_0 \frac{1}{\sqrt{c^2\omega_f^2 + (k - m\omega_f^2)^2}} \left\{ \left[c\omega_f (k - m\omega_f^2) - kc\omega_f \right] \right\}$$

$$= -cm\omega_f^3 x_0 \frac{1}{\sqrt{c^2\omega_f^2 + (k - m\omega_f^2)^2}}$$

Since

$$x_0 = \frac{f_0}{k}\beta = \frac{f_0}{k} \frac{1}{\sqrt{(1 - r^2)^2 + (2\xi r)^2}} = \frac{f_0}{\sqrt{(k - m\omega_f^2)^2 + (c\omega_f)^2}}$$

$$a_1 = \frac{-cm\omega_f^3 f_0}{(k - m\omega_f^2)^2 + (c\omega_f)^2} \tag{2.68}$$

and

$$b_1 = x_0 \frac{1}{\sqrt{c^2\omega_f^2 + (k - m\omega_f^2)^2}} \left[(c\omega_f)^2 + k(k - m\omega_f^2) \right]$$

$$= \frac{f_0 \left[c^2\omega_f^2 + k(k - m\omega_f^2) \right]}{c^2\omega_f^2 + (k - m\omega_f^2)^2} \tag{2.69}$$

If the damping force can be represented by $a\cos(\omega_f t)$ and the displacement can be represented by $b\sin(\omega_f t)$, then from the discussion of parametric equations in Chapter 1, the energy dissipation can be calculated through

$$E_d = \pi ab \tag{2.70}$$

When $\omega_f = \omega_n$, E_d becomes the damping capacity, and the damping ratio can be calculated through

$$\xi = \frac{E_d}{4\pi E_p} = \frac{\pi a}{2\pi kb} \tag{2.71}$$

Now, the coefficient a can be obtained by letting $t = 0$ and assuming $\omega_f = \omega_n$, that is,

$$f_d(0) = a = b_1\big|_{\omega_f=\omega_n} = \frac{f_0\left[c^2\omega_f^2 + k(0)\right]}{c^2\omega_f^2 + (0)^2} = f_0 \tag{2.72}$$

In addition, $b = x_0$. Now, the value of $E_d/4\pi E_p$ is examined,

$$\frac{E_d}{4\pi E_p} = \frac{\pi f_0 x_0}{2\pi k x_0^2} = \frac{f_0}{2kx_0}$$
$$= \frac{f_0}{2k\left[\dfrac{f_0}{\sqrt{(0)^2 + (c\omega_n)^2}}\right]} = \frac{c\omega_n}{2k} = \frac{2\xi\omega_n m\omega_n}{2k} = \xi \tag{2.73}$$

The structural force can be seen as the summation of the damping force $f_d = c\dot{x}(t)$ and the restoring force $f_r = kx(t)$. Following the above discussion, (also see Equation 1.220), denote the damping force and the displacement as the following Fourier series,

$$f_d(t) = f_1\cos(\omega_f t) + f_2\sin(\omega_f t) \tag{2.74a}$$

$$x(t) = x_1\cos(\omega_f t) + x_2\sin(\omega_f t) \tag{2.74b}$$

where

$$f_1 = c\omega_f x_0\cos(\phi) \text{ and } x_1 = x_0\sin(\phi) \tag{2.75}$$

$$f_2 = -c\omega_f x_0\sin(\phi) \text{ and } x_2 = x_0\cos(\phi) \tag{2.76}$$

In this case, again using Equation 2.70, the following can be directly obtained:

$$E_d = \pi c\omega_f x_0^2 \tag{2.77}$$

The above discussion on using the Fourier series to represent the linear damping and restoring force seems trivial. However, when the damping and restoring forces are nonlinear, as long as the total structural force can be represented by their summation, the operation of finding the corresponding Fourier series is linear. Suppose,

$$f_S(n) = c\dot{x}(n) + f_{dN}(n) + kx(n) + f_{rN}(n) \tag{2.78}$$

Here, for convenience, the discrete time series is used to represent the temporal signals. On the right side of Equation 2.78, $c\dot{x}(n)$ and $kx(n)$ are the linear viscous damping force and linear restoring force. The terms $f_{dN}(n)$ and $f_{rN}(n)$, respectively, are the nonlinear viscous damping force and nonlinear restoring force.

Thus,

$$\mathcal{F}\left[f_S(n)\right] = \mathcal{F}\left[c\dot{x}(n) + f_{dN}(n) + kx(n) + f_{rN}(n)\right]$$
$$= c\mathcal{F}\left[\dot{x}(n)\right] + \mathcal{F}\left[f_{dN}(n)\right] + k\mathcal{F}\left[x(n)\right] + \mathcal{F}\left[f_{rN}(n)\right] \tag{2.79}$$

Therefore, when the conditions described in Equations 2.59 and 2.60 hold, namely, when the system with arbitrary damping is tested by the harmonic forcing function, the Fourier series can be used to represent the structural force. Particularly, when the restoring force is linear with stiffness k and the driving frequency is equal to $\sqrt{k/m}$, only the Fourier coefficients that correspond to the effect of energy dissipation are counted.

In the following discussion, two examples of the types of damping forces primarily used in practice are examined.

2.1.4.3 Dry Friction Damping

First, consider dry friction (Coulomb friction) damping. The friction damping force can be expressed as

$$f_d(t) = \mu N \operatorname{sgn}(\dot{x}) \tag{2.80}$$

Note that

$$\operatorname{sgn}(\dot{x}) = \operatorname{sgn}\left[\cos\omega_f t\right] = \operatorname{sgn}\left[\cos(-\omega_f t)\right] \tag{2.81}$$

Thus, from Equations 2.80 and 2.81, it is seen that,

$$f_d(t) = f_d(-t) \tag{2.82}$$

Therefore, the damping force generated by dry friction damping is an even function of time t. It is understandable that an even function, f(t), will generate an even product with even-valued $\cos(n\omega_f t)$. In this case, the corresponding integration between the region (–T/2, 0) will be identical to that in the region (0, T/2).

However, the product with $\sin(n\omega_f t)$, which is odd, will also be odd. Therefore, the corresponding integration between the region (–T/2, 0) will be opposite to that in the region (0, T/2). In other words, the integrations in the two regions will cancel each other and the resulting Fourier series will no longer have sine terms. That is,

$$a_n = \frac{2}{T}\int_{-T/2}^{T/2} f(t)\cos n\omega_f t\,dt = \frac{8}{T}\int_0^{T/4} f(t)\cos n\omega_f t\,dt \tag{2.83}$$

and

$$b_n = \frac{2}{T}\int_{-T/2}^{T/2} f(t)\sin n\omega_f t\,dt = 0 \tag{2.84}$$

In addition, from Equation 2.80, it is seen that the mean value of the damping force is zero, that is,

$$a_0 = 0 \tag{2.85}$$

In the case $\omega_f = \omega_n$, the Fourier coefficient a_n can be written as

$$a_n = \frac{8}{T}\mu N\int_0^{T/4} \cos n\omega_n t\,dt = \frac{8\mu N}{Tn\omega_n}\sin n\omega_n t\Big|_0^{\pi/2\omega_n} = \frac{4\mu N}{n\pi}\sin\frac{n\pi}{2} \tag{2.86}$$

Since the damping force is an even function,

$$b_n = \frac{2}{T}\int_{-T/2}^{T/2} f(t)\sin n\omega_n t\,dt = 0 \tag{2.87}$$

Generally,

$$f_d(t) = \mu N \operatorname{sgn}(\dot{x}) = \frac{4\mu N}{\pi}\left(\cos\omega_f t - \frac{1}{3}\cos 3\omega_f t + \frac{1}{5}\cos 5\omega_f t - \cdots\right) \tag{2.88}$$

In Figure 2.7a, a normalized friction damping force and the corresponding Fourier series with the first Fourier term, the first four terms, and the first nine terms are plotted. It is seen that the more Fourier terms used, the closer the truncated series is to the original signal. However, as shown in Figure 2.7a, an overshoot is still in the corner, which is caused by the discontinuity of the signal and is referred to as the *Gibbs phenomenon*. It can be proven that when sufficient Fourier terms are used, the amplitude of this overshoot will be in the range of about 9%.

If only the first term is used to approximate the damping force, that is, if

$$f_d(t) \approx \frac{4\mu N}{\pi}(\cos\omega_f t) \tag{2.89}$$

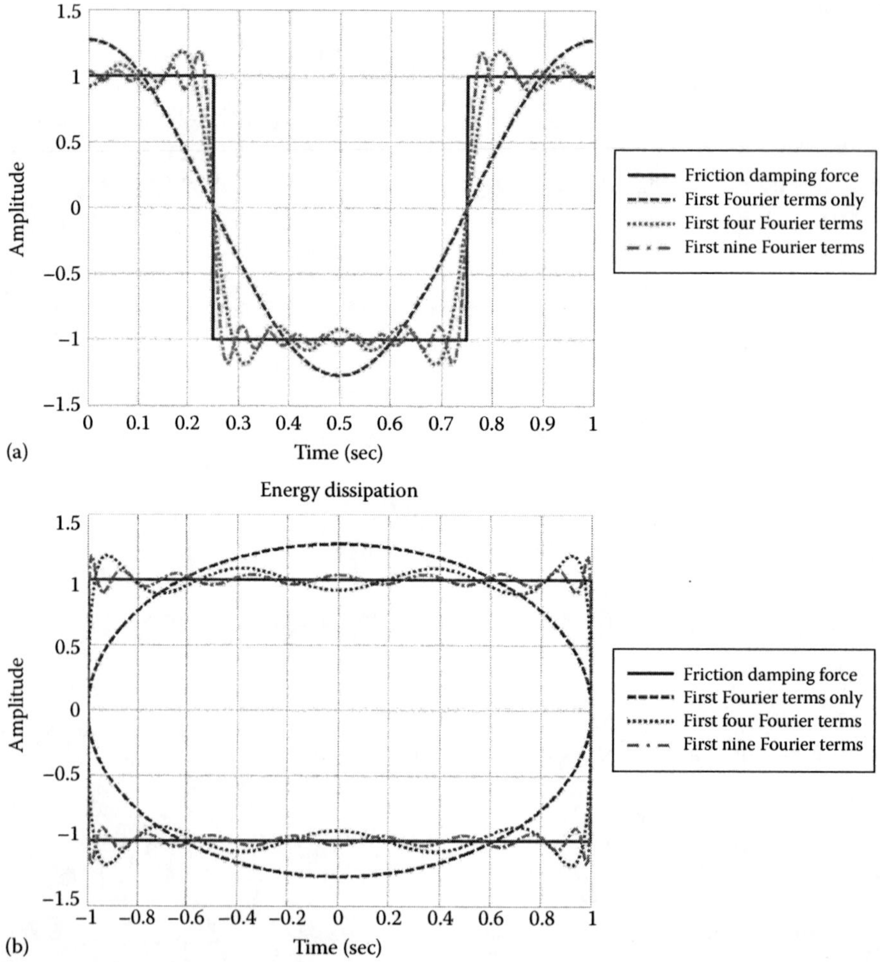

FIGURE 2.7 Normalized dry friction damping force and its Fourier series: (a) time history and (b) energy dissipation loop.

then from Chapter 1 it is seen that during one cycle of $x = \sin(\omega_f t)$, namely, a displacement with unit amplitude, the dissipated energy is

$$E_d \approx \pi\left(\frac{4\mu N}{\pi}\right)(1) = 4\mu N \tag{2.90}$$

However, it has already been seen that the energy dissipation by friction force during a cycle is

$$E_d = 4\mu N \tag{2.91}$$

Note that, since only a damper but not the vibration system is considered. E_d is only the dissipated energy but not the damping capacity.

In Figure 2.7b, the energy dissipations are shown by the enclosed areas. Each area of the corresponding damping force, as well as the representation by the Fourier series, is marked in this figure.

Comparing Equations 2.90 and 2.91, it is seen that if only the first term of the Fourier series is used, the calculated energy dissipation is exactly equal to 4 μN.

In fact, since

$$E_d = \int_{-T/2}^{T/2} f_d(t)\,dx(t) = \int_{-T/2}^{T/2} f_d \cos\omega_f t\, d\omega_f t \tag{2.92}$$

in this case,

$$\begin{aligned} E_d &= \int_{-T/2}^{T/2} \frac{4\mu N}{\pi}\left(\cos\omega_f t - \frac{1}{3}\cos 3\omega_f t + \frac{1}{5}\cos 5\omega_f t - \cdots\right)\cos\omega_f t\, d\omega_f t \\ &= \frac{4\mu N}{\pi}\int_{-T/2}^{T/2}(\cos\omega_f t)\cos\omega_f t\, d\omega_f t - \frac{4\mu N}{3\pi}\int_{-T/2}^{T/2}\cos 3\omega_f t\cos\omega_f t\, d\omega_f t \\ &\quad + \frac{4\mu N}{5\pi}\int_{-T/2}^{T/2}\cos 5\omega_f t\cos\omega_f t\, d\omega_f t - \cdots \end{aligned} \tag{2.93}$$

In Equation 2.93, the first term on the right side is

$$\frac{4\mu N}{\pi}\pi = 4\mu N$$

which agrees with Equation 2.90.

Note that, due to the aforementioned orthogonality of cosine functions,

$$\int_{-T/2}^{T/2} \cos(2n+1)\omega_f t \cos\omega_f t\, d\omega_f t = 0 \tag{2.94}$$

Therefore, the terms other than the first one in Equation 2.93 are zero. That is, any Fourier components other than the first term in Equation 2.88 will not contribute to the energy dissipation. This observation will simplify the computation of dissipative energy, and it is not limited to dry friction force.

Furthermore, when a friction damper is combined with certain linear restoring forces, the following combination is possible:

$$f_d(t) = \mu N \,\text{sgn}(\dot{x}) + kx \tag{2.95}$$

In this case, a special bilinear damping occurs, which is plotted in Figure 2.8a and b, by assuming that the stiffness $k = 1$ and $\mu N = 1$.

As previously discussed (see Equation 2.62), to find the Fourier series to represent this bilinear force,

$$\mathcal{F}\left[f_d(n)\right] = \mu N \mathcal{F}\left[\text{sgn}(\dot{x})\right] + k\mathcal{F}\left[x(n)\right] \tag{2.96}$$

Equation 2.96 shows that the Fourier series of the bilinear damping force can be derived by separately considering the term $\mu N \,\text{sgn}(\dot{x})$ and kx. It can be seen that the energy dissipation by kx is zero,

$$\int_{-T/2}^{T/2} kx \cos \omega_f t \, d\omega_f t = kx_0 \int_{-T/2}^{T/2} \sin \omega_f t \cos \omega_f t \, d\omega_f t = 0 \tag{2.97}$$

Therefore, the energy dissipation is calculated by the term $\mu N \,\text{sgn}(\dot{x})$ only, which was discussed earlier. The corresponding energy dissipation can be seen by using Figure 2.8b, where for convenience $k = 1$ and $\mu N = 1$. From this plot, the area is $2 \times 2 = 4$.

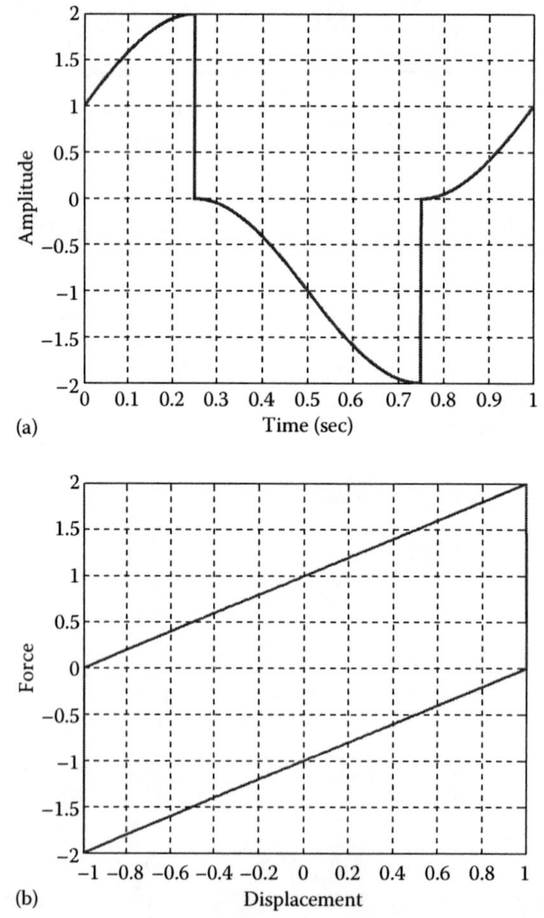

FIGURE 2.8 Bilinear damping force: (a) time history and (b) energy dissipation loop.

2.1.4.4 General Nonlinear Viscous Damping

In the literature, the general nonlinear viscous damping force is often written as

$$f_d(t) = c|\dot{x}|^{\beta} \operatorname{sgn}(\dot{x}) \tag{2.98}$$

where β is called the damping exponent. Under the test condition described by Equations 2.55 and 2.56,

$$\begin{aligned} f_d(t) &= c(x_0\omega_f)^{\beta} |\cos(\omega_f t)|^{\beta} \operatorname{sgn}[\cos(\omega_f t)] \\ &= \begin{cases} c(x_0\omega_f)^{\beta} \cos(\omega_f t)^{\beta}, & -\pi/2\,\omega_f \le t < \pi/2\,\omega_f \\ -c(x_0\omega_f)^{\beta} [-\cos(\omega_f t)]^{\beta}, & -\pi/\omega_f \le t < -\pi/2\,\omega_f, \text{ and } \pi/2\,\omega_f \le t < \pi/\omega_f \end{cases} \end{aligned} \tag{2.99}$$

Since

$$f_d(-t) = c(x_0\omega_f)^{\beta} |\cos(-\omega_f t)|^{\beta} \operatorname{sgn}[\cos(-\omega_f t)] = f_d(t) \tag{2.100}$$

the damping force is also an even function. Thus, the computation of the Fourier coefficient a_n is considered as

$$a_n = \frac{2}{T} c(x_0\omega_f)^{\beta} \int_{-T/2}^{T/2} \cos(\omega_f t)^{\beta} \cos n\omega_f t \, dt \tag{2.101}$$

Here, the period T is exactly the period of ω_f, the driving frequency. In this case, Equation 2.101 can be rewritten as

$$\begin{aligned} a_n &= \frac{c(\omega_f x_0)^{\beta}}{\pi} \int_{-T/2}^{T/2} |\cos(\omega_f t)|^{\beta} \operatorname{sgn}[\cos(\omega_f t)] \cos n\omega_f t \, d\omega_f t \\ &= \frac{c(\omega_f x_0)^{\beta}}{\pi} \int_0^{\frac{\pi}{2\omega_f}} 4 \cos(\omega_f t)^{\beta} \cos n\omega_f t \, d\omega_f t \end{aligned} \tag{2.102}$$

First, consider the Fourier coefficient when n = 1. The integral in Equation 2.102 can be denoted and evaluated as

$$A_{\beta} = 4\int_0^{\frac{\pi}{2\omega_f}} \cos^{\beta+1}(\omega_f t) \, d\omega_f t = \frac{2\sqrt{\pi}\,\Gamma\left(\frac{\beta+2}{2}\right)}{\Gamma\left(\frac{\beta+3}{2}\right)} \tag{2.103}$$

where Γ(.) is the Gamma function.

Thus,

$$a_1 = c\,(x_0\omega_f)^{\beta} / \pi A_{\beta} \tag{2.104}$$

Similar to the case of the dry friction force, it can be proven that

$$a_n = 0, \quad n = \text{even} \tag{2.105a}$$

and

$$b_n = 0, \quad n = 1, 2, \ldots \tag{2.105b}$$

Therefore consider the Fourier terms when n = 3, 5, 7, 9, …. Note that

$$\cos(3\omega_f t) = 4\cos(\omega_f t)^3 - 3\cos(\omega_f t) \tag{2.106a}$$

Furthermore,

$$\cos(5\omega_f t) = 16\cos(\omega_f t)^5 - 20\cos(\omega_f t)^3 + 5\cos(\omega_f t) \tag{2.106b}$$

$$\cos(7\omega_f t) = 64\cos(\omega_f t)^7 - 112\cos(\omega_f t)^5 + 56\cos(\omega_f t)^3 - 7\cos(\omega_f t) \tag{2.106c}$$

$$\cos(9\omega_f t) = 256\cos(\omega_f t)^9 - 576\cos(\omega_f t)^7 + 432\cos(\omega_f t)^5 - 120\cos(\omega_f t)^3 + 9\cos(\omega_f t) \tag{2.106d}$$

and so on.

The Gamma function satisfies the following property:

$$\Gamma(\sigma+1) = \sigma\Gamma(\sigma),\ \sigma > 0 \tag{2.107}$$

Thus,

$$\begin{aligned} a_3 &= \frac{c(\omega_f x_0)^\beta}{\pi}\int_0^{\frac{\pi}{2\omega_f}} 4\left[4\cos(\omega_f t)^{\beta+3} - 3\cos(\omega_f t)^{\beta+1}\right] d\omega_f t \\ &= c(x_0\,\omega_f)^\beta/\pi(\beta-1)/(\beta+3)A_\beta = (\beta-1)/(\beta+3)a_1 \end{aligned} \tag{2.108a}$$

Furthermore, it can be proven

$$a_5 = (\beta-1)(\beta-3)/(\beta+3)/(\beta+5)a_1 \tag{2.108b}$$

$$a_7 = (\beta-1)(\beta-3)(\beta-5)/(\beta+3)/(\beta+5)(\beta+7)a_1 \tag{2.108c}$$

$$a_9 = (\beta-1)(\beta-3)(\beta-5)(\beta-7)/(\beta+3)/(\beta+5)(\beta+7)(\beta+9)a_1 \tag{2.108d}$$

and so on. In Figure 2.9, these Fourier coefficients are plotted with special values of damping exponents $\beta = 0$, 0.5 and 1. In Figure 2.9, the plots are normalized by letting $c = \omega_f = 0 = 1$;

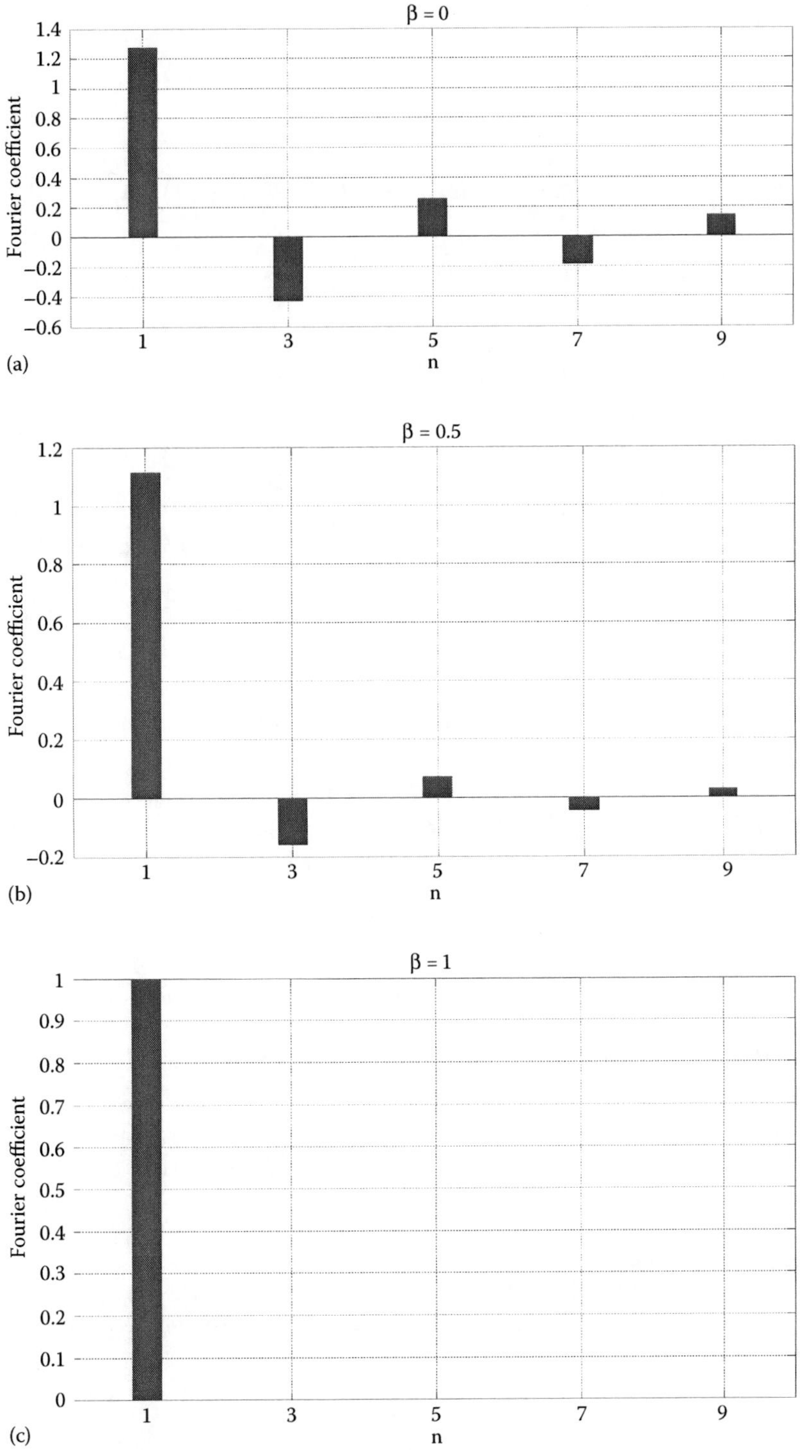

FIGURE 2.9 Fourier coefficients: (a) $\beta = 0$, (b) $\beta = 0.5$, and (c) $\beta = 1$.

Now, consider the energy dissipation.

$$E_d = \int_{-T/2}^{T/2} f_d\, x_0 \cos(\omega_f t)\, d\omega_f t$$

$$= \int_{-T/2}^{T/2} \left[a_1 \cos(\omega_f t) + a_3 \cos(3\omega_f t) + a_5 \cos(5\omega_f t) \cos(\omega_f t) + \ldots\right] x_0 d\omega_f t \quad (2.109)$$

Due to the orthogonality of cosine functions, Equation 2.109 is reduced to

$$E_d = \int_{-T/2}^{T/2} \left[a_1 \cos(\omega_f t)\right] x_0 \cos(\omega_f t) d(\omega_f t)$$

$$= c x_0^{\beta+1} \omega_f^{\beta} / \pi A_\beta \int_{-T/2}^{T/2} \cos(\omega_f t)^2 d(\omega_f t) = c x_0^{\beta+1} \omega_f^{\beta} A_\beta \quad (2.110)$$

Therefore, the energy dissipation during a cycle is only contributed by the first Fourier term of the damping force.

Consider the special case when $\beta = 0$, the nonlinear viscous damping is reduced to the dry friction damping. It is seen that $A_\beta = 4$; and since $c = \mu N$

$$a_1 = c(x_0 \omega_f)^0 / \pi A_\beta = 4\mu N/\pi \quad (2.111)$$

which is exactly the quantity given in Equation 2.86 for the friction damper.

Consider another special case when $\beta = 1$, the viscous damping becomes linear. In this case, the first term of the Fourier coefficient described in Equation 2.17 is

$$a_1 = c(x_0 \omega_f)^1 / \pi A_\beta = c x_0 \omega_f \quad (2.112)$$

where $A_\beta = \pi$. And, only the first Fourier term exists. In this case, the value of the amplitude of the linear viscous damping force is recovered.

From Equation 2.112, when $\beta = 1$ and using $\omega_f = \omega_n$ for a vibration system, the energy dissipation E_d becomes damping capacity, and

$$E_d = c(x_0 \omega_f)^1 / \pi A_\beta \int_{-T/2}^{T/2} \cos(\omega_f t) \cos(\omega_f t) d(\omega_f t) = c\pi \omega_f x_0^2 \quad (2.113)$$

which agrees with Equation 1.172.

From the above discussion, it is seen that, when the displacement and the damping force can be described by Equations 2.59 and 2.60, only the first term of the damping force represented by the Fourier series, denoted by $f_{d1}(t)$, contributes to the dissipative energy. That is,

$$f_{d1} \equiv a_1 \cos(\omega_f t) \quad (2.114)$$

and

$$E_d = \int_{-T/2}^{T/2} a_1 \cos(\omega_f t)\, x_0 d(\omega_f t) = \pi a_1 x_0 \quad (2.115)$$

In Chapter 1, it was seen that the energy dissipation or the damping capacity E_d would play an important role in defining the effective damping ratio. Equation 2.115 provides a simple formula to calculate E_d, when the first term in the Fourier series of the damping force is known.

Note that in the above equations, the symbol c is used to denote the damping coefficient of a general viscous damper, which was originally used to denote the linear viscous damping (see Equation 1.1). In Chapter 5, it will be seen that for nonlinear damping, the parameters of an individual damper or parameters of the entire system need to be distinguished. In the former case, the symbol c_{eq} is used to replace c and in the latter case, c_{eff} is used. Also in Chapter 5, the nonlinear damping force will be used to compute the energy dissipation and further model the effective damping ratios.

2.1.5 Response to Periodic Excitations

2.1.5.1 General Response

Next, a linear m-c-k system excited by a periodic forcing function f(t) of period T, with initial conditions x_0 and v_0, is considered. That is,

$$\begin{cases} m\ddot{x} + c\dot{x} + kx = f(t) \\ x(0) = x_0 \\ \dot{x}(0) = v_0 \end{cases} \tag{2.116}$$

Since the excitation is periodic, a Fourier series with a basic frequency of $\omega_T = 2\pi/T$ is used to represent the forcing functions. Suppose f(t) can be represented by the Fourier series

$$f(t) = \frac{f_{A0}}{2} + \sum_{n=1}^{N} \left[f_{An} \cos(n\omega_T t) + f_{Bn} \sin(n\omega_T t) \right] \tag{2.117}$$

where f_{A0}, f_{An}, and f_{Bn} are Fourier coefficients. Since the system is linear, the responses are first considered individually due to the forcing function

$$f_A = \frac{f_{A0}}{2} \tag{2.118a}$$

with the initial conditions denoted by $x_0(t)$ and the steady-state responses due to the forcing functions

$$f_{an}(t) = f_{An} \cos(n\omega_T t) \tag{2.118b}$$

$$f_{bn}(t) = f_{Bn} \sin(n\omega_T t) \tag{2.118c}$$

denoted by $x_{an}(t)$ and $x_{bn}(t)$.

The total response can then be seen as the summation of $x_0(t)$ and all $x_{an}(t)$ and $x_{bn}(t)$.

Thus,

$$x(t) = x_0(t) + \sum_{n=1}^{N} \left[x_{an}(t) + x_{bn}(t) \right] = x_0(t) + \sum_{n=1}^{N} \left[x_n(t) \right] \tag{2.119}$$

where

$$x_n(t) = x_{an}(t) + x_{bn}(t) \tag{2.120a}$$

2.1.5.2 The nth Steady-State Response

The steady-state response corresponding to the nth excitation component described by Equation 2.120a can also be written as

$$x_n(t) = \frac{f_N}{k}\beta_n \sin(n\omega_T t + \phi_n) \tag{2.120b}$$

Here,

$$f_N = \sqrt{f_{An}^2 + f_{Bn}^2} \tag{2.121}$$

is the amplitude of the nth forcing function, and the dynamic magnification β_n as well as phase angle ϕ_n are

$$\beta_n = \frac{1}{\sqrt{\left(1 - \frac{n^2\omega_T^2}{\omega_n^2}\right)^2 + \left(2\xi\frac{n\omega_T}{\omega_n}\right)^2}} = \frac{1}{\sqrt{(1-n^2r^2)^2 + (2\xi nr)^2}} \tag{2.122}$$

$$\phi_n = \tan^{-1}\left[\frac{2\xi nr}{n^2r^2 - 1}\right] + \tan^{-1}\left(\frac{f_{An}}{f_{Bn}}\right) + (h_\phi + h_\Phi)\pi \tag{2.123}$$

where

$$h_\phi = \begin{cases} 0 & n^2r^2 > 1 \\ -1 & n^2r^2 \le 1 \end{cases} \tag{2.124a}$$

and

$$h_\Phi = \begin{cases} 0 & f_{Bn} \ge 0 \\ 1 & f_{Bn} < 0 \end{cases} \tag{2.124b}$$

In Equations 2.122, 2.123, and 2.124a, r is the frequency ratio and

$$r = \frac{\omega_T}{\omega_n} \tag{2.125}$$

2.1.5.3 Transient Response

Assume that the transient response due to the initial condition and the force f_A, described by Equation 2.116 is

$$x_0(t) = e^{-\xi\omega_n t}\left[A\cos(\omega_d t) + B\sin(\omega_d t)\right] + \frac{f_A}{k} \tag{2.126}$$

The first term in Equation 2.126 is mainly generated by the initial conditions, which are discussed in Chapter 1, and the second term is a particular solution due to the step input described in Equation 2.118a. In the next section on transient excitation, more details of this solution are shown.

It can be proven that in Equation 2.126, the coefficients A and B are

$$A = x_0 - \frac{f_A}{k} - \sum_{n=1}^{N} \frac{f_N}{k} \beta_n \sin\phi_n \tag{2.128}$$

$$B = \frac{1}{\omega_d}\left(v_0 + A\xi\omega_n - \sum_{n=1}^{N} \frac{f_N}{k} \beta_n n\omega_T \cos\phi_n \right) \tag{2.129}$$

2.2 TRANSIENT EXCITATIONS

2.2.1 Transient Signals

Due to periodic excitations, the responses can be periodic signals; in general, there can be signals without a period.

For example, the function

$$f(t) = \sin t + \sin \pi t \tag{2.130}$$

does not have a period. This is because $1/\pi$ cannot be represented by a ratio of two integers n/m. Alternatively, it can be said that the period of $\sin t + \sin \pi t$ is infinitely long, denoted by

$$T \to \infty \tag{2.131}$$

In this circumstance, the Fourier series cannot be used, except for the case of DFTs. In fact, when the period extends toward infinity, the frequency ω_T becomes infinitesimally small, and alternative mathematical tools in the form of the Fourier integral and Fourier transform will need to be used. Furthermore, in many cases, the Fourier integral may not even exist; thus, an improved tool, namely, the Laplace transform is used instead.

2.2.2 Fourier Transform

Definition

Substituting Equation 2.37 into Equation 2.38 yields

$$f(t) = \frac{1}{T} \sum_{n=-\infty}^{\infty} \left[\int_{-T/2}^{T/2} f(t) e^{-j\omega_n t} dt \right] e^{j\omega_n t} \tag{2.132}$$

The condition of Equation 2.131 is examined next, for which Equation 2.132 can be rewritten as

$$f(t) = \lim_{T\to\infty} \frac{1}{T} \sum_{n=-\infty}^{\infty} \left[\int_{-T/2}^{T/2} f(t) e^{-j\omega_n t} dt \right] e^{j\omega_n t} \tag{2.133}$$

From Equation 2.18,

$$\frac{1}{T} = \frac{\Delta\omega}{2\pi} \tag{2.134}$$

Substituting Equation 2.134 into Equation 2.133 further yields

$$f(t) = \lim_{\substack{T\to\infty \\ \Delta\omega\to 0}} \frac{\Delta\omega}{2\pi} \sum_{n=-\infty}^{\infty} \left[\int_{-T/2}^{T/2} f(t) e^{-j\omega_n t} dt \right] e^{j\omega_n t} = \lim_{\substack{T\to\infty \\ \Delta\omega\to 0}} \frac{1}{2\pi} \sum_{n=-\infty}^{\infty} \left[\int_{-T/2}^{T/2} f(t) e^{-j\omega_n t} dt \right] e^{j\omega_n t} \Delta\omega$$

Note that when $T \to \infty$, $\Delta\omega \to d\omega$ and the summation transforms into an integration process. Thus, ω is used to replace ω_n and

$$f(t) = \frac{1}{2\pi} \int_{-\infty}^{\infty} \left[\int_{-T/2}^{T/2} f(t) e^{-j\omega t} dt \right] e^{j\omega t} d\omega \tag{2.135}$$

In other words, the Fourier series becomes an integral, the *Fourier integral*. Meanwhile, the Fourier coefficients c_n, as expressed in Equation 2.37, inherits a new form as

$$F(\omega) = \int_{-\infty}^{\infty} f(t) e^{-j\omega t} dt \tag{2.136}$$

which is now referred to as the *Fourier transform* of the function f(t). This may be denoted in a simple way as

$$F(\omega) = \mathcal{F}\left[f(t)\right] \tag{2.137}$$

Substituting Equation 2.136 into Equation 2.135 yields

$$f(t) = \frac{1}{2\pi} \int_{-\infty}^{\infty} F(\omega) e^{j\omega t} d\omega \tag{2.138}$$

From Equation 2.84, it is seen that if f(t) is given, its Fourier transform F(ω) can be uniquely determined, provided that the integral in Equation 2.136 exists; that is, provided the aforementioned absolutely integrable condition, given by

$$\left| \int_{-\infty}^{\infty} f(t) e^{-j\omega t} dt \right| < \infty \tag{2.139}$$

Furthermore, from Equation 2.138, if F(ω) is available, then the original function f(t) can be obtained. The operation described by Equation 2.138 is often called the *inverse Fourier transformation,* which can be denoted by

$$f(t) = \mathcal{F}^{-1}\left[F(\omega)\right] \tag{2.140}$$

In the following discussion, the functions f(t) and F(ω) are said to be a *Fourier pair* and the relationship between f(t) and F(ω) is denoted as

$$f(t) \leftrightarrow F(\omega) \tag{2.141}$$

To better understand vibration damping for damper design, the following features of the Fourier transform are important.

Mathematically speaking, for a signal f(t) to have the corresponding Fourier transform, both the conditions described in Equation 2.139 and the Dirichlet conditions must be satisfied. However, in the following discussion, the Dirichlet conditions will not be checked often, unless specifically necessary.

2.2.2.1 Important Features of the Fourier Transform: A Summary

In the following discussion, assume that f(t), $f_1(t)$, and $f_2(t)$ are real functions, which satisfy the conditions for the existence of their Fourier transforms and that a and b are constants.

2.2.2.1.1 Linearity

The Fourier transform of $f(t) = af_1(t) + bf_2(t)$ is

$$\mathcal{F}\left[f(t)\right] = \mathcal{F}\left[af_1(t) + bf_2(t)\right] = a\mathcal{F}\left[f_1(t)\right] + b\mathcal{F}\left[f_2(t)\right] \tag{2.142}$$

2.2.2.1.2 Theorem of Time Shift

The Fourier transform of $f(t + \tau)$ is

$$\mathcal{F}\left[f(t+\tau)\right] = e^{j\omega\tau}\mathcal{F}\left[f(t)\right] = e^{j\omega\tau}F(\omega) \tag{2.143}$$

This theorem can be proven as follows:

$$\mathcal{F}\left[f(t+\tau)\right] = \int_{-\infty}^{\infty} f(t+\tau)e^{-j\omega t}\,dt \tag{2.144}$$

Let $t + \tau = u$, and note that $dt = du$. Thus,

$$\begin{aligned}\int_{-\infty}^{\infty} f(u)e^{-j\omega(u-\tau)}\,du &= \int_{-\infty}^{\infty} f(u)e^{-j\omega u}\,e^{j\omega\tau}\,du \\ &= e^{j\omega\tau}\int_{-\infty}^{\infty} f(u)e^{-j\omega u}\,du \\ &= e^{j\omega\tau}F(\omega)\end{aligned} \tag{2.145}$$

2.2.2.1.3 Theorem of Multiplication

$$\begin{aligned}\int_{-\infty}^{\infty} f_1(t)\,f_2(t)\,dt &= \frac{1}{2\pi}\int_{-\infty}^{\infty} F_1^*(\omega)F_2(\omega)\,d\omega \\ &= \frac{1}{2\pi}\int_{-\infty}^{\infty} F_1(\omega)F_2^*(\omega)\,d\omega\end{aligned} \tag{2.146}$$

This can be seen as follows:

$$\int_{-\infty}^{\infty} f_1(t) f_2(t)\,dt = \int_{-\infty}^{\infty} f_1(t)\left[\frac{1}{2\pi}\int_{-\infty}^{\infty} F_2(\omega) e^{j\omega t}\,d\omega\right]dt$$

$$= \frac{1}{2\pi}\int_{-\infty}^{\infty} F_2(\omega)\int_{-\infty}^{\infty} f_1(t) e^{j\omega t} dt\,d\omega \tag{2.147}$$

$$= \frac{1}{2\pi}\int_{-\infty}^{\infty} F_2(\omega) F_1^*(\omega)\,d\omega$$

$$= \frac{1}{2\pi}\int_{-\infty}^{\infty} F_1^*(\omega) F_2(\omega)\,d\omega$$

2.2.2.1.4 Differentiation of Fourier Transform

$$\mathcal{F}\left[\dot{f}(t)\right] = j\omega F(\omega) \tag{2.148}$$

This theorem can be proven in the following way through the process of integration by parts:

$$\mathcal{F}\left[\dot{f}(t)\right] = \int_{-\infty}^{\infty} \dot{f}(t) e^{-j\omega t}\,dt = f(t) e^{-j\omega t}\Big|_{-\infty}^{\infty} + j\omega\int_{-\infty}^{\infty} f(t) e^{-j\omega t}\,dt \tag{2.149}$$

To ensure Equation 2.139,

$$f(t)\Big|_{|t|\to\infty} = 0 \tag{2.150}$$

so that Equation 2.148 holds.

2.2.3 Laplace Transform

Definition

In the above discussion, all the useful features must be based on the condition described in Equation 2.139. However, the integrals in certain cases can become infinitely large. In other words, the Fourier integral does not exist. One solution is to force the value contributed by function f(t) to vanish when the temporal variable t reaches infinity. This may be accomplished by multiplying the integrand with an exponential function of t as in $e^{-\upsilon t}$. In many cases, by just letting

$$\upsilon > 0 \tag{2.151}$$

the following can result:

$$\int_0^{\infty} f(t) e^{-(\upsilon + j\omega)t}\,dt < \infty \tag{2.152}$$

In this way, a new integral transform can be defined, the *Laplace transform*, written as

$$\mathcal{L}[f(t)] = F(s) = \int_0^{\infty} f(t) e^{-st}\,dt \tag{2.153}$$

where s is the Laplace variable with

$$s = \upsilon + j\omega \tag{2.154}$$

Similar to the case of the Fourier transform, the inverse Laplace transformation is

$$f(t) = \mathscr{L}^{-1}[F(s)] \tag{2.155}$$

and the *Laplace transform pair* can be denoted as

$$f(t) \leftrightarrow F(s) \tag{2.156}$$

2.2.3.1 Important Features of the Laplace Transform: A Summary

Here, it is assumed that f(t), $f_1(t)$, and $f_2(t)$ are real functions and that a and b are constants.

2.2.3.1.1 Linearity

The Laplace transform of $f(t) = af_1(t) + bf_2(t)$ is

$$\mathscr{L}[f(t)] = \mathscr{L}[af_1(t) + bf_2(t)] = a\mathscr{L}[f_1(t)] + b\mathscr{L}[f_2(t)] \tag{2.157}$$

2.2.3.1.2 Differentiation of Laplace Transform

If

$$\mathscr{L}[f(t)] = F(s) \tag{2.158}$$

then

$$\mathscr{L}[\dot{f}(t)] = sF(s) - f(0) \tag{2.159}$$

This theorem can be proven in the following way. First, write

$$\mathscr{L}[\dot{f}(t)] = \int_0^\infty \dot{f}(t)e^{-st}\,dt = f(t)e^{-st}\Big|_0^\infty + s\int_0^\infty f(t)e^{-st}\,dt \tag{2.160}$$

Here, the terms on the right side of the equation are obtained through the process of integration by parts. Thus,

$$f(t)e^{-st}\Big|_0^\infty + s\int_0^\infty f(t)e^{-st}dt = -f(0) + sF(s) = sF(s) - f(0) \tag{2.161}$$

which provides the desired result. From Equation 2.159, it can be further written that

$$\mathscr{L}[\ddot{f}(t)] = s^2F(s) - sf(0) - \dot{f}(0) \tag{2.162}$$

Example 2.4

Find the Fourier and Laplace transforms of f(t), which is the summation of inertia, damping, and spring forces of an SDOF vibration system given by

$$\begin{cases} f(t) = m\ddot{x} + c\dot{x} + kx \\ x(0) = 0 \\ \dot{x}(0) = 0 \end{cases}$$

It is seen that f(t) can be viewed as an external force of the vibration system. Taking the Fourier transform on both sides of the first equation in the above results in

$$\mathscr{F}[f(t)] = F(\omega) = \mathscr{F}[m\ddot{x} + c\dot{x} + kx] = (-m\omega^2 + j\omega c + k)X(\omega)$$

On the other hand, taking the Laplace transform on both sides of that equation, since the initial displacement and velocity are all zero, results in

$$\mathscr{L}[f(t)] = F(s) = \mathscr{L}[m\ddot{x} + c\dot{x} + kx] = (ms^2 + sc + k)X(s)$$

Here, $X(\omega) \leftrightarrow x(t)$ and $X(s) \leftrightarrow x(t)$.

2.2.4 Impulse Response

Besides harmonic excitation, an impulse is one of the simplest forcing functions. Practically, when an SDOF system is only subjected to a single impact force with sufficiently short duration, it will then have a free-decay vibration. This phenomenon is a foundation to explore further generally forced vibrations.

To examine the impact response, the idealized impulse force f(t) is considered first, which can be written as

$$f(t) = \begin{cases} I_0 \lim_{\varepsilon \to 0} \left(\dfrac{1}{2\varepsilon} \right), & \tau - \varepsilon \le t \le \tau + \varepsilon \\ 0, & \text{elsewhere} \end{cases} \tag{2.163}$$

where I_0 is the amplitude of the impulse function, τ is the time delay, and ε is a small time interval.

The rectangular-shaped impact force is shown in Figure 2.10, plotted with a solid line, which approximates the actual time history of force, plotted with a dotted line.

Equation 2.163 can be rewritten as

$$f(t) = I_0 \delta(t) \tag{2.164}$$

Here $\delta(t)$ is a special function, the *delta function*. Figure 2.10 shows a special case of time shift $t - \tau$ and the corresponding impulse function $\delta(t - \tau)$, which can be seen as

$$\delta(t - \tau) = \begin{cases} \lim_{\varepsilon \to 0} \left(\dfrac{1}{2\varepsilon} \right), & \tau - \varepsilon \le t \le \tau + \varepsilon \\ 0, & \text{elsewhere} \end{cases} \tag{2.165}$$

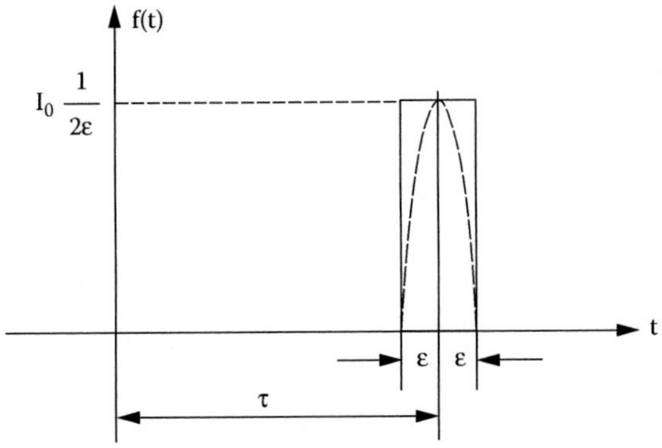

FIGURE 2.10 Impulse.

and

$$\int_{-\infty}^{\infty} \delta(t-\tau)\, dt = \int_{\tau-\varepsilon}^{\tau+\varepsilon} \delta(t-\tau) dt = \lim_{\varepsilon \to 0} \frac{1}{2\varepsilon} 2\varepsilon = 1 \tag{2.166a}$$

Thus,

$$\int_{-\infty}^{\infty} \delta(t) dt = 1 \tag{2.166b}$$

Therefore,

$$\int_{-\infty}^{\infty} f(t) dt = \int_{-\infty}^{\infty} I_0 \delta(t) dt = I_0 \tag{2.167}$$

When the initial displacement is zero, the mass is considered to be at rest just shortly prior to the application of the impulse I_0. At the moment when I_0 is applied, the momentum of the system gains mv_0. That is,

$$I_0 = f(t)\Delta t = mv_0 - 0 \tag{2.168}$$

Thus,

$$v_0 = \frac{f(t)\Delta t}{m} = \frac{I_0}{m} \tag{2.169}$$

Therefore, the effect of an impulse applied to the SDOF m-c-k system is identical to the case of a free vibration with zero initial displacement and initial velocity equal to that described in Equation 2.169.

From the discussion on the free vibration decay in Chapter 1, it is known that when $I_0 = 1$, the response is

$$x(t) = \frac{1}{m\omega_d}\left[e^{-\xi\omega_n t} \sin(\omega_d t)\right]$$

This expression is quite important. A special notation is used to represent this unit impulse response. Thus, let

$$h(t) = \frac{1}{m\omega_d}\left[e^{-\xi\omega_n t}\sin(\omega_d t)\right] \tag{2.170}$$

where the quantity h(t) is known as the *unit impulse response function.*

For an impulse with general amplitude I_0,

$$x(t) = I_0 h(t) = \frac{I_0}{m\omega_d}\left[e^{-\xi\omega_n t}\sin(\omega_d t)\right] \tag{2.171}$$

2.2.5 General Force and the Duhamel Integral

2.2.5.1 Convolution Integral

In the above discussion, the cause of the free decay is due to an impulse. Mathematically speaking, the special function called the *Dirac delta function*, δ(t), can be used to represent the impulse. In somewhat imprecise terms,

$$\delta(t-\tau) = \begin{cases} \infty & t = \tau \\ 0 & t \neq \tau \end{cases} \tag{2.172}$$

and

$$\int_0^\infty \delta(t-\tau)dt = 1, \quad \text{for } 0 < \tau < \infty \tag{2.173}$$

The δ function is seen to have a special effect on sampling. That is, if a continuous function f(t) is multiplied by δ(t – τ), and then integrated over the entire time axis, the contribution to the integral becomes zero except at t = τ, where the Dirac delta function operates on f(t) to sift out the value of f(τ) as the integral. This process implies that the Dirac delta function δ(t – τ) just samples the signal f(t) at time τ and the value of that sample is simply f(τ).

The δ function is plotted in Figure 2.11a. An arbitrary forcing function, such as an earthquake, can be seen as a combination of a series of impulse functions at time point t – τ. Particularly, at time point τ, the forcing function exists as an impulse f(τ)Δt. The corresponding responses due to the impulse can be denoted as Δx(t), and

$$\Delta x(\tau) = f(\tau)\Delta t\, h(t-\tau) \tag{2.174}$$

That is, this response due to the specific impulse response function is the product of the unit impulse response function h(t – τ) and the amplitude of the impulse f(τ)Δt. Figure 2.11b conceptually shows this treatment of an impulse response at time point τ.

Since the system is linear, the total response can be seen as the summation of Δx(t), which starts from time 0 up to time t, and the specific response at each particular interval $t - t_i$ is due to the corresponding impulse. Here, t_i is used to replace t to denote the particular instant. Therefore at $t = t_n$, the resulting summation is

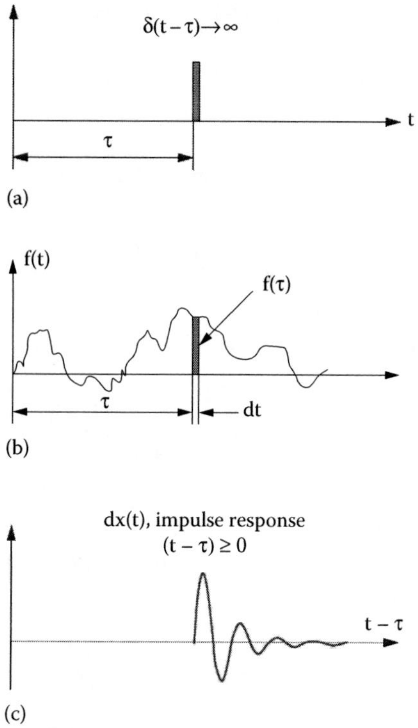

FIGURE 2.11 (a–c) Duhamel integral.

$$x(n) = \sum_{0}^{n} f(t_i)\Delta t\, h(t - t_i) \tag{2.175}$$

With the help of the above notations and concepts, the structural response under the arbitrary excitation f(t) can be seen as a combination of all the impulse responses, from time 0 to t, representing any given time. When the time interval Δt is minimized toward infinitesimally small dt, the summation described in Equation 2.175 is essentially an integration. Thus,

$$x(t) = \int_0^t h(t - \tau)\, f(\tau)d\tau \tag{2.176}$$

The integral shown in Equation 2.176 is referred to as a *convolution integral*, in this case, between the functions f(t) and h(t).

It is seen that by letting $u = t - \tau$, du = dt results. Therefore, by replacing the variable h in the above integral by $t - \tau$,

$$x(t) = -\int_t^0 h(u)\, f(t - u)du = \int_0^t h(u)\, f(t - u)du \tag{2.177}$$

Thus,

$$\int_0^t f(\tau)h(t - \tau)d\tau = \int_0^t h(\tau)\, f(t - \tau)d\tau \tag{2.178}$$

For convenience, the convolution of two functions, h(t) and f(t), can be denoted by

$$x(t) = f(t) * h(t) = h(t) * f(t) \tag{2.179}$$

Since the convolution is only taken between the starting point t = 0 and final time point t, for any time point beyond t, the integral will be equal to zero. Thus,

$$x(t) = \int_0^t h(t-\tau) f(\tau) d\tau + 0 = \int_0^t h(t-\tau) f(\tau) d\tau + \int_t^\infty h(t-\tau) f(\tau) d\tau$$

and h(t) = 0 for t < 0, therefore,

$$x(t) = \int_0^\infty h(t-\tau) f(\tau) d\tau = \int_{-\infty}^\infty h(t-\tau) f(\tau) d\tau \tag{2.180}$$

Equation 2.180 is also a popular form of the convolution.

Now, substituting Equation 2.170, the expression of the unit impulse response, into Equation 2.176 yields

$$x(t) = \frac{1}{m\omega_d} e^{-\xi\omega_n t} \int_0^t f(\tau) e^{-\xi\omega_n \tau} \sin \omega_d (t-\tau) d\tau \tag{2.181}$$

Thus, the response of an SDOF linear system under an arbitrary excitation f(t) with zero initial conditions is obtained. Such a formulation using the above-mentioned convolution to calculate the response is referred to as the *Duhamel integral*, as given by Equation 2.181.

2.2.6 Transfer Function of Unit Impulse Response

A linear SDOF m-c-k system subjected to unit impulse has the governing equation with zero initial condition defined as

$$\begin{cases} m\ddot{x} + c\dot{x} + kx = \delta(t) \\ x(0) = 0 \\ \dot{x}(0) = 0 \end{cases} \tag{2.182}$$

Taking the Laplace transform on both sides of the first equation in Equation 2.182, due to a zero initial condition, results in

$$\mathscr{L}[m\ddot{x} + c\dot{x} + kx] = \mathscr{L}[\delta(t)] \tag{2.183}$$

The left side of the Laplace transform has been obtained through Example 2.4. The term on the right side is considered next.

$$\mathscr{L}[\delta(t)] = \int_0^\infty \delta(t) e^{-st} dt = e^{-st}\big|_{t=0} = 1 \tag{2.184}$$

The above holds due to the sampling effect of the δ function, which makes all the values of e^{-st} vanish, except for the value at t = 0. Thus,

$$\left(ms^2 + sc + k\right)X(s) = 1 \tag{2.185}$$

Note that in this circumstance, X(s) is the particular Laplace transform of the unit impulse response, that is,

$$X(s) = \mathscr{L}\left[h(t)\right] \tag{2.186}$$

Substituting Equation 2.186 into Equation 2.185 yields

$$\mathscr{L}\left[h(t)\right] = \frac{1}{ms^2 + cs + k} \tag{2.187}$$

Equation 2.187 implies that the Laplace transform of the unit impulse response is one over the term ($ms^2 + cs + k$). This is another important relationship. This Laplace transform is denoted as H(s). That is, let

$$H(s) = \frac{1}{ms^2 + cs + k} \tag{2.188}$$

The Fourier transform of the unit impulse response is also important. This Fourier transform is denoted as H(jω). That is,

$$\mathscr{F}\left[h(t)\right] = H(j\omega) = \frac{1}{m\omega^2 + jc\omega + k} \tag{2.189}$$

In addition, consider a more general case defined as

$$\begin{cases} m\ddot{x}(t) + c\dot{x}(t) + kx(t) = f(t) \\ x(0) = 0 \\ \dot{x}(0) = 0 \end{cases} \tag{2.190}$$

Taking the Laplace transform on both sides of the first equation in Equation 2.190, because of zero initial condition, results in

$$\left(ms^2 + cs + k\right)X(s) = \mathscr{L}\left[f(t)\right] = F(s) \tag{2.191}$$

From Equation 2.191, with the help of Equation 2.188,

$$\frac{X(s)}{F(s)} = \frac{1}{ms^2 + cs + k} = H(s) \tag{2.192}$$

Equation 2.192 implies that the ratio of the Laplace transforms of the response and the forcing functions is exactly equal to the Laplace transform of the unit impulse response function. Here, this ratio is defined as the *transfer function*. Therefore,

$$X(s) = H(s)F(s) \tag{2.193}$$

Equation 2.193 implies that the force in the Laplace domain F(s), namely, the Laplace transform of the force, is transferred into the response in the Laplace domain X(s), by the effect of the transfer function H(s). In other words, F(s) transfers into X(s) through the "bridge" H(s). In the literature, Equation 2.193 is referred to as *Borel's theorem*.

By letting $s = j\omega$, Equation 2.193 can be rewritten as

$$X(j\omega) = H(j\omega)F(j\omega) \tag{2.194}$$

Similarly, Equation 2.194 implies that the force in the Fourier domain $F(j\omega)$, namely, the Fourier transform of the force, is transferred into the response in the Fourier domain X(s), by the effect of the function $H(j\omega)$. Here, the transfer function of $H(j\omega)$ has a special name—the *frequency response function*. That is, $F(j\omega)$ transfers into $X(j\omega)$ through the "bridge" $H(j\omega)$.

2.2.7 Integral Transform of Convolution

Comparing the convolution of h(t) and f(t) described in Equation 2.180 with the multiplication of H(s)F(s) as well as $H(j\omega)F(j\omega)$, both the Laplace transform and the Fourier transform of the convolution result in the multiplication of the integral transforms. Namely,

$$\mathscr{L}\left[h(t)*f(t)\right] = H(s)F(s) \tag{2.195}$$

$$\mathscr{F}\left[h(t)*f(t)\right] = H(j\omega)F(j\omega) \tag{2.196}$$

Such relationships are not a coincidence, in fact, it can be proven that for a general form of convolution y(t)*z(t),

$$\mathscr{L}\left[y(t)*z(t)\right] = Y(s)Z(s) \tag{2.197}$$

and

$$\mathscr{F}\left[y(t)*z(t)\right] = Y(j\omega)Z(j\omega) \tag{2.198}$$

Equation 2.198 can be proven by considering

$$\begin{aligned}\mathscr{F}\left[y(t)*z(t)\right] &= \int_{-\infty}^{\infty}\left[\int_{-\infty}^{\infty} y(\tau)z(t-\tau)d\tau\right]e^{-j\omega t}dt\\ &= \int_{-\infty}^{\infty}\int_{-\infty}^{\infty} y(\tau)e^{-j\omega\tau}z(t-\tau)e^{-j\omega(t-\tau)}d\tau dt\\ &= \int_{-\infty}^{\infty}\int_{-\infty}^{\infty} y(\tau)e^{-j\omega\tau}d\tau z(t-\tau)e^{-j\omega(t-\tau)}dt = Y(j\omega)Z(j\omega)\end{aligned} \tag{2.199}$$

Here, in the third step, the identity was inserted

$$1 = e^{-j\omega\tau}e^{-j\omega(-\tau)} \tag{2.200}$$

2.3 RANDOM EXCITATIONS

The responses of SDOF systems subjected to harmonic excitation were discussed in Chapter 1. In addition, the periodic and transient excitations were discussed in the previous sections of this chapter. Although these three types of excitations are very distinct, they have one common feature: they are all deterministic. Thus, for deterministic structures, the response is also deterministic. The earthquake excitations dealt with for damper design are random. As a deterministic signal, which is a function of time, if the time point is located, then the value of the forcing function will be known exactly. However, as a random signal, at a specific time point, the value is unknown.

For deterministic excitations, the response is automatically deterministic. This means that if the physical parameters of a system are given, at any time point, the value of the response can be calculated precisely. However, for random excitations, the value of the response at a specific time cannot be determined.

In engineering applications, the time history of the response does not necessarily need to be known. What is often needed is to locate the maximum value so that the structure to be designed and constructed can withstand the corresponding peak deformation and/or peak force. In earthquake engineering, often only the peak value of the random ground motions is needed to find the peak value of response, called the *response bound*. Standards based on an earthquake return period are set up and these standards are used to classify the input level of earthquakes. In this circumstance, although the detailed time history of a ground motion is unknown, the amplitude of the excitation is considered to be deterministic.

However, under general random excitations, such as earthquakes, the peak value of the structural response cannot be calculated, even when the input bound is specified. In other words, one of the important features of earthquake response is that bounded input does not yield bounded output.

A common practice to deal with such difficulties is to employ statistical studies. That is, sufficient earthquake records are collected, including artificial records, and then averages are taken. In this way, particular characteristics of the average response can be discovered. For example, on average, the most probable amplitude of the response is of interest, as is the trend of the structural response when supplemental damping is used, etc. In this section, the basic concept of random variable and random process is introduced, as well as the approach of averaging correlation analysis and power density spectrum analysis. Based on this preknowledge, the earthquake response spectrum and corresponding design spectrum are considered. The main approach for damper design described in this book is based on spectral analysis.

2.3.1 Random Variables

Suppose an SDOF vibration system is subjected to random excitations, such as earthquake ground motions. The amplitude of the response is considered at exactly 1 (s) after the excitation starts. It is understandable that, due to different earthquake excitations, the response displacement can be positive or negative; the amplitude can be large or small. In other words, the exact value of the displacement cannot be determined.

Consider the six different earthquake ground motion signals shown in Figure 2.12a. The first through the sixth records are, respectively, the El Centro earthquake measured at 0°, the El Centro earthquake measured at 90°, the Northridge earthquake measured at 0°, the Northridge earthquake measured at 90°, the Mexico earthquake, and the Taft earthquake. The amplitudes are accelerations with units in (cm/s^2). Figure 2.12b shows the amplitudes of these earthquake records at $t = 1$ (s), while Figure 2.12c displays the amplitude of the displacements due to these earthquakes at that same instant of time. From this group of data, it is seen that the amplitudes are somewhat random and it seems that no obvious regularity can be found by direct observation. Note that to obtain the displacement responses, an SDOF system with mass equal to 1, damping coefficient equal to 0.4π, and stiffness equal to $(2\pi)^2$ is assumed. (Therefore, this system has a damping ratio equal to 0.1 and a natural frequency equal to 1 (Hz).)

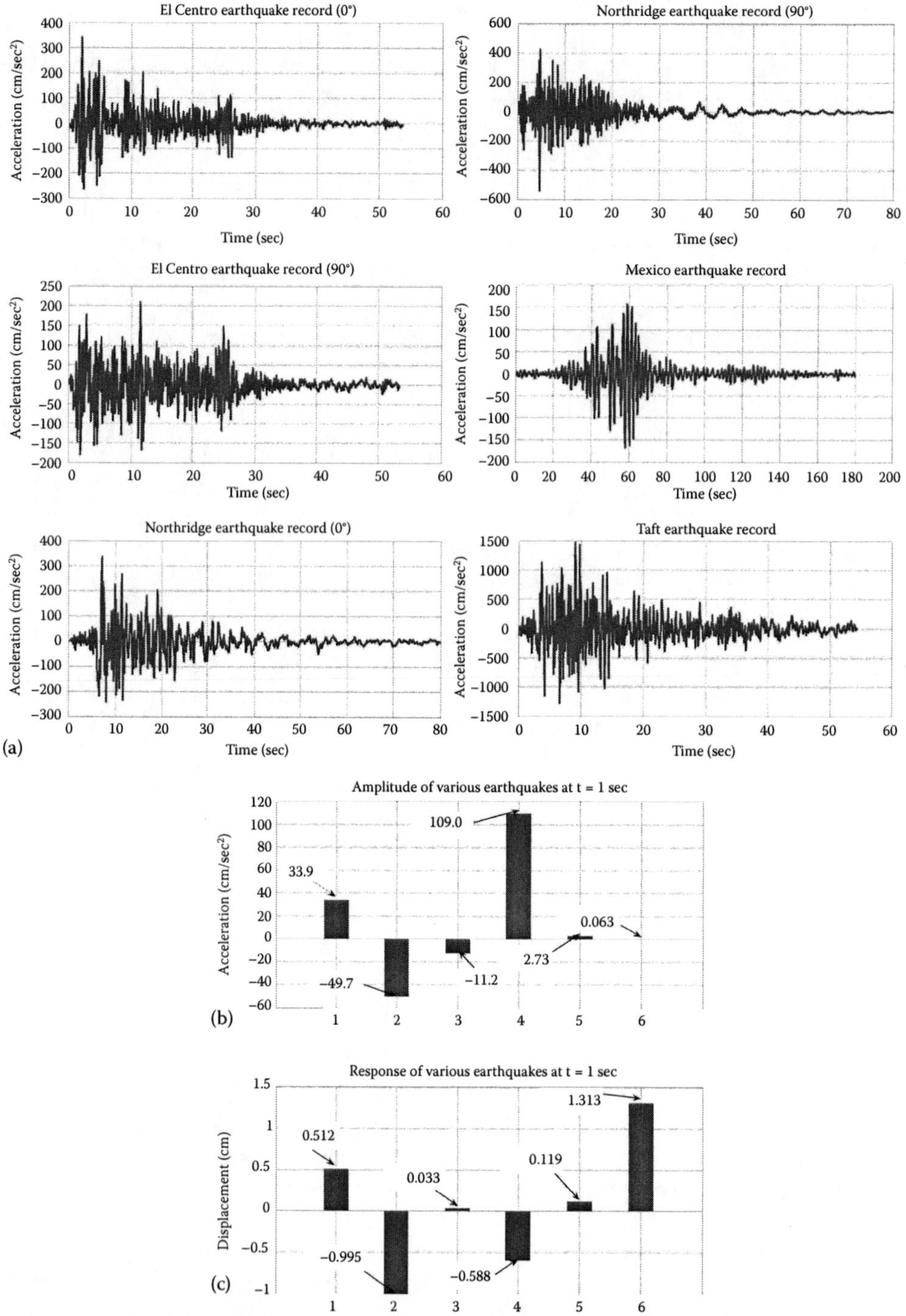

FIGURE 2.12 (a) Earthquake ground accelerations as random excitations (1)–(6). (b) Value of selected earthquake ground motions and corresponding response displacements of an assumed SDOF system at t = 1 (s). (c) Value of selected earthquake ground motions and corresponding response displacements of an assumed SDOF system at t = 1 (s).

By examining the responses at t = 1 (s), similar to the case of random excitation, it is seen that the responses are also somewhat random.

To study the random data, as many values as possible of the ground acceleration or response displacement or any kind of random data are collected by placing them inside a special vector, such that

$$\mathbf{x}_1 = \left[x_{11}, x_{12}, \ldots, x_{1n}\right] \quad (2.201)$$

Here, subscript 1 of the vector stands for the data collected at t = 1 (s). The first subscript of x_{1i}, which stands for an individual random data value, also denotes that the data are taken from the time 1 (s). The second subscript i of x_{1i} simply denotes that it is the ith data, which is often called the *samples*. In most cases, a point is not specified, whether t = 1 or not. Therefore, these data are simplified by rewriting Equation 2.201 as

$$\mathbf{x} = \left[x_1, x_2, \ldots, x_n\right] \quad (2.202)$$

In Equation 2.202, all the samples already taken or measured are known to be deterministic. However, when a sample x_i is arbitrarily chosen, the value of this sample is uncertain beforehand. Therefore, **x** is a *random* data set and x_i is a *random variable*. Whenever a variable x_i is chosen for the set **x**, its value is not predictable.

When the exact value is unknown, an alternative approach is chosen (mean value, variance, and standard deviation) to find a pattern by exploring if something is common among the values, such as displacements due to different earthquake excitations. To do so, several definitions will be reviewed first.

2.3.1.1 Mean Value, Mathematic Expectation

The *mean value* of **x** is defined as

$$\bar{x} = \frac{1}{n}\sum_{i=1}^{n} x_i \quad (2.203)$$

which provides the average of the group of data **x**.

In the case of an infinite number of samples, both summations and integrations will be present. Particularly, if the average of the total time history is considered, such as the ground excitation or the response, then the time history is denoted as x(t) and

$$\bar{x} = \lim_{T\to\infty} \frac{1}{T}\int_0^T x(t)dt \quad (2.204)$$

Here, T is the period of the time history. For example, in the first two plots in Figure 2.12, the ground motion records of the El Centro earthquake have T = 53.76 (s).

Equation 2.204 is rewritten as

$$\bar{x} = \lim_{T\to\infty} \int_0^T \left[\frac{1}{T}x(t)\right]dt \quad (2.205)$$

It is seen that the term in brackets implies that each value x(t) at time t has an equal chance 1/T to be counted. However, from all the ground motion records shown in Figure 2.13, it is realized that this is not true for earthquakes. Instead, at the very beginning of the earthquake, the amplitudes of

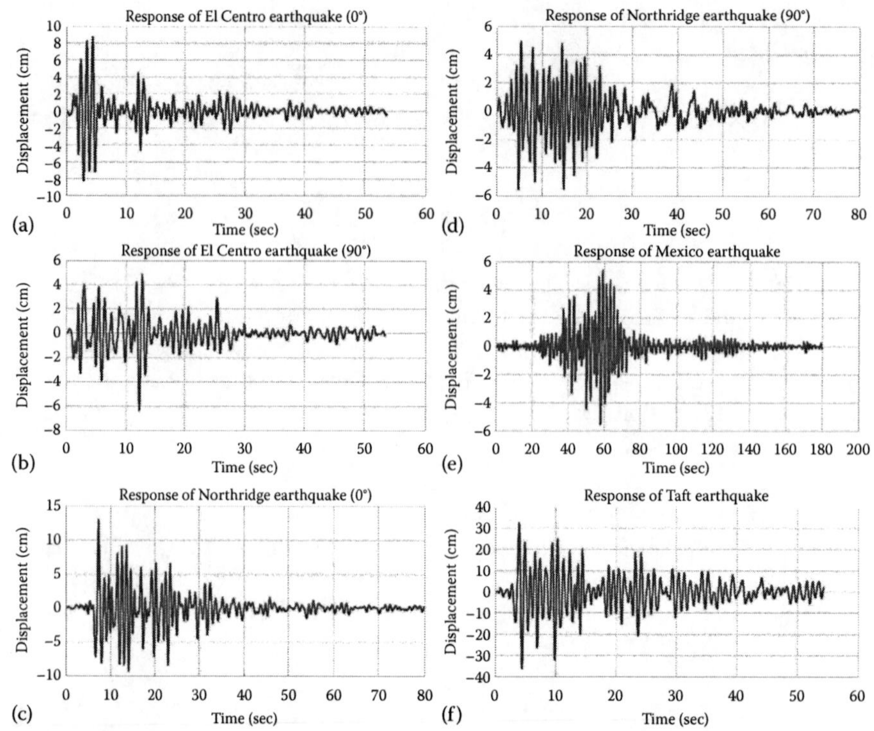

FIGURE 2.13 (a–f) Earthquake responses.

the ground motion are always rather small. After a while, the ground motions begin to increase and after several seconds, the amplitudes will gradually die out. This means that, at different times t, the chance of the amplitude being large or small is not equal. In certain circumstances, due to this unequal chance, the data can be counted with different weight factors, and this weight factor can often be taken as the probability of the appearance of x(t). Denoting such a probability as p(x) results in a more rigorous concept of average, the *mathematical expectation*, defined as

$$E[x] = \int_{-\infty}^{\infty} p(x)\,x\,dx \tag{2.206}$$

Note that in Equation 2.206, x is no longer treated as a function of time, but instead is taken as an individual value with the chance or probability density function p(x) to be counted. In this sense, E[x] can be called the *ensemble average*. As a comparison, the average expressed in Equation 2.205 is the average over time (with the period of T), called the *temporal average*.

Example 2.5

Let **x** be the collection of ground accelerations at t = 1 (s), taken from the earthquake records shown in Figure 2.13a, that is,

$$\mathbf{x}_1 = [33.9000 \quad -49.7000 \quad -12.0560 \quad 109.3820 \quad 2.7300 \quad 0.0530] \ (\text{cm/s}^2)$$

By using Equation 2.203,

$$\bar{x} = 14.05 \ (\text{cm/s}^2)$$

2.3.1.2 Variance and Mean Square Value

If sufficient data are taken to perform the averaging described in Equation 2.203, it is likely that the mean value will be very close to zero. This will happen at virtually any time point. However, from the ground motion plots shown in Figure 2.13, it is realized that at different time points, the peak amplitude can vary significantly. As mentioned above, at the very beginning of the recorded ground motion, there is always a small amplitude, but after a while, the amplitudes become larger, and so on. Therefore, a measure that avoids the positive and negative quantities canceling each other out is needed, so that the scattering of data can be represented. The simple way to treat the data is to take the square of the individual entries, so that all the negative signs are ignored. In this way, the *mean square value* is

$$\overline{x^2} = \frac{1}{n}\sum_{i=1}^{n}\left[x_i(t)\right]^2 \tag{2.207}$$

In the literature, this value is also called the "variance." The mean square value can be used to measure the magnitude of the fluctuation of the data x(t). For instance, in Example 2.5, the set of data **x** and the mean square value can be calculated as

$$\overline{x^2} = 2622.8 \quad \left(\mathrm{cm/s^2}\right)^2 \tag{2.208}$$

Now, it is seen that at t = 0.04 (s), the same group of earthquake records will have the following amplitude:

$$\mathbf{x}_{04} = \begin{bmatrix}-8.8000 & 2.9000 & 3.9130 & 13.4730 & 0.9430 & -0.0041\end{bmatrix} \quad \left(\mathrm{cm/s^2}\right)$$

Therefore, the corresponding mean square value is

$$\overline{x^2} = 47.3 \quad \left(\mathrm{cm/s^2}\right)^2 \tag{2.209}$$

Comparing Equation 2.208 and Equation 2.209, it is seen that at t = 1 (s), the earthquake may be more violent than at t = 0.04 (s). However, if the fluctuation that occurs from the average point instead of from the zero point is considered, it can be seen that such a measure around the average value may be more appropriate. In this case, the value known as *variance* can be defined as

$$\sigma^2 = \frac{1}{n}\sum_{i=1}^{n}\left[x_i(t) - \bar{x}\right]^2 \tag{2.210}$$

Note that in practical statistics, n – 1 instead of n is often used to calculate the variance as

$$\sigma^2 = \frac{1}{n-1}\sum_{i=1}^{n}\left[x_i(t) - \bar{x}\right]^2 \tag{2.211}$$

Similar to the case of average, for situations where the number of samples increases toward infinity, an integral to replace the summation to calculate the mean square value is

$$\overline{x^2} = \lim_{T\to\infty}\frac{1}{T}\int_0^T\left[x(t)\right]^2 dt \tag{2.212}$$

If the probability of each individual quantity of $x(t)^2$ must be considered and for ensemble averaging, the expected value is

$$\overline{x^2} = \int_{-\infty}^{\infty} [x]^2 p(x) dx \tag{2.213}$$

and for the variance,

$$\sigma^2 = \lim_{T\to\infty} \frac{1}{T} \int_0^T \left[x(t) - \overline{x}\right]^2 dt \tag{2.214}$$

The corresponding ensemble average is

$$\sigma^2 = \int_{-\infty}^{\infty} \left[x - \overline{x}\right]^2 p(x) dx \tag{2.215}$$

2.3.1.3 Standard Deviation and Root Mean Square Value

From Equations 2.208 and 2.209, it was found that both the variance and the mean square value are expressed in units of the square of the original units for the random data set. This is sometimes inconvenient. Thus, the *root means square value* (rms) is defined as

$$x_{rms} = \sqrt{\overline{x^2}} = \sqrt{\frac{1}{n-1} \sum_{i=1}^{n} \left[x_i(t)\right]^2} \tag{2.216}$$

or

$$x_{rms} = \sqrt{\overline{x^2}} = \sqrt{\lim_{T\to\infty} \frac{1}{T} \int_0^T \left[x(t)\right]^2 dt} \tag{2.217}$$

Furthermore,

$$x_{rms} = \sqrt{\overline{x^2}} = \sqrt{\int_{-\infty}^{\infty} [x]^2 p(x) dx} \tag{2.218}$$

Finally, the *standard deviation is defined* as

$$\sigma = \sqrt{\sigma^2} = \sqrt{\frac{1}{n-1} \sum_{i=1}^{n} \left[x_i(t) - \overline{x}\right]^2} \tag{2.219}$$

and

$$\sigma = \sqrt{\sigma^2} = \sqrt{\lim_{T\to\infty} \frac{1}{T} \int_0^T \left[x(t) - \overline{x}\right]^2 dt} \tag{2.220}$$

Also,

$$\sigma = \sqrt{\sigma^2} = \sqrt{\int_{-\infty}^{\infty} \left[x - \bar{x}\right]^2 p(x)\,dx} \tag{2.221}$$

The term standard deviation indicates the degree of data scattering around the mean value, as the variance does. However, it is expressed in the identical units as the original data, so that it is often more convenient to use.

Example 2.6

Using the formula described by Equation 2.211, the variance of $\mathbf{x}_1$ and the variance of $\mathbf{x}_{04}$ are calculated to be 2910.4 and 51.5686, respectively. The corresponding standard deviations are 53.946 (m/s²) and 7.1811 (m/s²), respectively.

2.3.2 Random Process

2.3.2.1 Random Time Histories

In the previous subsection, random variables (see Equation 2.202) were described; e.g., the ground acceleration of earthquake records at t = 0.04 (s) and t = 1 (s), denoted as $\mathbf{x}_{04}$ and $\mathbf{x}_1$. From these two sets of data, it was immediately found that using only random variables is insufficient to express the time histories. This is because $\mathbf{x}_{04}$ and $\mathbf{x}_1$ are only two time points. For the entire process of the time history, hundreds, even thousands, of time points that form a sequence should be considered. Thus, it will be necessary to introduce a new concept of random process, which is a function of time and, therefore, will be able to describe the time histories.

More specifically, a random process is a time process denoted by x(t), which is a collection of several sets of time histories, such as those ground motions plotted in Figure 2.12a. Although each time history is a specific sample set, so that it is deterministic, however, at a certain time, e.g., at t = 1 (s), the values taken from these sets of time histories are random, such as the collection described in Equation 2.201. For convenience, these sample time histories are denoted as $x_j(t)$. The entire collection or ensemble collection of all these time histories, namely, the $x_j(t)$, is referred to as a *random process*. It is also called a *stochastic process* in the literature. In Figure 2.13, the response of the system used to generate the data in Figure 2.12c is plotted, which was subjected to the earthquake excitations shown in Figure 2.12a.

From these plots, it is realized that due to random excitations, the response is also random. More specifically, when the excitation is from a random process, so are the responses. That is, for any given set of earthquake records, a deterministic response can be calculated. However, if one of the records is chosen arbitrarily, the type of response time history available beforehand is uncertain.

2.3.2.2 Statistical Averaging

To find possible patterns from a random process, the aforementioned averaging can certainly be used. Therefore, the mean and/or expected value, the mean square value and/or variance, and the root mean square and/or standard deviation can be used, when the collection of random time histories are seen as a collection of all random variables. In this sense, the phrase "statistics" actually means average.

On the other hand, to treat these random variables as functions of time, the above-mentioned averaging will not be sufficient. It can be easily realized that all the resulting quantities are only single-valued scalars, instead of temporal functions.

Generally, the ensemble average should not be replaced by the time average, unless the group of data is *ergodic*. If, with a different time period, the averages taken are identical, then the group of data is said to be *stationary*. The condition for random data or a random process to be ergodic is stronger than stationary. That is, if the random set x_R is not stationary, x_R is not ergodic. However, if x_R is stationary, then this does not necessarily mean that x_R is ergodic. It is easy to see that earthquake records are not stationary, because the averages taken from different time points t_1 and t_2 will likely be different. Therefore, mathematically speaking, the time average cannot be used to replace the ensemble average. Although, in practice, this substitution is often made, particular attention should be given to the validity of this assumption. For a more detailed discussion of stationary and ergodic data, readers may consult Bendat and Piersol (1971). In the following discussion, only the basic concepts are explained.

Consider the mean value at an arbitrary fixed time point t_1 as

$$\bar{x}(t_1) = \lim_{n\to\infty} \frac{1}{n} \sum_{i=1}^{n} x_i(t_1) \tag{2.222}$$

Here, $x_i(t)$ is the random time history, which represents the random process measured for the ith case. Suppose there are *n* times of such measurements. Then, apparently, Equation 2.222 denotes the ensemble averaging. Generally speaking, for a random process x(t), the value of the average $\bar{x}(t_1)$ depends on the time point t_1. In the case that the mean value remains constant, that is,

$$\bar{x}(t_1) = \bar{x} = \text{const} \tag{2.223a}$$

$$\sigma^2(t) = \sigma^2 = \text{const} \tag{2.223b}$$

the process x(t) can be referred to as *weakly stationary*. Note that for a more rigorous definition of a random process being weakly stationary, additional conditions of correlation functions are needed, which will be discussed in the next subsection. When all the possible averages are independent of time, say t_1, the process is said to be *strictly (strongly) stationary*. For engineering signals, there is no distinction between these two cases, and the signal is considered to be stationary.

Now, consider the time average of one of the above-mentioned time histories, say, the ith time history. Thus,

$$\bar{x}_i = \lim_{T\to\infty} \frac{1}{T} \int_0^T x_i(t)\,dt \tag{2.224}$$

Equation 2.224 is essentially identical to Equation 2.204, but rather distinct from the ensemble average described in Equation 2.222. For the case of an ergodic process, a single piece of time history has the value of the mean value, as well as the same value of the soon to be discussed correlation function.

However, from Equation 2.222, it is seen that to obtain the ensemble average, a large number of measurements are needed, which may not be available all the time. Thus, in many cases, engineers as well as researchers use the time average to replace the ensemble average, without carefully checking the condition of stationary and ergodic signals. Much needs to be explored in this area in the application of a probability-based approach in the design of structures to resist extreme hazard events.

Example 2.7

In order to compare ensemble and temporal averages, consider a numerical example by using MATLAB® simulations of an ergodic process $x_R(t)$ and a nonstationary process $y_R(t)$, which are plotted in Figures 2.14a and b, respectively. Here, both $x_R(t)$ and $y_R(t)$ consist of 2,500 time points, and both $x_R(t)$ and $y_R(t)$ contain 2,000 pieces of random time histories.

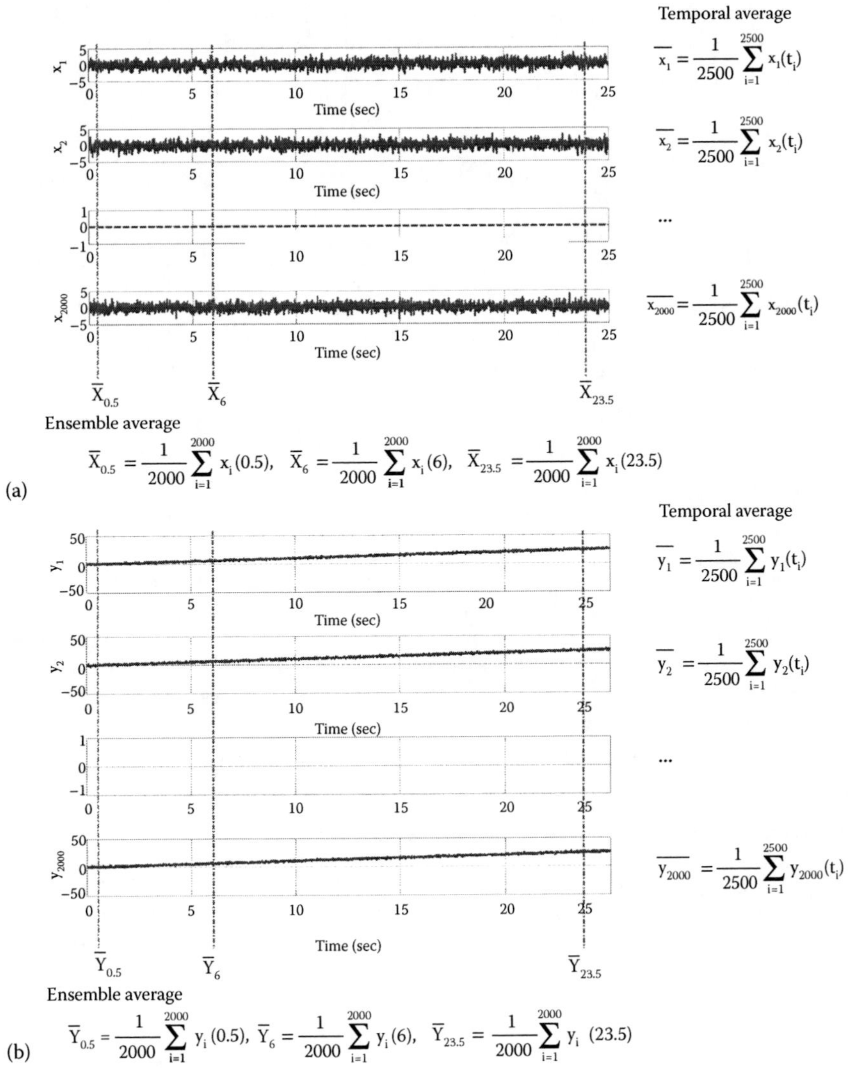

FIGURE 2.14 (a) Time histories of $x_R(t)$ and the temporal and ensemble averages. (b) Time histories of $y_R(t)$ and the temporal and ensemble averages.

The ensemble average of $x_R(t)$ at time 0.5 (s) is denoted by $\overline{X}_{0.5}$ and so on. The ensemble average of $y_R(t)$ at time 0.5 (s) is denoted by $\overline{Y}_{0.5}$ and so on.

The temporal average of $x_1(t)$ over the duration from 0 to 25 (s), totaling 2,500 time points is denoted by $\overline{x_i}$ and so on. The temporal average of $y_1(t)$ from 0 to 25 (s), totaling 2,500 time points is denoted by $\overline{y_i}$ and so on.

Table 2.1A lists the percentage differences among ensemble averages. It is seen that at different time points, namely, 0.5, 6, and 23.5 (s), the variation of $x_R(t)$ is smaller than 1.5%. Although the variation is not absolutely zero, for engineering signals, the averages of the ensemble averages can be treated as roughly identical. Note that if $\overline{X}(\bullet)$ are identical at any time point (•), and the corresponding variances at any time point (•) are also identical, then $x_R(t)$ is weakly stationary. In fact, an ergodic random signal must be stationary. On the other hand, it is seen that the variation of $y_R(t)$ is fairly large, which indicates that $y_R(t)$ is a nonstationary process.

TABLE 2.1A
Percentage Difference Among Ensemble Averages

$\frac{\bar{X}_{0.5}}{\max[x_i(0.5)]}$	$\frac{\bar{X}_6}{\max[x_i(6)]}$	$\frac{\bar{X}_{23.5}}{\max[x_i(23.5)]}$	$\frac{\bar{Y}_{0.5}}{\max[y_i(0.5)]}$	$\frac{\bar{Y}_6}{\max[y_i(6)]}$	$\frac{\bar{Y}_{23.5}}{\max[y_i(23.5)]}$
0.74	0.80	0.28	14.04	63.47	86.48

TABLE 2.1B
Percentage Difference Among Temporal Averages

$\frac{\bar{X}_{0.5}}{\max[x_i(0.5)]}$	$\frac{\bar{X}_6}{\max[x_i(6)]}$	$\frac{\bar{X}_{23.5}}{\max[x_i(23.5)]}$	$\frac{\bar{Y}_{0.5}}{\max[y_i(0.5)]}$	$\frac{\bar{Y}_6}{\max[y_i(6)]}$	$\frac{\bar{Y}_{23.5}}{\max[y_i(23.5)]}$
0.45	0.06	0.36	45.65	45.05	46.38

Furthermore, Table 2.1B lists the percentage differences among temporal averages. It is seen that for the averages for different pieces of time histories, say, x_1, x_2, and x_{2000}, the variation of $x_R(t)$ is smaller than 1.0%. Although the variation is not absolutely zero, for engineering signals, the averages of the temporal averages can be treated as roughly identical to the ensemble averages. In fact, for an ergodic random signal, the temporal average must be identical to the ensemble averages. Note that this statement is a necessary condition of an ergodic signal, but is not yet sufficient. On the other hand, it is seen that the variation of $y_R(t)$ is fairly large, which indicates that $y_R(t)$ is a nonergodic process.

2.3.3 Correlation Functions and Power Spectral Density Functions

To account for random processes, regardless of which concept or method (ensemble or temporal average) is used, the primary technology is still averaging. In the following discussion, the correlation functions and their Fourier transforms are described. Thus, correlation analysis and spectral analysis are considered.

2.3.3.1 Correlation Analysis

Suppose there are two time histories, $x_i(t)$ and $x_j(t)$. The $x_i(t)$ is a forcing function taken from location i and $x_j(t)$ represents the corresponding structural response taken from location j. Similar to the computation of mean value, for a general random process, the cross-correlation function of $x_i(t)$ and $x_j(t)$ is defined by the ensemble average, written as

$$R_{x_ix_j}(t_1, t_1+\tau) = \lim_{n\to\infty}\frac{1}{n}\sum_{i=1}^{n} x_i(t_1)\,x_j(t_1+\tau) \tag{2.225}$$

and the autocorrelation function is defined by the ensemble average

$$R_{x_ix_i}(t_1, t_1+\tau) = \lim_{n\to\infty}\frac{1}{n}\sum_{i=1}^{n} x_i(t_1)\,x_i(t_1+\tau) \tag{2.226}$$

Similar to the case of mean value computation, to investigate the statistical properties of the random process, Equations 2.225 and 2.226 require quite a collection of time histories x_i and x_j. Thus, in many cases, the temporal average is used instead of an ensemble average.

Suppose two pieces of sampled time histories are measured from a random process and are denoted as $x_1(t)$ and $x_2(t)$. The method used to correlate these two time histories is correlated based on

$$R_{12}(\tau) = \lim_{T\to\infty} \frac{1}{T}\int_0^T x_1(t)x_2(t+\tau)dt \tag{2.227}$$

Here, for simplicity, R_{12} is used instead of $R_{x_1x_2}$, to denote the cross-correlation function.

Mathematically, Equation 2.225 can be rewritten as

$$R_{12}(\tau) = \lim_{T\to\infty} \frac{1}{T}\int_{-T/2}^{T/2} x_1(t)x_2(t+\tau)\,dt = \lim_{T\to\infty}\int_{-\infty}^{\infty} \frac{1}{T}x_1(t)x_2(t+\tau)dt \tag{2.228a}$$

For more general cases, $1/T$ is replaced by a more general term of probability, $p_{12}(x_1, x_2, t, \tau)$, that is,

$$R_{12}(\tau) = \int_{-\infty}^{\infty} p_{12}(x_1,x_2,t,\tau)x_1(t)x_2(t+\tau)dt \tag{2.228b}$$

Here, $p_{12}(x_1, x_2, t, \tau)$ is the probability of the occurrence of the production of x_1 taken at time t and x_2 taken at $t + \tau$.

Since the above equations describe the relationship between the time histories of $x_1(t)$ and $x_2(t)$, it is referred to as the *cross-correlation function*.

In this particular case, when the time history $x_2(t)$ is chosen to be $x_1(t)$, the result is the *autocorrelation function*, which can be written as

$$R_{11}(\tau) = \lim_{T\to\infty} \frac{1}{T}\int_{-T/2}^{T/2} x_1(t)x_1(t+\tau)dt = \lim_{T\to\infty}\int_{-\infty}^{\infty} \frac{1}{T}x_1(t)x_1(t+\tau)dt \tag{2.229a}$$

The more general form is

$$R_{11}(\tau) = \int_{-\infty}^{\infty} p(t,\tau)x_1(t)x_1(t+\tau)dt \tag{2.229b}$$

Here, $p(t, \tau)$ is the probability of the occurrence of the production of x_1 taken at time t and $t + \tau$.

From both the cross- and autocorrelation functions, it is seen that the resulting quantities from the integrations are functions of τ. The variable τ in the correlation domain is similar to the temporal variable t in the real-world time domain. When τ is chosen from zero to a rational value, the resulting integrals $R_{12}(\tau)$ or $R_{11}(\tau)$ represent a deterministic process.

Note that in the above equations, the time histories $x_1(t)$ and $x_2(t)$ can be selected from a random process. After they are chosen, $x_1(t)$ and $x_2(t)$ are deterministic. The time histories $x_1(t)$ and $x_2(t)$ can also be known as periodic or transient signals, which are, of course, deterministic. Thus, the correlation analysis is not limited to random processes. However, in the following discussion, it will be seen that using the correlation functions for random processes will have additional limitations. Further, when the time average is used to define the correlation functions, care should be taken to check whether the signals are ergodic.

In Figure 2.15a, the autocorrelation function of $\sin(\omega t)$ is plotted. Here, $\omega = 2\pi$(rad), which is the natural frequency of the system mentioned above. In Figure 2.15b, the cross-correlation function of $\sin(\omega t)$, which is used as harmonic excitation, and the response displacement is plotted for the

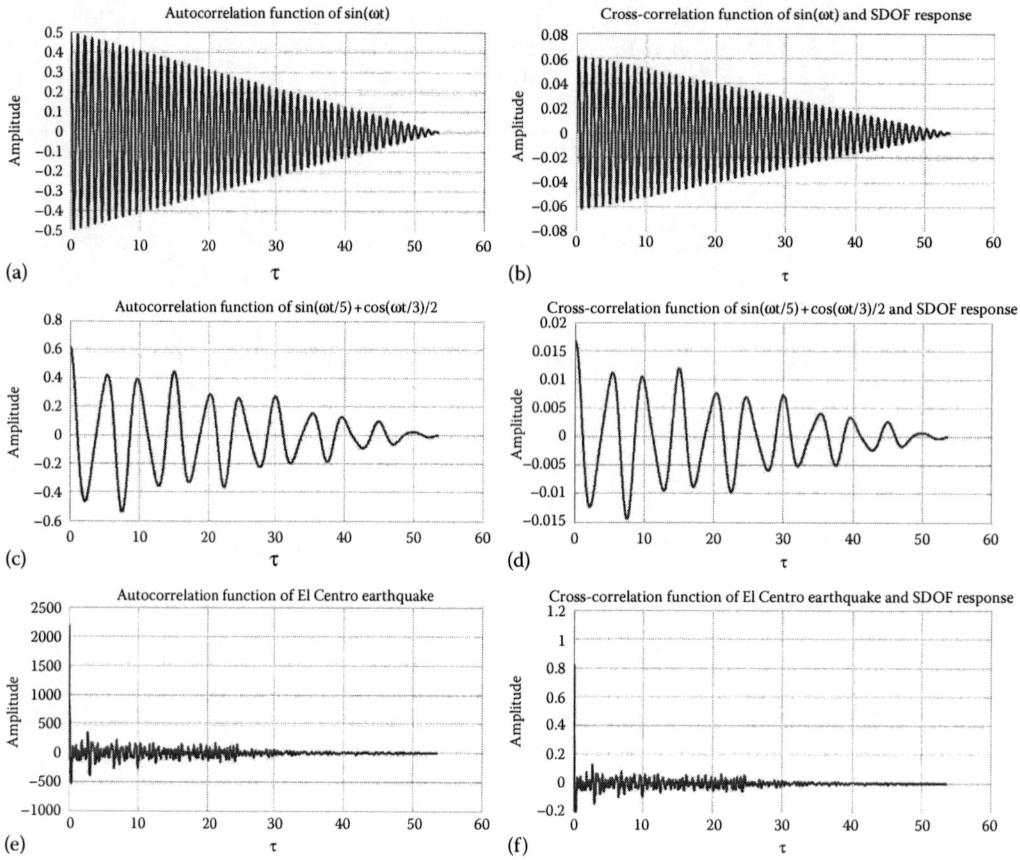

FIGURE 2.15 (a–f) Correlation functions.

SDOF system mentioned above. Furthermore, Figure 2.15c presents the autocorrelation function of [sin(ωt/5) + cos(ωt/3)/2]. Additionally, Figure 2.15d presents the cross-correlation function of [sin(ωt/5) + cos(ωt/3)/2], which is used as harmonic excitation, and the corresponding response displacement of the same SDOF system.

In addition, in Figure 2.15e, the autocorrelation function of the El Centro earthquake record is plotted, while Figure 2.15f provides the cross-correlation function of the El Centro earthquake, which is used as ground excitation, and the corresponding response displacement of the SDOF system.

From Figure 2.15, all these correlation functions look like free-decay time histories. That is, when the temporal variable t becomes quite large, the correlation functions tend toward zero. It is seen that for harmonic functions, the correlation functions decay slowly. For random excitation, the correlation function will decay rather fast. In fact, it can be proven that for any time histories $x_1(t)$ and $x_2(t)$,

$$\left|R_{11}(\tau)\right| < R_{11}(0) = E\left[x_1(t)^2\right] = \overline{x_1^2} \tag{2.230}$$

In the following discussion, it is seen that $R_{11}(0)$ actually stands for the average power of the process $x_R(t)$. Furthermore,

$$\left|R_{12}(\tau)\right|^2 < R_{11}(0)R_{22}(0) \tag{2.231}$$

Correlation functions are very helpful tools in analyzing the random processes. For convenience, these functions are denoted as

$$R_{11}(\tau) = \langle x_1, x_1 \rangle \tag{2.232}$$

and

$$R_{12}(\tau) = \langle x_1, x_1 \rangle \tag{2.233}$$

It can also be proven that the autocorrelation function is an even function of τ. That is,

$$R_{11}(-\tau) = R_{11}(\tau) \tag{2.234}$$

On the other hand, the cross-correlation function is neither an even nor an odd function. However,

$$R_{12}(-\tau) = R_{21}(\tau) \tag{2.235}$$

Because of the nature of the correlation function described in Equations 2.234 and 2.235, in practical computations, only the values of the correlation functions are calculated when

$$\tau \ge 0 \tag{2.236}$$

Suppose a signal contains several frequency components, which can be expressed as trigonometric functions, say, the frequencies are ω_1, ω_2, and ω_3. Also, suppose the signal has some random terms. From the definition of the autocorrelation function in Equation 2.229, it is realized that by using integration, the frequency components ω_1, ω_2, and ω_3 will remain in the correlation function, and the random terms will vanish due to the orthogonality of the trigonometric functions. Thus, by using the autocorrelation function, these particular frequency components can be emphasized and random terms can be eliminated. Suppose two signals share the same frequency components, which can be expressed as trigonometric terms, say, the frequencies are ω_1, ω_2, and ω_3. Also, suppose the signals have different frequency components and some random terms. From the definition of the cross-correlation function in Equation 2.228, it can also be realized that by using integration, the frequency components ω_1, ω_2, and ω_3 will remain in the cross-correlation function, and the frequency terms that belong to individual signals and the random terms will also vanish, due to the orthogonality of trigonometric functions. Thus, by using the cross-correlation function, these particular frequency components shared by both signals can be emphasized and the frequency terms of the individual signals and the random terms can be eliminated.

The above points can be realized by examining the plots of the earthquake responses given in Figure 2.15.

2.3.3.2 Power Spectral Density Function

Random signals may not always have Fourier transforms. Therefore, it is difficult to study the corresponding frequency spectra directly from the integral transformation. In this case, the Fourier transforms of their correlation functions are used, which exist for most engineering signals. Therefore, the frequency components of random signals can be studied.

2.3.3.2.1 Definition of Power Spectral Density Function

The Fourier transform of the autocorrelation function defines the *auto-power spectral density function*, which can be written as

$$S_{11}(\omega) = \int_{-\infty}^{\infty} R_{11}(\tau) e^{-j\omega\tau} d\tau \tag{2.237}$$

Thus,

$$R_{11}(\tau) \leftrightarrow S_{11}(\omega) \tag{2.238}$$

Furthermore, the Fourier transform of the cross-correlation function defines the *cross-power spectral density function*. This can be written as

$$S_{12}(\omega) = \int_{-\infty}^{\infty} R_{12}(\tau) e^{-j\omega\tau} d\tau \tag{2.239}$$

Thus,

$$R_{12}(\tau) \leftrightarrow S_{12}(\omega) \tag{2.240}$$

Equations 2.238 and 2.240 are referred to as Wiener–Khintchine equations (Silva 2007).

2.3.3.2.2 Properties of Auto-Power Spectral Function

It can be proven that if the signal x(t) is the response obtained through the convolution of h(t)∗f(t), then

$$S_{xx}(\omega) = |H(\omega)|^2 S_{ff}(\omega) \tag{2.241}$$

Here, S_{xx} is the autopower spectral density function of the response x(t) and S_{ff} is the autopower spectral density function of the forcing function f(t). From Equations 2.240 and 2.241,

$$R_{xx}(\tau)\big|_{\tau=0} = \frac{1}{2\pi} \int_{-\infty}^{\infty} |H(\omega)|^2 S_{ff}(\omega) d\omega = E\left[x^2\right] = \overline{X^2} \tag{2.242}$$

In addition to Equations 2.241 and 2.242, other important properties of the Fourier pair $R_{xx}(\tau)$ and $S_{xx}(\omega)$ can exist, which are further explored in a later subsection. Here, the proof of Equation 2.241 is obtained, which is very important in the analysis of random signals.

To prove Equation 2.241,

$$S_{xx}(\omega) = \int_{-\infty}^{\infty} R_{xx}(\tau) e^{-j\omega\tau} d\tau = \int_{-\infty\,(\tau)}^{\infty} \left[\int_{-\infty\,(\sigma)}^{\infty} x(\sigma) x(\sigma+\tau) d\sigma \right] e^{-j\omega\tau} d\tau \tag{2.243}$$

Here, in order to clearly denote the integrals, the integration symbols are followed by subscripts (τ) and (σ), which stand for the integrations for variables τ and σ, respectively. In the following discussion, similar notations are used to indicate the specific integrations.

Note that x(σ) is the response of time variable σ, which can be expressed as the convolution of the forcing function f(σ) and the unit impulse response function h(σ), that is,

$$x(\sigma) = \int_{-\infty}^{\infty} f(\sigma-\mu) h(\mu) d\mu \tag{2.244}$$

Similarly,

$$x(\sigma+\tau) = \int_{-\infty}^{\infty} f(\sigma-\mu+\tau) h(\mu) d\mu \tag{2.245}$$

Substituting Equations 2.244 and 2.245 into Equation 2.243 yields

$$S_{XX}(\omega)=\int_{-\infty\,(\tau)}^{\infty}\left\{\int_{-\infty\,(\sigma)}^{\infty}\left[\int_{-\infty\,(\mu)}^{\infty} f(\sigma-\mu)h(\mu)d\mu\right]\left[\int_{-\infty\,(\mu)}^{\infty} f(\sigma+\tau-\mu)h(\mu)d\mu\right]d\sigma\right\}e^{-j\omega\tau}d\tau \quad (2.246)$$

In order to denote clearly the two convolutions described in Equations 2.244 and 2.245, different variables are used. Namely, for the first convolution integral, variable μ is used, whereas in the second convolution integral, variable ν replaces μ. Such a replacement is only for, convenience of notation; in fact,

$$\mu=\nu \quad (2.247)$$

Now, using Equation 2.247 and realizing the functions $[f(\sigma-\mu)h(\mu)]$ and $[f(\sigma+\tau-\nu)h(\nu)]$ are mutually independent,

$$\begin{aligned}
&\int_{-\infty\,(\tau)}^{\infty}\left\{\int_{-\infty\,(\sigma)}^{\infty}\left[\int_{-\infty\,(\mu)}^{\infty} f(\sigma-\mu)h(\mu)d\mu\right]\left[\int_{-\infty\,(\nu)}^{\infty} f(\sigma+\tau-\nu)h(\nu)d\nu\right]d\sigma\right\}e^{-j\omega\tau}d\tau\\
&=\int_{-\infty\,(\tau)}^{\infty}\left\{\int_{-\infty\,(\sigma)}^{\infty}\left[\int_{-\infty\,(\mu)}^{\infty}\int_{-\infty\,(\nu)}^{\infty} f(\sigma-\mu)f(\sigma+\tau-\nu)h(\mu)h(\nu)d\mu\, d\nu\right]d\sigma\right\}e^{-j\omega\tau}d\tau\\
&=\int_{-\infty\,(\tau)}^{\infty}\int_{-\infty\,(\mu)}^{\infty}\int_{-\infty\,(\nu)}^{\infty}\int_{-\infty\,(\sigma)}^{\infty} f(\sigma-\mu)f(\sigma+\tau-\nu)h(\mu)h(\nu)d\sigma\, d\nu\, d\mu\, e^{-j\omega\tau}d\tau\\
&=\int_{-\infty\,(\tau)}^{\infty}\int_{-\infty\,(\mu)}^{\infty}\int_{-\infty\,(\nu)}^{\infty} h(\mu)h(\nu)\left\{\int_{-\infty\,(\sigma)}^{\infty} f(\sigma-\mu)f(\sigma+\tau-\nu)d\sigma\right\}d\nu\, d\mu\, e^{-j\omega\tau}d\tau
\end{aligned} \quad (2.248)$$

Because $\mu=\nu$ from Equation 2.247, the terms within the braces in Equation 2.248 are rewritten as

$$\begin{aligned}
\int_{-\infty}^{\infty} f(\sigma-\mu)f(\sigma+\tau-\nu)d\tau &= \int_{-\infty}^{\infty} f(\sigma-\mu)f(\sigma-\nu+\tau)d\tau\\
&=\int_{-\infty}^{\infty} f(\sigma-\mu)f(\sigma-\mu+\tau)d\tau\\
&=R_{ff}(\tau)
\end{aligned} \quad (2.249)$$

Substituting Equation 2.249 into Equation 2.248 yields

$$\begin{aligned}
S_{XX}(\omega)&=\frac{1}{2\pi}\int_{-\infty\,(\tau)}^{\infty}\int_{-\infty\,(\mu)}^{\infty}\int_{-\infty\,(\nu)}^{\infty} h(\mu)h(\nu)\{R_{ff}(\tau)\}d\nu\, d\mu\, e^{-j\omega\tau}d\tau\\
&=\int_{-\infty\,(\tau)}^{\infty} R_{ff}(\tau)\int_{-\infty\,(\mu)}^{\infty}\int_{-\infty\,(\nu)}^{\infty} h(\mu)h(\nu)d\nu\, d\mu\, e^{-j\omega\tau}d\tau
\end{aligned} \quad (2.250)$$

Inserting $1 = e^{-j\omega\mu}e^{j\omega\mu}$ between $h(\mu)$ and $h(v)$, Equation 2.250 can be written as

$$\int_{-\infty(\tau)}^{\infty} R_{ff}(\tau) \int_{-\infty(\mu)}^{\infty} \int_{-\infty(v)}^{\infty} h(\mu)e^{-j\omega\mu}\, e^{j\omega\mu} h(v)\, dv\, d\mu\, e^{-j\omega\tau} d\tau$$

$$= \int_{-\infty(\tau)}^{\infty} R_{ff}(\tau) \int_{-\infty(\mu)}^{\infty} h(\mu)e^{-j\omega\mu}\, d\mu \int_{-\infty(v)}^{\infty} h(v)e^{j\omega\mu}\, dv\, e^{-j\omega\tau} d\tau \tag{2.251}$$

Note that

$$H(\omega) = \int_{-\infty(\mu)}^{\infty} h(\mu)e^{-j\omega\mu} d\mu \tag{2.252}$$

and

$$H^*(\omega) = \int_{-\infty(v)}^{\infty} h(v)e^{j\omega\mu} dv \tag{2.253}$$

Therefore

$$S_{xx}(\omega) = \int_{-\infty(\tau)}^{\infty} R_{ff}(\tau)H(\omega)H^*(\omega)e^{-j\omega\tau} d\tau \tag{2.254}$$

Furthermore,

$$H(\omega)H^*(\omega) = |H(\omega)|^2 \tag{2.255}$$

Thus,

$$S_{xx}(\omega) = |H(\omega)|^2 \int_{-\infty}^{\infty} R_{ff}(\tau)e^{-j\omega\tau} d\tau \tag{2.256}$$

and Equation 2.241 is established. Thus,

$$S_{xx}(\omega) = |H(\omega)|^2 S_{ff}(\omega) \tag{2.257}$$

Next, since the autocorrelation function is an even function of τ (see Equation 2.234), the autopower spectral density function must be real valued, although in general the Fourier transform of a signal x(t) is complex valued.

Furthermore, from Equation 2.234, it is proven that the autopower spectral density function must be an even function of ω. This is because

$$S_{xx}(\omega) = \int_{-\infty}^{\infty} R_{xx}(\tau)\, e^{-j\omega\tau} d\tau = \int_{-\infty}^{\infty} R_{xx}(-\tau)e^{-j\omega\tau} d\tau$$

$$= \int_{\infty}^{-\infty} R_{xx}(-\tau)e^{j\omega(-\tau)} d(-\tau) = S_{xx}(-\omega) \tag{2.258}$$

Finally, from the above derivations, it is also realized that the autopower spectral density function must always be greater than zero. That is,

$$S_{XX}(\omega) \geq 0 \tag{2.259}$$

2.3.3.2.3 *Physical Meaning of Autopower Spectral Density Function*

In the above, the power spectral function was defined as the Fourier transform of the correlation function. Then, several important properties of the function were examined to emphasize the relationship between the correlation functions and the power spectral density functions. In the following discussion, the physical meaning of the power spectral density function is considered in more detail.

Recall Section 2.2, where it was noted that the existence of the Fourier transform of signal f(t) needs certain conditions. Recall the absolutely integrable condition,

$$\int_{-\infty}^{\infty} |f(t)| dt < \infty \tag{2.260}$$

Then, the time domain signal f(t) can have its Fourier transform pair F(ω).

Now, from the theorem of multiplication, in Equation 2.146, let $f_2 = f_1 = f$. Then,

$$\int_{-\infty}^{\infty} f^2(t)dt = \frac{1}{2\pi}\int_{-\infty}^{\infty} F^*(\omega)F(\omega)d\omega = \frac{1}{2\pi}\int_{-\infty}^{\infty} |F(\omega)|^2 d\omega \tag{2.261}$$

Equation 2.261 is referred to as the *Parseval equation*. The left side of the equation stands for the total energy within the range of (−∞, +∞). On the right side of the equation, the term $|F(\omega)|^2$ is called the *energy spectrum*. In this sense, the Parseval theorem can be seen as the energy equation of signal f(t).

It is understandable that many engineering signals within the range of (−∞, +∞) have an infinite amount of energy; in addition, the condition in Equation 2.260 may often be violated. The sine function is a simple example of such signals. Random signals also often fall into such a category. To study the frequency components, an alternative approach must be used. For example, instead of the energy, the average power can be written as

$$\lim_{T\to\infty} \frac{1}{2T}\int_{-T}^{T} f^2(t)dt \tag{2.262}$$

Now, assume that the signal $f(t)|_{-T\leq t\leq T}$ satisfies the Dirichlet conditions, so that it has the Fourier transform F(ω,T) written as

$$F(\omega,T) = \int_{-\infty}^{\infty} f(t)\big|_{-T\leq t\leq T}\, e^{-j\omega t}dt = \int_{-T}^{T} f(t)e^{-j\omega t}dt \tag{2.263}$$

The corresponding Fourier pair satisfies the Parseval energy equation, that is,

$$\int_{-\infty}^{\infty} f^2(t)\big|_{-T\leq t\leq T}\, dt = \int_{-T}^{T} f^2(t)\big|_{-T\leq t\leq T}\, dt = \frac{1}{2\pi}\int_{-\infty}^{\infty} |F(\omega,T)|^2 d\omega \tag{2.264}$$

Dividing by 2T on both sides of Equation 2.264, and letting $T \to \infty$, results in

$$\lim_{T\to\infty} \frac{1}{2T}\int_{-T}^{T} f^2(t)\,dt = \frac{1}{2\pi}\int_{-\infty}^{\infty} \lim_{T\to\infty} \frac{1}{2T}\left|F(\omega,T)\right|^2 d\omega \tag{2.265}$$

It is seen that Equation 2.265 represents the energy density. In fact, the term inside the integration sign on the right side is the average power density function of the transient signal f(t), denoted by

$$S_{ff}(\omega) = \lim_{T\to\infty} \frac{1}{2T}\left|F(\omega,T)\right|^2 \tag{2.266}$$

In Equation 2.266, the same symbol $S_{(\cdot)(\cdot)}$ is used and the identical name of "power density function" is used as in Equation 2.237, where the concept of the power density function was introduced by using the Fourier transform of autocorrelation functions. To see the logic in this approach, the essence of Equation 2.266 is examined by further considering random signals, instead of transient signals. From now on, assume that the average power always exists, and for convenience, use $x_R(t)$ instead of f(t) in the remaining analysis, where $x_R(t)$ represents a stationary random process. As mentioned before, $x_R(t)$ is a collection of random signals $x_i(t)$. Thus, Equations 2.263 and 2.265 are written as follows:

$$X(\omega,T) = \int_{-T}^{T} x_R(t)e^{-j\omega t}dt \tag{2.267}$$

$$\lim_{T\to\infty} \frac{1}{2T}\int_{-T}^{T} x_R^2(t)\,dt = \frac{1}{2\pi}\int_{-\infty}^{\infty} \frac{1}{2T}\left|X(\omega,T)\right|^2 d\omega \tag{2.268}$$

where $X(\omega, T)$ is the Fourier transform of the random process $x_R(t)|_{-T\le t\le T}$.

Note that the integrations in Equations 2.267 and 2.268 are random. The limit of the mathematical expectation of the left term in Equation 2.267 is considered, which is the average power of the stationary process $x_R(t)$, written as

$$\lim_{T\to\infty}\left\{E\left[\frac{1}{2T}\int_{-T}^{T} x_R^2(t)\,dt\right]\right\} = \lim_{T\to\infty}\left\{\frac{1}{2T}\int_{-T}^{T} E\left[x_R^2(t)\right]dt\right\} \tag{2.269}$$

Equation 2.269 is compared with the mean square value of the transient signal defined in Equation 2.212. It is seen that in Equation 2.212, x(t) is a deterministic signal, whereas in Equation 2.269, $x_R(t)$ is a random process. Therefore, to study the power of $x_R(t)$, the statistical average is needed and the mathematical operation of expectation $E[(\cdot)]$ is used. In addition, the integration range (0, T) in Equation 2.212 and (−T, T) in Equation 2.268 do not actually make an essential difference, especially for engineering signals. Therefore, Equation 2.269 implies that the average power of a stationary process equals the mean square value of that process. Additionally, the following notation is possible:

$$\lim_{T\to\infty}\left\{E\left[\frac{1}{2T}\int_{-T}^{T} x_R^2(t)\,dt\right]\right\} = \overline{X^2} \tag{2.270}$$

Furthermore, exchange the order of operations for the term on the right in Equation 2.268 and consider Equations 2.269 and 2.270. Then,

$$\overline{X^2} = \frac{1}{2\pi}\int_{-\infty}^{\infty} \lim_{T\to\infty} \frac{1}{2T} E\left[\left|X(\omega,T)\right|^2\right] d\omega \tag{2.271}$$

Examining the integral function in Equation 2.271, and comparing it with Equation 2.266,

$$S_{XX}(\omega) = \lim_{T\to\infty} \frac{1}{2T} E\left[\left|X(\omega,T)\right|^2\right] \tag{2.272}$$

where $S_{XX}(\omega)$ is the autopower spectral density function. Apparently, from Equation 2.272, $S_{XX}(\omega)$ is a real-valued, nonnegative function for the term $|X(\omega, T)|^2$, similar to the integrand in the integration operation of $E[(\cdot)]$ that is real valued and nonnegative. However, up to now, the term defined in Equation 2.272 as well as in Equation 2.266 has been the power spectral density, which has been used to name the Fourier transform of the autocorrelation function (see the Wiener–Khintchine Equation 2.237).

To see the term defined in Equation 2.272 as the Fourier transform of the autocorrelation function, substitute Equation 2.267 into Equation 2.272. That is,

$$S_{XX}(\omega) = \lim_{T\to\infty} \frac{1}{2T} E\left[\int_{-T}^{T} x_R(t_1) e^{j\omega t_1} dt_1 \int_{-T}^{T} x_R(t_2) e^{-j\omega t_2} dt_2\right]$$

$$= \lim_{T\to\infty} \frac{1}{2T} \int_{-T}^{T}\int_{-T}^{T} E\left[x_R(t_1) x_R(t_2)\right] e^{-j\omega(t_2-t_1)} dt_1\, dt_2 \tag{2.273}$$

Note that,

$$E\left[x_R(t_1) x_R(t_2)\right] = R_{XX}(t_2 - t_1) \tag{2.274}$$

Substituting Equation 2.274 into Equation 2.273 yields

$$S_{XX}(\omega) = \lim_{T\to\infty} \frac{1}{2T}\left[\int_{-T}^{T}\int_{-T}^{T} R_{XX}(t_2 - t_1) e^{-j\omega(t_2-t_1)} dt_1\, dt_2\right] \tag{2.275}$$

In Equation 2.275, let $\tau = t_2 - t_1$. Then, finally,

$$S_{XX}(\omega) = \int_{-\infty}^{\infty} R_{XX}(\tau) e^{-j\omega\tau} d\tau \tag{2.276}$$

Comparing Equation 2.276 with Equation 2.237, it is realized that the newly defined power spectral function is indeed the autopower spectral function. In addition, the reason that the power spectral density actually describes the statistics of the distribution of the frequency components of the stationary process $x_R(t)$ can be understood.

2.3.3.2.4 *Cross-Power Spectral Density Function*

Similar to the definition of the autopower spectral density function, the cross-power density spectral function had been defined in Equation 2.239. Particularly, $R_{12}(t)$ can be seen as the cross-correlation function of the response x(t) and the forcing function f(t), both of which are random signals. Additionally, similar to the explanation of the physical meaning of the autopower spectral density function,

$$S_{xf}(\omega) = \lim_{T\to\infty} \frac{1}{2T} E\left[X(-\omega,T)F(\omega,T)\right] = \lim_{T\to\infty} \frac{1}{2T} E\left[X^*(\omega,T)F(\omega,T)\right] \tag{2.277}$$

Now, some important properties of the cross-power spectral density function are noted as follows:

i. $S_{xf}(\omega) = X^*F = X^*_{fx}(\omega)$ (2.278)

ii. $\mathrm{Re}\left[S_{xf}(\omega)\right] = \mathrm{Re}\left[S_{xf}(-\omega)\right]$ (2.279)

and

$$\mathrm{Im}\left[S_{xf}(\omega)\right] = -\mathrm{Im}\left[S_{xf}(-\omega)\right] \tag{2.280}$$

iii. $\left|S_{xf}(\omega)\right| \le S_{xx}(\omega)S_{ff}(\omega)$ (2.281)

In addition, the autopower and cross-power spectral functions can be used to extract the transfer function from a random excitation F(t) or random response $x_R(t)$ process. Using Equation 2.278 and replacing F by X, results in $S_{xx}(\omega) = X^*X$ and it can be proven that

$$H(\omega) = \frac{S_{fx}(\omega)}{S_{ff}(\omega)} = \frac{S_{xx}(\omega)}{S_{xf}(\omega)} \tag{2.282}$$

Suppose that the random forcing functions and responses are measured. Denote $F_i(\omega)$ and $X_i(\omega)$ as the Fourier transforms of the ith random forcing function $f_i(t)$ and response function $x_i(t)$, respectively. (Note that $F_i(\omega)$ and $X_i(\omega)$ always exist, because once $f_i(t)$ and $x_i(t)$ are measured, these become deterministic and have finite time duration.) Then, the calculated transfer function is

$$H_i(\omega) = \frac{X_i(\omega)}{F_i(\omega)} \tag{2.283}$$

By multiplying $X_i^*(\omega)$ in both the numerator and the denominator on the right side of Equation 2.283,

$$H_i(\omega) = \frac{X_i(\omega)X_i^*(\omega)}{F_i(\omega)X_i^*(\omega)} = \frac{S_{x_ix_i}(\omega)}{S_{x_if_i}(\omega)} \tag{2.284}$$

Suppose that the average of the power density functions $S_{x_ix_i}$ and $S_{x_if_i}$ of the n sets of the measured response and forcing functions can approximate the power density function S_{xx} and S_{xf}. That is,

$$S_{xx} \approx \frac{1}{n}\sum_{i=1}^{n} S_{x_ix_i} \tag{2.285}$$

and

$$S_{xf} \approx \frac{1}{n}\sum_{i=1}^{n} S_{x_i f_i} \tag{2.286}$$

The result is the second part of Equation 2.282. The first part of Equation 2.283 can be proven via a similar methodology.

Example 2.8

Suppose there is significant noise mixed with the measurement of the forcing function, whereas when the signal of the response is picked up, the noise can be ignored. This case can be shown graphically, as in Figure 2.16a.

When the autopower spectral density function of the measured force is used, it is realized that

$$\begin{aligned}\left[F(\omega)+N(\omega)\right]\left[F(\omega)+N(\omega)\right]^* &= F(\omega)F^*(\omega)+N(\omega)F^*(\omega)+F(\omega)+N^*(\omega)+N(\omega)N^*(\omega)\\ &= S_{ff}(\omega)+S_{nf}(\omega)+S_{fn}(\omega)+S_{nn}(\omega)\end{aligned}$$

Here, $S_{ff}(\omega)$ and $S_{nn}(\omega)$ are, respectively, the autocorrelation functions of the force and the noise, while $S_{nf}(\omega)$ and $S_{fn}(\omega)$ are, respectively, the cross-correlation functions of the force and the noise.

In many cases, the noise $N(\omega)$ is not correlated with the forcing function and is not correlated with itself. Therefore, roughly,

$$S_{nf}(\omega) = S_{fn}(\omega) = S_{nn}(\omega) = 0$$

In this way, if the first part of Equation 2.282 is used to calculate the transfer function, the noise from the input sides can be eliminated, that is, with

$$H_1(\omega) = \frac{S_{fx}(\omega)}{S_{ff}(\omega)} \tag{2.287a}$$

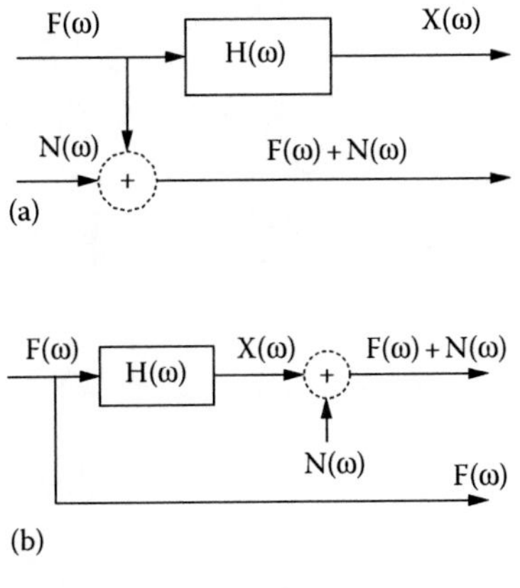

FIGURE 2.16 Noises and transfer function calculation: (a) input with significant noise and (b) output with significant noise.

If the measured response is contaminated with significant noise but the noise is relatively small from the input side, as in Figure 2.16b, then the second part of Equation 2.282 should be used to calculate the transfer function. Namely,

$$H_2(\omega) = \frac{S_{xx}(\omega)}{S_{xf}(\omega)} \quad (2.287b)$$

Similar to the first case of very large input noise, it can be proven that the noise from the output side will be greatly reduced by using $[X(\omega) + N(\omega)][X(\omega) + N(\omega)]^*$ to obtain $S_{xx}(\omega)$.

The above two equations represent the transfer functions. Subscripts 1 and 2 are used because in the literature, the computation of a transfer function that reduces input noise (see Figure 2.16a) is referred to as an H_1 transfer function. On the other hand, the computation of a transfer function that reduces the output noise (see Figure 2.16b) is referred to as an H_2 transfer function.

2.3.4 Correlation between Forcing Function and Impulse Response Function

In this subsection, the relationship between the Laplace transform of a temporal signal, which can be seen as a forcing function, is introduced, and the correlation functions between the forcing function and the unit impulse response function are examined. In this way, the essence of the Laplace variables and Laplace transform can be better understood, and the effect of damping can be further examined.

In Equation 2.154, the variable $\upsilon > 0$ in many cases, which can be denoted as

$$\upsilon = \xi\omega_n \quad (2.288)$$

And the variable ω is now denoted as

$$\omega = -\omega_d \quad (2.289)$$

In Equations 2.288 and 2.289, ξ, ω_n and ω_d can be seen as the damping ratio, natural and damped frequencies of a SDOF system. Thus, the Laplace transform of a temporal function that is seen as a general forcing function f(t) can be written as

$$\mathcal{L}[f(t)] = F(s) = \int_0^\infty f(t)e^{-st}dt = \int_0^\infty f(t)e^{(-\xi\omega_n + j\omega_d)t}dt \quad (2.290)$$

Multiplying by $(e^{-s\tau})/(m\omega_d)T$ on both sides of Equation 2.290 yields

$$\frac{e^{-s\tau}}{m\omega_d T}F(s) = \frac{e^{-s\tau}}{m\omega_d T}\int_0^\infty f(t)e^{-st}dt = \frac{1}{T}\int_0^\infty f(t)\frac{e^{-s(t+\tau)}}{m\omega_d}dt \quad (2.291)$$

This operation is equivalent to making a time shift τ of the forcing function f(t). In addition, the operation in Equation 2.290 also divides the Laplace transform F(s) by the factor $m\omega_d$.

Next, taking the complex conjugate of the above equation results in

$$\frac{e^{-s^*\tau}}{m\omega_d T}F(s^*) = \frac{1}{T}\int_0^\infty f(t)\frac{e^{-s^*(t+\tau)}}{m\omega_d}dt \quad (2.292)$$

Dividing both sides of Equations 2.291 and 2.292 by 2j and subtracting the resulting second equation from the first one yields

$$\frac{e^{-s\tau}}{2jm\omega_d T}F(s)-\frac{e^{-s^*\tau}}{2jm\omega_d T}F(s^*)=\frac{1}{T}\left[\int_0^\infty f(t)\frac{e^{-s(t+\tau)}}{2jm\omega_d}dt-\int_0^\infty f(t)\frac{e^{-s^*(t+\tau)}}{2jm\omega_d}dt\right]$$

$$=\frac{1}{T}\int_0^\infty f(t)\frac{e^{-\xi\omega_n(t+\tau)}}{m\omega_d}\left(\frac{e^{j\omega_d(t+\tau)}-e^{-j\omega_d(t+\tau)}}{2j}\right)dt \tag{2.293}$$

$$=\frac{1}{T}\int_0^\infty f(t)\left\{\frac{e^{-\xi\omega_n(t+\tau)}}{m\omega_d}\sin\left[\omega_d(t+\tau)\right]\right\}dt$$

From the previous section, it is known that the unit impulse response function with time shift τ, $h(t+\tau)$ of a linear SDOF vibration system (m-c-k) system can be written as,

$$h(t+\tau)=\frac{e^{-\xi\omega_n(t+\tau)}}{m\omega_d}\sin\omega_d(t+\tau)$$

In Equation 2.293, the operation of subtraction causes the left side term in Equation 2.292 to be real valued. Thus, Equation 2.293, can be rewritten as

$$\frac{e^{-s\tau}}{2jm\omega_d T}F(s)-\frac{e^{-s^*\tau}}{2jm\omega_d T}F(s^*)=\frac{1}{T}\int_0^\infty f(t)\{h(t+\tau)\}dt \tag{2.294}$$

Note that the term on the right side of Equation 2.293 is the cross-correlation function of f(t) and $h(t+\tau)$, denoted by $R_{fh}(\tau)$, Thus,

$$R_{fh}(\tau)=\frac{e^{-s^*\tau}}{2jm\omega_d T}F(s^*)-\frac{e^{-s\tau}}{2jm\omega_d T}F(s) \tag{2.295}$$

Denote

$$\alpha=\frac{1}{2jm\omega_d T} \tag{2.296}$$

And note that, similar to the theorem of time shift for Fourier transform (see Equation 2.143), multiplying F(s) by $e^{-s\tau}$ gives the Laplace transform of $f(t-\tau)$. Namely,

$$\mathcal{L}\left[\alpha f(t-\tau)\right]+\mathcal{L}\left[\alpha f(t-\tau)\right]^*=R_{fh}(\tau)=R_{hf}(-\tau) \tag{2.297}$$

Furthermore, denote

$$F(s)=F_0e^{j\theta} \tag{2.298}$$

Here F_0 and θ are the real-valued amplitude and phase angle of the Laplace transform F(s). Both are functions of s, namely, the function of the eigenvalue λ of the SDOF vibration system described in Equation 1.2, namely,

$$-s = \lambda = -\xi\omega_n \pm j\omega_d$$

and

$$F_0 = F_0(\lambda)$$

$$\theta = \theta(\lambda)$$

With the help of Equation 2.298, the right side of Equation 2.297 becomes

$$\frac{F_0 e^{-\xi\omega_n\tau}}{m\omega_d T}\frac{e^{j\omega_d\tau}e^{j\theta} - e^{-j\omega_d\tau}e^{-j\theta}}{2j} = \frac{F_0 e^{-\xi\omega_n\tau}}{m\omega_d T}\sin(\omega_d t + \theta) \tag{2.299}$$

Therefore,

$$R_{fh}(\tau) = \frac{F_0 e^{-\xi\omega_n\tau}}{m\omega_d T}\sin(\omega_d t + \theta) \tag{2.300}$$

Equation 2.300 implies that the cross-correlation function equals the amplitude of the corresponding Laplace transform F(s) times a function, which has the identical form of the unit impulse response h(t), except with a phase shift θ. Here, θ is the phase of F(s).

It is also seen that such a correlation function, obtained through the above manipulation of the Laplace transform of the forcing function, is the product of two functions. The function $\sin(\omega_d\tau + \theta)$ is a sinusoidal function with the damped natural frequency ω_d and the phase shift θ. The function $e^{-\xi\omega_n\tau}$ makes a decaying envelope with respect to time τ. It is seen that as the damping ratio ξ increases, the decay occurs more quickly.

It is also known that when the Laplace variable s is chosen, there is always a corresponding SDOF system with the solution λ of its characteristic to be $-s = -\xi\omega_n \pm j\omega_d$. The impulse response function used above is exactly determined through this SDOF system.

The above discussion implies that the sum of the Laplace transform and its complex conjugate of a forcing function with time delay τ, denoted by $f(t - \tau)$ and a proportional factor α equals the cross correlation of an impulse response function h(t) of a SDOF system and the function of $f(t - \tau)$. Due to the aforementioned orthogonality of sine and cosine functions, it is seen that the correlation integration will eliminate all uncorrelated frequency components in f(t) and only the frequency equal to the natural frequency ω_n of the SDOF system remains and the damping ratio of this system is ξ. Therefore, the operation of this correlation integral can be seen as a result of filtering by a SDOF system with eigenvalue λ. Thus, Equation 2.297 further implies that the Laplace transform of a signal f(t) can be seen as the filtered result through the corresponding SDOF systems, which act as a series of mechanical filters.

Example 2.9

Let the forcing function be a unit delta function $\delta(t)$, then

$$\mathscr{L}[\delta(t)] = 1$$

Thus,

$$F_0 = 1$$

and

$$\theta = 0$$

From Equation 2.300,

$$R_{fh}(\tau) = \frac{e^{-\xi\omega_n\tau}}{m\omega_d T}\sin(\omega_d\tau) = h(\tau)/T$$

This equation implies that the cross-correlation function of the unit impulse with a unit impulse response function is proportional to the unit impulse response function itself.

Example 2.10

Let the forcing function be a unit step u(t) and

$$u(t) = \begin{cases} 0, & t < 0 \\ 1, & t > 1 \end{cases}$$

Then

$$\mathcal{L}[u(t)] = \frac{1}{s}$$

and

$$\left|\frac{1}{s}\right| = \frac{1}{\omega_n}$$

$$\theta = -\tan^{-1}\frac{\sqrt{1-\xi^2}}{\xi}$$

Therefore,

$$\frac{e^{-\xi\omega_n\tau}}{m\omega_d\omega_n T}\sin(\omega_d\tau+\theta) = \frac{e^{-\xi\omega_n\tau}}{m\omega_n^2\sqrt{1-\xi^2}\,T}\sin(\omega_d\tau+\theta) = \frac{e^{-\xi\omega_n\tau}}{k\sqrt{1-\xi^2}\,T}\sin(\omega_d\tau+\theta)$$

The above should be equal to $R_{fh}(\tau)$, which can be proven to have the following form:

$$R_{fh}(\tau) = \frac{1}{T}\int_0^\infty u(t)h(t+\tau)dt = \frac{1}{T}\int_0^\infty h(t+\tau)dt$$

$$= \int_0^\infty \frac{e^{-\xi\omega_n(t+\tau)}}{m\omega_d T}\sin[\omega_d(t+\tau)]dt$$

$$= \frac{e^{-\xi\omega_n\tau}}{k\sqrt{1-\xi^2}\,T}\sin(\omega_d\tau+\theta)$$

2.3.5 Basic Approach to Dealing with Random Vibrations

As a brief summary, the basic approach to dealing with random vibrations is averaging with proper methods. For example, both mean value and standard deviation are arithmetic averages in dealing with random variables; the correlation functions are averages of the products of two sets of signals taken from certain time points.

The main effort in averaging is the summation for discrete signals, or integration for continuous signals, which helps distinguish specific frequency components based on the orthogonality of the vibration signals. The summation and/or integration will also help eliminate unwanted noises and uncertainties and indicate the major trends of certain random sets.

2.4 EARTHQUAKE RESPONSES OF SDOF LINEAR SYSTEMS

As mentioned before, structural vibrations are excited by three types of forces, namely, periodic, transient, and random forcing functions. In this section, a special random excitation is discussed, the earthquake ground motion. Although this issue seems not exactly parallel to the three basic types of excitations, the purpose is to emphasize the importance of studying earthquake-induced vibration. The corresponding level of structural vibration is indeed the target of the control by using the supplemental dampers.

Earthquake ground motions, as well as the corresponding responses of structures, are random processes, so their amplitudes are not deterministic. In addition, the excitation of earthquake ground motion is nonstationary random. Thus, even if the magnitude of the peak ground acceleration (PGA) is specified, it is rather difficult to have a deterministic value of the structural responses. As mentioned before, due to the random nature of earthquakes, the response bound of a structure with bounded excitations cannot be determined.

However, for engineering designs and constructions, deterministic values are needed. For example, the maximum possible amplitudes of earthquake-induced structural responses, such as forces and displacements, need to be known.

As mentioned in Section 2.3, the basic method to deal with random sets and random processes is averaging. And the average must follow certain rigorous rules. Through averages, uncertainties and/or noises can be effectively rejected and special patterns embedded in those random signals can be found. Therefore, the rule of thumb in dealing with random signals, namely, averaging, should be used to find patterns of earthquake responses.

The introduction of the response spectrum is an ingenious method of averaging, which provides the response bounds with given excitation bounds. Note that since earthquake responses are a nonstationary, nonergodic process, temporal averaging cannot be used to find the response patterns.

The construction of the response spectrum is a special type of ensemble average. It is not an average at a specific time point as mentioned in Section 2.3, but rather it is the averages at specific frequency (period) points. The ensemble averages discussed in Section 2.3 are, in general, a function of time. Now, the new ensemble averages are a function of period, which is often specifically called a spectrum.

In this section, the special ensemble average is introduced and its applications in earthquake engineering are discussed.

2.4.1 Response Spectrum

In Chapter 1, it is shown that the load and the response can have a simple proportional relationship by using the concept of dynamic stiffness. However, in that case, the excitation is sinusoidal. In the above section, the case in which the excitation or forcing function is deterministic was discussed. In this case, the convolution integral can be used to determine the responses. For the topic of random excitation mentioned in this chapter, mathematical tools, such as correlation functions and power spectral density function, were used to analyze the force and the response. Obtaining the structural responses has not yet been discussed.

Under earthquake loading, there cannot be an analytical relationship between the loading and the response, as in the case of harmonic excitation. Although the convolution integral can be used to obtain the response, the forcing function of the earthquake ground motion is unknown.

Furthermore, even under a given amplitude of earthquake loading, the amplitude of the response, in most cases, cannot be completely determined. Yet, to seek the deterministic relationship between the input level and the structural response is a necessity in earthquake engineering. Consequently, the concept of the response spectrum has been developed and is widely used (e.g., see Chopra 2006).

In earthquake engineering, for design convenience, the input levels are often treated as deterministic, although the seismic ground motion is random. This is because although earthquake excitation is random, it is expected that the peak amplitude will not exceed a certain level under some statistical rules. Therefore, a table that lists the deterministic level is possible by having prior knowledge of the regional history of earthquakes and the site conditions. A more sophisticated table will also consider the importance of the structure to be designed. The specifics of how to obtain such a table is beyond the scope of this book. However, with this prior knowledge, it is possible to conclude that the earthquake considered during an aseismic design is bounded. The upper bound is the design earthquake level, e.g., 0.4 (g). By using the idea of a bounded forcing function, it can be stated that one of the major characteristics of aseismic design, especially for an earthquake protective system, is that under bounded input excitation, the output bound for the response of the system is very difficult to determine.

In the case of earthquake response of an SDOF system,

$$m\ddot{x} + c\dot{x} + kx = f = -m\ddot{x}_g \tag{2.301}$$

Note that x, $\dot{x}$, and $\ddot{x}$ are relative displacement, velocity, and acceleration with units of (m), (m/s), and (m/s^2), respectively, and $\ddot{x}_g$ is the ground acceleration with units of meters per square second, as mentioned previously.

Mathematically speaking, the amplitude, or the bound of the forcing function f, is taken as $|m\ddot{x}_g|$ (N or kN) in Equation 2.301, where f is the earthquake loading. The bound of the response x, or the amplitude of x_0, cannot be specified if f is random. Under any given earthquake record, it is not difficult to solve Equation 2.301 for the peak response. However, for any given earthquake record, the forcing function is no longer random. Therefore, even if all the available records are used as input, Equation 2.301 will not give the peak response because the "next" earthquake ground motion is unknown.

However, for a practical design of the structural capacity of earthquake resistance, engineers need to have deterministic numbers, not only in terms of the input level, but more importantly, in terms of the response amplitude. To meet this requirement, a methodology for using the design response spectrum was introduced by using existing records of various earthquakes to establish the upper bound of an SDOF system. Therefore, it can provide the needed force for design, if the structure to be built can be approximated by an SDOF system.

Specifically, the design response spectrum is generated in several steps. First, a proper group of earthquake records is selected. The selection is based on certain criteria, e.g., high peak value of ground acceleration (PGA) and/or velocity (PGV) and/or displacement (PGD). Other criteria include the duration of an earthquake, the distance of signal pickup to earthquake epicenter, specific site conditions, etc. For convenience, let the number of these records be N.

Second, these records are normalized or scaled to a standard level. For example, all of the records will have the same level of PGA. Since each record has different peak amplitude, unscaled records will not have uniform input levels. This is the main reason for record scaling. In earlier days, PGA was often used for the scaling standard. However, many papers reported that PGA might not be a good factor, because peak acceleration might not directly relate to structural damage. In damping design, this statement must be carefully evaluated. In many cases, the use of additional dampers is intended to avoid serious damage to the structure. Note that structural damage may not only occur when large floor-drift happens, but may also occur in several other combinations of different responses. In this sense, using damage to a structure (which may occur during an earthquake with/without supplemental damping) may not be a proper criterion.

Third, the scaled records are used as forcing functions to excite a series of SDOF systems. Each of the systems will have the governing equation described in Equation 2.301 with a fixed damping ratio and the natural frequency or period will be varied. Therefore, with any period T_i, the convolution integral can be used to compute the corresponding responses. In so doing, all the numbers of *peak values* of the responses, say, the displacements denoted by x_{ij}, can be obtained. Here, the first subscript i stands for the period T_i, while the second subscript j means that the response is calculated by using the jth record. By collecting all the data and plotting the results in Cartesian coordinates with the X-axis as the period and the Y-axis as the peak amplitudes, a group of response spectra can be obtained. Figure 2.17 shows such spectra under 11 records, whose amplitudes are all scaled to be 0.4 (g) (Naeim and Kelly 1999, for these 11 records).

Fourth, at each period T_i, the mean value $\bar{x}_i$ and the standard deviation σ_i of these peak responses are computed. Here, subscript i stands for the statistics taken in accordance with the ith period T_i. The sum of $\bar{x}_i + \sigma_i$ is taken as the raw data for the statistical response spectral value x_i (m). That is,

$$x_i = \bar{x}_i + \sigma_i = \frac{1}{N}\sum_{j=1}^{N} x_{ij} + \frac{1}{N}\sum_{j=1}^{N}\left[x_{ij}^2 - \left(\frac{1}{N}\sum_{j=1}^{N} x_{ij}\right)^2\right]^{1/2} \tag{2.302}$$

In Equation 2.302, N stands for the number of records used. In this way, all x_i are positive, which are the functions of T_i, with reasonable resolution of T_i, and the *response spectrum* is obtained. Here, T_i is the ith natural period. Note that for convenience, subscript n is omitted from T_{ni} here and in the following text.

Fifth, since all of the quantities x_i form a nonsmooth curve, this result is not convenient to use. Therefore, further measurements are taken to smooth the curve, which may be the envelope of all x_i or other measures, which are referred to as the *design spectra*.

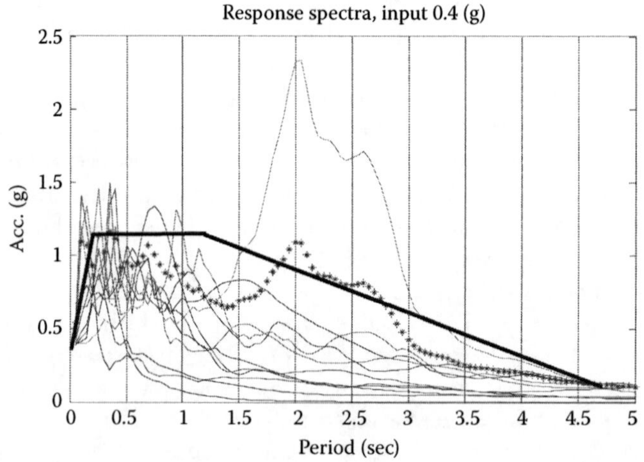

FIGURE 2.17 Earthquake response spectra, $\xi = 5\%$.

Recall the concept of ensemble average introduced in Section 2.3, which is an average of the random process at a specific time. The reason for using the ensemble average is that the random process is not ergodic. Now, it is seen that the earthquake responses are nonergodic so that the ensemble average is used instead of the temporal average. From Equation 2.302, index i implies that the average is taken when there is a fixed period T_i. That is, Equation 2.302 is actually an ensemble average of period, instead of time.

2.4.2 Design Spectra

The dark line in Figure 2.17 is the design displacement response spectrum and is denoted by S_D. The operation is rewritten to determine S_D through a series of SDOF oscillators by statistical means in Equation 2.302 with appropriate earthquake records using the following notation:

$$x_{max_i}\left(\sum \ddot{x}_{g_i}\right) \rightarrow S_D(T_i) \quad (m) \tag{2.303}$$

Here x_{max_i} with the symbol "→" denotes that the spectral value S_D is obtained through the maximum response of x_i. The notation $\ddot{x}_{g_i}$ in the parentheses of x_{max_i} indicates that the maximum response x_{max_i} is obtained through many appropriate earthquake records, $\ddot{x}_{g_i}$; T_i denotes that the value x_i is taken from an SDOF system with this natural period, where

$$T_i = \frac{2\pi}{\omega_{ni}} \tag{2.304}$$

where subscript i stands for the ith individual period.

In the notation for Equation 2.303, it is recognized that the displacement x_{max_i} is obtained with a certain uniformly scaled method, e.g., by using the same amplitude of PGA. Since each earthquake record is scaled with units of (g), this operation is denoted as x_{max_i}/g.

The notation defined in Equation 2.303 is significant. It solves a mathematically unsolvable problem. That is, mathematically, it is impossible to determine the peak value of the response under a given level of earthquake excitations. In the following discussion, the spectral values of the design spectrum are often used, e.g., S_D, to replace the statistical maxima x_{max}.

When the damping ratio is small, which is true for most structures without added dampers, the damping force is small when compared to the initial and restoring forces. Therefore, in Equation 2.301, the damping force can be ignored and reduced to

$$m\ddot{x} + kx = -m\ddot{x}_g \tag{2.305}$$

Note that x and $\ddot{x}$ in Equation 2.305 are relative displacement and acceleration, respectively. Meanwhile, the absolute acceleration is the sum of the relative and ground accelerations, that is,

$$\ddot{x}_A = \ddot{x} + \ddot{x}_g \tag{2.306}$$

Therefore, Equation 2.305 can be rewritten as

$$m\ddot{x}_A = -kx \tag{2.307}$$

or

$$\ddot{x}_A = -\frac{k}{m}x = -\omega_n^2 x \tag{2.308}$$

Using Equation 2.308, the following relationship between the spectrum of absolute acceleration, denoted by S_A, and the spectrum of the relative displacement, S_D, can be obtained, that is,

$$S_A = \frac{\omega_n^2 S_D}{g} \tag{2.309}$$

or

$$S_A = \frac{4\pi^2}{T^2} \cdot \frac{S_D}{g} \tag{2.310}$$

Note that S_A is dimensionless. This is seen from Equation 2.310, where the input records are scaled with their amplitude and with units in (g). From Equation 2.310, if the spectrum of relative displacement is already calculated, the spectrum of absolute acceleration, S_A, can be generated. In the procedure to obtain the acceleration spectrum, it is assumed that the damping force is zero, but it actually is not. A spectrum generated in this manner is called the *pseudo spectrum.* As a comparison, if the peak accelerations are computed directly from Equation 2.301, the *real response spectrum* can be obtained.

The variation of period in computing the response mentioned above can be seen as choosing different SDOF oscillators. Each has a different natural frequency or period. Figure 2.18b shows a series of pendulums used to represent this series of SDOF structures with these natural periods. The upper bound of the responses of this SDOF oscillator subjected to various earthquake ground motions with the uniform amplitude, say, 0.4 (g), is modified according to certain criteria and plotted in Figure 2.18a, shown as three thick lines. The upper thick line represents the response spectrum with a 5% damping ratio for each oscillator. The middle one denotes a damping ratio of 7%, while the lower one denotes a damping ratio of 15%.

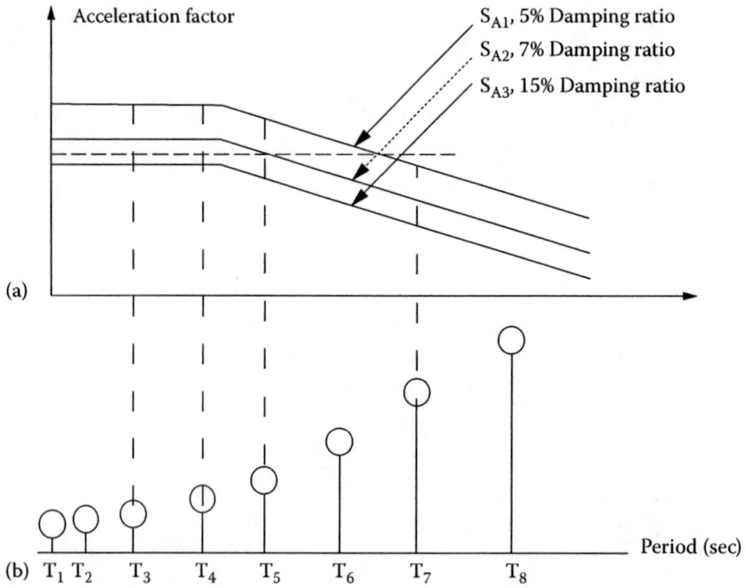

FIGURE 2.18 (a, b) Response spectrum.

From Figure 2.17, it is seen that there are two methods to lower the acceleration. The first is to shift the period from shorter values to longer ones. From Equation 2.308, it is seen that to make the period longer, the static stiffness should be decreased. However, the resulting acceleration is thus smaller, which means that the dynamic stiffness of the acceleration becomes larger instead of smaller.

The second method is to increase the damping. From Figure 2.18, the three curves conceptually with damping ratios at 5%, 7%, and 15%, respectively, indicate that at any period, larger damping will provide a lower acceleration level.

In practical applications, the design response spectrum is used to determine, design accelerations in a special way. For example, in NEHRP 2003 and/or 2009 (BSSC 2003/2009), several design spectral response acceleration parameters were specified as follows:

First, the maximum considered earthquake spectral response acceleration for short periods, S_{MS}, and that at 1 (s), S_{M1}, are defined in Equations 2.311 and 2.312, respectively:

$$S_{MS} = F_a S_S \tag{2.311}$$

and

$$S_{M1} = F_v S_1 \tag{2.312}$$

where S_S and S_1 are the maximum considered earthquakes, 5% damped, spectral response accelerations at a short period and at a period of 1 (s), respectively (i.e., see BSSC/NEHRP 2009, also see Figure 2.19). Here, F_a and F_v are site coefficients, which are defined in Tables 2.2A and B, respectively.

Note that the values of S_S and S_1 may not be listed in Tables 2.2A and B. Straight-line interpolations are used for intermediate values. Also note that the symbol * denotes that site-specific geotechnical investigations and dynamic site response analysis should be performed.

Next, the design spectral response acceleration parameters at short period S_{DS} and at 1 (s) period S_{D1} (also refer to Figure 2.18) can be determined by

$$S_{DS} = \frac{2}{3} S_{MS} \tag{2.313a}$$

$$S_{D1} = \frac{2}{3} S_{M1} \tag{2.313b}$$

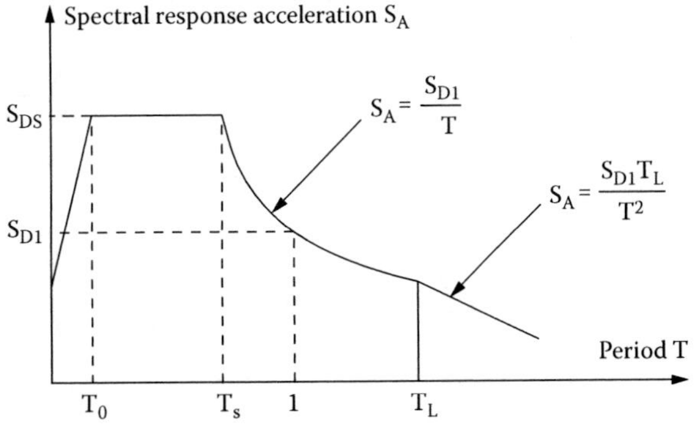

FIGURE 2.19 Design response spectrum.

TABLE 2.2A
Values of F_a as a Function of Site Class and Mapped Short-Period Maximum Considered Earthquake Spectral Acceleration

Site Class	$S_S \leq 0.25$	$S_S = 0.50$	$S_S = 0.75$	$S_S = 1.00$	$S_S \geq 1.25$
A	0.8	0.8	0.8	0.8	0.8
B	1.0	1.0	1.0	1.0	1.0
C	1.2	1.2	1.1	1.0	1.0
D	1.6	1.4	1.2	1.1	1.0
E	2.5	1.7	1.2	0.9	0.9
F	*	*	*	*	*

Source: NEHRP 2009.
Note: Use straight-line interpolation for intermediate values of S_S.
*See Section 11.4.7 of NEHRP 2009

When a design response spectrum is required by a design code (e.g., NEHRP 2009) and the site-specific procedures are not used, the design response spectrum curve shall be developed as indicated in Figure 2.19 and as follows:

1. For periods less than or equal to T_0, the design spectral response acceleration, S_A, shall be taken as given by

$$S_A = 0.6\frac{S_{DS}}{T_0}T + 0.4S_{DS} \tag{2.314}$$

Here, T_0 is the period at which the increasing acceleration and constant acceleration regions of the response spectrum intersect.
2. For periods greater than or equal to T_0 and less than T_S, the design spectral response acceleration, S_A, shall be taken as equal to S_{DS}. Here, T_S is the period at which the constant acceleration and constant velocity regions of the response spectrum intersect.
3. For periods greater than T_S, but less than T_L, the design spectral response acceleration, S_A, shall be taken as given by

TABLE 2.2B
Values of F_v as a Function of Site Class and Mapped 1 Sec Maximum Considered Earthquake Spectral Acceleration

Site Class	$S_1 \leq 0.1$	$S_1 = 0.2$	$S_1 = 0.3$	$S_1 = 0.4$	$S_1 \geq 0.5$
A	0.8	0.8	0.8	0.8	0.8
B	1.0	1.0	1.0	1.0	1.0
C	1.7	1.6	1.5	1.4	1.3
D	2.4	2.0	1.8	1.6	1.5
E	3.5	3.2	2.8	2.4	2.4
F	*	*	*	*	*

Source: NEHRP 2009.
Note: Use straight-line interpolation for intermediate values of S_1.
*See Section 11.4.7 of NEHRP 2009

$$S_A = \frac{S_{D1}}{T} \tag{2.315}$$

and T_L is the period to denote the log-period transition (see Figure 2.19), which is determined through a regional approach (NEHRP 2009).

4. For periods greater than T_L, the design spectral response acceleration, S_A, shall be taken as given by

$$S_A = \frac{S_{D1} T_L}{T^2} \tag{2.316}$$

Here

$$T_0 = 0.2 \frac{S_{D1}}{S_{DS}} \quad (s) \tag{2.317}$$

and

$$T_S = \frac{S_{D1}}{S_{DS}} \quad (s) \tag{2.318}$$

In the above, T, T_0, T_S, and T_L are all expressed in units of (s). Equations 2.315 and 2.316 as well as Figure 2.19 provide the value of S_A, which can be used to replace S_{DS} and S_{D1} (NEHRP 2009) whenever the detailed response acceleration vs. structural period is necessary for more accurate designs.

Note that the design response spectrum is based on the earthquake responses of a series of linear SDOF systems. Therefore, it can be directly used for the damper design of linear SDOF systems. In addition, many codes use several SDOF systems to represent a multi-degree-of-freedom (MDOF) system. In this circumstance, understanding the basic requirement of SDOF systems becomes the foundation of damper design. Problems and design errors of using linear or linearized SDOF systems to approximate general MDOF systems will be discussed in the next several chapters.

In the National Earthquake Hazards Reduction Program (NEHRP) recommended provisions for seismic regulations for new buildings and other structures (BSSC 2000, 2003, and/or 2009 editions), the SDOF approach is classified into the category of an equivalent lateral force analysis procedure, although the multiple-story-single-period approach is also in the same category. It states that a structure to be designed with the equivalent lateral force analysis procedure should be subjected to the following limitations:

1. In the direction of interest, the damping system has at least two damping devices in each story, configured to resist torsion.
2. The total effective damping of the fundamental mode, ξ, of the structure in the direction of interest is not greater than 35% of the critical damping.
3. The seismic-force-resisting system does not have a vertical irregularity of Type 1a, 1b, 2, or 3 (see Table 2.3A) or a plan irregularity of Type 1a or 1b (see Table 2.3B).
4. Floor diaphragms are rigid.
5. The height of the structure above the base does not exceed 30 m.
6. Peak dynamic response of the structure and elements of the damping system are confirmed by nonlinear time history analysis, when it is required that the structure be located at a site with S_1 greater than 0.60.

TABLE 2.3A
Vertical Structural Irregularities

Irregularity Type and Description	Reference Section[a]	Seismic[b] Design Category Application
1a Stiffness Irregularity—soft story A soft story is one in which the lateral stiffness is less than 70% of that in the story above or less than 80% of the average stiffness of the three stories above	4.4.1	D, E, and F
1b Stiffness Irregularity—extreme soft story An extreme soft story is one in which the lateral stiffness is less than 60% of that in the story above or less than 70% of the average stiffness of the three stories above	4.3.1.5.1 4.4.1	E and F D
2 Weight (Mass) Irregularity Mass irregularity shall be considered to exist where the effective mass of any story is more than 150% of the effective mass of an adjacent story. A roof that is lighter than the floor below need not be considered	4.4.1	D, E, and F
3 Vertical Geometric Irregularity Vertical geometric irregularity shall be considered to exist where the horizontal dimension of the later-force-resisting system in any story is more than 130% of that in an adjacent story	4.4.1	D, E, and F
4 In-Plan Discontinuity in Vertical Lateral-force-resisting Elements An in-plan offset of the lateral-force-resisting elements greater than the length of those elements or a reduction in stiffness of the resisting elements in the story below	4.6.1.7 4.6.3.2	B, C, D, E, and F D, E, and F
5 Discontinuity in Capacity—weak story A weak story is one in which the story lateral strength is less than 80% of that in the story above. The story strength is the total strength of all seismic-resisting elements shearing the story shear for the direction under consideration	4.3.1.5.1 4.6.1.6	E and F B, C, and D

Source: NEHRP 2003/2009.

[a] The reference sections are referred to as in NEHRP 2003/2009.

[b] See Table 2.4A, the site class definitions.

Note that in Table 2.4A, v_s stands for the average shear wave velocity in the top 30 (m), s_u stands for the average undrained shear strength in the top 30 (m), N is the average field standard penetration test for the top 30 (m), PI is the plastic index (ASTM D4318-93), w is the percentage moisture content, F_a is the acceleration-based site coefficient at 0.3 (s), F_v is the velocity-based site coefficient at 1.0 (s), and H is the thickness of the soil.

The above is the qualitative description of how to select the design procedures. In a later section on damper design for multiple-story-single-period systems, certain quantitative criteria for the selection of design procedures will be provided.

In addition, Table 2.3A and B only lists the irregularities associated with the weight (mass) and stiffness, which are sufficient for structural aseismic design, when the damping is comparatively very small. Therefore, they can be referred to as mass-stiffness irregularities. However, when a large amount of damping is added to the structure, the damping irregularity should also be considered. A detailed discussion is provided in Chapters 5 and 6.

NEHRP 2003 and 2009 have virtually identical tables as NEHRP 2000.

TABLE 2.3B
Plan Structural Irregularities

Irregularity Type and Description	Reference Section[a]	Seismic[b] Design Category Application
1a **Torsional Irregularity—to be considered when diaphragms are not flexible**	4.4.1	D, E, and F
	4.6.3.2	D, E, and F
Torsional irregularity shall be considered to exist when the maximum story drift, computed including accidental torsion, at one end of the structure transverse to an axis is more than 1.2 times the average of the story drifts at two ends of the structure	5.2.4.3 and 5.2.6.1	C, D, E, and F
1b **Extreme Tortional Irregularity—to be considered when diaphragms are not flexible**	4.3.1.5.1	E and F
	4.4.1	D
Extreme torsional irregularity shall be considered to exist when the maximum story drift, computed including accidental torsion, at one end of the structure transverse to an axis is more than 1.4 times the average of the story drifts at two ends of the structure	4.6.3.2	D
	5.2.4.3 and 5.2.6.1	C, D, E, and F
2 **Reentrant Corners** Plan configuration of a structure and its lateral-force-resisting system contain reentrant corners where both projections of the structure beyond a reentrant corner are greater than 15% of the plan dimension of the structure in the given direction	4.6.3.2	D, E, and F
3 **Diaphragm Discontinuity** Diaphragms with abrupt discontinuities or variations in stiffness including those having cutout or open areas greater than 50% of the gross enclosed diaphragm area or changes in effective diaphragm stiffness of more than 50% from one story to the next	4.6.3.2	D, E, and F
4 **Out-of-Plan Offsets**	4.6.1.7	B, C, D, E, and F
Discontinuities in a lateral force resistance path such as out-of-plan offsets of the vertical elements	4.6.3.2	D, E, and F
5 **Nonparallel Systems** The vertical lateral-force-resisting element is not parallel to or symmetric about the major orthogonal axes of the lateral-force-resisting system	4.4.2.2	C, D, E, and F

Source: NEHRP 2003/2009.
[a] The reference sections are referred to as in NEHRP 2003/2009.
[b] See Table 2.4A, the site class definition.

2.4.3 Control Factor for Damper Design: The Base Shear

For an SDOF system, the basic parameter to be controlled is the seismic base shear V, which should, however, not be less than the minimum base shear V_{min}, where V_{min} is determined as the greater of the following values:

$$V_{min} = \frac{V}{B} \quad (kN) \tag{2.319}$$

TABLE 2.4A
Site Class Definitions

Site	Descriptions
A	Hard rock with measured shear wave velocity vs > 1,500 (m/s)
B	Rock with 760 (m/s) < $v_s \le$ 1,500 (m/s)
C	Very dense soil and soft rock with 360 (m/s) < $v_s \le$ 760 (m/s) or with either N > 50 or s_u > 100 (kPa)
D	Stiff soil with 180 (m/s) < $v_s \le$ 360 (m/s) or with either 15 ≤ N ≤ 50 or 50 (kPa) ≤ $s_u \le$ 100 (kPa)
E	A soil profile with v_s < 180 (m/s) or with either N < 15, $s_u \le$ 50 (kPa) or any profile with more than 3 (m) of soft clay defined as soil with PI > 20, w ≥ 40%, and $s_u \le$ 25 (kPa)
F	Soil requiring site-specific evaluations: 1. Soil vulnerable to potential failure or collapse under seismic loading, such as liquefiable soil, quick and highly sensitive clays, and collapsible, weakly cemented soils. Exception: For structures having fundamental periods of vibration equal to or less than 0.5 (s), site-specific evaluations are not required to determine spectral accelerations for liquefiable soils. Rather, the site class may be determined in accordance with Section 3.5.2[a] and the corresponding values of F_a and F_v determined from Tables 3.3-1[a] and 3.3-2[a] 2. Peats and/or highly organic clays (H > 3 (m) of peat and/or highly organic clay with H = thickness of soil) 3. Very high plasticity clays (H > 8 (m) with PI > 75) 4. Very thick soft/medium stiff clays (H > 36 (m) with s_u < 50 (kPa))

Source: NEHRP 2003/2009.
[a] Referred to NEHRP 2003/2009.

$$V_{min} = 0.75V \quad (kN) \tag{2.320}$$

Here, B is the *numerical damping coefficient*, which is discussed in detail in Chapter 7. The base shear V, with units of (kN), in a given direction of the SDOF system is determined as follows:

$$V = C_s w \quad (kN) \tag{2.321}$$

Here, C_s is called the *elastic seismic coefficient*, which is also called the *seismic response coefficient*. This is discussed in detail in a later paragraph. Furthermore, w is the total vertical load, that is,

$$w = DL + \kappa LL \quad (kN) \tag{2.322}$$

Here LL is the live load; κ is a proportional coefficient of LL; and the total dead load DL is

$$DL = mg \quad (kN) \tag{2.323}$$

where m is the total mass of the structure with units in (t), and applicable portions of other loads, κLL, are listed below (BSSC/NEHRP 2003):

i. In areas used for storage, a minimum of 25% of the floor live load shall be applicable. Floor live load in public garages and open parking structures is not applicable.
ii. Where an allowance for partition load is included in the floor load design, the actual partition weight or a minimum weight of 0.5 (kPa) of floor area, whichever is greater, shall be applicable.

iii. Total operating weight of permanent equipment.
iv. In areas where the design flat roof snow load does not exceed 1.4 (kPa), the effective snow load is permitted to be taken as zero. In areas where the design snow load is greater than 1.4 (kPa) and where sitting and load duration conditions warrant and when approved by the authority having jurisdiction, the effective snow load is permitted to be reduced to not less than 20% of the design snow load.

Note that an MDOF system with order n can often be decomposed into n-SDOF systems. Therefore, the modal response for each mode may be computed by using the SDOF approach. In this way, Equation 2.322 can be rewritten as follows:

$$w_j = DL_j + \kappa LL_j \quad (kN) \tag{2.324}$$

where w_j is the load of the jth floor, with units of (kN), which is the combination of the total dead load DL_j for the jth floor,

$$DL_j = m_j g \quad (kN) \tag{2.325}$$

and applicable portions of other loads, κLL_j (see above description of the live load LL). g is the gravity (g = 9.8 (m/s^2)) and m_j is the jth lumped mass of the structure. In many cases, the modal mass, m_j, can be calculated by the following relationship:

$$m_j = w_j/g \quad (t) \tag{2.326}$$

In Equation 2.321, the units of the base shear V and the weight w are identical, e.g., (kN). In this case, the term C_s is dimensionless. In the literature, the base shear can also be represented as

$$V = S_A mg \quad (kN) \tag{2.327}$$

Note that again from Equation 2.326 in Equation 2.327, the spectral S_A is dimensionless, that is,

$$S_A = C_s \tag{2.328}$$

In the following, both S_A and C_s will be used. For the sake of simplicity, both will be referred to as the spectral accelerations.

Note that in Equation 2.322, C_s, the seismic response coefficient, shall be determined by

$$C_s = \frac{S_{DS}}{R} I \tag{2.329}$$

where S_{DS} is as defined previously, R is the response modification factor ranging from 1.25 to 8 (BSSC/NEHRP 2009), and I is the occupancy importance factor ranging from 1.0 to 1.5 (BSSC/NEHRP 2009).

The values of C_s need not exceed the following:

$$C_s = \frac{S_{DI}}{TR} I \tag{2.330}$$

where S_{DI} and T have been defined previously.

The values of C_s should also not be less than

$$C_s = 0.044 I S_{DS} \tag{2.331}$$

In the USA, for structures in seismic design categories E and F, the value of C_s should not be less than

$$C_s = 0.5 \frac{S_1}{R} I \tag{2.332}$$

where S_1 is as defined previously. For a regular structure with five stories or less in height and having a period T being 0.5 (s) or less, the value of C_s shall be permitted to use values of 1.5 and 0.6, respectively, for the mapped maximum considered earthquake spectral response accelerations, S_S and S_1.

Using Equations 2.329 through 2.332, one of the key issues is to determine the period. In many cases, the period of an SDOF system is less than T_S and greater than T_0, that is, the design response spectral acceleration S_{DS} is constant, as is the base shear V. In this case, the design criterion recognizes that by adding damping to a structure, the base shear V can be reduced. With the base shear as the only design parameter, it is seen that as more damping is added, the response of the base shear becomes smaller.

In the *Guide Specifications for Seismic Isolation Design* (AASHTO Interim 2000), the elastic seismic coefficient, C_s, has a simplified formula. This is because for base isolations, the "effective" period, T_{eff}, of the system is always longer than T_S. To make the AASHTO formula consistent within this context, it is written as

$$C_s = \frac{0.4 A S_i}{B T_{eff}} \tag{2.333}$$

However, if the damping effect cannot be ignored, then using the above approach may yield significant errors. Therefore, if the damping is large, Equation 2.333 should be modified by multiplying by a modification factor, $\sqrt{1+4\xi^2}$ as

$$C_s = \sqrt{1+4\xi^2}\, \frac{0.4 A S_i}{B T_{eff}} \tag{2.334}$$

This additional factor will be notably greater than unity when the damping ratio is large.

In Equation 2.333, A is the acceleration coefficient, which can be found in building codes such as NEHRP 2009 (BSSC 2009), where

$$A = \frac{S_{DS}}{0.4} \tag{2.335}$$

For example, if the amplitude of the ground excitation S_{DS} at short period is 0.4, A = 1, while if the amplitude of the ground excitation is 0.8, then A = 2.

Also in Equation 2.333, S_i is the site coefficient. For example, S_i specified by AASHTO is listed in Table 2.4B.

TABLE 2.4B
Site Coefficient

	Soil Profile Type			
	I	II	III	IV
S_i	1.0	1.5	2.0	2.7

Source: AASHTO 2000.

2.4.3.1 Spectral Accelerations and Displacement

In the above discussion, it is known that in earthquake engineering, the input force is random rather than sinusoidal. If only the amplitude, or the peak level of the input, is known then it is mathematically impossible to pinpoint the amplitude of the output. Therefore, to overcome this difficulty, the method involving design response spectrum is used. By means of statistical averaging, a sufficient number of random inputs are used, mostly real earthquake records, to find the peak value of a system under seismic excitation. In so doing, if the natural period and the damping of a structure are known, then the amplitude of the absolute acceleration can be written as follows:

$$S_A = C_s(T,\xi) \tag{2.336}$$

and the relative displacement can be determined from

$$S_D = C_s(T,\xi)\, g\frac{T^2}{4\pi^2} \quad (m) \tag{2.337}$$

Comparing Equations 2.336 and 2.337, it is seen that the relationship between the acceleration and the displacement is based on the zero-damping assumption described in Equations 2.309 and 2.310.

In the following chapters, Equation 2.336 is improved and the necessity of using the factor $\sqrt{1+4\xi^2}$ is shown.

Peak dynamic response of the structure and elements of the damping system are confirmed by nonlinear time history analysis, when it is required that the structure be located at a site with S_1 greater than 0.60.

2.5 SUMMARY

In this chapter, the focus was on forced vibration of SDOF systems under arbitrary loading, which can be classified as periodic, transient, and random excitations. In addition, earthquake ground motion is discussed as a special case of random dynamic loading. Understanding the concepts of dynamic responses of structures under random excitations is the foundation in considering the role of damping in vibration control. In the next two chapters, these concepts will be expanded to MDOF systems.

REFERENCES

AASHTO. 2000. *Guide Specifications for Seismic Isolation Design*. Washington, DC: American Association of State Highway Officials.

ASTM. 1994. Test method for liquid limit, plastic limit, and plasticity index of soils. ASTM Standard No. D 4318-93, *Annual Book of ASTM Standards*. Philadelphia, PA: ASTM.

Bendat, J.S. and Piersol, A.G. 1971. *Random Data: Analysis and Measurement Procedures*. NY: John Wiley and Sons.

Building Seismic Safety Council (BSSC). 2000. *NEHRP Recommended Provisions for Seismic Regulation for New Buildings and Other Structures, 2000 Edition*, Report Nos. FEMA 368, Federal Emergency Management Agency, Washington, DC.

———. 2003. *NEHRP Recommended Provisions for Seismic Regulation for New Buildings and Other Structures, 2003 Edition*, Report Nos. FEMA 450, Federal Emergency Management Agency, Washington, DC.

———. 2009. *NEHRP Recommended Provisions for Seismic Regulation for New Buildings and Other Structures, 2009 Edition*, Report Nos. FEMA P-750, Federal Emergency Management Agency, Washington, DC.

Chopra, A.K. 2006. *Dynamics of Structures: Theory and Applications to Earthquake Engineering*. 3rd ed. Upper Saddle River, NJ: Prentice Hall.

Naeim, F. and Kelly, J.M. 1999. *Design of Seismic Isolated Structures*. NY: John Wiley and Sons.

Silva, C.W. de (ed.). 2007. *Vibration Monitoring, Testing, and Instrumentation*. NY: Taylor & Francis.

3 Linear Proportionally Damped Multi-Degree-of-Freedom Systems

Practically speaking, most structures should be modeled as multi-degree-of-freedom (MDOF) systems. The basic idea to deal with an MDOF system is to decouple it into several SDOF systems. When the response is linear and the damping is either absent or proportional, which roughly means that the distribution of damping forces coincides with the restoring forces, the decoupling can be conducted in an n-dimensional modal space. Although such a treatment is not always practical, the study of proportionally damped systems can provide insight on how to handle general MDOF systems. In addition, most codes use this approach for aseismic designs. Furthermore, this chapter also offers important background information for the design of damped MDOF systems.

3.1 UNDAMPED MDOF SYSTEMS

To study the decoupling procedure for an MDOF system, an undamped system is considered, where the damping matrix is null. The basic idea in decoupling an MDOF system is to perform a linear transformation on the coefficient matrices to make them diagonal.

3.1.1 Eigen-Parameters of Linear Undamped Systems

3.1.1.1 Governing Equations

Similar to the SDOF systems discussed in Chapters 1 and 2, an MDOF system is also modeled by a set of second-order differential equations known as governing equations. A linear undamped MDOF system can have the following governing equation in matrix form:

$$\begin{bmatrix} m_{11} & m_{12} & \dots & m_{1n} \\ m_{21} & m_{22} & \dots & m_{2n} \\ & & \dots & \\ m_{n1} & m_{n2} & \dots & m_{nn} \end{bmatrix} \begin{Bmatrix} \ddot{x}_1(t) \\ \ddot{x}_2(t) \\ \dots \\ \ddot{x}_n(t) \end{Bmatrix} + \begin{bmatrix} k_{11} & k_{12} & \dots & k_{1n} \\ k_{21} & k_{22} & \dots & k_{2n} \\ & & \dots & \\ k_{n1} & k_{n2} & \dots & k_{nn} \end{bmatrix} \begin{Bmatrix} x_1(t) \\ x_2(t) \\ \dots \\ x_n(t) \end{Bmatrix} = 0 \tag{3.1}$$

or, simply

$$\mathbf{M}\ddot{\mathbf{x}}(t) + \mathbf{K}\mathbf{x}(t) = \mathbf{0} \tag{3.2}$$

Here **M** is the mass coefficient matrix and it must be symmetric, where

$$\mathbf{M} = \begin{bmatrix} m_{11} & m_{12} & \dots & m_{1n} \\ m_{21} & m_{22} & \dots & m_{2n} \\ & & \dots & \\ m_{n1} & m_{n2} & \dots & m_{nn} \end{bmatrix} \tag{3.3}$$

and m_{ij} is the ij[th] entry in matrix **M**; and n is the order of the MDOF system. In most cases, **M** is taken to be diagonal, instead of as described in Equation 3.3. Additionally, **K** is the stiffness coefficient matrix, with

$$\mathbf{K} = \begin{bmatrix} k_{11} & k_{12} & \dots & k_{1n} \\ k_{21} & k_{22} & \dots & k_{2n} \\ & & \dots & \\ k_{n1} & k_{n2} & \dots & k_{nn} \end{bmatrix} \tag{3.4}$$

where k_{ij} is the ij[th] entry in matrix **K**. The stiffness matrix must also be symmetric, which can be guaranteed by the Maxwell reciprocal theorem.

In Equation 3.2, **x** is the displacement vector

$$\mathbf{x} = \begin{Bmatrix} x_1 \\ x_2 \\ \dots \\ x_n \end{Bmatrix} = [x_1, x_2, \dots, x_n]^T \tag{3.5}$$

Here, x_i denotes the displacement at the ith location; the superscript T denotes the vector or matrix transpose. Note that **x** and x_i are time-dependent variables. However, the symbol (t) can be omitted, as in Equation 3.5, for convenience. In addition, $\ddot{\mathbf{x}}$ is the acceleration vector, which is the second derivative of **x** with respect to time t.

Similar, to SDOF systems, it is assumed that

$$\mathbf{x}(t) = \boldsymbol{u} e^{j\omega t} \tag{3.6}$$

and Equation 3.6 is substituted into Equation 3.2. In so doing, if Equation 3.6 is indeed one of the solutions of Equation 3.2, the natural frequencies of the MDOF systems can be found. In Equation 3.6, $\boldsymbol{u}$ is an $n \times 1$ vector. Thus,

$$-\omega^2 \mathbf{M} \boldsymbol{u} e^{j\omega t} + \mathbf{K} \boldsymbol{u} e^{j\omega t} = 0 \tag{3.7a}$$

or

$$\left(-\omega^2 \mathbf{M} + \mathbf{K}\right) \boldsymbol{u} = 0 \tag{3.7b}$$

since the term $e^{j\omega t}$ cannot be zero to allow both sides of Equation 3.7a to be divided by $e^{j\omega t}$, Equation 3.7b is obtained. Note that in Equation 3.7a, the temporal variable $e^{j\omega t}$ is present, whereas in Equation 3.7b all terms are time invariant. Furthermore, it is realized that vector $\boldsymbol{u}$ denotes the amplitude of spatial displacement. Thus, the operation from Equation 3.7a into Equation 3.7b is often referred to as the separation of the temporal and spatial variables.

The resulting Equation 3.7b is a linear matrix equation. To ensure $\boldsymbol{u}$ is one of the possible nonzero solutions, the determinant of matrix $(-\omega^2\mathbf{M} + \mathbf{K})$ needs to be zero. That is,

$$\det\left(-\omega^2 \mathbf{M} + \mathbf{K}\right) = 0 \tag{3.8}$$

In Equation 3.8, **M** and **K** are known. Equation 3.8 offers a unique solution in general, and there can be n values of ω^2. Thus, for each individual term ω_i^2,

$$\left(-\omega_i^2\mathbf{M}+\mathbf{K}\right)\boldsymbol{u}_i=\mathbf{0} \tag{3.9}$$

where the subscript i stands for the ith pair of vector $\boldsymbol{u}_i$ and scalar ω_i^2.

3.1.1.2 Modal Response of Free Vibrations

The vector $\boldsymbol{u}_i$ can be determined by solving Equation 3.9. On having the terms of ω^2_i and $\boldsymbol{u}_i$, the solution **x** can be determined. As mentioned above, subscript i is used to denote the ith possible solutions, that is, Equation 3.6 can be rewritten as

$$\mathbf{x}_i=\boldsymbol{u}_i e^{j\omega_i t} \tag{3.10}$$

From Equation 3.10, it is immediately seen that if Equation 3.10 is one of the possible solutions, then its complex conjugate

$$\mathbf{x}_i^*=\boldsymbol{u}_i e^{-j\omega_i t} \tag{3.11}$$

must also be a possible solution. Thus, associated with ω_i^2 and $\boldsymbol{u}_i$, Equations 3.10 and 3.11 can be combined as

$$\mathbf{x}_i=a_i\boldsymbol{u}_i e^{j\omega_i t}+b_i\boldsymbol{u}_i e^{-j\omega_i t} \tag{3.12}$$

Here a_i and b_i are complex-valued scalars. Suppose a_i is written as

$$a_i=\alpha_i+j\beta_i \tag{3.13}$$

where α_i and β_i are real-valued scalars, then it can be seen that

$$b_i=a_i^*=\alpha_i-j\beta_i \tag{3.14}$$

Now, using Euler's formula,

$$\mathbf{x}_i=\left[(a_i+b_i)\cos(\omega_i t)+j(a_i-b_i)\sin(\omega_i t)\right]\boldsymbol{u}_i=A_i\sin(\omega_i t+\varphi_i)\boldsymbol{u}_i \tag{3.15}$$

where A_i is the scalar amplitude

$$A_i=2\sqrt{a_i b_i}=2\sqrt{\alpha_i^2+\beta_i^2} \tag{3.16}$$

and φ_i is the phase angle

$$\varphi_i=\tan^{-1}\frac{-j(a_i+b_i)}{a_i-b_i}+h_{\varphi_i}\pi=\tan^{-1}\left(-\frac{\alpha_i}{\beta_i}\right)+h_{\varphi_i}\pi \tag{3.17}$$

where h_{φ_i} is a Heaviside function.

From Equation 3.15, it is seen that the ith vibration response is a combination of two portions. The first part is $A_i\sin(\omega_i t+\varphi_i)$, which is a temporal function and looks like an SDOF response. In Chapter 1, due to space constraints, when introducing free vibration of the SDOF system, the undamped response was not discussed. Here, this topic can be seen as a complementary item for the SDOF vibration. In a later section, it is shown that an MDOF system can indeed be represented by SDOF subsystems and $A_i\sin(\omega_i t+\varphi_i)$ defines the vibration status of the subsystem.

Each of the subsystems expressed by $A_i\sin(\omega_i t+\varphi_i)$ contains the amplitude A_i, frequency ω_i, and phase angle φ_i. Among these three parameters, ω_i is very important. The term ω_i indicates that this subsystem will vibrate at this specific frequency, which is only determined by the mass **M** and stiffness **K**, regardless of other factors such as external driving frequencies. Based on the concepts introduced in Chapter 1, such a frequency is nothing but the *angular natural frequency* or simply the *natural frequency* of this specific subsystem. Thus, ω_i is relabeled as ω_{ni} in the following and this subsystem is referred as a *vibration mode.*

The second part of Equation 3.15 is $\boldsymbol{u}_i$, which provides information on how the vibration is distributed spatially. For example, suppose there is a 2-DOF system such that

$$\mathbf{x}_i(t) = \begin{Bmatrix} x_{1i}(t) \\ x_{2i}(t) \end{Bmatrix} \quad i = 1,2,3,\ldots \tag{3.18}$$

and

$$\boldsymbol{u}_i = \begin{Bmatrix} u_{1i} \\ u_{2i} \end{Bmatrix} \quad i = 1,2,3,\ldots \tag{3.19}$$

Here, x_{1i} and x_{2i} are the displacement time history of the first and the second mass, while u_{1i} and u_{2i} are the first and the second amplitude of the displacement vector $\boldsymbol{u}_i$.

Using Equations 3.18 and 3.19, the vibration denoted by the temporal variable $A_i\sin(\omega_{ni}t + \varphi_i)$ is distributed to the first and second mass as

$$x_{1i}(t) = A_i \sin(\omega_{ni}t + \varphi_i)u_{1i} \quad i = 1,2,3,\ldots \tag{3.20}$$

and

$$x_{2i}(t) = A_i \sin(\omega_{ni}t + \varphi_i)u_{2i} \quad i = 1,2,3,\ldots \tag{3.21}$$

Since the response $\mathbf{x}_i(t)$ is only associated with ω_{ni}^2 and $\boldsymbol{u}_i$, it can be referred to as the vibration of the ith *modal response*. Here, ω_{ni}^2 and $\boldsymbol{u}_i$ are called the ith *modal parameters* or specifically, the ith *natural frequency* and *mode shape*, respectively. In addition, the amplitude A_i is called the ith *modal participating factor.*

As a brief summary, the pair $<\omega_{ni}, \boldsymbol{u}_i>$ denote the ith *vibration mode*. Here, the way to obtain information about the ith mode is through the semidefinite method by first assuming Equation 3.6 to be one of the solutions of Equation 3.2 and by further separating the spatial and temporal variables described in Equations 3.7.

In Equations 3.20 and 3.21, the amplitude A_i and the phase angle φ_i should be determined by initial conditions. This issue is addressed in Section 1.1.2 on the topic of damped free-decay vibrations.

Note that since Equations 3.1 and 3.2 are linear, the total response can be obtained by summing over all the modal responses as

$$\mathbf{x}(t) = \sum_{i=1}^{n} \mathbf{x}_i(t) = \sum_{i=1}^{n} A_i(\omega_{ni}t + \varphi_i)\boldsymbol{u}_i \tag{3.22}$$

which is referred to as *modal superposition.*

3.1.1.3 General Eigen-Parameters

In the above discussion, it is seen that the modal response to calculate the total response is needed. To determine the modal response, the natural frequency and the mode shape must be known. In addition to the semidefinite method described in Equation 3.6, a more systematic method to obtain these two modal parameters through eigen-decompositions is introduced.

In Equation 3.2, **M** and **K** are full rank matrices. Therefore, $\mathbf{M}^{-1}$, the inverse of **M**, exists. In this way, premultiplying $\mathbf{M}^{-1}$ on both sides of Equation 3.9 and rearranging these terms, results in

$$\omega_{ni}^2 \boldsymbol{u}_i = \left(\mathbf{M}^{-1}\mathbf{K}\right)\boldsymbol{u}_i \tag{3.23}$$

On the left side of Equation 3.23, a scalar ω_{ni}^2 times a vector $\boldsymbol{u}_i$ is present. On the right side of Equation 3.23, an n × n matrix $\mathbf{M}^{-1}\mathbf{K}$ times the same vector $\boldsymbol{u}_i$ exists. The resulting products must be equal, so this amounts to the so-called *eigen-problem*. That is, given an n × n square matrix, n pair of scalars and vectors can be found that satisfy Equation 3.10. The corresponding scalar is called the *eigenvalue*, while the vector is called the *eigenvector*. Writing these eigen-pairs for matrix $\mathbf{M}^{-1}\mathbf{K}$ in Equation 3.24, results in

$$\left[\omega_{n1}^2 \boldsymbol{u}_1, \omega_{n2}^2 \boldsymbol{u}_2, \ldots, \omega_{nn}^2 \boldsymbol{u}_n\right] = \left[\left(\mathbf{M}^{-1}\mathbf{K}\right)\boldsymbol{u}_1, \left(\mathbf{M}^{-1}\mathbf{K}\right)\boldsymbol{u}_2, \ldots, \left(\mathbf{M}^{-1}\mathbf{K}\right)\boldsymbol{u}_n\right] \tag{3.24}$$

which is rewritten as

$$\mathbf{U}\boldsymbol{\Omega}^2 = \left(\mathbf{M}^{-1}\mathbf{K}\right)\mathbf{U} \tag{3.25}$$

Equation 3.25 is the eigen-problem in matrix form, where

$$\boldsymbol{\Omega}^2 = \begin{bmatrix} \omega_{n1}^2 & 0 & \ldots & 0 \\ 0 & \omega_{n2}^2 & \ldots & 0 \\ & & \ldots & \\ 0 & 0 & \ldots & \omega_{nn}^2 \end{bmatrix} = \mathrm{diag}\left(\left[\omega_{n1}^2, \omega_{n2}^2, \ldots, \omega_{nn}^2\right]\right) \tag{3.26}$$

is the *eigenvalue matrix* of $\mathbf{M}^{-1}\mathbf{K}$. The square root of the ith entry of the eigenvalue matrix, ω_{ni}, is the ith natural frequency of this system.

Furthermore,

$$\mathbf{U} = \left[\boldsymbol{u}_1, \boldsymbol{u}_2, \ldots, \boldsymbol{u}_n\right] \tag{3.27}$$

is the *eigenvector matrix* of $\mathbf{M}^{-1}\mathbf{K}$.

Note that matrix $\mathbf{M}^{-1}\mathbf{K}$ is asymmetric in general, unless **M** = **I**. An asymmetric matrix can have complex-valued eigenvalues and eigenvectors. For comparison, a symmetric matrix will always have real-valued eigen-decompositions. However, the eigenvalues and eigenvectors of the specific matrix $\mathbf{M}^{-1}\mathbf{K}$ must be real valued.

It can be proven that all the column vectors, namely, the eigenvectors of **U**, can be linearly independent, so that **U** is a full rank matrix. In the next subsection, the concept of linear independence is discussed in detail. Here, this property is only used to obtain the inverse of **U**. Postmultiplying $\mathbf{U}^{-1}$ on both sides of Equation 3.26 results in

$$\mathbf{U}\boldsymbol{\Omega}^2\mathbf{U}^{-1} = \mathbf{M}^{-1}\mathbf{K} \tag{3.28}$$

Equation 3.28 implies that the square matrix $\mathbf{M}^{-1}\mathbf{K}$ is now decomposed into three matrices, namely, the eigenvector matrix **U**, the eigenvalue matrix Ω^2, and the inverse of the eigenvector matrix, $\mathbf{U}^{-1}$. Thus, Equation 3.28 is referred to as the *eigen-decomposition* of $\mathbf{M}^{-1}\mathbf{K}$.

Example 3.1

Suppose a structure has the following mass and stiffness matrices. Find the eigenvalues and eigenvectors of the system.

$$\mathbf{M} = 1000\ \text{diag}([1.1,\ 1.2,\ 1.3,\ 1.4])\quad (\text{t})$$

$$\mathbf{K} = 1000000\begin{bmatrix} 1 & -1 & 0 & 0 \\ -1 & 2 & -1 & 0 \\ 0 & -1 & 2 & -1 \\ 0 & 0 & -1 & 2 \end{bmatrix}\quad (\text{kN/m})$$

The eigenvalue matrix is calculated as
$\mathbf{\Omega}^2 = 1000\ \text{diag}([0.1016,\ 0.7914,\ 1.8446,\ 2.8051])$
The eigenvector matrix is calculated as

$$\mathbf{U} = \begin{bmatrix} 0.6503 & -0.6084 & -0.4734 & -0.3212 \\ 0.5776 & -0.0788 & 0.4872 & 0.6700 \\ 0.4344 & 0.5257 & 0.3694 & -0.5941 \\ 0.2339 & 0.5893 & -0.6342 & 0.0383 \end{bmatrix}$$

The angular natural frequency is the square root of the eigenvalue matrix, that is,
$\mathbf{\Omega} = \text{diag}([10.0821,\ 28.1318,\ 42.9489,\ 52.9635])$ (rad/s)
The period matrix **T** is
$\mathbf{T} = \text{diag}([0.6232,\ 0.2233,\ 0.1463,\ 0.1186])$ (s)

3.1.2 Brief Discussion of Vectors and Matrices

The matrix form of the governing equations was given in the previous subsection, where the eigen-decomposition of matrices to calculate the natural frequencies and mode shapes of the vibration system were discussed. The physical meaning and essence of these mathematical tools are addressed in this subsection. The logic of the discussion dictates that the vector and vector space be considered first, and then the concepts of linear independence and orthonormal vectors and matrices are introduced. Finally, the Rayleigh quotient, which provides the foundation for modal analysis and the physical meaning of the eigen-parameters, is discussed.

3.1.2.1 Vector

Suppose there is a nonzero n × 1 column vector $\boldsymbol{v}$ and

$$\boldsymbol{v} = \begin{Bmatrix} v_1 \\ v_2 \\ \dots \\ v_n \end{Bmatrix} \tag{3.29a}$$

where v_i is the ith element. The transpose of $\boldsymbol{v}$, $\boldsymbol{v}^T$, is another type of vector, an n × 1 row vector. The superscript T denotes the vector or matrix transpose,

$$\boldsymbol{v}^T = [v_1, v_2, \dots, v_n] \tag{3.29b}$$

3.1.2.2 Vector Norm

A scalar α can always be found such that

$$\boldsymbol{u} = \alpha \boldsymbol{v} \tag{3.30}$$

and

$$\boldsymbol{u}^{\mathrm{T}}\boldsymbol{u} = 1 \tag{3.31}$$

In Equation 3.30, the scalar α can be calculated as

$$\alpha = \frac{1}{\sqrt{\boldsymbol{v}^{\mathrm{T}}\boldsymbol{v}}} = \frac{1}{\sqrt{[\mathrm{v}_1, \ldots, \mathrm{v}_n]\begin{Bmatrix} \mathrm{v}_1 \\ \ldots \\ \mathrm{v}_n \end{Bmatrix}}} = \frac{1}{\sqrt{\mathrm{v}_1^2 +, \ldots, +\mathrm{v}_n^2}} \tag{3.32}$$

This can be seen by substituting Equations 3.32 and 3.30 into $\boldsymbol{u}^{\mathrm{T}}\boldsymbol{u}$. That is,

$$\boldsymbol{u}^{\mathrm{T}}\boldsymbol{u} = \alpha^2 \boldsymbol{v}^{\mathrm{T}}\boldsymbol{v} = \frac{\boldsymbol{v}^{\mathrm{T}}\boldsymbol{v}}{\boldsymbol{v}^{\mathrm{T}}\boldsymbol{v}} = 1 \tag{3.33}$$

When case $\boldsymbol{v}$ is used as one of the eigenvectors of an MDOF vibration system, $\alpha\boldsymbol{v}$ will also be an eigenvector associated with the identical eigenvalue. That is, the eigenvector $\boldsymbol{v}$ can be normalized by multiplying an arbitrary scalar α. Or, it can be said that the normalization is arbitrary. However, Equation 3.32 defines a special normalization, such that the vector satisfies Equation 3.31. In the literature, α is called the *Euclidian norm* or *2-norm* of vector $\boldsymbol{v}$, which is a measurement of the "length" of the vector. Thus, α is also referred to as the *Euclidian length* or simply the *length* of vector $\boldsymbol{v}$. Figure 3.1 shows the geometric meaning of such a normalization for a 2-DOF vibration system, in this case the dimension of $\boldsymbol{v}$ is 2, so that

$$\boldsymbol{v} = \begin{Bmatrix} \mathrm{v}_1 \\ \mathrm{v}_2 \end{Bmatrix} \tag{3.34}$$

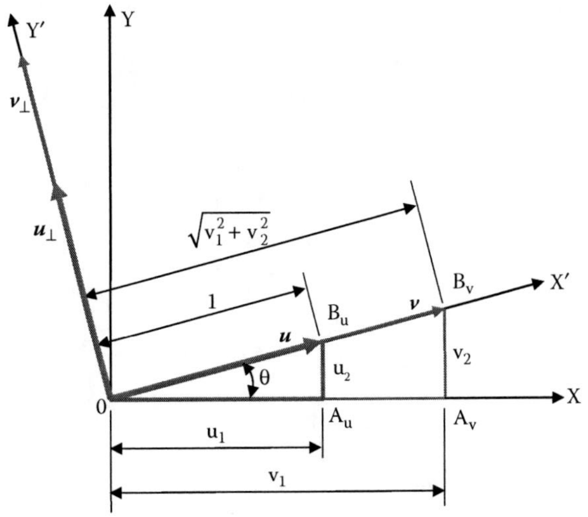

FIGURE 3.1 Vector and vector length.

From Figure 3.1, it is seen that if

$$\boldsymbol{u} = \alpha \boldsymbol{v} = \begin{Bmatrix} u_1 \\ u_2 \end{Bmatrix} \tag{3.35}$$

then

$$u_1 = \frac{v_1}{\sqrt{v_1^2 + v_2^2}} = \cos\theta \tag{3.36a}$$

and

$$u_2 = \frac{v_2}{\sqrt{v_1^2 + v_2^2}} = \sin\theta \tag{3.36b}$$

In Figure 3.1, vector $\boldsymbol{v}$ is represented by 0-B_v with the horizontal component v_1(0-A_v) and vertical component v_2(A_v-B_v). The vector has a magnitude $\sqrt{v_1^2 + v_2^2}$, which can be represented by the length (0-A_v). When the vector is divided by $\sqrt{v_1^2 + v_2^2}$, a new vector $\boldsymbol{u}$ is formed, which has the same direction as $\boldsymbol{v}$, but the length becomes 1. It has a horizontal component u_1 and a vertical component u_2, which can be calculated through Equations 3.36a and b. In addition,

$$\theta = \tan^{-1}\frac{v_1}{v_2} \tag{3.37}$$

3.1.2.3 Orthonormal Vectors

Two vectors $\boldsymbol{w}_i$ and $\boldsymbol{w}_j$ are said to be *orthogonal* if their inner product is zero. That is,

$$\boldsymbol{w}_i^T \boldsymbol{w}_j = 0 \tag{3.38}$$

Two vectors, $\boldsymbol{v}_i$ and $\boldsymbol{v}_j$, are said to be *orthonormal*, if

$$\boldsymbol{v}_i^T \boldsymbol{v}_j = \begin{cases} 1, & i = j \\ 0, & i \neq j \end{cases} \tag{3.39}$$

In Figure 3.1, $\boldsymbol{v}$ and $\boldsymbol{v}_\perp$ are such orthonormal vectors, which means that $\boldsymbol{v}$ is perpendicular to $\boldsymbol{v}_\perp$ and both vectors have a unit length.

A square matrix **W** is said to be *orthogonal*, if any two of its columns satisfy Equation 3.38. Furthermore, a square matrix **V** is said to be orthonormal if any two of its columns satisfy Equation 3.39.

It is seen that

$$\mathbf{V}^T\mathbf{V} = \mathbf{V}\mathbf{V}^T = \mathbf{I} \tag{3.40}$$

and

$$\mathbf{V}^T = \mathbf{V}^{-1} \tag{3.41}$$

3.1.2.4 Unit Vector

An n × 1 *unit vector* $\boldsymbol{e}_i$ has all its elements equal to zero except the ith element, which is equal to i^{th} unity. That is,

$$\boldsymbol{e}_i^T = \Big[0, 0...., 0, \underset{i}{1}, 0,..., 0\Big]^T \tag{3.42}$$

An *identity matrix* **I** has its columns (rows) arranged as a combination of $n \times 1$ ($1 \times n$) unit vectors as

$$\mathbf{I} = [\boldsymbol{e}_1, \boldsymbol{e}_2, \ldots, \boldsymbol{e}_n] \tag{3.43}$$

For an arbitrary nonzero vector $\boldsymbol{v}$, an orthonormal matrix $\mathbf{V}$ can always be found such that

$$\mathbf{V}^T \boldsymbol{v} = \alpha \boldsymbol{e}_i \tag{3.44}$$

where α is the norm of vector $\boldsymbol{v}$.

Example 3.2

For $n = 2$, $\boldsymbol{v}$ can be described by Equation 3.34. Thus,

$$\mathbf{V} = \begin{bmatrix} u_1 & -u_2 \\ u_2 & u_1 \end{bmatrix} = \begin{bmatrix} \frac{v_1}{\sqrt{v_1^2 + v_2^2}} & -\frac{v_2}{\sqrt{v_1^2 + v_2^2}} \\ \frac{v_2}{\sqrt{v_1^2 + v_2^2}} & \frac{v_1}{\sqrt{v_1^2 + v_2^2}} \end{bmatrix} = \begin{bmatrix} \cos\theta & -\sin\theta \\ \sin\theta & \cos\theta \end{bmatrix} \tag{3.45}$$

Equation 3.44 becomes

$$\mathbf{V}^T \boldsymbol{v} = \begin{Bmatrix} \frac{v_1}{\sqrt{v_1^2 + v_2^2}} v_1 & + \frac{v_2}{\sqrt{v_1^2 + v_2^2}} v_2 \\ -\frac{v_2}{\sqrt{v_1^2 + v_2^2}} v_1 & + \frac{v_1}{\sqrt{v_1^2 + v_2^2}} v_2 \end{Bmatrix} = \alpha \begin{Bmatrix} 1 \\ 0 \end{Bmatrix} = \alpha \mathbf{e}_1 \tag{3.46}$$

It is also seen that

$$\begin{bmatrix} \mathbf{V}_1 & \mathbf{V}_2 \\ -\mathbf{V}_2 & \mathbf{V}_1 \end{bmatrix}^T \begin{Bmatrix} v_1 \\ v_2 \end{Bmatrix} = \begin{Bmatrix} \frac{v_1}{\sqrt{v_1^2 + v_2^2}} v_1 & -\frac{v_2}{\sqrt{v_1^2 + v_2^2}} v_2 \\ \frac{v_2}{\sqrt{v_1^2 + v_2^2}} v_1 & +\frac{v_1}{\sqrt{v_1^2 + v_2^2}} v_2 \end{Bmatrix} = \alpha \begin{Bmatrix} 0 \\ 1 \end{Bmatrix} = \alpha \mathbf{e}_2 \tag{3.47}$$

From Figure 3.1, it is realized that premultiplying the transpose of the orthonormal matrix, namely, $\mathbf{V}^T$ on $\boldsymbol{v}$, implies the rotation of the Cartesian coordinate x-y into x′-y′, that is,

$$\begin{Bmatrix} x' \\ y' \end{Bmatrix} = \begin{bmatrix} \cos\theta & \sin\theta \\ -\sin\theta & \cos\theta \end{bmatrix} \begin{Bmatrix} x \\ y \end{Bmatrix} \tag{3.48}$$

Furthermore, it is seen that any $n \times 1$ vector $\boldsymbol{v} = [v_1, v_2, \ldots, v_n]^T$ can be represented by a linear combination of the unit vectors $\mathbf{e}_1, \mathbf{e}_2, \ldots, \mathbf{e}_n$. That is,

$$\boldsymbol{v} = a_1\mathbf{e}_1 + a_2\mathbf{e}_2 + \ldots + a_n\mathbf{e}_n \tag{3.49}$$

Equation 3.44 is valid by simply letting $a_1 = v_1$, $a_2 = v_2, \ldots, a_n = v_n$, in Equation 3.49.

3.1.2.5 Vector Space

In Equation 3.33, there are n vectors $\boldsymbol{e}_i$ that are used to represent vector $\boldsymbol{v}$, which raises a question of how many vectors are actually needed to represent an $n \times 1$ vector. Alternatively, the question may be posed from a different angle, i.e., how many eigenvectors are sufficient to represent the structural response of an n-DOF system. To answer this question, the set of all possible $n \times 1$ real-valued vectors are considered and the set is referred to as a *real-valued vector space*. In addition, all the complex-valued $n \times 1$ vectors are included, as a generalized *complex-valued vector space*. Apparently, the latter space covers the former. It is seen that the dimension of the vector space is n. Note that n is a limited number.

For scalars a and b, it is seen that if $\boldsymbol{u}$ and $\boldsymbol{v}$ are vectors in such a vector space, then

$$a(\boldsymbol{u} + \boldsymbol{v}) = a\boldsymbol{u} + a\boldsymbol{v} \tag{3.50}$$

$$(a + b)\boldsymbol{u} = a\boldsymbol{u} + b\boldsymbol{u} \tag{3.51}$$

$$a(b\boldsymbol{u}) = (ab)\boldsymbol{u} \tag{3.52}$$

and

$$1\boldsymbol{u} = \boldsymbol{u} \tag{3.53}$$

The $n \times 1$ vector space can be seen as the space that contains all the possible solutions of $\boldsymbol{u}_i$ for the homogeneous Equation 3.9 or all the possible eigenvectors in Equation 3.23 for arbitrary **M** and **K**.

3.1.2.6 Linear Independence

Suppose there are two $n \times 1$ vectors $\boldsymbol{v}$ and $\boldsymbol{u}$. If

$$\boldsymbol{v} = b\boldsymbol{u} \tag{3.54}$$

where b is a nonzero scalar, then $\boldsymbol{v}$ and $\boldsymbol{u}$ are linearly dependent. In other words, $\boldsymbol{v}$ can be represented by $\boldsymbol{u}$. If Equation 3.54 does not hold for $\boldsymbol{v}$ and $\boldsymbol{u}$, these two vectors are linearly independent. For example, if $n = 2$, and Equation 3.54 holds, geometrically, $\boldsymbol{v}$ and $\boldsymbol{u}$ are parallel. Otherwise, they are not parallel, in which case, if an orthonormal matrix **V** operates on $\boldsymbol{v}$ such that $\mathbf{V}^T\boldsymbol{v} = b\boldsymbol{e}_i$, then the same matrix cannot make "$\mathbf{V}^T\boldsymbol{u} = a\boldsymbol{e}_i$", where a is a scalar. This fact can easily be seen from Equation 3.54.

A set of $n \times 1$ vectors $\boldsymbol{u}_i$ are *linearly dependent* if n scalars $b_1, b_2, \ldots, b_n$ exist and are not all zero, such that

$$b_1\boldsymbol{u}_1 + b_2\boldsymbol{u}_2 + \cdots b_n\boldsymbol{u}_n = \mathbf{0} \tag{3.55}$$

where $\mathbf{0}$ is a null $n \times 1$ vector. If Equation 3.55 does not hold, then the set of $\boldsymbol{u}_i$ are *linearly independent*. In this case, the linear combination of all $\boldsymbol{u}_i$ will be a nonzero vector $\boldsymbol{v}$, that is,

$$b_1\boldsymbol{u}_1 + b_2\boldsymbol{u}_2 + \cdots b_n\boldsymbol{u}_n = \boldsymbol{v} \tag{3.56}$$

Equation 3.56 implies that the vector $\boldsymbol{v}$ is now represented by at most n linearly independent vectors. Compared with Equation 3.49, where $\boldsymbol{v}$ is represented by n unit vectors e_i, it can be seen that the n unit vectors $\boldsymbol{e}_i$ are linearly independent. Thus, a set of scalars b_i, not all zero, cannot be found to invalidate the relationship:

$$b_1\boldsymbol{e}_1 + b_2\boldsymbol{e}_2 + \cdots b_n\boldsymbol{e}_n \neq 0 \tag{3.57}$$

In fact, Equation 3.57 is equivalent to

$$[\boldsymbol{e}_1, \boldsymbol{e}_2, \ldots, \boldsymbol{e}_n]\begin{Bmatrix} b_1 \\ b_2 \\ \ldots \\ b_n \end{Bmatrix} = \mathbf{I}\begin{Bmatrix} b_1 \\ b_2 \\ \ldots \\ b_n \end{Bmatrix} = \begin{Bmatrix} b_1 \\ b_2 \\ \ldots \\ b_n \end{Bmatrix} \neq \mathbf{0} \tag{3.58}$$

Since the n unit vectors $\boldsymbol{e}_i$ can represent any $n \times 1$ vectors with proper linear combinations, they are called the *base* in the n-dimensional vector space. Of course, the set of base vectors is not unique. If Equation 3.55 holds, the n unit vectors $\boldsymbol{u}_i$ can also be used to represent any $n \times 1$ vectors with proper linear combinations. Thus, $\boldsymbol{u}_i$ also forms a base in the n-dimensional vector space.

It can be proven that if a square matrix **A** has all its column vectors independent, then this matrix is full rank, or

$$\operatorname{rank}(\mathbf{A}) = n \tag{3.59}$$

which is equivalent to

$$\det(\mathbf{A}) \neq 0 \tag{3.60}$$

In this case, **A** is nonsingular and $\mathbf{A}^{-1}$ exists.

Solving the following linear equations, the set of scalar b_i can be found to represent an $n \times 1$ vector $\boldsymbol{v} = [v_1, v_2, \ldots, v_n]^T$. The linear equations can be written as

$$[\boldsymbol{u}_1, \boldsymbol{u}_2, \ldots, \boldsymbol{u}_n]\begin{Bmatrix} b_1 \\ b_2 \\ \ldots \\ b_n \end{Bmatrix} = \begin{Bmatrix} v_1 \\ v_2 \\ \ldots \\ v_n \end{Bmatrix} \tag{3.61}$$

That is, since these $\boldsymbol{u}_i$ are linearly independent, $[\boldsymbol{u}_1, \boldsymbol{u}_2, \ldots, \boldsymbol{u}_n]$ has an inverse matrix $[\boldsymbol{u}_1, \boldsymbol{u}_2, \ldots, \boldsymbol{u}_n]^{-1}$. Thus, $[\boldsymbol{u}_1, \boldsymbol{u}_2, \ldots, \boldsymbol{u}_n]^{-1}$ can be multiplied on both sides of Equation 3.61, such that

$$\begin{Bmatrix} b_1 \\ b_2 \\ \ldots \\ b_n \end{Bmatrix} = [\boldsymbol{u}_1, \boldsymbol{u}_2, \ldots, \boldsymbol{u}_n]^{-1}\begin{Bmatrix} v_1 \\ v_2 \\ \ldots \\ v_n \end{Bmatrix} \tag{3.62}$$

Thus, $\boldsymbol{u}_1, \boldsymbol{u}_2, \ldots, \boldsymbol{u}_n$ can also be used as the base. Comparing these vectors with $\boldsymbol{e}_1, \boldsymbol{e}_2, \ldots, \boldsymbol{e}_n$, it is seen that $\boldsymbol{e}_i$ are more convenient to use, because $\boldsymbol{e}_i$ vectors are orthonormal. Note that a set of n orthogonal vectors are not necessarily $\boldsymbol{e}_i$. For example, in the 2-dimensional vector space, the two columns of matrix **V** expressed in Equation 3.45 are also orthonomal.

3.1.3 Symmetric Matrix and Rayleigh Quotient

3.1.3.1 Eigen-Parameters of Symmetric Matrices

In the eigen-problem described in Equation 3.23, if all the possible eigenvectors are normalized so that their Euclidian length is unity, then this set may or may not be orthogonal. This is because the eigenvectors are not necessarily perpendicular to each other. For example, two eigenvectors $\boldsymbol{u}_1$ and $\boldsymbol{u}_2$ are written as

$$\boldsymbol{u}_1 = a_1\boldsymbol{e}_1 + b_1\boldsymbol{e}_2 \tag{3.63}$$

and

$$\boldsymbol{u}_2 = a_2\boldsymbol{e}_1 + c_3\boldsymbol{e}_3 \tag{3.64}$$

where a_1, a_2, b_1, and c_3 are nonzero scalars.

It is apparent that

$$\boldsymbol{u}_1 \neq a\boldsymbol{u}_2 \tag{3.65}$$

Therefore, $\boldsymbol{u}_1$ and $\boldsymbol{u}_2$ are linearly independent. However,

$$\boldsymbol{u}_1^T\boldsymbol{u}_2 = a_1a_2 \neq 0 \tag{3.66}$$

Thus, they are not orthogonal. It is inconvenient to use linearly independent but nonorthogonal vectors, since the simple criterion described in Equation 3.38 cannot be used. The matrix product $\mathbf{M}^{-1}\mathbf{K}$ is often asymmetric and its eigenvector matrix, from now on denoted specifically by $\mathbf{U}$, is not orthogonal in general. Thus, $\mathbf{U}$ cannot be made orthonormal in general.

On the other hand, if a square matrix $\mathbf{K}$ is symmetric, that is

$$\mathbf{K} = \mathbf{K}^T \tag{3.67}$$

then all of its eigenvectors will be orthogonal to each other. In addition, all of its eigenvalues will be real valued. The symbols $\mathbf{V}_K$ and Λ_K are used to denote the corresponding eigenvector and eigenvalue matrix, respectively. That is,

$$\mathbf{K} = \mathbf{V}_K\Lambda_K\mathbf{V}_K^T \tag{3.68}$$

or

$$\mathbf{V}_K^T\mathbf{K}\mathbf{V}_K = \Lambda_K \tag{3.69}$$

Here, $\mathbf{V}_K$ is the orthonormal eigenvector matrix, which satisfies Equation 3.40 and 3.41. The condition of forming a symmetric matrix is not very difficult to realize for the structural dynamics problem. This is because both the mass and the stiffness matrices are symmetric. However, since the eigenvector matrix $\mathbf{U}$ of $\mathbf{M}^{-1}\mathbf{K}$ is often not orthogonal in general, a method to construct a symmetric matrix that provides the identical eigenvalues as the matrix $\mathbf{M}^{-1}\mathbf{K}$ and orthonormal eigenvectors needs to be found for convenience.

Before further exploring the eigen-problem of a symmetric matrix, a notation for $\mathbf{M}^{1/2}$ is introduced. Note that the square matrix $\mathbf{M}$ can have an eigen-decomposition as

$$\mathbf{M} = \mathbf{V}_M\Lambda_M\mathbf{V}_M^T \tag{3.70}$$

Here, $\mathbf{V}_M$ is the orthonormal eigenvector matrix, which satisfies Equation 3.40 as well as Equation 3.41. Matrix Λ_M provides the eigenvalues of **M**, where

$$\Lambda_M = \text{diag}\left([\lambda_{M1}, \lambda_{M2},...,\lambda_{Mn}]\right) \tag{3.71}$$

in which λ_{Mi} is the ith eigenvalue of **M**. Now, $\mathbf{M}^{1/2}$ can be calculated as follows:

$$\mathbf{M}^{1/2} = \mathbf{V}_M\Lambda_M^{1/2}\mathbf{V}_M^T \tag{3.72}$$

Since Λ_M is diagonal,

$$\Lambda_M^{1/2} = \text{diag}\left(\left[\sqrt{\lambda_{M1}}, \sqrt{\lambda_{M2}},...,\sqrt{\lambda_{Mn}}\right]\right) \tag{3.73}$$

Furthermore,

$$\mathbf{M}^{-1/2} = \mathbf{V}_M\Lambda_M^{-1/2}\mathbf{V}_M^T \tag{3.74}$$

and

$$\mathbf{M}^{1/2}\mathbf{M}^{-1/2} = \mathbf{I} \tag{3.75}$$

Note also that

$$\mathbf{M}^{1/2}\mathbf{M}^{1/2} = \mathbf{M}$$

Now, premultiplying $\mathbf{M}^{1/2}$ and postmultiplying $\mathbf{M}^{-1/2}$ on both sides of Equation 3.28, the following form results:

$$\mathbf{M}^{1/2}\mathbf{U}\mathbf{\Omega}^2\mathbf{U}^{-1}\,\mathbf{M}^{-1/2} = \mathbf{M}^{1/2}\mathbf{M}^{-1}\mathbf{K}\mathbf{M}^{-1/2} = \mathbf{M}^{-1/2}\mathbf{K}\mathbf{M}^{-1/2} \tag{3.76}$$

Denote

$$\mathbf{V} = \mathbf{M}^{1/2}\mathbf{U} \tag{3.77}$$

so that

$$\mathbf{V}^{-1} = \mathbf{U}^{-1}\mathbf{M}^{-1/2} \tag{3.78}$$

Therefore,

$$\mathbf{V}\mathbf{\Omega}^2\mathbf{V}^{-1} = \mathbf{M}^{-1/2}\mathbf{K}\mathbf{M}^{-1/2} \tag{3.79}$$

Equation 3.79 implies that matrix $\mathbf{M}^{-1/2}\,\mathbf{K}\,\mathbf{M}^{-1/2}$ can have an eigen-decomposition with eigenvector and eigenvalue matrices **V** and $\mathbf{\Omega}^2$, respectively. Since **K** is symmetric, it is easy to see that

$$(\mathbf{M}^{-1/2}\mathbf{K}\mathbf{M}^{-1/2})^{\mathrm{T}} = \mathbf{M}^{-1/2}\mathbf{K}\mathbf{M}^{-1/2} \tag{3.80}$$

That is, matrix $\mathbf{M}^{-1/2}\mathbf{K}\mathbf{M}^{-1/2}$ is symmetric, so the eigenvector matrix $\mathbf{V}$ is orthogonal and can be normalized to be orthonormal. In the following discussion, the notation $\tilde{\mathbf{K}}$ is used, where

$$\tilde{\mathbf{K}} = \mathbf{M}^{-1/2}\mathbf{K}\mathbf{M}^{-1/2} \tag{3.81}$$

to denote this symmetric matrix and is called the *generalized stiffness matrix*. In the following, $\mathbf{V}$ is used to denote the orthonomal eigenvector matrix. Thus,

$$\mathbf{V}^{-1} = \mathbf{U}^{-1}\mathbf{M}^{-1/2} = \mathbf{V}^{\mathrm{T}} = \mathbf{U}^{\mathrm{T}}\mathbf{M}^{1/2} \tag{3.82}$$

3.1.3.2 Rayleigh Quotient

The famous Rayleigh quotient is discussed next, which is the foundation of modal analysis that, in turn, is currently the basic tool used in damper design for civil engineering structures.

First, since the eigenvector $\mathbf{V}$ described in Equation 3.82 is orthonormal, thus satisfies $\mathbf{V}^{\mathrm{T}}\,\mathbf{V} = \mathbf{I}$,

$$\mathbf{U}^{\mathrm{T}}\mathbf{M}^{1/2}\mathbf{M}^{1/2}\mathbf{U} = \mathbf{U}^{\mathrm{T}}\mathbf{M}\mathbf{U} = \mathbf{I} \tag{3.83}$$

Furthermore,

$$\mathbf{U}^{\mathrm{T}}\mathbf{K}\mathbf{U} = \boldsymbol{\Omega}^2 \tag{3.84}$$

Note that in Equation 3.82, the eigenvector matrix $\mathbf{V}$ is assumed to be orthonormal. In fact, the eigenvector can indeed be made orthonormal with proper normalization by letting the Euclidian length of each column vector be unity. On the other hand, each column of the eigenvector matrix can be multiplied by an arbitrary nonzero scalar, so that the resulting eigenvector matrix is no longer orthonormal though it is still orthogonal. In this case, according to Equation 3.77, the eigenvector matrix $\mathbf{U}$ of $\mathbf{M}^{-1}\mathbf{K}$ will be affected, and thus Equations 3.83 and 3.84 will not hold. In this circumstance, there will be more general formulas.

$$\mathbf{U}^{\mathrm{T}}\mathbf{M}\mathbf{U} = \mathrm{diag}\left(\left[\mu_1, \mu_2, \ldots, \mu_n\right]\right) \tag{3.85}$$

and

$$\mathbf{U}^{\mathrm{T}}\mathbf{K}\mathbf{U} = \mathrm{diag}\left(\left[\kappa_1, \kappa_2, \ldots, \kappa_n\right]\right) \tag{3.86}$$

where μ_i and κ_i are modal mass and modal stiffness quantities, respectively, with their values altered by different normalizations of the eigenvectors in $\mathbf{U}$. However, their ratio is always fixed, such that

$$\frac{\kappa_i}{\mu_i} = \omega_{ni}^2 \quad i = 1, 2, \ldots, n \tag{3.87}$$

Now, Equations 3.85 through 3.87 are rewritten by considering each eigenvalue–eigenvector pair one by one. That is, since

$$\mathbf{U} = \left[\boldsymbol{u}_1, \boldsymbol{u}_2, \ldots, \boldsymbol{u}_n\right] \tag{3.88}$$

results in

$$\boldsymbol{u}_i^T \mathbf{M} \boldsymbol{u}_i = \mu_i, \quad i = 1, 2, \ldots, n \tag{3.89}$$

$$\boldsymbol{u}_i^T \mathbf{K} \boldsymbol{u}_i = \kappa_i, \quad i = 1, 2, \ldots, n \tag{3.90}$$

and

$$\frac{\boldsymbol{u}_i^T \mathbf{K} \boldsymbol{u}_i}{\boldsymbol{u}_i^T \mathbf{M} \boldsymbol{u}_i} = \frac{\kappa_i}{\mu_i} = \omega_{ni}^2 \tag{3.91}$$

Equation 3.91 represents a quotient of two vector-matrix products. The value of the quotient is a scalar, which is one of the eigenvalues of the **M-K** system. Suppose the vector in Equation 3.91 is an arbitrary vector $\boldsymbol{v}$ instead of one of the eigenvectors $\boldsymbol{u}_i$. A scalar quotient still exists, denoted by $R(\boldsymbol{v})$, which is named the *Rayleigh quotient*. That is,

$$R(\boldsymbol{v}) = \frac{\boldsymbol{v}^T \mathbf{K} \boldsymbol{v}}{\boldsymbol{v}^T \mathbf{M} \boldsymbol{v}} \tag{3.92}$$

It can be seen that when the vector $\boldsymbol{v}$ is taken to be one of the eigenvectors, the Rayleigh quotient will become the corresponding eigenvalue.

The ratios of the Rayleigh quotient and eigenvalues of a 2-DOF system are plotted. Figure 3.2a shows a comparison with the first eigenvalue, while Figure 3.2b shows the second eigenvalue. The X-axis represents an index of selection for the vectors. Each selection is the corresponding eigenvector with length 1, plus a random vector. At x = 50, the random vector is null. The length of the random vectors selected then increases as one moves away from x = 50, such that finally at x = 1 and x = 100, the random component will have a length of 1/2 of the eigenvector. In addition, since the selection is random, Equation 3.92 is evaluated 100 times for both plots in Figure 3.2a and b and the average of the resulting Rayleigh quotient is obtained. From Figure 3.2a, it is seen that the closer $\boldsymbol{v}$ is to the actual eigenvector, the smaller the value of the Rayleigh quotient will be, until the ratio reaches 1, which is the point at which the vector equals the first eigenvector. Thus, it is seen that in this example, the Rayleigh quotient is the minimum, which coincides with the first eigenvalue. On the other hand, from Figure 3.2b, the Rayleigh quotient will be the maximum, which equals the second eigenvalue.

This example is not a coincidence. In fact, it can be shown that for an n-DOF system, when the vector is chosen to be the eigenvector of the first mode, the corresponding Rayleigh quotient becomes equal to the first eigenvalue and will be the minimum among all the possible quotients. On the other hand, when the vector is chosen to be the eigenvector of the highest mode, the corresponding Rayleigh quotient becomes the largest eigenvalue and will be the maximum among all the possible quotients. In addition, in each instance when an eigenvector is chosen, the Rayleigh quotient will have a stationary value, which is exactly the same as the corresponding eigenvalue.

Example 3.3

Prove that the Rayleigh quotient

$$R(\boldsymbol{v}) = \frac{\boldsymbol{v}^T \tilde{\mathbf{K}} \boldsymbol{v}}{\boldsymbol{v}^T \boldsymbol{v}} \tag{3.93}$$

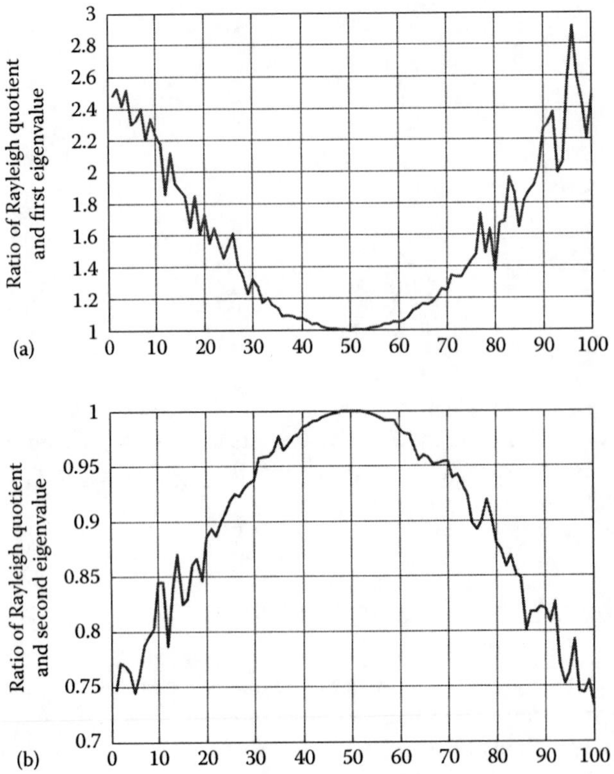

FIGURE 3.2 Ratio of Rayleigh quotient and eigenvalues (a) first order and (b) second order.

reaches a stationary value when $\boldsymbol{v}$ becomes one of the eigenvalues of the symmetric matrix $\tilde{\mathbf{K}}$ and the value of the Rayleigh quotient is the corresponding eigenvalue.

Taking the partial derivative of $R(\boldsymbol{v})$ with respect to each element of $\boldsymbol{v} = [v_1, v_2, \ldots, v_n]^T$ yields

$$\frac{\partial R(\boldsymbol{v})}{\partial v_i} = \frac{\left(\boldsymbol{v}^T\boldsymbol{v}\right)(2\tilde{K}_i\boldsymbol{v}) - (\boldsymbol{v}^T\tilde{\mathbf{K}}\boldsymbol{v})2v_i}{\left(\boldsymbol{v}^T\boldsymbol{v}\right)^2} \quad i = 1, 2, \ldots, n \tag{3.94}$$

Here

$$\tilde{\mathbf{K}}_i = \left[k_{i1}, k_{in}, \ldots k_{in}\right] \tag{3.95}$$

and k_{ij} is the ijth entry of matrix $\tilde{\mathbf{K}}$. Letting

$$\frac{\partial R(\boldsymbol{v})}{\partial v_i} = \mathbf{0}, \quad i = 1, 2, \ldots, n \tag{3.96}$$

and rewriting Equation 3.94 into matrix form results in

$$\tilde{\mathbf{K}}\boldsymbol{v} - R(\boldsymbol{v})\boldsymbol{v} = \mathbf{0} \tag{3.97}$$

or

$$R(\boldsymbol{v})\boldsymbol{v} = \tilde{\mathbf{K}}\boldsymbol{v} \tag{3.98}$$

From Equation 3.97, it is clear that $R(\boldsymbol{v})$ and $\boldsymbol{v}$ are one of the eigenvalue–eigenvector pairs of matrix $\tilde{\mathbf{K}}$.

3.2 PROPORTIONALLY DAMPED MDOF SYSTEMS

In this section, the Rayleigh quotient is considered as a means of decoupling the proportionally damped MDOF systems, which is often referred to as classic modal analysis or normal modal analysis.

3.2.1 Modal Analysis and Decoupling Procedure

3.2.1.1 Governing Equation of Damped MDOF Systems

Linear proportionally damped MDOF systems are often used to model a structure for damper design. Using *normal mode* analysis, the n-DOF system can be decoupled into n-SDOF subsystems, which are much easier to handle and more familiar to structural engineers. The decoupling procedure is examined in this section. In the previous section, a monic form with symmetric damping and stiffness matrices was used to study the eigen-problem of the MDOF systems. Under earthquake excitation, more generally, the governing equation of motion for MDOF is rewritten as

$$\mathbf{M}\ddot{\mathbf{x}}(t)+\mathbf{C}\dot{\mathbf{x}}(t)+\mathbf{K}\mathbf{x}(t) = -\mathbf{M}\mathbf{J}\ddot{\mathbf{x}}_g(t) \tag{3.99}$$

That is, similar to Equation 1.59, the force in Equation 3.99 is written as

$$\mathbf{f}(t) = -\mathbf{M}\mathbf{J}\ \ddot{x}_g(t) \tag{3.100}$$

where

$$\mathbf{J} = \{1\}_{n\times 1} \tag{3.101}$$

is a unit vector, known as the *input vector.*

It can be proven that the damping is proportional, if and only if the following defined *Caughey criterion* holds:

$$\mathbf{C}\mathbf{M}^{-1}\mathbf{K} = \mathbf{K}\mathbf{M}^{-1}\mathbf{C} \tag{3.102}$$

In other words, the matrices $\mathbf{M}^{-1}\mathbf{K}$ and $\mathbf{M}^{-1}\mathbf{C}$ share the identical eigenvector matrix **U**. Therefore, for a proportionally damped system, the eigenvector matrix **U** is necessary to make the matrices **M**, **C**, and **K** simultaneously diagonal. Similar to the case described in Equations 3.85 and 3.86, damping must be considered. That is,

$$\mathbf{U}^T\mathbf{M}\mathbf{U} = \text{diag}(\mu_i), \quad i = 1, 2, \ldots, n \tag{3.103}$$

$$\mathbf{U}^T\mathbf{C}\mathbf{U} = \text{diag}(\chi_i), \quad i = 1, 2, \ldots, n \tag{3.104}$$

and

$$\mathbf{U}^T\mathbf{K}\mathbf{U} = \mathrm{diag}(\kappa_i), \quad i = 1, 2, \ldots, n \tag{3.105}$$

The eigenvector **U** defined in Equations 3.103 through 3.105 is also the *mode shape matrix* of the proportionally damped system. Similar to the undamped case, matrix **U** is now said to be *orthogonal with respect to the weighting matrices* **M**, **C**, *and* **K**, for

$$\boldsymbol{u}_i^T\mathbf{M}\boldsymbol{u}_i = \mu_i, \quad i = 1, 2, \ldots, n \tag{3.106}$$

$$\boldsymbol{u}_i^T\mathbf{C}\boldsymbol{u}_i = \chi_i, \quad i = 1, 2, \ldots, n \tag{3.107}$$

and

$$\boldsymbol{u}_i^T\mathbf{K}\boldsymbol{u}_i = \kappa_i, \quad i = 1, 2, \ldots, n \tag{3.108}$$

Here, μ_i, χ_i, and κ_i are referred to as the ith *modal mass*, *modal damping*, and *modal stiffness coefficient*, respectively. From Equations 3.106 through 3.108, it is realized that

$$\mu_i > 0 \tag{3.109}$$

$$\kappa_i > 0 \tag{3.110}$$

and

$$\chi_i \geq 0 \tag{3.111}$$

Note that the mode shape vector is not unique. That is, if $\boldsymbol{u}_i$ is the ith mode shape of a system, then an alternative mode shape $\bar{\boldsymbol{u}}_i$ exists, where

$$\bar{\boldsymbol{u}}_i = \alpha_i\, \boldsymbol{u}_i \tag{3.112}$$

Here α_i is the generic normalization factor or proportionality factor, such that

$$\bar{\boldsymbol{u}}_i^{\,T}\mathbf{M}\bar{\boldsymbol{u}}_i = \mathbf{1} \tag{3.113}$$

In addition

$$\bar{\boldsymbol{u}}_i = \mathbf{M}^{-1/2}\boldsymbol{v}_i \tag{3.114}$$

where $\boldsymbol{v}_i$ is the ith orthonormal eigenvector of $\mathbf{M}^{-1/2}\mathbf{K}\mathbf{M}^{1/2}$ described in Equation 3.79.

3.2.1.2 Decoupling by Means of Rayleigh Quotient

The modal mass generated by the eigenvector $\boldsymbol{u}_i$ described in Equation 3.112 can be shown to be

$$\mu_i = \alpha_i^{-2} \tag{3.115}$$

Therefore, in Equations 3.106 through 3.108, μ_i, χ_i, and κ_i are not unique, because the vector $\boldsymbol{u}_i$ is not unique. However, their ratios are exactly the aforementioned Rayleigh quotient, when the vector $\boldsymbol{u}_i$ is taken to be one of the eigenvectors. Those ratios can be written as follows:

$$\chi_i/\mu_i = \frac{\boldsymbol{u}_i^T\mathbf{C}\boldsymbol{u}_i}{\boldsymbol{u}_i^T\mathbf{M}\boldsymbol{u}_i} = R_c(\boldsymbol{u}_i), \quad i = 1, 2, \ldots, n \tag{3.116}$$

and

$$\kappa_i/\mu_i = \frac{\boldsymbol{u}_i^T\mathbf{K}\boldsymbol{u}_i}{\boldsymbol{u}_i^T\mathbf{M}\boldsymbol{u}_i} = R_k(\boldsymbol{u}_i), \quad i = 1, 2, \ldots, n \tag{3.117}$$

Here, $R_c(\boldsymbol{u}_i)$ and $R_k(\boldsymbol{u}_i)$ denote the corresponding Rayleigh quotient, thus

$$R_c(\boldsymbol{u}_i) = \chi_i/\mu_i = 2\xi_i\omega_{ni}\alpha^2/\alpha^2 = 2\xi_i\omega_{ni}, \quad i = 1, 2, \ldots, n \tag{3.118}$$

$$R_k(\boldsymbol{u}_i) = \kappa_i/\mu_i = \omega_{ni}^2\alpha^2/\alpha^2 = \omega_{ni}^2, \quad i = 1, 2, \ldots, n \tag{3.119}$$

Here, ξ_i, ω_{ni}, and $\boldsymbol{u}_i$ are referred to as the *modal damping ratio, modal natural frequency*, and *mode shape* of the ith mode, respectively. All these are also referred to as the modal parameters.

Using the above notations, it can be assumed that

$$\mathbf{x}(t) = y_i(t)\boldsymbol{u}_i \tag{3.120}$$

is one of the possible solutions of Equation 3.99, where y(t) is the ith modal response. Therefore, substituting Equation 3.120 into Equation 3.99 and premultiplying $\boldsymbol{u}_i^T$ on both sides of the resulting equation yield

$$\boldsymbol{u}_i^T\mathbf{M}\boldsymbol{u}_i\ddot{y}_i(t) + \boldsymbol{u}_i^T\mathbf{C}\boldsymbol{u}_i\dot{y}_i(t) + \boldsymbol{u}_i^T\mathbf{K}\boldsymbol{u}_i y_i(t) = -\boldsymbol{u}_i^T\mathbf{M}\mathbf{J}\,\ddot{x}_g(t) \tag{3.121}$$

Now, dividing both sides of Equation 3.121 by the scalar $\boldsymbol{u}_i^T\mathbf{M}\boldsymbol{u}_i$ results in the relation:

$$\ddot{y}_i(t) + \frac{\boldsymbol{u}_i^T\mathbf{C}\boldsymbol{u}_i}{\boldsymbol{u}_i^T\mathbf{M}\boldsymbol{u}_i}\dot{y}_i(t) + \frac{\boldsymbol{u}_i^T\mathbf{K}\boldsymbol{u}_i}{\boldsymbol{u}_i^T\mathbf{M}\boldsymbol{u}_i}y_i(t) = -\frac{\boldsymbol{u}_i^T\mathbf{M}\mathbf{J}}{\boldsymbol{u}_i^T\mathbf{M}\boldsymbol{u}_i}\ddot{x}_g(t) \tag{3.122}$$

Equation 3.122 is a scalar equation, implying an SDOF vibration system, which was discussed in Chapters 1 and 2.

From the discussion of eigen-decomposition of $n \times n$ matrices, it is known that there will be n eigenvectors. Therefore, there can be n equations described by Equation 3.122. Since these n eigenvectors are linearly independent, they can represent any $n \times 1$ vector in the $n \times n$ vector space. Thus, $y_i(t)\,\boldsymbol{u}_i$ can be seen as one of the bases in the space. Since $\mathbf{x}(t)$ is also an $n \times 1$ vector, it can be represented by the linear combination of these $(y_i(t)\,\boldsymbol{u}_i)$. In this way, Equation 3.99 is solved by individually calculating $(y_i(t)\,\boldsymbol{u}_i)$ through Equation 3.122 and adding the individual solutions together to obtain the total solution. Similar to underdamped cases, the computation procedure to obtain

Equation 3.121 is referred to as *modal decoupling* or *modal analysis*, while the summation of (y_i (t) $\boldsymbol{u}_i$) is referred to as *modal superposition*. That is,

$$\mathbf{x}(t) = \sum_{i=1}^{n} y_i(t)\boldsymbol{u}_i \tag{3.123}$$

3.2.1.3 Decoupling by Means of Linear Transformation

The above modal decoupling superposition can also be reached by using a linear transformation method based on the eigenvector matrix. That is, premultiplying by matrix $\mathbf{U}^T$ on both sides of Equation 3.99 and using a new variable $\mathbf{y}$(t), such that

$$\mathbf{x}(t) = \mathbf{U}\mathbf{y}(t) \tag{3.124}$$

where

$$\mathbf{y}(t) = \begin{Bmatrix} y_1(t) \\ y_2(t) \\ \cdots \\ y_n(t) \end{Bmatrix} \tag{3.125}$$

results in

$$\mathbf{U}^T\mathbf{M}\mathbf{U}\ddot{\mathbf{y}}(t) + \mathbf{U}^T\mathbf{C}\mathbf{U}\dot{\mathbf{y}}(t) + \mathbf{U}^T\mathbf{K}\mathbf{U}\mathbf{y}(t) = -\mathbf{U}^T\mathbf{M}\mathbf{J}\ddot{x}_g(t) \tag{3.126}$$

Substituting Equations 3.103 through 3.105 into Equation 3.126 results in

$$\text{diag}(\mu_i)\ddot{\mathbf{y}}(t) + \text{diag}(\chi_i)\dot{\mathbf{y}}(t) + \text{diag}(\kappa_i)\mathbf{y}(t) = -\mathbf{U}^T\mathbf{M}\mathbf{J}\ddot{x}_g(t) \tag{3.127}$$

Equation 3.127 is thus a decoupled set of n-SDOF equations. The ith equation is

$$\mu_i\ddot{y}_i(t) + \chi_i\dot{y}(t) + \kappa_i y(t) = -\boldsymbol{u}_i^T\mathbf{M}\mathbf{J}\ddot{x}_g(t), \quad i = 1, 2, \ldots, n \tag{3.128}$$

Equation 3.128 is the same as described by Equation 3.121, which denotes vibrations of SDOF systems.

In engineering applications, for the set of SDOF systems, it may be preferable to have a modal mass rather than the monic form. However, the modal mass μ_i is not unique, and a uniquely defined method for normalization of matrix **U** is needed. In NEHRP 2000 (BSSC 2000) and many other codes, the mode shape, which is a column vector of **U**, namely, $\boldsymbol{u}_i$, is normalized as

$$\bar{\boldsymbol{u}}_i = \frac{\begin{Bmatrix} u_{1i} \\ u_{2i} \\ \cdots \\ u_{ni} \end{Bmatrix}}{u_{ni}} = \frac{\boldsymbol{u}_i}{\boldsymbol{u}_{ni}} \equiv \{\bar{u}_{ji}\}, \quad i, j = 1, 2, \ldots, n \tag{3.129}$$

where u_{ni} is the ith modal displacement at location n, which is located at the roof, unless specified otherwise.

Note that from now on, $\bar{\boldsymbol{u}}_i$ is no longer an arbitrary vector, but is used to denote special normalizations; in the following of this chapter, it stands for the normalization described by Equation 3.129. Using the notation in Equation 3.129, the uniquely defined approach for the *normalized modal mass coefficient* m_i for the ith mode is obtained as follows:

$$m_i = \bar{\boldsymbol{u}}_i^T \mathbf{M} \bar{\boldsymbol{u}}_i \tag{3.130}$$

Also denote

$$c_i = \bar{\boldsymbol{u}}_i^T \mathbf{C} \bar{\boldsymbol{u}}_i \tag{3.131}$$

$$k_i = \bar{\boldsymbol{u}}_i^T \mathbf{K} \bar{\boldsymbol{u}}_i \tag{3.132}$$

which are the *normalized modal damping* and *stiffness coefficients*, respectively.

Equations 3.130 through 3.132 describe the physical meaning of the product $\bar{\boldsymbol{u}}_i^T \mathbf{M} \bar{\boldsymbol{u}}_i$ and so on. In practical applications, a more convenient method to compute the modal responses is used, which is discussed in the next two sections.

Comparing Equations 3.130 through 3.132 with Equations 3.106 through 3.108, respectively, it is realized that the physical essence of these two groups of equations is identical, except that the normalizations of the mode shape vectors are different.

From Equations 3.130 through 3.132, it is further seen that

$$\mathbf{M}^{-1}\mathbf{K}\bar{\boldsymbol{u}}_i = \omega_{ni}^2 \bar{\boldsymbol{u}}_i \tag{3.133}$$

or

$$\mathbf{K}^{-1}\mathbf{M}\bar{\boldsymbol{u}}_i = 1/\omega_{ni}^2 \bar{\boldsymbol{u}}_i \tag{3.134}$$

and

$$\mathbf{M}^{-1}\mathbf{C}\bar{\boldsymbol{u}}_i = 2\xi_i \omega_{ni} \bar{\boldsymbol{u}}_i \tag{3.135}$$

Now, once all the modal responses y_i are solved using the notation in Equation 3.128, the modal response vector $\mathbf{y}(t)$ can be obtained. The total responses $\mathbf{x}(t)$ can then be written as the linear transformation $\mathbf{U}\mathbf{y}(t)$. This operation is described by Equation 3.124. Except in this case, $\bar{\mathbf{U}}$ is used instead of $\mathbf{U}$, where $\bar{\mathbf{U}} = [\bar{\boldsymbol{u}}_1, \bar{\boldsymbol{u}}_2, \ldots, \bar{\boldsymbol{u}}_n]$. In other words, the modal shape matrix $\bar{\mathbf{U}}$ transfers the responses $\bar{y}_i$ in the modal space into the physical space. Thus, the response $\bar{\mathbf{y}}(t)$ in the modal space can be mapped into the physical domain $\mathbf{x}(t)$ through mapping matrix $\bar{\mathbf{U}}$. This operation is also referred to as a linear transformation. In the next subsection, using Equation 3.153, the term $\bar{\mathbf{y}}(t) = \{\bar{y}_i\}$ will be formally defined.

The jth element of $\mathbf{x}(t)$ in Equation 3.99 can be written in the form of a *modal superposition* as (Clough and Penzien 1993; Chopra 2006)

$$x_j(t) = \sum_{i=1}^{n} \bar{u}_{ji} \bar{y}_i(t) \tag{3.136}$$

and the response vector is rewritten as

$$\mathbf{x}(t) = \sum_{i=1}^{n} \bar{\boldsymbol{u}}_i \bar{y}_i(t) \tag{3.137}$$

3.2.2 Free-Decay Vibration

If the MDOF system is only subjected to initial conditions without any excitation afterward, then the forcing function in Equation 3.99 is null. Correspondingly, Equation 3.122 can be further written as

$$\ddot{y}_i(t) + 2\xi_i\omega_{ni}\dot{y}_i(t) + \omega_{ni}^2 y_i(t) = 0, \quad i = 1, 2, \ldots, n \tag{3.138}$$

which is the free-decay equation for a typical SDOF vibration system with linear viscous damping. From Equation 1.30, it is known that the solution can be written as

$$y_i(t) = y_{i0}e^{-\xi_i\omega_{ni}t}\sin(\omega_{di}t + \varphi_i), \quad i = 1, 2, \ldots, n \tag{3.139}$$

Here, y_{i0} is the amplitude and the damped natural frequency is defined as $\omega_{di} = \sqrt{1-\xi_i^2}\ \omega_{ni}$, as previously discussed. To determine the amplitude y_{i0} and the phase φ_i angle in the solution, the initial conditions are needed. From Equation 3.124,

$$\mathbf{y}(t) = \mathbf{U}^{-1}\mathbf{x}(t) \tag{3.140}$$

and

$$\dot{\mathbf{y}}(t) = \mathbf{U}^{-1}\dot{\mathbf{x}}(t) \tag{3.141}$$

Therefore, if the initial conditions $\mathbf{x}(0)$ and $\dot{\mathbf{x}}(0)$ are given, then the modal initial conditions

$$\mathbf{y}(0) = \mathbf{U}^{-1}\mathbf{x}(0) \tag{3.142}$$

and

$$\dot{\mathbf{y}}(0) = \mathbf{U}^{-1}\dot{\mathbf{x}}(0) \tag{3.143}$$

In order to not compute the inverse of the eigenvector matrix, the following equations are chosen first:

$$\mathbf{U}^{-1} = \mathbf{V}^{\mathrm{T}}\mathbf{M}^{1/2} \tag{3.144}$$

Therefore,

$$y_i(0) = \boldsymbol{u}_i^{\mathrm{T}}\mathbf{M}^{1/2}\mathbf{x}(0) \tag{3.145}$$

and

$$\dot{y}_i(0) = u_i^T \mathbf{M}^{1/2} \dot{\mathbf{x}}(0) \tag{3.146}$$

When the mode $\bar{\boldsymbol{u}}_i$ is used instead of the generic term $\boldsymbol{u}_i$, the normalization constant α (see Equation 3.129) should be considered as

$$\alpha_i = 1/u_{ni} = 1/\mathbf{m}_n \boldsymbol{v}_i \tag{3.147}$$

Here, $\mathbf{m}_n$ is the nth row of matrix $\mathbf{M}^{-1/2}$. To see the validity of Equation 3.147, first examine

$$\mathbf{U} = \mathbf{M}^{-1/2}\mathbf{V} = \begin{bmatrix} u_{11} & u_{12} & \cdots & u_{1n} \\ u_{21} & u_{22} & \cdots & u_{2n} \\ & & \cdots & \\ u_{n1} & u_{n2} & \cdots & u_{nn} \end{bmatrix} \tag{3.148}$$

Therefore,

$$\bar{\mathbf{U}} = \begin{bmatrix} \frac{u_{11}}{u_{n1}} & \frac{u_{12}}{u_{n2}} & \cdots & \frac{u_{1n}}{u_{nn}} \\ \frac{u_{21}}{u_{n1}} & \frac{u_{22}}{u_{n2}} & \cdots & \frac{u_{2n}}{u_{nn}} \\ & & \cdots & \\ 1 & 1 & \cdots & 1 \end{bmatrix} = \mathbf{U} \begin{bmatrix} \frac{1}{u_{n1}} & & & \\ & \frac{1}{u_{n2}} & & \\ & & \cdots & \\ & & & \frac{1}{u_{nn}} \end{bmatrix} = \mathbf{U}\mathbf{A}_P \tag{3.149}$$

where $\mathbf{A}_p$ is the normalization matrix and

$$\mathbf{A}_P = \begin{bmatrix} \frac{1}{u_{n1}} & & & \\ & \frac{1}{u_{n2}} & & \\ & & \cdots & \\ & & & \frac{1}{u_{nn}} \end{bmatrix} \tag{3.150}$$

In Equation 3.149, the inverse of the modified mode shape matrix $\bar{\mathbf{U}}$ is given by

$$\bar{\mathbf{U}}^{-1} = \mathbf{A}_P^{-1}\mathbf{U}^{-1} = \mathbf{A}_P^{-1}\mathbf{V}^T\mathbf{M}^{1/2} \tag{3.151}$$

Thus,

$$\mathbf{x}(t) = \mathbf{U}\mathbf{y}(t) = \mathbf{U}\mathbf{A}_P\mathbf{A}_P^{-1}\mathbf{y}(t) = \bar{\mathbf{U}}\bar{\mathbf{y}}(t) \tag{3.152}$$

where $\bar{\mathbf{y}}(t)$ is the modal displacement vector and

$$\bar{\mathbf{y}}(t) = \mathbf{A}_P^{-1}\mathbf{y}(t) \tag{3.153}$$

From Equation (3.152),

$$\bar{\mathbf{y}}(t) = \bar{\mathbf{U}}^{-1}\mathbf{x}(t) = \mathbf{A}_P^{-1}\mathbf{V}^T\mathbf{M}^{1/2}\mathbf{x}(t) \tag{3.154}$$

Thus, for the initial displacement and velocity

$$\bar{\mathbf{y}}(0) = \bar{\mathbf{U}}^{-1}\mathbf{x}(0) = \mathbf{A}_P^{-1}\mathbf{V}^T\mathbf{M}^{1/2}\mathbf{x}(0) \tag{3.155}$$

and

$$\dot{\bar{\mathbf{y}}}(0) = \bar{\mathbf{U}}^{-1}\dot{\mathbf{x}}(0) = \mathbf{A}_P^{-1}\mathbf{V}^T\mathbf{M}^{1/2}\dot{\mathbf{x}}(0) \tag{3.156}$$

Equations 3.155 and 3.156 can be rewritten mode by mode as

$$\bar{y}_i(0) = u_{ni}\boldsymbol{v}_i^T\mathbf{M}^{1/2}\mathbf{x}(0) \tag{3.157}$$

and

$$\dot{\bar{y}}_i(0) = u_{ni}\boldsymbol{v}_i^T\,\mathbf{M}^{1/2}\dot{\mathbf{x}}(0) \tag{3.158}$$

Here, $\bar{y}_i$ is the ith element of the vector $\bar{\mathbf{y}}(t)$ and $\boldsymbol{v}_i$ is the ith column of the orthonormal eigenvector matrix **V**.

The corresponding equation for the newly defined variable $\bar{y}_i(t)$ can be obtained through Equation 3.138. That is,

$$\alpha_i\ddot{\bar{y}}_i(t) + 2\xi_i\omega_{ni}\alpha_i\dot{\bar{y}}_i(t) + \omega_{ni}^2\alpha_i\bar{y}_i(t) = 0, \quad i = 1, 2,\ldots, n$$

or

$$\frac{1}{u_{ni}}\ddot{\bar{y}}_i(t) + 2\xi_i\omega_{ni}\frac{1}{u_{ni}}\dot{\bar{y}}_i(t) + \omega_{ni}^2\frac{1}{u_{ni}}\bar{y}_i(t) = 0, \quad i = 1, 2,\ldots, n \tag{3.159}$$

Equation 3.159 is also the free-decay equation for a typical SDOF vibration system with linear viscous damping. However, as a difference from Equation 3.138, the modal mass is obtained in Equation 3.159.

$$\alpha_i = \frac{1}{u_{ni}}$$

and the corresponding solution can be written as

$$\bar{y}_i(t) = \bar{y}_{i0}e^{-\xi_i\omega_{ni}t}\sin(\omega_{di}t + \varphi_i), \quad i = 1, 2,\ldots, n \tag{3.160}$$

where $\bar{y}_{i0}$ is the amplitude of the modal displacement. Therefore,

$$\bar{y}_i(0) = (\mathbf{m}_n\boldsymbol{\nu}_i)\boldsymbol{\nu}_i^T\mathbf{M}^{1/2}\mathbf{x}(0) \tag{3.161}$$

and

$$\dot{\bar{y}}_i(0) = (\mathbf{m}_n\boldsymbol{\nu}_i)\boldsymbol{\nu}_i^T\mathbf{M}^{1/2}\,\dot{\mathbf{x}}(0) \tag{3.162}$$

Here, $\mathbf{m}_n$ is defined in Equation 3.147.

Now, using Equations 1.33 and 1.34, the formulas for the amplitude $\bar{y}_{i0}$ and the phase φ_i can be written as

$$\bar{y}_{i0} = \frac{1}{\omega_{di}}\sqrt{\left[(\mathbf{m}_n\boldsymbol{\nu}_i)\boldsymbol{\nu}_i^T\mathbf{M}^{1/2}\,\dot{\mathbf{x}}(0) + \xi_i\omega_{ni}(\mathbf{m}_n\boldsymbol{\nu}_i)\boldsymbol{\nu}_i^T\mathbf{M}^{1/2}\mathbf{x}(0)\right]^2 + \left[\omega_{di}(\mathbf{m}_n\boldsymbol{\nu}_i)\boldsymbol{\nu}_i^T\mathbf{M}^{1/2}\mathbf{x}(0)\right]^2} \tag{3.163}$$

$$\varphi_i = \tan^{-1}\left[\frac{\omega_{di}(\mathbf{m}_n\boldsymbol{\nu}_i)\boldsymbol{\nu}_i^T\mathbf{M}^{1/2}\mathbf{x}(0)}{(\mathbf{m}_n\boldsymbol{\nu}_i)\boldsymbol{\nu}_i^T\mathbf{M}^{1/2}\dot{\mathbf{x}}(0) + \xi_i\omega_{ni}(\mathbf{m}_n\boldsymbol{\nu}_i)\boldsymbol{\nu}_i^T\mathbf{M}^{1/2}\mathbf{x}(0)}\right] + h_{\varphi i}\pi \tag{3.164}$$

where, when the arctangent is calculated by limiting the values from $-\pi/2$ to $+\pi/2$, $h_{\varphi i}$ is defined as

$$h_{\varphi i} = \begin{cases} 0, & (\mathbf{m}_n\boldsymbol{\nu}_i)\boldsymbol{\nu}_i^T\mathbf{M}^{1/2}\dot{\mathbf{x}}(0) + \xi_i\omega_{ni}(\mathbf{m}_n\boldsymbol{\nu}_i)\boldsymbol{\nu}_i^T\mathbf{M}^{1/2}\mathbf{x}(0) \geq 0 \\ 1, & (\mathbf{m}_n\boldsymbol{\nu}_i)\boldsymbol{\nu}_i^T\mathbf{M}^{1/2}\dot{\mathbf{x}}(0) + \xi_i\omega_{ni}(\mathbf{m}_n\boldsymbol{\nu}_i)\boldsymbol{\nu}_i^T\mathbf{M}^{1/2}\mathbf{x}(0) < 0 \end{cases} \tag{3.165}$$

Note that if the mode shape $\mathbf{U}$ is used instead of $\bar{\mathbf{U}}$, the modal equation will be reduced to Equation 3.138, and according to Equations 3.145 and 3.146

$$y_{i0} = \frac{1}{\omega_{di}}\sqrt{\left[\boldsymbol{\nu}_i^T\mathbf{M}^{1/2}\dot{\mathbf{x}}(0) + \xi_i\omega_{ni}\boldsymbol{\nu}_i^T\mathbf{M}^{1/2}\mathbf{x}(0)\right]^2 + \left[\omega_{di}\boldsymbol{\nu}_i^T\mathbf{M}^{1/2}\mathbf{x}(0)\right]^2} \tag{3.166}$$

and

$$\varphi_i = \tan^{-1}\left[\frac{\omega_{di}\boldsymbol{\nu}_i^T\mathbf{M}^{1/2}\mathbf{x}(0)}{\boldsymbol{\nu}_i^T\mathbf{M}^{1/2}\dot{\mathbf{x}}(0) + \xi_i\omega_{ni}\boldsymbol{\nu}_i^T\mathbf{M}^{1/2}\mathbf{x}(0)}\right] + h_{\varphi i}\pi \tag{3.167}$$

Example 3.4

Suppose there is an **M-C-K** system with

$$\mathbf{M} = \begin{bmatrix} 0.3333 & 0.1667 \\ 0.1667 & 0.1667 \end{bmatrix}, \quad \mathbf{C} = \begin{bmatrix} 0.3667 & -0.0167 \\ -0.0167 & 0.1833 \end{bmatrix}, \quad \mathbf{K} = \begin{bmatrix} 20 & -10 \\ -10 & 10 \end{bmatrix}$$

and initial conditions:

$$\mathbf{x}(0) = \begin{Bmatrix} 1 \\ 0 \end{Bmatrix}, \quad \dot{\mathbf{x}}(0) = \begin{Bmatrix} 0 \\ -1 \end{Bmatrix}.$$

The quantities $\mathbf{M}^{1/2}$, $\mathbf{M}^{-1/2}$, $\tilde{\mathbf{K}}$, $\mathbf{V}$, $\mathbf{\Omega}^2$, $\mathbf{U}$, and $\bar{\mathbf{U}}$ can be calculated as

$$\mathbf{M}^{1/2} = \begin{bmatrix} 0.5477 & 0.1826 \\ 0.1826 & 0.3652 \end{bmatrix}, \quad \mathbf{M}^{-1/2} = \begin{bmatrix} 2.1912 & -1.0958 \\ -1.0958 & 3.2864 \end{bmatrix}, \quad \tilde{\mathbf{K}} = \begin{bmatrix} 156 & -168 \\ -168 & 204 \end{bmatrix}$$

$$\mathbf{V} = \begin{bmatrix} 0.7555 & -0.6552 \\ 0.6552 & 0.7555 \end{bmatrix}, \quad \Omega^2 = \mathrm{diag}([10.2944 \;\; 349.7056]), \quad \mathbf{U} = \begin{bmatrix} 0.9374 & 2.2630 \\ 1.3257 & -3.2004 \end{bmatrix}$$

$$\bar{\mathbf{U}} = \begin{bmatrix} 0.7071 & -0.7071 \\ 1.0 & 1.0 \end{bmatrix}$$

Therefore, $\omega_{n1} = 3.2085$ (rad/s) and $\omega_{n2} = 18.7004$ (rad/s), while $u_{21} = 1.3257$ and $u_{22} = 3.2004$. Furthermore, the damping ratios can be found as $\xi_1 = 0.094$ and $\xi_2 = 0.1069$, so that $\omega_{d1} = 3.1943$ (rad/s) and $\omega_{d2} = 18.5933$ (rad/s). The amplitudes are $\bar{y}_{10} = 0.7415$; $\bar{y}_{20} = 0.7088$, while the phase angles are $\varphi_1 = 1.2650$ and $\varphi_2 = 4.6430$.

The corresponding modal displacements are plotted in Figure 3.3a, where the solid line is the displacement of the first mode and the broken line is that of the second mode.

By using modal superposition, the response displacements can be calculated in the physical domain. In Figure 3.3b, the solid line is the displacement of the first mass and the broken line is that of the second mass.

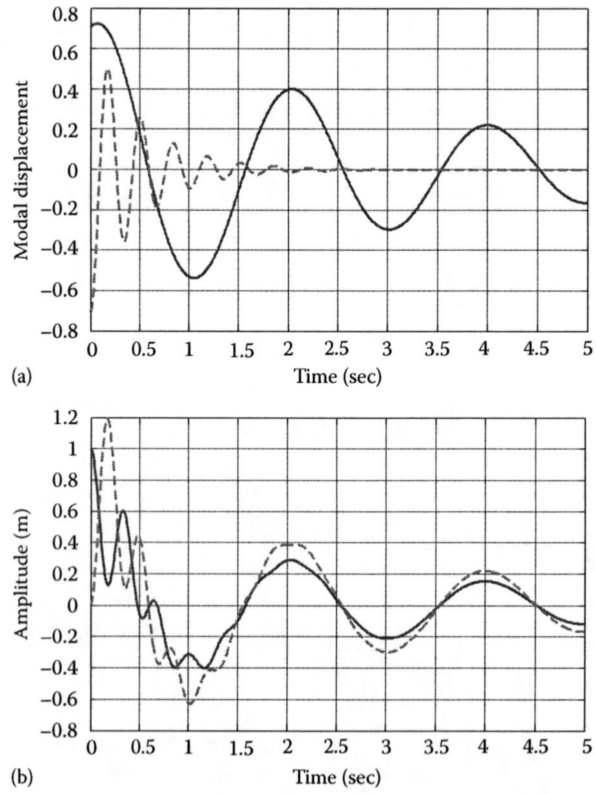

FIGURE 3.3 (a, b) Modal and physical responses.

3.3 MODAL PARTICIPATION AND TRUNCATION

3.3.1 Modal Participation Factor

Practically speaking, for an n-DOF system, all n modes do not have to be included in response computations, as described in Equation 3.137. Usually, the first several modes will provide sufficient contributions for the responses. That is, the first S modes are often used as follows:

$$x_j(t) = \sum_{i=1}^{S} \bar{u}_{ji} y_i(t) \tag{3.168}$$

by truncating the modal superposition after S terms. Usually, for large systems, the number of S modes can be considerably smaller than the number of total n modes, that is,

$$S << n \tag{3.169}$$

In order to find out exactly how many modes are needed, methods to quantitatively account for the modal truncation are considered. In this section, the concepts of modal participation and modal contribution are used.

Through the use of vector normalization, a quantity is denoted as follows:

$$\Gamma_i = \frac{\bar{\boldsymbol{u}}_i^T \mathbf{MJ}}{\bar{\boldsymbol{u}}_i^T \mathbf{M} \bar{\boldsymbol{u}}_i} \tag{3.170}$$

In Equation 3.170, the term $\Gamma_i \cdot \ddot{x}_g(t)$ is defined as the *modal participation factor for the* i^{th} *mode*, defined by M_{pfi}, while Γ_i is the *unit acceleration load for the* i^{th} *mode.* In the following discussion, for convenience, Γ_i are also called the modal participation factors.

Using the normalized modal mass, damping, and stiffness coefficients described in Equations 3.130 through 3.132, and also using the concept of modal participation described in Equation 3.170, the nonmonic form of the set of n-SDOF equations is

$$m_i \ddot{y}_i(t) + c_i \dot{y}_i(t) + k_i y_i(t) = -\Gamma_i m_i \ddot{x}_g(t), \quad i = 1, 2, \ldots, n \tag{3.171}$$

Equation 3.168 implies that in practical applications, the first S modes usually contribute most to the response. Therefore, all the modal responses described in Equation 3.171 do not have to be calculated. Instead, only the first S modal responses are used. That is, the displacement vector of the structure can be written as

$$\mathbf{x}(t)_{n\times 1} \approx [\bar{u}_1, \bar{u}_2, \ldots, \bar{u}_S]_{nxS} \begin{Bmatrix} y_1(t) \\ y_2(t) \\ \ldots \\ y_S(t) \end{Bmatrix}_{Sx1} \quad S < n \tag{3.172}$$

and the acceleration vector of the structure is

$$\ddot{\mathbf{x}}(t)_{n\times 1} \approx [\bar{u}_1, \bar{u}_2, \ldots, \bar{u}_S]_{nxS} \begin{Bmatrix} \ddot{y}_1(t) \\ \ddot{y}_2(t) \\ \ldots \\ \ddot{y}_S(t) \end{Bmatrix}_{Sx1} \quad S < n \tag{3.173}$$

In Equations 3.172 and 3.173, since only the first S modes are used, the procedure is referred to as *modal truncation;* the subscript $n \times S$ and $S \times 1$ are used to denote the corresponding dimensions of matrices and vectors. Here, for simplicity, $\mathbf{x}$ and $\ddot{\mathbf{x}}$ are used to denote the modally truncated displacement and the acceleration without any further specific notation.

Equation 3.172 can be written in matrix form as

$$\mathbf{x}(t) \approx \bar{\mathbf{U}}_C \bar{\mathbf{y}}_C \tag{3.174}$$

Here $\bar{\mathbf{U}}_C = [\bar{\boldsymbol{u}}_1, \bar{\boldsymbol{u}}_2, \ldots, \bar{\boldsymbol{u}}_S]_{nxS}$ is the truncated mode shape matrix and $\bar{\mathbf{y}}_C$ is the truncated modal response, such that

$$\bar{\mathbf{y}}_C = \begin{Bmatrix} \bar{y}_1(t) \\ \bar{y}_2(t) \\ \cdots \\ \bar{y}_S(t) \end{Bmatrix}_{S\times 1} \tag{3.175}$$

In many cases, only the first modal response is used, which is called the fundamental modal response to represent the displacement, that is,

$$\mathbf{x}(t) \approx \bar{\boldsymbol{u}}_1 \bar{y}_1(t) \tag{3.176}$$

and,

$$\ddot{\mathbf{x}}(t) \approx \bar{\boldsymbol{u}}_1 \ddot{\bar{y}}_1(t) \tag{3.177}$$

3.3.2 Modal Contribution Indicator

In order to determine if a specific mode should be used, criteria that are generally referred to as modal participation indicators or *modal contribution indicators* are needed. That is, these indicators will be used to establish if a specific peak response of a structure is sufficiently accurate when truncated modal superposition is used. The number of truncated modes is determined by the values of the indicators. For this purpose, Wilson (2004) suggested the modal mass ratio, which is helpful for proportionally damped systems. Chopra (2006) suggested another indicator called the modal contribution factor. In the following discussion, practical approaches that will be valid for both proportionally and nonproportionally damped systems are explored.

Before the discussion on the selection of various parameters that can be used as the modal contribution indicators, the criterion of selection is first described. There are two basic criteria for selection of the indicators.

3.3.2.1 Theory of the Indicator

The existence and application of an indicator should be scientifically sound, which means that the possible candidates should be mathematically consistent and legitimate.

Chopra (2006) states that the modal contribution indicators should have the following three properties:

i. The indicators should be dimensionless.
ii. The indicators should be independent of how the mode shapes are normalized.
iii. The sum of the modal contribution indicators over all modes should be unity.

As a matter of fact, these three properties have the same necessary essence. Namely, the indicators should be referenced by a given standard and the most convenient standard is unity. For example, the quantity of the modal participation factor cannot be directly used as the indicator. This is because it cannot satisfy the above conditions. Thus, it is necessary to find alternative quantities to determine the number of modes for the modal truncations.

However, the quantity used as an indicator of a specific mode should also be an amount of the percentage quantitatively describing the magnitude of the contribution. That is, if a specific mode has a larger contribution, the indicator should be proportionally larger. It is understood that the structural responses under earthquake excitations are dynamic quantities, which are the result of the convolutions of the ground excitations and the imposed response functions of structures.

Practically speaking, several additional modes are often acquired to improve the accuracy of the response computation, namely,

$$S = S_P + S_f << n \tag{3.178}$$

where S is the number of truncated modes in the response computation, S_p is the number calculated from various modal participation indicators, and S_f is the number of a few additional modes. Usually,

$$S_f \geq 1 \tag{3.179}$$

The greater the irregularity of a structure, the larger the S_f that should be considered. The concept of structural irregularity will be discussed later.

Using Equation 3.179, the burden of considering the influence of the excitations is removed. Thus, a single equation can be used to cover the essence of conditions (i) to (iii) listed above. Note that is, if the ith generic modal participation indicator is denoted as γ_i, then

$$\sum_{i=1}^{n} \gamma_i = 1 \tag{3.180}$$

Using the quantity γ_i, the idea that the larger the value of γ_i, the greater the contribution should be, is explored, and practically, individual modes may not be counted; instead, the summation of the first several indicators can be compared and a preset value G can be used as the criterion, that is, if

$$\sum_{i=1}^{S_P} \gamma_i \geq G \tag{3.181}$$

then the number S_p is specified.

Yet, the equations or the above-mentioned three requirements are not sufficient. In addition, it is best that the indicators are all nonnegative numbers. That is,

$$\gamma_i \geq 0, \quad i = 1, 2, \ldots, S \tag{3.182}$$

First, this is because the modal participations or contributions are counted by using the concept that the larger the value of γ_i, the greater the contribution. This concept therefore implies the use of absolute values. Secondly, if some of the indicators become negative, then the summation of the first

S_p modal participation indicators will not monotonically increase, and will not be convenient to use. That is, if there are two numbers for the required modes, namely, S_2 and S_1, it may be required that

$$\sum_{i=1}^{S_2} \gamma_i \geq \sum_{i=1}^{S_1} \gamma_i, \quad S_2 > S_1 \tag{3.183}$$

3.3.2.2 Realization of Indicators

The second aspect for the selection of indicators is related to using the indicator in practical design. First, it is noticed that to compute different types of responses, different numbers of modes may be needed to guarantee the accuracy of modal truncation.

Secondly, it is understood that computing responses at different locations may require different numbers of modes to guarantee the accuracy of modal truncation.

It is also noted that the consideration of accuracy of modal truncation is related to the dynamic behavior of a structure. The response due to the Duhamel convolution contains two factors, namely, the structure itself and the external excitation. In this case, any indicator that does not involve the factors of the ground excitation cannot be absolutely precise. Instead of seeking more accurate modal contributions by including earthquake, as well as other excitations, Equation 3.178 is used and several additional modes are included.

In addition, to calculate the modal participation indicator, the fewer pieces of information needed, the easier it will be to obtain the quantity.

3.3.2.2.1 Modal Mass Ratio

Having discussed the criteria for selecting the modal contribution factors, one of the oldest parameters, the concept of *modal mass ratio* (Wilson 2004), is considered. It is defined as

$$\gamma_{mi} = \frac{\left[\bar{\boldsymbol{u}}_i^T \mathbf{MJ}\right]^2}{\bar{\boldsymbol{u}}_i^T \mathbf{M} \bar{\boldsymbol{u}}_i \sum_{j=1}^{n} m_j} = \frac{\Gamma_i^2 \, \bar{\boldsymbol{u}}_i^T \mathbf{M} \bar{\boldsymbol{u}}_i}{M_\Sigma} = \frac{m_{effi}}{M_\Sigma} \tag{3.184}$$

where

$$M_\Sigma = \sum_{j=1}^{n} m_j \tag{3.185}$$

is the *total mass* of the structure. Here, it is noted that the mass matrix is taken to be diagonal (see Equation 3.187). Let

$$m_{effi} = \frac{\left[\bar{\boldsymbol{u}}_i^T \mathbf{MJ}\right]^2}{\bar{\boldsymbol{u}}_i^T \mathbf{M} \bar{\boldsymbol{u}}_i} \tag{3.186}$$

which is called the *effective mass* of the ith mode. Note that if the generic mode shape $\boldsymbol{u}_i$ is used to replace $\bar{\boldsymbol{u}}_i$, Equation 3.186 is still valid.

It can be seen that the modal mass ratio will satisfy all four conditions described in Equations 3.180 through 3.183. Plus, to obtain the modal mass ratio, only the mass matrix and the mode shapes of the first few modes are needed. Thus, the modal mass ratio can be a good index to indicate the proportion of the contribution of the corresponding mode. It is also easy to use for practical engineers.

In later chapters on practical damper designs, a preset value to compare the first S summations of the modal mass ratios is provided. The idea is that if the modal mass ratio is large enough, which implies that this particular mode will contribute significantly to the total responses, the mode should be considered. Otherwise, it can be dismissed.

In more general cases, there can also be complex-valued mode shapes or complex modes, as well as overdamped pseudo modes. The corresponding modal participation factor and modal mass ratio will be more complicated. However, Equation 3.180 will still hold, although conditions described in Equations 3.181 through 3.183 may be violated.

Note that in practice, it is often assumed that the mass matrix is diagonal, that is,

$$\mathbf{M} = \mathrm{diag}\left(m_j\right) \quad j = 1, 2, \ldots, n \tag{3.187}$$

Equation 3.187 is used in NEHRP 2000 (BSSC 2000) and many other codes. Mathematically, however, Equation 3.187 is not a necessary condition to use in defining the modal participation factor and modal mass ratio. Thus, in the following derivations, a diagonal mass matrix is not required, unless specifically stated.

3.3.2.2.2 Other Indicators

In addition to the modal mass ratio, several other kinds of indicators exist. For example, the modal contribution factor γ_{Ci} is given by

$$\gamma_{Ci} = \frac{r_n^{st}}{r^{st}} \tag{3.188}$$

where r_n^{st} and r^{st} are respectively the static response of the ith mode and of the total external force, (Chopra, 2006).

As a brief summary, modal truncation can save significant computational time and provide sufficiently accurate response estimations, with the proper number of selected modes, which can be determined by modal contribution indicators. The modal mass ratio is a comparatively better and simpler criterion; this indicator is used in the practical damper designs discussed in Chapters 7 and 8.

3.3.3 Response Computation of Truncated Modal Superposition

In this section, examples are used to demonstrate the procedure of modal truncation as well as to compare the modal contribution indicators described above.

3.3.3.1 Computation Procedure

Suppose the number S is obtained from Equation 3.178 through a certain quantity of the modal participation indicators. It is now possible to compute the structural responses by the truncated S modes. In the following discussion, a method to carry out the truncated modal superposition is explained in detail. The response computation is also used to compare the above-mentioned modal participation indicators to examine their accuracy.

Note that the modal truncation discussed here is for proportionally damped systems only. In other words, all the modes of concern are normal modes. The system with complex modes and/or overdamped subsystems will be discussed in Chapter 4.

The procedure of modal decoupling for proportionally damped systems was theoretically explained in the previous section. It was shown that, generally, the decoupled systems, namely, the individual modes, can be used to compute the modal responses, such as described in Equation

3.122. After these modal responses are computed, the modal superposition described in Equation 3.168 can be used.

To carry out the modal decomposition in practice, the first step is to find the modal parameters of interest. To compute the modal participation indicators, the number of modes that should be considered needs to be estimated. Usually, a few more modes are estimated to meet the requirement in Equation 3.178. There are various procedures of modal decomposition or modal analysis. Although they look quite different, the essence of the modal parameters yielded is essentially the same. In the following discussion, one of the popular procedures is used to illustrate this concept.

1. Make sure the system is proportionally damped. This can be done by checking to see if the Caughey criterion (Equation 3.102) is satisfied.
2. If Equation 3.102 holds, then Equation 3.25 is used to compute the corresponding eigenvalues and mode shapes of matrix $\mathbf{M}^{-1}\mathbf{K}$. Note that the eigenvalues are the squares of the natural frequencies. Namely, Equation 3.25 is rewritten as follows:

$$\mathbf{M}^{-1}\mathbf{K}\bar{\boldsymbol{u}}_i = \omega_{ni}^2\bar{\boldsymbol{u}}_i \quad 1 \le i \le S$$

In earthquake engineering, there is another criterion for determining the order of S. That is, the natural frequency $f_{ni} = \omega_{ni}/2\pi$ is less than 33(Hz), namely,

$$f_{ni} \le 33 \quad (\text{Hz})$$

Many commercially available computational software programs can provide the modal parameters, such as the natural frequencies ω_{ni}, damping ratios ξ_i, and mode shapes $\bar{\boldsymbol{u}}_i$.
3. Use the proper modal contribution indicators to determine the number of modes, S. For this, the modal mass ratio described in Equation 3.184 can be used.

Note that to determine the modal contribution indicators, the mode shape $\mathbf{p}_i$ is not necessarily normalized with respect to special standards at this time. To obtain the number S_p, the preset value G is often required. For regular structures, it can be taken as 85%–90%. For irregular structures, G should preferably be greater than 95%. That is, Equation 3.181 may be rewritten as

$$\sum_{i=1}^{S_p} \gamma_{mi} \ge G = 95\%$$

Once the number S_p is obtained, Equations 3.178 and 3.179 are used to add a few more modes.
4. Equation 3.113 is used to normalize the mode shape. Note that for general computation, this step is not necessary.
5. Equation 3.188 is used to find the modal damping coefficient $2\xi_i \cdot \omega_{ni}$, that is,

$$2\xi_i\,\omega_{ni} = \frac{\bar{\boldsymbol{u}}_i^T\left(\mathbf{M}^{-1}\mathbf{C}\right)\bar{\boldsymbol{u}}_i}{\bar{\boldsymbol{u}}_i^T\bar{\boldsymbol{u}}_i}, \quad i = 1, 2, \ldots, S \tag{3.189}$$

6. Two methods can be used to compute the modal responses. First, the following method can be used to compute the response mode by mode. That is, rewrite Equation 3.122 as

$$\ddot{y}_i(t) + 2\xi_i\omega_{ni}\dot{y}_i(t) + \omega_{ni}^2 y_i(t) = -\frac{\boldsymbol{u}_i^T\mathbf{MJ}}{\boldsymbol{u}_i^T\mathbf{M}\boldsymbol{u}_i}\ddot{x}_g(t), \quad i = 1, 2, \ldots, S \tag{3.190}$$

By solving Equation 3.190, all the modal responses $\bar{y}_i(t)$ of interest can be obtained. Note that in many codes, the damping ratio ξ_i of a structure before adding any dampers is assumed to be 5%. After adding damping, ξ_i is increased. For damper design, ξ_i is often specified with a given value, e.g., 15%. In this case, this value can be used along with Equation 3.189 to compute the modal responses. For the problem when damping parameters other than the damping ratio are given, and it is necessary to find ξ_i, Equation 3.189 can be used to determine the value of the damping ratio.

Secondly, state equations can be used to simultaneously compute the first S modal responses. For the truncated mode approach, the state matrix with dimension 2S × 2S is needed, which can be constructed as follows:

$$\mathcal{A}_C = \begin{bmatrix} -\text{diag}(2\xi_i\omega_{ni})_{S\times S} & -\text{diag}(\omega_{ni}^2)_{S\times S} \\ \mathbf{I} & \mathbf{0} \end{bmatrix}_{2S\times 2S} \tag{3.191}$$

Here **I** and **0** are the identity and null matrices with dimension S × S.

The input matrix $\mathcal{B}_C$ is also necessaDry, where

$$\mathcal{B}_C = -\begin{bmatrix} \left(\bar{\mathbf{U}}_C^T\mathbf{M}\bar{\mathbf{U}}_C\right)^{-1} \bar{\mathbf{U}}_C^T\mathbf{MJ} \\ \mathbf{0} \end{bmatrix} \tag{3.192a}$$

where $\bar{\mathbf{U}}_C$ is the truncated mode shape matrix with dimension n × S mentioned previously. Note that in Equation 3.192a, it is not necessary to compute the inverse of $\bar{\mathbf{U}}_C^T\mathbf{M}\bar{\mathbf{U}}_C$, which increases the computational burden. The following formula can be used:

$$\mathcal{B}_C = -\begin{bmatrix} \dfrac{\bar{\boldsymbol{u}}_1^T\mathbf{MJ}}{\bar{\boldsymbol{u}}_1^T\mathbf{M}\bar{\boldsymbol{u}}_1} \\ \dfrac{\bar{\boldsymbol{u}}_2^T\mathbf{MJ}}{\bar{\boldsymbol{u}}_2^T\mathbf{M}\bar{\boldsymbol{u}}_2} \\ \cdots \\ \dfrac{\bar{\boldsymbol{u}}_S^T\mathbf{MJ}}{\bar{\boldsymbol{u}}_S^T\mathbf{M}\bar{\boldsymbol{u}}_S} \\ 0 \\ 0 \\ \cdots \\ 0 \end{bmatrix}_{2S\times 1} \tag{3.192b}$$

Using the state and the input matrices, the following state equations can be obtained:

$$\dot{\bar{\boldsymbol{Y}}} = \mathcal{A}_C\bar{\boldsymbol{Y}} + \mathcal{B}_C\,\ddot{x}_g \tag{3.193}$$

where $\bar{\boldsymbol{Y}}$ is a state variable:

$$\bar{\boldsymbol{Y}} = \begin{Bmatrix} \dot{\bar{\mathbf{y}}}_C(t) \\ \bar{\mathbf{y}}_C(t) \end{Bmatrix}_{2S\times 1} \tag{3.194}$$

and $\bar{\mathbf{y}}_C$ is defined in Equation 3.175.

Note that most commercially available software uses the second method to solve for all the modal responses of interest.

7. Equation 3.168 is rewritten as follows to carry out the truncated modal superposition:

$$\mathbf{x}_C(t) = \bar{\mathbf{U}}_C \bar{\mathbf{y}}_C(t) = \sum_{i=1}^{S} \bar{\boldsymbol{u}}_i \bar{y}_i(t) \tag{3.195}$$

Note that in this case, the dimension of the modal shape matrix $\bar{\mathbf{U}}_C$ is $n \times S$.

In the above discussion, it was shown how to use truncated modal superposition to calculate the approximate time histories. In the following discussion, examples are used to examine the quantities to determine the modal contribution introduced above, by comparing their values and the percentages of the peak responses of the truncated and the total modal superposition.

Example 3.5

A 10-story and one-bay shear frame of a building model is used as a simulation example to examine the modal mass ratio and the modal contribution factor. The model structure is shown in Figure 3.4. The material masses of the columns are ignored, the rotation DOFs around the axis Y are restrained, and the floors are assumed to be rigid, so that only X-transverse DOFs are available. In addition, the shear effect is restrained.

Let the mass of the first floor be 800 (t), the second be 700 (t), the ninth be 600 (t), and the rest floors be 500 (t). That is,

$$\mathbf{M} = \text{diag}[800,\ 700,\ 500,\ \ldots\ 500,\ 600,\ 500]\ \ (t)$$

The interfloor stiffness along the height is identical, which means $k_1 = k_2 = ,\ldots, k_{10} = k$. Let k = 1.5e6 (kN/m), so that the corresponding stiffness is

$$\mathbf{K} = 1.5\times10^6 \begin{bmatrix} 2 & -1 & \cdots & 0 \\ -1 & 2 & \cdots & 0 \\ & \cdots & & \\ 0 & \cdots & -1 & 1 \end{bmatrix}_{10\times10} \quad (\text{kN/m})$$

From the mass and stiffness matrices, the natural frequencies ω_{ni} and mode shapes $\mathbf{U}$ are calculated, which are not affected by the damping matrix. To compute the response, however, the damping matrix is needed. Assume that there is a 2% damping ratio for every mode. The corresponding damping matrix can be constructed as

$$\mathbf{C} = \mathbf{MU}\text{diag}(2\xi_i\omega_{ni})\mathbf{U}^{-1} = \mathbf{MU}\text{diag}(0.1\omega_{ni})\mathbf{U}^{-1} \quad (\text{kN-s/m}) \tag{3.196}$$

The natural frequencies of this structure are listed in Table 3.1.

Based on Equation 3.184, the modal mass ratios are summarized in Table 3.2.

From Table 3.2, it is seen that all the modal mass ratios and the static modal energy ratios are positive values, but smaller than unity. The modal mass ratio of the first mode is less than 82%. The summation of the first two ratios is 93.3%. The summation of the first three ratios is 97.40%. If a preset criterion requires the summation to be not less than 90%, then only the first two modes can sufficiently represent the modal contribution. On the other hand, if

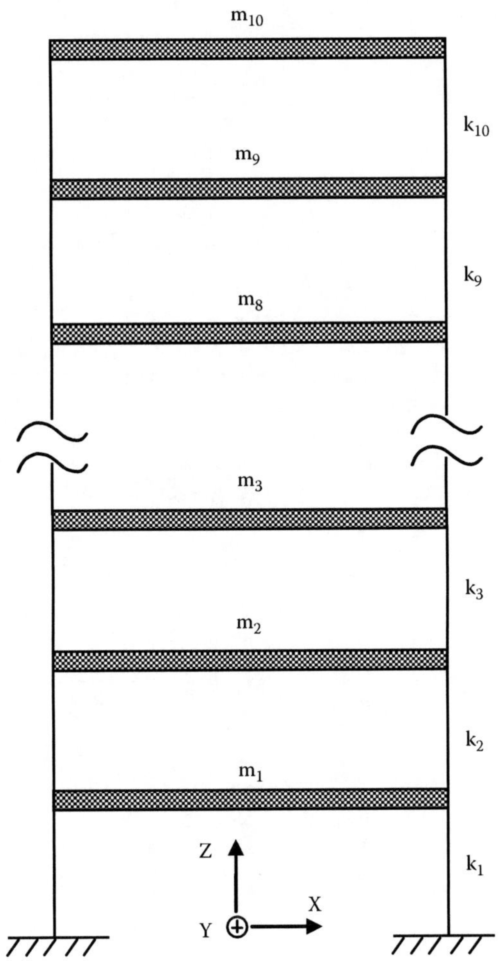

FIGURE 3.4 Example of a 10-story building.

a preset criterion requires the summation to be not less than 95%, then the first three modes are needed.

The summation of the modal mass ratio vs. the number of modes selected in the order from the lower to higher modes is plotted in Figure 3.5. It is seen that the curve increases monotonically.

The ground motion record of the El Centro (1940) earthquake is used to compute the floor displacement relative to the ground and the data are used to examine how the above modal participation indicators work. The responses are computed by using the aforementioned procedure

TABLE 3.1
Natural Frequencies

Mode	1st	2nd	3rd	4th	5th	6th	7th	8th	9th	10th
(Hz)	1.2742	3.7012	5.9520	8.2220	10.2916	11.8749	13.3296	14.8394	16.1904	17.1109

TABLE 3.2
Modal Mass Ratio

Mode	1st	2nd	3rd	4th	5th	6th	7th	8th	9th	10th
γ_{mi}	0.8151	0.1179	0.0409	0.0143	0.0074	0.0035	0.0007	0.0001	0.0000	0.0000
$\Sigma\gamma_{mi}$	0.8151	0.9330	0.9739	0.9882	0.9956	0.9991	0.9998	0.9999	1.0000	1.0000

from steps 1 through 7. The ratios of the maximum acceleration of the truncated response and the precise response of each floor are listed in Table 3.3. From the second column onward, the ratios of calculated acceleration are compared with the participation of the first mode, the first two modes up to the fifth modes. It is seen that the more modes counted, the closer the response will be to the true values.

The modal mass ratio of the aforementioned cases is checked, and the results are listed in Table 3.4, where the smallest acceleration and displacement ratios are also provided. From Table 3.4, it is seen that the modal mass ratio provides acceptable estimates.

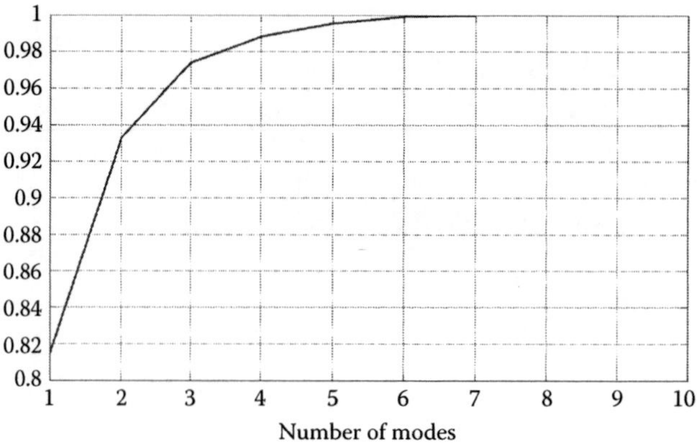

FIGURE 3.5 Summation of modal mass ratio for a regular 10-story building.

TABLE 3.3
Ratio of Floor Acceleration of Truncated Response

	First Mode Only	First 2 Modes	First 3 Modes	First 4 Modes	First 5 Modes
1st	0.7902	0.9881	0.9277	0.9651	0.9827
2nd	0.6618	1.1232	1.0464	0.9769	0.9767
3rd	0.6477	0.9833	0.9701	1.0227	1.0221
4th	0.7015	0.8939	0.8972	0.9965	0.9963
5th	0.8550	0.9056	0.9987	1.0192	1.0016
6th	0.9415	0.8807	1.0063	0.9974	1.0001
7th	0.8678	0.8735	0.9147	0.9862	0.9895
8th	0.8310	0.9524	0.9484	0.9952	0.9976
9th	0.7865	0.9876	0.9964	0.9971	0.9934
10th	0.7070	0.9247	0.9499	0.9844	0.9909

TABLE 3.4
Comparisons of Smallest Acceleration and Displacement Ratio vs. Modal Mass Ratio

	First Mode Only	First 2 Modes	First 3 Modes
Smallest acc. ratio	0.6477	0.8807	0.8972
Smallest disp. ratio	0.9843	0.9961	0.9982
Modal mass ratio	0.8151	0.9331	0.9740

Note that in the above example, the model structure is not very irregular. That is, the structure is symmetric and the distributions of both the mass and the stiffness are rather even. In practical design situations, irregular structures are common. Chapter 4 provides an explanation of structural irregularities, which are referred to as vertical and plan irregularities, as well as weight distribution irregularities. In addition, adding dampers may also add damping irregularities. This will cause the structure to be nonproportionally and/or overly damped, which is discussed in the last section as well as in Chapter 4. The behavior of the modal participation indicators, such as the modal mass ratios of an irregular structure, may be of interest. In Figure 3.6, the summations of the modal mass ratios are plotted for another 10-story building with significant plan and vertical irregularities. As a comparison, the summations of modal contribution factors γ_{Ci} are also plotted in Figure 3.7.

Figure 3.6 shows that the summation of the modal mass ratio increases faster than that of more regular buildings. This means that to have the same amount of modal contribution or to have the same computational accuracy, fewer modes may be needed to participate. That is, to have the modal mass ratio greater than 95%, only two modes are needed. However, this phenomenon may not be true for other irregular structures. Generally speaking, the more irregular the structure, the more modes are needed to have the same computational accuracy.

From Figure 3.7, it is seen that for irregular structures, the summation of the modal contribution factors is to converge to unity more slowly. For example, the modal contribution factors for the displacement responses of the first and third floors are greater than unity up to the seventh mode. For the fourth floor, even the ninth mode is needed to compute the displacement.

A comparison of modal contribution factors and the displacement ratio is given in Table 3.5 and the values of the modal mass ratio are listed in Table 3.6. In Table 3.6, for comparison, the largest percentage error of the displacement is also provided. Using all 10 modes should provide the exact

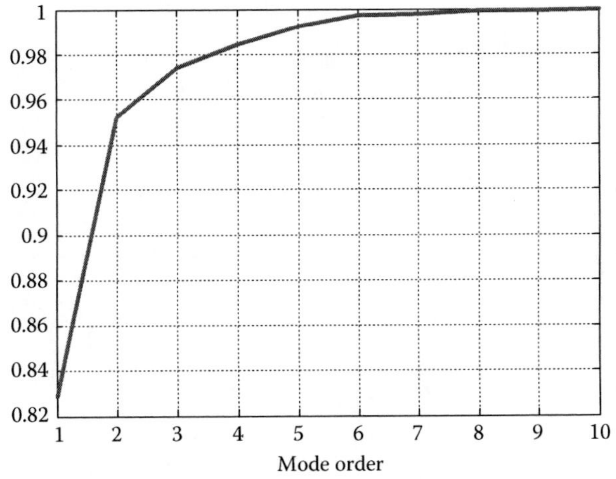

FIGURE 3.6 Summation of modal mass ratio for an irregular 10-story building.

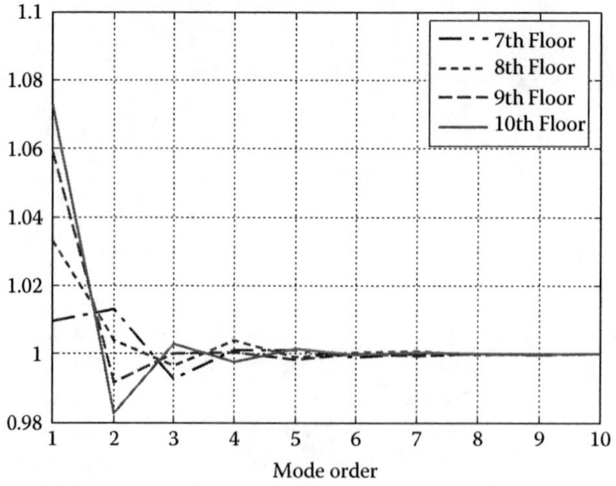

FIGURE 3.7 Summation of modal contribution factors.

response of the displacement. Using the first several modes will result in some error in the computed displacement. The largest error is taken among all the errors of the 10 displacements. It is seen in Table 3.5 that the data do not quite agree with each other, as those given in Table 3.3. The modal contribution factors are used to examine the structural responses in detail at each location. Since the displacement ratio at each floor does not agree with the modal contribution factor, it again implies that the modal mass ratio factor may be a better indicator for use in selecting the number of modes in modal truncations.

TABLE 3.5
Floor Displacement Ratio of Truncated Response

	First Mode Disp. Ratio	First Mode γ_{ci}	First 2 Modes	First 2 Modes γ_{ci}	First 3 Modes	First 3 Modes γ_{ci}
1st	1.0154	1.0730	0.9923	0.9827	1.0045	1.0030
2nd	1.0153	1.0593	0.9979	0.9918	1.0028	1.0000
3rd	1.0136	1.0333	1.0059	1.0040	1.0012	0.9966
4th	1.0124	1.0096	1.0133	1.0128	1.0005	0.9928
5th	0.9946	0.9721	1.0085	1.0223	1.0031	0.9971
6th	0.9764	0.9327	1.0078	1.0238	1.0053	1.0065
7th	0.9567	0.8836	1.0058	1.0085	1.0066	1.0141
8th	0.9170	0.7942	1.0016	0.9811	1.0066	1.0115
9th	0.8931	0.7299	0.9944	0.9371	1.0026	0.9845
10th	0.8017	0.6254	0.9302	0.8701	0.9548	0.9271

TABLE 3.6
Comparisons of Smallest Displacement Ratio vs. Modal Mass

	First Mode	First 2 Modes	First 3 Modes
Percentage error	0.8017	0.9302	0.9548
Modal mass ratio	0.8287	0.9524	0.9739

However, it is seen that in Table 3.6 the values of the modal mass ratio are always greater than the errors, which means that if only the modal mass ratio is used without adding additional modes, say S_f modes, then the result will be unsafe.

3.3.4 Peak Responses

At the preliminary design phase using the modal contribution factors, it is not necessary to check the detailed responses at specific places in a structure. However, their values still need to be estimated. This task can be done through the quantity of peak level of the lateral force applied at level i of the structure. The local lateral force can be seen as the product of mass at level i, denoted by m_i, and the amplitude of the ith absolute acceleration, $x_i(t)+\ddot{x}_g(t)$.

Namely, in aseismic design, the peak responses are a common quantity. By using the modal participation factor, the peak response can be computed mode by mode.

Compare an SDOF system described by the governing equation as

$$m\ddot{x}(t)+c\dot{x}(t)+kx(t)=-m\ddot{x}_g(t) \tag{3.197}$$

with the ith mode of the MDOF system described by the governing equation as

$$m_i\ddot{y}_i(t)+c_i\dot{y}_i(t)+k_iy_i(t)=-\Gamma_i m_i\ddot{x}_g(t) \tag{3.198}$$

It is seen that Equation 3.198 can be used to determine the modal responses. On the other hand, from Equations 2.334 and 2.337, it was also seen that by using the design response spectrum, the peak pseudo acceleration and the peak relative displacement of the SDOF system can be determined, too. Repeated as follows, specific terms a_s, a_a, and d_D are used to denote the spectral acceleration and displacement:

$$a_s = C_s(T,\xi)g \quad (m/s^2) \tag{3.199}$$

where a_s is the pseudo acceleration used to approximate the absolute acceleration a_a.

As previously mentioned, when damping is large, Equation 3.199 should be modified as

$$a_a = \sqrt{1+4\xi^2}C_s(T,\xi)g \quad (m/s^2) \tag{3.200}$$

Furthermore,

$$d_D = C_s(T,\xi)\frac{T^2}{4\pi^2}g \quad (m) \tag{3.201}$$

Note that the unit of acceleration is often taken to be 1 (g) = 9.8 (m/s^2), and the unit of displacement is often taken to be (m). With these, the peak pseudo acceleration of the ith mode of the MDOF system can be written as

$$a_{si} \approx \Gamma_i C_s(T_i,\xi_i) \quad (g) \tag{3.202}$$

Note that when damping is large, Equation 3.202 should be modified as

$$a_{ai} \approx \sqrt{1+4\xi^2}\Gamma_i C_s(T_i,\xi_i) \quad (g) \tag{3.203}$$

Furthermore, the peak relative displacement of the ith mode of the MDOF system is

$$d_{iD} \approx \Gamma_i C_s(T_i, \xi_i) \frac{T_i^2}{4\pi^2} g \quad (m) \tag{3.204}$$

In the above, a_{ai} and d_{iD} can be seen as the spectral acceleration and displacement.

It is noted that Equation 3.203 can only provide the approximate value of the absolute acceleration for the ith mode. Similarly, Equation 3.204 is a rough estimate of the displacement. In fact, the relationship of

$$d_{iD} \approx a_{ai} \frac{T_i^2}{4\pi^2} \tag{3.205}$$

is obtained by assuming that the damping force is negligible. With added damping, this assumption may no longer be valid, and the effect of the factor $\sqrt{1+4\xi^2}$ should be considered. Note that Equations 3.199, 3.202, and 3.205 are widely used for the aseismic method of using the design response spectrum in the earthquake engineering community. In the following, for simplicity, the factor $\sqrt{1+4\xi^2}$ will not be written together with these conventionally used expressions for response.

Once the amplitudes of the modes of interest, which are now taken to be the spectral values, are determined, methods for the summation of these peak accelerations can be applied, such as the method of the *square-root-of-the-sum-of-squares* (SRSS) and/or the *complete quadratic combination* (CQC) (e.g., see Clough and Penzien 1993), which are introduced later.

As a comparison, a more accurate value may be obtained from the specific response history analysis. Thus, the peak level of the total absolute acceleration can be written as

$$\ddot{\mathbf{x}}_A(t) = \left\| [\bar{\boldsymbol{u}}_1, \quad \bar{\boldsymbol{u}}_2, \quad \ldots, \quad \bar{\boldsymbol{u}}_S] \begin{Bmatrix} \ddot{\bar{y}}_{a1}(t) \\ \ddot{\bar{y}}_{a2}(t) \\ \ldots \\ \ddot{\bar{y}}_{aS}(t) \end{Bmatrix} \right\| \tag{3.206}$$

where the term $\ddot{\bar{y}}_{ai}(t)$ is the absolute acceleration of the ith mode with proper normalization. The peak level of the total relative displacement can be written as

$$\mathbf{x}_{(t)} = \left\| [\bar{\boldsymbol{u}}_1, \quad \bar{\boldsymbol{u}}_2, \quad \ldots, \quad \bar{\boldsymbol{u}}_S] \begin{Bmatrix} \bar{y}_1(t) \\ \bar{y}_2(t) \\ \ldots \\ \bar{y}_S(t) \end{Bmatrix} \right\| \tag{3.207}$$

where the term $\bar{y}_i(t)$ is the relative displacement of the ith mode with proper normalization.

3.4 BASE SHEAR AND LATERAL FORCE

If the peak acceleration level is already known, the amplitudes of the lateral force can be written as

$$\mathbf{f}_L = \mathbf{M}\mathbf{a}_a \quad (N) \tag{3.208}$$

where $\mathbf{f}_L$ is a vector whose jth element is the story lateral force of the jth level of the structure. Note that the unit of lateral force is often taken to be (N). In particular, when only the first mode is used, the peak value of the absolute acceleration at the jth level can be written as

$$a_{aj1} = \boldsymbol{e}_j^T \bar{\boldsymbol{u}}_1 \Gamma_1 C_s(T_1, \xi_1) g \quad (m/s^2) \tag{3.209}$$

while the peak value of the relative displacement at the jth level is written as

$$d_{1j} = \boldsymbol{e}_j^T \bar{\boldsymbol{u}}_1 \Gamma_1 C_s(T_1, \xi_1) \frac{T_1^2}{4\pi^2} g \quad (m) \tag{3.210}$$

where T_1 and ξ_1 are the period and the damping ratio of the first mode, and $\boldsymbol{e}_j$, defined in Equation 3.42, is the selection vector with all elements equals to zero, except the jth element, which equals 1. That is, $\boldsymbol{e}_j^T \bar{\boldsymbol{u}}_1 = \bar{u}_{1j}$.

The peak level of the story lateral force at the jth level contributed only by the fundamental mode can then be written as

$$f_{Lj1} = m_j a_{aj1} = m_j \boldsymbol{e}_j^T \bar{\boldsymbol{u}}_1 \Gamma_1 C_s(T_1, \xi_1) g \quad (N) \tag{3.211}$$

The total base shear v(t) is a time vatiable. From Equation 3.208, it is seen that the amplitude of v(t), denoted by V, which is the summation of the amplitude of the total lateral force contributed by all modes, can be written as

$$V = \mathbf{J}^T \mathbf{f}_L \approx \mathbf{J}^T \mathbf{M}\, \mathbf{a}_a \quad (N) \tag{3.212}$$

where $\mathbf{J}$ is the unit vector defined before. Generally speaking, the value $\mathbf{a}_a$ is often obtained through spectral analysis of $\mathbf{a}_{ai}$ with SRSS and/or or CQC methods, which are approximations. In addition, the peak value of accelerations often occurs at different time. Therefore, the computation of V by using the acceleration in Equation 3.212 is only an approximation.

Considering Equations 3.206 and 3.212, results in

$$V = \left\| v(t) \right\| = \mathbf{J}^T \mathbf{M} \left\| \bar{\mathbf{U}}\, \ddot{y}(t) \right\| \quad (N) \tag{3.213}$$

which is based on the time history analysis of the total base shear. Comparing Equations 3.212 and 3.213, the time history analysis provides exact base shear but not the maximum peak value in general cases. In practical design, both methods are used. In the following, the spectral analysis is discussed in a more detailed way.

If only the first mode is considered, the total base shear contributed by the fundamental mode only can be obtained, and is denoted by V_1. Using the spectral analysis,

$$V_1 = \mathbf{J}^T \mathbf{M} \bar{\boldsymbol{u}}_1 \Gamma_1 C_s(T_1, \xi_1) g = \bar{\boldsymbol{u}}_1^T \mathbf{M} \mathbf{J} \Gamma_1 C_s(T_1, \xi_1) g \quad (N) \tag{3.214}$$

Furthermore, the total base shear contributed by the modal response of the ith mode is

$$V_i = \mathbf{J}^T \mathbf{M} \bar{\boldsymbol{u}}_i \Gamma_i C_s(T_i, \xi_i) g = \bar{\boldsymbol{u}}_i^T \mathbf{M} \mathbf{J} \Gamma_i C_s(T_i, \xi_i) g \quad (N) \tag{3.215}$$

Now, the effective mass of the system of the ith mode can be written further as

$$m_{effi} = \frac{\left(\bar{\boldsymbol{u}}_i^T \mathbf{MJ}\right)^2}{\bar{\boldsymbol{u}}_i^T \mathbf{M} \bar{\boldsymbol{u}}_i} = \Gamma_i^2 m_i \quad (\text{kg}) \tag{3.216}$$

With the two notations defined in Equations 3.215 and 3.216, the expression of the story lateral force at the jth level contributed by the fundamental mode only can be rewritten as:

$$f_{L_{1j}} = m_j \bar{u}_{1j} \Gamma_1 C_s\left(T_1, \xi_1\right) \frac{\bar{\boldsymbol{u}}_1^T \mathbf{MJ}}{1} \frac{1}{\bar{\boldsymbol{u}}_1^T \mathbf{MJ}} \frac{1}{\bar{\boldsymbol{u}}_1^T \mathbf{MJ}} \frac{\bar{\boldsymbol{u}}_1^T \mathbf{M} \bar{\boldsymbol{u}}_1}{1} \frac{1}{\bar{\boldsymbol{u}}_1^T \mathbf{M} \bar{\boldsymbol{u}}_1} \frac{\bar{\boldsymbol{u}}_1^T \mathbf{MJ}}{1} g \quad (\text{N}) \tag{3.217}$$

Therefore,

$$f_{L_{1j}} = m_j \bar{u}_{1j} \left[\Gamma_1 C_s\left(T_1, \xi_1\right) \bar{\boldsymbol{u}}_1^T \mathbf{MJ}\right] \left[\frac{\bar{\boldsymbol{u}}_1^T \mathbf{M} \bar{\boldsymbol{u}}_1}{\left(\bar{\boldsymbol{u}}_1^T \mathbf{MJ}\right)^2}\right] \left[\frac{\bar{\boldsymbol{u}}_1^T \mathbf{MJ}}{\bar{\boldsymbol{u}}_1^T \mathbf{M} \bar{\boldsymbol{u}}_1}\right] g \quad (\text{N}) \tag{3.218}$$

In Equation 3.218, the term in the first bracket is the amplitude of the total base shear V_1, the term in the second bracket is the inverse of the effective mass of the first mode, m_{eff1}, and the term in the third bracket is the unit acceleration load for the first mode Γ_1. Therefore, the story lateral force at the jth level contributed by the fundamental mode becomes

$$f_{L_{ij}} = m_j \bar{u}_{1j} \frac{\Gamma_1}{m_{eff1}} V_1 = m_j \bar{u}_{1j} \Gamma_1 C_s g \quad (\text{N}) \tag{3.219}$$

Similarly, the lateral force at the jth level contributed by the ith mode can be written as

$$f_{L_{ij}} = m_j \bar{u}_{ij} \frac{\Gamma_i}{m_{effi}} V_i = m_j \bar{u}_{ij} \Gamma_i C_s g \quad (\text{N}) \tag{3.220}$$

and the total base shear contributed by the ith mode can be rewritten as

$$V_i = \mathbf{J}^T \mathbf{M} \bar{\boldsymbol{u}}_i \Gamma_i C_s\left(T_i, \xi_i\right) g = \bar{\boldsymbol{u}}_i^T \mathbf{MJ} \frac{\bar{\boldsymbol{u}}_i^T \mathbf{MJ}}{\bar{\boldsymbol{u}}_i^T \mathbf{M} \bar{\boldsymbol{u}}_i} C_s\left(T_i, \xi_i\right) g = m_{effi} C_s\left(T_i, \xi_i\right) g \quad (\text{N}) \tag{3.221}$$

If the first S modes must be considered, then the SRSS method can be used to find the modal combination. For simplicity, V and $\mathbf{f}_L = \{f_{Lj}\}$ are used to denote the modally truncated total base shear and story lateral forces, that is,

$$V = \sqrt{\sum_{i=1}^{S} V_i^2} = \sqrt{\sum_{i=1}^{S} m_{effi}^2 C_s^2\left(T_i, \xi_i\right)}\ g \quad (\text{N}) \tag{3.222}$$

and f_{Lj} is the modally truncated story lateral force applied on the jth level,

$$f_{Lj} = \sqrt{\sum_{i=1}^{S} f_{ij}^2} = m_j \sqrt{\sum_{i=1}^{S} \left(\bar{u}_{ij}\Gamma_i\right)^2 C_s^2(T_i,\xi_i)}\ g \quad (N) \tag{3.223}$$

Furthermore, the modally truncated displacement is denoted as $\mathbf{d} = \{d_j\}$, where d_j is the displacement at the jth level of the structure, which can be written as

$$d_j = \sqrt{\sum_{i=1}^{S} d_{ij}^2} = \sqrt{\sum_{i=1}^{S} \left[\frac{T_i^2 C_s(T_i,\xi_i)}{4\pi^2}\right]^2}\ g \quad (m) \tag{3.224}$$

Equations 3.222 and 3.223 can be used to find the number of modes, S. In fact, the concept of modal truncation was previously introduced. To explain the inequality, an example of the case where S = 2 is discussed. From Equation 3.222, in this case, the total base shear is expressed by

$$V = \sqrt{m_{eff1}^2 C_s^2(T_1,\xi_1) + m_{eff2}^2 C_s^2(T_2,\xi_2)}\ g \quad (N) \tag{3.225}$$

Assuming that the spectral amplitude remains constant,

$$V = C_s g m_{eff1}\sqrt{1 + \frac{m_{eff2}^2}{m_{eff1}^2}} \approx C_s g m_{eff1}\left(1 + \frac{m_{eff2}}{2m_{eff1}}\right) = C_s g m_{eff1}\left(1 + \frac{\gamma_{m2}}{2\gamma_{m1}}\right) \quad (N) \tag{3.226}$$

If $\gamma_{m1} = 0.8$, then the value in the last parenthesis in Equation 3.226 is 1.125, which means that if $G = 0.8$, there will be an error of one minus one over 1.125, which is about 11%. In this case, it can be approximated that if the value of G is taken to be 0.8, the result is a fairly good estimation of the modally truncated responses.

Note that, again, if a structure is nonproportionally damped, the above formulation should be modified, which is discussed in Chapter 4.

3.5 NATURAL FREQUENCY AND MODE SHAPE ESTIMATION

3.5.1 Natural Frequency

For damper design, one important parameter is the natural period or its inverse, the natural frequency. In the above discussion, it was seen that the natural frequency of an SDOF system can be simply calculated using Equation 1.12, if the mass and stiffness are known. For MDOF systems, Equation 3.25 can be used. However, practically speaking, in the preliminary design stage, the natural frequency may have to be estimated, since important information may not be available in the earlier stage. Or, for a simplified design, information such as the stiffness matrix may never be used. Therefore, a method to estimate the natural frequencies is needed. The discussion begins with an SDOF system.

As shown in Figure 3.8a, an applied force mg can cause a structural drift p:

$$p = mg/k \tag{3.227}$$

Here, p is used to denote the specific value. In the following discussion, regular letter p are used to denote more general values.

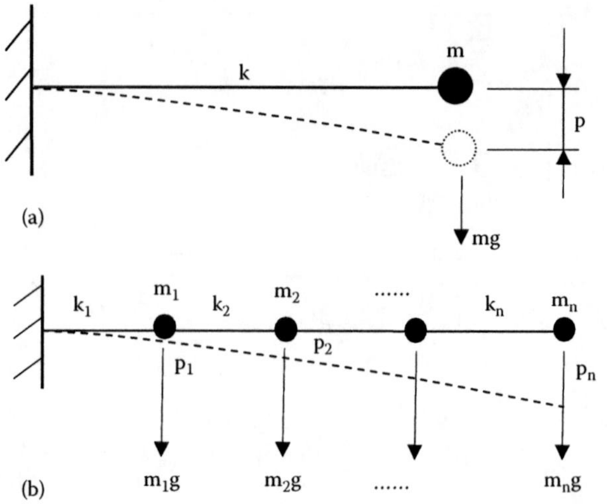

FIGURE 3.8 (a, b) Natural frequency estimation SDOF.

If p happens to be the maximum displacement of mass when the system vibrates in its own natural frequency, then the maximum potential energy stored by k can be expressed as

$$E_p = 1/2\,mgp \tag{3.228}$$

The maximum kinetic energy stored by m is

$$E_k = 1/2\,m(\omega_n p)^2 \tag{3.229}$$

Because $E_p = E_k$,

$$\omega_n = \sqrt{\frac{g}{p}} \tag{3.230}$$

The mass and stiffness matrices **M** and **K** are present in the case of an MDOF system. Since the structure will not have rigid body motion, the structural flexibility matrix, denoted by **S**, can be expressed as

$$\mathbf{S} = \mathbf{K}^{-1} \tag{3.231}$$

Now, a series of lateral forces are applied to the system, as shown in Figure 3.8b. The applied force vector:

$$\mathbf{f} = \{f\} = g\mathbf{MJ} \tag{3.232}$$

will cause the drift vector:

$$\boldsymbol{u} = \{u_i\} = g\mathbf{SMJ} \tag{3.233}$$

From Equation 3.232,

$$\mathbf{J} = \mathbf{M}^{-1}\mathbf{K}\boldsymbol{u}/g \tag{3.234}$$

Assuming that the structural system vibrates approximately with the structural first modal frequency ω_{n1} and modal shape $\boldsymbol{u}$, the maximum potential energy and kinetic energy of the system can be approximated as

$$E_p = 1/2\,\mathbf{f}^T\boldsymbol{u} = 1/2\,g\boldsymbol{u}^T\mathbf{MJ} \tag{3.235}$$

and

$$E_k = 1/2\,\omega_{n1}^2\boldsymbol{u}^T\mathbf{M}\boldsymbol{u} \tag{3.236}$$

Similar to the approach used for an SDOF system, since $E_p = E_k$,

$$1/2\,g\boldsymbol{u}^T\mathbf{MJ} = 1/2\,\omega_{n1}^2\boldsymbol{u}^T\mathbf{M}\boldsymbol{u} \tag{3.237}$$

Therefore,

$$\omega_{n1}^2 = \frac{g\boldsymbol{u}^T\mathbf{MJ}}{\boldsymbol{u}^T\mathbf{M}\boldsymbol{u}} \tag{3.238}$$

and

$$\omega_{n1} = \sqrt{\frac{g\boldsymbol{u}^T\mathbf{MJ}}{\boldsymbol{u}^T\mathbf{M}\boldsymbol{u}}} \quad (\text{rad/s}) \tag{3.239}$$

Equation 3.239 has only the mass matrix, which is comparatively easier to obtain than the stiffness or the flexibility matrices. However, when this equation is derived, the displacement caused by the series of lateral forces is unique. Therefore, the vector $\boldsymbol{u}$ cannot be used as a general mode shape function, which can be normalized with respect to many standards and therefore is not unique. Thus, an italic letter is used to denote this vector. In Chapter 8, a method to determine $\boldsymbol{u}$ in detail is introduced.

If the stiffness matrix **K** is available, then a more accurate formula to estimate the natural frequency using the Rayleigh quotient can be developed. That is,

$$\omega_{n1} = \sqrt{\frac{\boldsymbol{u}_1^T\mathbf{K}\boldsymbol{u}_1}{\boldsymbol{u}_1^T\mathbf{M}\boldsymbol{u}_1}} \quad (\text{rad/s}) \tag{3.240}$$

where instead of the specific deformation vector $\boldsymbol{u}$, $\boldsymbol{u}_1$ is specifically the first mode shape normalized with respect to any standard. From Equation 3.240,

$$\boldsymbol{u}_1^T\mathbf{K}\boldsymbol{u}_1 = \omega_{n1}^2\boldsymbol{u}_1^T\mathbf{M}\boldsymbol{u}_1 \tag{3.241}$$

3.5.2 Mode Shape Estimation

In damper design, another important factor that must be considered is the mode shape. For a simplified approach to modeling the structure and/or for preliminary design, the modal shape can be estimated with a good degree of accuracy.

If a structure has no notable plan and vertical as well as damping irregularities, which are discussed later in more detail, then it can be treated as a simple cantilever beam with evenly distributed mass and stiffness. Therefore, the ith natural period can be described by

$$T_i = \frac{T_1}{2i-1}, \quad i = 1, 2, \ldots, S \tag{3.242}$$

where S is the total number of modes of interest. The jth element of the ith mode shape, u_{ji}, can be described as

$$u_{ji} = (-1)^{i+1} \sin\left[\frac{(2i-1)(N+1-j)\pi}{2N}\right], \quad i = 1, 2, \ldots, S, \quad j = 1, 2, \ldots, N \tag{3.243}$$

Here, N is the number of the stories. Note that in Equation 3.243, the mode shape is normalized by allowing the roof level modal displacement to be equal to unity.

The mode shape vector for the entire ith mode becomes

$$\boldsymbol{u}_i = \{u_{ji}\}, \quad i = 1, 2, \ldots, m, \quad j = 1, 2, \ldots, n \tag{3.244}$$

Example 3.6

An example is given to show the results of estimated mode shapes. With this example, the first modal mass ratio is calculated by using the triangular mode shape, which is 0.7925. This value is slightly smaller than the above-mentioned criterion, so the second mode shape is determined by using Equation 3.239. For comparison purposes, the third and fourth mode shapes are calculated as well.

In Figures 3.9a through d, the actual and estimated first, second, third, and fourth modes shapes, respectively, are shown for the 10-story structure. The stiffness coefficients are evenly distributed, shown in Figure 3.4, and the mass matrix has some irregularities. That is,

$$\mathbf{M} = \begin{bmatrix} 439.9 & & & & & & & & & \\ & 499 & & & & & & & & \\ & & 492.2 & & & & & & & \\ & & & 419.80 & & & & & & \\ & & & & 512.9 & & & & & \\ & & & & & 447.2 & & & & \\ & & & & & & 570.8 & & & \\ & & & & & & & 459.7 & & \\ & & & & & & & & 526.4 & \\ & & & & & & & & & 511.0 \end{bmatrix} \text{(t)}$$

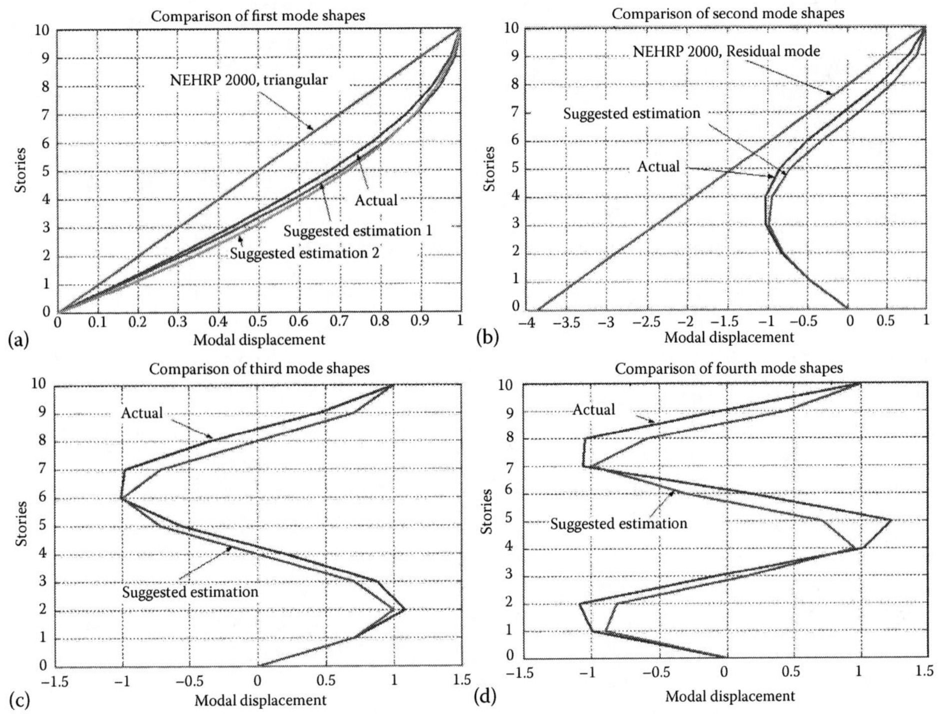

FIGURE 3.9 Mode shapes (a) mode #1, (b) mode #2, (c) mode #3, and (d) mode #4.

It is seen that the maximum variation is about 30%.

In Figure 3.9a, the suggested estimation method 1 is described by Equation 3.243, whereas the suggested estimation method 2 is described later in Chapter 8. In the rest of the plots, the suggested estimations are all described by Equation 3.243. Figure 3.9 shows that the estimated mode shapes using Equation 3.243 can provide good approximations with the actual mode shapes. As a comparison, the mode shapes are plotted with the triangular approach and the corresponding residual modes as suggested by NEHRP 2003/2009. It can be seen that a large error may occur with the triangular approach.

Next, the natural periods are checked. Table 3.7 lists the actual values and those estimated by Equation 3.243. From Table 3.7, it is seen that once the fundamental period is correctly estimated, using Equation 3.243 can provide good results.

By using the estimated mode shapes, the modal participation factors can be computed, which are listed in Table 3.8. From Table 3.8, it is seen that by using the estimated mode shape, except for the fundamental mode, greater errors result. This is one of the limitations of using the simplified approach.

The fundamental modal mass ratios, calculated with the actual and estimated mode shapes, results in $\gamma_{m1} = 85.17\%$ and $\gamma_{m1} = 85.95\%$, respectively. First, these values are close. Second, both values are greater than 80%, whereas using the triangular mode shape, γ_{m1} is less than 80%. The remaining modal participation factors for the "residual mode" obtained by using the actual and

TABLE 3.7
Comparison of Natural Periods

Mode	I	II	III	IV
Actual value (sec)	0.5945	0.1966	0.1213	0.0879
Estimated (sec)	0.5945	0.1982	0.1189	0.0849
Error (%)	0.00	0.79	−2.02	−3.50

TABLE 3.8
Comparison of Modal Participation Factor

Mode	I	II	III	IV
Actual value	1.2590	−0.3981	0.2170	−0.1242
Estimated	1.2354	−0.2427	0.3034	−0.0662
Error (%)	1.87	39.02	−39.85	46.66

estimated mode shapes are, respectively, 14.85% and 14.83%, which are fairly close. However, with the triangular mode shape, the value jumps to 20.75%.

Note that the above example is only used to show the accuracy of the mode shape and natural period estimations. However, the results show qualitatively the trend of using Equations 3.233 and 3.234. Namely, if a structure does not have large plan and vertical or weight irregularities, these equations can provide a fairly good estimate of the mode shapes and natural periods for the first few modes. However, care must be taken in the calculation of modal participation factors, which may yield large errors, except in the fundamental mode. Fortunately, the modal participation factors of higher modes only contribute slightly to the error.

Also note that in Figure 3.9, only the closeness of the estimated mode shapes and the mode shapes of normal modes are shown. Practically speaking, normal modes are rare. Therefore, when an estimated mode shape is not close to a normal mode, it may not necessarily be a good approach. In the next chapter, this issue is discussed in detail.

3.6 COEFFICIENT MATRIX FOR PROPORTIONAL DAMPING

3.6.1 Rayleigh Damping

Several equations represent proportional damping. The simplest one is the *Rayleigh damping*, which expresses that the damping coefficient matrix **C** can be represented by a linear combination of the mass and stiffness matrices. That is,

$$\mathbf{C} = \alpha_0\mathbf{M} + \alpha_1\mathbf{K} \tag{3.245}$$

where α_0 and α_1 are scalars.

In certain circumstances, the proportionality parameters a_0 and a_1 need to be identified, which can be carried out as follows by using the Rayleigh quotient. Substituting Equation 3.245 into Equation 3.116 yields

$$\frac{\boldsymbol{u}_i^T(\alpha_0\mathbf{M} + \alpha_1\mathbf{K})\boldsymbol{u}_i}{\boldsymbol{u}_i^T\mathbf{M}\boldsymbol{u}_i} = (\alpha_0 + \alpha_1\omega_{ni}^2) = 2\xi_i\omega_{ni} \tag{3.246}$$

Equation 3.246 implies that if all the damping ratios and natural frequencies can be located, Equation 3.247, based on the least squares method, can be used to calculate the proportionality coefficients α_0 and α_1. That is,

$$\begin{bmatrix} 1 & \omega_{n1}^2 \\ 1 & \omega_{n2}^2 \\ \dots & \dots \\ 1 & \omega_{nn}^2 \end{bmatrix} \begin{Bmatrix} \alpha_0 \\ \alpha_1 \end{Bmatrix} = 2\begin{Bmatrix} \xi_1\omega_{n1} \\ \xi_2\omega_{n2} \\ \dots \\ \xi_n\omega_{nn} \end{Bmatrix} \tag{3.247}$$

or

$$\begin{Bmatrix} \alpha_0 \\ \alpha_1 \end{Bmatrix} = 2 \begin{bmatrix} 1 & \omega_{n1}^2 \\ 1 & \omega_{n2}^2 \\ \dots & \dots \\ 1 & \omega_{nn}^2 \end{bmatrix}^{+} \begin{Bmatrix} \xi_1\omega_{n1} \\ \xi_2\omega_{n2} \\ \dots \\ \xi_n\omega_{nn} \end{Bmatrix} \tag{3.248}$$

Here, + stands for the pseudo inverse. For a rectangular matrix **A** with dimension m×n, the pseudo inverse of **A** can be calculated as

$$\mathbf{A}^{+} = \left(\mathbf{A}^{T}\mathbf{A}\right)^{-1}\mathbf{A}^{T} \quad \text{for } m > n \tag{3.249}$$

or

$$\mathbf{A}^{+} = \mathbf{A}^{T}\left(\mathbf{A}\,\mathbf{A}^{T}\right)^{-1} \quad \text{for } m < n \tag{3.250}$$

3.6.2 Caughey Damping

Alternatively, Caughey proposed a more general formula for proportional damping as

$$\mathbf{C} = \sum_{s=0}^{n-1} \alpha_s \mathbf{M}\left(\mathbf{M}^{-1}\mathbf{K}\right)^{s} = \alpha_0\mathbf{M} + \alpha_1\mathbf{K} + \alpha_2\mathbf{K}\mathbf{M}^{-1}\mathbf{K} + \cdots \tag{3.251}$$

Note that

$$\frac{\boldsymbol{u}_i^T\left[\mathbf{M}(\mathbf{M}^{-1}\mathbf{K})^{s}\right]\boldsymbol{u}_i}{\boldsymbol{u}_i^T\mathbf{M}\boldsymbol{u}_i} = \left(\omega_{ni}^2\right)^{s} \tag{3.252}$$

Therefore, substituting Equation 3.251 into Equation 3.116 results in

$$\frac{\boldsymbol{u}_i^T\left(\alpha_0\mathbf{M} + \alpha_1\mathbf{K} + \alpha_2\mathbf{K}\mathbf{M}^{-1}\mathbf{K} + \cdots\right)\boldsymbol{u}_i}{\boldsymbol{u}_i^T\mathbf{M}\boldsymbol{u}_i} = \left(\alpha_0 + \alpha_1\omega_{ni}^2 + \alpha_2\omega_{ni}^4 + \cdots\right) = 2\xi_i\omega_{ni} \tag{3.253}$$

Similarly,

$$\begin{bmatrix} 1 \ \omega_{n1}^2 \ \dots \ \left(\omega_{n1}^2\right)^{n-1} \\ 1 \ \omega_{n2}^2 \ \dots \ \left(\omega_{n2}^2\right)^{n-1} \\ \dots \\ 1 \ \omega_{nn}^2 \ \dots \ \left(\omega_{nn}^2\right)^{n-1} \end{bmatrix} \begin{Bmatrix} \alpha_0 \\ \alpha_1 \\ \dots \\ \alpha_{n-1} \end{Bmatrix} = 2 \begin{Bmatrix} \xi_1\omega_{n1} \\ \xi_2\omega_{n2} \\ \dots \\ \xi_n\omega_{nn} \end{Bmatrix} \tag{3.254}$$

Therefore, the n coefficients α_i can be calculated as

$$\begin{Bmatrix} \alpha_0 \\ \alpha_1 \\ \cdots \\ \alpha_{n-1} \end{Bmatrix} = 2\begin{bmatrix} 1 \ \omega_{n1}^2 \ \ldots \ \left(\omega_{n1}^2\right)^{n-1} \\ 1 \ \omega_{n2}^2 \ \ldots \ \left(\omega_{n2}^2\right)^{n-1} \\ \ldots \\ 1 \ \omega_{nn}^2 \ \ldots \ \left(\omega_{nn}^2\right)^{n-1} \end{bmatrix}_{n\times n}^{-1} \begin{Bmatrix} \xi_1\omega_{n1} \\ \xi_2\omega_{n2} \\ \cdots \\ \xi_n\omega_{nn} \end{Bmatrix} \tag{3.255}$$

In many cases, the modal parameters do not have to be used for all modes, just the first S modes. In this case, the first S terms in the series are used to represent the damping matrix. That is, Equation 3.251 can be written as

$$\mathbf{C} \approx \sum_{s=0}^{S-1} \alpha_s \mathbf{M}\left(\mathbf{M}^{-1}\mathbf{K}\right)^s \tag{3.256}$$

Therefore,

$$\begin{Bmatrix} \alpha_0 \\ \alpha_1 \\ \cdots \\ \alpha_{S-1} \end{Bmatrix} = 2\begin{bmatrix} 1 \ \omega_{n1}^2 \ \ldots \ \left(\omega_{n1}^2\right)^{S-1} \\ 1 \ \omega_{n2}^2 \ \ldots \ \left(\omega_{n2}^2\right)^{S-1} \\ \ldots \\ 1 \ \omega_{nS}^2 \ \ldots \ \left(\omega_{nS}^2\right)^{S-1} \end{bmatrix}_{S\times S}^{-1} \begin{Bmatrix} \xi_1\omega_{n1} \\ \xi_2\omega_{n2} \\ \cdots \\ \xi_S\omega_{nS} \end{Bmatrix}_{S\times 1} \tag{3.257}$$

3.6.3 Modification of Caughey Damping

The use of Caughey damping to compute the modal damping can be written as

$$\sum_{s=0}^{n-1} \alpha_s \left(\omega_{ns}^2\right)^s = 2\xi_i\omega_{ni} \tag{3.258}$$

It is understandable that when the exponent s becomes a large number, the value of $(\omega_{ns}^2)^s$ will be very large, especially for higher modes. Accordingly, the coefficient α_s will become rather small, easily smaller than the lowest limit of computer digits. Therefore, mathematically, these coefficients can exist; but computationally, quite a few of these coefficients will become computational zero. To improve the situation, an alternative formula can be used, such as

$$\mathbf{C} = \sum_{s=0}^{n-1} \alpha_s \mathbf{M}\left(\mathbf{M}^{-1}\mathbf{K}\right)^{b_s} \tag{3.259}$$

Here, b_s can be a series of arbitrary nonzero numbers, not necessarily the integers 0, 1, 2, 3…, as long as $b_i \neq b_j$. However, in order to use higher modal frequencies, b_s should not be very large. Suppose the maximum possible value of the frequency is denoted as ω_{nS}. The corresponding coefficient α_s must be greater than the lowest value of the common computational digit ε. It is suggested that for the largest number among all the exponents, denoted by b_s,

$$\left\|\left(\mathbf{M}^{-1}\mathbf{K}\right)^{b_s}\right\| < 10^5 \left\|\mathbf{M}^{-1}\mathbf{K}\right\| \tag{3.260}$$

where $\|\bullet\|$ stands for a norm of matrix (.).

Accordingly,

$$\left(\omega_{nS}\right)^{2b_s} < 10^4 \tag{3.261}$$

or

$$b_s < \left[4 - \log_{10}(\varepsilon)\right] / \left[2 \log_{10}(\omega_{nS})\right] \tag{3.262}$$

For example, MATLAB® for PCs has the digit $\varepsilon = 10^{-16}$. An alternative formula can be written as

$$b_s < 10 / \left[2 \log_{10}(\omega_{nS})\right] \tag{3.263}$$

Now, there are S terms in Equation 3.256, the smallest is 0 and the largest is b_s, and the ith exponent b_l can be written as

$$b_i = \left(\frac{i-1}{S-1}\right) b_s \tag{3.264}$$

Note that the choices of b_s in inequalities 3.262 and 3.263 as well as b_l in Equation 3.264 are only a suggestion.

3.6.4 Overdamped Modes

When a structure is supplied with dampers, it is highly possible that certain modes are overdamped. The overdamped case is considered using the Rayleigh quotient. It is known that, since the stiffness matrix is symmetric and positive definite, the Rayleigh quotient must be a positive scalar. That is,

$$\frac{\boldsymbol{u}_i^T \mathbf{K} \boldsymbol{u}_i}{\boldsymbol{u}_i^T \mathbf{M} \boldsymbol{u}_i} = \lambda_{ki} > 0 \tag{3.265}$$

Here, for simplicity, the ith eigenvector $\boldsymbol{u}_i$ is used directly, instead of an arbitrary vector in Equation 3.265. Since overdamped cases may occur, λ_{ki} is used instead of ω_{ni}^2 in Equation 3.275 to denote the general case. Similarly, for a positive semidefinite damping matrix,

$$\frac{\boldsymbol{u}_i^T \mathbf{C} \boldsymbol{u}_i}{\boldsymbol{u}_i^T \mathbf{M} \boldsymbol{u}_i} = \lambda_{ci} \geq 0 \tag{3.266}$$

Here, λ_{ci} is used instead of $2\xi_i \cdot \omega_{ni}$ in Equation 3.266. Note that in practice, there is always a positive Rayleigh quotient in Equation 3.266.

To determine if the ith mode is overdamped, check if

$$\chi_i^2 - 4\mu_i\kappa_i > 0 \tag{3.267}$$

or

$$\chi_i / 2(\mu_i\kappa_i)^{1/2} > 1 \tag{3.268}$$

That is equivalent to checking whether

$$\lambda_{ci} / 2(\lambda_{ki})^{1/2} > 1 \tag{3.269}$$

or

$$\frac{\boldsymbol{u}_i^T \mathbf{C} \boldsymbol{u}_i}{2\sqrt{(\boldsymbol{u}_i^T \mathbf{M} \boldsymbol{u}_i)(\boldsymbol{u}_i^T \mathbf{K} \boldsymbol{u}_i)}} > 1 \tag{3.270}$$

3.6.5 Alternative Expression of Proportional Damping

Suppose the mass and stiffness matrices **M** and **K** and all the damping ratios of an n-DOF system are given. The damping ratios can be arranged in a diagonal matrix as

$$\Xi = \text{diag}([\xi_1, \xi_2, \ldots, \xi_n]) \tag{3.271}$$

From the matrices **M** and **K**, the eigenvalue and eigenvector matrices can be calculated and denoted by Ω^2 and **U**. The square root of Ω^2 is further calculated as Ω, which is the diagonal matrix containing all the natural frequencies. That is,

$$\Omega = \text{diag}([\omega_1, \omega_2, \ldots, \omega_n]) \tag{3.272}$$

The corresponding damping matrix **C** can be expressed as

$$\mathbf{C} = 2\mathbf{M}\mathbf{U}\boldsymbol{\Xi}\,\boldsymbol{\Omega}\mathbf{U}^{-1} \tag{3.273}$$

Example 3.7

Suppose the mass and stiffness matrices are as given in Example 3.1. In addition, the damping ratio matrix is

$$\boldsymbol{\Xi} = \text{diag}([0.05, 0.05, 0.04, 0.06])$$

By using the formula described in Equation 3.248, $\alpha_0 = 0.9016$ and $\alpha_1 = 0.0018$, such that

$$\mathbf{C} = 1000\begin{bmatrix} 2.7971 & -1.8053 & 0 & 0 \\ -1.8053 & 4.6926 & -1.8053 & 0 \\ 0 & -1.8053 & 4.7827 & -1.8053 \\ 0 & 0 & -1.8053 & 4.8729 \end{bmatrix} \ (\text{kN-s/m})$$

Using this damping matrix, the damping ratios are calculated as 0.0538, 0.0414, 0.0493, and 0.0563 for the first through the fourth mode, respectively. It is seen that using the approach of Rayleigh damping, computational errors can exist.

By using the formula described in Equation 3.257, $\alpha_0 = 0.5182$, $\alpha_1 = 0.0052$, $\alpha_2 = -3.5702\text{e-}6$, and $\alpha_3 = 8.7961\text{e-}10$. The corresponding damping matrix is calculated to be

$$\mathbf{C} = 1000\begin{bmatrix} 2.8050 & -1.7117 & 0.0406 & -0.5639 \\ -1.7117 & 4.6536 & -2.2257 & 0.3889 \\ 0.0406 & -2.2257 & 4.9837 & -1.4389 \\ -0.5639 & 0.3889 & -1.4389 & 4.6919 \end{bmatrix} \ (\text{kN-s/m})$$

Using this damping matrix, the damping ratio is calculated as 0.05, 0.05, 0.04, and 0.06 for the first through the fourth mode, respectively. It is seen that Caughey damping is a fairly good approach.

Finally, by using the formula described in Equation 3.273,

$$\mathbf{C} = 1000\begin{bmatrix} 2.8050 & -1.7117 & 0.0406 & -0.5639 \\ -1.7117 & 4.6536 & -2.2257 & 0.3889 \\ 0.0406 & -2.2257 & 4.9837 & -1.4389 \\ -0.5639 & 0.3889 & -1.4389 & 4.6919 \end{bmatrix} \ (\text{kN-s/m})$$

which is identical to the Caughey damping matrix.

3.6.5.1 Generalized Symmetric Damping Matrix

In Equation 3.81, the concept of a generalized stiffness matrix $\tilde{\mathbf{K}}$ is introduced. It can be seen that the eigenvalue matrix of $\tilde{\mathbf{K}}$ is Ω^2. The advantage of the generalized stiffness matrix is that it is symmetric so that the corresponding eigenvector matrix can be orthonormal. In addition, the same treatment of premultiplying and postmultiplying $\mathbf{M}^{-1/2}$ on both sides of the damping matrix $\mathbf{C}$ to obtain the generalized damping matrix $\tilde{\mathbf{C}}$ can be used. That is,

$$\tilde{\mathbf{C}} = \mathbf{M}^{-1/2}\mathbf{C}\mathbf{M}^{-1/2} \tag{3.274}$$

Similarly, matrix $\tilde{\mathbf{C}}$ is also symmetric. When the system is proportionally damped, the generalized damping matrix $\tilde{\mathbf{C}}$ will share the identical eigenvector matrix $\mathbf{V}$ with $\tilde{\mathbf{K}}$.

Therefore $\tilde{\mathbf{K}}$ and $\tilde{\mathbf{C}}$ will commute, thus

$$\tilde{\mathbf{K}}\tilde{\mathbf{C}} = \tilde{\mathbf{C}}\tilde{\mathbf{K}} \tag{3.275}$$

Equation 3.275 is an alternative form of the Caughey criterion. Furthermore, it can be realized that for the proportionally damped system, matrix $\tilde{\mathbf{C}}$ will have the following eigen-decomposition:

$$\tilde{\mathbf{C}} = \mathbf{V}\boldsymbol{\Lambda}_C\mathbf{V}^T \tag{3.276}$$

or

$$\mathbf{V}\tilde{\mathbf{C}}\mathbf{V}^T = \boldsymbol{\Lambda}_C \tag{3.277}$$

where the eigenvalue matrix $\boldsymbol{\Lambda}_C$ is

$$\boldsymbol{\Lambda}_C = \text{diag}\left(2\xi_i\omega_{ni}\right) \tag{3.278}$$

By using the generalized damping and stiffness matrices, Equation 3.99 can be replaced by

$$\mathbf{I}\ddot{\mathbf{z}}(t) + \tilde{\mathbf{C}}\dot{\mathbf{z}}(t) + \tilde{\mathbf{K}}\mathbf{z}(t) = -\mathbf{M}^{-1/2}\mathbf{J}\,\ddot{x}_g(t) \tag{3.279}$$

through premultiplying $\mathbf{M}^{1/2}$ on both sides of Equation 3.99.
Also,

$$\mathbf{z}(t) = \mathbf{M}^{-1/2}\mathbf{x}(t) \tag{3.280}$$

By using Equation 3.279, the similar form of modal Equation 3.138 can be directly obtained from the eigen-decomposition of $\tilde{\mathbf{C}}$ and $\tilde{\mathbf{K}}$.

3.7 SUMMARY

In this chapter, the concept of proportionally damped MDOF systems is introduced. To understand the fundamental difference between SDOF and MDOF systems, an undamped structure is first discussed. The Rayleigh quotient and matrix algebra are introduced as the basic tools to describe the MDOF vibratory systems. The principal approach to dealing with an MDOF system is to decouple the system into n-SDOF systems, namely, n-modes, so that the responses can be obtained mode by mode. In the next chapter, generally damped MDOF systems will be discussed.

REFERENCES

Building Seismic Safety Council (BSSC). 2000. *NEHRP Recommended Provisions for Seismic Regulation for New Buildings and Other Structures, 2000 Edition*, Report Nos. FEMA 368, Federal Emergency Management Agency, Washington, DC.

Chopra, A.K. 2006. *Dynamics of Structures: Theory and Applications to Earthquake Engineering*. 3rd ed. Upper Saddle River, NJ: Prentice Hall.

Clough, R.W. and Penzien, J. 1993. *Dynamics of Structures*. 2nd ed. NY: McGraw-Hill.

Wilson, E.L. 2004. *Static and Dynamic Analysis of Structures*. 4th ed. Berkeley, CA: Computers and Structures.

4 Multi-Degree-of-Freedom Systems with General Damping

In Chapter 3, a multi-degree-of-freedom (MDOF) system with proportional damping was examined. When large supplemental damping is added to a structure, the total system is likely to be nonproportionally damped. In this case, the system will have complex-valued mode shapes, which are referred to as *complex modes*. Furthermore, the system may have certain "modes" with a damping ratio greater than 1, which will make the corresponding mode reduce to two overdamped subsystems. In many cases, the contributions of both the complex modes and the overdamped subsystems to the total structural response may not be ignored or treated as normal modes. These phenomena may be thought of as the results of *damping irregularity*. In current building codes, neither the complex modes nor the overdamped subsystems are required to be considered. In structures with added dampers, damping irregularity exists. The question is the degree of irregularity. The complex modal responses, as well as overdamped cases, are quantitatively discussed in this chapter, so that a determination can be made whether a pair of specific complex modes or an overdamped subsystem should be considered in damper design. Note that although installing dampers in a structure will likely make the damping of the system nonproportional, if the total damping is still not very large, then the proportional damping approach can still be used.

Since adding dampers to a structure may also reduce a mode that was originally underdamped and can be expressed as a second-order system to two first-order overdamped subsystems. In this case, using the conventional method based on the theory of vibration systems will have limitations. Therefore, the overdamped problem is also discussed in this section. In the following paragraphs, systems with nonproportional damping and/or overdamped subsystems are referred to as *generally damped* systems. The overdamped subsystems are referred to as *pseudo modes*.

The phrases normal mode, complex mode, and pseudo mode are used to represent the cases listed in Figure 4.1.

4.1 STATE EQUATIONS AND CONVENTIONAL TREATMENT

Conventionally, nonproportionally damped systems with n-degrees-of-freedom are treated in the 2n state space. State equations and the state matrix are the major tools used in the approach. Similar to the system with proportional damping, eigen-decomposition is used to examine the modal responses, though for a nonproportionally damped system, the eigen-decomposition is in the 2n-complex modal space. In addition, the state equation is very useful in numerical simulations in carrying out the time history analysis.

In the following subsections, a nonproportionally damped system with every mode underdamped is considered first. It is also assumed that this system does not have repeated eigenvalues in general.

4.1.1 State Matrix and Eigen-Decomposition

4.1.1.1 State Equations

As mentioned in Chapter 3, a proportionally damped linear MDOF system can be decoupled into n single-degree-of-freedom (SDOF) subsystems or modes in the n-dimensional modal domain. This is because matrices $\mathbf{M}^{-1}\mathbf{C}$ and $\mathbf{M}^{-1}\mathbf{K}$ share the identical eigenvector matrix, which is then used

$$\text{Linear MDOF system}\begin{cases}\text{Proportionally damped} & \begin{cases}\text{Underdamped} \rightarrow \text{Normalmode}\\ \text{Overdamped} \rightarrow \text{Pseudomode}\end{cases}\\ \text{Nonproportionally damped} & \begin{cases}\text{Underdamped} \rightarrow \text{Complexmode}\\ \text{Overdamped} \rightarrow \text{Pseudomode}\end{cases}\end{cases}$$

FIGURE 4.1 General linear MDOF systems.

to transform both $\mathbf{M}^{-1}\mathbf{C}$ and $\mathbf{M}^{-1}\mathbf{K}$ into diagonal matrices. However, for a generally damped system, the matrices $\mathbf{M}^{-1}\mathbf{C}$ and $\mathbf{M}^{-1}\mathbf{K}$ do not have to share identical eigenvectors. In other words, the Caughey criterion described in Equation 3.102 will not hold. Thus, an alternative way to decouple the entire system needs to be found in order to obtain the responses of such a system, mode by mode.

Consider the governing equation of a linear MDOF system introduced in Chapter 3 (see Equation 3.99). Here, for a more general case, the forcing vector F(t) is used, that is,

$$\mathbf{M}\ddot{\mathbf{x}}(t)+\mathbf{C}\dot{\mathbf{x}}(t)+\mathbf{K}\mathbf{x}(t)=\mathbf{f}(t) \tag{4.1a}$$

In Chapter 3, Equation 3.202, the state equations were introduced as an alternative description of vibration systems. Now, Equation 3.202 is repeated and more useful properties of the state equations, which can be used as mathematical tools to decouple generally damped systems, are further explored. That is,

$$\begin{Bmatrix}\ddot{\mathbf{x}}\\ \dot{\mathbf{x}}\end{Bmatrix}=\begin{bmatrix}-\mathbf{M}^{-1}\mathbf{C} & -\mathbf{M}^{-1}\mathbf{K}\\ \mathbf{I} & \mathbf{0}\end{bmatrix}\begin{Bmatrix}\dot{\mathbf{x}}\\ \mathbf{x}\end{Bmatrix}+\begin{Bmatrix}\mathbf{M}^{-1}\mathbf{f}\\ \mathbf{0}\end{Bmatrix} \tag{4.1b}$$

Furthermore, using the state and the input matrices, the following state equations can be obtained by rewriting Equation 4.1b as

$$\dot{\boldsymbol{X}}=\boldsymbol{\mathcal{A}X}+\boldsymbol{\mathcal{B}F}(t) \tag{4.2}$$

In contrast to Equation 3.202, here the dimension of the vector $\boldsymbol{X}$ and $\boldsymbol{F}$ is $2n \times 1$, denoted in Equations 4.3a and 4.3b as a subscript, that is,

$$\boldsymbol{X}=\begin{Bmatrix}\dot{\mathbf{x}}(t)\\ \mathbf{x}(t)\end{Bmatrix}_{2n\times 1} \tag{4.3a}$$

$$\boldsymbol{F}=\begin{Bmatrix}\mathbf{M}^{-1}\mathbf{f}(t)\\ \mathbf{0}\end{Bmatrix}_{2n\times 1} \tag{4.3b}$$

and note that the dimension of the vector $\mathbf{x}$ is $n \times 1$, denoted in Equation 4.4 as a subscript,

$$\mathbf{x}=\begin{Bmatrix}x_1(t)\\ x_2(t)\\ \dots\\ x_n(t)\end{Bmatrix}_{n\times 1} \tag{4.4}$$

The state matrix is written as

$$\mathcal{A}=\begin{bmatrix}\left[-\mathbf{M}^{-1}\mathbf{C}\right]_{n\times n} & \left[-\mathbf{M}^{-1}\mathbf{K}\right]_{n\times n}\\ \mathbf{I}_{n\times n} & \mathbf{0}_{n\times n}\end{bmatrix}_{2n\times 2n} \tag{4.5}$$

Here **I** and **0** are the identity and null matrices, respectively, with dimension n × n.

The input matrix $\mathcal{B}$ is

$$\mathcal{B} = \begin{bmatrix} \mathbf{M}^{-1}_{n\times n} \\ \mathbf{0}_{n\times n} \end{bmatrix}_{2n\times n} \tag{4.6}$$

4.1.1.2 Eigen-Decomposition

As with undamped systems mentioned in Chapter 3, the homogeneous form of Equation 4.4 is used,

$$\dot{\boldsymbol{X}} = \boldsymbol{\mathcal{A}}\boldsymbol{X}(t) \tag{4.7}$$

to introduce the eigen-decomposition of the state matrix $\mathcal{A}$. Note that for the undamped system, the matrix to be decomposed is $\mathbf{M}^{-1}$K, which is an n × n matrix, whereas the state matrix $\mathcal{A}$ is 2n × 2n. To evaluate the eigen-properties of these systems, it is first assumed that

$$\boldsymbol{X} = \boldsymbol{P}e^{\lambda t} \tag{4.8}$$

is a solution of the homogeneous Equation 4.7. That is,

$$\lambda\boldsymbol{P}e^{\lambda t} = \boldsymbol{\mathcal{A}}\boldsymbol{P}e^{\lambda t} \tag{4.9}$$

or

$$\lambda\boldsymbol{P} = \boldsymbol{\mathcal{A}}\boldsymbol{P} \tag{4.10}$$

In Equations 4.8 through 4.10, λ is a scalar and $\boldsymbol{P}$ is a 2n × 1 vector.

If $\boldsymbol{X} = \boldsymbol{P}e^{\lambda t}$ is a solution of Equation 4.7, then Equation 4.10 should hold. Also, if Equation 4.10 holds, then $\lambda\boldsymbol{P}e^{\lambda t} = \boldsymbol{\mathcal{A}}\boldsymbol{P}e^{\lambda t}$ will be a solution of Equation 4.7. Furthermore, from the theory of linear algebra and the theory of vibration systems, it can be proven that these necessary and sufficient conditions hold.

Now, further assume that the system does not have large damping, that is, the entire system is underdamped. Since $\mathcal{A}$ is an asymmetric matrix, the complex-valued λ and $\boldsymbol{P}$ will generally exist. Taking the complex conjugate of Equation 4.10, denoted by $()^*$,

$$\lambda^*\boldsymbol{P}^* = \boldsymbol{\mathcal{A}}\boldsymbol{P}^* \tag{4.11}$$

Equations 4.10 and 4.11 form the typical eigen-problem. That is, if the pair <λ, $\boldsymbol{P}$> does make $\boldsymbol{X} = \boldsymbol{P}e^{\lambda t}$ a solution of Equation 4.8, then Equation 4.10 implies that λ is one of the eigenvalues of the matrix $\mathcal{A}$ and $\boldsymbol{P}$ is the corresponding eigenvector. Suppose the system has n-DOFs. Thus, there are n pairs of eigenvalues and eigenvectors in the complex conjugates. That is, Equations 4.10 and 4.11 can be further expanded as

$$\lambda_i\boldsymbol{P}_i = \boldsymbol{\mathcal{A}}\boldsymbol{P}_i \quad i = 1,2,\ldots,n \tag{4.12}$$

Taking the complex conjugate of both sides of Equation 4.12,

$$\lambda_i^*\boldsymbol{P}_i^* = \boldsymbol{\mathcal{A}}\boldsymbol{P}_i^* \quad i = 1,2,\ldots,n \tag{4.13}$$

It is known that the eigen-problem described in Equations 4.12 and 4.13 implies that all the eigenvectors, $\boldsymbol{P}_i$ and $\boldsymbol{P}_i^*$, are linearly independent. Furthermore, each eigenvector is only associated with a unique eigenvalue λ_i (or λ_i^*), which is defined in Chapter 1 and is represented as follows:

$$\lambda_i = -\xi_i\omega_i + j\sqrt{1-\xi_i^2}\,\omega_i, \quad i = 1, 2, \ldots, n \tag{4.14}$$

Note that in the case of a nonproportionally damped system, both the natural frequency ω_i and the damping ratio ξ_i are derived from Equation 4.14. Thus,

$$\omega_i = |\lambda_i|, \quad i = 1, 2, \ldots, n \tag{4.15}$$

and

$$\xi_i = -\mathrm{Re}(\lambda_i)/\omega_i \tag{4.16}$$

In previous chapters, the natural frequency (or angular natural frequency) is denoted as ω_{ni} with subscript n standing for "natural." This natural frequency is obtained through the square roots of stiffness k over m or κ_i over μ_i. In generally damped systems, however, the natural frequency cannot be obtained in this way, but rather through Equation 4.14. In this case, to distinguish it from the previously defined quantities, subscript n is omitted.

Note that an eigenvalue λ_i (or λ_i^*) can have an infinite number of eigenvectors. That is, if $\boldsymbol{P}_i$ is one of the corresponding eigenvectors, then vector $\boldsymbol{R}_i$,

$$\boldsymbol{R}_i = \alpha\boldsymbol{P}_i \tag{4.17}$$

will also be the eigenvector associated with that λ_i, where α is an arbitrary nonzero scalar. Since $\boldsymbol{P}_i$ is a $2n \times 1$ vector, it is seen that through the assumption described in Equation 4.8,

$$\boldsymbol{P}_i = \begin{Bmatrix} \lambda_i \boldsymbol{p}_i \\ \boldsymbol{p}_i \end{Bmatrix} \tag{4.18}$$

where $\boldsymbol{p}_i$ is an $n \times 1$ vector. Since the system is linear, the solution can have all the linear combinations of $\boldsymbol{p}_i$ and $\boldsymbol{p}_i^*$ as follows:

$$\begin{aligned} \mathbf{x}(t) &= \boldsymbol{p}_1 e^{\lambda_1 t} + \boldsymbol{p}_2 e^{\lambda_2 t} + \cdots + \boldsymbol{p}_n e^{\lambda_n t} + \boldsymbol{p}_1^* e^{\lambda_1^* t} + \boldsymbol{p}_2^* e^{\lambda_2^* t} + \cdots + \boldsymbol{p}_n^* e^{\lambda_n^* t} \\ &= [\boldsymbol{p}_1, \boldsymbol{p}_2, \ldots, \boldsymbol{p}_n] \begin{Bmatrix} e^{\lambda_1 t} \\ e^{\lambda_2 t} \\ \cdots \\ e^{\lambda_n t} \end{Bmatrix} + [\boldsymbol{p}_1^*, \boldsymbol{p}_2^*, \ldots, \boldsymbol{p}_n^*] \begin{Bmatrix} e^{\lambda_1^* t} \\ e^{\lambda_2^* t} \\ \cdots \\ e^{\lambda_n^* t} \end{Bmatrix} \end{aligned} \tag{4.19}$$

Denote

$$\mathbf{P} = [\boldsymbol{p}_1, \boldsymbol{p}_2, \ldots, \boldsymbol{p}_n] \tag{4.20}$$

and

$$e(t) = \begin{Bmatrix} e^{\lambda_1 t} \\ e^{\lambda_2 t} \\ \dots \\ e^{\lambda_n t} \end{Bmatrix} \tag{4.21}$$

the result is

$$\mathbf{x}(t) = \mathbf{P}e(t) + \mathbf{P}^* e^*(t) \tag{4.22}$$

$$\dot{\mathbf{x}}(t) = \mathbf{P\Lambda}e(t) + \mathbf{P}^*\mathbf{\Lambda}^* e^*(t) \tag{4.23}$$

and

$$\ddot{\mathbf{x}}(t) = \mathbf{P\Lambda}^2 e(t) + \mathbf{P}^*\mathbf{\Lambda}^{*2} e^*(t) \tag{4.24}$$

or

$$\begin{Bmatrix} \dot{\mathbf{x}}(t) \\ \mathbf{x}(t) \end{Bmatrix} = \begin{bmatrix} \mathbf{P\Lambda} & \mathbf{P}^*\mathbf{\Lambda}^* \\ \mathbf{P} & \mathbf{P}^* \end{bmatrix} \begin{Bmatrix} e(t) \\ e^*(t) \end{Bmatrix} \tag{4.25}$$

and

$$\begin{Bmatrix} \ddot{\mathbf{x}}(t) \\ \dot{\mathbf{x}}(t) \end{Bmatrix} = \begin{bmatrix} \mathbf{P\Lambda} & \mathbf{P}^*\mathbf{\Lambda}^* \\ \mathbf{P} & \mathbf{P}^* \end{bmatrix} \begin{bmatrix} \mathbf{\Lambda} & \mathbf{0} \\ \mathbf{0} & \mathbf{\Lambda}^* \end{bmatrix} \begin{Bmatrix} e(t) \\ e^*(t) \end{Bmatrix} \tag{4.26}$$

Here $\mathbf{\Lambda}$ is defined as the diagonal $n \times n$ matrix, which contains all the n-sets of eigenvalues, as follows:

$$\mathbf{\Lambda} = \mathrm{diag}(\lambda_i) = \mathrm{diag}\left(-\xi_i\omega_i + j\sqrt{1-\xi_i^2}\,\omega_i\right) = \begin{bmatrix} \lambda_1 & & & \\ & \lambda_2 & & \\ & & \dots & \\ & & & \lambda_n \end{bmatrix}_{n\times n} \tag{4.27}$$

where λ_i is defined in Equation 4.14, and $\mathbf{0}$ stands for a null matrix with proper dimension. In the following discussion, for simplicity, the off-diagonal submatrix $\mathbf{0}$ will not appear.

Substituting Equation 4.26 into Equation 4.7 with assistance from Equation 4.3 results in

$$\begin{bmatrix} \mathbf{P\Lambda} & \mathbf{P}^*\mathbf{\Lambda}^* \\ \mathbf{P} & \mathbf{P}^* \end{bmatrix} \begin{bmatrix} \mathbf{\Lambda} & \\ & \mathbf{\Lambda}^* \end{bmatrix} \begin{Bmatrix} e(t) \\ e^*(t) \end{Bmatrix} = \mathcal{A} \begin{bmatrix} \mathbf{P\Lambda} & \mathbf{P}^*\mathbf{\Lambda}^* \\ \mathbf{P} & \mathbf{P}^* \end{bmatrix} \begin{Bmatrix} e(t) \\ e^*(t) \end{Bmatrix} \tag{4.28}$$

There can further be a $2n \times u$ matrix $\mathcal{E}$, defined as

$$\mathcal{E} = \left[\begin{Bmatrix} e(t) \\ e^*(t) \end{Bmatrix}, \begin{Bmatrix} e(t+\Delta t) \\ e^*(t+\Delta t) \end{Bmatrix}, \dots, \begin{Bmatrix} e[t+(u-1)\Delta t] \\ e^*[t+(u-1)\,\Delta t] \end{Bmatrix} \right], \quad u \geq 2n \tag{4.29}$$

which can be shown to have full rank 2n. Therefore,

$$\begin{bmatrix} \mathbf{P\Lambda} & \mathbf{P}^*\mathbf{\Lambda}^* \\ \mathbf{P} & \mathbf{P}^* \end{bmatrix} \begin{bmatrix} \mathbf{\Lambda} & \\ & \mathbf{\Lambda}^* \end{bmatrix} \mathcal{E} = \mathcal{A} \begin{bmatrix} \mathbf{P\Lambda} & \mathbf{P}^*\mathbf{\Lambda}^* \\ \mathbf{P} & \mathbf{P}^* \end{bmatrix} \mathcal{E} \tag{4.30}$$

Postmultiplying by $\mathcal{E}^+$ on both sides of Equation 4.30, since

$$\mathcal{E}\mathcal{E}^+ = \mathbf{I}_{2n\times 2n} \tag{4.31}$$

where again superscript + stands for the pseudo inverse, results in

$$\begin{bmatrix} \mathbf{P\Lambda} & \mathbf{P}^*\mathbf{\Lambda}^* \\ \mathbf{P} & \mathbf{P}^* \end{bmatrix} \begin{bmatrix} \mathbf{\Lambda} & \\ & \mathbf{\Lambda}^* \end{bmatrix} = \mathcal{A} \begin{bmatrix} \mathbf{P\Lambda} & \mathbf{P}^*\mathbf{\Lambda}^* \\ \mathbf{P} & \mathbf{P}^* \end{bmatrix} \tag{4.32}$$

Equation 4.32 indicates that the state matrix $\mathcal{A}$ can be decomposed into the eigenvalue matrix:

$$\mathbf{\Delta} = \begin{bmatrix} \mathbf{\Lambda} & \\ & \mathbf{\Lambda}^* \end{bmatrix} \tag{4.33}$$

and eigenvector matrix $\mathcal{P}$:

$$\mathcal{P} = \begin{bmatrix} \mathbf{P\Lambda} & \mathbf{P}^*\mathbf{\Lambda}^* \\ \mathbf{P} & \mathbf{P}^* \end{bmatrix} = [\boldsymbol{P}_1, \boldsymbol{P}_2, \ldots, \boldsymbol{P}_{2n}] \tag{4.34}$$

Note that the eigenvector matrix is now arranged to have the form of a complex conjugate pair:

$$\begin{bmatrix} \mathbf{P\Lambda} \\ \mathbf{P} \end{bmatrix} \text{and} \begin{bmatrix} \mathbf{P\Lambda} \\ \mathbf{P} \end{bmatrix}^* \tag{4.35}$$

Thus,

$$\mathcal{P} = \left[[\boldsymbol{P}_1, \boldsymbol{P}_2, \ldots, \boldsymbol{P}_n], [\boldsymbol{P}_1, \boldsymbol{P}_2, \ldots, \boldsymbol{P}_n]^* \right] \tag{4.36}$$

and

$$\begin{bmatrix} \mathbf{P\Lambda} \\ \mathbf{P} \end{bmatrix}^* = \begin{bmatrix} \mathbf{P}^*\mathbf{\Lambda}^* \\ \mathbf{P}^* \end{bmatrix} \tag{4.37}$$

That is,

$$\mathcal{A} = \mathcal{P}\, \mathbf{\Delta}\, \mathcal{P}^{-1} \tag{4.38}$$

or

$$\mathbf{\Delta} = \mathcal{P}^{-1} \mathcal{A} \mathcal{P} \tag{4.39}$$

Note that, the method of using the state equation also works for proportionally damped systems.

Matrix **Δ** in Equation 4.39 maintains exactly the same eigenvalue format as that for proportionally damped systems. However, the eigenvector $\mathscr{P}$ can have a different form from the proportionally damped case due to the nonuniqueness of $\boldsymbol{P}_i$ discussed above.

Equations 4.38 and 4.39 can be used to define modal analysis in the 2n complex modal domain, since the submatrix $\boldsymbol{P}$ in the eigenvector matrix $\mathscr{P}$ is generally complex valued. Here $\boldsymbol{P}$ is called the *mode shape matrix.* Note that $\boldsymbol{P}$ contains n vectors expressed in Equation 4.20.

In this case, there is an n set of triples $<\boldsymbol{p}_i, \xi_i, \omega_i>$ and another n set of its complex conjugates. It is apparent that the damping ratio ξ_i and the natural frequency ω_i can be obtained through Equation 4.27. In this circumstance, the triple $<\boldsymbol{p}_i, \xi_i, \omega_i>$ and its complex conjugate define the ith complex mode.

Example 4.1

Suppose an **M-C-K** system with

$$\mathbf{M} = \begin{bmatrix} 2 & \\ & 2.5 \end{bmatrix}(\mathrm{t}),\ \mathbf{C} = \begin{bmatrix} 10 & -5 \\ -5 & 15 \end{bmatrix}(\mathrm{kN\text{-}s/m}),\ \mathbf{K} = \begin{bmatrix} 1500 & -750 \\ -750 & 750 \end{bmatrix} \quad (\mathrm{kN/m})$$

First, let us check the Caughey criterion. It is found that

$$\mathbf{CM}^{-1}\mathbf{K} = \begin{bmatrix} 9000 & -5250 \\ -8250 & 6375 \end{bmatrix}$$

and

$$\mathbf{KM}^{-1}\mathbf{C} = \begin{bmatrix} 9000 & -8250 \\ -5250 & 6375 \end{bmatrix}$$

Therefore, $\mathbf{CM}^{-1}\mathbf{K} \neq \mathbf{KM}^{-1}\mathbf{C}$, so the Caughey criterion is not satisfied. In other words, the eigenvector matrix of **K** cannot be used to decouple **C** as mentioned in Chapter 3, but rather the state matrix is used to find the eigenvalues and mode shapes.

The state matrix can be written as

$$\mathcal{A} = \begin{bmatrix} -5 & 2.5 & -750 & 375 \\ 2 & -6 & 300 & -300 \\ 1 & 0 & 0 & 0 \\ 0 & 1 & 0 & 0 \end{bmatrix}$$

The corresponding eigen-decomposition will result in

$$\boldsymbol{\Delta} = \mathrm{diag}\big([-1.9668 + 10.8474\mathrm{j}, -3.5332 + 30.2190\mathrm{j}, -1.9668 - 10.8474\mathrm{j}, -3.5332 - 30.2190\mathrm{j}]\big)$$

and

$$\mathscr{P} = \begin{bmatrix} 0.5068 + 0.0277\mathrm{j} & -0.9009 & 0.5068 - 0.0277\mathrm{j} & -0.9009 \\ 0.8569 & 0.4278 + 0.0654\mathrm{j} & 0.8569 & 0.4278 - 0.0654\mathrm{j} \\ -0.0057 - 0.0457\mathrm{j} & 0.0034 + 0.0294\mathrm{j} & -0.0057 + 0.0457\mathrm{j} & 0.0034 - 0.0294\mathrm{j} \\ -0.0139 - 0.0765\mathrm{j} & 0.0005 - 0.0142\mathrm{j} & -0.0139 - 0.0765\mathrm{j} & 0.0005 + 0.0142\mathrm{j} \end{bmatrix}$$

So,

$$\Lambda = \text{diag}\left(\left[-1.9668+10.8474j, -3.5332+30.2190j\right]\right)$$

and the mode shape matrix is

$$\mathbf{P} = \begin{bmatrix} -0.0057-0.0457j & 0.0034+0.0294j \\ -0.0139-0.0765j & 0.0005-0.0142j \end{bmatrix}$$

In Section 4.1.4, additional measurements to quantify the damping nonproportionality are introduced.

4.1.2 Accompanist Matrix of Mode Shapes

For convenience, in this section, another matrix $\mathbf{Q}$ is introduced, which is closely related to the mode shape matrix $\mathbf{P}$. The matrix $\mathbf{Q}$ will play an important role in analyzing the non-proportionally damped systems.

Note that the eigenvector matrix $\mathcal{P}$ is of full rank. Thus, it must have an inverse matrix, denoted as

$$\mathcal{Q} = \mathcal{P}^{-1} \tag{4.40}$$

It can be proven that matrix $\mathcal{Q}$ has the following structure with complex conjugate pair:

$$\mathcal{Q} = \begin{bmatrix} \mathbf{Q} & \mathbf{R} \\ \mathbf{Q}^* & \mathbf{R}^* \end{bmatrix} \tag{4.41}$$

The upper diagonal submatrix $\mathbf{Q}$ in Equation 4.41 can be written as

$$\mathbf{Q} = -\mathbf{P}^{-1}\left(\mathbf{P}^*\Lambda^*\mathbf{P}^{*-1} - \mathbf{P}\,\Lambda\,\mathbf{P}^{-1}\right)^{-1} \tag{4.42}$$

while the upper off-diagonal submatrix $\mathbf{R}$ is

$$\mathbf{R} = \mathbf{P}^{-1} - \mathbf{Q}\,\mathbf{P}\,\Lambda\,\mathbf{P}^{-1} \tag{4.43}$$

or

$$\mathbf{R} = -\Lambda^{-1}\mathbf{Q}\,\mathbf{M}^{-1}\mathbf{K} \tag{4.44}$$

Matrix $\mathbf{Q}$ plays an important role in system analysis. It is called the *accompanist matrix* of the mode shape $\mathbf{P}$.

It is seen that the state matrix $\mathcal{A}$ can be written as a product of two symmetric matrices $\mathcal{R}$ and S, that is,

$$\mathcal{A} = \mathcal{R}S \tag{4.45}$$

where

$$\mathcal{R} = \begin{bmatrix} -\mathbf{M}^{-1}\mathbf{C}\mathbf{M}^{-1} & \mathbf{M}^{-1} \\ \mathbf{M}^{-1} & \mathbf{O} \end{bmatrix} \tag{4.46}$$

and

$$\mathcal{S} = \begin{bmatrix} \mathbf{M} & \mathbf{O} \\ \mathbf{O} & -\mathbf{K} \end{bmatrix} \tag{4.47}$$

Therefore,

$$\mathcal{R}\,\mathcal{S}\,\mathcal{P} = \mathcal{P}\boldsymbol{\Delta} \tag{4.48}$$

Premultiplying $\mathcal{Q}$ on both sides, and inserting a 2n × 2n identity $I = \mathcal{Q}^T\,\mathcal{P}^T$ in between $\mathcal{R}$ and $\mathcal{S}$,

$$\mathcal{Q}\,\mathcal{R}\left(\mathcal{Q}^T\,\mathcal{P}^T\right)\mathcal{S}\,\mathcal{P} = \boldsymbol{\Delta} \tag{4.49}$$

or

$$\left(\mathcal{Q}\,\mathcal{R}\,\mathcal{Q}^T\right)\left(\mathcal{P}^T\,\mathcal{S}\,\mathcal{P}\right) = \boldsymbol{\Delta} \tag{4.50}$$

Since $\mathcal{R}$ and $\mathcal{S}$ are symmetric, $\mathcal{Q}\,\mathcal{R}\,\mathcal{Q}^T$ and $\mathcal{P}^T\,\mathcal{S}\,\mathcal{P}$ will also be symmetric. That is, we can have

$$\left(\mathcal{Q}\,\mathcal{R}\,\mathcal{Q}^T\right)^T = \mathcal{Q}\,\mathcal{R}\,\mathcal{Q}^T \tag{4.51}$$

$$\left(\mathcal{P}^T\mathcal{S}\,\mathcal{P}\right)^T = \mathcal{P}^T\,\mathcal{S}\,\mathcal{P} \tag{4.52}$$

Thus, taking the transpose on both sides of Equation 4.50,

$$\mathcal{P}^T\,\mathcal{S}^T\,\mathcal{P}\,\mathcal{Q}\,\mathcal{R}^T\,\mathcal{Q}^T = \boldsymbol{\Delta}^T = \boldsymbol{\Delta} \tag{4.53}$$

or

$$\left(\mathcal{P}^T\,\mathcal{S}\,\mathcal{P}\right)\left(\mathcal{Q}\,\mathcal{R}\,\mathcal{Q}^T\right) = \boldsymbol{\Delta}^T = \boldsymbol{\Delta} \tag{4.54}$$

Comparing Equations 4.50 and 4.54, it is seen that matrices ($\mathcal{Q}\,\mathcal{R}\,\mathcal{Q}^T$) and ($\mathcal{P}^T\,\mathcal{S}\,\mathcal{P}$) can commute, which implies that both of them have the identical eigenvector matrix. Furthermore, since ($\mathcal{Q}\,\mathcal{R}\,\mathcal{Q}^T$) and ($\mathcal{P}^T\,\mathcal{S}\,\mathcal{P}$) are symmetric, the associated eigenvector matrix can be orthonormal, which is denoted by $\boldsymbol{\Phi}$. Thus,

$$\mathcal{Q}\,\mathcal{R}\,\mathcal{Q}^T = \boldsymbol{\Phi}\,\mathrm{diag}\left(r_i\right)\boldsymbol{\Phi}^T \tag{4.55}$$

and

$$\mathcal{P}^T\,\mathcal{S}\,\mathcal{P} = \boldsymbol{\Phi}\,\mathrm{diag}\left(s_i\right)\boldsymbol{\Phi}^T \tag{4.56}$$

where r_i and s_i are corresponding eigenvalues of $\mathbf{Q}\mathcal{R}\mathbf{Q}^T$ and $\mathbf{P}\mathcal{S}\mathcal{P}^T$; and

$$\boldsymbol{\Phi}\boldsymbol{\Phi}^T = I \tag{4.57}$$

Substituting Equations 4.55 and 4.56 into Equation 4.54 results in

$$\boldsymbol{\Phi}\ \mathrm{diag}(r_i)\ \boldsymbol{\Phi}^T\boldsymbol{\Phi}\ \mathrm{diag}(s_i)\ \boldsymbol{\Phi}^T = \boldsymbol{\Delta} \tag{4.58}$$

Premultiplying and postmultiplying $\boldsymbol{\Phi}^T$ and $\boldsymbol{\Phi}$ on both sides of Equation 4.58 respectively and considering Equation 4.57,

$$\mathrm{diag}(r_i)\ \mathrm{diag}(s_i) = \boldsymbol{\Phi}^T\boldsymbol{\Delta}\boldsymbol{\Phi} \tag{4.59}$$

Therefore,

$$\mathrm{diag}(r_i s_i) = \boldsymbol{\Phi}^T\boldsymbol{\Delta}\boldsymbol{\Phi} \tag{4.60}$$

Equation 4.60 implies that

$$\boldsymbol{\Phi}^T\ \boldsymbol{\Delta}\boldsymbol{\Phi} = \text{diagonal} \tag{4.61}$$

Thus,

$$\mathcal{Q}\ \mathcal{R}\ \mathcal{Q}^T = \mathrm{diag}(r_i) \tag{4.62}$$

and

$$\mathcal{P}^T\ \boldsymbol{S}\ \mathcal{P} = \mathrm{diag}(s_i) \tag{4.63}$$

as well as

$$r_i\, s_i = \lambda_i, \quad i = 1, 2, \ldots, n \tag{4.64}$$

and

$$r_i^*\, s_i^* = \lambda_i^*, \quad i = 1, 2, \ldots, n \tag{4.65}$$

Note that, in the above discussion, the condition of nonrepeated eigenvalues are not necessary.

Example 4.2

Reviewing the **M-C-K** system mentioned in Example 4.1, the inverse matrix of $\mathcal{P}$ is calculated as

$$\mathcal{Q} = \begin{bmatrix} 0.2156+0.0459j & 0.4596+0.0719j & 0.7635+2.3339j & 0.5342+5.1603j \\ -0.4351-0.0215j & 0.2563+0.0522j & -0.5014-13.2852j & 0.1416+7.9275j \\ 0.2156-0.0459j & 0.4596-0.0719j & 0.7635-2.3339j & 0.5342-5.1603j \\ -0.4351+0.0215j & 0.2563-0.0522j & -0.5014+13.2852j & 0.1416-7.9275j \end{bmatrix}$$

Therefore, the accompanist matrix **Q** of the mode shape **P** is

$$\mathbf{Q} = \begin{bmatrix} 0.2156+0.0459j & 0.4596+0.0719j \\ -0.4351-0.0215j & 0.2563+0.0522j \end{bmatrix}$$

In addition, it can be seen that

$$\mathbf{Q} = -\mathbf{P}^{-1}\left(\mathbf{P}^{*}\boldsymbol{\Lambda}^{*}\mathbf{P}^{*-1} - \mathbf{P}\,\boldsymbol{\Lambda}\,\mathbf{P}^{-1}\right)^{-1}$$

Futhermore,

$$\mathcal{R} = \begin{bmatrix} -2.5 & 1 & 0.5 & 0 \\ 1 & -2.4 & 0 & 0.4 \\ 0.5 & 0 & 0 & 0 \\ 0 & 0.4 & 0 & 0 \end{bmatrix} \text{ and } \boldsymbol{S} = \begin{bmatrix} 2 & 0 & 0 & 0 \\ 0 & 2.5 & 0 & 0 \\ 0 & 0 & -1500 & 750 \\ 0 & 0 & 750 & 750 \end{bmatrix}$$

It is seen that, $\mathcal{Q}\,\mathcal{R}\,\mathcal{Q}^{T} = \mathrm{diag}\left([-0.7862+2.2610j, -1.2137+7.2554j, -0.7862-2.2610j, -1.2137-7.2554j]\right)$ and $\mathcal{P}^{T}\,\boldsymbol{S}\mathcal{P} = \mathrm{diag}\left([4.5499-0.7122i, 4.1309-0.2040i, 4.5499+0.7122i, 4.1309+0.2040i]\right)$.

4.1.3 Linear Independency and Orthogonality Conditions

In Chapter 3, the orthogonality of normal modes was introduced without further explanation as to why the modes were orthogonal (see also Clough and Penzien 1993).

Note that Equations 4.62 through 4.65 can also be obtained by examining the generalized eigen-decomposition of symmetric matrices $\mathcal{R}$ and $\boldsymbol{S}$ by using orthogonal conditions. For example,

$$\boldsymbol{Q}_i\,\mathcal{R}\,\boldsymbol{Q}_j^{T} = \begin{cases} r_i & i = j_i \\ 0 & i \neq j \end{cases} \tag{4.66}$$

$$\boldsymbol{P}_i^{T}\,\boldsymbol{S}\,\boldsymbol{P}_j = \begin{cases} s_i & i = j_i \\ 0 & i \neq j \end{cases} \tag{4.67}$$

That is, vectors $\boldsymbol{Q}_i$ and $\boldsymbol{P}_i$ are weighted orthogonal with the weighting function $\mathcal{R}$ and $\boldsymbol{S}$, respectively. Note that $\boldsymbol{Q}_i$ and $\boldsymbol{P}_i$ are the ith row and column of matrices $\mathcal{Q}$ and $\mathcal{P}$, and their dimensions are $1 \times 2n$ and $2n \times 1$, respectively.

Although the orthogonal properties described in Equations 4.66 and 4.67 have been proven in Equations 4.62 and 4.63, the orthogonality is checked in an alternative way as follows (Ewins 1984):

Premultiplying $\mathcal{R}^{-1}$ on both sides of Equation 4.48, results in

$$\boldsymbol{S}\,\mathcal{P} = \mathcal{R}^{-1}\,\mathcal{P}\,\boldsymbol{\Delta} \tag{4.68}$$

Therefore, for the ith and j^{th} eigenvalues and eigenvectors,

$$\boldsymbol{S}\,\boldsymbol{P}_i = \lambda_i\,\mathcal{R}^{-1}\,\boldsymbol{P}_i \tag{4.69}$$

and

$$\boldsymbol{S}\boldsymbol{P}_j = \lambda_j \mathcal{R}^{-1} \boldsymbol{P}_j \tag{4.70}$$

Premultiplying $\boldsymbol{P}_j^T$ on both sides of Equation 4.69, results in

$$\boldsymbol{P}_j^T \boldsymbol{S}\boldsymbol{P}_i = \lambda_i \boldsymbol{P}_j^T \mathcal{R}^{-1} \boldsymbol{P}_i \tag{4.71}$$

Because both $\mathcal{R}$ and $\boldsymbol{S}$ are symmetric, it follows that $\mathcal{R}^{-1}$ will also be symmetric. Postmultiplying $\boldsymbol{P}_i$ on both sides of the transpose of Equation 4.70, results in

$$\boldsymbol{P}_j^T \boldsymbol{S}\boldsymbol{P}_i = \lambda_j \boldsymbol{P}_j^T \mathcal{R}^{-1} \boldsymbol{P}_i \tag{4.72}$$

Subtracting Equation 4.72 from Equation 4.71 yields

$$\mathbf{0} = \boldsymbol{P}_j^T \mathcal{R}^{-1} \boldsymbol{P}_i (\lambda_i - \lambda_j) \tag{4.73}$$

Equation 4.73 implies that when $\lambda_i \neq \lambda_j$, the matrix production $\boldsymbol{P}_j^T \mathcal{R}^{-1} \boldsymbol{P}_i$ must be zero, since both λ_i and λ_j are nonzero values. Therefore, $\boldsymbol{P}_j$ and $\boldsymbol{P}_i$ are orthogonal with a weighting matrix of $\mathcal{R}^{-1}$.

Furthermore, it is easy to see that $\boldsymbol{P}_j$ and $\boldsymbol{P}_i$ are also orthogonal with weighting matrix $\mathcal{R}$. The inverse of both sides of Equation 4.48 can be taken and $\mathcal{R}$ can be postmultiplied on both sides of the resulting equation. Since $\mathcal{Q} = \mathcal{P}^{-1}$,

$$\mathcal{Q}\, \boldsymbol{S}^{-1} = \boldsymbol{\Delta}^{-1}\, \mathcal{Q}\, \mathcal{R} \tag{4.74}$$

which implies

$$\boldsymbol{Q}_i \boldsymbol{S}^{-1} = 1/\lambda_i \boldsymbol{Q}_i \mathcal{R} \tag{4.75}$$

and

$$\boldsymbol{Q}_j \boldsymbol{S}^{-1} = 1/\lambda_j \boldsymbol{Q}_j \mathcal{R} \tag{4.76}$$

Note that it is assumed that the system does not have repeated eigenvalues. Therefore, similarly,

$$\mathbf{0} = \boldsymbol{Q}_i \mathcal{R} \boldsymbol{Q}_j^T (1/\lambda_i - 1/\lambda_j) \tag{4.77}$$

Equation 4.76 implies that when $1/\lambda_i \neq 1/\lambda_j$, the matrix production $\boldsymbol{Q}_i \mathcal{R} \boldsymbol{Q}_j^T$ must be zero, since both $1/\lambda_i$ and $1/\lambda_j$ are nonzero values. Therefore, $\boldsymbol{Q}_j$ and $\boldsymbol{Q}_i$ are orthogonal with weighting matrix $\mathcal{R}$. Similarly,

$$\mathcal{R}^{-1}\, \mathcal{P} = \boldsymbol{S}\, \mathcal{P}\, \boldsymbol{\Delta}^{-1} \tag{4.78}$$

As well as

$$\mathcal{R}^{-1} \boldsymbol{P}_i = \boldsymbol{S}\boldsymbol{P}_i/\lambda_i \tag{4.79}$$

and

$$\mathcal{R}^{-1}\boldsymbol{P}_j = \boldsymbol{S}\boldsymbol{P}_j/\lambda_j \tag{4.80}$$

By the same procedure, it can be proven that $\boldsymbol{P}_j$ and $\boldsymbol{P}_i$ are also orthogonal with weighting matrix $\boldsymbol{S}$ by generating

$$\boldsymbol{P}_j^T\boldsymbol{S}\boldsymbol{P}_i\ (1/\lambda_i - 1/\lambda_j) = 0 \tag{4.81}$$

The orthogonal conditions described by Equations 4.66 and 4.67 are good properties to use in the study of state equations, and to prove these the assumption of $\lambda_i \neq \lambda_j$ when $i \neq j$ is required. However, it should be understood that the essence of Equations 4.62 and 4.63 is not necessarily the orthogonal conditions, but the linear independency of all the eigenvectors $\boldsymbol{P}_i$ in matrix $\mathcal{P}$. This is the first reason not to use the orthogonal conditions for further studies.

The second reason not to use the orthogonal condition is that the value of r_i is not unique, due to the nonuniqueness of the eigenvector $\boldsymbol{P}_i$ for all modes. In order to use the values of r_i and s_i for all modes, the method of normalization needs to be specified, which may be different from those provided by most building codes.

The third reason not to use the orthogonal approach is that it is preferable to use a minimum number of equations in practical designs.

A method to obtain the relationship between $\boldsymbol{Q}_i$ and $\boldsymbol{P}_i$ without using the orthogonal conditions is explained below. Note that $\boldsymbol{Q}_i$ and $\boldsymbol{P}_i$ are further used in a later section to introduce the modal participation indicators, etc.

Postmultiplying $\mathcal{P}^T$ on both sides of Equation 4.62 yields

$$\mathcal{Q}\,\mathcal{R} = \text{diag}(r_i)\ \mathcal{P}^T \tag{4.82}$$

Now, consider the ith row of the above matrix equation. The ith row of matrix $\mathcal{Q}$ is denoted as

$$\boldsymbol{Q}_i = [\boldsymbol{q}_i, \boldsymbol{r}_i] \tag{4.83}$$

which results in

$$[\boldsymbol{q}_i, \boldsymbol{r}_i]\mathcal{R} = r_i\left[\lambda_i\boldsymbol{p}_i^T,\ \boldsymbol{p}_i^T\right] \tag{4.84}$$

where $\boldsymbol{r}_i$ is the ith row of matrix $\mathbf{R}$ defined in Equation 4.41.

Note that $\boldsymbol{q}_i$ contains the first n elements of $\mathbf{Q}$, and $\boldsymbol{p}_i$ contains the last n elements of $\mathbf{P}$. Therefore, the dimensions of $\boldsymbol{q}_i$ and $\boldsymbol{p}_i$ are, respectively, $1 \times n$ and $n \times 1$.

Using Equation 4.46, Equation 4.84 can further be written as

$$\left[-\boldsymbol{q}_i\mathbf{M}^{-1}\mathbf{C}\mathbf{M}^{-1} + \boldsymbol{r}_i\mathbf{M}^{-1},\ \boldsymbol{q}_i\mathbf{M}^{-1}\right] = \left[r_i\lambda_i\boldsymbol{p}_i^T,\ r_i\boldsymbol{p}_i^T\right] \tag{4.85}$$

Therefore, from the second element of Equation 4.85,

$$\boldsymbol{q}_i = r_i\boldsymbol{p}_i^T\mathbf{M} \tag{4.86}$$

Combining all n of $\boldsymbol{q}_i$ in Equation 4.86, it is seen that

$$\mathbf{Q} = \boldsymbol{\Lambda}_R \mathbf{P}^T \mathbf{M} \tag{4.87}$$

where

$$\boldsymbol{\Lambda}_R = \mathrm{diag}(r_i), \quad i = 1, 2, \ldots, n \tag{4.88}$$

There are n conjugate pairs of r_i. Note that, those real-valued r_i associated with overdamped subsystems can also be treated as pairs. In Equation 4.88, r_i is arranged as the first portion of the conjugate pair. Next, all values of r_i are examined, in order to further represent vector $\boldsymbol{q}_i$.

Taking the inverse of both sides of Equation 4.62 and using the relationship described in Equation 4.88,

$$\mathcal{P}^T \mathcal{R}^{-1} \mathcal{P} = \mathrm{diag}(1/r_i) \tag{4.89}$$

or

$$\boldsymbol{P}_i^T \mathcal{R}^{-1} \boldsymbol{P}_i = 1/r_i \tag{4.90}$$

It can be seen that

$$\mathcal{R}^{-1} = \begin{bmatrix} \mathbf{0} & \mathbf{M} \\ \mathbf{M} & \mathbf{C} \end{bmatrix} \tag{4.91}$$

Thus,

$$\boldsymbol{p}_i^T \mathrm{M} \boldsymbol{p}_i \lambda_i + \lambda_i \boldsymbol{p}_i^T \mathrm{M} \boldsymbol{p}_i + \boldsymbol{p}_i^T \mathbf{C} \boldsymbol{p}_i = 1/r_i \tag{4.92}$$

Therefore,

$$r_i = \frac{1}{\boldsymbol{p}_i^T (2\lambda_i \mathbf{M} + \mathbf{C}) \boldsymbol{p}_i} = \frac{\lambda_i}{\boldsymbol{p}_i^T (\lambda_i^2 \mathbf{M} - \mathbf{K}) \boldsymbol{p}_i} \tag{4.93}$$

Similarly,

$$s_i = \boldsymbol{p}_i^T (\lambda_i^2 \mathbf{M} - \mathbf{K}) \boldsymbol{p}_i \tag{4.94}$$

Furthermore, the ith row vector of matrix $\mathbf{Q}$ can be written as

$$\boldsymbol{q}_i = \frac{\boldsymbol{p}_i^T \mathbf{M}}{\boldsymbol{p}_i^T (2\lambda_i \mathbf{M} + \mathbf{C}) \boldsymbol{p}_i} = \frac{\lambda_i \boldsymbol{p}_i^T \mathbf{M}}{\boldsymbol{p}_i^T (\lambda_i^2 \mathbf{M} - \mathbf{K}) \boldsymbol{p}_i} \tag{4.95}$$

Note that

$$\xi_i = \frac{1}{2\omega_i} \frac{\boldsymbol{p}_i^H \mathbf{C} \boldsymbol{p}_i}{\boldsymbol{p}_i^H \mathbf{M} \boldsymbol{p}_i} \tag{4.96}$$

and

$$\omega_i^2 = \frac{p_i^H K p_i}{p_i^H M p_i} \tag{4.97}$$

Here, superscript H stands for Hermitian transpose.

Using Equation 4.86, it can be proven that the following equations hold:

$$\xi_i = \frac{1}{2\omega_i}\frac{q_i^* M^{-1} C p_i}{q_i^* p_i} \tag{4.98}$$

and

$$\omega_i^2 = \frac{q_i^* M^{-1} K p_i}{q_i^* p_i} \tag{4.99}$$

Example 4.3

Example 4.3 continues with a review of Example 4.1. Through Equation 4.93, it is calculated that

$$r_1 = \frac{1}{2\lambda_1(p_1^T M p_1) + p_1^T C p_1} = \frac{1}{\lambda_1(p_1^T M p_1) - (p_1^T K p_1)/\lambda_1} = -0.7862 + 2.2610j$$

$$r_2 = \frac{1}{2\lambda_2(p_2^T M p_2) + p_2^T C p_2} = \frac{1}{\lambda_2(p_2^T M p_2) - (p_2^T K p_2)/\lambda_2} = -1.2137 + 7.2554i$$

and

$$s_1 = \lambda_1^2 p_1^T M p_1 - p_1^T K p_1 = 4.5499 - 0.7122j$$

$$s_2 = \lambda_2^2 p_2^T M p_2 - p_2^T K p_2 = 4.1309 - 0.2040j$$

Furthermore,

$$\omega_1^2 = \frac{q_1^* M^{-1} K p_1}{q_1^* p_1} = 121.5336$$

so that $\omega_1 = 11.0242$, and

$$\omega_2^2 = \frac{q_2^* M^{-1} K p_2}{q_2^* p_2} = 925.6701$$

thus, $\omega_2 = 30.4248$. In addition,

$$\xi_1 = \frac{1}{2\omega_1}\frac{q_1^* M^{-1} C p_1}{q_1^* p_1} = 0.1784$$

and

$$\xi_2 = \frac{1}{2\omega_2}\frac{q_2^* M^{-1} C p_2}{q_2^* p_2} = 0.1161$$

4.1.4 Approach of Complex Damping

From the above discussion, it is known that nonproportionally damped systems have unique characteristics that are different from proportionally damped systems. Generally speaking, because the distribution of the damping is irregular with respect to the distribution of the stiffness, the function of damping will not only dissipate energy, but also transfer it. A proportionally damped system can be decoupled in an n-dimensional space, whereas a nonproportionally damped system cannot. A nonproportionally damped system can be realized as a system that has evolved from a proportionally damped system by gradually adding nonproportional damping. It is understandable that, at the very beginning, when the amount of damping is sufficiently small, the behavior of a nonproportionally damped system is very similar to the original proportionally damped system. However, to accurately represent the nonproportionally damped systems modally, complex modes must be used. In this case, the set of complex modes can be thought of as those that have evolved from the original set of normal modes. The reason for the evolution is the energy transfer due to nonproportional damping among the original set of normal modes.

It is understandable that either a normal mode or a complex mode will always be an independent "energy island," which will have no energy exchange with other modes. However, if the complex mode is seen as the result of an evolutionary process, it can further be understood that this new "energy island" is formed by energy transfer among the original set of modes. In this situation, it is seen that the function of the nonproportional damping will not only dissipate energy, but also transfer energy. However, the function of proportional damping is only energy dissipation. In order to quantitatively express the energy dissipation, the concept of a damping ratio has already been introduced. The concept of the modal energy transfer ratio (ETR) in order to quantitatively express the energy transfer has been discussed in Liang and Lee (1991). For convenience, some important issues are repeated here to form a foundation for damper design in the following discussion.

In regular structural design against earthquake excitations, many building codes employ the concept of structural irregularity, such as plan and vertical irregularity. In damper design, the concept of structural irregularity should be considered as stiffness, mass, and damping irregularity. The concept of complex mode and modal ETR can help engineers further understand the damping irregularity, both qualitatively and quantitatively.

To begin this discussion, the *model energy transfer ratio* of the ith mode, denoted by ζ_i, is considered as (Liang and Lee 1991)

$$\zeta_i = \ln\frac{\omega_i}{\omega_{ni}} \tag{4.100}$$

Here, ω_i is the ith natural frequency of a nonproportionally damped system, whereas ω_{ni} is the ith natural frequency of the corresponding proportionally damped system, which can be seen as an undamped system (C = 0) whose mass and stiffness matrices are, respectively, identical to the nonproportionally damped system. Since both the undamped and the proportionally damped systems have normal modes, the undamped system can be seen as a special case of the proportionally damped system, as described in Chapter 3. In the following discussion, it can be seen that the quantity ζ_i, defined in Equation 4.100, is related to modal energy transfer.

The ETR ζ_i can be examined from a different perspective. In the following paragraph, it is seen that if the ith mode is truly complex, then the modal natural frequency ω_i must be different from the ith normal mode of the corresponding proportionally damped system.

To begin this discussion, consider the homogeneous form of Equation 3.279. That is,

$$\mathbf{I}\ddot{z}(t)+\tilde{\mathbf{C}}\dot{z}(t)+\tilde{\mathbf{K}}z(t)=0 \tag{4.101}$$

Note that the system described by Equation 4.101 has monic mass or the mass matrix is an identity matrix. In Equation 4.101, generalized stiffness and damping matrices $\tilde{\mathbf{K}}$ and $\tilde{\mathbf{C}}$ are defined in Equations 3.81 and 3.274, respectively. In Chapter 3, proportionally damped systems were primarily discussed, but in this chapter, nonproportional damping is introduced. Therefore, the Caughey criterion described in Equation 3.285 may no longer hold.

This monic (**I**-$\tilde{\mathbf{C}}$-$\tilde{\mathbf{K}}$) system will generally have a different modal space from the original (**M-C-K**) system. The newly formed (**I**-$\tilde{\mathbf{C}}$-$\tilde{\mathbf{K}}$) system is only used for simplicity, because both $\tilde{\mathbf{C}}$ and $\tilde{\mathbf{K}}$ are symmetric (see Equations 3.284 and 3.81). For convenience, the modal space of the monic (**I**-$\tilde{\mathbf{C}}$-$\tilde{\mathbf{K}}$) system is called the *monic modal space*, and the corresponding mode is called the *monic mode.* The modal space of the generic original (**M-C-K**) system is called the *general modal space*, and the corresponding mode is called the *mode* if necessary. In Section 4.5.4, the method of dual modes is introduced to solve a *generic* practical problem in damper design when the damping ratios for a structure with added dampers must be calculated more accurately and the stiffness matrix is not available. In that section, the monic mode will again be used.

Now, one of the conjugate pair of the spatial portion of Equation 4.101 can be further written as follows:

$$\mathbf{V}^{\mathrm{T}}\tilde{\mathbf{P}}\Lambda^2+\mathbf{V}^{\mathrm{T}}\tilde{\mathbf{C}}\tilde{\mathbf{P}}\Lambda+\mathbf{\Omega}^2\tilde{\mathbf{P}}=0 \tag{4.102}$$

Equation 4.64 is in matrix form. Here **V** and $\tilde{\mathbf{P}}$ are the eigenvector matrix of the stiffness $\tilde{\mathbf{K}}$ and the mode shape matrix of the system defined previously, respectively. That is,

$$\mathbf{V}=[\boldsymbol{v}_1,\boldsymbol{v}_2,\ldots,\boldsymbol{v}_n]_{n\times n} \tag{4.103}$$

and

$$\tilde{\mathbf{P}}=[\tilde{\boldsymbol{p}}_1,\tilde{\boldsymbol{p}}_2,\ldots,\tilde{\boldsymbol{p}}_n]_{n\times n} \tag{4.104}$$

Note that again

$$\mathbf{\Omega}^2=\mathrm{diag}\left(\omega_{ni}^2\right) \tag{4.105}$$

is the eigenvalue matrix of stiffness $\tilde{\mathbf{K}}$. Note that if the system is proportionally damped, then

$$\omega_{ni}=\omega_i \tag{4.106}$$

which is the ith natural frequency of the system.

With the notations defined in Equation 4.105, the ii^{th} entry of each matrix of Equation 4.102 forms

$$\omega_i^2\boldsymbol{v}_i^{\mathrm{T}}\tilde{\boldsymbol{p}}_i+\omega_i\boldsymbol{v}_i^{\mathrm{T}}\tilde{\mathbf{C}}\tilde{\boldsymbol{p}}_i+\omega_{ni}^2\boldsymbol{v}_i^{\mathrm{T}}\tilde{\boldsymbol{p}}_i=0$$

or

$$\omega_i^2 + \omega_i\left(\boldsymbol{v}_i^T\tilde{\mathbf{C}}\tilde{\boldsymbol{p}}_i\right)/\left(\boldsymbol{v}_i^T\tilde{\boldsymbol{p}}_i\right) + \omega_{ni}^2 = 0 \tag{4.107}$$

Now, consider the generalized Rayleigh quotient in Equation 4.107 and note that subscript i denotes the ith mode.

Since the ith mode shape $\boldsymbol{p}_i$ is, in general, complex valued, so is the generalized Rayleigh quotient. Therefore,

$$\frac{\boldsymbol{v}_i^T\tilde{\mathbf{C}}\tilde{\boldsymbol{p}}_i}{\boldsymbol{v}_i^T\tilde{\boldsymbol{p}}_i} = \alpha_i + j\beta_i \tag{4.108}$$

where α_i and β_i are real numbers. When the damping of the system is not sufficiently large, Equation 4.108 is rewritten as

$$\frac{1}{2\omega_{ni}}\frac{\boldsymbol{v}_i^T\tilde{\mathbf{C}}\tilde{\boldsymbol{p}}_i}{\boldsymbol{v}_i^T\tilde{\boldsymbol{p}}_i} \approx \xi_i + j\zeta_i \tag{4.109}$$

That is, the real part on the right side of Equation 4.109 is approximately the conventionally defined damping ratio ξ_i and the imaginary part is identical to that defined in Equation 4.100 (Liang and Lee 1991).

Equation 4.109 can also be rewritten as

$$\frac{1}{2\omega_{ni}}\frac{\boldsymbol{u}_i^T\mathbf{C}\boldsymbol{p}_i}{\boldsymbol{u}_i^T\mathbf{M}\boldsymbol{p}_i} \approx \xi_i + j\zeta_i \tag{4.110}$$

where $\boldsymbol{u}_i$ and $\boldsymbol{p}_i$ are, respectively, the ith mode shapes of the **M-zero-K** and the **M-C-K** system. Namely, $\boldsymbol{u}_i$ is the ith eigenvector of matrix $\mathbf{M}^{-1}\mathbf{K}$.

The imaginary part ζ_i is not zero valued unless the ith mode is normal and only if the ith mode is normal, $\zeta_i = 0$. In other words, all the imaginary parts

$$\zeta_i = 0, \quad i = 1,2,\ldots,n \tag{4.111}$$

satisfy the *sufficient and necessary condition of a proportional damping system.*

The damping ratio here is still the ratio between the dissipated energy E_{di} during a vibration cycle and the total energy E_{pi} of the ith mode in the case of steady-state vibration, that is, Timoshenko damping still holds (see Equation 1.217), which is rewritten as

$$\xi_i = E_{di}/4\pi E_{pi}$$

In addition, ζ_i is the ratio between the transferred energy E_{Ti} during a vibration cycle and the total energy E_{pi} of the ith mode,

$$\zeta_i = E_{Ti}/4\pi E_{pi} \tag{4.112}$$

That is, the term ζ_i is the ratio between the energy E_{Ti} transferred during a vibration cycle and the total energy E_{pi} before the cycle. In this regard, ζ_i is the ith modal ETR. Here the energy E_{Ti} transferred is defined as work done by imaginary damping force $2\zeta_i\omega_{ni}\dot{x}_i(t)$ during a complete cycle of $2\pi/\omega_{ni}$ (Liang and Lee 1991).

Furthermore, it can be shown that

$$\omega_i = \omega_{ni} e^{\zeta_i} \tag{4.113}$$

Equation 4.113 (also see Equation 4.100) states that if the ith mode is complex, the natural frequency is changed by the factor e^{ζ_i}. Note that the natural frequency ω_i (nonproportionally damped) and/or ω_{ni} (proportionally damped) is actually the square root of the generalized modal energy. Therefore, Equation 4.113 implies that, for a nonproportionally damped system, there must be a certain amount of energy transferred among modes and the ETR is the quantity ζ_i. From this perspective, the modal ETR ζ_i changes the modal energy quantitatively represented by the relationship given by Equation 4.113. It follows that the real part ξ_i, the damping ratio, is the ratio of energy dissipation.

In the above derivation, no assumptions or restrictions are placed on the physical parameters of mass, damping, and stiffness. If and only if the system is proportionally damped, then all the modal ETRs are equal to zero. This implies that the modal ETR is the parameter of the system itself and, from Equation 4.110, it is shown that the modal ETR cannot be represented by the conventionally defined modal parameters (the natural frequency, damping ratio, and mode shape).

Generally speaking, if the norm of the damping matrix is comparatively small, both the damping ratio and the ETR will be small. As in conventional design, when the damping ratio is taken at about 5%, the ETR may be smaller than 1%. When damping is purposely increased for vibration control, such as in damper-added and/or base isolated structures, both the damping ratio and the ETR can be significant.

The responses of the structure can be further magnified under two-directional ground motion inputs. In base isolation, often a greater than 15% damping ratio is incorporated into the design, where the damping of the base isolator is often large, but that of the superstructure can be very small. Because of the unbalanced nature of damping distribution, the structure can be heavily nonproportionally damped.

Example 4.4

Consider the **M-C-K** system mentioned in Example 4.1. The eigenvector matrix $\boldsymbol{V}$ is calculated by the eigen-decomposition of $\tilde{\mathbf{K}} = \mathbf{M}^{-1/2}\mathbf{K}\mathbf{M}^{-1/2}$, and the eigenvector matrix $\boldsymbol{U}$ is obtained by the eigen-decomposition of $\mathbf{M}^{-1}\mathbf{K}$, which are, respectively,

$$\mathbf{V} = \begin{bmatrix} 0.4706 & -0.8824 \\ 0.8824 & 0.4706 \end{bmatrix}, \quad \mathbf{U} = \begin{bmatrix} 0.5122 & 0.9026 \\ 0.8589 & -0.4306 \end{bmatrix}$$

The mode shape matrix of the corresponding monic ($\mathbf{I}$-$\tilde{\mathbf{C}}$-$\tilde{\mathbf{K}}$) system is

$$\tilde{\mathbf{P}} = \begin{bmatrix} 0.0053+0.0042j & 0.0034+0.0287j \\ 0.0142+0.0785j & 0.0005-0.0155j \end{bmatrix}$$

Note that the mode shape $\tilde{\mathbf{P}}$ matrix of the **M-C-K** system was obtained in Example 4.1. Thus,

$$\frac{1}{2\omega_{n1}} \frac{\boldsymbol{v}_1^T \tilde{\mathbf{C}}\, \tilde{\boldsymbol{p}}_1}{\boldsymbol{v}_1^T\, \tilde{\boldsymbol{p}}_1} = \frac{1}{2\omega_{n1}} \frac{\boldsymbol{u}_1^T \mathbf{C}\, \boldsymbol{p}_1}{\boldsymbol{u}_1^T \mathbf{M}\, \boldsymbol{p}_1} \approx \xi_1 + j\zeta_1 = 0.1784 - 0.0017j$$

$$\frac{1}{2\omega_{n2}} \frac{\boldsymbol{v}_2^T \tilde{\mathbf{C}}\, \tilde{\boldsymbol{p}}_2}{\boldsymbol{v}_2^T\, \tilde{\boldsymbol{p}}_2} = \frac{1}{2\omega_{n2}} \frac{\boldsymbol{u}_2^T \mathbf{C}\, \boldsymbol{p}_2}{\boldsymbol{u}_2^T \mathbf{M}\, \boldsymbol{p}_2} \approx \xi_2 + j\zeta_2 = 0.1161 + 0.0017j$$

Furthermore, it is seen that the ETRs are 0.0017 and –0.0017 for the first and second modes. Note that,

$$\zeta_1 + \zeta_2 = 0$$

If the damping is proportional, the corresponding natural frequency will be the square roots of the eigenvalues of matrix $\tilde{\mathbf{K}}$, which are 11.0051 and 30.4777, respectively. Further comparing ω_1 and ω_2, which are 11.0242 and 30.4248 rad/sec, respectively, it is seen that,

$$\ln(11.0242/11.0051) = 0.0017$$

and

$$\ln(30.4248/30.4777) = -0.0017$$

The results agree with the calculated ETRs.

Example 4.5

A specially selected 5-DOF system is used to illustrate the effect of nonproportional damping, which results from adding certain damping matrices. Suppose, originally, $\mathbf{M} = \mathbf{I}$ (t); $\mathbf{C}_0 = 142.9\ \mathbf{M}$ (kN-s/m), and

$$\mathbf{K} = \begin{bmatrix} 338.4783 & -308.4909 & 116.2266 & 43.8386 & -42.3444 \\ -308.4909 & 454.7049 & -264.6522 & 73.8822 & 1.4942 \\ 116.2266 & -264.6522 & 412.3605 & -306.9967 & 117.7208 \\ 43.8386 & 73.8822 & -306.9967 & 456.1991 & -190.7701 \\ -42.3444 & 1.4942 & 117.7208 & -190.7701 & 147.7082 \end{bmatrix} \times 1000 \quad (\text{kN/m})$$

Since $\mathbf{C}$ is a Rayleigh damping matrix, the system is proportionally damped. The natural frequencies are listed in the first row of Table 4.1.

With added damping matrix $\mathbf{C}_a$,

$$\mathbf{C}_a = \begin{bmatrix} 0.6369 & -0.2676 & 0.3728 & -0.3699 & 0.5741 \\ -0.2676 & 0.1134 & -0.2100 & 0.1287 & -0.2636 \\ 0.3728 & -0.2100 & 3.0939 & 1.2200 & 1.5422 \\ -0.3699 & 0.1287 & 1.2200 & 0.9324 & 0.2691 \\ 0.5741 & -0.2636 & 1.5422 & 0.2691 & 1.0234 \end{bmatrix} \times 1000 \quad (\text{kN-s/m})$$

the system will become nonproportionally damped; the corresponding natural frequencies are listed in the first and second rows of Table 4.1 and the ETRs and damping ratios are listed in the third and fourth rows, of Table 4.1, respectively.

The mode shapes of the proportionally and nonproportionally damped systems are, respectively,

TABLE 4.1
Modal Parameters of the Proportionally and Newly Formed Systems

	1st Mode	2nd Mode	3rd Mode	4th Mode	5th Mode
ω_{ni}	1.1370	1.1500	1.4895	4.0000	5.0000
ω_i	1.1500	1.1569	1.4661	4.0000	4.9923
ζ_i	0.0120	0.0063	−0.0158	0.0028	−0.0015
ξ_i	0.0118	0.2194	0.0625	0.0028	0.0292

$$\begin{bmatrix} 1.0000 & 1.0000 & 1.0000 & 1.0000 & 1.0000 \\ 1.9190 & 1.3097 & 0.2846 & -0.8308 & -1.6825 \\ 2.6825 & 0.7154 & -0.9190 & -0.3097 & 1.8308 \\ 3.2287 & -0.3728 & -0.5462 & 1.0882 & -1.3979 \\ 3.5133 & -1.2036 & 0.7635 & -0.5944 & 0.5211 \end{bmatrix}$$

and

$$\begin{bmatrix} 1.0000 & 1.0000 & 1.0000 & 1.0000 & 1.0000 \\ 1.3097+0.0000j & 1.6347+1.3214j & 0.3603+0.1022j & -0.8308 & -1.6782-0.1070j \\ 0.7154+0.0000j & 1.9435+3.0877j & -0.7668+0.2365j & -0.3097 & 1.8178-0.0246j \\ -0.3728+0.0000j & 2.5134+3.1437j & -0.3790+0.2420j & 1.0882 & -1.3996-0.1370j \\ -1.2036+0.0000j & 2.9862+2.2994j & 0.8846+0.1768j & -0.5944 & 0.5186-0.0885j \end{bmatrix}$$

Examining the above mode shapes, it is seen that by adding damping $\mathbf{C}_a$, the mode shapes of the nonproportionally damped system generally become complex valued. The resulting first mode evolves from the second normal mode of the original proportionally damped system. It does not have a large imaginary value because the corresponding ETR is comparatively small (0.59%). The second complex mode evolves from the first normal mode. The positions of the third, fourth, and fifth modes do not change. The twist between the first and second modes can be illustrated by multiplying a factor on $\mathbf{C}_a$ from 0% to 100%. The variation of the first and second natural frequencies is plotted in Figure 4.2.

Note that the fourth mode remains normal because the corresponding ETR is exactly zero, which implies that no energy is transferred into or out of the fourth mode. Note that by using

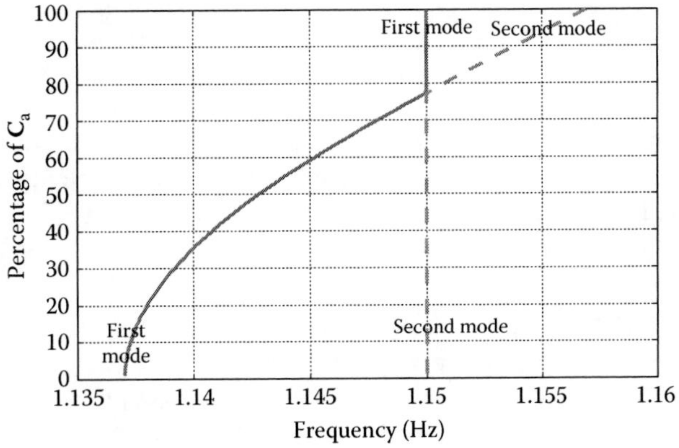

FIGURE 4.2 Twist of natural frequencies.

TABLE 4.2
Total Base Shears (1000 kN)

Original system	4.6155
Proportionally damped	4.2067
Nonproportionally damped	1.7918

the Caughey criterion, the normal mode in nonproportionally damped systems cannot be identified.

Furthermore, the damping ratios ξ_i and ETRs ζ_i of the nonproportionally damped system are listed in the third row of Table 4.1 in the form of $\xi_i + j\zeta_i$. Note that the sum of all ETRs is zero, which means that by considering the modal energy transfers only, no energy is dissipated.

$$\sum \zeta_i = [(1.20+0.63)+(-1.58-0.15)]\% = [1.83-1.83]\% = 0$$

Now, if the following damping matrix, $\mathbf{C}_p$, is used, a proportionally damped system with exactly the same damping ratios as the nonproportionally damped system can be obtained.

$$\mathbf{C}_p = \begin{bmatrix} 1.3289 & 0.6395 & 0.4508 & -0.6902 & -0.3971 \\ 0.6395 & 1.7796 & -0.0507 & 0.0536 & -1.0873 \\ 0.4508 & -0.0507 & 1.3825 & -0.4479 & -0.6366 \\ -0.6902 & 0.0536 & -0.4479 & 0.6922 & 0.0028 \\ -0.3971 & -1.0873 & -0.6366 & 0.0028 & 1.3316 \end{bmatrix} \times 1000 \quad \text{(kN-s/m)}$$

The corresponding damping ratios are listed in the fourth row of Table 4.1.

By using the El Centro earthquake records, the base shears of the proportionally and nonproportionally damped systems can be calculated. The peak values are listed in Table 4.2. Note that Table 4.2 contains only the results of using a single earthquake record. It does not necessarily mean that with the same damping ratios, a nonproportionally damped system will always have better results. The purpose of the numerical simulations is to illustrate the differences between the proportionally and nonproportionally damped systems, which may imply that damper design of a nonproportionally damped system, using a simplified proportional damping method, may result in errors.

4.1.5 Solutions in 2N Modal Space

In this section, the solution of forced response of a generally damped system is discussed. For a generally damped system, 2n-state space is often used and the mode shape function that is employed to decouple the state equation is often complex valued. However, with a specific arrangement, real-valued variables can still be used to represent the solutions.

4.1.5.1 Mode Decoupling

The solutions of Equation 4.2 are considered here. Conventionally, using the eigen-decomposition in Equation 4.38 and rearranging Equation 4.2, $\mathcal{Q} = \mathcal{P}^{-1}$ is premultiplied on both sides of the rearranged equation, and the following is obtained:

$$\mathcal{Q}\dot{X} - \Delta\mathcal{Q}X = \mathcal{Q}\begin{Bmatrix} \mathbf{M}^{-1}\mathbf{f} \\ \mathbf{0} \end{Bmatrix} \tag{4.114}$$

Denoting

$$\mathbf{Z} = \mathcal{Q}\mathbf{X} = \begin{Bmatrix} z_1 \\ z_2 \\ \dots \\ z_n \\ z_1^* \\ \dots \\ z_n^* \end{Bmatrix} \text{ and } \mathcal{F} = \mathcal{Q}\begin{Bmatrix} \mathbf{M}^{-1}\mathbf{f} \\ \mathbf{0} \end{Bmatrix} = \begin{Bmatrix} f_1 \\ f_2 \\ \dots \\ f_n \\ f_1^* \\ \dots \\ f_n^* \end{Bmatrix} \tag{4.115}$$

thus,

$$\dot{\mathbf{Z}} - \mathbf{\Delta Z} = \mathcal{F} \tag{4.116}$$

Equation 4.116 has the following decoupled form:

$$\dot{z}_i - \lambda_i z = f_i, \quad i = 1,2,\dots,n \tag{4.117}$$

Equation 4.117 is the complex-valued, first-order differential equation. Therefore,

$$\dot{z}_i^* - \lambda_i^* z^* = f_i^*, \quad i = 1,2,\dots,n \tag{4.118}$$

Equation 4.118 can be solved by direct integration. However, most computational software cannot directly handle complex integration. In the following discussion, an alternative way to solve Equations 4.117 and 4.118 is introduced.

Under earthquake excitation, the forcing function is

$$\mathbf{f} = -\mathbf{MJ}x_{g''} \tag{4.119}$$

Substituting Equation 4.119 into Equation 4.114 yields

$$\mathcal{Q}\dot{\mathbf{X}} - \mathbf{\Delta}\mathcal{Q}\mathbf{X} = -\mathcal{Q}\begin{Bmatrix} \mathbf{J}x_{g''} \\ \mathbf{0} \end{Bmatrix} \tag{4.120}$$

Thus, using the notations described in Equations 4.115 and 4.116, Equations 4.117 and 4.118 can be rewritten as

$$\dot{z}_i - \lambda_i z_i = -\boldsymbol{q}_i \mathbf{J} x_{g''} \tag{4.121}$$

and

$$\dot{z}_i^* - \lambda_i^* z^* = -\boldsymbol{q}_i^* \mathbf{J} x_{g''}^* \tag{4.122}$$

Here, $\boldsymbol{q}_i$ is defined in Equation 4.86.

4.1.5.2 Alternative Computation Method

Furthermore, Equation 4.121 and its complex conjugate Equation 4.122 can be combined to yield the following:

$$\begin{Bmatrix} \dot{z}_i \\ \dot{z}_i^* \end{Bmatrix} = \begin{bmatrix} \lambda_i & 0 \\ 0 & \lambda_i^* \end{bmatrix} \begin{Bmatrix} z_i \\ z_i^* \end{Bmatrix} - \begin{Bmatrix} \boldsymbol{q}_i \mathbf{J} \\ \boldsymbol{q}_i^* \mathbf{J} \end{Bmatrix} x_{g''} \tag{4.123}$$

For convenience, denote

$$z_i = z_{Ri} + j z_{Ii} \tag{4.124}$$

and let

$$\boldsymbol{q}_i \mathbf{J} = q_{Ri} + j q_{Ii} \tag{4.125}$$

where z_{Ri}, z_{Ii}, q_{Ri}, and q_{Ii} are real-valued scalars. Thus, the above equations separate the variable z_i and the forcing function $\boldsymbol{q}_i \mathbf{J}$ into real and imaginary parts.

Also denote a full rank matrix **G**:

$$\mathbf{G} = \frac{1}{2} \begin{bmatrix} 1 & 1 \\ -j & j \end{bmatrix} \tag{4.126}$$

It is seen that

$$\mathbf{G}^{-1} = \begin{bmatrix} 1 & j \\ 1 & -j \end{bmatrix} \tag{4.127}$$

Therefore,

$$\mathbf{G} \begin{Bmatrix} z_i \\ z_i^* \end{Bmatrix} = \begin{Bmatrix} z_{Ri} \\ z_{Ii} \end{Bmatrix} \tag{4.128}$$

and

$$\mathbf{G} \begin{Bmatrix} \boldsymbol{q}_i \mathbf{J} \\ \boldsymbol{q}_i^* \mathbf{J} \end{Bmatrix} = \begin{Bmatrix} q_{Ri} \\ q_{Ii} \end{Bmatrix} \tag{4.129}$$

Using Equations 4.128 and 4.129, premultiplying **G** on both sides of Equation 4.123, and inserting the identity

$$\mathbf{I} = \mathbf{G}^{-1}\mathbf{G} \tag{4.130}$$

in between $\begin{bmatrix} \lambda_i & \\ & \lambda_i^* \end{bmatrix}$ and $\begin{Bmatrix} z_i \\ z_i^* \end{Bmatrix}$ results in

$$\begin{Bmatrix} \dot{z}_{Ri} \\ \dot{z}_{Ii} \end{Bmatrix} = \begin{bmatrix} -\xi_i \omega_i & -\sqrt{1-\xi_i^2}\,\omega_i \\ \sqrt{1-\xi_i^2}\,\omega_i & -\xi_i \omega_i \end{bmatrix} \begin{Bmatrix} z_{Ri} \\ z_{Ii} \end{Bmatrix} - \begin{Bmatrix} q_{Ri} \\ q_{Ii} \end{Bmatrix} x_{g''} \quad i = 1, 2, \ldots, n \tag{4.131}$$

Equation 4.131 can be rewritten as

$$\begin{Bmatrix} \dot{z}_{Ri} \\ \dot{z}_{Ii} \end{Bmatrix} = \omega_i \begin{bmatrix} -\xi_i & -\sqrt{1-\xi_i^2} \\ \sqrt{1-\xi_i^2} & -\xi_i \end{bmatrix} \begin{Bmatrix} z_{Ri} \\ z_{Ii} \end{Bmatrix} - \begin{Bmatrix} q_{Ri} \\ q_{Ii} \end{Bmatrix} x_{g''} \quad i = 1,2,\ldots,n \tag{4.132}$$

or

$$\begin{Bmatrix} \dot{z}_{Ri} \\ \dot{z}_{Ii} \end{Bmatrix} = \omega_i \mathbf{L} \begin{Bmatrix} z_{Ri} \\ z_{Ii} \end{Bmatrix} - \begin{Bmatrix} q_{Ri} \\ q_{Ii} \end{Bmatrix} x_{g''} \quad i = 1,2,\ldots,n \tag{4.133}$$

Here matrix

$$\mathbf{L} = \begin{bmatrix} -\xi_i & -\sqrt{1-\xi_i^2} \\ \sqrt{1-\xi_i^2} & -\xi_i \end{bmatrix} = \begin{bmatrix} \cos(\gamma) & -\sin(\gamma) \\ \sin(\gamma) & \cos(\gamma) \end{bmatrix} \tag{4.134}$$

is seen to be an orthonormal matrix, where γ is an angle and

$$\gamma = \cos^{-1}(-\xi_i) \tag{4.135}$$

When damping is small, $\cos(\gamma) \approx 0$, and $\gamma \approx -90°$. Furthermore, in this case, the homogeneous form of Equation 4.133 can be written as

$$\begin{Bmatrix} \dot{z}_{Ri} \\ \dot{z}_{Ii} \end{Bmatrix} \approx \omega_i \begin{bmatrix} \cos(-90°) & -\sin(-90°) \\ \sin(-90°) & \cos(-90°) \end{bmatrix} \begin{Bmatrix} z_{Ri} \\ z_{Ii} \end{Bmatrix} \tag{4.136}$$

Equation 4.136 implies that the vector $\begin{Bmatrix} \dot{z}_{Ri} \\ \dot{z}_{Ii} \end{Bmatrix}$ can be seen as a rotation to nearly 90° from vector $\begin{Bmatrix} z_{Ri} \\ z_{Ii} \end{Bmatrix}$ with a uniform multiplication ω_i. In this case, vector $\begin{Bmatrix} z_{Ri} \\ z_{Ii} \end{Bmatrix}$ can be called the pseudo displacement and vector $\begin{Bmatrix} \dot{z}_{Ri} \\ \dot{z}_{Ii} \end{Bmatrix}$ can be called the pseudo velocity; from Equation 4.136, the following can result:

$$\max\left(\left|\dot{z}_{Ri}\right|\right) \approx \max\left(\omega_i \left|z_{Ii}\right|\right) \tag{4.137}$$

and

$$\max\left(\left|\dot{z}_{Ii}\right|\right) \approx \max\left(\omega_i \left|z_{Ri}\right|\right) \tag{4.138}$$

Note that Equations 4.127 and 4.138 are obtained by ignoring the forcing function. Under earthquake excitations, they will no longer hold.

Equation 4.133 turns the complex-valued Equation 4.117 and its complex conjugate Equation 4.118 into a real-valued state equation, which can be numerically calculated by most commercially available software. Thus, by using Equation 4.133, the modal response $z_i = z_{Ri} + j \cdot z_{Ii}$ for the ith mode (and its complex conjugate) can be computed, which is useful in practical designs if a particular mode is specially considered.

Consider two cases of forcing functions separately as

$$f_{Ii} = -\begin{Bmatrix} 0 \\ q_{Ii} \end{Bmatrix} x_{g''} \tag{4.139}$$

and

$$f_{Ri} = -\begin{Bmatrix} q_{Ri} \\ 0 \end{Bmatrix} x_{g''} \tag{4.140}$$

The first case described in Equation 4.139 has only the imaginary part of the forcing vector and the second case described in Equation 4.140 has only the real part of the forcing vector. Since Equation 4.131 is a linear equation, the cases of the above excitations can also be considered separately and later combined to compute the solutions of Equation 4.131.

Consider the first excitation with the imaginary part of the forcing function.

$$\begin{Bmatrix} \dot{z}_{RIi} \\ \dot{z}_{IIi} \end{Bmatrix} = \begin{bmatrix} -\xi_i\,\omega_i & -\sqrt{1-\xi_i^2}\,\omega_i \\ \sqrt{1-\xi_i^2}\,\omega_i & -\xi_i\omega_i \end{bmatrix} \begin{Bmatrix} z_{RIi} \\ z_{IIi} \end{Bmatrix} - \begin{Bmatrix} 0 \\ q_{Ii} \end{Bmatrix} x_{g''} \tag{4.141a}$$

Here, for convenience, the variables under this forcing function are denoted by z_{RIi} and z_{IIi}. The second subscript I stands for the case when only the imaginary part of the excitation is considered; the third subscript i stands for the ith mode. Similarly, with the second excitation with the real part,

$$\begin{Bmatrix} \dot{z}_{RRi} \\ \dot{z}_{IRi} \end{Bmatrix} = \begin{bmatrix} -\xi_i\,\omega_i & -\sqrt{1-\xi_i^2}\,\omega_i \\ \sqrt{1-\xi_i^2}\,\omega_i & -\xi_i\,\omega_i \end{bmatrix} \begin{Bmatrix} z_{RRi} \\ z_{IRi} \end{Bmatrix} - \begin{Bmatrix} q_{Ri} \\ 0 \end{Bmatrix} x_{g''} \tag{4.141b}$$

Under the forcing function used in Equation 4.139, the corresponding response is denoted by z_{Ii}. It is seen that

$$z_{Ii} = z_{RIi} + z_{IIi} \tag{4.142}$$

Similarly, the response under the forcing function Equation 4.140 can be written as

$$z_{Ri} = z_{RRi} + z_{IRi} \tag{4.143}$$

The variables under this forcing function are denoted as z_{RRi} and z_{IRi}. The second subscript R stands for the case when only the real part of the excitation is considered; the third subscript i also stands for the ith mode.

From the first row of Equation 4.141a,

$$z_{IIi} = -\frac{\dot{z}_{RIi} + \xi_i\,\omega_i\,z_{RIi}}{\sqrt{1-\xi_i^2}\,\omega_i} \tag{4.144}$$

and

$$\dot{z}_{IIi} = -\frac{\ddot{z}_{RIi} + \xi_i\omega_i\dot{z}_{RIi}}{\sqrt{1-\xi_i^2}\,\omega_i} \tag{4.145}$$

Substituting Equations 4.144 and 4.145 into the second row of Equation 4.141a yields

$$\ddot{z}_{RIi} + 2\xi_i\omega_i\dot{z}_{RIi} + \omega_i^2 z_{RIi} = -\sqrt{1-\xi_i^2}\ \omega_i q_{Ii} x_{g''} \tag{4.146a}$$

Similarly,

$$\ddot{z}_{IRi} + 2\xi_i\omega_i\dot{z}_{IRi} + \omega_i^2 z_{IRi} = -\sqrt{1-\xi_i^2}\ \omega_i q_{Ri} x_{g''} \tag{4.146b}$$

Now, the case when the system is proportionally damped is considered first.

$$\boldsymbol{q}_i = \frac{\boldsymbol{p}_i^T \mathbf{M}}{2j\,\boldsymbol{p}_i^T \mathbf{M}\,\boldsymbol{p}_i\,\sqrt{1-\xi_i^2}\ \omega_i} \tag{4.147}$$

Since for the proportionally damped system the mode shape $\boldsymbol{p}_i$ is real valued, it is seen that $\boldsymbol{q}_i$ has only the imaginary part. That is, the first case is sufficient to cover the seismic ground excitation and

$$q_{Ii} = \mathrm{Im}(\boldsymbol{q}_i\mathbf{J}) = \frac{\boldsymbol{p}_i^T \mathbf{M}\ \mathbf{J}}{2\,\boldsymbol{p}_i^T \mathbf{M}\,\boldsymbol{p}_i\sqrt{1-\xi_i^2}\,\omega_i} \tag{4.148}$$

Using Equation 4.148 results in

$$\ddot{z}_{RIi} + 2\xi_i\omega_i\dot{z}_{RIi} + \omega_i^2 z_{RIi} = \frac{1}{2}\ \Gamma_i x_{g''} \tag{4.149}$$

Here Γ_i is the modal participating factor, defined in Equation 3.170. From Equation 4.149, the equation for the SDOF forced vibration that was discussed in Chapter 2, z_{RIi} can be solved. Furthermore, through Equation 4.144, z_{IIi} can be calculated. Also, using Equation 4.142, z_{Ii} will be determined. Note that the ith complex conjugate of Equation 4.149 for the proportionally damped system is exactly the same. Similarly, z_{Ri} can be determined as well.

The combination of the complex conjugate parts of the variables z_{Ii} and z_{Ii}^*, denoted as y_{Ii}, can be written as

$$y_{Ii} = \left[\boldsymbol{p}_i, \boldsymbol{p}_i^*\right]\begin{Bmatrix} z_{Ii} \\ z_{Ii}^* \end{Bmatrix} \tag{4.150a}$$

Note that the response can also be calculated from the combination of the complex conjugate parts of the variables z_{Ri} and z_{Ri}^*, and denoted as y_{Ri}.

$$y_{Ri} = \left[\boldsymbol{p}_i, \boldsymbol{p}_i^*\right]\begin{Bmatrix} z_{Ri} \\ z_{Ri}^* \end{Bmatrix} \tag{4.150b}$$

By further combining y_{Ii} and y_{Ri}, the response due to the ith mode can be determined.

Since in the case of the proportionally damped system, $\boldsymbol{p}_i = \boldsymbol{p}_i^*$, the term z_{IIi} will be canceled and it is seen that the equation in the 2n space for the general underdamped system reduces to a familiar form for the normal modes. That is, the response by using the forcing function described in Equation 4.139, denoted as y_{Ii}, is sufficient to represent the total response of the ith mode. That is,

$$y_{Ii} = 2\boldsymbol{p}_i z_{RIi} \tag{4.151}$$

and it is seen that the maximum absolute value of the response y_{Ii} can be replaced by the approximation of the spectral value y_{Di}.

When the system is nonproportionally damped,

$$\boldsymbol{p}_i \neq \boldsymbol{p}_i^* \tag{4.152}$$

The quantity z_{IIi} must be considered. This is an important difference between the proportionally and nonproportionally damped systems.

The above method is an alternative way to calculate the response of the ith mode by solving the second-order differential equation that was discussed in Chapter 2. If the modal responses z_i are computed, their complex conjugate z_i^* can easily be found. Furthermore, once all the modal responses are calculated, the response vector $\boldsymbol{X}$ can be obtained by the following linear transformation:

$$\boldsymbol{X} = \mathcal{P}\boldsymbol{Z} \tag{4.153}$$

Note that Equation 4.117 can also be solved directly to find z_i and Equation 4.153 can be used to compute the response $\boldsymbol{X}$, as well. Furthermore, the displacement $\mathbf{x}$ as well as the velocity $\dot{\mathbf{x}}$ can be determined.

If m time points are collected in the response of $\boldsymbol{Z}$ and arranged in a $2n \times m$ matrix $\mathcal{Z}_m$,

$$\mathcal{Z}_m = \left[\boldsymbol{Z}(1), \boldsymbol{Z}(2), \ldots, \boldsymbol{Z}(m)\right] \tag{4.154}$$

Then,

$$\mathcal{Z}_m = \begin{bmatrix} \mathbf{z}_{1m} \\ \mathbf{z}_{2m} \\ \ldots \\ \mathbf{z}_{nm} \\ \mathbf{z}_{1m}^* \\ \mathbf{z}_{2m}^* \\ \ldots \\ \mathbf{z}_{nm}^* \end{bmatrix}_{2n\times m} \tag{4.155}$$

where

$$\mathbf{z}_{im}(t) = \left[z_i(1), z_i(2), \ldots, z_i(m)\right] \tag{4.156}$$

Here, subscript m stands for the total time points; from Equation 4.34,

$$\mathcal{P} = \begin{bmatrix} \lambda_1 \boldsymbol{p}_1 & \ldots & \lambda_n \boldsymbol{p}_n & \lambda_1^* \boldsymbol{p}_1^* & \ldots & \lambda_n^* \boldsymbol{p}_n^* \\ \boldsymbol{p}_1 & \ldots & \boldsymbol{p}_n & \boldsymbol{p}_1^* & \ldots & \boldsymbol{p}_n^* \end{bmatrix} \tag{4.157}$$

Using Equations 4.155 and 4.157, and considering m time points, Equation 4.153 can also be written as

$$\mathcal{X}_m = \mathcal{X}_m(t)_{2n\times m} = \sum_i^n \left\{ \begin{bmatrix} \lambda_i \boldsymbol{p}_i \\ \boldsymbol{p}_i \end{bmatrix} \mathbf{z}_{im}(t) + \begin{bmatrix} \lambda_i^* \boldsymbol{p}_i^* \\ \boldsymbol{p}_i^* \end{bmatrix} \mathbf{z}_{im}^*(t) \right\} \tag{4.158}$$

Here, subscript m stands for m time points.

Equation 4.158 is a general form of the solution of Equation 4.2. Since $\mathcal{X}_m$ contains both the velocity $\dot{\mathbf{x}}$ and the displacement $\mathbf{x}$, once $\mathcal{X}_m$ is determined, the solution $\mathbf{x}(t)$ can be obtained for Equation 4.1. Generally, to calculate the responses of nonproportionally damped systems, four times larger computer memories are required. For proportionally damped systems, n-physical space and the response calculation are likely to involve the design response spectrum. Conventionally, n-physical space is decoupled into an n-SDOF subspace that can then be linked with the design spectrum that was originally associated with the SDOF systems. The nonproportionally damped system needs 2n modal space for its solutions. Similar to Chapter 3, where the response computation through the modal superposition without involving all modes was discussed, for the nonproportionally damped system, the same modal truncation approach is used.

4.2 DAMPER DESIGN FOR NONPROPORTIONALLY DAMPED SYSTEMS

Since a nonproportionally damped system cannot be decoupled in n-dimensional modal space, and adding dampers to a structure will likely introduce nonproportional damping, this issue must be considered in damper design for nonproportionally damped systems. Theoretically, the problems come from the fact that for a nonproportionally damped system, the solutions contain both the displacement and the velocity, whereas for a proportionally damped system, only the term of displacement is involved.

4.2.1 Issues with General Damper Design

In Chapter 3, issues with damper design for proportionally damped cases were briefly discussed. Some additional issues with general damper design are explored in this section.

First, the computation of the solution for generally damped systems is in the 2n complex modal domain with the first-order differential equations, whereas in earthquake engineering, the n-dimensional normal mode domain with second-order equations is more common. In particular, the design response spectrum is obtained by the following second-order differential equation:

$$m_i \ddot{y}_i + c\dot{y}_i + ky_i = -m_i \Gamma_i \, x_{g''}(t) \tag{4.159}$$

where Γ_i is the ith modal participation factor. For generally damping system, the modal participation factor will be discussed in the next section. As discussed earlier, Equation 4.159 can be rewritten as the following equation by dividing m_i on both sides to obtain the monic equation as:

$$\ddot{y}_i + 2\xi_i \omega_i \dot{y}_i + \omega_i^2 y_i = -\Gamma_i x_{g''}(t) \tag{4.160}$$

As previously mentioned, the statistical maxima of the responses x(t)'s of a series of the SDOF oscillators described by Equation 4.160 with different damping ratio ξ_i and natural frequency ω_i (natural period $T_i = 2\pi/\omega_i$) can be represented by the spectral value $S_D(\xi_i,T_i)$ through the design response spectrum introduced in the previous section. For convenience, Equation 2.303 of this operation is rewritten as

$$y_{max_i}(\xi_i,T_i)/g \rightarrow S_D(\xi_i,T_i) \tag{4.161}$$

where the subscript i of ξ_i and T_i stand for the ith damping ratio and the natural period. Symbol $\rightarrow$ stands for the statistical operation again.

In Equation 4.161, it is seen that to represent the maximum response of a vibration system with the damping ratio ξ_i and the natural period T_i, the second-order differential Equation 4.160 should be used instead of the first-order Equations 4.117 and 4.118.

Therefore, in order to use the design spectrum, the two complex conjugate pairs need to be combined (see Equations 4.117 and 4.118). Namely combine,

$$\dot{z}_i - \lambda_i z = f_i, \quad i = 1, 2, \ldots, n \tag{4.162}$$

and

$$\dot{z}_i^* - \lambda_i^* z^* = f_i^*, \quad i = 1, 2, \ldots, n \tag{4.163}$$

into the form of the second-order equation as generally described in Equation 4.159.

Secondly, in damper design as well as in the general computation of the structural responses, all the modal responses do not have to be calculated. Instead, only the first several modes can be used.

For proportionally damped systems, the modal mass ratio γ_{mi} can be used as the modal participation indicator to determine if a specific mode must be considered, as described in Equation 3.184 and repeated as

$$\gamma_{mi} = \frac{\left[\boldsymbol{p}_i^T \mathbf{MJ}\right]^2}{\boldsymbol{p}_i^T \mathbf{M} \boldsymbol{p}_i \sum_{j=1}^{n} m_j} = \frac{\Gamma_i^2\, \boldsymbol{p}_i^T \mathbf{M} \boldsymbol{p}_i}{M_\Sigma} = \frac{m_{effi}}{M_\Sigma} \tag{4.164}$$

However, for the complex mode, this modal mass ratio is no longer valid. A new criterion is needed to determine if a specific mode must be considered. Note that in Equation 4.164, $\boldsymbol{p}_i$ is used to denote more general cases.

In addition, if a mode reduces to two overdamped first-order subsystems, the question is whether the overdamped subsystems will still be capable of contributing to the total responses. If some of them will, then criteria are needed to determine whether or not the first order should be included.

Thirdly, criteria are needed to determine if the nonproportional damping approach is really needed. On the one hand, because of the complexity of nonproportional damping, this approach should be avoided if possible. On the other hand, if the simplified approach of using proportional damping yields unacceptable errors, a more complicated approach will be required. The judgment of whether nonproportional damping is used becomes critical in damper design.

4.2.2 Essence of the Solution of Nonproportionally Damped System

The damper design issues for nonproportionally damped system mentioned above are examined mathematically in this section.

Taking the Laplace transformation on both sides of Equation 4.1b and assuming the zero initial conditions of both the displacements and the velocities,

$$\mathscr{L}\left[\begin{Bmatrix} \ddot{\mathbf{x}}(t) \\ \dot{\mathbf{x}}(t) \end{Bmatrix} - \boldsymbol{a} \begin{Bmatrix} \dot{\mathbf{x}}(t) \\ \mathbf{x}(t) \end{Bmatrix}\right] = \mathscr{L}\left[\begin{Bmatrix} \mathbf{M}^{-1}\mathbf{f}(t) \\ \mathbf{0} \end{Bmatrix}\right] \tag{4.165}$$

which is further written as

$$s\begin{Bmatrix} s\mathbf{X}(s) \\ \mathbf{X}(s) \end{Bmatrix} - \boldsymbol{a} \begin{Bmatrix} s\mathbf{X}(s) \\ \mathbf{X}(s) \end{Bmatrix} = \begin{Bmatrix} \mathbf{M}^{-1}\mathbf{F}(s) \\ \mathbf{0} \end{Bmatrix} \tag{4.166}$$

where s is the Laplace variable and $\mathbf{X}(s)$ and $\mathbf{F}(s)$ are the Laplace transformation of $\mathbf{x}(t)$ and $\mathbf{f}(t)$, respectively. Then

$$\begin{Bmatrix} s\mathbf{X}(s) \\ \mathbf{X}(s) \end{Bmatrix} = (s\mathbf{I} - \boldsymbol{\mathcal{A}})^{-1} \begin{bmatrix} \mathbf{M}^{-1} & \\ & \mathbf{0} \end{bmatrix} \begin{Bmatrix} \mathbf{F}(s) \\ \mathbf{0} \end{Bmatrix} \tag{4.167}$$

Here, $(s\mathrm{I} - \boldsymbol{\mathcal{A}})^{-1} \begin{bmatrix} \mathbf{M}^{-1} & \\ & \mathbf{0} \end{bmatrix}$ is called the *transfer function matrix* denoted by **H** with dimension $2n \times 2n$, that is,

$$\begin{aligned} \mathbf{H}(s) = (s\mathbf{I} - \boldsymbol{\mathcal{A}})^{-1} \begin{bmatrix} \mathbf{M}^{-1} & \\ & \mathbf{0} \end{bmatrix} &= \mathcal{P}\,\mathrm{diag}\left(\frac{1}{s-\lambda_i}\right) \mathcal{P}^{-1} \begin{bmatrix} \mathbf{M}^{-1} & \\ & \mathbf{0} \end{bmatrix} \\ &= \mathcal{P}\,\mathrm{diag}\left(\frac{1}{s-\lambda_i}\right) \mathcal{Q} \begin{bmatrix} \mathbf{M}^{-1} & \\ & \mathbf{0} \end{bmatrix} \end{aligned} \tag{4.168}$$

Substituting Equations 4.34 and 4.41 into Equation 4.168 yields

$$\begin{aligned} \mathbf{H} &= \begin{bmatrix} \mathbf{P}\boldsymbol{\Lambda} & \mathbf{P}^*\boldsymbol{\Lambda}^* \\ \mathbf{P} & \mathbf{P}^* \end{bmatrix} \begin{bmatrix} \mathrm{diag}\left(\frac{1}{s-\lambda_i}\right) & \\ & \mathrm{diag}\left(\frac{1}{s-\lambda_i^*}\right) \end{bmatrix} \begin{bmatrix} \mathbf{Q} & \mathbf{R} \\ \mathbf{Q}^* & \mathbf{R}^* \end{bmatrix} \begin{bmatrix} \mathbf{M}^{-1} & \\ & \mathbf{0} \end{bmatrix} \\ &= \begin{bmatrix} \boldsymbol{\Xi} & \boldsymbol{\Psi} \\ \mathbf{P}\,\mathrm{diag}\left(\frac{1}{s-\lambda_i}\right)\mathbf{Q}\mathbf{M}^{-1} & \mathbf{Z} \end{bmatrix} + \begin{bmatrix} \boldsymbol{\Xi}^* & \boldsymbol{\Psi}^* \\ \mathbf{P}^*\mathrm{diag}\left(\frac{1}{s-\lambda_i^*}\right)\mathbf{Q}^*\mathbf{M}^{-1} & \mathbf{Z}^* \end{bmatrix} \end{aligned} \tag{4.169}$$

Here, only the submatrix at the lower left corner is of interest, so the symbols $\boldsymbol{\Xi}$, $\boldsymbol{\Psi}$, and $\mathbf{Z}$ are used to represent the rest of the submatrices.

With the help of Equations 4.169 and 4.167,

$$\begin{aligned} \mathbf{X}(s) &= \left\{ \mathbf{P}\,\mathrm{diag}\left(\frac{1}{s-\lambda_i}\right)\mathbf{Q}\mathbf{M}^{-1} + \mathbf{P}^*\mathrm{diag}\left(\frac{1}{s-\lambda_i^*}\right)\mathbf{Q}^*\mathbf{M}^{-1} \right\} \mathbf{F}(s) \\ &= \left\{ \sum_{i=1}^{n} \left[\frac{\boldsymbol{p}_i\boldsymbol{q}_i\mathbf{M}^{-1}}{s-\lambda_i} + \frac{(\boldsymbol{p}_i\boldsymbol{q}_i)^*\,\mathbf{M}^{-1}}{s-\lambda_i^*} \right] \right\} \mathbf{F}(s) \end{aligned} \tag{4.170}$$

The ith term with dimension $n \times n$ in brackets is the transfer function matrix between **X**(s) and **F**(s) and is denoted by

$$\mathbf{H}_i(s) = \frac{\boldsymbol{p}_i\boldsymbol{q}_i\mathbf{M}^{-1}}{s-\lambda_i} + \frac{(\boldsymbol{p}_i\boldsymbol{q}_i)^*\,\mathbf{M}^{-1}}{s-\lambda_i^*} \tag{4.171}$$

Denote $\mathcal{L}\left[x_{g''}(t)\right] = x_{g''}(s)$. Thus

$$\mathbf{F}(s) = -\mathbf{MJ}x_{g''}(s) \tag{4.172}$$

therefore, for convenience, the location vector **J** is included inside the transfer function vector between **X**(s) and $-x_{g''}(s)$ and the resulting term is denoted by

$$\boldsymbol{h}_i(s) = \frac{\mathbf{X}(s)}{x_{g''}(s)} = -\left[\frac{\boldsymbol{p}_i\boldsymbol{q}_i\mathbf{J}}{s-\lambda_i} + \frac{(\boldsymbol{p}_i\boldsymbol{q}_i\mathbf{J})^*}{s-\lambda_i^*}\right] \tag{4.173}$$

Note that the dimension of $\boldsymbol{h}_i(s)$ is now reduced to $n \times 1$.

The product of $\boldsymbol{p}_i\boldsymbol{q}_i\mathbf{J}$ is generally complex valued, so vector $\boldsymbol{p}_i\boldsymbol{q}_i\mathbf{J}$ can be denoted as

$$\boldsymbol{p}_i\boldsymbol{q}_i\mathbf{J} = \boldsymbol{\Phi}_i + j\boldsymbol{\Psi}_i \tag{4.174}$$

and its complex conjugate is

$$(\boldsymbol{p}_i\boldsymbol{q}_i\mathbf{J})^* = \boldsymbol{\Phi}_i - j\boldsymbol{\Psi}_i \tag{4.175}$$

where the real part, vector

$$\boldsymbol{\Phi}_i = \{\phi_{ij}\} \tag{4.176a}$$

and the imaginary part, vector

$$\boldsymbol{\Psi}_i = \{\psi_{ij}\} \tag{4.176b}$$

are real valued.

Substituting Equation 4.95 into Equation 4.174 yields

$$\boldsymbol{p}_i\boldsymbol{q}_i\mathbf{J} = \frac{\boldsymbol{p}_i(\boldsymbol{p}_i^T\mathbf{MJ})}{2\lambda_i(\boldsymbol{p}_i^T\mathbf{M}\boldsymbol{p}_i) + \boldsymbol{p}_i^T\mathbf{C}\boldsymbol{p}_i} = \boldsymbol{\Phi}_i \pm j\boldsymbol{\Psi}_i, \quad i = 1, 2, \ldots, n \tag{4.177}$$

From Equation 4.177, it is realized that the values of $\boldsymbol{\Phi}_i$ and $\boldsymbol{\Psi}_i$ are unique. In other words, they are independent from various normalizations. This is because Equation 4.17, namely ($\alpha\boldsymbol{p}_i$), can be used to replace $\boldsymbol{p}_i$ in Equation 4.177 and the factor α will be canceled.

Substituting Equations 4.174 and 4.175 into Equation 4.173 and considering Equation 4.14,

$$\boldsymbol{h}_i(s) = -2\omega_i\frac{-\xi_i\boldsymbol{\Phi}_i + \sqrt{1-\xi_i^2}\,\boldsymbol{\Psi}_i}{s^2 + 2\xi_i\omega_i s + \omega_i^2} + 2s\frac{\boldsymbol{\Phi}_i}{s^2 + 2\xi_i\omega_i s + \omega_i^2} \tag{4.178}$$

The response of the ith mode $\mathbf{x}_i(t)$ can be obtained by the inverse of the Laplace transform of the product of the modified transfer function $\boldsymbol{h}_i(s)$ and the Laplace transform of the ground excitation $-x_{g''}(s)$. Note that the location vector J has been considered and included inside the transfer function. Thus,

$$\begin{aligned}\mathbf{x}_i(t) &= \mathcal{L}^{-1}\{-\boldsymbol{h}_i(s)\, x_{g''}(s)\}\\ &= \mathcal{L}^{-1}\left\{-2\omega_i\left(-\xi_i\boldsymbol{\Phi}_i + \sqrt{1-\xi_i^2}\,\boldsymbol{\Psi}_i\right)\frac{x_{g''}(s)}{s^2 + 2\xi_i\omega_i s + \omega_i^2} - 2\boldsymbol{\Phi}_i\frac{s x_{g''}(s)}{s^2 + 2\xi_i\omega_i s + \omega_i^2}\right\}\end{aligned} \tag{4.179}$$

It is seen that $\mathbf{x}_i(t)$ is the vector of time variables and contains only the ith mode. That is, it is the *modal response*.

Denote

$$y_{Di}(t) = \mathscr{L}^{-1}\left\{\frac{x_{g''}(s)}{s^2 + 2\xi_i\omega_i s + \omega_i^2}\right\} \tag{4.180}$$

and

$$y_{Vi}(t) = \mathscr{L}^{-1}\left\{\frac{s x_{g''}(s)}{s^2 + 2\xi_i\omega_i s + \omega_i^2}\right\} \tag{4.181}$$

Here, $y_{Di}(t)$ and $y_{Vi}(t)$ can be seen as the responses of the displacement and the velocity under normalized unit input $x_{g''}(t)$ and

$$y_{Vi}(t) = \frac{d}{dt} y_{Di}(t) \tag{4.182}$$

Thus, the vector equation can be written as

$$\mathbf{x}_i(t) = -2\omega_i\left(\xi_i\mathbf{\Phi}_i - \sqrt{1-\xi_i^2}\,\mathbf{\Psi}_i\right)y_{Di}(t) + 2\mathbf{\Phi}_i y_{Vi}(t) \tag{4.183}$$

Equation 4.183 expresses the solution of the ith complex mode, which is already in the time domain. The total response contributed by all the modes is simply the linear combinations, that is,

$$\mathbf{x}(t) = \sum_{i=1}^{n} \mathbf{x}_i(t) = -\sum_{i=1}^{n}\left\{2\omega_i\left(\xi_i\mathbf{\Phi}_i - \sqrt{1-\xi_i^2}\,\mathbf{\Psi}_i\right)y_{Di}(t) + 2\mathbf{\Phi}_i y_{Vi}(t)\right\} \tag{4.184}$$

Note that Equation 4.184 can be used to reduce the computational burden to find the exact solution of Equation 4.2. This is because the computation is carried out in the n-dimensional space, instead of a 2n space and many commercially available computational programs can simultaneously provide the displacement and the velocity for SDOF systems, since the displacement y_{Di} and the velocity y_{Vi} are calculated from the identical SDOF system. Note that when the system is proportionally damped, $\mathbf{\Phi}_i = 0$, Equation 4.184 reduces to the case described by the normal mode.

Example 4.6

Consider the **M-C-K** system mentioned in Example 4.1.

$$p_1 q_1 J = \mathbf{\Phi}_1 + j\mathbf{\Psi}_1 = \begin{Bmatrix} -0.0015 \\ 0.0003 \end{Bmatrix} + j\begin{Bmatrix} 0.0315 \\ 0.0533 \end{Bmatrix}$$

$$p_2 q_2 J = \mathbf{\Phi}_2 + j\mathbf{\Psi}_2 = \begin{Bmatrix} 0.0015 \\ -0.0003 \end{Bmatrix} + j\begin{Bmatrix} 0.0052 \\ -0.0026 \end{Bmatrix}$$

Therefore

$$\mathbf{x}(t) = \sum_{i=1}^{n} \mathbf{x}_i(t) = -\sum_{i=1}^{n}\left\{2\omega_i\left(\xi_i\mathbf{\Phi}_i - \sqrt{1-\xi_i^2}\,\mathbf{\Psi}_i\right)y_{Di}(t) + 2\mathbf{\Phi}_i y_{Vi}(t)\right\}$$

$$= -\left[\begin{Bmatrix} -0.6892 \\ -1.1543 \end{Bmatrix} y_{D1} + \begin{Bmatrix} -0.0030 \\ 0.0007 \end{Bmatrix} y_{V1}\right] + \left[\begin{Bmatrix} -0.3008 \\ 0.1521 \end{Bmatrix} y_{D2} + \begin{Bmatrix} 0.0030 \\ -0.0007 \end{Bmatrix} y_{V1}\right]$$

4.2.3 Modal Truncations for Nonproportionally Damped Systems

As in proportionally damped systems, in most cases, all the modes do not have to be considered if the first S modes can be dominant; that is, the modal truncation can be obtained as follows:

$$\mathcal{X}_C \approx \mathcal{P}_C\, \mathcal{Z}_C \tag{4.185}$$

Here $\mathcal{Z}_C$ is the modal time history matrix with the dimension 2S × m and again m is the total number of time points, namely,

$$\mathcal{Z}_C = \begin{bmatrix} \mathbf{z}_{1m} \\ \mathbf{z}_{2m} \\ \cdots \\ \mathbf{z}_{Sm} \\ \mathbf{z}_{1m}^* \\ \mathbf{z}_{2m}^* \\ \cdots \\ \mathbf{z}_{Sm}^* \end{bmatrix}_{2S\times m} \tag{4.186}$$

where $z_{im}(t)$ is defined in Equation 4.156, and $\mathcal{P}_C$ is a truncated eigenvector matrix,

$$\mathcal{P}_C = \begin{bmatrix} \lambda_1 \boldsymbol{p}_1 & \cdots & \lambda_S \boldsymbol{p}_S & \lambda_1^* \boldsymbol{p}_1^* & \cdots & \lambda_S^* \boldsymbol{p}_S^* \\ \boldsymbol{p}_1 & \cdots & \boldsymbol{p}_S & \boldsymbol{p}_1^* & \cdots & \boldsymbol{p}_S^* \end{bmatrix}_{2n\times 2S} \tag{4.187}$$

Using Equations 4.186 and 4.187, the response $\mathcal{X}_C$ in Equation 4.185, which now contains m time points, can also be written as

$$\mathcal{X}_C = \mathcal{X}_C(t)_{2n\times m} = \sum_i^S \left\{ \begin{bmatrix} \lambda_i \boldsymbol{p}_i \\ \boldsymbol{p}_i \end{bmatrix} z_i(t) + \begin{bmatrix} \lambda_i^* \boldsymbol{p}_i^* \\ \boldsymbol{p}_i^* \end{bmatrix} z_i^*(t) \right\} \tag{4.188}$$

Equations 4.185 and 4.188 indicate that in many cases, only the first S modal responses are needed. Therefore,

$$\mathbf{x}_C(t) \approx \sum_{i=1}^S \mathbf{x}_i(t) = \sum_{i=1}^S \left\{ 2\omega_i \left(\xi_i \mathbf{\Phi}_i - \sqrt{1-\xi_i^2}\, \mathbf{\Psi}_i \right) y_{Di}(t) + 2\mathbf{\Phi}_i y_{Vi}(t) \right\} \tag{4.189}$$

That is, the time history of $\mathbf{x}_C(t)$ can be obtained by the modal superposition described in Equation 4.189. However, it is often the case that the peak value of $\mathbf{x}_C(t)$ and not the time history itself is of interest during aseismic design. Although the total peak response can be found through the maximum responses of each individual mode, the maximum modal responses cannot simply be summarized. The summation of the peak value can be carried out through the complete quadratic combination (CQC) and the square-root-of-the-sum-of-squares (SRSS) method (Song et al. 2008).

Now the remaining questions are, first, what if certain modes reduce to overdamped? and second, how can the number S be determined?

4.3 OVERDAMPED SUBSYSTEMS

Very often, in attempting to achieve the desired damping ratio of certain modes by increasing the damping either proportionally or nonproportionally, some other "modes" will become overdamped.

That is, the corresponding damping ratios will be greater than 1. In 2n-scape, a pair of complex conjugate modes will become independent. Both their eigenvalues and eigenvectors will become real valued. For convenience, these overdamped subsystems are referred to as *pseudo modes*. However, certain overdamped subsystems still make notable contributions to the total responses. Ignoring such contributions will introduce design errors.

4.3.1 Concept of Overdamped System

Note that when large damping is added to a structure, more modes will become critically damped and will then vanish to form two overdamped first-order subsystems. To realize this point in detail, a monic homogeneous equation of motion for an SDOF system is considered first.

$$\ddot{y}(t)+2\xi\omega_n\dot{y}(t)+\omega_n^2 y(t)=0 \tag{4.190}$$

Here the subscript i is omitted for simplicity.

Equation 4.190 can represent a pure SDOF system. It can also be used to describe a decoupled MDOF system. It is known that the corresponding characteristic equation of Equation 4.190 can be written as (see Equation 1.19)

$$\lambda^2+2\xi\omega_n\lambda+\omega_n^2=0 \tag{4.191}$$

The roots of Equation 4.191 were discussed in Chapter 1 (see Equation 1.20) and are repeated again as

$$\lambda_{1,2}=\left(-\xi\pm\sqrt{\xi^2-1}\right)\omega_n \tag{4.192}$$

It has already been shown that, if

$$\xi^2<1,$$

or

$$\xi<1 \tag{4.193}$$

then the underdamped system will exist. In this case, the roots are complex conjugates.

For a free vibration with certain initial conditions, the system will conceptually have a solution of (see Equation 1.30)

$$y(t)=1/2(y_0e^{\lambda t}+y_0e^{\lambda^* t})=y_0e^{-\xi\omega_n t}[\sin(\omega_n t+\varphi)] \tag{4.194}$$

The cosine term in brackets on the right side of Equation 4.194 implies that the mass will move back and forth around its equilibrium position. Thus, it is a vibration system. It is seen that to have these trigonometric time variables, the imaginary part in Equation 4.194 needs to be nonzero, that is,

$$\lambda_{1,2}=\left(-\xi\pm j\sqrt{1-\xi^2}\right)\omega_n \tag{4.195}$$

Namely, the term $\sqrt{1-\xi^2}$ in Equation 4.195 must be real and nonzero valued. In this case, Equation 4.193 is the necessary and sufficient condition for a free-decay system to have vibration.

The SDOF system can also represent one of the normal modes of an MDOF system. The corresponding mode is called the *underdamped mode* and ξ is the damping ratio.

When damping is gradually added to an underdamped system, the damping ratio gradually increases until the critical level is reached. The SDOF system can also include the case in which one of the underdamped normal modes of an MDOF system becomes critically damped by gradually adding damping, as

$$\xi = 1 \tag{4.196}$$

The corresponding mode at this point becomes two identical subsystems, with real-valued eigenvalues and eigenvectors. Thus, the roots will become real valued as

$$\lambda_1 = \lambda_2 = -\xi\omega_n \tag{4.197}$$

which is referred to as the *critical damping*. In this case, the SDOF system reduces to two identical subsystems. For convenience, they are referred to as the pseudo modes.

When damping continues to increase to the above system, the damping ratio of either the MDOF system or one of the modes of the MDOF system will be greater than 1, which is the above-mentioned critical damping; that is

$$\xi_i > 1 \tag{4.198}$$

From Equation 4.192, it is realized that the roots will become two unequal real numbers. Thus, two *overdamped subsystems* will exist. For convenience, they are also referred to as the pseudo modes. In this case, as described by Equation 4.198, subscript i represents these specific pseudo mode.

It is thus seen that once Inequality 4.193 is violated, there will no longer be free vibrations for the ith pair of subsystems.

If the SDOF system is not monic, then the damping ratio is defined as (see Equation 1.27)

$$\xi = \frac{c}{2\sqrt{mk}}$$

It is seen that when the system is overdamped,

$$c > 2\sqrt{mk} \tag{4.199}$$

Suppose the system is under a sinusoidal excitation with a driving frequency equal to ω_n. From Equations 1.5 and 1.6, it is realized that the peak value of the damping force f_d and restoring force f_r, when the steady state is reached, can have the following forms:

$$|f_d| = |c\,\dot{x}| = 2\xi\omega_n^2 m x_0 \tag{4.200a}$$

and

$$|f_r| = |k\,x| = \omega_n^2 m x_0 \tag{4.200b}$$

The corresponding ratio is

$$\frac{|f_d|}{|f_r|} = 2\xi \tag{4.201}$$

That is, when the damping is 50%, the amplitude of the damping force will be equal to that of the restoring force. When the system is critically damped, the amplitude of the damping force will be twice that of the restoring force.

Note that this is only true for the linear SDOF at the resonant point. When a system is generally damped, the relationship as described by Equation 4.201 does not occur. However, for convenience, a system whose damping force is more than twice that of the restoring force is still referred to as overdamped.

Now, a nonproportionally damped system cannot be decoupled in the n-dimensional space. Therefore, there must be certain modes that cannot be described by Equation 4.190. However, the eigenvalues of the system can still be described by Equation 4.195, if it is underdamped.

Similar to the case of a proportionally damped system, if damping is continuously added, sooner or later the eigenvalues of the ith mode will reach a point when the imaginary part is equal to zero as the condition of Equation 4.196 is reached. Thus, the critical damping for the specific mode is obtained and the mode becomes two identical subsystems. When damping is continuously increased, these two subsystems will have different real-valued eigenvalues. They are exactly like the case of a proportionally damped system.

Therefore, it does not have to be distinguished if a system is proportionally damped. When adding damping to the system so that it possesses two real-valued eigenvalues, which originally come from the ith mode, they are denoted using superscript R as

$$\lambda_{i1}^{R} < 0 \text{ and } \lambda_{i2}^{R} < 0 \tag{4.202}$$

The corresponding eigenvectors will also reduce to real-valued

$$\boldsymbol{p}_{i1}^{R} \text{ and } \boldsymbol{p}_{i2}^{R} = \text{real} \tag{4.203}$$

The corresponding first-order differential equations will become

$$\dot{z}_{i1}^{R} - \lambda_{i1}^{R} \cdot z_{i1}^{R} = f_{i1} \tag{4.204}$$

and

$$\dot{z}_{i2}^{R} - \lambda_{i2}^{R} \cdot z_{i2}^{R} = f_{i2} \tag{4.205}$$

Usually,

$$\lambda_{i1}^{R} \neq \lambda_{i2}^{R} \tag{4.206}$$

Therefore,

$$z_{i1}^{R} \neq z_{i2}^{R} \tag{4.207}$$

Note that even certain modes become overdamped, and by using the state equations, the total solutions can still be obtained. This is the conventional way to deal with the overdamped case and the solution is accurate.

From the conclusion that in nonproportionally underdamped systems the eigen-pair $<\lambda_i, \boldsymbol{p}_i>$ must have its complex conjugate, it is seen that if there is an n-DOF system with n_o modes reducing to overdamped subsystems, and the remaining n_u modes are underdamped,

$$n_o + n_u = n \tag{4.208}$$

Secondly, in a 2n space, there will be $2n_o$ overdamped subsystems or pseudo modes. In this case, $\mathbf{P}$ and $\boldsymbol{\Lambda}$ are still used to denote the complex-valued eigenvector and eigenvalue matrices, and $\mathbf{P}^R$ and $\boldsymbol{\Lambda}^R$ are used to denote the set of $2n_o$ real-valued eigenvector and eigenvalue matrices, where

$$\mathbf{P}^R = \left[\boldsymbol{p}_1^R, \boldsymbol{p}_2^R, \ldots, \boldsymbol{p}_{2no}^R\right] \tag{4.209}$$

$$\boldsymbol{\Lambda}^R = \text{diag}\left(\lambda_i^R\right), \quad i = 1, 2, \ldots, 2n_o \tag{4.210}$$

Here, superscript R represents real-valued parameters for the pseudo modes. Using Equations 4.209 and 4.210, the eigenvalue matrix $\boldsymbol{\Delta}$ is rewritten as

$$\boldsymbol{\Delta} = \begin{bmatrix} \boldsymbol{\Lambda} & & \\ & \boldsymbol{\Lambda}^* & \\ & & \boldsymbol{\Lambda}^R \end{bmatrix} \tag{4.211}$$

and the eigenvector matrix is

$$\mathcal{P} = \begin{bmatrix} \mathbf{P}\boldsymbol{\Lambda} & \mathbf{P}^*\boldsymbol{\Lambda}^* & \mathbf{P}^R\boldsymbol{\Lambda}^R \\ \mathbf{P} & \mathbf{P}^* & \mathbf{P}^R \end{bmatrix} \tag{4.212}$$

The inverse of the eigenvector matrix, still denoted by $\mathcal{Q}$ will become

$$\mathcal{Q} = \mathcal{P}^{-1} = \begin{bmatrix} \mathbf{Q} & \mathbf{R} \\ \mathbf{Q}^* & \mathbf{R}^* \\ \mathbf{Q}^R & \mathbf{R}^R \end{bmatrix} \tag{4.213}$$

It can be proven that the submatrices $\mathbf{Q}^R$ and $\mathbf{R}^R$ in the third row are real valued. Similar to the process of derivation of the representation of $\mathbf{X}(s)$ with the total underdamped modes, the following expression for the case where some of the modes have been reduced to overdamped cases can be used. That is,

$$\mathbf{X}(s) = \left\{\left[\mathbf{P}\ \text{diag}\left(\frac{1}{s-\lambda_i}\right)\mathbf{Q}\right] + \left[\mathbf{P}^*\ \text{diag}\left(\frac{1}{s-\lambda_i^*}\right)\mathbf{Q}^*\right] + \left[\mathbf{P}^R\ \text{diag}\left(\frac{1}{s-\lambda_j^R}\right)\mathbf{Q}^R\right]\right\}\mathbf{M}^{-1}\mathbf{F}(s)$$

$$= \left[\sum_{i=1}^{n_u}\left\{\frac{\boldsymbol{p}_i\boldsymbol{q}_i}{s-\lambda_i} + \frac{(\boldsymbol{p}_i\boldsymbol{q}_i)^*}{s-\lambda_i^*}\right\} + \sum_{j=1}^{2n_o}\frac{\boldsymbol{p}_j^R\boldsymbol{q}_j^R}{s-\lambda_j^R}\right]\mathbf{M}^{-1}\mathbf{F}(s) = \sum_{i=1}^{n_u}\mathbf{X}_i^C(s) + \sum_{j=1}^{2n_o}\mathbf{X}_j^R(s) \tag{4.214}$$

where $\boldsymbol{p}_j^R$ and $\boldsymbol{q}_j^R$ are, respectively, the jth column and row in matrices $\mathbf{P}^R$ and $\mathcal{Q}^R$ and the normal character $\mathbf{X}_j^C(s)$ is used to represent the ith underdamped modal response. Superscript C is used because the corresponding eigenvalue is complex valued. However, in certain formulas, in order to be consistent with conventional equations that only account for the underdamped system, the superscript can be omitted. The term $X_j^R(s)$ is used to represent the jth overdamped response. Note that the underdamped response can be expressed as the case of the system with total modes underdamped.

Now, consider the jth overdamped subsystem first, as follows:

$$\mathbf{X}_j^R(s) = \frac{\boldsymbol{p}_j^R \boldsymbol{q}_j^R \mathbf{M}^{-1}}{s-\lambda_j^R}\mathbf{F}(s) = -\frac{\boldsymbol{p}_j^R \boldsymbol{q}_j^R \mathbf{J}}{s-\lambda_j^R} x_{g''}(s) = \left(\frac{\boldsymbol{r}_j^R}{s-\lambda_j^R}\right) x_{g''}(s) \tag{4.215}$$

Here, since $\boldsymbol{p}_j^R$ and $\boldsymbol{q}_j^R$ are real valued, the real-valued vector $\boldsymbol{r}_j^R$ is used to represent the product $-\boldsymbol{p}_j^R \boldsymbol{q}_j^R \mathbf{J}$, that is,

$$\boldsymbol{r}_j^R = -\boldsymbol{p}_j^R \boldsymbol{q}_j^R \mathbf{J} \tag{4.216}$$

Note that in Equation 4.215, in order to distinguish the underdamped modes and the overdamped subsystems, i and j are used to denote them, respectively. In the following discussion, for convenience, both i and j are used to denote the pseudo modes.

In Equation 4.215, the term in brackets on the right side is the transfer function. By letting s = j ω, the corresponding frequency response function of the jth pseudo mode can be written as

$$\boldsymbol{h}_j^R(j\omega) = \frac{\boldsymbol{r}_j^R}{j\omega - \lambda_j^R} \tag{4.217}$$

Taking the inverse of the Laplace transform on both sides of Equation 4.215,

$$\mathbf{x}_j^R(t) = \mathscr{L}^{-1}\left[\mathbf{X}_j^R(s)\right] = \boldsymbol{r}_j^R \int_0^t e^{\lambda_j^R(t-\tau)} x_{g''}(\tau)\, d\tau = \boldsymbol{r}_j^R y_{Dj}^R(t) \tag{4.218}$$

where

$$y_{Dj}^R(t) = \int_0^t e^{\lambda_j^R(t-\tau)} x_{g''}(\tau)\, d\tau \tag{4.219}$$

is the unit response of the pseudo modes.
Using Equation 4.219, the total overdamped responses $\mathbf{x}^R(t)$ can be written as

$$\mathbf{x}^R(t) = \sum_{j=1}^{2n_o} \mathbf{x}_j^R(t) = \sum_{j=1}^{2n_o} \boldsymbol{r}_j^R y_{Dj}^R(t) \tag{4.220}$$

and the total exact response can be written as

$$\begin{aligned} \mathbf{x}(t) &= \sum_{i=1'}^{n} \mathbf{x}_i(t) \\ &= \sum_{i=1'}^{n_u} \left\{ 2\omega_i \left(\xi_i \boldsymbol{\Phi}_i - \sqrt{1-\xi_i^2}\, \boldsymbol{\Psi}_i \right) y_{Di}(t) + 2\boldsymbol{\Phi}_i y_{Vi}(t) \right\} + \sum_{j=1}^{2n_o} \boldsymbol{r}_j^R y_{Dj}^R(t) \end{aligned} \tag{4.221}$$

Using statistical measures, earthquake records can be used to calculate all the corresponding $y_{Dj}^R(t)$ to find their maxima, similar to the procedure for determining the design response spectra.

Example 4.7

Assume that the **M-C-K** system mentioned in Example 4.1 is a two-story structure and a damper with the coefficient 60 (kN-s/m) is installed in between the first and the second stories. The supplemental damping matrix can be written as

$$\mathbf{C}_a = \begin{bmatrix} 60 & -60 \\ -60 & 60 \end{bmatrix} \quad (\text{kN-s/m})$$

The corresponding state matrix is

$$\mathcal{A} = \begin{bmatrix} -35 & 32.5 & -750 & 375 \\ 26 & -30 & 300 & -300 \\ 1 & 0 & 0 & 0 \\ 0 & 1 & 0 & 0 \end{bmatrix}$$

From this state matrix,

$$\mathbf{\Lambda} = \text{diag}\left([-3.0801+11.9335i, \; -3.0801-11.9335i, \; -18.24444, \; -40.5954]\right)$$

and

$$\mathcal{P} = \begin{bmatrix} 0.5707+0.2148j & 0.5707-0.2148j & 0.4397 & -0.6947 \\ 0.7884 & 0.7884 & -0.8965 & 0.7189 \\ 0.0053-0.0492j & 0.0053+0.0492j & -0.0241 & 0.0171 \\ -0.0160-0.0619j & -0.0160+0.0619j & 0.0491 & -0.0177 \end{bmatrix}$$

Therefore,

$$\mathbf{P}^R = \begin{bmatrix} -0.0241 & 0.0171 \\ 0.0491 & -0.0177 \end{bmatrix}$$

$$\mathbf{\Lambda}^R = \text{diag}(-18.2444, \; -40.5954)$$

$$\mathcal{Q} = \begin{bmatrix} 0.3191+0.0754j & 0.5256-0.0676j & -4.3804+7.9715j & 4.5850+2.0028j \\ 0.3191-0.0754j & 0.5256+0.0676j & -4.3804-7.9715j & 4.5850-2.0028j \\ -0.4264 & 1.0865 & -35.3919 & 26.6285 \\ -1.2316 & 1.5930 & -34.5272 & 23.1498 \end{bmatrix}$$

$$\boldsymbol{r}_1^R = -\boldsymbol{p}_1^R\boldsymbol{q}_1^R\mathbf{J} = -\begin{Bmatrix} -0.0241 \\ 0.0491 \end{Bmatrix}[-0.4264 \;\; 1.0865]\mathbf{J} = \begin{Bmatrix} 0.0159 \\ -0.0324 \end{Bmatrix}$$

Similarly,

$$\boldsymbol{r}_2^R = -\boldsymbol{p}_2^R\boldsymbol{q}_2^R\mathbf{J} = \begin{Bmatrix} -0.0062 \\ 0.0064 \end{Bmatrix}$$

4.3.2 Design Response Spectra for Overdamped Subsystem

4.3.2.1 Spectral Value

From Chapter 2, the design response spectrum was shown to be very useful for simplifying damper design. By using a statistical survey, the design response spectrum can provide the acceleration if the period and the damping are known. However, the currently used design spectrum is obtained through underdamped SDOF systems. A new type of design response spectrum is necessary to account for the overdamped system.

The statistical maximum response due to a given peak ground acceleration, denoted by S_{Do}, is the only function of the real eigenvalue λ_j^R. That is,

$$S_{Do} = f\left(\lambda_j^R\right) \tag{4.222}$$

Compared with the statistical maximum response of an underdamped system, which is often referred to as the spectral value of the design response spectrum introduced in Chapter 2, Equation 4.222 can be rewritten using the following relationship:

$$\lambda_j^R = -2\pi/T_j^R \tag{4.223}$$

or

$$T_j^R = -2\pi/\lambda_j^R \tag{4.224}$$

where the quantity T_j^R is actually a time constant with the unit "second" for the subsystem with a real eigenvalue. This is referred to as a pseudo period and superscript R again stands for the eigenvalue of the overdamped subsystem or the pseudo mode is real valued. The reason for introducing this quantity is that for a pseudo mode, there is no commonly defined period or natural frequency. However, for convenience in comparing it with regular systems and eventually to use the concept of period in order to employ the design spectra, an equivalent quantity with the same dimension as the period is needed.

Thus, using Equation 4.224, Equation 4.222 is rewritten as

$$S_{Do} = f\left(T_j^R\right) \tag{4.225}$$

That is, although there is no commonly defined period for the overdamped system, Equation 4.225 is still used and the value S_{Do} is called the spectral value for the specific design response spectrum of the overdamped subsystems.

One way to determine the design response spectrum is to use the same procedure as the regular design spectrum by calculating the mean and standard deviation of the responses of overdamped subsystems with different values of λ_j^R or T_j^R through a sufficiently large number of ground excitations. However, in the following discussion, an alternative method to find the design spectrum using the current regular spectrum is provided.

4.3.2.2 Overdamping Constant

In Section 4.3.1, the modal superposition and modal combination methods for complex modes and overdamped subsystems are discussed. In the literature, one of the key issues for modal combination is to assume that the peak response of an underdamped system is proportional to the root mean square value of the response and the proportional factors are constant, despite the value of the modal frequency ω_i. Now, this issue is discussed in detail.

In the case of the overdamped subsystem, it is also assumed that the peak response of a pseudo mode is proportional to the root mean square value of the response and the proportional factor is constant, despite the value of the real eigenvalue λ_j^R. This assumption cannot be proven theoretically. Statistically, it can be seen that, even though the factor is not exactly constant, the errors are rather small. This proportional factor is denoted by α_R, which can be determined by a regression method based on a white noise input assumption.

In fact, the same assumption is made for underdamped systems (Clough and Penzien 1993). That is, the peak response of an underdamped mode is proportional to the root mean square value of the modal response. In addition, the proportion factor is constant, despite the natural frequencies and damping ratios, namely, eigenvalues. This proportion factor is denoted as α_C, which can also be determined by using the regression method based on a white noise input assumption.

In the above assumption, the ratio of α_R and α_C must also be constant. Through statistical studies, it was found that the following is approximately true:

$$\eta = \frac{\alpha_C}{\alpha_R} \approx 0.85 \tag{4.226}$$

For convenience, the above ratio η is called the *overdamping constant.*

In the following discussion, for convenience, d is used to denote the peak response of a nonhomogeneous form of Equation 4.190, which is excited by earthquake ground motions.

Referring to the formula developed for a general complete quadratic combination, the relationship of the peak response of a SDOF underdamped system to the spectral acceleration is as follows:

$$d \approx \frac{1}{\alpha_C}\sqrt{\frac{\pi S_A}{2\xi \omega_n^3}} \quad (m) \tag{4.227}$$

Here S_A is the value of design spectrum of acceleration, (see Equation 2.309) and d is for all reasonable natural frequencies ω_n (or periods) and damping ratios ξ can be obtained from the standard response spectrum. Generally, the assumption of 5% damping ratio is used. For simplicity, the superscript C is omitted, although d is the response of underdamped modes. Equation 4.227 is given here (to be proven later). It is obtained from the square roots of the mean square responses y(t) through the inverse Fourier transform of the frequency response function of Y(jω).

From Equation 4.227, with proper dimensions,

$$S_A = \frac{0.1\alpha_C^2 \omega_n^3}{\pi} d^2 \tag{4.228}$$

Similarly, the relationship between the peak response and the spectral value is obtained as follows by recalling Equation 4.218, the frequency response function of $\mathbf{x}_j^R(t)$ of the jth, pseudo mode, that is,

$$d_R \approx \frac{1}{\alpha_R}\sqrt{\int_{-\infty}^{\infty} \frac{S_A}{\omega^2 + (\lambda^R)^2} d\omega} \tag{4.229}$$

Here, the subscript R again stands for the pseudo mode. For convenience, the subscript j of the eigenvalue λ_j^R is omitted; and furthermore, the amplitude of an element of $\mathbf{x}_j^R(t)$ and the spectral displacement, which is denoted as d_R, are considered.

During an earthquake, for both the mode with a complex eigenvalue and the pseudo mode with a real eigenvalue, the input must be the same. That is, S_A in Equations 4.228 and 4.229 are identical. Thus, substituting Equation 4.228 into Equation 4.229 results in,

$$d_R \approx \frac{\alpha_C}{\alpha_R}\sqrt{\frac{0.1}{\pi}\int_{-\infty}^{\infty} \frac{|\omega|^3 y_a^2}{\omega^2 + (\lambda^R)^2} d\omega} = \frac{1}{\eta}\sqrt{\frac{0.2}{\pi}\int_{0}^{\infty} \frac{|\omega|^3 y_a^2}{\omega^2 + (\lambda^R)^2} d\omega} \tag{4.230}$$

Note that the integration in Equation 4.230 should be carried out for all reasonable real-eigenvalue λ_i^R's. That is, d_R is a function of λ_i^R. Recall Equation 4.212, the statistical spectral value or the spectral value S_{Do} is used to replace d_R in Equation 4.220 (also see Equation 4.215). That is,

$$S_{Do} \approx \frac{\alpha_C}{\alpha_R}\sqrt{\frac{0.1}{\pi}\int_{-\infty}^{\infty} \frac{\omega^3 d^2}{\omega^2 + (\lambda^R)^2} d\omega} = \frac{1}{\eta}\sqrt{\frac{0.2}{\pi}\int_{0}^{\infty} \frac{\omega^3 d^2}{\omega^2 + (\lambda^R)^2} d\omega} \tag{4.231}$$

Again, for convenience, a variable substitution for Equation 4.231 is made by using Equation 4.223 as well as

$$\omega = 2\pi/T \tag{4.232}$$

that is,

$$S_{Do} \approx \frac{1}{\eta}\sqrt{0.8\pi\int_0^{\infty}\frac{\left(T^R\right)^2 d^2}{T^3\left[T^2+\left(T^R\right)^2\right]}dT} \tag{4.233}$$

Equation 4.233 can be obtained through both analytical and numerical approaches.

Note that in Equation 4.233, subscript j of the pseudo period T_j^R is omitted for convenience.

Equations 4.230 and 4.233 provide the analytical forms of the peak responses of the pseudo modes. They can also provide the spectral value for the design response spectra, which is compatible with underdamped systems. In other words, the numerical integration method does not have to be used statistically to obtain the response spectrum for the pseudo modes with real eigenvalues. Instead, the peak value of the pseudo mode response can therefore be used to generate the overdamped design spectra as

$$d_{Rj}\left(\lambda_j^R\right) \rightarrow S_{Do}\left(\lambda_j^R\right) \tag{4.234}$$

In Equation 4.234, the subscript j is replaced to denote the peak response of the jth pseudo mode. Or, by using the pseudo period, T_j^R,

$$d_{Rj}\left(T_j^R\right) \rightarrow S_{Do}\left(T_j^R\right) \tag{4.235}$$

Example 4.8

In the following example, numerical studies to verify Equations 4.230 and 4.235 are presented. Figures 4.3 and 4.4 show the simulation results. Figures 4.3a and 4.4a are the regular response spectra (one mean value plus one standard deviation) generated by 28 and 99 groups of earthquake ground motion accelerations with 0.4 (g) amplitudes, respectively. Figures 4.3a and b show the comparisons of exact and estimated response spectra for pseudo modes with real eigenvalues. The exact results (one mean value plus one standard deviation) are calculated directly by 28 and 99 groups of earthquake ground motion accelerations with 0.4 (g) amplitudes as input, respectively. The estimated curves are obtained through regular response spectra, as shown in Figures 4.3b and 4.4b and Equation 4.230. It is seen that the estimation results are acceptable for use in earthquake engineering practice.

Note that for convenience, in Figure 4.3b, the unit of X axes is frequency, which is actually the pseudo-frequency for the pseudo mode, in the sense of the reciprocal of the pseudo period.

From Equations 4.230 and 4.235 as well as Figures 4.3b and 4.4b, it is realized that when the absolute value of λ_j^R is relatively small, $S_{Do}(\lambda_j^R)$ can have a relatively large value. Therefore, it may not be safe to simply ignore the overdamped systems.

The maximum acceleration can also be obtained by using statistical measures, denoted by $S_{Ao}(\lambda_j^R)$. It is also seen that when the absolute value of λ_j^R is relatively small, $S_{Ao}(\lambda_j^R)$ can also have a relatively large value.

The term $S_{Do}(\lambda_j^R)$ is called the *design spectrum of an overdamped SDOF system* or the *overdamping displacement spectrum* and the term $S_{Vo}(\lambda_j^R)$ is called the *design acceleration spectrum of an overdamped SDOF system* or the *overdamping acceleration spectrum*. For convenience, the

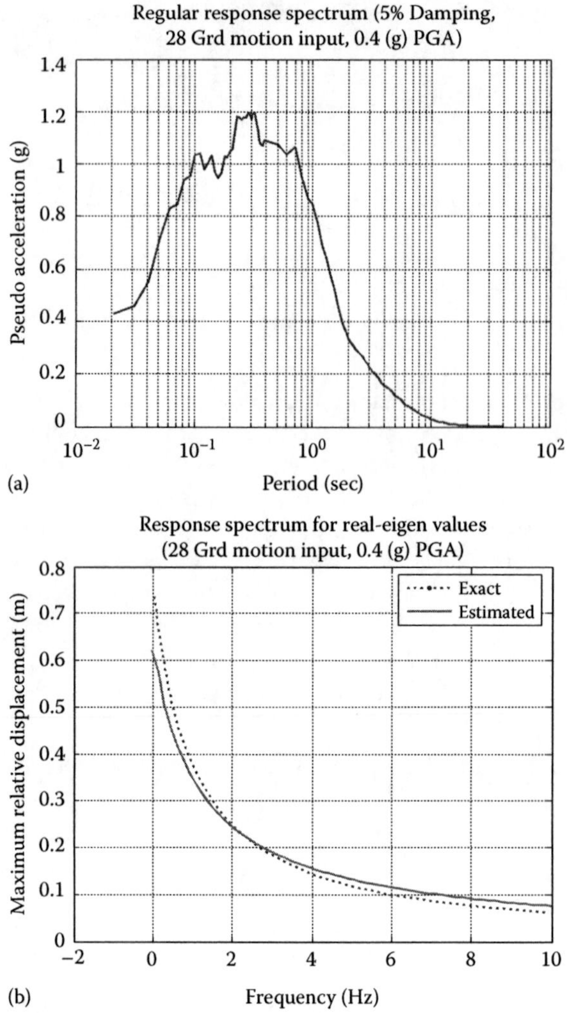

FIGURE 4.3 (a) Statistical response spectrum (5% damping ratio, 28 records). (b) Comparisons between statistically obtained ("exact") and that estimated through Equation 4.230.

real-valued eigenvalue $-\lambda_j^R$ is called the jth *overdamping factor.* Note that systems and/or subsystems with different damping can be more clearly shown in the following chart. In this book, we refer to a structure with all three cases of damping (proportional damping, nonproportional damping, and overdamping) and a system with unknown damping is referred to as a *generally damped* system. The phrases normal mode, complex mode, and pseudo mode are used to represent the cases already shown in Figure 4.1.

4.4 RESPONSES OF GENERALLY DAMPED SYSTEMS AND THE DESIGN SPECTRA

Conventionally, to obtain the responses of a generally damped system, the computation is carried out by decoupling in the 2n complex modal space. This is not convenient for engineers in design practice. They often use design spectra for SDOF systems and use the modal summation for the MDOF systems in n-dimensional space. In order to be compatible with currently used design

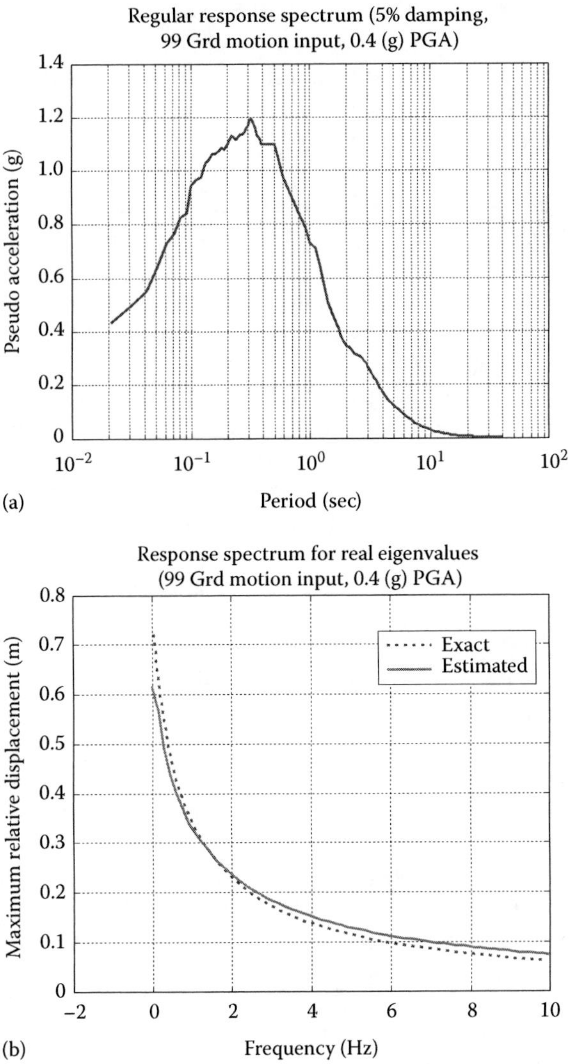

FIGURE 4.4 (a) Statistical response spectrum (5% damping ratio, 99 records). (b) Comparisons between statistically obtained ("exact") and that estimated through Equation 4.148.

response spectrum, a computation in n-dimensional space is introduced for a generally damped system by using the modal approaches with displacement only. In this way, existing design procedure and design spectra can be used directly.

4.4.1 Approach in 2n- and n-Space, Design Codes

The first-order differential Equations 4.117, 4.118, and/or 4.204 and 4.205 have been transferred into the form of second-order equations such as those described by Equation 4.184. This equation is further linked with the design spectra, which are primarily used in damper design. In order to pursue this approach more effectively, the responses are simplified by examining each mode. Note that the response of the jth pseudo mode is already known. Thus, in this section, we focus first on the ith underdamped mode. After all the interested responses are determined, we can use the method of modal summation to find the peak value of the total responses and relate these to the design spectra.

Thus, Equation 4.183 is considered instead of Equation 4.189. From Equation 4.183, it is realized that the response of the ith mode contains two parts: $y_{Di}(t)$ and $y_{Vi}(t)$, which are, respectively, the monic displacement and the monic velocity described in Equations 4.180 and 4.181, respectively. Here, for simplicity, superscript C is omitted, although these responses are taken from underdamped systems. The reason they are called unit responses is that the input $x_{g''}(t)$ is normalized with unit amplitude.

From the above analysis, it is seen that because of the existence of the velocity term, to find the modal solution, the 2n-complex modal space must be used. However, all the design codes use modal summation in the n-dimensional space.

That is, the design spectral value S_D (also S_V or S_A) can only be related with the sole displacement (or sole velocity, sole acceleration). Yet, the displacement $y_{Di}(t)$ cannot be separated from the velocity $y_{Vi}(t)$ from Equation 4.183 alone.

There are two possible methods to deal with the combination of $y_{Di}(t)$ and $y_{Vi}(t)$. First, $y_{Di}(t)$ and $y_{Vi}(t)$ can be treated as two different sets of responses, like two different modes. Then, they are linked separately with the design spectra, and finally combined using the common modal summation methods, such as the SRSS and/or the CQC. This task has been achieved by Veletsos and Ventura (1986), and Song et al. (2008), who pioneered the linking of responses of a nonproportionally damped system with the earthquake design spectra. However, since this method treats the displacement and the velocity of the same mode as two different ones, it is complicated and requires more computation.

The second is to find the relationship between the displacement $y_{Di}(t)$ and the velocity $y_{Vi}(t)$. Because both belong to the same mode, the responses should have the same natural frequency and damping ratio, which allow them to be combined. If these two responses can be combined together and treated as a single modal response, the design formula can be greatly simplified, more easily understood, and requires less computation.

Note that one of the goals is to find the maximum modal response from Equation 4.183. This goal is realized by using the design response spectrum of the displacement, S_D.

That is, the monic design spectrum is statistically obtained by using the mean pulse of the standard deviation of $y_{Di}(t)$ with the statistical measures using multiple ground motions. Therefore, $y_{Di}(t)$ can be replaced by S_D as the statistical maximum response of displacements. Similarly, $y_{Vi}(t)$ is treated as the statistical maximum response of velocities and the design response spectrum of the velocity, S_V, is statistically obtained by using the mean and the standard deviation of $y_{Vi}(t)$. It is also noted that $y_{Di}(t)$ and $y_{Vi}(t)$ do not provide the entire design spectra, but the spectral value at the modal frequency $\omega_i = 2\pi/T_i$ under normalized unit excitation $x_{g''}(t)$. Therefore, they can be replaced as $S_D(T_i)$ and $S_V(T_i)$, when the maximum value is calculated.

However, using both spectral values $S_D(T_i)$ and $S_V(T_i)$ is just as inconvenient as treating them as two different modes. This is because S_D and S_V cannot be used to replace the terms defined in Equation 4.103, because the maximum responses of $y_{Di}(t)$ and $y_{Vi}(t)$ will not occur at the same time. Instead, the following approaches can be used.

Figure 4.5a shows the response of the displacement and velocity of an SDOF structure with a natural frequency of $\omega_i = 14$(rad/s) (f = 2.2(Hz), T = 0.45(s)) under El Centro earthquake excitation. From the response plots, it can be seen that:

1. When the velocity reaches a very high level, the displacement is close to zero. Then, the velocity drops from its peak value, whereas the amplitude of the displacement increases. At the exact moment when the velocity reaches zero, the displacement reaches its peak value. Therefore, it can approximately be said that the velocity is leading the displacement at a 90° phase.
2. When the velocity and the displacement with close to a 90° phase difference reach their maximum peak value, both the velocity and the displacement are close to half sine waves.

Therefore, two sinusoidal functions can be used to approximate the peak velocity and the peak displacement.

3. Both sinusoidal functions have their periods close to $2\pi/\omega_i$. Therefore, the angle $\theta = \omega_i t$ can be used to construct the sinusoidal functions.

Therefore, in order to find the maximum response contributed by both $y_{Di}(t)$ and $y_{Vi}(t)$, the maximum response of both $y_{Di}(t)$ and $y_{Vi}(t)$ can be approximated by the half sine waves and are denoted as $\alpha\sin(\theta)$ and $\beta\cos(\theta)$. Their peak values are denoted as α and β, respectively.

Figure 4.5b shows the peak values of the displacement, each of which is purposely multiplied by its natural frequency ω_i, and the peak velocity. It is seen that these peak values can be represented by the half sine and cosine waves. Figure 4.5c shows these two trigonometric functions, where θ is chosen to be 90° or 180° by not considering the minus sign of $y_{Di}(t)$, since only the maximum value is of interest. Note that in Figure 4.5b, the peak responses are purposely chosen to be not very close to the trigonometric waves. In many cases, there can be noticeably better approximation than using

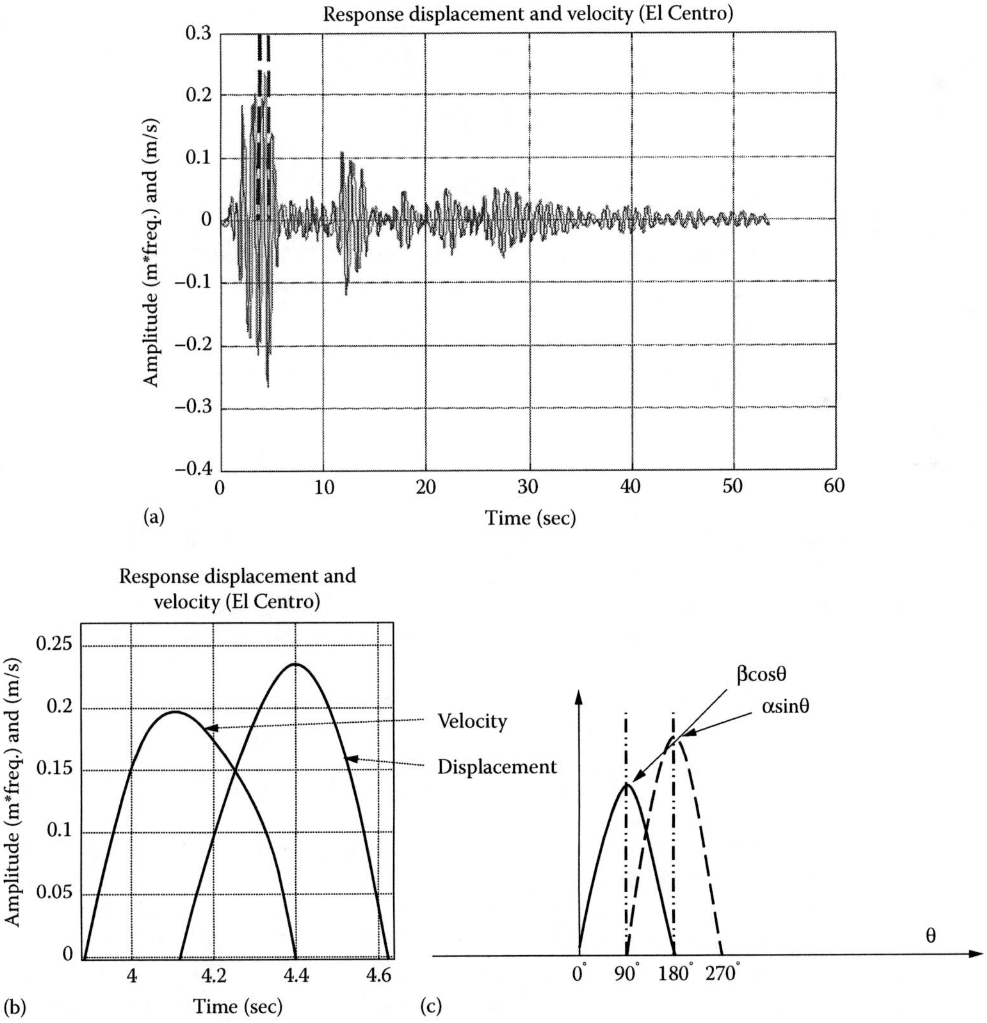

FIGURE 4.5 (a–c) Peak velocity and displacement.

the trigonometric functions. However, it is seen that it is already sufficiently close to the result that would be obtained by using the trigonometric functions. Also note that in Figure 4.5b, if the displacement is not amplified by the factor ω, its amplitude will be considerably smaller.

4.4.2 Modal Solution in n-Dimensional Space

To evaluate the coefficients α and β (see Figure 4.5c), the coefficients of $y_{Di}(t)$ and $y_{Vi}(t)$, namely, $2\omega_i\left(\xi_i\mathbf{\Phi}_i - \sqrt{1-\xi_i^2}\mathbf{\Psi}_i\right)$ and $2\mathbf{\Phi}_i$, are considered. The following relationship can be obtained:

$$\left\|y_{Vi}(t)\right\| \approx \left\|\omega_i y_{Di}(t)\right\|_{max} \tag{4.236}$$

where the symbol $\|(\bullet)\|$ specially stands for the maximum value of (•), which was defined previously.

That is, the amplitude of $y_{Vi}(t)$ is ω_i times the amplitude of y_{Di}. Thus, it is written as

$$\left\|y_{Vi}(t)\right\| \approx \omega_i \left\|y_{Di}(t)\right\| \tag{4.237}$$

There is another angle to describe the relationship of Equation 4.237, that is, the pseudo design spectra:

$$S_{PV} = \omega_i S_D \tag{4.238}$$

The pseudo spectral value of the velocity is ω_i times the value of the spectral displacement.

According to the approximation described in Equation 4.237, a notation that expresses the relationship of the peak amplitudes needs to be found, which is ω_i times in amplitude difference and 90° in phase difference. Specifically, the following equations can be used:

$$y_{Di}(t) \approx \sin(\theta)\, y_{Di}^{max} \tag{4.239}$$

and

$$y_{Vi}(t) \approx \omega_i \cos(\theta)\, y_{Di}^{max} \quad \text{with } 90° \le \theta \le 180° \tag{4.240}$$

where y_{Di}^{max} is the amplitude of the displacement $y_{Di}(t)$; and the variations of the displacement $y_{Di}(t)$ and velocity $y_{Vi}(t)$ can be described by the variation of angle θ.

Equations 4.239 and 4.240 imply that the responses $y_{Di}(t)$ and $y_{Vi}(t)$, under unit excitation, can be summarized together by only considering the phase difference of 90° or π/2, since their amplitude is known to be y_{Di}^{max} and $\omega_i y_{Di}^{max}$.

Using Equations 4.239 and 4.240, the maximum value of $\mathbf{x}_i(t)$ can be rewritten as

$$\mathbf{x}_{max\,i} = -2\omega_i\left\{\left(-\xi_i\mathbf{\Phi}_i + \sqrt{1-\xi_i^2}\mathbf{\Psi}_i\right)\sin(\theta_m) + \mathbf{\Phi}_i\cos(\theta_m)\right\} y_{Di}^{max} \tag{4.241}$$

where the angle θ is chosen so that the displacement $\mathbf{x}_i(t)$ reaches its peak value $\mathbf{x}_{max\,i}$ and the specific angle is denoted by θ_m. Note that θ_m is only used to justify the existence of the maximum value; it is not used in real-numbered computations. The displacement and the velocity are combined in Equation 4.241. Note that in Equation 4.241, the coefficients are vectors, which can be treated element by element in order to compare their amplitudes. In these cases, the angle θ_m can be different for different elements. Here, for simplicity, only one angle θ_m is used. For convenience, the coefficients of the jth elements sin(θ)and cos(θ) in Equation 4.241, respectively, are denoted by α and β; and for simplicity, subscript j is ignored. The problem now is how to determine the total value of α sin(θ) + β cos(θ) by examining the coefficients α and β.

First, assume $\beta < \alpha$ and the maximum value of $\alpha/\beta \sin(\theta) + \cos(\theta)$ is denoted as y and the ratio (α/β) is denoted as x. That is,

$$y = \alpha/\beta \sin(\theta) + \cos(\theta) \tag{4.242}$$

and

$$x = \alpha/\beta \tag{4.243}$$

The maximum values are obtained by varying the angle θ. Figure 4.6 plots the curve y vs. the ratio x marked by the solid line. In Figure 4.6, a regression curve marked by the dotted line is also plotted, which can be expressed as

$$y = 0.9986\,x + 0.4213 \tag{4.244}$$

Equation 4.242 can be further simplified as

$$y = x + 0.4 \tag{4.245}$$

Using Equation 4.161, the maximum absolute error will be around 0.2, which occurs at the point

$$\cos(\theta) = \sin(\theta) \tag{4.246}$$

The maximum percentage error is less than 15%.

Note that the case described by Equation 4.246 will occur when the order of the mode becomes sufficiently high. Since the higher mode will contribute considerably less to the entire responses, the overall error will be much smaller than 15%, usually smaller than 5%.

Equation 4.245 is obtained by assuming that

$$\beta < \alpha \tag{4.247}$$

However, if

$$\alpha \le \beta \tag{4.248}$$

the same relationship described by Equation 4.245 can still exist, except that in this case,

$$x = \beta/\alpha \tag{4.249}$$

Using Equation 4.245, the total value is obtained by multiplying either the factor β or the factor α back. The remaining task is to determine whether $\beta < \alpha$ or $\alpha \le \beta$ element by element of the two vectors $2\omega_i\left(\xi_i \mathbf{\Phi}_i - \sqrt{1-\xi_i^2}\,\mathbf{\Psi}_i\right) y_{Di}^{max}$ and $2\omega_i \mathbf{\Phi}_i y_{Di}^{max}$, or more simply $\left(\xi_i \mathbf{\Phi}_i - \sqrt{1-\xi_i^2}\,\mathbf{\Psi}_i\right)$ and $\mathbf{\Phi}_i$. First, assume that Equation 4.247 holds; that is, assume for the jth element,

$$\left(-\xi_i \varphi_{ij} + \sqrt{1-\xi_i^2}\,\psi_{ij}\right) > \varphi_{ij} \tag{4.250}$$

The plot of Figure 4.6 is thus the combined values of

$$\left[\left(-\xi_i \varphi_{ij} + \sqrt{1-\xi_i^2}\,\psi_{ij}\right)/\varphi_{ij}\right] \sin(\theta) + \cos(\theta) \tag{4.251}$$

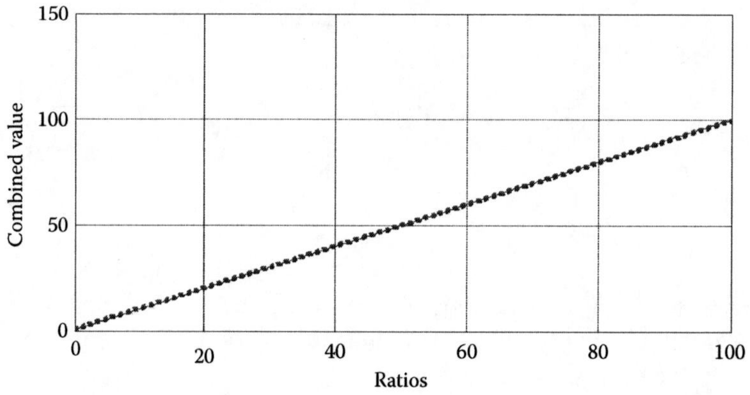

FIGURE 4.6 Combined value vs. the ratio.

vs. one of the element ratios taken from

$$\left(-\xi_i\varphi_{ij} + \sqrt{1-\xi_i^{\,2}}\,\psi_{ij}\right)\Big/\varphi_{ij} \tag{4.252}$$

Here the symbol $\boldsymbol{u}\bullet/\boldsymbol{v}$ stands for each element of vector $\boldsymbol{u}$, e.g., the ith element is divided by the corresponding i[th] element of vector $\boldsymbol{v}$. Further, $\boldsymbol{u}\bullet\boldsymbol{v}$ is used to denote that for each element of vector $\boldsymbol{u}$, such as the ith element, the corresponding i[th] element of vector $\boldsymbol{v}$ is multiplied. Furthermore, $\boldsymbol{u}^{\bullet b}$ is used to denote that each element of vector $\boldsymbol{u}$ will have the power of scalar b. That is,

$$\boldsymbol{u} = \{u_i\} \tag{4.253}$$

$$\boldsymbol{u} = \{v_i\} \tag{4.254}$$

$$\boldsymbol{u}\bullet/\boldsymbol{v} = \{u_i/v_i\} \tag{4.255}$$

$$\boldsymbol{u}\bullet\boldsymbol{v} = \{u_i \cdot v_i\} \tag{4.256}$$

$$\boldsymbol{u}^{\bullet b} = \{u_i^b\} \tag{4.257}$$

With the above notation,

$$x_{\max_{ij}} = -2\omega_i\varphi_{ij}\left\{\left[\left(-\xi_i\varphi_{ij} + \sqrt{1-\xi_i^{\,2}}\,\psi_{ij}\right)\Big/\varphi_{ij}\right]\sin(\theta_m) + \cos(\theta_m)\right\} y_{Di}^{\max} \tag{4.258}$$

and $y_{Di}^{\max}$ is used to denote that the displacement $y_{Di}(t)$ has reached its maximum value.

By using the approximation relationship expressed in Equation 4.245,

$$x_{\max_{ij}} = -2\omega_i\varphi_{ij}\{y\}\, y_{Di}^{\max} \tag{4.259}$$

or

$$x_{\max_{ij}} = -2\omega_i\varphi_{ij}\{x + 0.4\}\, y_{Di}^{\max} \tag{4.260}$$

Furthermore,

$$x_{\max_{ij}} = -2\omega_i\varphi_{ij}\{\alpha/\beta + 0.4\}\, y_{Di}^{\max} \tag{4.261}$$

and finally,

$$x_{max_{ij}} = -2\omega_i \left\{ \left(-\xi_i \varphi_{ij} + \sqrt{1-\xi_i^2}\,\psi_{ij} \right) + 0.4\,\varphi_{ij} \right\} y_{Di}^{max} \tag{4.262}$$

In addition, it is realized that the maximum value of the displacement, y_{Di}^{max}, can be statistically obtained by using the design response spectrum. That is,

$$y_{Di}^{max} \rightarrow S_D(T_i, \xi_i)\,g \tag{4.263}$$

Thus, Equation 4.261 is rewritten by replacing the maximum value of y_{Di}^{max} for $y_{Di}(t)$ and the maximum value for $y_{Vi}(t) = \omega_i y_{Di}(t)$ with $S_D(T_i, \xi_i)$ and $\omega_i\ S_D(T_i, \xi_i)$, also replacing $x_{max_{ij}}$ for $d_{max_{ij}}$ as follows:

$$d_{max_{ij}} = 2\omega_i \left| \left\{ \left[-\xi_i \varphi_{ij} + \sqrt{1-\xi_i^2}\,\psi_{ij} \right] + 0.4\varphi_{ij} \right\} \right| S_D g \tag{4.264}$$

Similarly, it can be proven that if

$$\left| -\xi_i \varphi_i + \sqrt{1-\xi_i^2}\,\psi_i \right|_j \le \left| \varphi_{ij} \right| \tag{4.265}$$

then

$$x_{max_{ij}} = -2\omega_i \left\{ 0.4\, \left| -\xi_i \varphi_{ij} + \sqrt{1-\xi_i^2}\,\psi_{ij} \right| + \left| \varphi_{ij} \right| \right\} y_{Di}^{max} \tag{4.266}$$

or

$$d_{max_{ij}} = -2\omega_i \left\{ 0.4 \left| -\xi_i \varphi_{ij} + \sqrt{1-\xi_i^2}\,\psi_{ij} \right| + \left| \varphi_{ij} \right| \right\} S_D g \tag{4.267}$$

The significance of Equations 4.264 and 4.267 is that the ith modal response at the jth location can be directly related to the spectral value S_D, whether or not the system is proportionally or non-proportionally damped.

Furthermore, the modal response vector can be obtained by considering the following shape function. That is, in order to have the general form of the maximum value of the ith modal displacement, the response coefficient vector $\mathbf{Y}_i$ is denoted as

$$\mathbf{Y}_i = \left\{ \upsilon_{ij} \right\} = \begin{cases} 2\omega_i \left\{ \left| -\xi_i \varphi_{ij} + \sqrt{1-\xi_i^2}\,\psi_{ij} \right| + 0.4 \left| \varphi_{ij} \right| \right\}, & \text{when } \left| -\xi_i \varphi_{ij} + \sqrt{1-\xi_i^2}\,\psi_{ij} \right| < \left| \varphi_{ij} \right| \\ 2\omega_i \left\{ 0.4 \left| -\xi_i \varphi_{ij} + \sqrt{1-\xi_i^2}\,\psi_{ij} \right| + \left| \varphi_{ij} \right| \right\}, & \text{when } \left| -\xi_i \varphi_{ij} + \sqrt{1-\xi_i^2}\,\psi_{ij} \right| \ge \left| \varphi_{ij} \right| \end{cases} \tag{4.268}$$

Thus,

$$\mathbf{d}_{max_i} = \mathbf{Y}_i S_D(T_i, \xi_i)\,g \tag{4.269}$$

Here, the spectral vector $\mathbf{d}_{max_i}$ is used to replace x_{max_i}.

Equation 4.269 defines the peak value of the ith modal response in terms of an $n \times 1$ real-valued vector and the spectral value $S_D(T_i, \xi_i)$ taken from the displacement design response spectrum. In practical applications, the pseudo acceleration design spectrum, $S_A(T_i, \xi_i)$, is often used and

$$S_D(T_i, \xi_i) = \frac{T_i^2}{4\pi^2} \cdot S_A(T_i, \xi_i) g \tag{4.270}$$

It can be written that

$$\mathbf{d}_{\max i} = \mathbf{Y}_i \frac{T_i^2}{4\pi^2} S_A(T_i, \xi_i) g \tag{4.271}$$

4.4.3 Modal Truncations for a Generally Damped System

Suppose a system contains both n_u underdamped modes and $2n_o$ overdamped subsystems.

According to Equation 4.221, the total response can be written as

$$\mathbf{x}(t) = \sum_{i=1}^{n} \mathbf{x}_i(t) = \sum_{i=1}^{n_u} \mathbf{x}_i^C(t) + \sum_{j=1}^{2n_o} \mathbf{x}_j^R(t) \tag{4.272}$$

Here, $\mathbf{x}_i^C(t)$ is the underdamped response and $\mathbf{x}_j^R(t)$ is the overdamped response, which was previously defined. Thus,

$$\mathbf{x}_i(t) = \mathbf{x}_i^C(t) + \mathbf{x}_j^R(t) \tag{4.273a}$$

Note that if a system does not contain overdamped subsystems,

$$\mathbf{x}_i(t) = \mathbf{x}_i^C(t) \tag{4.273b}$$

It was previously discussed that for the underdamped system, all the modes do not have to be used because the first several modes often contribute the greatest portions of the responses. Thus, the concept of modal truncation can be used. Here, because the concept is extended to include the overdamped cases and for convenience, it is still called modal truncation.

That is, if only the first S_U modes of the underdamped part and the first S_O subsystems of the overdamped part are of interest,

$$\mathbf{x}(t) = \sum_{i=1}^{S} \mathbf{x}_i(t) = \sum_{i=1}^{S_U} \mathbf{x}_i^C(t) + \sum_{j=1}^{S_O} \mathbf{x}_j^R(t) \tag{4.274}$$

where

$$S = S_U + S_O \tag{4.275}$$

The remaining questions are: (1) what if certain modes reduce to overdamped? (2) how can the number S be determined (this is discussed later)? and (3) how should the peak values be considered? By using Equation 4.272, the individual maximum value of the underdamped response can be obtained. To find the maximum value of the overdamped response, Equation 4.218 is repeated as follows:

$$\mathbf{x}_j^R(t) = \boldsymbol{r}_j^R \, y_{Dj}^R(t) \tag{4.276}$$

so that

$$\mathbf{x}_{\max j}^R = \left|\boldsymbol{r}_j^R\right| \, y_j^{R\max} \tag{4.277}$$

Here $\mathbf{x}_{\max j}^R$ is the maximum response of the jth overdamped subsystem with a real eigenvalue λ_j^R. Considering spectral values, Equation 4.277 can be rewritten as

$$\mathbf{d}_{\max j}^R = \left|\boldsymbol{r}_j^R\right| \, S_{Do}\left(\lambda_j^R\right) g \tag{4.278}$$

Note that, these maximum response values of each individual mode and/or subsystem cannot be directly summarized together. In the following discussion, in order to determine the number of S_U and S_O modes, the concept of modal participation is studied first. Then the truncated modal summation will be performed.

4.5 MODAL PARTICIPATION AND MODAL CRITERIA

The generally damped system is further examined in this section. Similar to a proportionally damped system, the concept of modal participation must be addressed by defining modal contribution indicators to determine if a mode or a subsystem should be included in the response computation. Different from a proportionally damped system, complex-valued mode shapes and real-valued eigenvalues may present.

4.5.1 Criteria on Complex Mode

From the above discussion, it is seen that although using the complex mode approach can increase the computational accuracy, it will make the problem more complicated; even when using the n-space method. However, on the other hand, there are certain cases where this approach must be used. Thus, in order to determine if complex modes should be used, criteria are needed. In the following discussion, specific indices to estimate whether the complex mode must be used in damper design are introduced.

4.5.1.1 Modal Energy Index

From Equation 4.113,

$$\omega_i \approx e^{\zeta_i} \omega_{ni} \approx \left(1 + \zeta_i\right) \omega_{ni} \tag{4.279}$$

Specifically, the difference between the nonproportionally and proportionally damped systems, in terms of the natural frequencies, can be denoted by the modal ETR ζ_i.

When

$$\zeta_i > 0 \tag{4.280}$$

then

$$\omega_i > \omega_{ni} \tag{4.281}$$

On the other hand, when

$$\zeta_i < 0 \tag{4.282}$$

then

$$\omega_i < \omega_{ni} \tag{4.283}$$

This phenomenon encourages the use of the ETR as the indicator to determine whether or not the mode of interest can be approximated by the simplified normal mode approach.

As mentioned above, it can be proven that

$$\zeta_i \neq 0 \tag{4.284}$$

is the necessary and sufficient condition for the ith mode to be complex. Namely, if the ith mode is complex, then the modal participation factor will be affected by the modal ETR ζ_i. Generally, if ζ_i is greater than zero, this particular mode is formed by receiving energy from other modes and vice versa for negative ζ_i.

For complex modes, a uniform modal participation factor no longer exists. Instead, in a later paragraph, the modal participation factor vector is introduced, which means that for the response at different locations, the modal participation indicator may vary. In this case, the response of a certain location may be amplified, whereas others may be reduced. The energy transfer is only an average measure of the entire mode. The resulting peak values at each location may be either larger or smaller. In this case, the ETR cannot be only used as a uniform measure to examine the magnification of the responses.

The damping ratio may or may not contribute to the value of the modal participation factor. For a proportionally damped system, it is seen that from Equation 3.170, the modal participation factor will have nothing to do with the damping in the normal mode. In this circumstance, the damping ratio will not affect the mode shape $\boldsymbol{P}_i$; therefore, the value of Γ_i will not be changed if the damping ratio varies. That is, unless the ith mode reduces to two pseudo modes, the modal participation factor will remain the same value for the proportionally damped system.

However, in the second case of nonproportional damping, it is seen that adding damping will affect the particular modal participation factor. That is, in an attempt to determine if a particular mode can contribute to the total response as well as if it can be simplified as a normal mode, we should consider both the damping ratio and the ETR.

For the above reasons, the ratio r_{Ei} is defined as follows:

$$r_{Ei} = \frac{|\zeta_i|}{\sqrt{\xi_i^2 + \zeta_i^2}} \tag{4.285}$$

To obtain an overall view of the degree of having the complex mode, the following term can be used:

$$r_E = \max_{i=1,S}\left(r_{Ei}\right) \tag{4.286}$$

From Equation 4.285, it is seen that

$$0 \leq r_E < 1 \tag{4.287}$$

In practical damper design, this ratio can be used to determine whether or not the ith mode can be treated as a normal mode. It is seen that when the ith mode of a system is normal, $r_E = 0$. When it become complex, the higher the degree of complexity, the larger the value of r_{Ei}. For convenience, this ratio is called the *modal energy transfer factor* or the *ETR factor.*
Practically speaking, a value of G can be preset so that when

$$r_{Ei} < G, i = 1,\ldots, S \tag{4.288a}$$

the particular i^{th} mode can be treated as a simplified normal mode. Furthermore, in practice, the first mode is checked for simplicity, that is, the criterion can be written as

$$r_{E1} < G \tag{4.288b}$$

Furthermore, note that one of the advantages of using the ETR factor is that the quantity of ETR of the first mode can be obtained by

$$\zeta_1 \approx \omega_i/\omega_{ni} - 1 \tag{4.289}$$

or

$$\zeta_1 \approx T_{n1}/T_1 - 1 \tag{4.290}$$

That is, given the natural frequencies for the nonproportionally damped system, ω_{n1}, and the corresponding undamped system, ω_1, the ETR can be calculated for the first mode with the simple formula described in Equation 4.289. Or, if the natural period for the nonproportionally damped system, T_{n1}, and the corresponding undamped system, T_1, can be obtained, the ETR can be calculated for the first mode by using the simple formula described in Equation 4.290.

4.5.1.2 Complex Modal Factor

In order to study the existence of the complex mode more generally, an alternative index based on a comparison between the responses of the complex mode and the normal mode is further considered.

From Equation 4.183, it was learned that the response of the ith complex mode can be represented by two parts, namely, the one contributed by displacement $y_{Di}(t)$ and the one contributed by velocity $y_{Vi}(t)$. Also, from Equation 4.236, it is known that the amplitude of velocity $y_{Vi}(t)$ is approximately ω_i times the amplitude of displacement $y_{Di}(t)$, that is,

$$\left\| y_{Vi}(t) \right\| \approx \omega_i \left\| y_{Di}(t) \right\| \tag{4.291}$$

Here "$\|(\bullet)\|$" again stands for the amplitude of $(\bullet)$.
In this way, the amplitude coefficient of the displacement is denoted as

$$\mathbf{c}_{Di} = 2\omega_i \left(\xi_i \boldsymbol{\Phi}_i - \sqrt{1-\xi_i^2}\, \boldsymbol{\Psi}_i \right) \tag{4.292}$$

and the amplitude coefficient of the velocity as

$$\mathbf{c}_{Vi} = 2\omega_i \boldsymbol{\Phi}_i \tag{4.293}$$

Here, the vectors $\boldsymbol{\Phi}_i$ and $\boldsymbol{\Psi}_i$ as before are defined as the real and imaginary parts of $\boldsymbol{p}_i\boldsymbol{q}_i\mathbf{J}$, which is used in the transfer function of the nonproportionally damped i^{th} mode described in

Equation 4.173. It is known that a formula such as Equation 4.173 can also be used for a proportionally damped system. Equation 4.173 is rewritten as follows:

$$\boldsymbol{h}_i(s) = -\left[\frac{\boldsymbol{p}_i\boldsymbol{q}_i\mathbf{J}}{s-\lambda_i} + \frac{(\boldsymbol{p}_i\boldsymbol{q}_i\mathbf{J})^*}{s-\lambda_i^*}\right] \tag{4.294}$$

Next, the difference between the transfer functions of the nonproportionally and proportionally damped systems is examined by considering the term $\boldsymbol{p}_i\boldsymbol{q}_i\mathbf{J}$. This is because the rest of the terms $s - \lambda_i$, etc., are identical for both systems. Now, substituting Equation 4.86 into $\boldsymbol{p}_i\boldsymbol{q}_i\mathbf{J}$ results in

$$\boldsymbol{p}_i\boldsymbol{q}_i\mathbf{J} = \boldsymbol{p}_i\left(r_i\boldsymbol{p}_i^T\mathbf{M}\right)\mathbf{J} \tag{4.295}$$

If the system is proportionally damped, $\boldsymbol{p}_i$ is real valued, that is,

$$\boldsymbol{p}_i = \text{real}\left(\text{Im}(\boldsymbol{p}_i) = 0\right)$$

Also, note that $\mathbf{M}$ and $\mathbf{J}$ are real valued.

The remaining term, r_i, is described in Equation 4.93. From Chapter 3, it was seen that the mode shape vectors $\boldsymbol{p}_i$ are orthogonal with the weighting matrix $\mathbf{M}$ and it is known that $\boldsymbol{p}_i^T\mathbf{M}\boldsymbol{p}_i = m_i$, $\boldsymbol{p}_i^T\mathbf{C}\boldsymbol{p}_i = c_i = 2\xi_i\omega_{ni}m_i$. Therefore, Equation 4.93 is rewritten for the proportionally damped system as

$$\begin{aligned} r_i &= \frac{1}{2\lambda_i\left(\boldsymbol{p}_i^T\mathbf{M}\boldsymbol{p}_i\right)+\boldsymbol{p}_i^T\mathbf{C}\boldsymbol{p}_i} \\ &= \frac{1}{2\lambda_i(m_i)+2\xi_i\omega_{ni}m_i} = \frac{1}{j\sqrt{1-\xi_i^2}\,\omega_{ni}m_i} \qquad i = 1, 2, \ldots, n \end{aligned} \tag{4.296}$$

Here, r_i only has an imaginary part.

From Equation 4.296, it is seen that the term $\boldsymbol{p}_i\boldsymbol{q}_i\mathbf{J}$ is a product of four pure real quantities and one pure imaginary quantity. Thus, it is concluded that

$$\boldsymbol{p}_i\boldsymbol{q}_i\mathbf{J} = j\text{Im}(\boldsymbol{p}_i\boldsymbol{q}_i\mathbf{J}) \tag{4.297}$$

Compared with generally damped systems, it is seen that the only difference between the nonproportionally and proportionally damped systems is that, for the latter,

$$\boldsymbol{p}_i\boldsymbol{q}_i\mathbf{J} = j\Psi_i \tag{4.298}$$

or

$$\boldsymbol{\Phi}_i = \mathbf{0} \tag{4.299}$$

Assuming small damping, substituting Equation 4.299 into Equations 4.292 and 4.293 yields, respectively,

$$\mathbf{c}_{Di} = 2\omega_i\sqrt{1-\xi_{Di}^2}\,\boldsymbol{\Psi}_i \approx 2\omega_i\boldsymbol{\Psi}_i \tag{4.300}$$

and

$$\mathbf{c}_{Vi} = \mathbf{0} \tag{4.301}$$

for the pure proportionally damped system.

It is understandable that when the degree of damping nonproportionality is lower, the value of the velocity will be smaller. Once the system reaches a pure proportionally damped case, the velocity term disappears and the vector $\mathbf{\Phi}_I$ will become null. This phenomenon implies that the amplitude of the coefficients c_{Di} and c_{Vi} can be seen as a new index for determining the complex mode, called the *complex mode factor*, r_{Vi}, and is defined as follows:

$$r_{Vi} = \sqrt{\frac{\mathbf{\Phi}_i^T \mathbf{\Phi}_i}{\mathbf{\Phi}_i^T \mathbf{\Phi}_i + \mathbf{\Psi}_i^T \mathbf{\Psi}_i}} \tag{4.302}$$

For an overall understanding of the degree of the significance having the complex mode, the following term can be used:

$$r_V = \max_{i=1,S} \left(r_{Vi} \right) \tag{4.303}$$

From Equation 4.302, it is seen that

$$0 \le r_{Vi} < 1 \tag{4.304}$$

In practice, the complex mode factor needs more information in order to be computed, and this is one of its limitations. However, it is suggested that both indices be considered to determine if the complex mode approach should be used, since using only one factor may lead to errors and/or incorrect decisions.

4.5.2 Modal Participation Factors

To compute the structural response efficiently without loss of accuracy, the modes that contribute the largest part of the total response need to be selected and the modes that only have minimal contribution need to be excluded. In other words, the truncated modal superposition must be sufficient to represent the total responses. To determine whether a mode needs to be included in our computation, we need the help of the concept of modal participation.

The complex mode is more accurate for modeling nonproportionally damped systems, but requires more computations. In order to determine whether or not to use it, a criterion is needed. This may help us simplify response estimations and less computation may be used without loss of design accuracy.

In order to consider another specific case of overdamped subsystems, which is often ignored by conventional approaches but may affect the total response computation, we need additional criterion. This criterion allows the overdamped subsystems to be considered quantitatively. If an overdamped subsystem is to be included, although the computation is unfamiliar to most practical engineers, this criterion should be considered.

In this section, these criteria are examined based on an explanation of the dynamic behavior of generally damped systems.

In the above discussion, Equation 4.271 was obtained to link the maximum response of the ith mode and/or subsystem to the design spectral value. In the following, the modal participation factor for the generally damped systems will be further defined, which will be convenient for use in damper design.

4.5.2.1 Nonproportionally Damped Modes

The first step is to make the equation compatible with building codes, such as NEHRP 2003 (BSSC 2003), which use a special form of mode shape, and is normalized with the method described by Equation 3.129. Note that the mode shape used in the code is real valued, whereas the one described here thus far is complex valued. Therefore, a real-valued mode shape $\boldsymbol{p}_i$ is used to approximate the exact mode shape as

$$\bar{\boldsymbol{p}}_i = \left\{\frac{|\boldsymbol{p}_i|}{\mathrm{p}_{ni}}\right\} \bullet \mathrm{sgn}(\boldsymbol{p}_{ni}) = \{\bar{\mathrm{p}}_{ij}\}, \quad j = 1, 2, \ldots, n \tag{4.305}$$

Here, the symbol • is described in Equation 4.256. The vector $\boldsymbol{p}_{ni}$ is the ith mode shape of a corresponding normal mode system with mass and stiffness matrices identical to the nonproportionally damped system and a null damping matrix. And, when $-90° < \angle(\mathrm{p}_{ij}) < 90°$, $\sin(\mathrm{p}_{ij}) = 1$. Otherwise, $\sin(\mathrm{p}_{ij}) = -1$.

Using Equation 4.305, the modal participation factor of the ith mode at the jth location is further defined as

$$\Gamma_{ij} = \frac{\psi_{ij}}{\bar{\mathrm{p}}_{ij}} = \begin{cases} 2\omega_i\left\{\left|-\xi_i\varphi_{ij} + \sqrt{1-\xi_i^2}\,\psi_{ij}\right| + 0.4|\varphi_{ij}|\right\}\Big/\mathrm{p}_{ij}, & \text{when } \left|-\xi_i\varphi_{ij} + \sqrt{1-\xi_i^2}\,\psi_{ij}\right| > |\varphi_{ij}| \\ 2\omega_i\left\{0.4\left|-\xi_i\varphi_{ij} + \sqrt{1-\xi_i^2}\,\psi_{ij}\right| + |\varphi_{ij}|\right\}\Big/\mathrm{p}_{ij}, & \text{when } \left|-\xi_i\varphi_i + \sqrt{1-\xi_i^2}\,\psi_{ij}\right| \le |\varphi_{ij}| \end{cases} \tag{4.306}$$

Therefore, for the nonproportionally damped system, there is no longer a uniform modal participation factor as in the proportionally damped system described in Equation 3.170. Instead, the modal participation factors are written in the following vector form:

$$\boldsymbol{\Gamma}_i = \{\Gamma_{ji}\} = \begin{Bmatrix} \Gamma_{1i} \\ \Gamma_{2i} \\ \cdots \\ \Gamma_{ni} \end{Bmatrix} \tag{4.307}$$

where the braces indicate that the modal participation factors are in vector form. In this case, Equation 4.269 is rewritten as

$$\mathbf{x}_{\max i} = \bar{\mathbf{P}}_i \bullet \boldsymbol{\Gamma}_i S_D(T_i, \xi_i) g \tag{4.308}$$

Note that, Γ_i is a positive vector, which is used for design convenience.

4.5.2.2 Proportionally Damped Modes

If the system is proportionally damped with all modes underdamped, then the definition of Equation 4.173 can still be used; that is,

$$\boldsymbol{h}_i(s) = -\left[\frac{\boldsymbol{p}_i\boldsymbol{q}_i\mathbf{J}}{s-\lambda_i} + \frac{(\boldsymbol{p}_i\boldsymbol{q}_i\mathbf{J})^*}{s-\lambda_i^*}\right]$$

However, it was proven that the product of $\boldsymbol{p}_i\boldsymbol{q}_i\boldsymbol{J}$ has only imaginary parts; that is,

$$\boldsymbol{p}_i\boldsymbol{q}_i\mathbf{J} = j\boldsymbol{\Psi}_i$$

and its complex conjugate is

$$(\boldsymbol{p}_i\boldsymbol{q}_i\mathbf{J})^* = -j\boldsymbol{\Psi}_i$$

Therefore, in this case, $\boldsymbol{h}_i(s)$ reduces to

$$\boldsymbol{h}_i(s) = \frac{2\sqrt{1-\zeta_i^2}\,\omega_{ni}\boldsymbol{\Psi}_i}{s^2 + 2\zeta_i\omega_{ni}s + \omega_{ni}^2}$$

The corresponding response is then reduced to

$$\mathbf{x}_i(t) = \mathscr{L}^{-1}\left\{-\boldsymbol{h}_i(s)x_{g''}(s)\right\} = \mathscr{L}^{-1}\left\{\left(-2\omega_{ni}\sqrt{1-\xi_i^2}\,\boldsymbol{\Psi}_i\right)\frac{x_{g''}(s)}{s^2+2\xi_i\omega_{ni}s+\omega_{ni}^2}\right\} \tag{4.309}$$

Thus, the vector equation is written as

$$\mathbf{x}_i(t) = -2\omega_{ni}\sqrt{1-\xi_i^2}\,\boldsymbol{\Psi}_i y_{Di}(t) \tag{4.310}$$

Furthermore,

$$\mathbf{x}_{\max_i} = \left\{x_{\max_i}\right\} = 2\omega_{ni}\sqrt{1-\xi_i^2}\,\boldsymbol{\Psi}_i y_{Di}^{\max} \tag{4.311}$$

That is, the maximum value, denoted by vector $\mathbf{x}_{\max_i}$, is only the function of $\boldsymbol{\Psi}_i$. Next, denote

$$\bar{\boldsymbol{p}}_{ni} = \left\{\frac{\boldsymbol{\Psi}_i}{\psi_{in}}\right\} = \left\{\bar{p}_{ij}\right\}, \quad j = 1, 2, \ldots, n \tag{4.312}$$

Using Equation 4.312, as well as Equation 4.311, results in a similar form to the nonproportionally damped system:

$$\mathbf{x}_{\max_i} = \bar{\boldsymbol{p}}_{ni}\Gamma_i S_D(T_i, \xi_i)g \tag{4.313}$$

except that here, Γ_i, the modal participation factor, is no longer a vector; and

$$\Gamma_i = \left|2\omega_{di}\psi_{i1}\right| \tag{4.314a}$$

Therefore, one of the major differences of the modal participation factor between nonproportionally and proportionally damped systems is that the latter is a scalar, which applies to all the jth elements. However, for nonproportionally damped systems, the modal participation factor for each complex mode will vary when the elements are different.

It can be proven that Equation 4.314a yields the same absolute value of the modal participation factor as in Equation 3.170. Again, for design convenience, the sign of the factor is omitted, that is,

$$\Gamma_i = \left|\frac{\bar{\boldsymbol{p}}_i^T\mathbf{MJ}}{\bar{\boldsymbol{p}}_i^T\mathbf{M}\bar{\boldsymbol{p}}_i}\right| \tag{4.314b}$$

Note that, in this case,

$$\bar{\boldsymbol{p}}_{ni} = \bar{\boldsymbol{u}}_i$$

Here, $\bar{\boldsymbol{u}}_i$ is the ith mode shape of the proportionally damped system, and so will be $\bar{\boldsymbol{p}}_{ni}$.

4.5.2.3 Overdamped Subsystems

For the overdamped subsystems, the vector $\boldsymbol{r}_j^R$ was used, which is defined in Equation 4.216 and repeated as follows:

$$\boldsymbol{r}_j^R = -\boldsymbol{p}_j^R \boldsymbol{q}_j^R \mathbf{J}$$

Thus, the modal participation factor, denoted by Γ_j^R, is defined for the jth overdamped subsystem as

$$\Gamma_j^R = \left| \boldsymbol{q}_j^R \mathbf{J} \right| \tag{4.315}$$

In this case, using Γ_i^R to replace $\boldsymbol{q}_j^R \mathbf{J}$ in Equation 4.278,

$$d_{\max j}^R = \boldsymbol{p}_j^R \Gamma_j^R S_{Do}\left(\lambda_i^R\right) g \tag{4.316}$$

4.5.3 Modal Contribution Indicators

To attempt to determine the number of modes that should be included to compute the total responses in a nonproportionally damped system, modal contribution indicators are still needed. In this section, the indicators that can potentially be used to choose the number of truncated modes are examined.

4.5.3.1 Modal Mass Ratio

It has been established that for proportionally damped systems, to determine if the contribution of a mode needs to be included when computing the total structural responses, the concept of a modal mass ratio can be applied (see Equation 3.184).

$$\gamma_{mi} = \frac{m_{effi}}{M_\Sigma} = \frac{\left(\bar{\boldsymbol{u}}_i^T \mathbf{M} \mathbf{J}\right)^2}{\bar{\boldsymbol{u}}_i^T \mathbf{M} \bar{\boldsymbol{u}}_i M_\Sigma} = \Gamma_i^2 \frac{m_{mi}}{M_\Sigma}, \quad i = 1, 2, \ldots, n$$

The favorable properties of the modal mass ratio are that, first, the summation of the modal mass ratio is unity (see Equation 3.180) and, secondly, $\Sigma\gamma_{mi}$ will monotonically increase when the order of the mode becomes larger. In other words, γ_{mi} is a good index or indicator of the modal contributions in the total structural responses.

For nonproportionally damped systems as well as systems containing overdamped subsystems, the *modal mass ratio* needs to be redefined. For a generally damped system, the eigenvalue λ_i, and the mode shape $\boldsymbol{p}_i$ as well as its accompanist vector $\boldsymbol{q}_i$, can be obtained from one of the following cases:

1. They represent one of the identical pairs of a proportionally damped system, namely, the normal mode.
2. They represent one of the identical pairs of a nonproportionally damped system, namely, the complex mode.
3. They represent one of the overdamped subsystems, namely, the pseudo mode.

In the following discussion, the modified modal mass ratio is introduced for generally damped systems. Then, the viability of using this parameter is proven. In a later section, other indicators will be introduced and compared using numerical simulations.

If there is no distinction between the above three cases, but the eigenvalue λ_i, the mode shape $\boldsymbol{p}_i$, and as its accompanist vector $\boldsymbol{q}_i$ are treated as a subsystem,

$$\gamma_{mi} = \begin{cases} \dfrac{\left(\boldsymbol{p}_i^T\mathbf{MJ}\right)^2}{\boldsymbol{p}_i^T\mathbf{M}\boldsymbol{p}_i\sum_{j=1}^{n} m_j}, & \text{underdamped normal mode shape,} \quad i = 1,\ldots, n_{Ru} \\ \dfrac{2\mathrm{Re}\left\{\lambda_i\left(\boldsymbol{q}_i\mathbf{J}\right)\left(\boldsymbol{p}_i^T\mathbf{MJ}\right)\right\}}{\sum_{j=1}^{n} m_j}, & \text{underdamped complex mode shape,} \quad i = 1,\ldots, n_{Cu} \\ \dfrac{\lambda_i^R\left(\boldsymbol{q}_i^R\,\mathbf{J}\right)\left(\boldsymbol{p}_i^{R^T}\mathbf{MJ}\right)}{\sum_{j=1}^{n} m_j}, & \text{overdamped pseudo mode,} \quad i = 1,\ldots, 2n_{oO} \end{cases} = \frac{m_{effi}}{M_\Sigma} \tag{4.317}$$

Here, n_{Ru} and n_{Cu} are respectively the number of normal and complex modes; $n_{Ru} + n_{Cu} = n_u$ and the total mass M_Σ is defined in Equation 3.185, and the ith effective mass is redefined as

$$m_{effi} = 2\mathrm{Re}\left\{\left(\lambda_i\boldsymbol{q}_i\mathbf{J}\boldsymbol{p}_i^T\mathbf{MJ}\right)\right\} \tag{4.318}$$

This newly defined modal mass ratio, similar to the one defined in Equation 3.184, is used to determine if a specific mode needs to be included in the total structural response computations. Therefore, such a quantity has to satisfy the following requirements.

First, the summation of all these quantities must be a fixed value. For convenience, this fixed value can be taken as 1. This requirement provides a reference, e.g., a unity, so that the modal contribution can be quantitatively determined.

Second, if the contribution of a specific mode to the total structural responses is significant, the corresponding modal mass ratio must be sufficiently large. In fact, the larger the contribution, the larger this quantity should be in a proportional sense. This requirement guarantees that the quantity must be a monotonic function of the modal contribution, so that it can be used as an index or indicator of the modal contribution.

Now, the first requirement is proven, namely,

$$\sum_{i=1}^{n} m_{effi} = M_\Sigma \tag{4.319a}$$

or

$$\frac{\sum_{i=1}^{n} m_{effi}}{M_\Sigma} = 1 \tag{4.319b}$$

or

$$\sum_{i=1}^{n} \gamma_{mi} = 1 \tag{4.319c}$$

The term $\left(\lambda_i \boldsymbol{q}_i \mathbf{J} \boldsymbol{p}_i^{T} \mathbf{MJ}\right)$ is examined next. For convenience, there is no distinction regarding whether the ith subsystem is the complex-conjugate pair of a nonproportionally damped system, or one of the identical pairs of a proportionally damped system, or a pseudo mode of structure. Then, the Equation 4.319 can be written as

$$\sum_{i=1}^{2n} \lambda_i (\boldsymbol{q}_i \mathbf{J})(\boldsymbol{p}_i^{T} \mathbf{MJ}) = \mathbf{J}^{T} \boldsymbol{Q}'^{T} \boldsymbol{\Delta} \boldsymbol{P}'^{T} \mathbf{MJ} \tag{4.320}$$

Here, $\boldsymbol{P}'_{n\times 2n}$ is used to represent all $\mathbf{P}$, $\mathbf{P}^*$, and $\mathbf{P}^R$ in n-dimensional space, which are defined in Equation 4.212. That is,

$$\boldsymbol{P}' = \left[\mathbf{P},\ \mathbf{P}^*,\ \mathbf{P}^R\right] \tag{4.321a}$$

Also $\boldsymbol{Q}'_{n\times 2n}$ is used to represent all $\boldsymbol{Q}$, $\boldsymbol{Q}^*$, and $\boldsymbol{Q}^R$, which are defined in Equation 4.213, that is,

$$\boldsymbol{Q}' = \begin{bmatrix} \mathbf{Q} \\ \mathbf{Q}^* \\ \mathbf{Q}^R \end{bmatrix} \tag{4.321b}$$

It is known that

$$\mathcal{P} \boldsymbol{\Delta} \mathcal{Q} = \mathcal{A} = \begin{bmatrix} -\mathbf{M}^{-1}\mathbf{C} & -\mathbf{M}^{-1}\mathbf{K} \\ \mathbf{I} & \mathbf{0} \end{bmatrix} \tag{4.322}$$

where $\boldsymbol{\Delta}$ was defined previously.

Substituting Equations 4.321a and 4.321b into Equation 4.322 and considering the lower left submatrix,

$$\boldsymbol{P}' \boldsymbol{\Delta} \boldsymbol{Q}' = \mathbf{I} = \boldsymbol{Q}'^{T} \boldsymbol{\Delta} \boldsymbol{P}'^{T} \tag{4.323}$$

Substituting the right part of Equation 4.323 into Equation 4.320 results in

$$\sum_{i=1}^{2n} \left\{\left(\lambda_i \boldsymbol{q}_i \mathbf{J} \boldsymbol{p}_i^{T} \mathbf{MJ}\right)\right\} = \mathbf{J}^{T} \mathbf{IMJ} = \mathbf{J}^{T} \mathbf{MJ} = \mathrm{M}_{\Sigma} \tag{4.324}$$

Therefore,

$$\frac{\sum_{i=1}^{2n} \left\{\left(\lambda_i \boldsymbol{q}_i \mathbf{J} \boldsymbol{p}_i^{T} \mathbf{MJ}\right)\right\}}{\mathrm{M}_{\Sigma}} = 1 \tag{4.325}$$

Thus, the requirement of Equation 4.319 is proven.

In the following, underdamped system with proportional damping is further examined. In this circumstance, it is known that the ith mode shape $\boldsymbol{p}_i$ can be real valued and reduced to $\boldsymbol{u}_i$ defined in Equation 3.88. It is also known that $\boldsymbol{p}_i$ can be normalized with many respects. For example, $\boldsymbol{p}_i$ can be a special vector $\bar{\boldsymbol{u}}_i$, such that

$$\bar{\boldsymbol{u}}_i^{\mathrm{T}}\mathbf{M}\bar{\boldsymbol{u}}_i = 1 \tag{4.326}$$

In this case, let

$$\boldsymbol{p}_i^{\mathrm{T}}\mathbf{M}\boldsymbol{p}_i = m_i \tag{4.327}$$

and

$$\boldsymbol{p}_i^{\mathrm{T}}\mathbf{C}\boldsymbol{p}_i = 2\xi_i\omega_{ni}m_i \tag{4.328}$$

Substituting Equations 4.237 and 4.238 into Equation 4.93 results in

$$r_i = \frac{1}{m_i\left(2\lambda_i + 2\xi_i\omega_{ni}\right)} = \frac{1}{2jm_i\sqrt{1-\xi_i^2}\,\omega_{ni}} \tag{4.329}$$

Furthermore,

$$r_i = \frac{1}{2j\,\boldsymbol{p}_i^{\mathrm{T}}\mathbf{M}\,\boldsymbol{p}_i\sqrt{1-\xi_i^2}\,\omega_{ni}} \tag{4.330}$$

Therefore, the following is correct:

$$\boldsymbol{q}_i = r_i\boldsymbol{p}_i^{\mathrm{T}}\mathbf{M} = \frac{\boldsymbol{p}_i^{\mathrm{T}}\mathbf{M}}{2j\boldsymbol{p}_i^{\mathrm{T}}\mathbf{M}\boldsymbol{p}_i\sqrt{1-\xi_i^2}\,\omega_{ni}} \tag{4.331}$$

Substituting Equation 4.331 into Equation 4.317 results in

$$\gamma_{mi} = 2\mathrm{Re}\left[\frac{\lambda_i\left(\boldsymbol{p}_i^{\mathrm{T}}\mathbf{MJ}\right)\left(\boldsymbol{p}_i^{\mathrm{T}}\mathbf{MJ}\right)}{2j\boldsymbol{p}_i^{\mathrm{T}}\mathbf{M}\boldsymbol{p}_i\sqrt{1-\xi_i^2}\,\omega_{ni}\sum_{j=1}^{n}m_j}\right] = \frac{\left(\boldsymbol{p}_i^{\mathrm{T}}\mathbf{MJ}\right)^2}{\boldsymbol{p}_i^{\mathrm{T}}\mathbf{M}\boldsymbol{p}_i\sum_{j=1}^{n}m_j} \tag{4.332}$$

That is, the modal mass ratio reduces to Equation 3.184, which was introduced for the case of an underdamped system with proportional damping only.

Now, consider the overdamped subsystems. In this case,

$$\boldsymbol{q}_i^{\mathrm{R}} = \mathbf{r}_i^{\mathrm{R}}\boldsymbol{p}_i^{\mathrm{R}^{\mathrm{T}}}\mathbf{M} \tag{4.333}$$

$$r_i^{\mathrm{R}} = \frac{1}{\boldsymbol{p}_i^{\mathrm{R}^{\mathrm{T}}}\left(2\lambda_i^{\mathrm{R}}\mathbf{M} + \mathbf{C}\right)\boldsymbol{p}_i^{\mathrm{R}}} \tag{4.334}$$

thus,

$$q_i^R = \frac{p_i^{R^T} \mathbf{M}}{p_i^{R^T} \left(2\lambda_i^R \mathbf{M} + \mathbf{C}\right) p_i^R} \tag{4.335}$$

For the overdamped subsystems,

$$\gamma_{mi} = \frac{2\left\{\lambda_i^R \left(q_i^{R^T} \mathbf{J}\right) \left(q_i^{R^T} \mathbf{MJ}\right)\right\}}{\sum_{j=1}^{n} m_j} = \frac{\lambda_i^R \left(q_i^{R^T} \mathbf{J}\right) \left(p_i^{R^T} \mathbf{MJ}\right)}{\sum_{j=1}^{n} m_j} \tag{4.336}$$

In generally damped systems, both proportionally damped modes, namely, the normal modes, and the nonproportionally damped modes, namely, the complex modes can exist. In addition, the overdamped subsystems can also exist. In this case, both underdamped and pseudo modes should be considered, and the formula described by Equation 3.184 should be modified.

It is seen that Equations 4.319b and 4.319c can also be proven. Here, for convenience, the notation m_{effi} is used to denote both underdamped and overdamped pseudo modes or subsystems.

It is known that the modal mass ratio is used to determine whether or not the ith mode should be considered. That is, a criterion *G* can be preset. If the summation of a modal mass ratio of the first S modes is greater than this value, the corresponding mode should be considered when the structural responses are calculated. This requirement was described in Equation 3.181 and is repeated as follows:

$$\sum_{i=1}^{S} \gamma_{mi} \geq G \tag{4.337}$$

In Chapter 3, it was mentioned that it was best that the summation for the first S modes monotonically increases with respect to the number S.

Unfortunately, the modal mass ratio defined above for the nonproportionally damped system does not satisfy these requirements. That is, the ratios of certain modes can be negative and the summation of those for the first S modes can be greater than unity. In the worst cases, the ratio of the first mode can be negative. This fact may mislead a damper designer to make an incorrect decision when choosing modes, although it can be compensated by using larger value of S_f in Equation 3.178. Therefore, a better method of choosing the modes may be needed to ensure the accuracy of the modal truncation.

4.5.3.2 Static Modal Energy Ratio

Consider a generally damped system that consists of a structure with added damping, which may contain the complex modes and overdamped subsystems, described by Equation 4.338, a system with real modes only can be generated as described by Equation 4.339, which can also have overdamped subsystems. In addition, an underdamped system with proportional damping only can also be generated. That is,

$$\mathbf{M}\ddot{\mathbf{x}} + \mathbf{C}_g\dot{\mathbf{x}} + \mathbf{K}\mathbf{x} = -\mathbf{MJ}x_{g''} \tag{4.338}$$

where $\mathbf{C}_g$ is a general damping, which may cause complex modes as well as overdamped subsystems;

$$\mathbf{M}\ddot{\mathbf{x}} + \mathbf{C}_r\dot{\mathbf{x}} + \mathbf{K}\mathbf{x} = -\mathbf{M}\mathbf{J}x_{g''} \tag{4.339}$$

where $\mathbf{C}_r$ is the damping that causes the system to only have a real-valued mode shape; and

$$\mathbf{M}\ddot{\mathbf{x}} + a\mathbf{C}_r\dot{\mathbf{x}} + \mathbf{K}\mathbf{x} = -\mathbf{M}\mathbf{J}x_{g''} \tag{4.340}$$

where a is a scalar that reduces the amount of damping such that the resulting system is underdamped.

Note that from the above-mentioned assumption, if the original system described by Equation 4.340 is used, then by adding damping that is proportional, the system becomes the one described by Equation 4.339. By adding general damping devices, the system becomes the one described by Equation 4.338. Now, suppose the system is as described by Equation 4.338. The next step is to find the systems described by Equations 4.339 and 4.340 reduced from the generally damped system.

First of all, matrix $\mathbf{C}_r$ can be obtained as follows:

$$\mathbf{C}_r = \mathbf{M}\mathbf{U}\mathrm{diag}\left(2\xi_i\omega_{ni}\right)\mathbf{U}^{-1} \tag{4.341}$$

Here, the damping ratios ξ_i are obtained from the system Equation 4.338 and ω_{ni} as well the eigenvector matrix $\mathbf{U}$ are obtained through eigen-decomposition of matrix $\mathbf{M}^{-1}\mathbf{K}$, that is,

$$\mathbf{M}^{-1}\mathbf{K}\mathbf{U} = \mathbf{U}\mathrm{diag}\left(\omega_{ni}^2\right) \tag{4.342}$$

Note that the system may contain overdamped subsystems, with the largest damping ratio denoted as ξ_i,

$$\xi_i > 1 \tag{4.343}$$

This fact will not affect using Equation 4.341 to generate matrix $\mathbf{C}_r$. To generate the corresponding underdamped system described by Equation 4.340, the factor used in Equation 4.340 can approximately be calculated by

$$a < 1/\xi_i \tag{4.344}$$

Further assume that when the responses of system Equations 4.348 and 4.340 are calculated by choosing the same number of modes for both systems, the same percentage errors can exist for each system. That is, the modal contribution indicator of system Equation 4.340 can be used to choose the number of modes for system Equation 4.338.

Practically speaking, this is not always true. However, by choosing a few more modes as described in Equation 3.178 and repeated as follows:

$$S = S_p + S_f \tag{4.345}$$

In so doing, the computational accuracy is ensured. The benefit of using this method is that the requirements described by both Equations 3.182 and 3.183 are satisfied.

Therefore, consider the *modified static modal energy ratio*, which can be obtained by the following procedure.

First, calculate the mode shape matrix **V**, which is expressed in Equation 3.77 and repeated as follows:

$$\mathbf{V} = \mathbf{M}^{1/2}\mathbf{U} = \left[\boldsymbol{\nu}_1, \boldsymbol{\nu}_2, \ldots, \boldsymbol{\nu}_s\right]_{n\times s} \tag{4.346}$$

Furthermore, a matrix $\mathbf{R}_I$ containing an incomplete set of normalized orthogonal vectors $\mathbf{r}_{Ii}$ is given by

$$\mathbf{R}_I = \mathbf{V}\left[\text{diag}\left(\mathbf{V}^+\{\text{a}\mathbf{I}\,\mathbf{J}\}\right)\right]_{n\times s} = \left[\boldsymbol{r}_{I1}, \boldsymbol{r}_{I2}, \ldots, \boldsymbol{r}_{Is}\right]_{n\times s} \tag{4.347}$$

The static modal energy ratio for the general damped system is then denoted as

$$\gamma_{Ii} = \left(\boldsymbol{r}_{Ii}^{\ T}\boldsymbol{r}_{Ii}\right)/\text{na} \tag{4.348}$$

Similar to Equation 4.337, the criterion can be written as

$$\sum_{i=1}^{s} \gamma_{Ii} \geq G \tag{4.349}$$

Practically speaking, for both the modal mass ratio and the static modal energy ratio, the value of G can be chosen as

$$G \geq 85\% - 90\% \tag{4.350}$$

Example 4.9

In order to compare the performance of these modal contribution indicators, recall the example of the 10-story building (see Example 3.5). This time, suppose two dampers are installed in the first and second stories. It is also assumed that before any damper is added, the original damping ratio is 2% for all 10 modes. Using the dampers with a damping coefficient of 3.0×10^4 (kN-s/m), the system is nonproportionally damped.

In Figure 4.7, two curves are plotted, which are marked in the legend. This figure shows that the summation of both the static modal energy ratio and the modal mass ratio increases monotonically vs. the numbers of modes used, whereas the summation of the modal mass ratio becomes greater than unity when five or six modes are used. However, it can be shown that, when the damping coefficieint is chosen to be 4.5×10^4 (kN-s/m), the summation of the modal mass ratio becomes greater than unity when six modes are used. In this case, although the summation of the modal mass ratio eventually becomes unity, it is realized that the nonmonotonic increase will make the determination of the number of modes to be included difficult.

In Figure 4.8, the curve of the modified static modal energy ratio as well as the largest percentage error is plotted. This error is calculated by comparing the responses obtained by all 10 modes and the truncated modal superposition, when the El Centro earthquake (1940) is used as the ground excitation. It is seen that although the modified static modal energy ratio is taken from the corresponding proportionally damped system by using Equation 4.260b, it can provide a good estimate for the modal truncation. For example, if $G = 85\%$ is chosen, and therefore, the first mode is chosen only to compute the response, the error is <3% (1%–97% = 3%).

Note that in Example 4.9, only one earthquake record is used in the computation for comparison purposes. The authors have conducted many other calculations with several types of structures as well as 99 earthquake records, with similar results.

4.5.4 Modal Reconstruction of Generally Damped System

To calculate the modal participation as well as the modal contribution and to determine if complex modes should be considered, mode shapes are needed. In later chapters on practical damper design, these mode shapes are also needed. In most circumstances, the entire mode shape matrix does not have to be used, and only the first few are needed. If a structure has pure proportional damping, several methods can be used to compute or estimate the mode shapes, which were discussed in

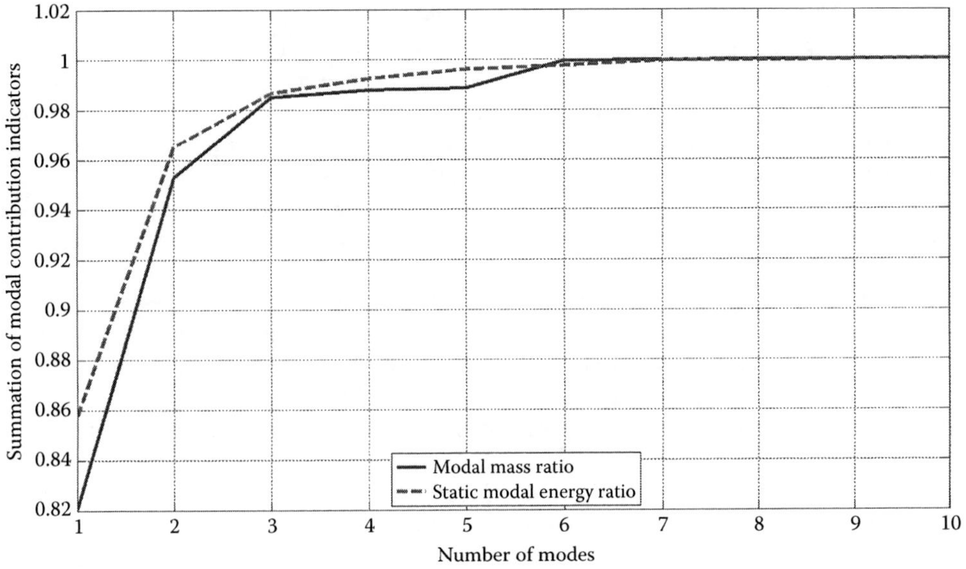

FIGURE 4.7 Comparisons of modal contribution indicators.

Chapter 3. However, as mentioned above, a structure with damping installed is likely to become generally damped; namely, it may have nonproportionally damped complex modes and/or overdamped subsystems. In this circumstance, the mode shape can be substantially different from the pure proportionally damped system and often the corresponding mode shape for the damper design is necessary.

In many cases, the dimension of mode shape can be very large. Previously, the concept of modal truncations based on n-dimensional normal modal space and in 2n-dimensional state space is introduced for reducing computational burdens. Modal reconstruction (also referred to as modal condensation) will be discussed in the following, not only for computations, but also for design convenience because of significant dimension reductions.

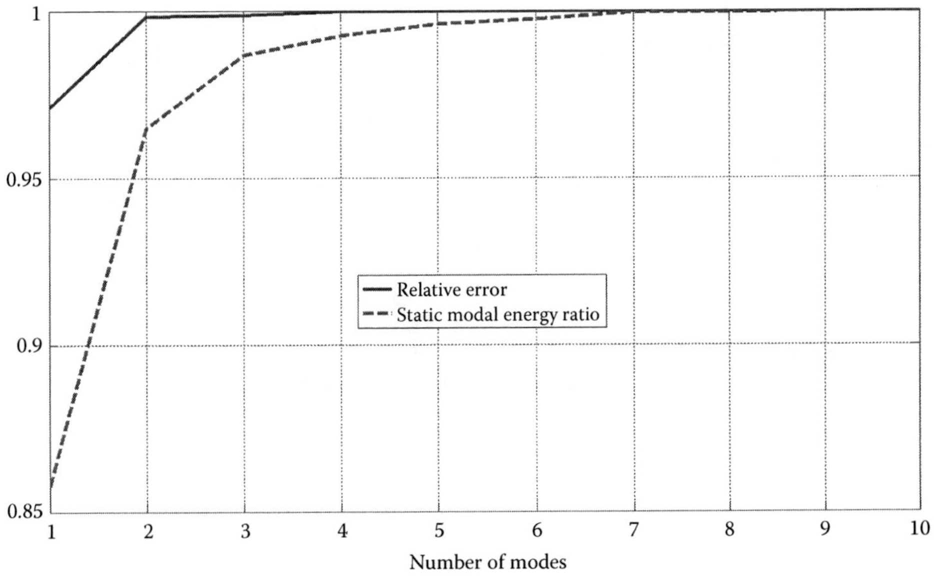

FIGURE 4.8 Static modal energy ratio and relative error of displacement computation.

4.5.4.1 Modal Reconstruction for Damper Design

To compute the mode shape, it is necessary to have the mass, damping, and stiffness matrices. As mentioned earlier, the mass matrix is comparatively easier to obtain. However, it is difficult to obtain the full damping and stiffness matrices. In this section, reconstructing a generally damped system to approximate the system when key pieces of information are missing is discussed, which mainly involves reconstruction of reduced-size stiffness and damping matrices.

In most cases, when an analytical model is extracted from a real-world building design, the stiffness matrix cannot be directly obtained. However, by using commercially available FEM software, such as SAP2000 and ABAQUS, a structural model can be established, a structural analysis performed, and real modal frequencies and modal shapes (not stiffness matrix) can be output. Based on these modal parameters, a modal reanalysis (modal subspace transformation) approach is applied to approximately calculate new complex modal parameters and further predict the structural dynamic characteristics after installing dampers in the structure. This procedure is referred to as modal reconstruction. That is, using the modally reconstructed system, the mode shapes can be obtained for further damper design.

Note that for the requirement of stiffness, a reduced-size matrix can be constructed by considering the natural frequencies of the modes of interest. For the requirement of damping, a reduced-size matrix can also be constructed. That is, a percentage of the damping ratio is assumed and a proportional damping matrix is constructed if the eigenvector matrix of the stiffness is known. Secondly, there is some knowledge of the added dampers, e.g., the damping coefficient of the dampers and the location of the dampers installed. Although there are limitations on where the added dampers may be placed, a reduced-size damping matrix can still be constructed.

Consequently, from a reduced-size space to the full-size space, a linear transformation is needed. As mentioned previously, this is done by using the modal superposition of the truncated modes and/or overdamped subsystems. In the following discussion, this method is introduced step by step.

Up to now, in the space of the normal modes, all the vectors are real valued. Further, in the 2n-dimensional state space, all the vectors are complex valued. Note that the complex-valued 2n state space does not necessarily contain only the complex modes; it is also valid for the normal modes. However, a set of complex modes of a system that is nonproportionally damped can only be decoupled in the 2n space.

In the n-dimensional vector space, there are n linearly independent vectors, denoted by **P**, which is called the base. That is, the n × n matrix containing all the base vector $\boldsymbol{p}_i$ is of full rank n:

$$\mathbf{P} = [\boldsymbol{p}_1, \boldsymbol{p}_2, \ldots, \boldsymbol{p}_n]_{n\times n} \tag{4.351}$$

If there can be all $\boldsymbol{p}_i$, then the complete set of bases is available. Otherwise, they are incomplete; e.g., the truncated modal space. In this case,

$$\mathbf{P}_C = [\boldsymbol{p}_1, \boldsymbol{p}_2, \ldots, \boldsymbol{p}_S]_{n\times S} \tag{4.352}$$

and the rank of the above matrix is S.

Furthermore, any vector in the n-dimensional space denoted by $\boldsymbol{u}$ can be represented by these bases. That is,

$$\boldsymbol{u} = \sum_{i=1}^{n} a_i \boldsymbol{p}_i \tag{4.353}$$

The cases of mode shapes that can be used as bases to represent vectors in damper design are summarized as follows:

4.5.4.1.1 Generic Normal Mode

The differential equations of a proportionally damped structural vibration system with n-DOFs was described by Equation 4.1a and with earthquake excitations, it is repeated again as follows:

$$\mathbf{M\ddot{x}} + \mathbf{C\dot{x}} + \mathbf{Kx} = -\mathbf{MJ}x_{g''} \tag{4.354}$$

The differential equations of an undamped structural vibration system with n-DOFs can be expressed as

$$\mathbf{M\ddot{x}} + \mathbf{Kx} = -\mathbf{MJ}x_{g''} \tag{4.355}$$

The mode shape of the above two cases is identical, that is,

$$\left(\mathbf{M}^{-1}\mathbf{K}\right)\boldsymbol{u}_i = \omega_{ni}^2 \boldsymbol{u}_i, \quad i = 1, 2, \ldots, n \tag{4.356}$$

4.5.4.1.2 Monic Normal Mode

Premultiplying $M^{-1/2}$ on both sides of the homogeneous form of Equation 4.355 and using the notation

$$z = \mathbf{M}^{1/2}\mathbf{x} \tag{4.357}$$

results in

$$\ddot{z} + \tilde{\mathbf{K}}z = \mathbf{0} \tag{4.358}$$

Here $\tilde{\mathbf{K}} = \mathbf{M}^{-1/2}\mathbf{K}\mathbf{M}^{-1/2}$ as defined previously.

It is known that the corresponding eigen-equation is

$$\tilde{\mathbf{K}}\boldsymbol{v}_i = \omega_{ni}^2 \boldsymbol{v}_i \quad i = 1, 2, \ldots, n \tag{4.359}$$

where ω_{ni} and $\boldsymbol{v}_i$ can be seen as the ith natural frequency and mode shape of the system. Note that the eigenvector $\mathbf{V}$ can have different normalizations. Here, for convenience, the orthonormal normalization described by Equation 3.40 is used.

It is seen that Equation 4.359 is also valid for the system described by Equation 4.101, which is repeated as

$$\mathbf{I}\ddot{\mathbf{z}}(t) + \tilde{\mathbf{C}}\dot{\mathbf{z}}(t) + \tilde{\mathbf{K}}\mathbf{z}(t) = \mathbf{0} \tag{4.360}$$

4.5.4.1.3 Generic Complex Mode

It is known that when the damping matrix in Equation 4.360 does not satisfy the aforementioned Caughey criterion, the 2n-dimensional state equation is employed and for generic cases, the homogeneous form of the state Equation 4.7 is obtained and is repeated as

$$\dot{X}(t) = \boldsymbol{a}X(t) \tag{4.361}$$

and the eigen-decomposition

$$\tilde{a}P_i = \lambda_i P_i = \lambda_i \begin{Bmatrix} \lambda_i p_i \\ p_i \end{Bmatrix}, \quad i = 1, 2, \ldots, 2n \tag{4.362}$$

can provide the complex mode shape p_i. Here, the generic state matrix is described in Equation 4.5.

4.5.4.1.4 Monic Complex Mode

From the monic Equation 4.360, an alternative state matrix $\tilde{a}$ is

$$\tilde{a} = \begin{bmatrix} -\tilde{\mathbf{C}} & -\tilde{\mathbf{K}} \\ \mathbf{I} & \mathbf{O} \end{bmatrix} \tag{4.363}$$

The eigen-decomposition of matrix $\tilde{a}$ is

$$\tilde{a}\tilde{P}_i = \lambda_i \tilde{P}_i = \lambda_i \begin{Bmatrix} \lambda_i \tilde{p}_i \\ \tilde{p}_i \end{Bmatrix}, \quad i = 1, 2, \ldots, 2n \tag{4.364}$$

It is also learned that the system Equation 4.362 and the system Equation 4.364 are *similar* since they both have identical eigenvalues. In fact,

$$\tilde{a} = \begin{bmatrix} \mathbf{M}^{1/2} & \\ & \mathbf{M}^{1/2} \end{bmatrix} a \begin{bmatrix} \mathbf{M}^{-1/2} & \\ & \mathbf{M}^{-1/2} \end{bmatrix} \tag{4.365}$$

or

$$a = \begin{bmatrix} \mathbf{M}^{-1/2} & \\ & \mathbf{M}^{-1/2} \end{bmatrix} \tilde{a} \begin{bmatrix} \mathbf{M}^{1/2} & \\ & \mathbf{M}^{1/2} \end{bmatrix} \tag{4.366}$$

So that, with proper normalization,

$$\tilde{P}_i = \begin{bmatrix} \mathbf{M}^{1/2} & \\ & \mathbf{M}^{1/2} \end{bmatrix} P_i \tag{4.367}$$

or

$$P_i = \begin{bmatrix} \mathbf{M}^{-1/2} & \\ & \mathbf{M}^{-1/2} \end{bmatrix} \tilde{P}_i \tag{4.368}$$

and

$$\tilde{p}_i = \mathbf{M}^{1/2} p_i \tag{4.369}$$

or

$$\boldsymbol{p}_{\mathrm{i}} = \mathbf{M}^{-1/2}\tilde{\boldsymbol{p}}_{\mathrm{i}} \tag{4.370}$$

4.5.4.1.5 Simple Monic Mode

Premultiplying $\mathbf{V}^{\mathrm{T}}$ on both sides of Equation 4.360 results in

$$\mathbf{V}^{\mathrm{T}}\ddot{z}(\mathrm{t}) + \mathbf{V}^{\mathrm{T}}\tilde{\mathbf{C}}\mathbf{V}\dot{z}(\mathrm{t}) + \mathbf{V}^{\mathrm{T}}\tilde{\mathrm{K}}\mathbf{V}z(\mathrm{t}) = \mathbf{0} \tag{4.371}$$

Denote

$$\mathbf{r} = \mathbf{V}^{\mathrm{T}}z \tag{4.372}$$

and

$$\mathbf{D} = \mathbf{V}^{\mathrm{T}}\tilde{\mathbf{C}}\mathbf{V} \tag{4.373}$$

Also notice that

$$\mathbf{V}^{\mathrm{T}}\tilde{\mathrm{K}}\mathbf{V} = \boldsymbol{\Omega}^2 = \mathrm{diag}(\omega_{\mathrm{ni}}^2), \quad \mathrm{i} = 1,\ 2, \ldots,\ \mathrm{n} \tag{4.374}$$

Then,

$$\ddot{\mathbf{r}}(\mathrm{t}) + \mathbf{D}\dot{\mathbf{r}}(\mathrm{t}) + \boldsymbol{\Omega}^2\mathbf{r}(\mathrm{t}) = \mathbf{0} \tag{4.375}$$

The corresponding state matrix can be written as

$$\hat{a} = \begin{bmatrix} -\mathbf{D} & -\boldsymbol{\Omega}^2 \\ \mathbf{I} & \mathbf{O} \end{bmatrix}$$

The system described by Equation 4.375 can be called the *simple monic system*. The eigen-decomposition of matrix $\hat{a}$ is

$$\hat{a}\hat{\boldsymbol{P}}_{\mathrm{i}} = \lambda_{\mathrm{i}}\hat{\boldsymbol{P}}_{\mathrm{i}} = \lambda_{\mathrm{i}}\begin{Bmatrix} \lambda_{\mathrm{i}}\hat{\boldsymbol{p}}_{\mathrm{i}} \\ \hat{\boldsymbol{p}}_{\mathrm{i}} \end{Bmatrix} \tag{4.376}$$

It can be seen that the systems of Equations 4.362, 4.364, and 4.376 are all *similar*, since they all have identical eigenvalues. In fact,

$$\begin{aligned} \hat{a} &= \begin{bmatrix} \mathbf{V}^{\mathrm{T}} & \\ & \mathbf{V}^{\mathrm{T}} \end{bmatrix} \tilde{a} \begin{bmatrix} \mathbf{V} & \\ & \mathbf{V} \end{bmatrix} \\ &= \begin{bmatrix} \mathbf{V}^{\mathrm{T}} & \\ & \mathbf{V}^{\mathrm{T}} \end{bmatrix} \begin{bmatrix} \mathbf{M}^{1/2} & \\ & \mathbf{M}^{1/2} \end{bmatrix} a \begin{bmatrix} \mathbf{M}^{-1/2} & \\ & \mathbf{M}^{-1/2} \end{bmatrix} \begin{bmatrix} \mathbf{V} & \\ & \mathbf{V} \end{bmatrix} \end{aligned} \tag{4.377}$$

or

$$a = \begin{bmatrix} \mathbf{M}^{-1/2} & \\ & \mathbf{M}^{-1/2} \end{bmatrix} \begin{bmatrix} \mathbf{V} & \\ & \mathbf{V} \end{bmatrix} \hat{a} \begin{bmatrix} \mathbf{V}^T & \\ & \mathbf{V}^T \end{bmatrix} \begin{bmatrix} \mathbf{M}^{1/2} & \\ & \mathbf{M}^{1/2} \end{bmatrix} \tag{4.378}$$

So that,

$$\hat{P}_i = \begin{bmatrix} \mathbf{V}^T & \\ & \mathbf{V}^T \end{bmatrix} \tilde{P}_i = \begin{bmatrix} \mathbf{V}^T & \\ & \mathbf{V}^T \end{bmatrix} \begin{bmatrix} \mathbf{M}^{1/2} & \\ & \mathbf{M}^{1/2} \end{bmatrix} P_i \tag{4.379}$$

or

$$P_i = \begin{bmatrix} \mathbf{M}^{-1/2} & \\ & \mathbf{M}^{-1/2} \end{bmatrix} \begin{bmatrix} \mathbf{V} & \\ & \mathbf{V} \end{bmatrix} \hat{P}_i \tag{4.380}$$

and

$$\hat{p}_i = \mathbf{V}^T \tilde{p}_i = \mathbf{V}^T \mathbf{M}^{1/2} p_i \tag{4.381}$$

or

$$p_i = \mathbf{M}^{-1/2} \mathbf{V} \hat{p}_i \tag{4.382}$$

4.5.4.2 Damper Design without Stiffness Matrix

In the following discussion, the theory and procedure of modal reanalysis are discussed, particularly for the damper design under the condition that the frequencies, damping ratios, and modal shapes are obtained, but the structural stiffness matrix is unknown. This treatment is often referred to as the *method of dual modal space* since the original system has a generic normal mode and the re-analyzed modal space contains truncated monic modes. This treatment is also referred to as *modal (or model) condensation* (Song et al. 2008).

4.5.4.2.1 Dual Modal Space

It is known that the differential equations of an undamped structural vibration system with n-DOFs can be expressed as

$$\mathbf{M}\ddot{\mathbf{x}}(t) + \mathbf{K}\mathbf{x}(t) = \mathbf{f}(t)$$

The corresponding eigen-decomposition has been written in Equation 4.356. After solving this equation, the lower S (S ≤ n) order reserved eigenvalues and eigenvectors (modal shapes) can be obtained:

$$\boldsymbol{\Omega}_C^2 = \text{diag}\left(\omega_{n1}^2, \omega_{n2}^2, \ldots, \omega_{nS}^2\right)_{S\times S} \tag{4.383}$$

$$\bar{\mathbf{U}}_C = \left[\bar{u}_1, \bar{u}_2, \ldots, \bar{u}_S\right]_{n\times S} \tag{4.384}$$

Here, as mentioned earlier, ω_{ni} is the ith natural frequency (i = 1, 2,…, S). Suppose all the modal shapes, namely $\boldsymbol{u}_i$, are normalized to the identical modal mass matrix, that is,

$$\bar{\mathbf{U}}_C \mathbf{M} \bar{\mathbf{U}}_C = \mathbf{I} \tag{4.385}$$

For the sake of simplicity, the overhead bar now stands for the normalization realized by the following (see Equation 3.113) instead of that in Equation 3.129,

$$\bar{\boldsymbol{u}}_i = \frac{\boldsymbol{u}_i}{\sqrt{\boldsymbol{u}_i^T \mathbf{M} \boldsymbol{u}_i}} \tag{4.386}$$

where $\boldsymbol{u}_i$ is the ith eigenvector obtained by solving the eigen-decomposition Equation 4.356.

As a consequence of the assumption in Equation 4.385, the following relationship exists:

$$\bar{\mathbf{U}}_C \mathbf{K} \bar{\mathbf{U}}_C = \boldsymbol{\Omega}_C^2 \tag{4.387}$$

Suppose that in the original structure, before supplemental dampers are installed, the damping is proportional. The equation for the proportionally damped system can be rewritten as

$$\mathbf{M}\ddot{\mathbf{x}}(t) + \mathbf{C}_0\dot{\mathbf{x}}(t) + \mathbf{K}\mathbf{x}(t) = \mathbf{f}(t) \tag{4.388}$$

where C_0 is the proportional damping coefficient matrix, which has the following property:

$$\bar{\mathbf{U}}_C \mathbf{C}_0 \bar{\mathbf{U}}_C = \operatorname{diag}\left(2\xi_{01}\omega_{n1}, 2\xi_{02}\omega_{n2}, \ldots, 2\xi_{0S}\omega_{nS}\right) \tag{4.389}$$

where ξ_{0i} is the ith damping ratio of the original system.

Now, when the supplemental dampers are added to the structure, an extra damping coefficient matrix $\mathbf{C}_a$ will be formed, which may or may not satisfy the Caughey criterion. Now a new and complete damping matrix should be established as

$$\mathbf{C} = \mathbf{C}_0 + \mathbf{C}_a \tag{4.390}$$

Thus, a new equation of motion can be obtained as

$$\mathbf{M}\ddot{\mathbf{x}}(t) + \mathbf{C}\dot{\mathbf{x}}(t) + \mathbf{K}\mathbf{x}(t) = \mathbf{f}(t) \tag{4.391}$$

Further, the coordinates of the transformation for Equation 4.391 can be determined from the following formula:

$$\mathbf{x}(t) = \bar{\mathbf{U}}_C \mathbf{y}_C(t) \tag{4.392}$$

Note that $\bar{\mathbf{U}}_C$ and $\mathbf{y}_C(t)$ are an $n \times S$ matrix and an S-dimension vector, respectively. Thus, Equation 4.392 is an incomplete and approximate linear space transformation when $S < n$, which will transform a time-variable vector from the n-dimension complete physical space to S-dimension incomplete monic modal subspace; in Equation 4.392, this point is seen clearly. Using these *dual modal spaces*, the following treatment of the modal truncation is called the *method of dual modal space*. In a later paragraph, the S-dimensional reduced modal space will be further extended into a 2S state space.

4.5.4.2.2 Truncated Modes

Now, suppose the modal number S is set to be sufficiently large, and S/2 order eigen reanalysis is carried out as a database for further earthquake ground motion responses analysis (time history analysis and/or response spectrum analysis). Experience from many other fields and numerical simulations in earthquake engineering indicate that satisfactory final results can be guaranteed. The detailed procedure continues as follows:

Substituting Equation 4.392 into Equation 4.391 and premultiplying $\bar{\mathbf{U}}_C^T$ to the consequent equation

$$\bar{\mathbf{U}}_C^T \mathbf{M} \bar{\mathbf{U}}_C \ddot{\mathbf{y}}_C(t) + \bar{\mathbf{U}}_C^T \mathbf{C} \bar{\mathbf{U}}_C \dot{\mathbf{y}}_C(t) + \bar{\mathbf{U}}_C^T \mathbf{K} \bar{\mathbf{U}}_C \mathbf{y}_C(t) = \bar{\mathbf{U}}_C^T \mathbf{f}(t) \tag{4.393}$$

Considering Equations 4.387, 4.388, and 4.393, it is further found that

$$\ddot{\mathbf{y}}_C(t) + \mathbf{D}_C\, \dot{\mathbf{y}}_C(t) + \boldsymbol{\Omega}_C^2\, \mathbf{y}_C(t) = \mathbf{f}_C(t) \tag{4.394}$$

Comparing Equation 4.394 with Equation 4.360, it is seen that the system described by Equation 4.394 is in the monic modal space.

Here

$$\mathbf{D}_C = \mathbf{D}_{C0} + \mathbf{D}_{CS} \tag{4.395}$$

In Equation 4.395,

$$\mathbf{D}_{C0} = \bar{\mathbf{U}}_C^T \mathbf{C}_0\, \bar{\mathbf{U}}_C = \text{diag}\left(2\xi_{01}\omega_{n1},\, 2\xi_{02}\omega_{n2}, \ldots,\, 2\xi_{0S}\omega_{nS}\right) \tag{4.396}$$

and

$$\mathbf{D}_{CS} = \bar{\mathbf{U}}_C^T \mathbf{C}_a\, \bar{\mathbf{U}}_C \tag{4.397}$$

Furthermore,

$$\mathbf{f}_C(t) = \bar{\mathbf{U}}_C^T\, \mathbf{f}(t) \tag{4.398}$$

If $\mathbf{C}_a$ satisfies the Caughey criterion, $\mathbf{D}_C$ can be a diagonal matrix and can be expressed as

$$\mathbf{D}_C = \text{diag}\left(2\xi_1\omega_{n1},\, 2\xi_2\omega_{n2}, \ldots,\, 2\xi_S\omega_{nS}\right) \tag{4.399}$$

which will make Equation 4.394 a group of L independent differential equations in the real modal space. In this case, real modal analysis is applied to solve the structural responses.

However, in most cases, the Caughey criterion cannot be satisfied for $\mathbf{C}_a$. Therefore, complex modal analysis must be considered. In this circumstance, rewriting Equation 4.394 as a state-space form results in

$$\hat{\mathcal{R}}_C^{-1} \dot{\boldsymbol{Y}}_C(t) + \hat{\mathcal{S}}_C \boldsymbol{Y}_C(t) = \boldsymbol{F}_C(t) \tag{4.400}$$

where

$$\hat{\mathcal{R}}_C^{-1} = \begin{bmatrix} -\mathbf{D}_C & \mathbf{I} \\ \mathbf{I} & \mathbf{O} \end{bmatrix}^{-1} = \begin{bmatrix} \mathbf{O} & \mathbf{I} \\ \mathbf{I} & \mathbf{D}_C \end{bmatrix}_{(2S\times 2S)} \tag{4.401}$$

$$\hat{\mathcal{S}}_C = \begin{bmatrix} \mathbf{I} & \mathbf{O} \\ \mathbf{O} & -\mathbf{\Omega}_L^2 \end{bmatrix}_{(2S\times 2S)} \tag{4.402}$$

$$\boldsymbol{Y}_C(t) = \begin{Bmatrix} \dot{\mathbf{y}}_C(t) \\ \mathbf{y}_C(t) \end{Bmatrix}_{2S\times 1} \tag{4.403}$$

and

$$\boldsymbol{F}_C(t) = \begin{Bmatrix} \mathbf{f}_C(t) \\ \mathbf{0} \end{Bmatrix}_{2S\times 1} \tag{4.404}$$

If the system is underdamped, the corresponding eigen-equations can be written as,

$$\lambda_i\ \hat{\mathcal{R}}_C^{-1}\hat{\boldsymbol{P}}_{Ci} - \hat{\mathcal{S}}_C\hat{\boldsymbol{P}}_{Ci} = \mathbf{0}, \quad i = 1, 2, \ldots, S \tag{4.405a}$$

and

$$\lambda_i^*\hat{\mathcal{R}}_C^{-1}\ \hat{\boldsymbol{P}}_{Ci}^* - \hat{\mathcal{S}}_C\ \hat{\boldsymbol{P}}_{Ci}^* = \mathbf{0}, \quad i = 1, 2, \ldots, S \tag{4.405b}$$

Note that the state Equation 4.400 is an alternative format to the one expressed in Equation 4.2. The reason using this form is to show that in addition to Equation 4.2, Equation 4.0 is also popularly used and the corresponding eigen-equations 4.405a and 4.405b are often referred to as the *generalized eigen-equation*. The dimension of this particular state space is 2L, which is reduced from 2n, the original state space. In this case, the reader can further realize the meaning of the method of the dual modal space.

In addition, Equation 4.405a can be rewritten as

$$\lambda_i\ \hat{\boldsymbol{P}}_{Ci} = \hat{\mathcal{A}}_C\hat{\boldsymbol{P}}_{Ci}, \quad i = 1, 2, \ldots, S \tag{4.406a}$$

and Equation 4.405b can be rewritten as

$$\lambda_i^*\hat{\boldsymbol{P}}_{Ci}^* = \hat{\mathcal{A}}_C\ \hat{\boldsymbol{P}}_{Ci}^*, \quad i = 1, 2, \ldots, S \tag{4.406b}$$

Note that if the system contains overdamped pseudo modes, then Equations 4.406a and 4.406b should be modified.

In Equations 4.406a and 4.406b,

$$\hat{\boldsymbol{P}}_{Ci} = \begin{Bmatrix} \lambda_i\ \hat{\boldsymbol{p}}_{Ci} \\ \hat{\boldsymbol{p}}_{Ci} \end{Bmatrix}_{2S\times 1} \quad i = 1, 2, \ldots, L \tag{4.407}$$

and

$$\hat{\mathcal{A}}_C = \hat{\mathcal{R}}_C\ \hat{\mathcal{S}}_C = \begin{bmatrix} -\mathbf{D}_C & -\mathbf{\Omega}_C^2 \\ \mathbf{I} & \mathbf{O} \end{bmatrix}_{2S\times 2S} \tag{4.408}$$

Note that the 2S × 2S matrices $\hat{\mathcal{R}}_C$ and $\hat{\mathcal{S}}_C$ have the same meaning as described in Equations 4.46 and 4.47, except here the mass, damping, and stiffness matrices become **I**, $\mathbf{D}_C$, and $\mathbf{\Omega}_L^2$, which are not physical coordinates, but are the reduced modal coordinates. Thus, Equation 4.406a is compatible with Equation 4.10 and so on.

In this way, a new state equation is established

$$\dot{Y}_C(t) = \widehat{\mathcal{A}}_C\, Y_C(t) + F_C(t) \tag{4.409a}$$

which can be decoupled by means of the above-mentioned eigen-decompositions,

$$\widehat{\mathcal{P}}_C^{-1}\dot{Y}_C(t) = \left(\widehat{\mathcal{P}}_C^{-1}\widehat{\mathcal{A}}_C\,\widehat{\mathcal{P}}_C\right)\widehat{\mathcal{P}}_C^{-1}Y_C(t) + \widehat{\mathcal{P}}_C^{-1}F_C(t) \tag{4.409b}$$

Thus,

$$\dot{z}_i(t) - \lambda_i z_i(t) = f_i(t), \quad i = 1,\ldots, 2S \tag{4.410}$$

similar to the aforementioned procedure in Equations 4.121 and 4.122. Here, $z_i(t)$ and $f_i(t)$ are the ith elements of vectors $\widehat{\mathcal{P}}_C^{-1}\dot{Y}_C(t)$ and $\widehat{\mathcal{P}}_C^{-1}F_C(t)$ respectively; and

$$\widehat{\mathcal{P}}_C = [\widehat{P}_{C1}, \widehat{P}_{C2}, \ldots, \widehat{P}_{CS}, \widehat{P}_{C1}^*, \widehat{P}_{C2}^*, \ldots, \widehat{P}_{CS}^*]_{2S\times 2S} \tag{4.411a}$$

where

$$\widehat{P}_i = \begin{Bmatrix} \lambda_i \widehat{p}_{Ci} \\ \widehat{p}_{Ci} \end{Bmatrix}_{2S\times 1} \quad i = 1,\ldots, 2S \tag{4.411b}$$

The last step to carry out is to transform modal shapes and response vectors back to the original physical space. That is, after solving Equation 4.407, 2S pairs of eigenvalues, eigenvectors, and complex modal shapes denoted as λ_i, $\widehat{P}_{Ci}$, and $\widehat{p}_{Ci}$ (i = 1,2,..., 2S) are obtained, in which λ_i is independent to the coordinate systems, while the mode shapes, $\widehat{p}_{Ci}$, are not, which should be obtained through the eigen-decomposition described in Equations 4.406a and 4.406b.

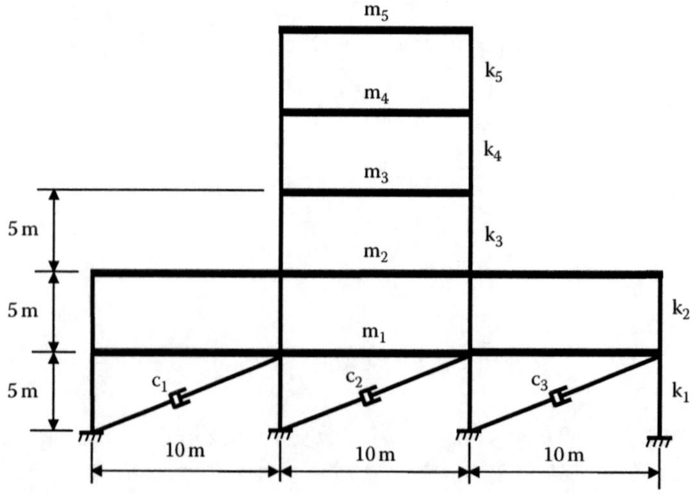

FIGURE 4.9 Five DOF structure with nonproportional damping.

TABLE 4.3
Natural Frequencies and Damping Ratios

	1st Mode	2nd Mode	3rd Mode	4th Mode	5th Mode
Circular frequency	7.1556	14.8819	27.2193	31.6228	40.6583
Damping ratio	0.0200	0.0204	0.0291	0.0327	0.0406

The n-dimensional mode shape, denoted as $\hat{\boldsymbol{p}}_i$, can be estimated as

$$\hat{\boldsymbol{p}}_i = \mathbf{U}_C \, \hat{\boldsymbol{p}}_{Ci} \tag{4.412}$$

The structural responses in the original physical space are finally obtained as

$$\boldsymbol{X}(t)_{2n\times 1} \approx \begin{bmatrix} \mathbf{U}_C & \\ & \mathbf{U}_C \end{bmatrix}_{2n\times 2S} \hat{\mathscr{P}}_C \begin{Bmatrix} z_1(t) \\ \cdots \\ z_{2S}(t) \end{Bmatrix} \tag{4.413}$$

From Equations 4.394 through 4.413 it can be seen that the reformed governing equation and the eigen-equation are not directly denoted by the matrices **M** and **K**. As mentioned above, these matrices are replaced by **I** and $\boldsymbol{\Omega}_L^2$, respectively.

The benefit of the cases where, S << n, yield noticeably smaller scales of eigen-solutions than these in the original physical space can be understood.

Example 4.10

A five-story building is shown in Figure 4.9. The story masses are $m_1 = m_2 = 3m_0$, $m_3 = m_4 = m_5 = m_0$, and $m_0 = 4.0 \times 10^5$ (kg). The inter-story lateral stiffnesses are $k_1 = k_2 = 2k_0$, $k_3 = k_4 = k_5 = k_0$, and $k_0 = 2.0 \times 10^8$ (N/m). Suppose that the original structure has proportional damping, which can be expressed as $\mathbf{C}_0 = 0.19\,\mathbf{M} + 0.00188$ (N-s/m). From the assumed **M**, **K**, and $\mathbf{C}_0$, all five orders of the normal modal parameters can be solved. The natural frequencies and damping ratios are listed in Table 4.3, and the modal shapes are provided in Table 4.4.

Three dampers are installed as shown and $c_1 = c_2 = c_3 = c_0 = 1.0 \times 10^7$ N-s/m, which forms an underdamped complex modal system. Here, the angle of the damper installation must be considered, which is discussed in detail in Chapter 5.

By using the truncated mode approach with the dual modal space, the resulting periods and damping ratios can be computed.

TABLE 4.4
Mode Shapes

1st Mode	2nd Mode	3rd Mode	4th Mode	5th Mode
1.0000	1.0000	1.0000	1.0000	1.0000
0.8976	0.5571	−0.4818	−1.0000	−2.3062
0.7033	−0.1326	−1.2497	−1.0000	2.0124
0.4369	−0.7636	−0.1658	1.0000	−0.3224
0.2366	−0.5717	0.7447	−1.0000	−0.1089

TABLE 4.5
Modal Periods and Percentage Error Comparison

Reserved Mode No.		1		2		3		4		5	
Mode #	Exact Value (sec)	Est. (sec)	Error	Est. (sec)	Error	Est. (sec)	Error	Est. (sec)	Error	Est. (sec)	Error
1	0.8562	0.8781	2.5556	0.8640	0.9133	0.8591	0.3434	0.8562	0.0019	0.8562	0.0000
2	0.4007			0.4291	7.0829	0.4128	3.0322	0.4008	0.0159	0.4007	0.0000
3	0.2354					0.2413	2.4915	0.2355	0.0315	0.2354	0.0000
4	0.2104							0.2104	0.0070	0.2104	0.0000
5	0.1546									0.1546	0.0000

For example, if only the first two modes are considered,

$$\mathbf{\Omega}_C^2 = \begin{bmatrix} 51.2030 & \\ & 221.4717 \end{bmatrix}$$

$$\bar{\mathbf{U}}_C = \begin{bmatrix} 0.9067 & 0.7849 \\ 0.8138 & 0.4373 \\ 0.6377 & -0.1041 \\ 0.3962 & -0.5994 \\ 0.2146 & -0.4488 \end{bmatrix} \times 10^{-3}$$

$$\mathbf{D}_{C0} = \bar{\mathbf{U}}_C^T \mathbf{C}_0 \bar{\mathbf{U}}_C = \begin{bmatrix} 0.2863 & \\ & 0.6064 \end{bmatrix}$$

The damping matrix $\mathbf{D}_a$ for the supplemental dampers can be formed and

$$\mathbf{D}_{Ca} = \bar{\mathbf{U}}_C^T \mathbf{C}_a \bar{\mathbf{U}}_C = \begin{bmatrix} 1.1049 & -2.3109 \\ -2.3109 & 4.8332 \end{bmatrix}$$

TABLE 4.6
Damping Ratios and Percentage Error Comparison

Reserved Mode No.		1		2		3		4		5	
Mode #	Exact Value (100%)	Est. (100%)	Error	Est. (100%)	Error	Est. (100%)	Error	Est. (100%)	Error	Est. (100%)	Error
1	9.132	9.7205	6.4469	9.5924	5.0448	9.3444	2.3293	9.1331	0.0148	9.132	0.0000
2	19.177			18.560	3.2209	19.654	2.4856	19.184	0.0348	19.177	0.0000
3	26.046					16.115	38.128	25.994	0.1969	26.046	0.0000
4	5.447							5.4448	0.0344	5.447	0.0000
5	4.101									4.101	0.0000

By using matrices $\mathbf{D}_C = \mathbf{D}_{C0} + \mathbf{D}_{Ca}$ and $\mathbf{\Omega}_C^2$, the corresponding modal parameters can be calculated. The first period is 0.8640 sec. The second period is 0.4291 sec. Compared to the exact values of 0.8562 and 0.4007, respectively, it is seen that the corresponding errors are 0.913% for the first mode and 7.083% for the second mode. In addition, the first modal damping ratio is 9.592%. The second one is 18.56%. Compared with the exact values of 9.132% and 19.177%, respectively, it is seen that the corresponding errors are 5.045% for the first mode and 3.221% for the second mode. That is, with the truncated first two modes, the simplified model is obtained by using the dual mode method and the corresponding errors are small.

In Tables 4.5 and 4.6, the accurate and approximately estimated values of the periods and the damping ratios are listed and compared. From these two tables, it is seen that by using the truncated modal approach, the percentage errors are not significant. Even using only the first mode, the error of the period is only about 2.6%. The error of the period estimation is about 6.5%. For more complex structures, larger errors using this method can be expected. However, through sufficient numerical simulations, it is seen that the percentage errors are acceptable in most cases. Since this method yields a much smaller scale of equations with reasonable error, it can be applied in practical design for simplifications.

4.6 SUMMARY

In this chapter, generally damped MDOF systems have been discussed. For this generalized consideration, the damping matrix is not necessarily proportional to the mass and the stiffness. Furthermore, the systems can also be overdamped. Although the basic approach remains the decoupling of an entire system into SDOF modes, the generally damped system can no longer be decoupled into n-normal modes. Instead, 2n-complex modes must be used.

Up until Chapter 4 in this book, the focus has been on fundamental vibration theories. In the next two chapters, Part II of this book, principles related to damper design and their application to different types of systems will be discussed.

REFERENCES

Building Seismic Safety Council (BSSC). 2003. *NEHRP Recommended Provisions for Seismic Regulation for New Buildings and Other Structures, 2003 Edition*, Report Nos. FEMA 450, Federal Emergency Management Agency, Washington, DC.

Clough, R.W. and Penzien, J. 1993. *Dynamics of Structures*. 2nd ed. NY: McGraw-Hill.

Ewins, B.J. 1984. *Modal Testing Theory and Practice*. England: Research Studies Press.

Liang, Z. and Lee, G.C. 1991. *Damping of Structures: Part 1—Theory of Complex Damping*, NCEER Report 91-0014 NCEER, University at Buffalo, Buffalo, NY.

Song, J., Chu Y., Liang, Z., and Lee, G.C. 2008. *Modal Analysis of Generally Damped Linear Structure Subjected to Seismic Excitations*, Technical Report MCEER-08-0005, SUNY Buffalo, Buffalo, NY.

Veletsos, A.S. and Ventura, C.E. 1986. Modal analysis of non-classically damped linear systems, *Earthquake Engineering and Structural Dynamics* 14:217–43.

Part II

Principles and Guidelines for Damping Control

In Part II, the role of damping, with a special emphasis on supplemental damping, is discussed. At the same time, the scope is extended to include nonlinearities of the dynamic response.

Nonlinear structural dynamics is governed by both the structural stiffness and damping. These two parameters have their own characteristics, which are described before consideration of the nonlinear response of structures.

For stiffness, two typical cases are addressed: linear and nonlinear.

- If the stiffness of a structure enters the inelastic range of the material, or becomes geometrically nonlinear, or a combination of both, and the degree-of-nonlinearity is large, using a linearization method for the nonlinear stiffness may yield many errors in response estimations. Pushover analysis is a commonly used method today. It is an iterative approach, and its initial curve may only exist for the initial supposition. The pushover method may not always converge, and the convergent range (or working condition) of this method is not well established. For given ground excitations, nonlinear time history analysis is the only way to achieve accurate response estimation. However, generalization is a challenge.
- If the degree of nonlinearity is not large, the use of a linearization method is generally workable for response analysis and damping design. However, choosing an appropriate method for linearization of the stiffness still requires attention. For example, using "secant stiffness" as a so-called "effective stiffness" may not result in reasonable response estimations.

For damping of structures, especially with added supplemental damping, the system is nonlinear. The typical approach in design is to use "effective damping" and "effective mode shape" in addition to "effective stiffness" obtained through linearization procedures.

- When the damping force is small (i.e., the effective damping ratio is < 5%), virtually any type of approximation can be used to account for nonlinear damping with acceptable design result.
- When the damping force is at an intermediate level (i.e., the effective damping ratio is less than 15%–25% but greater than 5%), proper linearization methods can be used to account for the damping nonlinearity. In this case, however, the choice of a proper modeling

approach can also be important. In Chapter 5, equations that use the Timoshenko damping approach are given for several supplemental damping technologies. The equations for a force-based effective damping approach to model these particular types of supplemental damping are also given. The accuracy of linearization also depends on correct treatment of the effective stiffness. Generally speaking, the Timoshenko damping approach tends to overestimate the value of damping, while the force-based effective damping approach is more conservative.

For an MDOF systems, problems associated with inaccurate modeling of the "proportional damping" can occur. That is, even when the linearization of the effective damping ratios for individual modes is carried out with acceptable errors, the damping can still be nonproportional for the overall system. Thus, criteria to determine if the system should be modeled as proportionally damped are needed. A method to account for nonproportional damping is also needed, which is discussed in Chapter 6.

- When the damping force is sufficiently large (i.e., the effective damping ratio is 25% or more), linearization of damping is only a measure for simplified design, which may often be used for initial estimations. This is because the linearization, no matter what type of approximation is employed, can still yield large inaccuracies in response estimation. Thus, for cases involving large damping, nonlinear time history analysis must be used.

5 Principles of Damper Design

The major topic of this chapter is the fundamental design considerations of supplemental dampers for structures. Before presenting the design procedures, the basic principles are first described. This can help to identify conditions under which an optimal design can be achieved and where special attention is needed. In addition, several types of design inaccuracies and inefficiencies can be avoided.

Various models of damping are briefly introduced, followed by a description of three basic models: linear viscous damping, bilinear damping, and sublinear damping.

The key damping design principles emphasized in this book are established by (1) interpreting the Lazan (1968) formulation of damping mechanisms within the context of structural engineering; (2) advanced structural dynamics, particularly with high damping; (3) observation from various experimental studies; and (4) interactions with the damper design industry. Although these principles are fundamentally rather theoretical, they are readily applicable to engineering practice.

Lazan (1968) classified damping by three standards: rate-dependency, quadratic/nonquadratic, and recoverable/nonrecoverable damping. His definition provides the most fundamental essence of damping. However, his idea has not yet been adapted in the design of earthquake protective systems, where added damping is the predominant measure for reducing the seismic vibration level.

Yet, Lazan did not pinpoint the application of added damping to a civil engineering structure; his suggestions are primarily for material damping. Interpretation of these concepts in the field of structural engineering is given in the first three subsections of this chapter.

For structural engineers, these three laws of Lazan alone are not sufficient in damping design; and, it is seen that sometimes these laws are mutually contradictory. Thus, a more uniform method is needed that is based on these laws, to guide engineering practice. This topic is also discussed in the following sections.

The fundamental principles discussed in this chapter are general and somewhat abstract. In the following chapters, design procedures based on these principles for specific dampers are discussed.

5.1 MODELING OF DAMPING

Mathematical modeling of damping is necessary for damper design. Modeling has two levels, the model of an individual damper and the model of the system including both the structure and the dampers. There are at least four approaches to model damping. The first is to model the damping coefficient, which is a sole parameter of an individual damping device. The second is to model the damping ratio, which is a system parameter. The third is to consider the damping force, which can be a quantity associated with a damper. However, the damping force is always related to the system velocity and/or displacement. The fourth approach is to calculate the dissipative energy, which is also related to the system response. A damper provides damping forces that dissipate energy and may also alter restoring forces. The primary modeling method discussed herein is the damping force.

Practically speaking, most commercially available dampers are nonlinear devices. In order to use design spectrum obtained through linear SDOF systems, the linearization of the nonlinear damping force is an important step, which is also a major task in damping modeling.

5.1.1 General Classifications of Damping

Damping must be classified to develop damper design guidelines. Generally speaking, the design spectrum for damper design can be employed to determine which models of effective damping ratio are used to linearize systems with nonlinear damping. Time history analysis can also be used by employing models of exact damping forces. In the following discussion, a uniform damping classification is used for design convenience.

5.1.1.1 Damping Ratios of Systems

The models of damping ratios are considered first.

In Chapter 1, Section 1.3.3, Timoshenko damping and force-based effective damping were used to linearize various nonlinear damping contributions. Timoshenko damping is based on the calculation of energy dissipation, whereas force-based effective damping is based on the computation of damping force.

Modeling of the damping ratio is important when using design response spectrum, because the corresponding design parameter is the damping ratio. This parameter describes the total system. Furthermore, the damping ratio parameter represents a very simple model. Generally,

$$\xi_{eff} = f\left(c_{eq}, m_{eq}, k_{eq}, I_g\right) \tag{5.1a}$$

Here, ξ_{eff}, c_{eq}, m_{eq}, and k_{eq} are respectively the effective damping ratio, equivalent damping coefficient, and mass and stiffness of an SDOF system or of a certain mode of MDOF system. Meanwhile, I_g is the input level (in the case of earthquake engineering, $I_g = AS_i$). Note that for sublinear damping in particular (discussed in Chapter 8), c_β replaces c_{eq}.

Equation 5.1a implies that the effective damping ratio is not only affected by the system parameters c_{eq}, m_{eq}, and k_{eq}, but also by the input level. To emphasize this phenomenon, x_0 and v_0 are used, which are, respectively, the amplitudes of the displacement and the velocity of the structure, to replace the input in Equation 5.1a. Then,

$$\xi_{eff} = f\left(c_{eq}, m_{eq}, k_{eq}, x_0, v_0\right) \tag{5.1b}$$

Note that in engineering practice, the purpose of determining ξ_{eff} is to calculate the responses, such as x_0 and v_0.

According to Lazan, if the energy dissipated by a damper is proportional to the square of the peak displacement, the damping is called *quadratic*. That is, quadratic damping dissipates energy as

$$E_d \propto x_0^2 \tag{5.2}$$

Since the conservative energy is also proportional to the square of the peak displacement, if the stiffness is linear, then from Equation 1.227 for Timoshenko damping, it is seen that when the damping ratio is independent of the displacement x_0, *linear damping* occurs. Otherwise, the damping is *nonlinear.*

Equation 5.1b implies that for nonlinear systems, the task to calculate the damping ratio will not be as easy as in linear systems. Often, iterative procedures should be used to determine the values of the response x_0 and v_0 and the damping ratio ξ_{eff}. More discussion on iteration and an alternative way to compute the response and the damping ratio is carried out in Chapter 8 for bilinear and sublinear damper designs.

Generally speaking, the phrase *effective* means that the capability of energy dissipation of a nonlinear and/or nonviscous damper equals that of a linear viscous damper, when the displacements of both dampers are identical. For a simplified damper design, the damping ratio or effective damping ratio is a major *control parameter.* Here, the effective damping ratio is mainly defined by Equation 1.227. With Timoshenko damping, the damping ratio can be used for any type of damper for a simplified design based on the design spectrum. Note that the design spectrum is determined through linear systems. Thus, to use the spectrum-based design, linear or linearized damping ratios and natural periods are needed. In this case, an effective damping ratio and effective natural period imply that the linearized parameter actually belongs to certain nonlinear systems.

On the other hand, the term *equivalent* means that such a quantity can be readily used. For an individual damper, its damping properties need to be described quantitatively. One important parameter is the equivalent damping coefficient, which is a fixed value when the damper is fabricated. The value of this parameter is not affected by the structure in which it is installed. In other words, an equivalent parameter is a *design parameter.*

When the structure is an SDOF system, the control and design parameters do not have distinct mathematical meaning. However, for an MDOF system, the control parameters are those in the modal domain, whereas the design parameters are in the physical domain and are more easily distinguished. A detailed discussion of control parameters and design parameters is provided in Section 5.4. Note that in MDOF systems, more often the first effective damping ratio is taken to be the basic control parameter, without introducing those of the higher modes.

Traditionally, the terms effective and equivalent are not always distinguished from each other. For example, an effective stiffness may be obtained from an inelastic system through certain linearization procedures. Therefore, we distinguish the terms effective and equivalent in this book only when describing dampers.

Since the effective damping ratio of a nonlinear system is generally a function of the amplitude of displacement x_0 and/or the velocity v_0 of the system, the effective damping ratio ξ_{eff} is used to obtain the response of the structure, i.e., x_0. The response can also be seen as a function of the damping ratio. Therefore, ξ_{eff} will not be determined through Equations 5.1a and b. Often, an iterative method will be used.

In this book, bilinear and sublinear dampings used to discuss dampers used in practice for spectra-based design. In both cases, explicit formulas are derived for the displacement x_0 as functions of physical damping parameters of the dampers and structures, which is called the direct method. This way, the iterative procedure can be avoided so that more accurate and reliable estimations of the displacement can be carried out with less computation effort. When a structure is designed to remain in its elastic range, the direct method is particularly helpful for the structural parameters, such that the periods T_i and the normalized mode shapes P_i are fixed, so that the explicit formulas are simple and easy to use. The disadvantage of the direct method is that the damping ratio will no longer be a control parameter and the corresponding design procedure is not familiar to designers. Due to limited space, the direct methods in this chapter are not discussed, instead they are introduced in Chapter 8 for bilinear damping and sublinear damping.

5.1.1.2 Damping Force of Systems

5.1.1.2.1 Modeling of Damping Force

When time history analysis is conducted, a different type of modeling that constitutes the relationship between the damping force and the displacement is often used. This type of damping model is independent of the structure. The force is determined as long as the velocity and displacement of the damper are given. However, if the structure-damper system is considered, the damping force is often written as

$$f_d = f_d(t) = f(v, x, \xi_{eff}, c_{eq}, m_{eq}, k_{eq}) \tag{5.3a}$$

and/or

$$f_d = f_d(t) = f(c_{eff}, \omega_{eff}, x) \tag{5.3b}$$

Here f_d is the time-varying damping force of the SDOF system, or of a certain mode of MDOF system, and ω_{eff} is the effective natural frequency.

From the viewpoint of damping force, when f_d is proportional to x_0 and/or v_0, linear damping occurs, that is,

$$f_d \propto x \tag{5.4a}$$

or

$$f_d \propto \dot{x} \tag{5.4b}$$

Otherwise, the damping is nonlinear.

Analyses of the response time history of structures with dampers are often carried out using various finite element programs. The modeling described in Equation 5.4 can be comparatively easier to use with commercially available computer programs.

5.1.1.2.2 Characteristics of Individual Dampers

A traditional way to model damping is to consider damping forces. Governing equations for a specific damper are often a nonlinear set of first-order differential or integral equations. These equations can be totally independent of the system. However, for nonlinear dampers, the structural responses must be considered. In many cases, closed-form analytical solutions to the constitutive relationships are not available and numerical solutions must be developed. Sometimes, it is difficult to attach models to commercially available finite element programs. In this case, even though these differential/integral equations may provide more accurate results, such an approach to modeling the damping force is not emphasized in this book. For those interested in precise damping forces, a number of models for individual dampers are available in the literature. For example, Constantinou et al. (1998) summarizes a number of the most popular damper models. More details can be found in the following original publications.

Metallic dampers:
Caughey (1960), Jennings (1964), Iwan (1979), Valanis (1971), Kelly et al. (1972), Dafalias and Popov (1975), Krieg (1975), Masri (1975), Skinner et al. (1975), Ozdemir (1976), Iawn and Gates (1979), Rivlin (1981), Capecchi and Vestroni (1985), Cofie and Krawinkler (1985), Su and Hanson (1990), Graesser and Cozzarelli (1991), Whittaker et al. (1991), Xia and Hanson (1992), Pong et al. (1994), Tsai and Tsai (1995), Dargush and Soong (1995), Sabelli (2001), Shuhaibar et al. (2002), Kim et al. (2004), and Christopoulos and Fillatrault (2006).

Friction dampers:
Mayes and Mowbray (1975), Keightley (1977), Pall et al. (1980), Pall and Marsh (1982), and Aiken and Kelly (1990).

Viscoelastic dampers:
Gemant (1936), Willam (1964), Ferry (1980), Bagley and Torvik (1983), Huffmann (1985), Arima et al. (1988), Zhang (1989), Makris and Constantinou (1990), Soong and Lai (1991), Kasai et al. (1993), Tsai and Lee (1993), Constantinou and Symans (1993), Makris et al. (1993, 1995), and Shen and Soong (1995).

Fluid and recentering dampers:
Graesser and Cozzarelli (1989), Richter et al. (1990), Nims et al. (1993), Tsopelas and Constantinou (1994), Pong et al. (1994), and Pekcan et al. (1995).

In addition to these references, Badrakhan (1985) classified damping into two categories, namely, material damping and structural damping. Under alternating load, the hysteretic phenomena between stress and strain dissipate energy. This represents material damping. Metallic damping and viscoelastic (VE) damping belong to this category. On the other hand, phenomena occurring between substructures or substances, such as connection surfaces, fluid orifices, sliding friction, fluid flows, shear deformations, etc., cause structural damping.

In the literature, there are also many other classifications. From the viewpoint of damper design, these classifications are not always beneficial. To focus on practical damper design, nonlinear damping devices are classified into two categories involving bilinear and sublinear dampings.

5.1.1.3 Bilinear Damping

In Chapter 2, friction damping was introduced, which can be classified as bilinear damping. This is because the force of Coulomb dry friction can be written as

$$f_d = -\eta(x)\operatorname{sgn}(\dot{x}) + \gamma(x) \tag{5.5}$$

Here, the variables $\eta(x)$ and $\gamma(x)$ are defined as

$$\eta(x) = \frac{1}{2}(f_+ - f_-) \tag{5.6a}$$

$$\gamma(x) = \frac{1}{2}(f_+ + f_-) \tag{5.6b}$$

where

$$f_+, f_- = \pm\mu N \operatorname{sgn}(\dot{x}) \tag{5.6c}$$

are, respectively, the forward and backward forces. Furthermore, N is the normal force, and μ is the coefficient of friction that was previously defined. Finally, the bilinear damping force can also be written as

$$f_d = \begin{cases} k_u x + k_u x_0 - f_{max} & \dot{x} \ge 0 \text{ and } -x_0 \le x < -x_c \\ k_d x - k_d x_0 + f_{max} & \dot{x} \ge 0 \text{ and } -x_c \le x < x_0 \\ k_u x - k_u x_0 + f_{max} & \dot{x} < 0 \text{ and } x_c < x \le x_0 \\ k_d x + k_d x_0 - f_{max} & \dot{x} < 0 \text{ and } -x_0 < x \le x_c \end{cases} \tag{5.7}$$

Here, k_u and k_d are the unloading and yielding stiffness as mentioned earlier, and f_{max} and x_0 are, respectively, the maximum force and displacement. Also, $-x_c$ and x_c are the displacements at points C_- and C_+, respectively, as indicated in Figure 5.1.

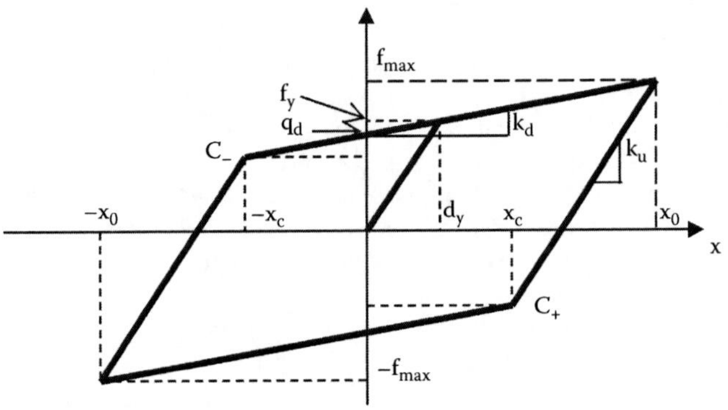

FIGURE 5.1 Bilinear model.

Many damping devices can have energy dissipation loops approximated by a parallelogram, namely, modeled by the bilinear damping. According to the classification by Lazan (1968), bilinear damping is rate independent.

Example 5.1

A special material has *complex modulus* k_C for the *linear hysteretic* or a special type of viscoelastic (VE) damping, described as

$$k_C = (j\eta + 1)k \tag{5.8a}$$

For the shear mode, the *complex shear modulus* G_C

$$G_C = (j\eta + 1)G'' \tag{5.8b}$$

Here k″ is the spring rate and G″ is the shear modulus. Closely related to this model, the loss factor η can exist, which is

$$\eta = k'/k'' \tag{5.9a}$$

and for the shear mode,

$$\eta = G'/G'' \tag{5.9b}$$

where k′ and k″ are, respectively, the loss and restoring modulus, while G′ and G″ are the loss and restoring shear modulus, respectively.

In the literature, complex modulus or stiffness is one of the most common models of VE damping (Sun and Lu 1995). Although there can be many other models, in this book, the term VE damping is used to specifically indicate such a damping model. Note that this type of linear hysteretic damping is also called structural damping in the mechanical engineering community (Inman 2008).

The concept of complex stiffness is useful for steady-state responses under harmonic excitation, but cannot be used directly for transient and/or random vibration. In the literature, an equation is suggested as

$$\xi = \eta/2 \tag{5.10}$$

The application of Equation 5.10 is limited, because the damping ratio is a parameter of a vibration system, whereas the loss factor is a parameter of an individual VE damper.

Equation 5.10 implies that the loss factor is independent of the system displacement. Based on the aforementioned judgment, the VE damping is linear. Furthermore, it must be rate dependent. For this linear rate-dependent damping, the complex stiffness formula cannot be used to calculate the system response. Additionally, care must be taken to use Equation 5.10 in order to have the effective damping ratio. These bring difficulties to the design of a VE damper. However, when the excitation level is given and/or the displacement is given, the energy dissipation loops of a VE damper are very close to the bilinear parallelogram. In this particular case, a VE damper may be treated as bilinear.

In fact, although the loss factor is independent of displacement, it is a function of driving frequency ω. In addition, it is also a function of temperature T. That is,

$$\xi = \xi(\omega, T) \tag{5.11}$$

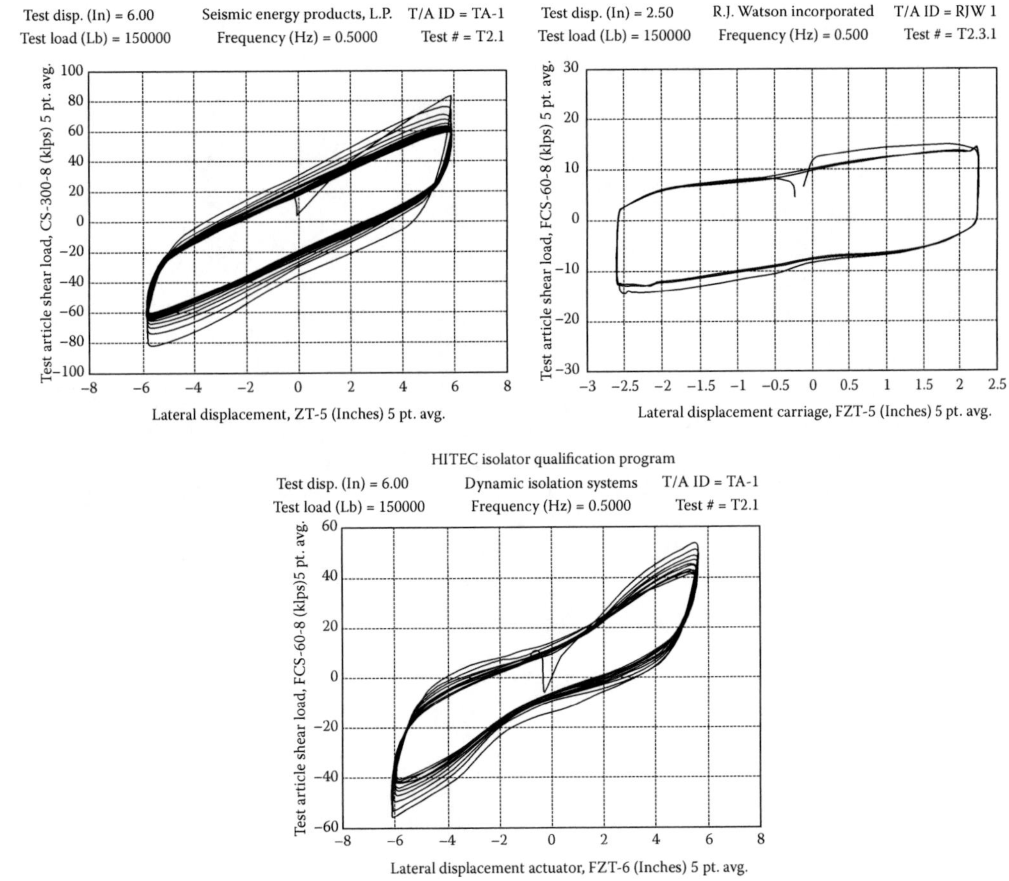

FIGURE 5.2 Examples of bilinear damping. (Photo courtesy of ASCE, Reston, VA.)

VE dampers can be applied for relatively fixed working conditions, such as displacement and temperature, with great efficiency. In such working conditions, the modeling of bilinear damping can be sufficiently accurate.

Figure 5.2 shows some test results for energy dissipation loops that can be approximated by bilinear damping.

5.1.1.4 Sublinear Damping

In Chapter 2, the concept of sublinear damping was introduced. Based on this concept, when the driving frequency is close to the natural frequency ω_n so that ω_n is the dominant frequency, the maximum damping force can be written as

$$f_{d\max} = cv_0^{\beta} \approx c\omega_n^{\beta} x_0^{\beta} \tag{5.12}$$

In this case, the damping force is not proportional to the displacement, nor is it proportional to the natural frequency ω_n. To further explore nonlinear damping, a more general form is used to represent the damping force,

$$f_d(t) \approx c_{eq}\omega_n^{\alpha} \left|x(t)\right|^{\beta} \operatorname{sgn}\left[\dot{x}(t)\right] \tag{5.13}$$

TABLE 5.1
Variations of Damping

	α	β
Friction damping	0	0
Special Viscoelastic damping	0	1
Viscous damping	1	1
Sublinear damping	$0 < \alpha < 1$	$0 < \beta < 1$
Superlinear damping	>1	>1

Here, the equivalent damping coefficient, c_{eq}, is used instead of c to denote more general damping, instead of viscous damping only. When the term c_{eq} denotes damping coefficients of nonlinear dampers, its dimensions or units will vary, according to the exponents α and β. As a comparison, the damping coefficient of a linear viscous damper has fixed units of newton seconds per meter in the SI system.

Table 5.1 lists several cases of different parameters of α and β as special cases described by Equation 5.13.

Example 5.2

In order to see the validity of Equation 5.12, the following example is considered. Suppose a displacement time history is given by

$$x(t) = \sum_{i=1}^{p} x_i \sin(\omega_i t + \theta_i) + N(t)$$

where N(t) is a zero mean random noise with the peak value of $\|N\|$, which cannot be expressed by a Fourier series such as the first term on the right side of the above equation. The steady-state amplitude of x(t) is denoted as

$$x_0 = \max|x(t)|$$

Note that the corresponding velocity is

$$\dot{x}(t) = \sum_{i=1}^{p} x_i \omega_i \cos(\omega_i t + \theta_i) + \frac{d}{dt} N(t)$$

Consider a normalized viscous damping force ($c_{eq} = 1$) written as

$$f_d(t) = |\dot{x}(t)|^{\beta} \operatorname{sgn}(\dot{x}(t))$$

Denote the amplitude of f_d as

$$f_{d\,max} = \max\left|\left[|\dot{x}(t)|^{\beta} \operatorname{sgn}(\dot{x}(t))\right]\right| = \max\left[|\dot{x}(t)|^{\beta}\right]$$

If ω_1 is the dominant frequency and $\|N\|$ is small, then $f_{d\max} \approx \omega_1^{\alpha} x_0^{\beta}$, and it is seen that

$$\alpha \approx \frac{\log\left(F_0/x_0^{\beta}\right)}{\log(\omega_1)}$$

Using a MATLAB® computational program, it is found that if $\|N\| = 0$, the exponent α is a function of β. That is, as long as the frequency series ω_i and the amplitude series x_i are given, the ratio of α/β is a constant. If $\|N\|$ is sufficiently smaller than $F_0/10$, then the ratio of α/β will vary in a small range. Furthermore, If $\|N\|$ is larger than $F_0/10$, then the ratio of α/β can vary with more than a 100% difference, which implies that when x(t) and $\dot{x}(t)$ cannot be expressed mainly as sums of a Fourier series, Equation 5.12 will no longer hold. Note that in a system with linear stiffness and a relatively small viscous damping force, the corresponding displacement and velocity can always be expressed as combinations of several frequency components, namely, certain Fourier series.

For example, let $[\omega_1, \omega_2, \omega_3, \omega_4, \omega_5] = 2\pi\,[1, 2, 3, 4, 5]$, $[x_1, x_2, x_3, x_4, x_5] = [5, 4, 3, 2, 1]$, $[\theta_1, \theta_2, \theta_3, \theta_4, \theta_5] = \pi/180\,[10, 20, 30, 40, 50]$. When $N = 0$, $\alpha \equiv 1.6197\beta$. When $N = 0.01$, mean(α) = 1.6190β and std(α) = 0.02%β, for β is chosen from 0.1 to 0.9.

Example 5.3

To better understand the relationship between energy dissipation E_d and damping exponent β, several examples are considered.

First, suppose the displacement $x_0 = 0.1$ (m), damping coefficient c = 40 (kN-s/m), and the natural frequency f = 0.8 (Hz). When $\beta = 0$, 0.3, 0.6, and 1, the damping force–displacement curves are shown in Figure 5.3. Note that when $\beta = 0$, the damper becomes a Coulomb friction damper, and when $\beta = 1$, the damper becomes a regular linear viscous damper. At these specific points, the corresponding values of A_β are 4 and π, respectively.

As a second example, let $x_0 = 0.1$ (m), c = 40 (kN-s/m), and f = 2 (Hz). When $\beta = 0$, 0.3, 0.6, and 1, the damping force–displacement curves are shown in Figure 5.4. In this case, the value of A_β is also fixed. Again, when the displacement is fixed, as the value of β becomes smaller, the damping force also decreases.

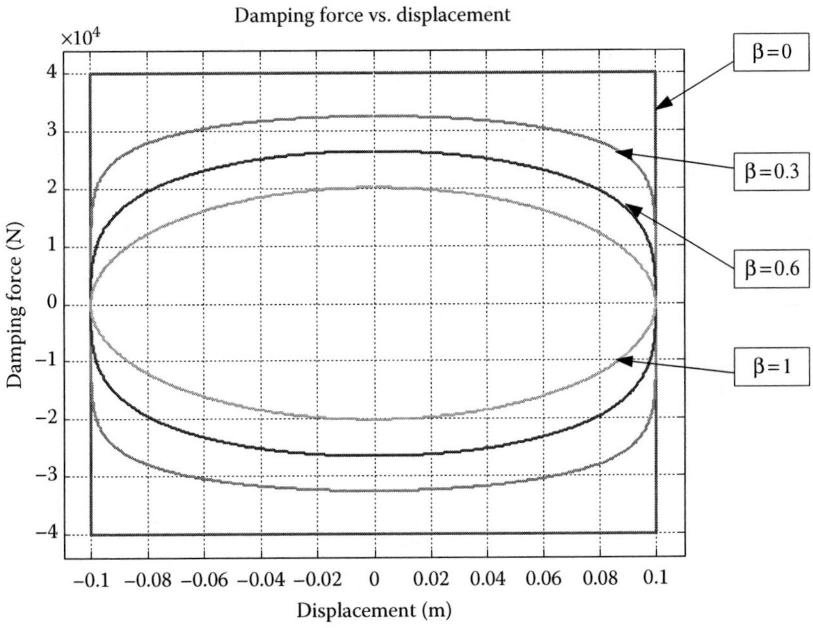

FIGURE 5.3 Energy dissipation and damping force vs. β for the first example.

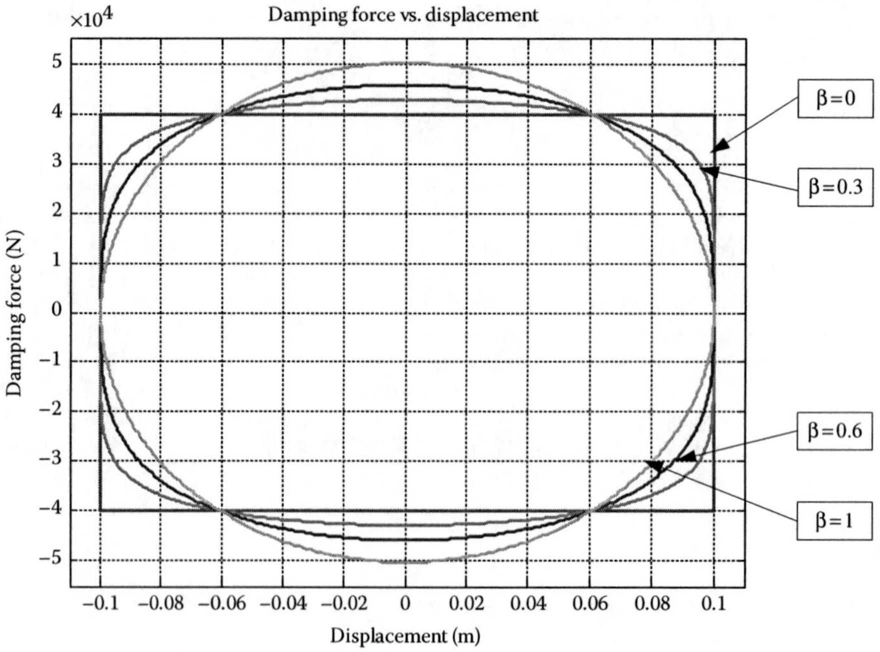

FIGURE 5.4 Energy dissipation and damping force vs. β for the second example.

5.1.2 Effective Damping Ratios for MDOF Systems

In damper design, the effective damping ratios for MDOF systems need to be determined, because most of the structures are MDOF systems. In Chapter 8, in the section on damping ratio realization, this effort is discussed in detail, and is based entirely on the aforementioned approach for SDOF systems.

One of the advantages of using Timoshenko damping is that it can be used for both SDOF and MDOF systems. This is because the number of degrees-of-freedom for the system does not matter, since the energy relationship is a scalar equation. The force-based damping ratio, on the other hand, simultaneously uses quite a few force equations to decide a single damping ratio, which is, in general, a difficult task. Therefore, except in special cases where these equations are all consistent, a virtual displacement vector is specified that will be used to multiply on both sides of the force equations in order to obtain a single scalar equation. This scalar equation will have the dimension of energy, so that the force-based effective damping will also have to adapt to the concept of energy, which is the basic idea of Timoshenko damping. In the special case of sublinear damping, however, the consistent force equations can be found, so the scalar equation will not be used.

5.1.2.1 Timoshenko Damping

5.1.2.1.1 General Viscous Damping

Using Equation 1.227, the expression for the effective damping ratio of a system with general viscous damping can be denoted as

$$\xi_{eff} = \frac{c_{eq} x_0^{\beta-1} \omega_{eff}^{\alpha} A_\beta}{2\pi m_{eff} \omega_{eff}^2} = \frac{c_{eq} x_0^{\beta-1} \omega_{eff}^{\alpha} A_\beta}{2\pi k_{eff}} \tag{5.14}$$

As mentioned before, for an MDOF system, if the stiffness remains linear and the damping force is comparatively small, the concept of effective mode is used. Thus, this relation is considered "mode" by "mode." In the case of the ith effective mode, the effective damping ratio can be rewritten as

$$\xi_{eff_i} = \frac{c_{eq_i} \omega_i^{\alpha_i} A_{\beta_i} \sum_{j=1}^{S} x_{ij}^{\beta_i + 1}}{2\pi \boldsymbol{p}_i^T K \boldsymbol{p}_i} \tag{5.15}$$

where it is supposed that a total of S dampers are installed and the proportionality between excitation and response approximately holds. Subscript i denotes the ith mode, while ω_i and x_i are the ith natural frequency or effective natural frequency and modal amplitude at the jth location, respectively, and

$$x_{ij} = G_{mj} \left| (p_{ji} - p_{j-1,i}) \right| \tag{5.16}$$

In Equations 5.15 and 5.16, $\boldsymbol{p}_i$ is the ith "effective" mode shape with "actual" modal displacement; p_{ji} and $p_{j-1,i}$ are the "actual" floor displacements of the ith mode which are the elements of $\boldsymbol{p}_i$, and G_{mj} is a geometrical magnification factor, which is defined in detail in Section 5.5.4.

Note that since this system is nonlinear, theoretically the linear modal analysis will not apply. However, to calculate the damping ratio, the "mode" shape function is needed. Generally, the shape functions are certain vectors denoting the displacement of the "effective" mode. Here, the "effective" mode and the corresponding mode shapes are often obtained in two different ways. The first is the exact mode shape of undamped M-K systems. In this case, suppose the original system without the supplemental dampers is linear, so the mode shape can be computed through the mass and the stiffness matrices **M** and **K**, respectively. Secondly, if the structure itself has been undergoing inelastic deformation, then it will no longer be linear. In this case, the concept of effective stiffness can be used so that the mode shape of the system with mass **M** and effective stiffness $\mathbf{K}_{eff}$ will be employed. Either way, the corresponding displacement shape function is referred to as the *effective mode shape*. In the following discussion, the term "effective" is used to denote the nonlinear system, in order to remind the reader of the theoretical nonexistence of modes in nonlinear systems. For simplicity, the double quotation marks may also be omitted.

To calculate the effective damping ratio ξ_{effi}, the maximum "effective modal" displacements x_{ij} at each of the jth stories are needed. Without knowing the damping ratio, x_{ij} cannot be determined. To solve this problem, the iteration process must be used.

Example 5.4

Use Timoshenko damping to derive Equation 5.14.

From Equation 1.227, it was shown that when the excitation frequency is ω_n, the Timoshenko damping formula for a general system, either linear or nonlinear, is as follows:

$$\xi_{eff} = \frac{E_d}{4\pi E_p}$$

Thus, as mentioned in Chapter 1, an effective linear system is defined, which has the identical energy dissipation as the nonlinear system, that is,

$$c_{eff} \pi x_0^2 \omega_{eff} = E_d$$

Furthermore, by using the expression of the damping force described in Equation 5.13, a more general equation than the one expressed by Equation 2.111 in Chapter 2 is obtained. That is,

$$E_d = 4\int_0^{\frac{\pi}{2}} f_d dx = 4c_{eq}\omega_n^{\alpha}\, x_0^{\beta+1}\int_0^{\frac{\pi}{2}} \cos^{\beta+1}(\omega t)\, d\omega t = c_{eq}\omega_n^{\alpha}\, x_0^{\beta+1} A_\beta$$

In the above equations, the symbol c_{eq} is the equivalent damping coefficient of the nonlinear viscous damper, whereas c_{eff} is the effective damping coefficient of the nonlinear system. By using the above two equations,

$$c_{eq} = \frac{c_{eff}\pi\omega_{eff}x_0^2}{x_0^{\beta+1}\omega_{eff}^{\alpha}A_\beta} = \pi c_{eff}x_0^{1-\beta}\omega_{eff}^{1-\alpha}A_\beta^{-1}$$

It is seen that if $\alpha = 1$ and $\beta = 1$, c_{eq} is simply equal to c, the linear viscous damping coefficient. Now, assume

$$c_{eff} = 2\xi_{eff}\,\omega_{eff}\,m_{eff} \tag{5.17}$$

and denote

$$\omega_{eff} = \sqrt{\frac{k_{eff}}{m_{eff}}} \tag{5.18}$$

Again, subscript "eff" is used to denote not only an effective damping ratio, but also effective stiffness and mass. Practically speaking, effective stiffness means linearized stiffness obtained from nonlinear systems and, generally, the quantity of mass, either for the linearized system or for the original nonlinear system, is often modeled identically. However, for the sake of generality, subscript "eff" is used for all parameters of mass, damping, and stiffness anyway, even when the stiffness is linear. In the following discussion, for the sake of simplicity, the nonlinear system is not distinguished in such a detailed way, but rather ω_n is used to denote generic natural frequencies in most cases.

Therefore, from Equation 5.18, the expression for the effective damping can be written as

$$\xi_{eff} = \frac{c_{eq}x_0^{\beta-1}\omega_{eff}^{\alpha}A_\beta}{2\pi m_{eff}\omega_{eff}^2} = \frac{c_{eq}x_0^{\beta-1}\omega_{eff}^{\alpha}A_\beta}{2\pi k_{eff}}$$

To obtain a general form of the energy dissipation, the integration for A_β is performed as shown in Equation 5.14.

In Equation 2.112,

$$A_\beta = \int_0^{\frac{\pi}{2}} \cos^{\beta+1}(\omega_f t)\, d\omega_f t = \frac{2\sqrt{\pi}\,\Gamma\left(\dfrac{\beta+2}{2}\right)}{\Gamma\left(\dfrac{\beta+3}{2}\right)}$$

However, since the expression involves the gamma function, it is not convenient to use. Thus, polynomials are used to represent the relationship between the value of integration and the parameter β.

The relationship A_β vs. β is close to a quadratic curve, so it may be regressed by a quadratic equation. The results are displayed in Figure 5.5.

Thus, the regressed equation can be written as

$$A_\beta = 0.298\beta^2 - 1.147\beta + 4 \tag{5.19}$$

The estimation standard deviation is 2.5223×10^{-3} and maximum estimation error is ≤0.143% ($0 \le \beta \le 1$).

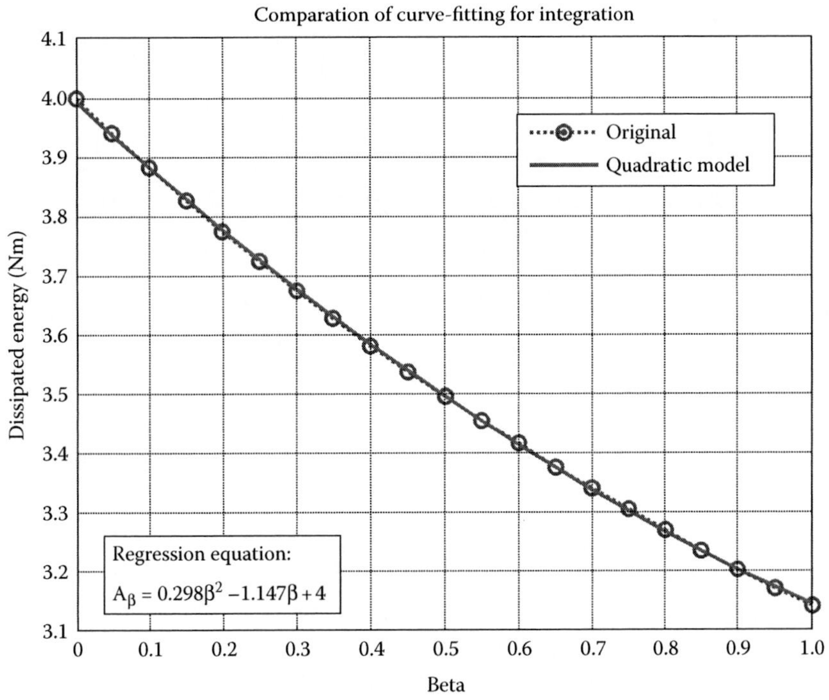

FIGURE 5.5 Quadratic regression.

5.1.2.1.2 Dry Friction Damping

Using Equation 1.227, the expression for the effective damping ratio of a system with friction damping is obtained as

$$\xi_{eff} = \frac{2c_{eq}}{\pi k x_0} \tag{5.20}$$

where

$$c_{eq} = \mu N \tag{5.21}$$

When the structure must be modeled as an MDOF system, assume that L dampers are installed. The effective damping ratio of the ith mode can be written as

$$\xi_{eff_i} = \frac{2\mu N \sum_{j=1}^{L} x_{ji}}{\pi \boldsymbol{p}_i^T \mathbf{K} \boldsymbol{p}_i} \tag{5.22}$$

where $\boldsymbol{p}_i$ is the ith mode shape defined earlier. Note that the number of dampers L may be less than the number of DOF. To calculate the numerator in Equation 5.22, the locations are ignored where no damper is installed.

5.1.2.1.3 Bilinear Damping

It shown in Chapter 6 that the maximum potential energy of a bilinear damper can be written as

$$E_{pd} = \frac{1}{2}\left[k_u d_y^2 + k_d\left(x_0 - d_y\right)^2\right]$$

Then, together with the maximum potential energy of the system with stiffness k, the total potential energy will be

$$E_p = \frac{1}{2}\left[k_u d_y^2 + k_d\left(x_0 - d_y\right)^2 + kx_0^2 0\right]$$

Using Equation 1.227, the expression for the effective damping ratio of an SDOF system with general viscous bilinear damping is denoted as

$$\xi_{eff} = \frac{2q_d\left(x_0 - d_y\right)}{\pi\left[k_u d_y^2 + k_d\left(x_0 - d_y\right)^2 + kx_0^2\right]} \tag{5.23}$$

(See Figure 5.1 for definitions of the terms used in Equation 5.23.)

When the structure must be modeled by an MDOF system, assume that L dampers are installed. The effective damping ratio of the ith mode can be approximately written as

$$\xi_{eff_i} = \frac{2q_d \sum_{j=1}^{L}\left(x_{ji} - d_y\right)}{\pi\left[\boldsymbol{p}_i^T \mathbf{K}\, \boldsymbol{p}_i + k_u d_y^2 + k_d \sum_{j=1}^{L}\left(x_{ji} - d_y\right)^2\right]} \tag{5.24}$$

Here, d_y is the yielding displacement of the bilinear damping as defined previously. In order to distinguish the maximum value of the modal displacement from the effective displacement used to model the damping force, subscript m is used to denote the maximum displacement. That is,

$$\boldsymbol{p}_i = \begin{Bmatrix} p_{m_{1i}} \\ p_{m_{2i}} \\ \cdots \\ p_{m_{ni}} \end{Bmatrix} \tag{5.25}$$

Since the potential energy of the system can be much higher than that stored in the dampers, the effective damping ratio of the ith mode can be written approximately as

$$\xi_{eff_i} = \frac{2q_d \sum_{j=1}^{L}\left(x_{ji} - d_y\right)}{\pi\, \boldsymbol{p}_i^T \mathbf{K}\, \boldsymbol{p}_i} \tag{5.26a}$$

Note that in most cases, it is not necessary to consider the small displacement d_y. Therefore, in this case,

$$\xi_{eff_i} = \frac{2q_d \sum_{j=1}^{L} (x_{ji})}{\pi \, \boldsymbol{p}_i^T \mathbf{K} \, \boldsymbol{p}_i} \tag{5.26b}$$

otherwise, the difference between $x_{ji} - d_y$ must be considered.

Note that the number of dampers L may be less than the number of DOF. To calculate the numerator in Equation 5.24, the locations where no damper is installed are simply ignored.

5.1.2.1.4 Viscoelastic Damping

Using Equation 1.227, the expression of the effective damping ratio of a system with VE damping is given by

$$\xi_{eff} = \frac{\pi\eta k'' x_0^2}{2\pi(k + k'')x_0^2} = \frac{k'}{2k + 2k''} \tag{5.27}$$

Or, in shear mode,

$$\xi_{eff} = \frac{\pi\eta G' \frac{A}{t} x_0^2}{2\pi\left(k + G''\sum_L \frac{A}{t}\right)x_0^2} = \frac{\beta}{2k + 2G''\sum_L \frac{A}{t}} \tag{5.28}$$

Here, k is the stiffness of the system, k′ and k″, and G′ and G″ are the properties of the damper (see example 5.1). Additionally, the terms A and t are, respectively, the working area and the thickness of the VE damper, and totally L dampers are used.

When the structure must be modeled by an MDOF system, the effective damping ratio of the ith mode can be written as

$$\xi_{eff_i} = \frac{\boldsymbol{p}_i^T \mathbf{B} \, \boldsymbol{p}_i}{2\boldsymbol{p}_i^T \mathbf{K} \, \boldsymbol{p}_i} \tag{5.29a}$$

Considering the contribution of the restoring modulus for storing additional potential energies, the following can be approximated:

$$\xi_{eff_i} = \frac{\boldsymbol{p}_i^T \mathbf{B} \, \boldsymbol{p}_i}{2(1.0-1.05)\boldsymbol{p}_i^T \mathbf{K} \, \boldsymbol{p}_i} \tag{5.29b}$$

where the factor (1.0–1.05) is used to accommodate the additional potential energy; when the term $G''\Sigma A/t$ has a larger value, the effective damping ratio takes a smaller value.

In Equation 5.29, $\boldsymbol{p}_i$ and **K** are defined as earlier. The undamped mode shape is exactly the same as that of a proportionally damped system with the same mass and stiffness, whatever the damping is, as long as it is proportional. Also, in Equation 5.29, the term **B** is called the loss-β matrix, where

$$\mathbf{B} = [\beta_{ji}] \tag{5.30}$$

and β_{ji} is defined by

$$\beta_{ji} = c_{con,ij}\beta_{ji} \tag{5.31}$$

where $c_{con,ij}$ is the entry of the configuration matrix, which is defined in detail in Chapter 7. Also, β_{ji} is the loss-β coefficient for the ith mode, which is rewritten as

$$\beta_{ji} = \frac{G'(\omega_i)A_j}{t_j} \tag{5.32}$$

Here, ω_i is the ith natural frequency, while A_j and t_j are, respectively, the working area and the thickness of the VE material.

In Equations 5.30 through 5.32, subscript j stands for the jth damper. If no damper is installed in between the location of the jth and (j – 1)[th] modal displacement, then

$$\beta_{ji} = 0 \tag{5.33}$$

5.1.2.2 Force-Based Effective Damping

Compared to the cases of Timoshenko damping, it is more difficult to write the formula for the MDOF system in the form of force only. Using the inner product of vectors of virtual relative displacements $\boldsymbol{p}_i$ and the damping forces $\mathbf{f}_d$ of the ith effective mode, the corresponding damping ratio can be represented by

$$\xi_{eff_i} = \frac{\boldsymbol{p}_i^T\mathbf{f}_d}{2\times(1.0\sim1.05)\,\boldsymbol{p}_i^T\mathbf{K}\,\boldsymbol{p}_i} \tag{5.34}$$

When the supplemental dampers can store larger potential energy, the effective damping ratio takes a smaller value.

In addition to Equation 5.34, in certain special cases, the effective damping ratio can be obtained directly from the damping force. Since the ith effective damping is a scalar, whereas the damping force is an n-dimensional vector, it can be very difficult to directly obtain the damping ratio, unless all the n equations are consistent. In Chapter 8, a special case of a sublinear damper is shown that will approximately satisfy this condition, so that the effective damping ratio can be computed. Note that the damping force vector $\mathbf{f}_d$ and the relative displacement vector $\boldsymbol{p}_i$ should be taken as the same with proper units.

5.1.2.2.1 Sublinear Damping

Assume that the equivalent damping coefficient remains constant under different frequencies ω_i and also assume that the damping exponent is constant. In the ith effective mode, $\omega_i^\beta x_{ij}$ can be used to represent the term $\dot{x}_{ij}^\beta$. For example, take a two-story building where each story is equipped with the same sublinear damper. The damping force can be written as

$$\mathbf{f}_d(t) = \begin{Bmatrix} c_{eq}|\dot{x}_1|^\beta \operatorname{sgn}(\dot{x}_1) + c_{eq}|\dot{x}_1 - \dot{x}_2|^\beta \operatorname{sgn}(\dot{x}_1 - \dot{x}_2) \\ c_{eq}|\dot{x}_2|^\beta \operatorname{sgn}(\dot{x}_2) \end{Bmatrix} \tag{5.35}$$

For the ith effective mode, the amplitude of the damping force is

$$\mathbf{f}_{d_i} = \begin{Bmatrix} c_{eq}\omega_i^\beta p_{1i}^\beta + c_{eq}\omega_i^\beta \left(|p_{1i} - p_{2i}|\right)^\beta \\ c_{eq}\omega_i^\beta p_{2i}^\beta \end{Bmatrix} \tag{5.36}$$

Based on the approach described conceptually in Equation 5.34, the effective damping coefficient of sublinear damping can be written as

$$\xi_{\text{eff}_i} = \frac{c_{eq_i}\omega_i^{\beta}\sum_{j=1}^{L} p_j x_{ij}^{\beta+1}}{2\times(1.0\sim1.05)\,\boldsymbol{p}_i^{\mathrm{T}}\mathbf{K}\,\boldsymbol{p}_i} \tag{5.37}$$

When the supplemental dampers can store larger potential energy, the effective damping ratio takes a smaller value.

As mentioned earlier, the effective damping ratio can also be obtained directly from the damping force. This is particularly true for sublinear damping. However, due to space limitations here, this equation is not discussed in detail.

Now, comparing the expression of sublinear damping with the alternative formula described in Equation 5.36 and Timoshenko damping described in Equation 5.15, it is found that the ratio of Timoshenko damping to the force-based effective damping is A_{β_i}/π. Thus, in most cases, the estimation of Timoshenko damping will be larger.

5.1.2.2.2 Bilinear Damping

The force-based effective damping ratio for bilinear damping can be written as

$$\xi_{\text{eff}_i} = \frac{q_d\sum_{j=1}^{L}\left(x_{ji}-d_y\right)}{2\times(1.0\sim1.05)\,\boldsymbol{p}_i^{\mathrm{T}}\mathbf{K}\,\boldsymbol{p}_i} \tag{5.38}$$

When the supplemental dampers can store larger potential energy, the force-based effective damping ratio takes a smaller value.

Comparing the Timoshenko damping described in Equation 5.23 and the force-based effective damping described in Equation 5.38, it is found that the Timoshenko damping is estimated to be $4/\pi$ times larger.

5.1.2.2.3 Dry Friction Damping

Since the maximum damping force of a friction damper can be written as

$$f_d = \mu N$$

the force-based effective damping ratio of the bilinear damping becomes

$$\xi_{\text{eff}_i} = \frac{\mu N\sum_{j=1}^{L}\left(x_{ji}\right)}{2\boldsymbol{p}_i^{\mathrm{T}}\mathbf{K}\,\boldsymbol{p}_i} \tag{5.39}$$

Comparing Equation 5.39 with Equation 5.22, again the Timoshenko damping is approximately $4/\pi$ times larger.

5.2 RECTANGULAR LAW, MAXIMUM ENERGY DISSIPATION PER DEVICE

The relationship between the damping force and the displacement of a damper is the constitutive loop of energy dissipation. Given the amplitude of the force and the displacement, the maximum energy dissipation per damper is a rectangular loop with an area four times the product of the force and displacement amplitudes. Furthermore, the relationship between the maximum seismic force and the displacement defines the maximum work done by the seismic force. Suppose the maximum accelerations of several systems are known to be identical. Then, the displacement of the system with a rectangular-shaped maximum seismic force vs. displacement is the smallest in the period range between 0.5 and 5 (s).

Installation of nonlinear dampers in structures, which operate either in the elastic or inelastic range, would have very different effects due to the nonlinearity of the damping. Therefore, it is helpful to study the maximum possible energy dissipation with given allowed amplitudes of the damping force and damper displacement. The maximum possible energy dissipation is closely related to the vibration reduction, though larger energy dissipation does not necessarily mean larger vibration reduction.

5.2.1 Maximum Energy Dissipation, Rectangular Law of Damping

Dampers, or damping devices, are used to dissipate vibration energy. Therefore, one of the basic measurements of a damper is its capacity for energy dissipation. No matter what type of damping mechanism the damper possesses, the dissipated energy can be represented by the loop area of the force vs. displacement. This is referred to as the force–displacement constitutive relationship for a damper, denoted by $f_d(\dot{x}, x)$ in scalar form or $\mathbf{f}_d(\mathbf{x}, \dot{\mathbf{x}})$ in vector form.

From the previous section, given the amplitude of the force $f_{d\max}$ applied on a damper and the amplitude of the relative displacement between the two ends of the damper, x_0, the maximum energy that can possibly be dissipated, denoted by $E_{d\max}$, is

$$E_{d\max} = 4 f_{d\max} x_0 \tag{5.40}$$

In Figure 5.6, the force and displacement of a damper are described. From Figure 5.7, it is easy to see that any energy dissipation loop other than the rectangular loop will have dissipated energy less than $E_{d\max}$. For example, the viscous damping, shown as an ellipse, has less energy dissipation than the rectangular $4f_{d\max}x_0$. The maple-seed-shaped damping shown in Figure 5.7 has even less energy dissipation than the viscous damping behavior.

Suppose the damping force $f_d(\dot{x}, x)$ can be modeled in the more general form as,

$$f_d(t) = \left[c_0 + c_1|\dot{x}|^{\beta} + c_2|\dot{x}|^{2\beta} + \cdots\right]\operatorname{sgn}(\dot{x}) \tag{5.41}$$

A polynomial of the frequency ω_n and displacement x_0 is then used to represent the amplitude of the damping force, $f_{d\max}$:

$$f_{d\max} = c_0 + c_1\omega_n^{\alpha}x_0^{\beta} + c_2\omega_n^{2\alpha}x_0^{2\beta} + \cdots \tag{5.42}$$

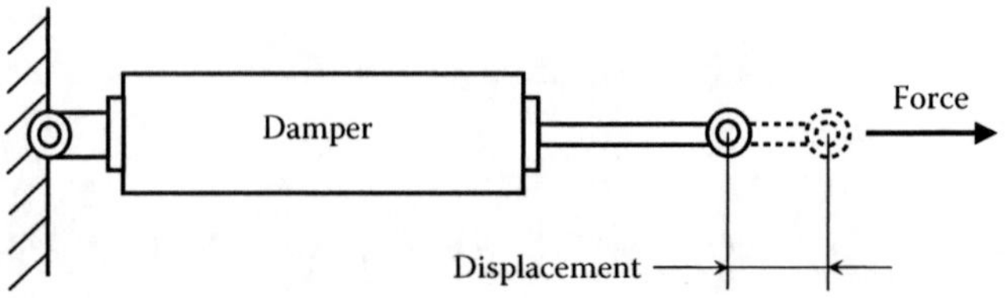

FIGURE 5.6 Force and displacement of a damper.

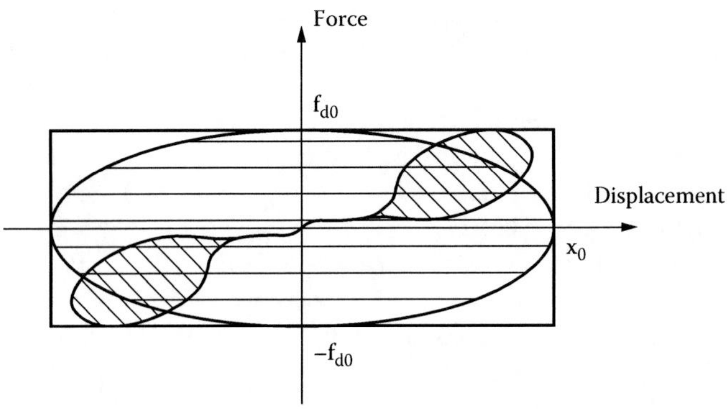

FIGURE 5.7 Maximum energy dissipation.

where c_k are coefficients and the first term involving c_0 can be written as $c_0\omega_n^0 x_0^0$. Note that in Equations 5.41 and 5.42, the exponential β has a different meaning than in Equation 5.14. For the sake of simplicity, the same symbol is used.

It can be seen that the smaller these powers α and β become, the closer the energy dissipation will be to a rectangular loop. Only in this case, for example, $f_{d\,max} \propto x_0^{1/2}$ can be better than $f_{d\,max} \propto x_0^2$.

The concept of rate dependency of dampers is defined as follows (Lazan 1968):

Rate-dependent damping will depend on the rate, the derivative of the force or displacement with respect to time; that is, the energy dissipated will be a function of either the displacement or the velocity. Rate-dependent damping may also depend on the amplitude of a cyclic load. *Rate-independent damping* will only depend on the amplitude of the cyclic loading, but will be independent of the rate of load or displacement.

Thus, it is seen that a damper with rectangular energy dissipation has rate-independent damping. On the other hand, rate-dependent damping consists of many more complicated subcategories. As a matter of fact, the amplitude of the load, the amplitude of the displacement, and the driving frequency may all affect the rate.

Within the context of the energy dissipation capacity, the Lazan classifications are of fundamental significance. His idea can be extended to the rate dependency of the damping force, and the damping force can be an important design criterion. It follows that the frequency dependence of the damping force is very important. For example, a viscous damper will not change its damping coefficient if the driving frequency changes but the velocity remains the same. The damping force will then remain constant. However, a VE damper will change its damping force under the same circumstances. Furthermore, if the supporting stiffness of a viscous damper is comparatively weak, the resulting damper assembly will tend to behave with frequency dependence.

5.2.2 Smallest Maximum Dissipation, Rectangular Law of Seismic Work

In Chapter 2, the concept of work done by maximum seismic force was introduced. The corresponding response spectra of the seismic work for linear SDOF systems with viscous damping were then plotted. It was learned that during an earthquake, the work done by the maximum seismic force is a deterministic value, in the sense of statistical averaging under a given set of earthquake excitations. It was also seen that the relationship of the maximum seismic work with the displacement of the linear SDOF system defines an elliptic loop.

5.2.2.1 Minimum Work Done by Maximum Seismic Force

In an idealized case, there can be a nonlinear system in which the relationship of the maximum seismic work with the displacement of the linear SDOF system has a rectangular loop. In the previous section, the energy dissipation loop is formed by the damping force vs. the displacement. Now, the loop is formed by the maximum seismic force. In this specific case, another rectangular law can be obtained through numerical simulations by comparing the displacement with the given level of the maximum seismic force or maximum absolute acceleration. Thus, the concept of the maximum seismic work can be used as a tool to examine the smallest possible displacement associated with earthquake responses.

To see this point, a numerical simulation is used to generate nonlinear displacements for an idealized structure that has rectangular maximum seismic work. The absolute acceleration of the earthquake responses is regulated during the simulation so that it is identical to linearly viscous damped systems. Namely, the responses of the linear systems are first computed by letting the periods be 0.10, 0.15, 0.20, 0.25, 0.30, 0.35, 0.4, 0.5, 0.6, 0.7, 0.8, 0.9, 1.0, 1.2, 1.4, 1.6, 1.8, 2.0, 2.2, 2.4, 2.6, 2.8, 3.0, 3.2, 3.4, 3.6, 3.8, 4.0, 4.2, 4.4, 4.6, 4.8, 5.0, 5.2, 5.4, 5.6, 5.8, 6.0, 7.0, 8.0, 9.0, and 10.0, (s) for a total of 42 different periods. The damping ratios are 0.02, 0.05, 0.10, 0.15, 0.20, 0.25, 0.3, 0.35, 0.40, 0.45, 0.50, 0.55, 0.60, 0.65, 0.70, 0.80, 0.90, 1.00, 1.25, and 2.00, for a total of 20 damping ratios. Therefore, the total number of selected linear systems is 42 × 20 = 840. In each case, the absolute acceleration and relative displacement are computed. Consequently, a nonlinear system with the rectangular maximum seismic work is generated in such a way that both systems have identical peak values for the acceleration. The corresponding displacement is then calculated.

In order to conduct a statistical survey, 99 earthquake records are used, which were first used in Chapter 4, Example 4.8, and will be discussed in detail in Sections 5.4 and 5.5. For each pair of linear and nonlinear systems, the corresponding mean values and standard deviations are calculated through the 99 pair responses.

It is found that when the period falls in the range of greater than 0.1 (s) and smaller than 7.5 (s), the nonlinear displacement is always smaller than that for the linear system. In Figure 5.8 and following, the symbol "d.r." denotes "damping ratio". The displacements in this plot are mean values plus one standard deviation, obtained statistically using the 99 earthquake records.

In order to visualize the detailed comparison, Figure 5.9 shows the nonlinear and linear displacements vs. the damping ratios, which are plotted separately in Figures 5.9a and b. For the nonlinear systems, the damping ratio is the corresponding value to the linear system.

From Figure 5.9, it can be realized that the nonlinear system has considerably smaller displacement. Furthermore, it is seen that when the corresponding damping ratios increase, the nonlinear

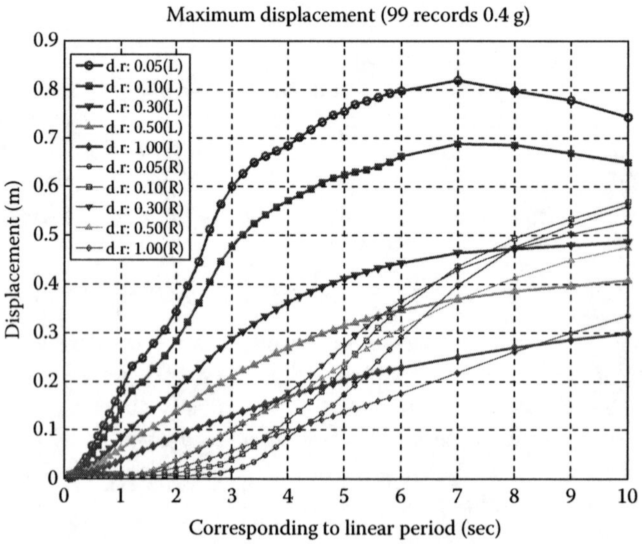

FIGURE 5.8 Comparison of displacement between linear and rectangular systems, (L)—Linear and (R)—Rectangular.

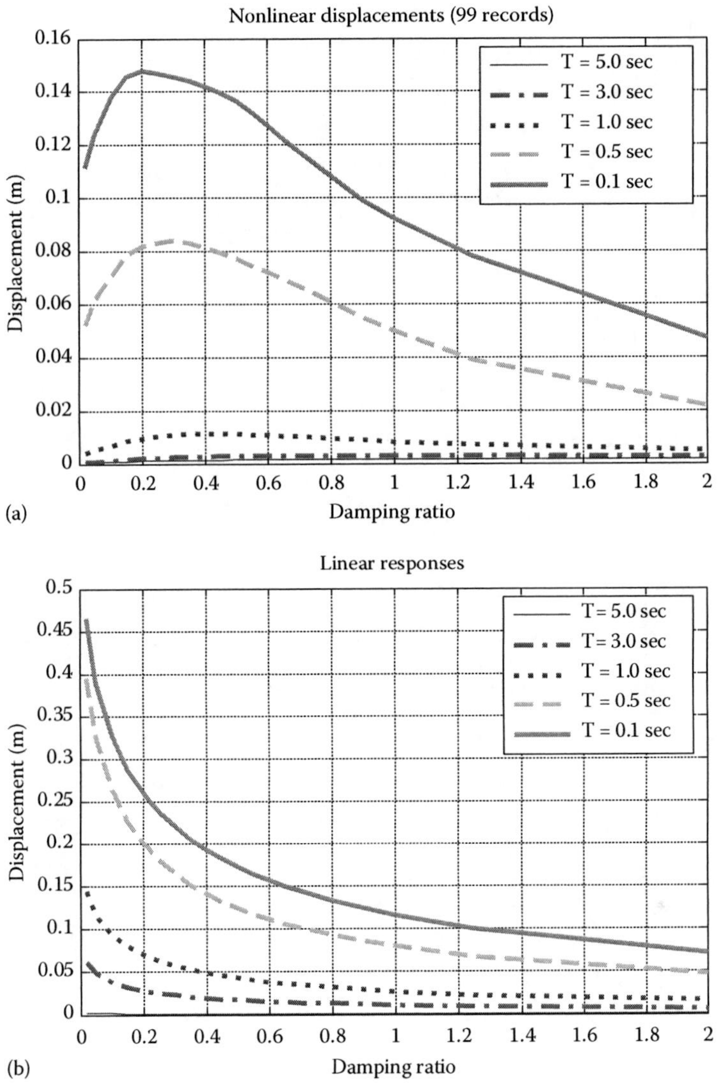

FIGURE 5.9 Comparisons of response vs. damping between linear and rectangular systems: (a) rectangular systems and (b) linear systems.

responses are first increased and then decreased. As a comparison, the linear displacements are monotonically decreased.

In order to visualize this further, Figure 5.10 shows the nonlinear and linear displacements vs. the periods, which are plotted separately in Figures 5.10a and b. For the nonlinear systems, the period is the corresponding value to that of the linear system.

From Figure 5.10, it is realized again that the nonlinear system has a considerably smaller displacement. Furthermore, it is seen that when the corresponding periods increase, the nonlinear responses are increased almost linearly. The linear displacements are also monotonically increased within the period range below 3 (s). However, when the periods become longer, the linear displacements will reach their peak values and then start to decrease.

It also noted that the corresponding period and damping terms do not have the conventionally defined effective period and damping ratio.

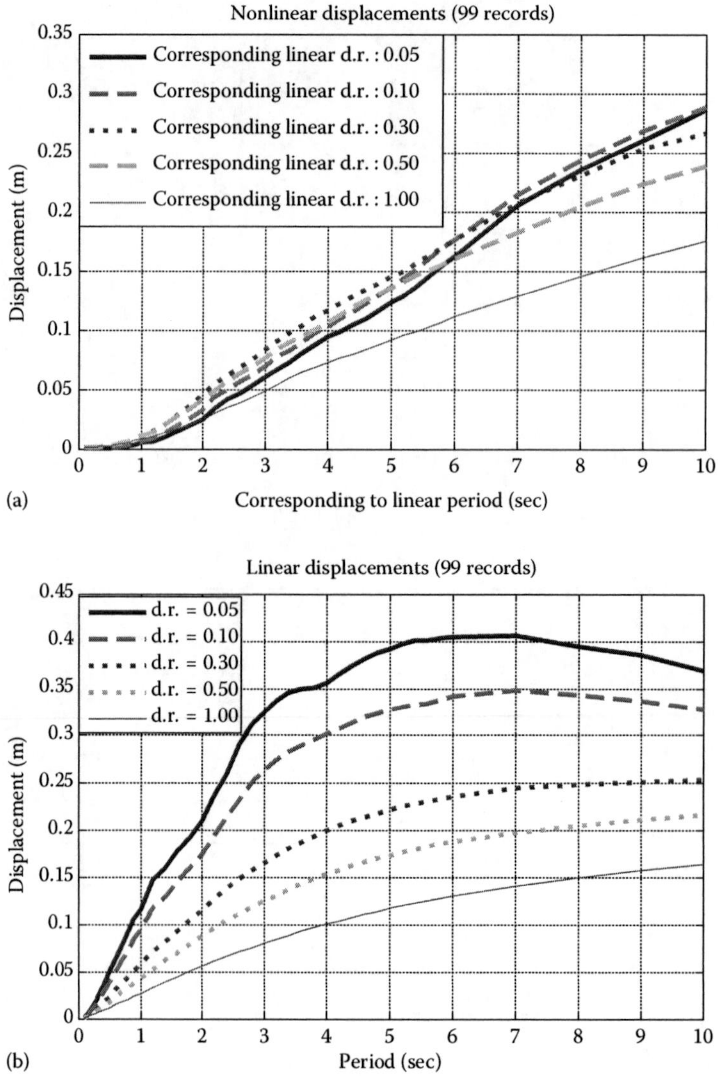

FIGURE 5.10 Comparisons of response vs. damping between linear and rectangular systems: (a) rectangular systems and (b) linear systems.

5.2.2.2 Linearity of Nonlinear Responses

Note that in Figure 5.10, and also in the response comparisons in the previous section, the simulations are conducted by specifying the input acceleration level as 0.4 (g). Since the response is nonlinear, it is necessary to study whether this conclusion holds for other input levels. For this purpose, other input levels are used to study the linearity of the nonlinear responses, which are 0.2, 0.6, 0.8, and 1.0 (g). In Figure 5.11, several selected groups of displacements for the rectangular systems are plotted. Group 1 provides the response corresponding to the linear system with a damping ratio equal to 0.05; Groups 2, 3, and 4 correspond to the linear systems with damping ratios 0.10, 0.30, and 0.50, respectively. Note that the X-axis is the "period," which is explained above.

In each group, the response under 0.2 (g) excitations is first plotted. Then, the response under 0.4, 0.6, 0.8 and 1.0 (g) excitations with multiplication factors 0.2/0.4 = 1/2, 0.2/0.6 = 1/3,

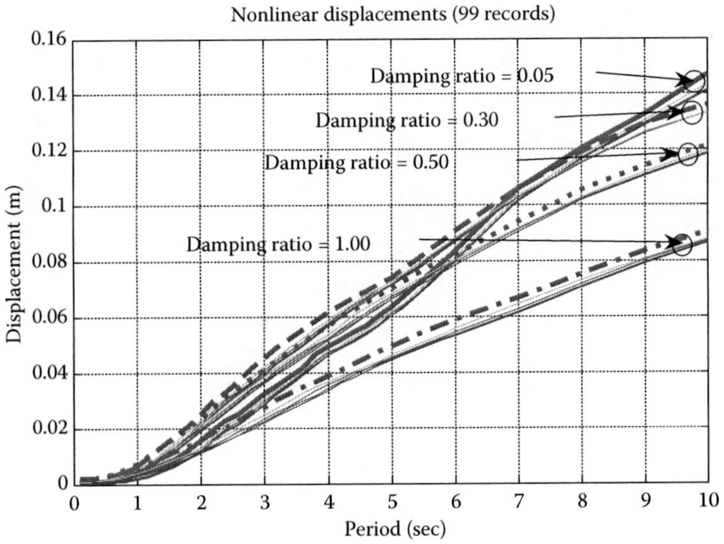

FIGURE 5.11 Examination of nonlinearity of rectangular seismic work systems.

0.2/0.8 = 1/4 and 0.2/1.0 = 1/5 are plotted respectively. Thus, in each group, five curves are included.

Using Figure 5.11, the nonlinearity can be examined. If the responses are linear, the five curves in each group should be exactly overlapped. Since the systems are nonlinear, there will be certain variations. From Figure 5.11, however, it is realized that the variation is rather small.

It is known that when the loop of the maximum seismic work is rectangular, the system must be highly nonlinear. However, the earthquake excitation is rather random. The averaging of random excitations can smooth the nonlinearity. Therefore, the results from excitations from 0.2(g) to 1.0(g) actually show good linearity, which can greatly simplify the damper design procedure. Thus, in practical design, the result obtained from a particular excitation, i.e., 0.4 (g), can be taken and modified by the factor n/0.4, if the required ground acceleration is n (g).

Furthermore, the same measures are applied to several parallelogram-shaped systems and the results are plotted in Figure 5.12, where the corresponding damping ratios are chosen to be 5%, 30%, 50%, and 100%, and the input levels are chosen to be 0.4, 0.6, and 1.0 (g). With the same treatment mentioned in Figure 5.12, the displacements are plotted in Figure 5.12a and the accelerations are plotted in Figure 5.12b. Similar to the pure rectangular system, the linearity of the nonlinear responses are very good in the displacement plots, where all three curves almost overlap. In the acceleration plots, when the periods are about 0.5 (s) or shorter, differences in the responses under different input levels in each group can be clearly seen. However, when the periods are sufficiently long, the three responses in each group nearly overlap, as well. In engineering applications, these can be treated as linear responses without noticeable error.

In this case, the above conclusion of the minimum work done by the maximum seismic force can be used for any possible ground input.

5.2.3 Quality Factor

In the above discussion of maximum energy dissipation, it is assumed that a damper can provide the idealized rectangular loop. In reality, this rarely happens. In fact, most dampers are modeled for their corresponding constitutive force–displacement relationship. However, they all barely provide the energy dissipation loop for the idealized loop.

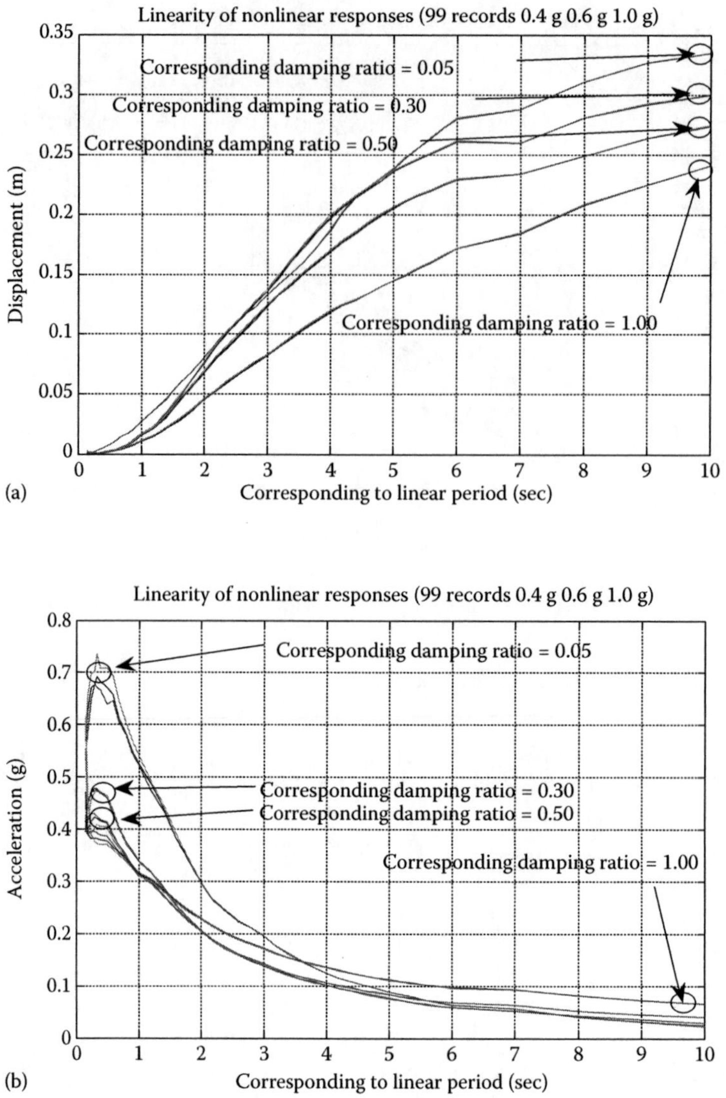

FIGURE 5.12 Examination of nonlinearity of parallelogram seismic work systems: (a) relative mean displacement and (b) absolute mean acceleration.

To handle this type of imperfection, a safety factor, defined as *quality factor* and denoted as q_H, can be used. This is explained in the following:

In NEHRP 2003 (BSSC 2003), the quality factor or the hysteresis loop adjustment factor is defined as the ratio of the actual area of the hysteresis loop and the area of the assumed elastoplastic representation of the loop. The hysteresis loop is formed by assuming that the base-roof displacement loop is converted to a special capacity form. In NEHRP 2003 (BSSC 2003), this factor is calculated by

$$0.5 < q_H = 0.67\frac{T_s}{T_1} \tag{5.43}$$

where T_s is the ratio of S_{D1}/S_{DS}, which was discussed in Chapter 2 with detailed definitions; and T_1 is the period of the first (or fundamental) mode. From Equation 5.43, this is a question of how closely the quality of an individual damper relates to the system. In fact, when a damper is evaluated, cyclic load by sinusoidal excitation is often used with fixed displacement. However, when a damper starts to work under earthquake excitation, the displacement is both nonfixed and nonsinusoidal. Therefore, the quality factor here is not as accurate.

Practically speaking, the quality factor should be provided by vendors who supply the damper. In most cases, each individual damper should be tested before it is installed in a building. The theoretical energy dissipation loops, denoted as E_d, which were briefly introduced earlier and are discussed in a later section, should be the base to evaluate individual dampers. The area of an actual loop of each specific damper, denoted as E_r, should be used under a standard testing procedure. Then,

$$q_H = \frac{E_r}{E_d} \tag{5.44}$$

5.2.4 Issues of Multiple Dynamic Equilibrium Positions

In Section 5.5.1, the issue of recoverable damping and self-centering systems is discussed. That is, for a linear vibration system, when the external force no longer activates, the system will always return to its original equilibrium position. However, for nonlinear systems, such as a system with rectangular damping, there is a chance that when the external force vanishes, the system has permanent displacement. Moreover, during excitations, some or all of the displacements will have multiple dynamic equilibrium positions, which will be discussed in detail in Chapter 6 (see Figures 6.29 and 6.30).

The cause of multiple equilibrium positions is complex. Intuitively speaking, if the required time needed for a system to return its original equilibrium posision from a certain displacement is too long, then before centering, the system is forced to have a further displacement in the same direction. A detailed explanation of these causes is beyond the scope of this book. It is seen that when a nonlinear system is excited by different earthquake ground motions, this phenomena may or may not occur all the time.

In Section 5.2.2, it was shown that statistically, a nonlinear system with rectangular energy dissipation had the smallest average displacement when compared to a linear system with the same amount of peak damping force. However, when the responses of a nonlinear system have multiple equilibrium positions, it is understood that the total displacement can become very large. Therefore, comparing a linear system and an individual nonlinear system that has a peak damping force identical to that of the linear system, the nonlinear system can have larger displacements, although on average, the rectangular law still applies. In order to simplify the explanation of rectangular law and its application in damper design, the case of the multiple equilibrium positions caused by nonlinear damping is excluded in the following discussion.

5.3 DAMPING ADAPTABILITY

Suppose the amplitude of the damping force of a damper is proportional to the amplitude of displacement to the power of β. When the displacement of a damper increases, while the force also increases, the damping can adapt as defined by the quantity β. The higher the damping adaptability, the better the chance that it will work in a wider range and dissipate more energy. A damper with zero adaptability can only work within a narrow range. Therefore, such a damper may not be suitable for an earthquake protective system. On the other hand, the damping force will not increase significantly for devices with low damping adaptability. Thus, the energy dissipation will be closer to rectangular. Properly designed supplemental damping can be more effective for response reduction. In addition, variable damping force affects the vibration shape function, whereas constant damping force does not, which results in different design strategies.

5.3.1 Concept of Damping Adaptability

When a damper is installed in a structure, the relative deformation of the damper is often determined by the corresponding displacement of the structure. Let, the relative deformation and velocity between the two ends of a damper be denoted by x and v, respectively, the values x and v cannot be arbitrarily adjusted by dampers. However, the force applied to the damper can be varied, depending on the damping characteristics.

It was seen that when only the constant term exists, that is, the damping force $f_{dmax}(\dot{x}, x) = c_0$, and $\alpha = \beta = 0$ in Equation 5.112, the relationship between force f_d and deformation x_0 is rectangular. However, since the damping force is constant, no matter how large the velocity or deformation, the area of energy dissipation cannot be increased accordingly. In this case, the damper has no *adaptability* to absorb more energy.

This concept can be better explained through Figure 5.13, where x_{01} stands for a smaller displacement, and x_{02} is a larger displacement. Suppose at displacement x_{01}, there is a square-shaped energy dissipation loop E_1 and an ellipse-shaped energy dissipation loop E_2. According to the above discussion, the damping mechanism E_1 is better than E_2.

However, when the displacement becomes larger than x_{02}, due to different damping mechanisms, the situation may change.

First, the square-shaped loop can become rectangular, as denoted by E_5. Examples of this type of damping are friction damping and metallic damping. The loop can also become a larger square, as denoted by E_4. In addition, the ellipse may become the one denoted by E_3 in Figure 5.13.

Quantitatively, the amplitude of the damping force, f_{dmax}, is used as follows,

$$f_{d\max} \approx c x_0^{\beta} \tag{5.45}$$

If the exponent β is larger than unity and $x_0 > 1$, then the force can be larger, thereby making the total energy dissipation larger, and vice versa. In this case, for example, it is seen that $f_{d\max} \propto x_0^{1/2}$ can be worse than $f_{d\max} \propto x_0^2$.

It is readily seen that this conclusion contradicts the one reached in the previous section. The correct way to select criterion depends on two basic factors: the value of the deformation and the dynamics of the entire structure.

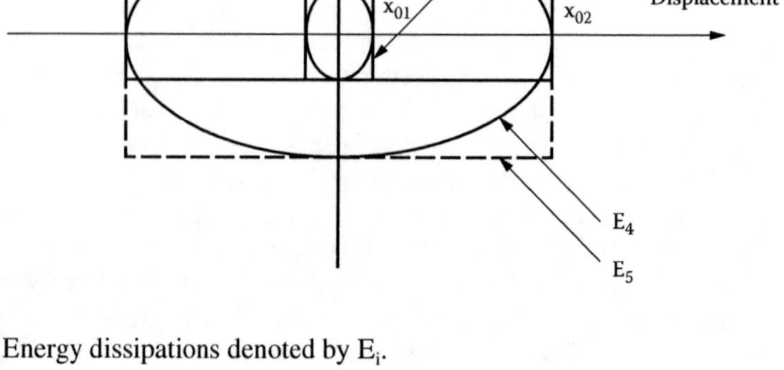

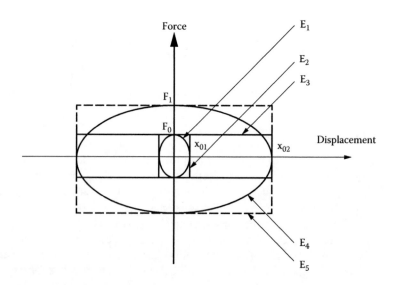

FIGURE 5.13 Energy dissipations denoted by E_i.

Generally speaking, if only the capability of energy dissipation by the dampers is taken into account, the following integration can be maximized in one vibration cycle:

$$E_d = \oint f_d dx \tag{5.46}$$

In order to quantitatively describe the energy dissipation, a quantitative definition of how the energy dissipation loop will be changed as the displacement changes is needed, which is referred to as the damping adaptability. This concept is actually a logical expansion of Lazan's second classification of damping. The first is the rate dependency, which was previously discussed. The second relates to the *quadratic or nonquadratic damping* form. Lazan (1968) suggested that the energy dissipation can be written as

$$E_d = \kappa x_0^{\gamma} \tag{5.47}$$

where κ is a proportionality coefficient and

$$\gamma = 1 + \beta \tag{5.48}$$

where β is the exponent defined previously. When $\gamma = 2$, quadratic damping exists, otherwise there is nonquadratic damping. For example, linear viscous damping is quadratic.

By using $f_{d\,max} = c_{eq}\omega_n^{\alpha} x_0^{\beta}$,

$$\kappa = c_{eq}\omega_n^{\alpha} A_{\beta} \tag{5.49}$$

As a comparison, for the case of linear viscous damping, $f_{dmax} = c\omega_n x_0$ for cycles with frequency equal to the natural frequency of the system ω_n, and

$$\kappa = c\pi\omega_n \tag{5.50}$$

E_q and E_d are denoted as the energy dissipation achieved by quadratic damping and other damping, respectively. Subscripts q and g are used to distinguish the proportional coefficients κ for quadratic and general damping. That is,

$$E_q = \kappa_q x_0^2 \propto x_0^2 \tag{5.51}$$

and

$$E_d = \kappa_g x_0^{1+\beta} \propto x_0^{1+\beta} \tag{5.52}$$

From Equations 5.51 and 5.52, the rate of energy dissipation by general sublinear damping will be smaller as the displacement increases.

For comparison purposes, suppose there can be a reference of fixed values for c, c_{eg}, ω_n, and α to make

$$\kappa_q = \kappa_g \tag{5.53}$$

In this situation, the ratio of energy dissipations of the general damping and the quadratic damping is considered:

$$E_d/E_q = x_0^{1-\beta} \tag{5.54}$$

It is seen that only if $\beta = 1$ will the general damping dissipate the same amount of energy as the quadratic damping.

Now, consider bilinear damping. The corresponding energy dissipation,

$$E_d = 4q_d x_0 - 4q_d d_y \tag{5.55}$$

If the small amount of $4q_d d_y$ is ignored:

$$E_d = 4q_d x_0 \propto x_0 \tag{5.56}$$

From Equation 5.56, it is seen that the rate of energy dissipation by bilinear damping will also be comparatively smaller than that of quadratic damping, when the displacement increases.

Based on the above discussion, the *damping adaptability factor* is defined as follows:

$$f_{adp} = \beta \tag{5.57}$$

It is important to note that the terms f_{adp} and β are no longer the characteristics of a damper per se, but rather global parameters of the entire integration of a structure and added energy dissipation devices. By comparison, Lazan's notation is for a damper only.

Note that when $\beta = 1$, the damping is quadratic:

$$f_{adp} = 1 \tag{5.58}$$

From Equation 5.54, it is seen that for sublinear damping,

$$f_{adp} = 1 - \frac{\ln(E_d/E_q)}{\ln(x_0)} \tag{5.59}$$

That is, using the logarithm of displacement x_0 as a reference, as the energy dissipated by the general damping becomes larger, the damping adaptability will also increase.

Now, as an example, the concept of using damping adaptability in damper design is examined. By employing the same subscripts q and g as previously defined, consider the damping ratio ξ. For quadratic damping,

$$\xi_q = \frac{\kappa_q x_0^2}{2\pi k x_0^2} = \frac{\kappa_q}{2\pi k} \tag{5.60}$$

while for general damping,

$$\xi_g = \frac{\kappa_g x_0^{1+\beta}}{2\pi k x_0^2} = \frac{\kappa_g}{2\pi k} x_0^{\beta-1} = d\xi_q x_0^{f_{adp}-1} \tag{5.61}$$

where d is a proportional factor.

From Equation 5.60, it is seen that the damping ratio is an important control factor in damping design and is independent of the displacement if the damping is quadratic, i.e., Newtonian viscous damping. In this case, no matter how large or small the vibration level, the damping ratio will remain constant. In addition, the units of κ_q and k, which are identical, do not need to be considered.

However, if the damping is nonquadratic, the situation is different. From Equation 5.61, it is seen that the damping ratio depends on the displacement. In this case, the amplitude of vibration level matters, and the damping ratio will no longer be constant. Furthermore, κ_g and k have different units so that the term ($\kappa_g/2\pi k$) denotes dimensionless. Therefore, the proportional factor will have a unit, e.g., $(m)^{1-\beta}$. When the unit meter is used for the displacement x_0 as a seismic response of a building, this amplitude is often smaller than unity. In this particular case, for convenience, the value of *d* is determined and smaller damping adaptability often results in a larger damping ratio. Therefore, in this case, smaller damping adaptability is beneficial.

On the other hand, for nonunity damping adaptability, the amplitudes of the structure have a nonlinear relationship with the input level, and this relationship is exponential. If the nonlinearity is larger, the system will become unstable and will yield rather large displacements. This situation must be considered in design practice, especially in earthquake zones where the dynamic range of ground excitation is relatively large. Based on limited numerical simulations, the following range of damping adaptability of a damper appears more beneficial:

$$0.1 \le f_{adp} \le 0.5 \quad (5.62)$$

5.3.2 Deformation Shape Function

Now, consider the effect of f_{adp} on the structural deformation shape functions.

When

$$f_{adp} = 1 \quad (5.63)$$

linear damping force occurs. In addition, when the damping force is relatively small, say the damping ratio is less than 10.0%, the corresponding mode shape functions will not be affected by displacement. Therefore, for structural response estimation, constant deformation shape functions can be used when changing the amount of supplemental damping. That is,

$$\boldsymbol{p}_i \approx \text{const.} \quad (5.64)$$

Here, $\boldsymbol{p}_i$ is the ith shape function. For a linear system, $\boldsymbol{p}_i$ is the ith mode shape. Note that for proportional damping, the mode shape functions remain identical as the damping matrix varies. For a nonproportionally damped system, Equation 5.64 will only hold for small damping force. This phenomenon will make the response estimation easier. Note that for a nonlinear system, $\boldsymbol{p}_i$ can be seen as a deformation shape function that evolved from the ith linear mode shape due to an increased amount of nonlinear damping force.

However, when sublinear dampers are used,

$$f_{adp} \neq 1 \quad (5.65)$$

the corresponding shape function will be affected by the level of input, which determines the value of the displacement. That is,

$$\boldsymbol{p}_i \neq \text{const.} \quad (5.66)$$

This fact increases the difficulty of the response estimation.

It is known that the variation of the deformation shape function is due to the different distribution of forces, both restoring force and damping force.

When bilinear damping is used,

$$f_{\text{adp}} = 0 \tag{5.67}$$

a constant damping force with respect to the level of displacement will exist. In this case, it is still assumed that Equation 5.64 holds. This is the main reason for using the different approaches of sublinear and bilinear dampings in practical design.

In the above discussion, the concepts of the rectangular law and damping adaptability are described. These two design principles seem to contradict each other. Namely, by using the rectangular law, a smaller damping exponent is desired, and the best possible case is zero exponents or dry friction damping. However, to gain damping adaptability, larger damping exponents are needed. The central issue lies in the proper choice of this parameter.

Generally speaking, if the excitation does not have a very broad dynamic range, the rectangular law should be considered. In most cases of earthquake excitation, the input dynamic range is comparatively small. Thus, the damping exponents should have lower values is also true for inelastic structures. In the previous section, 99 earthquake records were used to plot the response reduction, as well as the energy status. With many other groups of earthquake records, this phenomenon is illustrated.

5.4 DESIGN AND CONTROL PARAMETERS

For convenience, "control parameters" and "design parameters" will represent different concepts in this book. From a practical standpoint, control parameters are directly related to design spectra, whereas design parameters are used to describe real-world parameters, such as an equivalent damping coefficient for sublinear dampers.

5.4.1 Low Damping and High Damping Structures

5.4.1.1 Control Parameters and Design Parameters of Linear Systems

The fundamental approach of damper design discussed in this book is based on spectrum analysis. Regardless of whether the structure is a linear SDOF system or a nonlinear MDOF system, the computation of the structural responses is calculated based on the existing design spectrum, which is generated through linear SDOF systems. Many discussions are therefore devoted to the rationale and process of designing a nonlinear MDOF system using the design spectrum.

Therefore, it is helpful to analyze the differences between SDOF systems and MDOF systems; the difference among proportionally and nonproportionally damped MDOF systems, and nonlinearly damped, as well as overdamped, MDOF systems. This topic is first considered by analyzing the differences in the control and design parameters between the SDOF system and linear systems.

The control parameters are used to control the peak structural responses within a certain allowed level. If these parameters are changed, the structural responses will be either increased or decreased. The design parameters are the basic structural responses, which are to be decided or designed.

It is known that for a linear SDOF system, the primary control parameters are the natural period and the numerical damping coefficient. Since the numerical damping coefficient B (see Equation 2.319) can be directly determined through the damping ratio, there are only two primary control parameters: the natural period and the damping ratio. Other parameters, such as the damping coefficient c and damping exponent β of viscous dampers, are considered as the secondary control parameters. Once the periods and the damping ratios are determined, these secondary parameters can be determined as well. For linear systems, these two sets of parameters are directly related. For example, if the damping ratio and the natural period are known, the damping coefficient per unit mass can be calculated. When the damping coefficient and the natural period are known, the damping ratio is determined. For nonlinear systems, this bilateral relationship usually does not exist, because the quantities depend on their amplitudes.

For a linear SDOF system, the traditional primary design parameter is the total base shear, which is the product of the mass and the pseudo acceleration. Therefore, base shear can be replaced by the pseudo acceleration. In many codes, such as NEHRP 2009 (BSSC 2009), the design procedure of an aseismic structure is to first find the base shear, using the aforementioned control parameters. That is,

$$V = WS_A = V_L \quad (N) \tag{5.68}$$

where V is the maximum total base shear or simply called the base shear, W is the corresponding weight, S_A is dimensionless spectral value, and V_L is the amplitude of the lateral shear force of the column.

Equation 5.68 can be extended to linear MDOF systems. The total base shear is used to find the lateral force for each story, aided by the modal shape of the mode of interest, that is,

$$\mathbf{f}_L = \begin{Bmatrix} f_{L1} \\ f_{L2} \\ \dots \\ f_{Ln} \end{Bmatrix} = \begin{Bmatrix} p_1 \\ p_2 \\ \dots \\ p_n \end{Bmatrix} V \quad (N) \tag{5.69}$$

Here $\mathbf{f}_L$ is the lateral force vector for the ith mode and f_{Lj} is the amplitude of the lateral force of the jth floor for the ith mode. However, in Equation 5.69, for convenience, the indicator of the ith mode is not written. Additionally, in Equation 5.69, p_j is the ith modal displacement or the modal shape element at the jth floor. Here, the condition of the modal shape normalization is such that

$$\sum_{k=1}^{n} p_k = 1 \tag{5.70}$$

After the lateral force has been calculated, this quantity is used to define the floor displacement or drift and so on. For example, the column shear force of the jth story can be determined as

$$\mathbf{v}_L = \begin{Bmatrix} V_{L1} \\ V_{L2} \\ \dots \\ V_{Ln} \end{Bmatrix} = \begin{bmatrix} 1 & 1 & \dots & 1 \\ 0 & 1 & .. & 1 \\ & & \dots & \\ 0 & 0 & 0 & 1 \end{bmatrix} \begin{Bmatrix} f_{L1} \\ f_{L2} \\ \dots \\ f_{Ln} \end{Bmatrix} \quad (N) \tag{5.71}$$

Or, for the jth floor,

$$V_{Lj} = \sum_{k=1}^{j} f_{Lk} \quad (N) \tag{5.72}$$

Here, $\mathbf{v}_L$ is the shear force vector for the ith mode and V_{Lj} is the amplitude shear force of the jth floor for the ith mode. Similarly, in Equation 5.72, for convenience, the indicator of the ith mode is not written.

Substituting Equation 5.69 into Equation 5.71 results in

$$\mathbf{v}_L = \begin{bmatrix} 1 & 1 & \dots & 1 \\ 0 & 1 & .. & 1 \\ & & \dots & \\ 0 & 0 & 0 & 1 \end{bmatrix} \begin{Bmatrix} p_1 \\ p_2 \\ \dots \\ p_n \end{Bmatrix} V \ (N) \tag{5.73}$$

Therefore, it is realized that the total base shear is indeed the primary design parameter.

There is an alternative design procedure starting with the displacement. Either way, only one primary design parameter is sufficient for the entire design. Two control parameters and one design parameter are the essence of the design of a traditional linear SDOF system based on the design spectrum. On the surface, this nature of design is quite understandable. That is, one must have two control parameters to render the design spectrum. Then, only one parameter is sufficient to cover both the displacement and the acceleration, since the pseudo acceleration is proportional to the displacement, and the proportionality constant is simply the square of the natural frequency, whose information has already been given by the period. This design feature can be summarized by the following equations:

$$S_A = f(T, \xi) \tag{5.74a}$$

and

$$S_D = \left(\frac{T^2}{4\pi^2}\right) S_A \, g \ (m) \tag{5.74b}$$

When a linear MDOF system is to be designed, if it is proportionally damped, Equations 5.74a and b can be simply rewritten as follows, where i stands for the ith mode of interest:

$$S_{Ai} = f(T_i, \xi_i) \tag{5.75a}$$

and

$$S_{Di} = \left(\frac{T_i^2}{4\pi^2}\right) S_{Ai} \, g \ (m) \tag{5.75b}$$

Therefore, for each mode, the essence of the two control parameters and one design parameter remains unchanged.

Note that in Equation 2.337 of Chapter 2, a factor $\sqrt{1+4\xi^2}$ is suggested to modify the relationship between the spectral values of displacement and acceleration. This can be proven by using the approach of the square root of the sum of squares (SRSS) for SDOF systems. Adding the damping force to Equation 2.307 results in

$$m\ddot{x}_a = -c\dot{x} - kx$$

or

$$\ddot{x}_a = -2\xi\omega\dot{x} - \omega^2 x$$

The maximum value of the absolute acceleration can be seen as the SRSS value of the maximum values of the monic damping and stiffness forces. That is,

$$a_a = \left[4\xi^2 \frac{4\pi^2}{T^2} \left(\frac{2\pi}{T} d_{max} \right)^2 + \left(\frac{4\pi^2}{T^2} \right)^2 (d_{max})^2 \right]^{1/2} = \sqrt{1+4\xi^2} \frac{4\pi^2}{T^2} d_{max} \quad (m/s^2) \tag{5.76}$$

The maximum values of the absolute acceleration a_a and the relative displacement d_{max} are replaced by the spectral values S_A and S_D, respectively,

$$S_A = \sqrt{1+4\xi^2} \frac{4\pi^2}{T^2} S_D / g \tag{5.77a}$$

and for linear MDOF systems, we have

$$S_{Ai} = \sqrt{1+4\xi_i^2} \frac{4\pi^2}{T_i^2} S_{Di} / g \tag{5.77b}$$

Therefore,

$$S_D = \frac{1}{\sqrt{1+4\xi^2}} \left(\frac{T^2}{4\pi^2} \right) S_A \, g \quad (m) \tag{5.78a}$$

and for linear MDOF systems,

$$S_{Di} = \frac{1}{\sqrt{1+4\xi_i^2}} \left(\frac{T_i^2}{4\pi^2} \right) S_{Ai} \, g \quad (m) \tag{5.78b}$$

5.4.1.2 Necessity of Additional Design Parameters

The above-mentioned feature of design, however, will vary when a larger amount of damping is added to an MDOF structure. As mentioned previously, Equations 5.75 and 5.78 are used to simplify the expression of accelerations for design purposes, particularly when damping is small. As mentioned in Chapter 4, for structures with added damping, Equations 5.71 and 5.73 must be used with caution. That is,

$$\mathbf{v}_L = \begin{Bmatrix} V_{L1} \\ V_{L2} \\ \cdots \\ V_{Ln} \end{Bmatrix} \neq \begin{bmatrix} 1 & 1 & \cdots & 1 \\ 0 & 1 & \cdots & 1 \\ & & \cdots & \\ 0 & 0 & 0 & 1 \end{bmatrix} \begin{Bmatrix} f_{L1} \\ f_{L2} \\ \cdots \\ f_{Ln} \end{Bmatrix} \quad (N) \tag{5.79}$$

or

$$V_{Lj} \neq \sum_{k=1}^{j} f_{Lk} \tag{5.80}$$

Example 5.5

In order to see this point, the following simple linear example, with given mass and stiffness, is considered, as follows:

$$\mathbf{M} = 10^7\mathbf{I}(\text{kg}),$$

$$\mathbf{K} = 10^{10}\begin{bmatrix} 2 & -1 & 0 \\ -1 & 2 & -1 \\ 0 & -1 & 1 \end{bmatrix} \quad (\text{N/m})$$

To have a 1% damping of all the three modes, the proportional damping matrix is

$$\mathbf{C}_1 = 10^6\begin{bmatrix} 8.567 & -2.512 & -0.546 \\ -2.512 & 8.021 & -3.058 \\ -0.546 & -3.058 & 5.510 \end{bmatrix} \quad (\text{N-s/m}).$$

Note that when theoretically increasing the proportional damping matrix to have a 10% damping ratio per mode, and assuming that this new damping matrix $\mathbf{C}_2$ can be realized, it is given as

$$\mathbf{C}_2 = 10\mathbf{C}_1$$

Suppose a linear viscous damper with c = 1.606 × 10^8 (N-s/m) is installed between the first and third floors, which increases the total damping ratio of the first mode to 10%. The added damper forms the damping matrix C_a,

$$\mathbf{C}_a = 10^8\begin{bmatrix} 1.606 & 0 & -1.606 \\ 0 & 0 & 0 \\ -1.606 & 0 & 1.606 \end{bmatrix} \quad (\text{N-s/m}).$$

That is, the newly formed damping matrix $\mathbf{C}_3$ is

$$\mathbf{C}_3 = \mathbf{C}_1 + \mathbf{C}_a$$

If the value of the added damping $\mathbf{C}_a$ is doubled, then another newly formed damping matrix:

$$\mathbf{C}_4 = \mathbf{C}_1 + 2\mathbf{C}_a$$

will increase the damping ratio of the first mode to 14.67%.

Now, the total base shear and total interstory shears of the aforementioned four cases with damping matrices $\mathbf{C}_1$, $\mathbf{C}_2$, $\mathbf{C}_3$, and $\mathbf{C}_4$, respectively, are calculated through numerical simulations under El Centro earthquake excitation. As a comparison, Equations 5.79 and 5.74 are also used to compute the results. The comparison is given in Table 5.2.

In Table 5.2, the term "Exact" means the exact maximum shear force calculated through the time history analysis. The term "Assu." means the maximum shear force calculated through the time history at the time point where the total base shear reaches the maximum value. It is found that when using damping matrix $\mathbf{C}_1$, whose damping ratios are 1% and proportional, the shear force is in error at the second floor by –4.72% and at the third floor by –13.42%. Both are not considerably smaller than the exact values. When using damping matrix $\mathbf{C}_2$, whose damping ratios are 10% and still proportional, the shear force at the third floor is in error by –22.77%, which is

TABLE 5.2
Base Shear Comparisons

		1st Floor		2nd Floor		3rd Floor	
		V_{L1} (N)	Time (s)	V_{L2} (N)	Time (s)	V_{L3} (N)	Time (s)
C_1	Exact	4.33e8	5.03	3.38e8	5.70	2.01e8	5.70
	Assu.	4.33e8	5.03	3.22e8	5.06	1.74e8	5.06
	Error			–4.72%		–13.42%	
	Cal.	4.33e8		3.47e8		1.93e8	
	Error			2.78%		–3.86%	
C_2	Exact	1.87e8	5.06	1.44e8	2.34	0.88e8	2.35
	Assu.	1.87e8	5.06	1.38e8	5.06	0.68e8	5.06
	Error			–3.97%		–22.77%	
	Cal.	1.87e8		1.50e8		0.83e8	
	Error			4.17%		–5.37%	
C_3	Exact	1.85e8	2.34	1.43e8	2.34	0.84e8	2.35
	Assu.	1.85e8	2.34	1.43e8	2.34	0.83e8	2.34
	Error			–0.31%		–1.34%	
	Cal.	1.85e8		1.48e8		0.82e8	
	Error			3.26%		–2.67%	
C_4	Exact	1.71e8	2.34	1.24e8	2.34	0.69e8	2.35
	Assu.	1.71e8	2.34	1.23e8	2.34	0.66e8	2.34
	Error			–0.39%		–3.47%	
	Cal.	1.71e8		1.37e8		0.76e8	
	Error			10.31%		10.43%	

somewhat large. When using damping matrix $\mathbf{C}_3$, whose damping ratio is still 10% for the first mode but becomes nonproportional, the shear force error at the third floor is –1.34%. Finally, when using damping matrix $\mathbf{C}_4$, whose damping ratio is 14.67% for the first mode and remains nonproportional, the error of shear force at the third floor increases to –3.47%. This value is considerably smaller than the exact shear force.

In Table 5.2, "Cal" denotes the maximum shear force calculated by Equation 5.73, which is the conventional calculation suggested by building codes. It is found that when using damping matrix $\mathbf{C}_1$, the shear force error at the second floor is 2.78% and at the third floor is –3.86%. Both are negligible. When using damping matrix $\mathbf{C}_2$, the shear forces of both floors are still somewhat close to the exact values. When using damping matrix $\mathbf{C}_3$ with nonproportional damping, the errors in shear force at the second and third floors are 3.26% and –2.67%, respectively. When using damping matrix $\mathbf{C}_4$, the shear force errors at the second and third floors increase to 10.31% and 10.43%, respectively.

Note that this is only an example. In actual structures, the errors using Equation 5.73 may vary from one case to another. However, the tendency of increased damping to cause the exact maximum floor forces and shear forces to differ from the calculated value can be seen through statistical surveys. More importantly, when the damping becomes nonproportional and larger, this phenomenon becomes worse.

The reasons are as follows: First, Equation 5.73 means that the maximum shear force is the summation of the maximum lateral forces. Second, this equation actually implies that all the peak forces occur at exactly the same time. In most cases for an MDOF system, this assumption does not hold, especially for heavily damped structures, the maximum floor forces cannot be reached at the exact time point. This phenomenon can be clearly seen from the columns of time in Table 5.2, which indicates that the peak values can occur at different time points, even for proportionally damped systems. In addition, for nonproportionally damped system, even within the same mode, peak values may occur at different time points. In this circumstance, the summation of these forces to form the column shear loses its validity. Thus, when the damping ratio becomes high, the corresponding errors can easily be more than 50%.

In addition, by carefully analyzing the lateral forces of each floor, it is realized that these forces are not the product of the floor mass times the pseudo acceleration, but rather the real absolute acceleration. In Chapter 2, it was shown that the spectrum of real acceleration and the pseudo acceleration can be quite distinct when damping becomes significantly large.

In other words, when the damping becomes larger, an additional parameter is needed to describe the dynamic behavior of a system. In fact, in Chapter 4, this phenomenon was explained mathematically. In the case of engineering applications, it can be interpreted as the requirement for an additional design parameter. For example, when large damping is added to a structure, while the displacement always decreases, the absolute acceleration may increase. However, if the formula of pseudo acceleration is used as described in Equations 5.76 and 5.78, it can be found that the values of the pseudo acceleration will always be reducing.

Note that in the above example, the damping force has not yet been counted. Otherwise, when the damping and the restoring forces cannot be separated, the actual shear force will be even larger. When the damping is large, this problem becomes notably worse.

In addition, for the problem of damper design, if the damping is sufficiently high, the structure needs two design parameters per mode. In other words, two design spectra are needed: the displacement spectrum and the real acceleration spectrum. In Chapter 1, equations are presented to relate the displacement spectrum and the real acceleration spectrum. However, these equations were developed based on a specific group of earthquake records. This is why there is a notable difference between the equations generated through the 99 earthquake records, and Mohraz and Sadek's (2001) work. In practical applications, local earthquake histories should be considered in order to establish more accurate real spectra.

5.4.2 Issues of Damping Ratios

In linear systems, the damping ratio is a well-defined parameter, which is dimensionless and independent of other basic parameters, such as natural frequencies and mode shapes. The damping ratio is also not affected by the input level. Therefore, the damping ratio can be a solid design parameter. Note that when a damping coefficient is given, which is roughly fixed when damping devices are installed, the damping ratio will be affected by the physical parameters of the mass and stiffness. In a nonlinear system, the effective damping will be influenced by additional parameters, such as the effective natural frequencies, as well as input levels. In this case, further care must be taken when using the damping ratio as a design parameter.

The damping ratio is discussed further, as it is one of the most important control parameters. It is known that frequency-independent and frequency-dependent damping are also the characteristics of the entire system. When exponent α of ω_n is zero, both the damping force and the damping ratio will not be functions of the natural frequency. In this case, the variation mass of a structure will have no effect on the damping ratio. Otherwise, increasing the mass will decrease the value of the damping ratio.

In the previous section, it was shown that the damping ratio expressions for certain dampers do not contain the term for natural frequency, whereas they do for other dampers. For those that contain a natural frequency term, the corresponding damping ratio will be influenced by the frequency, and the degree of influence will depend on exponent α.

As the simplest example of a damping ratio that is not affected by the natural frequency, idealized friction damping is first considered by reexamining Equation 5.20. That is,

$$\xi_{eff} = \frac{2c_{eq}}{\pi k x_0} \tag{5.81}$$

in which the damping ratio is not a function of the natural frequency ω_n.

Note that $\omega_n = (k/m)^{1/2}$, thus the natural frequency is a function of mass; therefore, for the friction damping, it is realized that the damping ratio will not be affected by any change of the mass.

In real applications, it is well known that certain structures can have a significant amount of variable mass. For example, the mass of a bridge for a railway train can increase several times when a train is passing. Therefore, studying the influence of mass can be helpful.

The damping ratio from Equation 5.81 is not a function of mass. This is because the damping force, in fact, the friction force, is not a function of ω_n. That is,

$$f_{d\max} = c_{eq}\,\mathrm{sgn}(v) \quad (N) \tag{5.82}$$

Note that the amplitude of the damping force can be generally expressed as

$$f_{d\max} = c_{eq}\omega_n^{\alpha}x_0^{\beta} \quad (N) \tag{5.83}$$

in which exponent α plays an important role. If

$$\alpha = 0, \tag{5.84}$$

the corresponding damping ratio will not be a function of the mass.

Similarly, it is known that the damping forces of both the bilinear damping and roughly the VE damping are not functions of ω_n. That is, Equation 5.84 is satisfied by both cases. Thus, these damping types can be classified as *frequency-independent damping.*

It is also noticed that for all three of these classes of damping, the damping ratios are only functions of the stiffness. That is,

$$\xi_{eff} \propto \frac{1}{k} \tag{5.85}$$

When supplemental damping is installed, the total stiffness of a structure will not often be greatly affected. Therefore, no matter how many frequency-independent dampers are installed, the damping ratio will not be affected by the factor given by Equation 5.85.

It is worth mentioning that certain viscoelastic materials can have a large amount of restoring modulus, which will change the total stiffness. In this case, the damping ratio will be smaller than when the restoring modulus is small.

It is also worth noting that certain viscoelastic materials can be significantly influenced by working frequencies. In this case, Equation 5.83 will be applied and the corresponding dampers are no longer frequency independent.

As seen in Equation 5.83, if exponent α is not zero, the damping force will be a function of the natural frequency ω_n, as will the damping ratio. The corresponding damping will become *frequency-dependent damping.* In the following discussion, several cases of frequency-dependent damping are examined.

Linear viscous damping is considered first. In this case, the amplitude of the damping force is

$$f_{d\max} = c\omega_n x_0 = c x_0 \frac{\sqrt{k}}{\sqrt{m}} \quad (N) \tag{5.86}$$

and the damping ratio can be written as

$$\xi = \frac{c}{2\omega_n m} = \frac{c}{2\sqrt{mk}} \tag{5.87}$$

Therefore, the damping ratio is inversely proportional to the square root of the mass. That is, if the mass is doubled, then the damping ratio will be reduced by a factor of approximate 40%. For structures with variable mass, such as the aforementioned bridge or a movie theater with a near full load of people, care must be taken when dampers are used to reduce the seismic-induced vibration. Note that in this case, when the mass increases, the damping force decreases if the maximum displacement remains unchanged.

Now, for the general case of nonlinear viscous damping, the damping force is

$$f_{d\,max} = c\omega_n^{\alpha} x_0^{\beta} = c x_0^{\beta} k^{\alpha/2} m^{-\alpha/2} \quad (N) \tag{5.88}$$

From Equation 5.88, it is seen that the damping force is affected by the mass and exponent α. When the displacement is fixed, decreasing exponent α will increase the amplitude of the damping force. Increasing the mass will decrease the damping force.

The damping ratio can be written as

$$\xi_{eff} = \frac{c_{eq}\omega_n^{\alpha} x_0^{\beta-1} A_{\beta}}{2\pi k} = \frac{c_{eq} x_0^{\beta-1} A_{\beta}}{2\pi} k^{(\alpha/2)-1} m^{-\alpha/2} \tag{5.89}$$

From Equation 5.89, the damping ratio is also affected by the mass and exponent α. When the displacement is fixed, increasing exponent α will decrease the damping ratio. Increasing the mass will also decrease the damping ratio.

It is worth mentioning that the nonlinear hydraulic damper, which provides nonlinear viscous damping, has recently become more popular. Most vendors do not provide users with the exponent for the frequency. Instead, the exponent of velocity is used. For the reasons discussed in this chapter, it is suggested that vendors should measure exponents α and β separately, so that engineers can select the proper parameters and design their structure more safely.

5.5 DAMPING FORCE–RELATED ISSUES

The concept of recoverable damping from Lazan (1968) can be further described by two types of relationships between pure stiffness and pure damping. The first is that the stiffness and damping are in series. Thus, when a damper is installed in a structure, it needs a certain supporting stiffness, the value of which is usually considered to be finite. When damping is nonrecoverable, the supporting stiffness does not contribute to withstanding the static load. Finite supporting stiffness will limit the damping capability and can change a viscous damper into a VE damper. Therefore, this issue should not be overlooked. A special state matrix can be used to model the integration of a structure and its dampers with finite stiffness support. Roughly speaking, every 10% of the damping ratio designed into a structure requires a supporting stiffness of about 100% of the structural stiffness. The second type of relationship between stiffness and damping is related to the parallel configuration, as it defines the capability of a structure with added dampers to be self-centered.

5.5.1 Recoverable Damping and Self-Centering Systems

For civil engineering structures, the primary consideration is static load. Damping is mainly designed to withstand the dynamic force, but most dampers need supporting stiffness. In Figure 5.14, a damper c needs to be supported by a stiffness k_s, other than the structural stiffness k. Thus, it is necessary to know if the stiffness can withstand the dead load. A practical thought is, if the needed supporting stiffness must be very strong, why bother using dampers? Why not simply increase the stiffness? These questions will be answered in Section 5.5.3.

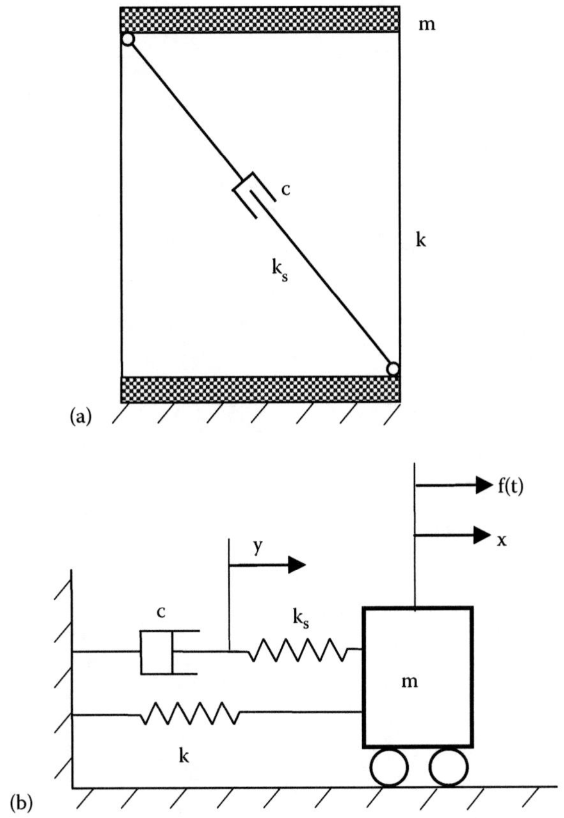

FIGURE 5.14 (a, b) Supporting stiffness for dampers.

5.5.1.1 Recoverable Damping

The third and last important classification of damping recommended by Lazan rates the damping on the basis of *recoverability*, which is shown in Figure 5.15. In the following, it is further noted that nonlinear damping makes the system nonrecoverable dynamically and nonlinear stiffness results in permanent displacement.

It is seen that with the same amplitudes of the force and deformation, nonrecoverable damping can have considerably larger energy dissipation. However, recoverable damping can contribute to system stiffness. This is because for recoverable damping, when the external force drops to zero, the corresponding displacement returns back to its center position. Whereas for nonrecoverable damping, there will be an offset, d_0. This implies that the nonrecoverable damping cannot contribute static force. In other words, when the driving frequency tends to zero, there will be no force. Neither damping force nor spring force will exist. Therefore, a nonrecoverable damping device cannot be used to support static load. It thus also indicates that all the supporting stiffness cannot be used to support static load.

On the other hand, recoverable damping contributes static stiffness and can withstand the dead load.

The most popular dampers in use today provide only nonrecoverable damping. The recoverable mechanism of the structure is to be centered, when the external force is canceled due to its stiffness. In this case, the integration of the structure and the dampers contributes recoverable damping.

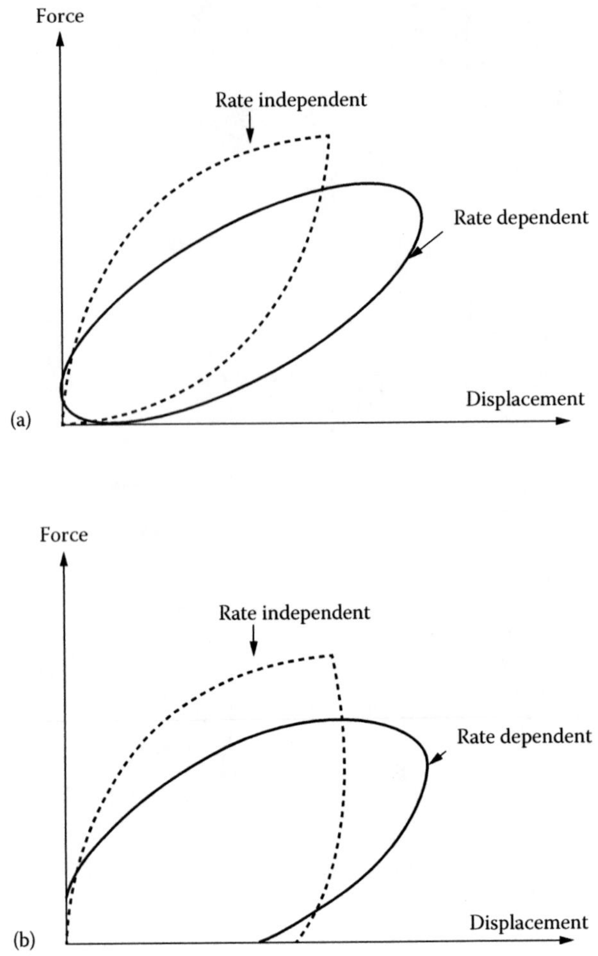

FIGURE 5.15 Damping classifications: (a) recoverable and (b) nonrecoverable.

From the above discussion, it is seen that the concept of recovery is important for damper and structure designers. However, another significant factor not explicitly included in the concept of damping recovery is the rate of recovery. For example, at a working frequency of, i.e., around 2 (Hz), the VE damper has a notable offset, d_0. However, at about zero driving frequency, it contributes a low recovery force.

Note that a fluid damper will never provide a recoverable (or more precisely, restoring) force. In this sense, at a different driving frequency, the VE damping may contribute some stiffness, but fluid dampers will not. This factor must be included in the damper design.

Furthermore, different types of rate-dependent dampers will have different energy dissipation levels at different driving frequencies. Together with its capacity for recovery, there should be a clear understanding of its supporting stiffness. For example, fluid dampers can be used to reduce wind-induced vibration. Since the driving frequency is low, a very small orifice of the oil path is necessary to create large damping. If the dampers are also used for earthquake response reduction, the supporting stiffness must be stronger than that required for wind engineering only, since the driving frequency of earthquake excitation is considerably higher. In Section 5.5.3, while discussing specific dampers, a more detailed explanation of how to choose the supporting stiffness is given.

5.5.1.2 Self-Centering Structures

Generally speaking, a civil engineering structure that did not completely collapse following an earthquake can have different configurations. If the vibration of the structure is always in the elastic range, then it will not have permanent deformation.

However, if the vibration of the structure enters the inelastic range, there will be a change in configuration associated with permanent deformation. In many cases, although a structure has inelastic deformations, the remaining stiffness is still sufficient to withstand the vertical and horizontal loads. In other words, the structure can still be in-service. However, if it has a large permanent deformation, the owner either has to consider a major repair or replacement of the structure. Thus, self-centering is an important consideration in the seismic design of structures.

From the viewpoint of structural restoring forces, the capability of structural recentering can be explained by noting that when the external force is decreased to zero, the deformation will also approach zero. To study the concept of self-centering, two situations should be examined.

First, when the structure reaches its center position, the restoring force must be equal to zero and, when the restoring force equals zero, the structure must return to its center position. That is,

$$\begin{aligned} x &\to 0 \\ f &\to 0 \end{aligned} \tag{5.90}$$

This condition can be graphically described by Figure 5.15a, namely, the case of recoverable damping. In practice, there can be various types of the recoverable damping, as shown in Figure 5.16.

In Figure 5.16a and d, the typical rate-dependent and rate-independent self-centering damping is shown, respectively. In Figure 5.16b and e, the damping with a symmetric stiffness is shown, while in Figure 5.16c, the plot indicates asymmetric damping and stiffness.

The second situation is that when the driving frequency is not close to zero, the above-mentioned criterion in Equation 5.80 is not satisfied. However, when the driving frequency ω_f approaches zero at the end of the earthquake excitation, the deformation also approaches zero. That is,

$$\begin{aligned} \omega_f &\to 0 \\ x &\to 0 \end{aligned} \tag{5.91}$$

In Figure 5.16f, typical viscous damping with symmetric stiffness is shown. When the velocity of the structure reaches zero, the damping force will also reach zero, thereby satisfying Equation 5.91. That is, this structure also has a self-centering capability. In a later section the issue of driving frequency is examined in detail.

5.5.1.3 Relation between Stiffness and Damping

In Figure 5.14b, the damping c has a relationship with two springs, k_s and k. It is seen that the spring k is *in series* with the damping, whereas the spring k_s is *in parallel* with the damping. The two types of relationships have a general meaning. In a later section, the case of damping in series with stiffness is discussed in more detail. Here, the case of damping and stiffness is considered in parallel.

When dampers are added to a structure, it is known that the additional damping is in parallel with the original stiffness of the structure.

In the case of inelastic deformation in vibration cycles, the restoring force and dissipative force can be treated separately. That is,

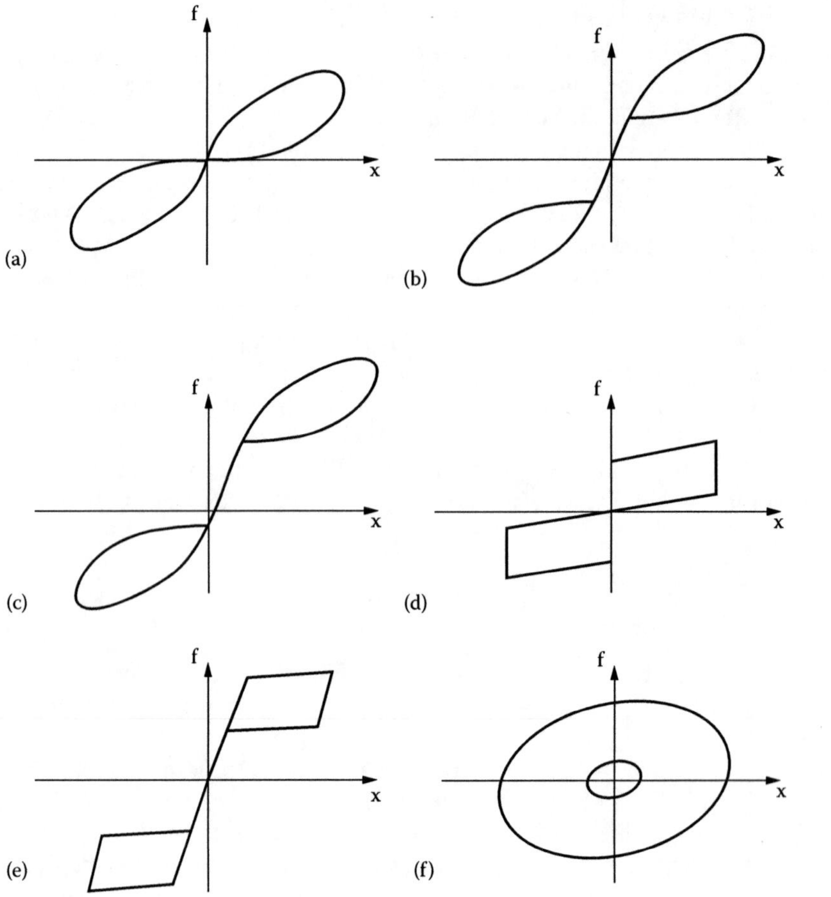

FIGURE 5.16 (a–f) Self-centered structures.

$$f(t) = f_d(t) + f_c(t) \tag{5.92}$$

Different from the situation that separates the conservative and dissipative force at the maximum displacement, Equation 5.92 describes the situation at any possible displacement. Thus, lowercase subscripts d and c are used to denote the general situation. Apparently, when the displacement is close to the center position, the force separation can be written as

$$f_0 = f_{0d} + f_{0c} \tag{5.93}$$

where subscript 0 stands for the close neighborhood of the center position. In practice, the word close can be quantified with certain values, e.g., one thousandth of the height of a structural story (see Figure 5.15a, the solid line for example).

Within this neighborhood, if the following criterion is satisfied, the structure can be treated as a self-centering system:

$$f_{0d} < f_{0c} \tag{5.94}$$

Therefore, this case can also be treated as a pure damping loop in parallel with a given stiffness.

5.5.2 Working Frequency and Temperature of Dampers

In many cases, the working frequency will affect the damping force, thereby influencing the issue of structural self-centering and affecting the capability of added dampers to reduce the vibration.

5.5.2.1 Effect on Self-Centering

Using Equation 5.91, the second criterion for a self-centering structure is provided. This criterion relates to the specific driving frequencies that appear close to the end of the earthquake. In Figure 5.16, although several types of self-centering structures are shown, the only type that relates to the driving frequency is the one with linear viscous damping. However, in practice, there are other nonlinear types of damping that can also ensure self-centering. Generally speaking, if the damping force can be expressed as

$$f_{SC} = f_{d\omega}(\omega) f_{dn} \tag{5.95}$$

then, after earthquake excitations, namely, when the driving frequency approaches zero, the added damping force will become zero and self-centering is ensured. Here, f_{SC} is the generic damping force and f_{dn} is the amplitude of the damping force that is not a function of frequency. Furthermore, $f_{d\omega}$ is the portion of the damping force that is affected by the frequency and is defined as

$$f_{d\omega}(\omega) = \begin{cases} f(\omega), & \omega > 0 \\ 0, & \omega = 0 \end{cases} \tag{5.96}$$

where f(ω) stands for any possible function of the frequency.

On the one hand, Equations 5.95 and 5.96 are equivalent to the conventional concept of velocity-dependent damping. When the earthquake excitation ends, the velocity of ground motion reaches zero, and the velocity-dependent damping will no longer exist. Then, the damping will become zero.

On the other hand, Equations 5.95 and 5.96 provide a different perspective with which to check the effect of damping on self-centering. In many cases, when the excitation ends, a damper can have a permanent displacement, which may or may not create a residual force, depending on whether the condition described in Equation 5.96 is satisfied.

5.5.2.2 Frequency-Dependent and Temperature-Dependent Damping

Many damping materials and/or damping mechanisms are frequency dependent. There are two types of frequency-dependent damping. The first one is when the damping itself is frequency dependent. Strictly speaking, most damping materials are frequency dependent. However, earthquake excitations and corresponding structural responses do not have a very large frequency band. The variation of the damping properties of many damping materials due to the change of working frequency may not be notably large within such a narrow band. In this case, the issue of frequency-dependent damping is often ignored. In addition, the function of the damping property vs. the driving frequency can be rather complex. For the same reason as the narrow frequency band, for many frequency-dependent damping materials, the approach described below can be used. That is, the damping force is assumed to be proportional to certain exponents of the frequency, such as

$$f_d \propto \omega_n^{\alpha} \tag{5.97}$$

Viscous fluid damping, due to the existence of a certain amount of air inside the fluid, often has the above behavior.

Other types of damping, such as VE damping, are often described by loss factors. Usually, the loss factor η can be expressed as concave curves due to the change in frequency. One of the simplest functions can be expressed as

$$\eta = a\omega^{\alpha} + b\omega + c \tag{5.98}$$

where a, b, c, and α are parameters to be determined by the specific type of damping.

For example, Figure 5.17 conceptually shows the typical frequency responses of viscoelastic materials. Here, the fine solid line is the storage modulus and the dotted line is the loss modulus, where both represent shear modes. Meanwhile, the thick solid line is the loss factor. It is seen that when the frequency increases, the storage modulus and the loss modulus will vary accordingly. Since the loss factor is the ratio between the two, this factor will also vary according to the working frequency.

Note that both the working frequency and the temperature can influence the storage and loss modulus of VE damping materials in a similar way as shown in Figure 5.17, where the thick, thin, solid, and dotted lines conceptually represent different types of curves. That is, when both the frequency and the temperature are varied, the loss factor will first increase and then decrease. In Figure 5.17, the curves of the frequency vs. the modulus and loss factor are recorded when the frequency increases. A similar shape of curves can be obtained when the temperature decreases, for a broad class of thermorheologically simple viscoelastic materials.

In this case, an alternative type of equation is used to express the relationship between the damping and the temperature. For example, for commonly used viscoelastic materials,

$$\ln(\beta) = aT + b \tag{5.99}$$

where a and b are parameters determined for specific materials and T is the temperature in degrees Celsius. For example, in the range between 21°C and 40°C, for a specific viscoelastic material (Shen and Soong 1995),

$$\ln(\beta) = -0.0561T + 1.218 \tag{5.100}$$

The second type of frequency-dependent damper is the situation when a damper needs to be supported and the supporting stiffness does not have an infinite value. In the next section, the concept and influence of supporting stiffness are discussed in more detail. And the fact that finite supporting

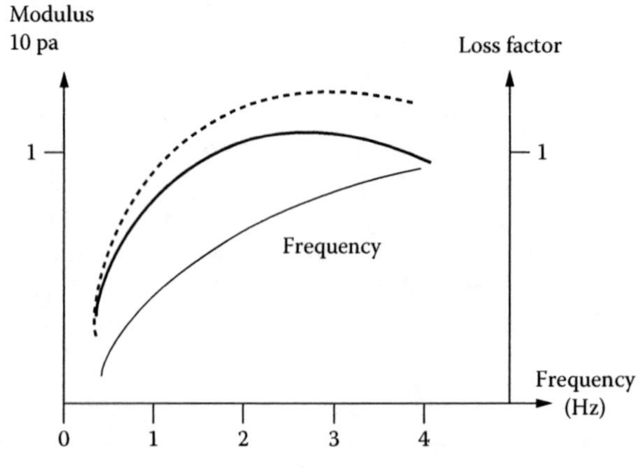

FIGURE 5.17 Example of frequency-dependent damping.

stiffness of a damper changes the damping and resonant frequency of a structure, although the damper itself may have a frequency-independent response, is addressed.

5.5.3 Damper Design Considering Supporting Stiffness

In the following discussion, two related issues are considered: modeling of the structure with limited supporting stiffness and the geometrical magnification factor for damper installations.

5.5.3.1 Modeling

The supporting or connecting stiffness of a damper plays an important role in damper design, which should not be ignored. In order to have a better design, modeling the damper with the supporting stiffness becomes critical.

First, with finite supporting stiffness, the effectiveness of a damper is more limited than with infinite supporting stiffness. In certain circumstances, e.g., when the supporting stiffness is decreased from a very large value to about five times the structural stiffness, even if the damping ratio continues to increase, the response will become greater, not smaller. Note that this relates only to an SDOF example. The situation for an MDOF system becomes more complex. The standard Laplace transform approach becomes difficult and a more useful modeling approach involves the state equation and the state matrix. A small mass of the supporter can be added for an additional degree-of-freedom to create the regular state matrix. However, this method increases the number of equations and sometimes the state matrix becomes stiff owing to the difference between the large mass of the structure and the small mass of the support. In the following discussion, different state variables are used to yield a smaller state matrix. In this case, the state variables become

$$\boldsymbol{X} = \begin{Bmatrix} \dot{x} \\ x \\ y \end{Bmatrix} \tag{5.101}$$

This results in

$$\dot{\boldsymbol{X}} = \mathcal{A}\boldsymbol{X} + \mathcal{B}\mathbf{f} \tag{5.102}$$

where

$$\mathcal{A} = \begin{bmatrix} 0 & -\frac{k+k_s}{m} & \frac{k_s}{m} \\ 1 & 0 & 0 \\ 0 & \frac{k_s}{c} & -\frac{k_s}{c} \end{bmatrix} \tag{5.103}$$

and

$$\mathcal{B} = \begin{bmatrix} 1, & 0, & 0 \end{bmatrix}^T \tag{5.104}$$

It is seen that the state matrix has an additional dimension. Other methods can be used to account for the stiffness instead of the state equations approach. That is, to further explore the issue of damper supporting stiffness, the discussion can be started from a single mass system. Now, consider the linear modeling of a damper with supporting stiffness, as shown in Figure 5.18.

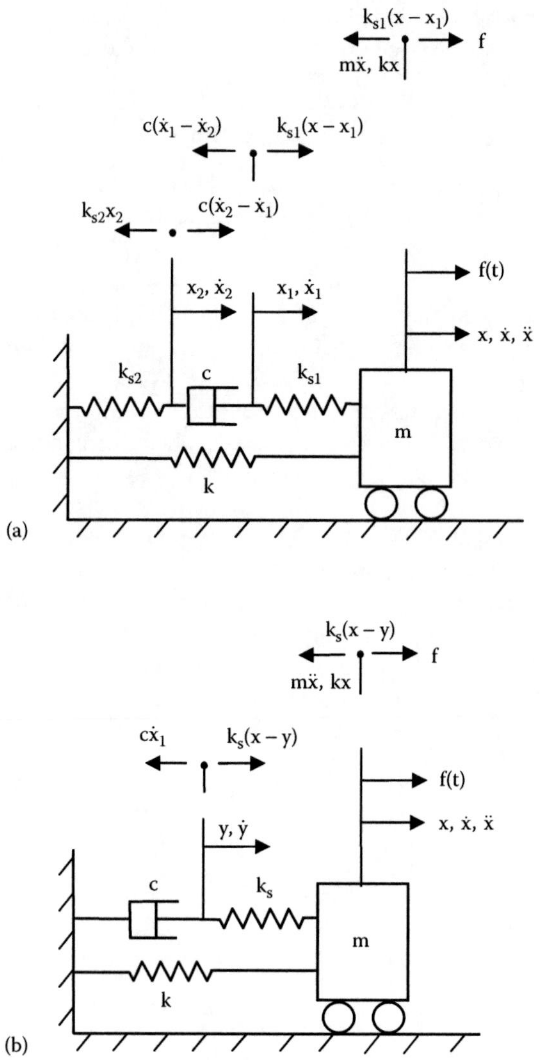

FIGURE 5.18 Modeling of linear system with finite damper supporting stiffness. (a) two supporting members. (b) one supporting member.

Based on the force relationship shown in Figure 5.18a, the following set of equations result:

$$k_{s2}x_2 = c\left(\dot{x}_1 - \dot{x}_2\right) \tag{5.105}$$

$$c\left(\dot{x}_1 - \dot{x}_2\right) = k_{s1}\left(x - x_1\right) \tag{5.106}$$

$$m\ddot{x} + k_{s1}\left(x - x_1\right) + kx = f \tag{5.107}$$

Here, k_{s1} and k_{s2} are the supporting stiffness.

Equations 5.105 and 5.106 can be rewritten as

$$x_2 = \frac{c}{k_{s2}}(\dot{x}_1 - \dot{x}_2) \tag{5.108}$$

and

$$x - x_1 = \frac{c}{k_{s1}}(\dot{x}_1 - \dot{x}_2) \tag{5.109}$$

Therefore,

$$x - (x_1 - x_2) = c\left(\frac{1}{k_{s1}} + \frac{1}{k_{s2}}\right)(\dot{x}_1 - \dot{x}_2) \tag{5.110}$$

Equation 5.177 can be rewritten as

$$m\ddot{x} + kx + k_s(x - y) = f \tag{5.111}$$

Denote $y = (x_1 - x_2)$ and $k_s = k_{s1}k_{s2}/(k_{s1} + k_{s2})$.
Equation 5.110 becomes:

$$k_s(y - x) + c\dot{y} = 0 \tag{5.112}$$

If $k_{s2} = \infty$, then $k_s = k_{s1}$, $x_2 = 0$ and $y = x_1$; or if $k_{s1} = \infty$, then $k_s = k_{s2}$, $x_1 = x$ and $y = (x - x_2)$. For either way, Equations 5.105, 5.106, and 5.107 can be reduced to Equations 5.111 and 5.112 and the two cases are equivalent. The equivalent physical model is shown in Figure 5.18(b).

Applying Laplace transform to Equations 5.111 and 5.112 with zero initial conditions assumption yields

$$s^2mX + kX + k_s(X - Y) = F \tag{5.113}$$

$$k_s(Y - X) + scY = 0 \tag{5.114}$$

where X, Y, and F are the Laplace transforms of x, y, and f, respectively.
In addition, letting the Laplace variable $s = j\omega$ (as in steady state) and denoting $X_0(j\omega) = X(s)|_{s=j\omega}$, $Y_0(j\omega) = Y(s)|_{s=j\omega}$ and $F_0(j\omega) = F(s)|_{s=j\omega}$. Equations 5.113 and 5.114 become, respectively,

$$X_0 = \frac{k_s + jc\omega}{-jmc\omega^3 - mk_s\omega^2 + j(k + k_s)c\omega + kk_s}F_0 \tag{5.115}$$

and

$$Y_0 = \frac{k_s}{k_s + jc\omega}X_0 \tag{5.116}$$

From Equation 5.115, it is seen that the system has an order of cubic. Only if $k_s = \infty$ is it reduced to the quadratic form:

$$X_0 = \frac{1}{m\left(-\omega^2 + j\frac{c}{m}\omega + \frac{k}{m}\right)} F_0 \tag{5.117}$$

which is the case of the commonly accepted formula to calculate the peak value of the displacement.

5.5.3.2 Approximation

In order to further analyze the details of the supporting stiffness in linear systems, the complete form of the frequency response function $H(j\omega) = X_0(j\omega)/F_0(j\omega)$ is considered. Based on Equation 5.115, $H(j\omega)$ can be expressed as

$$H(j\omega) = \frac{1}{-m\omega^2 + j\omega\frac{ck_s^2}{(c\omega)^2 + k_s^2} + \left[k + \frac{(c\omega)^2 k_s}{(c\omega)^2 + k_s^2}\right]} \tag{5.118}$$

Comparing Equation 5.118 with Equation 5.117, the equivalent stiffness k_{eq} and equivalent damping coefficient c_{eq} can be defined as

$$k_{eq} = k + \frac{(c\omega)^2 k_s}{(c\omega)^2 + k_s^2} \tag{5.119}$$

$$c_{eq} = \frac{ck_s^2}{(c\omega)^2 + k_s^2} \tag{5.120}$$

From the above equations, it is seen that both the equivalent stiffness and the equivalent damping are functions of the excitation frequency. If the structure is excited by a nonharmonic force signal, then the frequency ω will vary with time. Therefore, it can be quite difficult to use in practical damper design.

In engineering practice, however, to simplify the analysis, the equivalent structural natural circular frequency ω_{eq} can be estimated from Equation 5.119 first, and then c_{eq} and the corresponding equivalent damping ratio ξ_{eq} can be calculated. These parameters are then used in the simplified design and the "equivalent" quantities can be kept as time invariant in time response analysis. Simulation results indicate that the analysis errors are in the acceptable range. The procedure to determine these quantities is presented next.

First, dividing both sides of Equation 5.119 by m, and replacing ω by ω_{eff},

$$\frac{k_{eq}}{m} = \frac{k}{m} + \frac{\frac{(c\omega_{eq})^2}{m^2}\frac{k_s}{m}}{\frac{(c\omega_{eq})^2 + k_s^2}{m^2}} \tag{5.121}$$

In order to define the operations more completely, the following variables are denoted. The "original" natural frequency for the cubic equation is denoted as

$$\omega_0 = (k/m)^{1/2} \tag{5.122}$$

the original damping ratio for the cubic equation is represented by

$$\xi_0 = \frac{c}{2m\omega_0} \tag{5.123}$$

the equivalent natural frequency is

$$\omega_{eq} = \left(k_{eq}/m\right)^{1/2} \tag{5.124}$$

the equivalent damping ratio is

$$\xi_{eq} = \frac{c_{eq}}{2m\omega_{eq}} \tag{5.125}$$

the stiffness ratio is defined as

$$\gamma = k_s/k \tag{5.126}$$

the frequency ratio is

$$\lambda = \omega_{eq}/\omega_0 \tag{5.127}$$

while the ratio of the damping ratios is

$$\mu = \xi_{eq}/\xi_0 \tag{5.128}$$

The square of the frequency ratio is also denoted as

$$\nu = \lambda^2 \tag{5.129}$$

Using the above notations,

$$\omega_{eq}^2 = \omega_0^2 + \frac{\left(2\xi_0\omega_0\omega_{eq}\right)^2\left(\gamma\omega_0^2\right)}{\left(2\xi_0\omega_0\omega_{eq}\right)^2 + \left(\gamma\omega_0^2\right)^2} \tag{5.130}$$

Dividing both sides of Equation 5.130 by ω_0^2,

$$\left(\frac{\omega_{eq}}{\omega_0^2}\right)^2 = 1 + \frac{4\xi_0^2\gamma\left(\frac{\omega_{eq}}{\omega_0^2}\right)^2}{4\xi_0^2\gamma\left(\frac{\omega_{eq}}{\omega_0^2}\right)^2 + \gamma^2} \tag{5.131}$$

or write

$$\nu = 1 + \frac{4\xi_0^2\gamma\nu}{4\xi_0^2\nu + \gamma^2} \tag{5.132}$$

Equation 5.132 can be further written as a quadratic equation about ν,

$$\left(4\xi_0^2\right)\nu^2 + \left[\gamma^2 - 4\xi_0^2(1+\gamma)\right]\nu - \gamma^2 = 0 \tag{5.133}$$

The roots of Equation 5.133 can be expressed as

$$\nu_{1,2} = \frac{-\left(\gamma^2 - 4\xi_0^2(1+\gamma)\right) \pm \sqrt{\left(\gamma^2 - 4\xi_0^2(1+\gamma)\right)^2 + \left(4\xi_0\gamma\right)^2}}{8\xi_0^2} \quad (\xi_0 \neq 0) \tag{5.134}$$

Since υ must be larger than zero and

$$\sqrt{\left[\gamma^2 - 4\xi_0^2(1+\gamma)\right]^2 + \left(4\xi_0\gamma\right)^2} \geq \left[\gamma^2 - 4\xi_0^2(1+\gamma)\right] \tag{5.135}$$

there must be only one reasonable root, which is

$$\nu = \frac{-\left[\gamma^2 - 4\xi_0^2(1+\gamma)\right] + \sqrt{\left[\gamma^2 - 4\xi_0^2(1+\gamma)\right]^2 + \left(4\xi_0\gamma\right)^2}}{8\xi_0^2} \quad (\xi_0 \neq 0) \tag{5.136}$$

Note that if $\xi_0 = 0$, which means that there is no damper in the undamped system, then we have $k_{eq} = k$, which will be simplified to $\nu = 1$.

After solving for ν, λ as well as c_{eq} and ξ_{eq} can be calculated. From Equation 5.129,

$$\lambda = \sqrt{\nu} \tag{5.137}$$

and

$$k_{eq} = \omega_{eq}^2 m = \left(\lambda\omega_0\right)^2 m \tag{5.138}$$

From Equations 5.121 and 5.125, the equivalent damping ratio can be calculated by

$$\xi_{eq} = \frac{\dfrac{ck_s^2/m^2}{\left[(c\omega)^2 + k_s^2\right]/m^2}}{2m\omega_{eq}} = \frac{\gamma^2\xi_0}{\lambda\left[\left(2\xi_0\lambda\right)^2 + \gamma^2\right]} \tag{5.139}$$

and

$$c_{eq} = 2\xi_{eq}\omega_{eq}m = 2\xi_{eq}\lambda\omega_0 m \tag{5.140}$$

Based on Equations 5.136, 5.139, and 5.132, the curves of the natural frequency ratio vs. stiffness ratio, equivalent damping ratio vs. stiffness ratio, and the ratio of damping ratios vs. stiffness ratio, are plotted in Figures 5.19 through 5.21, respectively with constant value of λ. From Figure 5.19, it is seen that if the support stiffness of the damper is equal to zero, the equivalent natural frequency is identical to the original structural natural frequency, which indicates that the damper does not function. Along with the increase of the support stiffness, the equivalent natural frequency will also increase at first, it will approach a peak, and then decrease gradually and

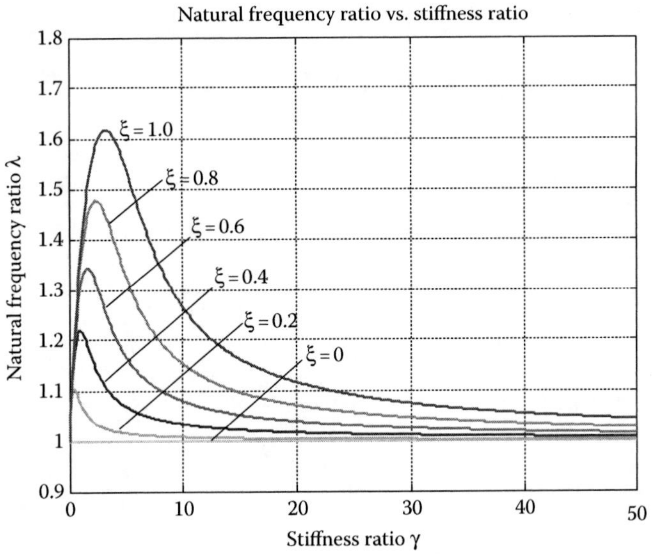

FIGURE 5.19 λ vs. γ.

tend to the original structural natural frequency. On the other hand, as the original damping ratio increases, the equivalent natural frequency will always increase.

From Figures 5.20 and 5.21, it is observed that as the support stiffness increases, the equivalent damping ratio will also increase from zero toward the original value. Additionally, the larger the original damping, the more slowly the equivalent value approaches the original one.

In order to examine the calculation accuracy, an SDOF vibration system is used as an example, based on Figure 5.18b. Consider the case with the following parameters: m = 20,000 (kg); k = 3.1583e + 006 (N/m); c = 3.0159e + 005 (N-s/m), and $k_1 = k$, from which we can obtain $f_0 = 2$ (Hz); $\omega_0 = 12.566$ (rad/s); $\xi_0 = 60\%$ and $\gamma = 1$, as well as $\omega_{eq} = 16.439$ (rad/s); $f_{eq} = 2.6163$ (Hz) and $\xi_{eq} = 13.239\%$.

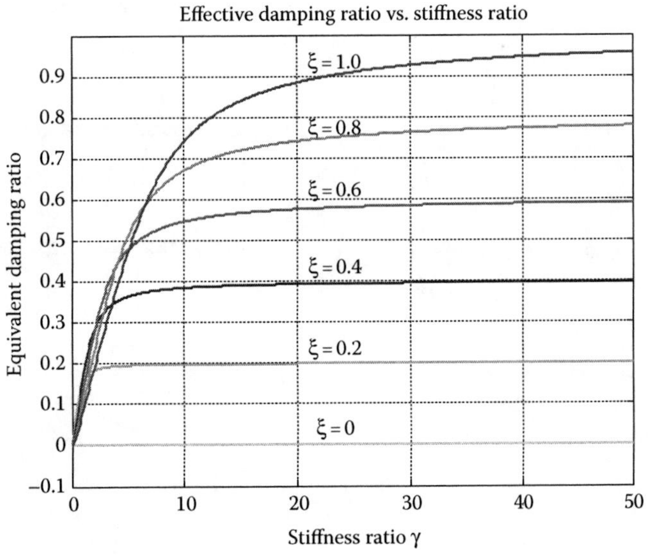

FIGURE 5.20 Effective damping ratio vs. stiffness ratio ($\gamma = 1.2$).

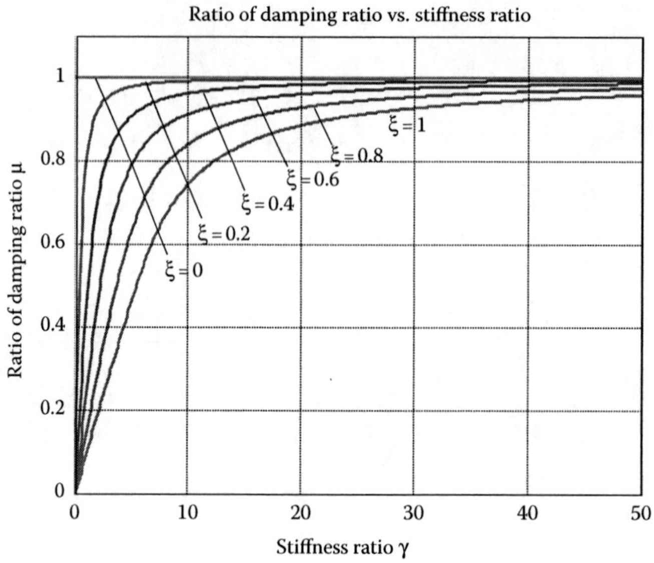

FIGURE 5.21 Ratio of damping ratio vs. stiffness ratio ($\gamma = 1.2$).

5.5.3.3 Generalized Supporting Stiffness

The concept of finite supporting stiffness can be extended to linear structures with the number of DOF different from the number of locations of installed dampers. Figure 5.22a and b show a two-story and a ten-story building, respectively. Both buildings have linear viscous dampers installed in the first story only. The damping of both cases is nonproportional. The frequencies and damping ratios of the corresponding modes are shown in Table 5.3A and B, respectively.

From Table 5.3, it is seen that with the dampers installed, the damping ratios of the first mode for both buildings are approximately 1%. A record from the Northridge earthquake with a peak acceleration equal to 0.36 (g) is used as an input to excite these buildings. Then, the value of the damping

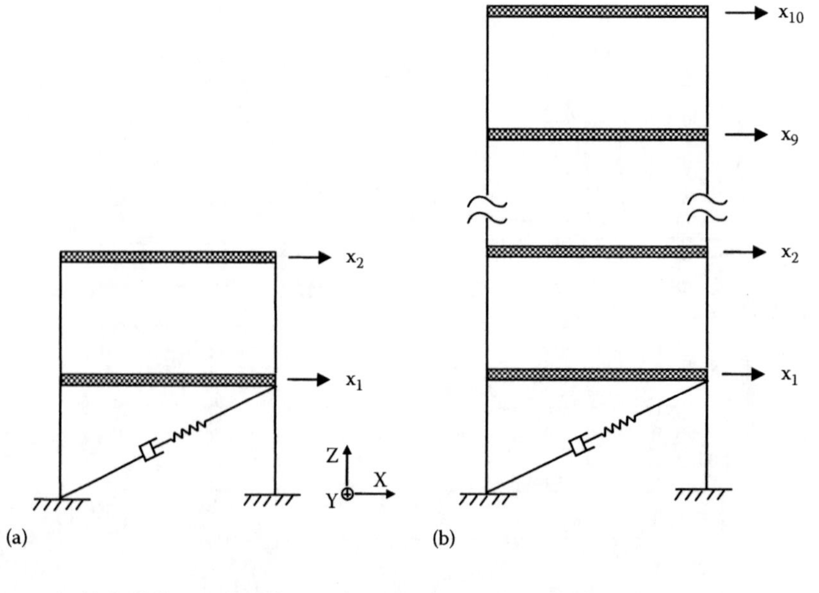

FIGURE 5.22 (a, b) Buildings with linear viscous damper installed in the first story.

TABLE 5.3A
Modal Parameters for the Two-Story Structure

Mode	1st	2nd
f_i (Hz)	3.11	8.14
ξ_i (%)	1.01	1.01

TABLE 5.3B
Modal Parameters for the Ten-Story Structure

Mode	1st	2nd	3rd	4th	5th	6th	7th	8th	9th	10th
f_i (Hz)	2.4	7.1	11.8	16.2	20.1	23.1	25.7	28.3	30.2	31.4
ξ_i (%)	0.98	2.69	3.93	4.96	6.18	6.14	3.21	1.42	0.54	0.13

coefficient is increased until it reaches 50 times the original values. At each time, the displacement response is calculated and the maximum values for the two-story and ten-story structures are plotted in Figures 5.22a and b, respectively.

In Figure 5.23a, the upper curve is the response of displacement for the second floor, x_2, and the lower one is for the first floor, x_1. It is seen that as the value of the damping coefficient increases, the responses will continuously decrease for both the first and second floors. This means that adding damping only at the first story will help reduce the responses of both the first and second floors.

In Figure 5.23b, from bottom to top, the curves represent the displacement response of the first to the tenth floors, namely, x_1 to x_{10}, respectively. From this latter figure, it is seen that, unlike the case of the two-story structure, the displacements will no longer decrease monotonically when the damping coefficients increase, except for those at the first and second stories.

The reason for the phenomena shown in Figure 5.23 can be qualitatively explained as follows. For both structures, since only the first story has dampers, the energy dissipation of the remaining stories will be carried out by the first story only. Here, the damping of the primary structure is ignored, since it is quite small in both cases. The energy dissipation of the remaining stories occurs through the stiffness that connects the rest of the stories to the first story. For the two-story structure, the second story is close to the first one and therefore the "connecting" stiffness is comparatively strong. This is similar to the above-mentioned supporting stiffness concepts. When the stiffness is strong, a large damping coefficient will still function to absorb energy.

However, for the ten-story structure, the "connecting" stiffness becomes much weaker than in the two-story structure. The function of energy dissipation of the damper for the remaining floors becomes weaker and weaker. At the very beginning, when the damping coefficient is not yet large, the connecting stiffness is relatively strong, allowing energy dissipation to take place. However, when the value of the damping coefficient becomes sufficiently large, only the second story, which is closest to the first, can have enough energy dissipation to reduce the vibration level. The responses of the other stories will begin to increase instead of decrease as the value of the damping coefficient gets larger and larger.

5.5.3.4 Design Considerations

In the above discussion, the effect on the supporting stiffness due to increasing damping from supplemental dampers and the corresponding response reductions was theoretically shown. Next, this issue is related to practical damper design.

From Equations 5.126 and 5.139,

$$\gamma = 2\lambda\xi_0\sqrt{\frac{\lambda\xi_{eq}}{\xi_0 - \lambda\xi_{eq}}} \tag{5.141}$$

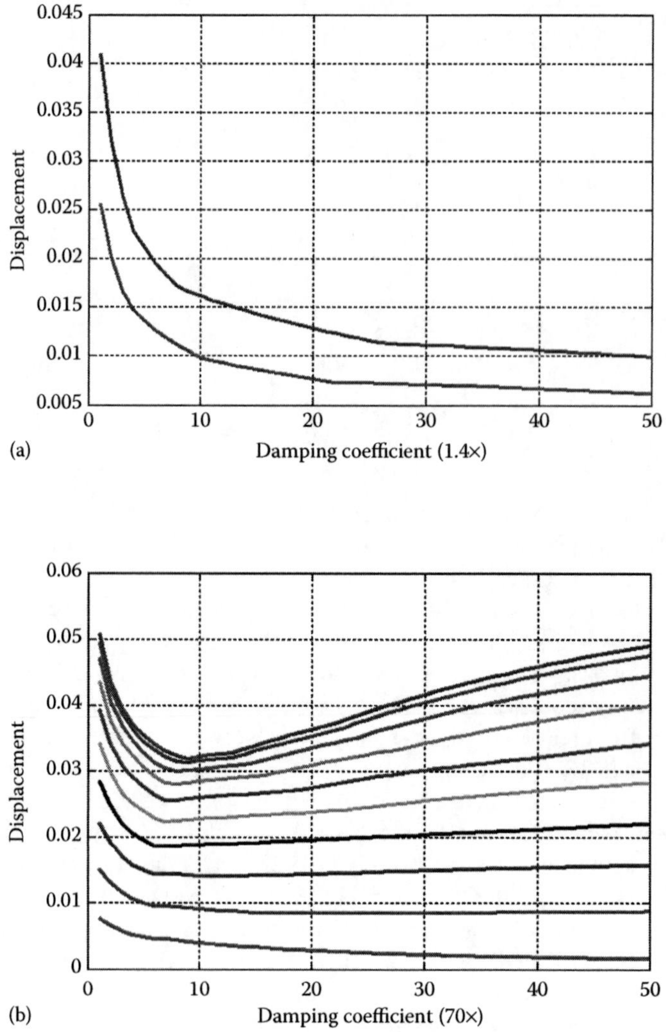

FIGURE 5.23 Structural responses vs. increase of damping coefficients: (a) response of 2-story structure and (b) response of 10-story structure.

If ξ_{eq} is required to not be 90% of ξ_0, then

$$\gamma = \frac{k_s}{k} \geq 10\xi_{eq}$$

That is, generally speaking, when the damping ratio is designed to be around 10%, the supporting stiffness k_s should be greater than k. When the damping ratio is designed to be around 20%, the supporting stiffness k_s should be greater than 2k. Therefore, for the case of a smaller damping ratio used in damper design, such as, less than 15%, an approximate equation to describe the requirement of supporting stiffness can be written as follows:

$$k_s \geq 10\xi k \tag{5.142}$$

where ξ is the generic term of damping ratio. This statement is quite rough. If more accuracy is needed, state equations and/or the modal condensed equations should be used.

In Chapter 3, it was learned that when higher modes are concerned, considerably higher damping is usually involved. In this case, Equation 5.141 becomes insufficient. By using the aforementioned 99 earthquake records, it can be found that if the reduction of responses of the acceleration and/or the displacement due to finite supporting stiffness is limited to within about 10%, then the following empirical formula can be used:

$$k_s \geq \left[10\ln(\xi - 6.4\%) + 23\right]\xi k = \gamma_k k \tag{5.143a}$$

This formula is obtained by a least squares approximation. The condition of using this form is that the design damping ratio should be larger than 7%. Here γ_k is the ratio of the supporting stiffness and the lateral stiffness in practical design, and

$$\gamma_k = \frac{k_s}{k} = \left[10\ln(\xi - 6.4\%) + 23\right]\xi \tag{5.143b}$$

Note that Equation 5.143 is obtained from the numerical simulation by using the specific group of 99 earthquake records. Obviously, by using another group of earthquake records, alternative parameters will be obtained in the empirical formula; for example, by using the 28 records mentioned by Naeim and Kelly (1999) due to the randomness of earthquake ground motions. However, based on the authors' limited experience, the differences are not significant. These 99 records are used as references of simulation in this book for more pieces of records are available.

Many ground motion records are available from the USGS Strong-Motion CD-ROM and some online strong-motion databases. Among them, 99 typical ground motion records are commonly used (Naeim and Anderson (1996), www.berkeley.edu/smcat, Lee, et al (2007)). Since they are corrected by different techniques, the peak values are slightly different from the records originally used by Naeim and Anderson, which are listed in Table 5.4. The soil site classifications according to FEMA-273 (FEMA 1997) are also indicated in Table 5.4.

Note that, in Equation 5.143, the factor of period T does not appear. However, at a different period to limit the reduction to within 10%, a different amount of supporting stiffness is required. In Equation 5.143, the period range is considered to be from 0.1 (s) to 6 (s). A more accurate formula can be obtained by explicitly considering the value of the period. Yet, in engineering applications, if sufficient supporting stiffness can be guaranteed, a very detailed formula does not need to be used.

Table 5.5 shows the required stiffness ratio, as well as the percentage increase of the responses due to the effect of finite supporting stiffness. For example, if a 50% damping ratio is designed and the supporting stiffness is approximately 9.4 times greater than the lateral stiffness, the acceleration and displacement will actually be larger than the case of theoretically assumed infinite supporting stiffness by factors of 9.87% and 6.68%, respectively.

From Table 5.5, it is also realized that the increase of the acceleration is always larger than the displacement. However, this phenomenon can only be observed in a linear system. If nonlinear damping is used, then the increase of the displacement is often found to be larger than the acceleration.

For larger damping nonlinearity, a safety factor S_n is introduced as

$$k_s \geq 10 S_n \xi k \tag{5.144}$$

Furthermore, when the supporting stiffness is included, the viscous damping becomes "visco-elastic" damping, and several degrees of phase shift may be observed. When the driving frequency becomes higher, this phenomenon can become more serious.

TABLE 5.4
99 earthquake records

No.	Year	Earthquake	Station	Deg.	Attribues
1	1940	El Centro	El Centro-Imp Vall Irr Dist	180	Duration
2	1940	El Centro	El Centro-Imp Vall Irr Dist	270	Duration
3	1949	Western Washington	Olympia Hwy Test Lab	356	Duration
4	1949	Western Washington	Olympia Hwy Test Lab	86	Duration
5	1952	Kern County	Taft	21	Duration
6	1952	Kern County	Taft	111	*
7	1966	Parkfield CA	Cholame Shandon Array 2	65	IV
8	1971	San Fernando	Pacoima Dam	164	PA, PV, PD, IV, ID, EPA, EPV
9	1971	San Fernando	Pacoima Dam	254	PA, PV, IV, EPA, EPV
10	1979	Coyote Lake, CA	Gilroy Array #6	230	EPV
11	1979	Coyote Lake, CA	Gilroy Array #6	320	*
12	1979	Imperial Valley, CA	Bonds Corner	230	PA, PV, IV, EPA, EPV
13	1979	Imperial Valley, CA	Bonds Corner	140	PA, PV, IV, EPA, EPV
14	1979	Imperial Valley, CA	Cerro Prieto	237	Duration
15	1979	Imperial Valley, CA	Cerro Prieto	147	Duration
16	1979	Imperial Valley, CA	Delta	352	Duration
17	1979	Imperial Valley, CA	Delta	262	Duration
18	1979	Imperial Valley, CA	El Centro, Array #4	230	PD, IV, ID
19	1979	Imperial Valley, CA	El Centro, Array #4	140	EPV
20	1979	Imperial Valley, CA	El Centro, Array #5	230	IV, ID
21	1979	Imperial Valley, CA	El Centro, Array #5	140	Duration
22	1979	Imperial Valley, CA	El Centro, Array #6	230	PD, IV
23	1979	Imperial Valley, CA	El Centro, Array #6	140	PD, ID
24	1979	Imperial Valley, CA	El Centro, Array #7	230	PD, IV, ID, EPV
25	1979	Imperial Valley, CA	El Centro, Array #7	140	EPV
26	1979	Imperial Valley, CA	El Centro Array #8	230	PD, ID
27	1979	Imperial Valley, CA	El Centro Array #8	140	PA, PV
28	1979	Imperial Valley, CA	El Centro, Array #10	50	PD
29	1979	Imperial Valley, CA	El Centro, Array #10	320	*
30	1979	Imperial Valley, CA	El Centro, Differential Array	270	PD, ID
31	1979	Imperial Valley, CA	El Centro, Differential Array	360	*
32	1979	Imperial Valley, CA	Meloland Overpass FF	0	PD, ID
33	1979	Imperial Valley, CA	Meloland Overpass FF	270	PD, IV, ID
34	1979	Imperial Valley, CA	Holtville Post Office	225	ID
35	1979	Imperial Valley, CA	Holtville Post Office	315	*
36	1980	Mammoth Lakes	Long Valley Dam (u.l.abut)	0	Duration
37	1980	Mammoth Lakes	Long Valley Dam (u.l.abut)	90	*

TABLE 5.4 (Continued)
99 earthquake records

No.	Year	Earthquake	Station	Deg.	Attribues
38	1983	Coalinga, CA	Parkfield Fault Zone 14	90	EPV
39	1983	Coalinga, CA	Parkfield Fault Zone 14	0	EPV
40	1983	Coalinga Aftershock	Pleasant Valley Pump Plant Yr	45	PA, PV
41	1983	Coalinga Aftershock	Pleasant Valley Pump Plant Yr	135	*
42	1984	Morgan Hill, CA	Coyote Lake Dam	285	PA, PV, IV, EPA, EPV
43	1984	Morgan Hill, CA	Coyote Lake Dam	195	PA, PV, IV, EPA, EPV
44	1985	Nahanni	Site 1	10	PA, PV, EPA
45	1985	Nahanni	Site 2	280	PA, PV, EPA
46	1989	Loma Prieta-October 17	Agnew-Agnews State Hospital	0	ID
47	1989	Loma Prieta-October 17	Agnew-Agnews State Hospital	90	*
48	1989	Loma Prieta-October 17	Corralitos-Eureka Canyon Rd.	0	PA, PV, IV, EPA, EPV
49	1989	Loma Prieta-October 17	Corralitos-Eureka Canyon Rd.	90	IV, EPV
50	1989	Loma Prieta-October 17	Gilroy #1-Gavilan College, Water Tank	90	EPV
51	1989	Loma Prieta-October 17	Gilroy #1-Gavilan College, Water Tank	0	*
52	1989	Loma Prieta-October 17	Gilroy #3-Gilroyh Sewage Plant	0	PA, PV, EPA
53	1989	Loma Prieta-October 17	Gilroy #3-Gilroyh Sewage Plant	90	*
54	1989	Loma Prieta-October 17	Hollister-South Street and Pine Drive	0	PD, IV, ID, EPV
55	1989	Loma Prieta-October 17	Hollister-South Street and Pine Drive	90	PD, ID
56	1989	Loma Prieta-October 17	Saratoga-Aloha Ave.	90	PD, ID
57	1989	Loma Prieta-October 17	Saratoga-Aloha Ave.	0	*
58	1991	Sierra Madre-June 28	Altadena-Eaton Canyon Park	0	EPA
59	1991	Sierra Madre-June 28	Altadena-Eaton Canyon Park	90	*
60	1992	Landers-June 28	Amboy	90	Duration
61	1992	Landers-June 28	Amboy	0	*
62	1992	Landers-June 28	Barstow-Vineyard&H St.	0	ID
63	1992	Landers-June 28	Barstow-Vineyard&H St.	90	*
64	1992	Landers-June 28	Desert Hot Springs	0	Duration
65	1992	Landers-June 28	Desert Hot Springs	90	Duration
66	1992	Landers-June 28	Joshua Tree-Fire Station	0	Duration
67	1992	Landers-June 28	Joshua Tree-Fire Station	90	Duration
68	1992	Landers-June 28	Lucerne Valley	T	ID
69	1992	Landers-June 28	Lucerne Valley	L	*

TABLE 5.4 (Continued)
99 earthquake records

No.	Year	Earthquake	Station	Deg.	Attribues
70	1992	Landers-June 28	Yermo-Fire Station	270	PD, ID
71	1992	Landers-June 28	Yermo-Fire Station	360	PD, ID
72	1992	Petrolia-April 25	Cape mendocino	0	PA, PV, PD, IV, ID, EPA, EPV
73	1992	Petrolia-April 25	Cape mendocino	90	PA, PV, EPA
74	1992	Petrolia-April 25	Fortuna-701 S. Fortuna Bl vd.	0	PD
75	1992	Petrolia-April 25	Fortuna-701 S. Fortuna Bl vd.	90	*
76	1992	Petrolia-April 25	Petrolia	0	PA, PV, IV, EPV
77	1992	Petrolia-April 25	Petrolia	90	PA, PV, PD, IV, EPV
78	1992	Petrolia-April 25	Shelter Cove-Airport	90	Duration
79	1992	Petrolia-April 25	Shelter Cove-Airport	0	*
80	1994	Northridge-January 17	Jensen Filtratioin Plant	22	PV, PD, IV, ID, EPV
81	1994	Northridge-January 17	Jensen Filtratioin Plant	292	PA, PV, PD, IV, EPV
82	1994	Northridge-January 17	Los Angeles, Sepulveda V.A. Hospital	360	PA, PV, IV, EPA, EPV
83	1994	Northridge-January 17	Los Angeles, Sepulveda V.A. Hospital	270	PA, PV, IV, EPA, EPV
84	1994	Northridge-January 17	Castaic-Old Ridge Route	360	IV
85	1994	Northridge-January 17	Castaic-Old Ridge Route	90	*
86	1994	Northridge-January 17	Newhall-La County Fire Station	90	PV, IV, EPA
87	1994	Northridge-January 17	Newhall-La County Fire Station	360	PV, PD, IV, EPA, EPV
88	1994	Northridge-January 17	Pacoima Dam-Upper Left Abutment	104	PA, IV, EPA, [D]
89	1994	Northridge-January 17	Pacoima Dam-Upper Left Abutment	194	PA, PV, IV, EPA, EPV
90	1994	Northridge-January 17	Santa Monica-City Hall Grounds	90	PA, EPA
91	1994	Northridge-January 17	Santa Monica-City Hall Grounds	0	*
92	1994	Northridge-January 17	Sylmar-County Hosp. Parking Lot	90	PA, PV, IV
93	1994	Northridge-January 17	Sylmar-County Hosp. Parking Lot	360	PA, PV, PD, IV, ID, EPA, EPV
94	1994	Northridge-January 17	Tarzana-Cedar Hill Nursery A	90	PA, PV, PD, IV, EPA, EPV
95	1994	Northridge-January 17	Tarzana-Cedar Hill Nursery A	360	PA, PV, PD, IV, EPA, EPV
96	1994	Northridge-January 17	17645 Saticoy St., Northridge, CA	180	IV
97	1994	Northridge-January 17	17645 Saticoy St., Northridge, CA	90	*
98	1994	Northridge-January 17	14145 Mulholland Dr., Beverly Hills, CA	9	EPV
99	1994	Northridge-January 17	14145 Mulholland Dr., Beverly Hills, CA	279	IV

PA, PV, PD (large peak acceleration, vecolcity and displacement); IV, ID (large incremental velocity, displacement), EPA, EPV (large effective peak acceleration, displacement); * Unknown

TABLE 5.5
Supporting Stiffness

Damping Ratio (%)	Increase of Acc. (%)	Increase of Disp. (%)	γ_k
10	10.27	4.83	0.86
20	8.72	4.62	2.87
30	9.21	5.51	5.02
40	9.61	6.15	7.30
50	9.87	6.68	9.39
60	9.92	7.32	12.17
70	9.47	7.78	14.72
80	9.54	8.33	17.33
90	9.4	8.69	20.00
100	9.28	9.07	22.21

5.5.4 Damper Installation

There are several methods typically used to install a damper in a structure. In the literature, damper installation is also referred to as damper configuration. In this book, the term configuration is used to denote the locations of dampers installed in a structure, and installation is used to denote the detailed manner of how a damper is installed.

In Figure 5.24, the most popular methods of damper installation are illustrated. A unified factor, called the geometrical magnification factor, is used to quantify the effect of the type of installation.

From Figure 5.24, it is seen that when level j have level j – 1 have displacements p_j and p_{j-1}, or velocities v_j and v_{j-1}, respectively, the two ends of the damper do not necessarily have the same displacements and/or velocities, except for the case shown in Figure 5.24c.

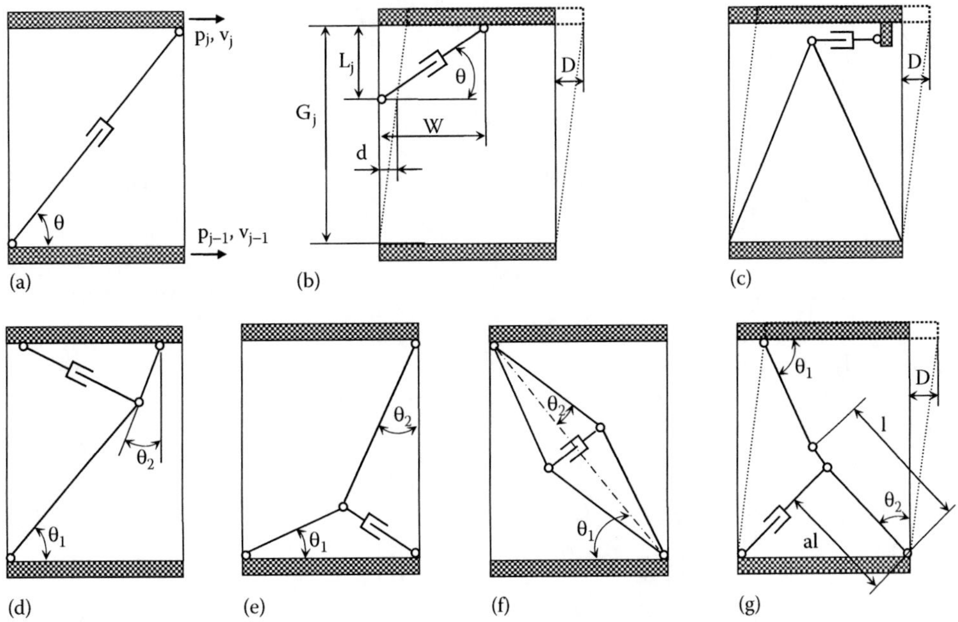

FIGURE 5.24 Damper installation.

The *geometrical magnification factor*, G_{mj}, is now used to unify all these cases. Denoting the relative displacement between the two ends of the damper j as d_{rj} and the relative displacement between the two levels as $p_j - p_{j-1}$, the *geometrical magnification factor*, G_{mj}, for the jth level is defined as

$$G_{mj} = \frac{d_{rj}}{p_j - p_{j-1}} \tag{5.145}$$

With the assumption of linear systems, if the modal displacement of the ith mode, denoted as d_{rji}, is present, then

$$G_{mj} = \frac{d_{rji}}{p_{ji} - p_{j-1i}} \tag{5.146}$$

Here, $(p_{ji} - p_{j\text{-}1i})$ is the relative displacement (floor drift) between the two levels of the ith mode. From Equation 5.146, it is seen that the value of this factor is not affected by the normalization of modes, which is only a function of the geometrical installation. It can be proven that for the installation shown in Figure 5.24a, the factor is

$$G_{mj} = \cos(\theta_j) \tag{5.147}$$

where θ_j is the installation angle.

Meanwhile, for the installation shown in Figure 5.24b, the magnification factor is

$$G_{mj} = \cos(\theta_j)\frac{L_j}{G_j} \tag{5.148}$$

where G_i and L_i are, respectively, the height of the story and the installation height.

The most basic damper installation is shown in Figure 5.24c, where the amplification factor is always

$$G_{mj} = 1 \tag{5.149}$$

For the installation depicted in Figure 5.24d, the magnification factor can be written as

$$G_{mj} = \sin(\theta_1) + \frac{\sin\theta_2}{\cos(\theta_1 + \theta_2)} \tag{5.150}$$

where θ_1 and θ_2 are the angles defined in the figure.

For the remaining three installations shown in Figures 5.24e through g, the amplification factors are

$$G_{mj} = \frac{\cos\theta_1}{\cos(\theta_1 + \theta_2)}, \tag{5.151}$$

$$G_{mj} = \frac{\sin\theta_1}{\tan\theta_2} \tag{5.152}$$

and

$$G_{mj} = \frac{a\cos\theta_1}{\cos(\theta_1+\theta_2)} - \cos\theta_2 \tag{5.153}$$

respectively. It can be seen that for the cases shown in Figures 5.24a through c, the geometrical magnification factors are equal to or less than 1. However, the cases shown in Figures 5.24d through g will have geometrical magnification factors that are considerably greater than 1.

On the one hand, when the factor is greater than 1, a broader range of choice is available to select various types of dampers. This is because the difference of the displacement $p_j - p_{j-1}$ is usually a very small number. Without the help of a magnification factor that is greater than 1, the relative displacement d_{rj} will also be quite small. Similarly, the relative velocity will be small. With such a small relative displacement or velocity, many types of dampers will not provide effective capability for energy dissipation. By using large values of G_m, however, these types of dampers can be used.

On the other hand, when the geometrical magnification factor is considerably greater than 1, the supporting stiffness becomes a more important issue. That is, if the supporting stiffness is not sufficiently strong, the effectiveness of the geometrical magnification will be greatly reduced.

Note that with th e presence of the geometrical magnification factor, the equation for damping ratio estimation is modified. Using the case of SDOF systems with linear viscous damping, e.g.,

$$\xi = \frac{c\omega_n (G_m x_0)^2}{2kx_0^2} = G_m^2 \frac{c}{2\sqrt{mk}}, \tag{5.154}$$

the factor G_m^2 can be a large number. For example, in the case shown in Figure 5.24f, if $\theta_1 = 30°$ and $\theta_2 = 4°$, then $G_m = 7.15$ and $G_m^2 = 51.13$.

Practically speaking, using a large value of G_m is not for large energy dissipation. This is because if a larger amount of energy needs to be dissipated, the value of the damping coefficient can simply be increased. It is not necessary to use the damping installations shown in Figures 5.24d through g, which need a more complex supporting mechanism and will thus increase the cost. In other words, using these installations is primarily to magnify the displacement.

However, Equation 5.154 indicates that the geometrical magnification factor will also magnify the damping force. In this case, the damping force can be magnified 7.15 times, which should be accommodated by the four bars. Such a large force may tend to buckle these bars, thereby greatly reducing the supporting stiffness.

In practical design, the corresponding supporting stiffness can be calculated on a case-by-case basis. However, the basic formula suggested in this section will apply in most cases.

5.6 SUMMARY

In this chapter, several basic design principles for practical damper design have been presented. They include general classification and modeling of damping based on practical considerations; the rectangular law to approach optimal design and to estimate the limit of damping control; the concept of damping adaptability used for damper selection; as well as issues related to the installation of damping devices. In the next chapter, application of these design principles to nonlinear and irregular structures will be considered.

REFERENCES

Aiken, I.D. and Kelly, J.M. 1990. *Earthquake Simulator Testing and Analytical Studies of Two Energy-Absorbing Systems for Multistory Structures*, Technical Report UCB/EERC-90/03, University of California, Berkeley, CA.

Arima, F., Miyazaki, M., Tanaka, H., and Yamazaki, Y. 1988. A study on buildings with large damping using viscous damping walls. Proceeding of the Nineth World Conference on Earthquake Engineering, Tokyo, V, 821–26.

Badrakhan, F. 1985. Separation and determination of combined dampings from free vibrations. *Journal of Sound and Vibration* 100 (2): 243–55.

Bagley, R.L. and Torvik, P.J. 1983. Fractional calculus — a different approach to the analysis of viscoelastically damped structures. *AIAA Journal* 21 (5): 742–48.

Building Seismic Safety Council (BSSC). 2003. *NEHRP Recommended Provisions for Seismic Regulation for New Buildings and Other Structures, 2003 Edition*, Report No. FEMA 450, Federal Emergency Management Agency, Washington, DC.

———. 2009. *NEHRP Recommended Provisions for Seismic Regulation for New Buildings and Other Structures, 2009 Edition*, Report No. FEMA P-750, Federal Emergency Management Agency, Washington, DC.

Capecchi, D. and Vestroni, F. 1985. Steady-state dynamic analysis of hysteretic systems. *Journal of Engineering Mechanics* 111 (12): 1515–31.

Caughey, T.K. 1960. Sinusoidal excitation of a system with bilinear hysteresis. *Journal of Applied Mechanics* 27 (4): 640–43.

Cofie, N.G. and Krawinkler, H. 1985. Uniaxial cyclic stress-strain behavior of structural steel. *Journal of Engineering Mechanics* 111 (9): 1105–20.

Constantinou, M.C. and Symans, M.D. 1993. Experimental study of seismic response of buildings with supplemental fluid dampers. *Structural Design of Tall Buildings* 2 (2): 93–132.

Constantinou, M.C., Soong, T.T., and Dargush, G.F. 1998. Passive energy dissipation systems for structural design and retrofit. Monograph No. 1, Multidisciplinary Center for Earthquake Engineering Research, University at Buffalo, NY.

Dafalias, Y.F. and Popov, E.P. 1975. A model of nonlinearity hardening materials for complex loading. *Acta Mechanica* 21:173–192.

Dargush, G.F. and Soong, T.T. 1995. Behavior of metallic plate dampers in seismic passive energy dissipation systems. *Earthquake Spectra* 11 (4): 545–68.

Federal Emergency Management Agency 1997. NEHRP Guidelines for the Seismic Rehabilitation of Buildings (FEMA-273).

Ferry, J.D. 1980. *Viscoelastic Properties of Polymers.* NY: John Wiley.

Gemant, A. 1936. A method of analyzing experimental results obtained from elastoviscous bodies. *Physics* 7 (8): 311–17.

Graesser, E.J. and Cozzarelli, F.A. 1989. *Multidimensional Models of Hysteretic Material Behavior for Vibration Analysis of Shape Memory Energy Absorbing Devices*, Technical Report NCEER-89-0018, National Center for Earthquake Engineering Research, Buffalo, NY.

———. 1991. Shape-memory alloys as new materials for aseismic isolation. *Journal of Engineering Mechanics* 117 (11): 2590–2608.

Huffmann, G.K. 1985. Full base isolation for earthquake protection by helical springs and viscodampers. *Nuclear Engineering Design* 84 (2): 331–38.

Inman, D.J. 2008. *Engineering Vibration.* 3rd ed. Upper Saddle River, NJ: Pearson Prentice Hall.

Iwan, W.D. 1965. The steady-state response of the double bilinear hysteretic model. *Journal of Applied Mechanics* 32:921–25.

Iwan, W.D. and Gates, N.C. 1979. Estimating earthquake response of simple hysteretic structures. *Journal of Applied Mechanics* 105 (EM3): 391–405.

Jennings, P.G. 1964. Periodic response of a general yielding structure. *Journal of the Engineering Mechanics Division* (EM2):131–66.

Kasai, K., Munshi, J.A., Lai, M.L., and Maison, B.F. 1993. Viscoelastic damper hysteretic model: Theory, experiment and application. Proceedings of ATC-17-1 on Seismic Isolation, Energy Dissipation, and Active Control, San Francisco, CA, 2, 521–32.

Keightley, W.O. 1977. Building damping by Coulomb friction. *Sixth World Conference on Earthquake Engineering*, New Delhi, India, 3043–48.

Kelly, J.M., Skinner, R.I., and Heine, A.J. 1972. Mechanisms of energy absorption in special devices for use in earthquake resistant structures. *Bulletin of the New Zealand National Society for Earthquake Engineering* 5:63–88.

Kim, H.J., Christopoulos, C., and Tremblay, R. 2004. *Experimental Characterization of Bolt-stressed Non-Asbestos Organic (NOA) Material-to-Steel Interfaces*, Report No. UT2004-3, Department of Civil Engineering, University of Toronto, Canada.

Krieg, R.D. 1975. A practical two surface plasticity theory. *Journal of Applied Mechanics* E42:641–46.

Lazan, B.J. 1968. *Damping of Materials and Members in Structural Mechanics.* NY: Pergamon Press.

Lee, G.C., Ou, Y.C., Liang, Z., Niu, T., and Song, J. 2007. Principles and performance of roller seismic isolation bearings for highway bridges. Technical Report MCEER-Report-07-0019, MCEER, University at Buffalo, Buffalo, NY.

Makris, N. and Constantinou, M.C. 1990. *Viscous Dampers: Testing, Modeling and Application in Vibration and Seismic Isolation*, Technical Report NCEER-90-0028, National Center for Earthquake Engineering Research, Buffalo, NY.

Makris, N., Constantinou, M.C., and Dargush, G.F. 1993. Analytical model of viscoelastic fluid dampers. *Journal of Structural Engineering* 119 (11): 3310–25.

Makris, N., Dargush, G.F., and Constantinou, M.C. 1993. Dynamic analysis of generalized viscoelastic fluids. *Journal of Engineering Mechanics* 119:1663–79.

———. 1995. Dynamic analysis of viscoelastic fluid dampers. *Journal of Engineering Mechanics* 121 (10): 1114–21.

Masri, S.F. 1975. Forced vibration of the damped bilinear hysteretic oscillator. *Journal of the Acoustic Society of America* 57 (1): 106–12.

Mayes, R.L. and Mowbray, N.A. 1975. The effect of Coulomb damping on multidegree of freedom elastic structures. *Earthquake Engineering and Structural Dynamics* 3:275–86.

Mohraz, B. and Sadek, F. 2001. Earthquake ground motion and response spectra. Chapter 2 in *The Seismic Design Handbook*, F. Naeim (ed.). 2nd ed. Boston: Kluwer Academic.

Naeim F. and Anderson, J.C. 1996. Design Classification of Horizontal and Vertical Earthquake Ground Motion, JAMA Report No. 7738.68-96.

Naeim, F. and Kelly, J.M. *Design of Seismic Isolated Structures, From Theory to Practice*, 1999, John Willy & Sons, INC, NY.

Nims, D.K., Richter, P.J., and Bachman, R.E. 1993. The use of the energy dissipating restraint for seismic hazard mitigation. *Earthquake Spectra* 9 (3): 467–89.

Ozdemir, H. 1976. Nonlinear transient dynamic analysis of yielding structures. PhD Thesis, University of California.

Pall, A.S. and Marsh, C. 1982. Response of friction damped braced frames. *Journal of the Structural Division* 108 (ST6): 1313–23.

Pall, A.S., Marsh, C., and Fazio, P. 1980. Friction joints for seismic control of large panel structures. *Journal of the Prestressed Concrete Institute* 25 (6): 38–61.

Pekcan, G., Mander, J.B., and Chen, S.S. 1995. The seismic response of a 1:3 scale model R.C. structure with elastomeric spring dampers. *Earthquake Spectra* 11 (2): 249–67.

Pong, W.S., Tsai, C.S., and Lee, G.C. 1994. *Seismic Study of Building Frames with Added Energy-absorbing Devices*, Technical Report NCEER-94-0016, National Center for Earthquake Engineering Research, Buffalo, NY.

Richter, P.J., Nims, D.K., Kelly, J.M., and Kallenbach, R.M. 1990. The EDR-energy dissipating restraint. A new device for mitigation of seismic effects. Proceedings of the 1990 Structural Engineers Association of California (SEAOC) Convention, Lake Tahoe, NV.

Rivlin, R.S. 1981. Some comments on the endochronic theory of plasticity. *International Journal of Solids and Structures* 17:231–48.

Sabelli, R. 2001. *Research on Improving the Design and Analysis of Earthquake-Resistant Steel Braced Frames*, EERI/FEMA NEHRP Fellowship Report, Earthquake Engineering Research Institute, Oakland, CA.

Shen, K.L. and Soong, T.T. 1995. Modeling of viscoelastic dampers for structural applications. *Journal of Engineering Mechanics* 121 (6): 694–701.

Shuhaibar, C., Lopez, W., and Sabelli, R. 2002. Buckling-restrained braced frames. ATC-17-2 Seminar on Response Modification Technologies for Performance-Based Seismic Design, Applied Technology Council, Redwood City, CA, 321–28.

Skinner, R.I., Kelly, K.M., and Heine, A.J. 1975. Hysteresis dampers for earthquake-resistant structures. *Earthquake Engineering and Structural Dynamics* 3:287–96.

Soong, T.T. and Lai, M.L. 1991. Correlation of experimental results with predictions of viscoelastic damping of a model structure. Proceedings of Damping 1991, San Diego, CA, FCB.1–FCB.9.

Su, Y.F. and Hanson, R.D. 1990. *Seismic Response of Building Structures with Mechanical Damping Devices*, Technical Report UMCE 90-02, The University of Michigan, Ann Arbor, MI.

Sun, C.T. and Lu, Y.P. 1995. *Vibration Damping of Structural Elements*. Englewood Cliffs, NJ: Prentice Hall PTR.

Tsai, C.S. and Lee, H.H. 1993. Application of viscoelastic dampers to high-rise buildings. *Journal of Structural Engineering* 119 (4): 1222–33.

Tsai, C.S. and Tsai, K.C. 1995. TPEA device as seismic damper for high-rise buildings. *Journal of Engineering Mechanics* 121 (10): 1075–81.

Tsopelas, P. and Constantinou, M.C. 1994. *NCEER-Taisei Corporation Research Program on Sliding Seismic Isolation Systems for Bridges: Experimental and Analytical Study of a System Consisting of Sliding Bearings and Fluid Restoring Force/Damping Devices*, Technical Report NCEER-94-0014, National Center for Earthquake Engineering Research, Buffalo, NY.

Valanis, K.C. 1971. A theory of viscoplasticity without a yield surface. *Archives of Mechanics* 23:517–34.

Whittaker, A.S., Bertero, V.V., Thompson, C.L., and Alonso, L.J. 1991. Seismic testing of steel plate energy dissipation devices. *Earthquake Spectra* 7 (4): 563–604.

Williams, M.L. 1964. Structural analysis of viscoelastic materials. *AIAA Journal* 2 (5): 785–808.

Xia, C. and Hanson, R.D. 1992. Influence of ADAS element parameters on building seismic response. *Journal of Structural Engineering* 118 (7): 1903–18.

Zhang, R.H., Soong, T.T., and Mahmoodi, P. 1989. Seismic response of steel frame structures with added viscoelastic dampers. *Earthquake Engineering and Structural Dynamics* 18:389–96.

6 System Nonlinearity and Damping of Irregular Structures

This chapter continues the discussion on available information of damping design principles and approaches for nonlinear and irregular structures. While this general area is a current frontier in earthquake engineering research with many challenges yet to be addressed, only the limited amount of knowledge currently available is discussed and summarized for possible damping design applications.

6.1 NONLINEAR SYSTEMS

Linear design spectra can achieve good results in the design of nonlinear dampers when the structure is simple and the effective damping ratio is low. However, most actual buildings and structures are complex in shape and cannot be modeled as SDOF systems, and when higher levels of damping are needed, improvements in the design procedure are required. First, Timoshenko damping may overdamped the effective damping ratio in these cases, one improvement often can be realized by using the force-based effective damping approach. Furthermore, special requirements for supporting stiffness should be taken into account. Two essentially different types of damping should be distinctly designed by considering the effective damping estimation, the required supporting stiffness, and the amplitude of the input level of ground excitations. Secondly, the design spectra approach based on linearized SDOF systems and used to compute the responses of multiple-story structures requires modification, both with regard to the formulation of damping and for the estimation of the effective period and effective modal shapes. Even with these modifications, the estimation of the structural response can still contain significant errors due to system nonlinearities and structural irregularities. Consequently, in many cases, time history analyses are needed to evaluate and/or finalize the corresponding damper designs.

6.1.1 Classifications of Nonlinear Damping

As noted in Chapter 5, damping was classified according to Lazan's theory (1968), which focuses on the total damping behavior of a structure with added damping systems. In practical design, damping needs to be classified, but the focus is on the damping devices. Currently, most codes classify the damping as either displacement dependent or velocity dependent. In Chapter 5, it was pointed out that such a classification does not clearly distinguish the concepts of linear and nonlinear damping. For example, the amplitude of the damping force of a linear viscously damped system will be proportional not only to the velocity, but also to the displacement and, for certain types of nonlinear damping, the damping force does not directly depend on the velocity or the displacement. In this book, this classification is not followed.

Alternative classifications are hysteretic damping, viscoelastic damping, and recentering damping (Constantinou, Soong, and Dargush 1998). These classifications are helpful for mathematical modeling. However, for practical design, often the boundaries among these damping categories are somewhat blurred. For example, a viscoelastic damper is not self-centered during a strong earthquake; however, it will have a self-centering capability after the vibration. In addition, the viscoelastic material may exhibit a good hysteretic loop for energy dissipation. In damper design, it is often necessary to clearly know the damping category in order to perform damper modeling and response computations.

One method of damping classification relating to practical design is to divide the dampers into linear and nonlinear categories. These classifications are helpful in understanding the nature of dampers and the functions of damping in a structure and are used in this book.

Linear dampers include the linear viscous damper, which is primarily a hydraulic damping mechanism with a damping force proportional to structural velocity. Most viscoelastic dampers also provides linear damping, whose damping force is proportional to structural displacement. This kind of viscoelastic damping is often obtained through viscoelastic materials and the most popular damper utilizes the shear mode. Note that some researchers also believe that the damping of a structure is better modeled with the damping force proportional to the displacement. Therefore, this type of damping with viscoelastic characteristics is sometimes called structural damping.

Nonlinear dampers include the *sublinear* viscous damper, with the amplitude of the damping force proportional to the amplitude of velocity raised to an exponent. These types of dampers are also often made by hydraulic devices. When the damping exponent is greater than 1, it is called a superlinear damper, which is less effective in earthquake vibration control, and thus it has a primarily theoretical meaning. Note that when the damping exponent is equal to 1, the damping becomes linear. Generally speaking, the amplitude of structural velocity contains two components, namely, the frequency and the amplitude of the displacement. The exponent of the frequency and the exponent of the displacement can be different. In practice, the slight difference can yield rather large design errors. Nonlinear damping can also be accomplished through the use of friction dampers. Once the maximum friction force is reached due to transverse vibration, the force stays virtually constant and the energy dissipation loop is close to a rectangular loop, which is a special case of the parallelogram-shaped force–displacement loop. In fact, when the stiffness of the structure is considered, the energy dissipation loop is a general parallelogram. Metallic dampers start to dissipate energy when the metallic material begins to yield. Thus, for these dampers, the energy dissipation loop is close to the *bilinear* parallelogram.

6.1.1.1 Control Parameters and Design Parameters of Nonlinear Systems

6.1.1.1.1 Linear and Nonlinear Systems

The concept of control parameters and design parameters of a nonlinear system is basically the same as that specified for linear systems. However, a linear system has two advantages over nonlinear systems in terms of analysis and design. The first is that only two independent control parameters, the period and the damping ratio, are needed for the entire damper design. The second advantage is that the level of any responses of the linear system is proportional to the level of the excitation. In other words, if the amplitude of the excitation is doubled, so are the responses. Therefore, if these two independent control parameters are determined, all the responses, including the control parameters, can be determined if the input level is specified.

Theoretically, a nonlinear system will not have the period and damping ratio of a linear system. Also, different nonlinear systems will have different parameters and the number of these independent parameters is greater than two. For example, a structure with sublinear viscous dampers can be characterized by the equivalent damping coefficient, the effective period, and the damping exponent. On the other hand, a structure with friction dampers can be described by the characteristic strength, the loading and unloading stiffness, etc.

Because the relationship between the input and the response is not exactly proportional, the responses of the nonlinear system cannot be determined with information from these design parameters and the input level. Furthermore, modal superposition will not apply. In other words, if the design parameters are given, the control parameters cannot be directly determined. Quite often, iterative estimation of the structural responses is necessary. This issue is discussed later in this chapter.

To simplify the design procedure, nonlinear systems may be linearized. As mentioned above, the resulting equivalent linear system has only two design parameters. Therefore, no matter what the characteristics of a nonlinear system are, the two basic control parameters will be involved.

6.1.1.1.2 *Design Procedure*

In the early years of damping systems development, it was often assumed that the total damping of a structure was not very large, so that the structure with dampers, linear or nonlinear, was treated as a linear system. In this case, the design procedure for a structure with added dampers is rather simple and straightforward. Furthermore, even if an initial design is not satisfactory, the damping ratio control parameter can simply be adjusted until the design parameters are optimal.

In nonlinear systems with large damping, even though they can sometimes be treated as equivalent linear systems, both design parameters, often need to be adjusted. The following two design procedures will help clarify the difference between these two approaches.

6.1.1.1.2.1 Common Design Procedure The first design procedure may be referred to as the control parameter-based design, or more specifically, the damping ratio based design, which contains the following steps:

1. The designer estimates the response of a structure without added dampers. If the design parameters exceed the required values, then the original control parameters, the period, and especially the damping ratio shall be examined.
2. If the damping ratio is found to be small and the corresponding value needs to be increased to control seismic-induced vibration, the desired control parameters, especially the damping ratio or the numerical damping coefficient, will be established.
3. According to the required control parameters, appropriate dampers are then selected.
4. Based on the specifications of the selected dampers as well as information taken from the original design of the structure, the control parameters of an equivalent linear system are computed.
5. According to the "effective" control parameters, the design parameters of the structure are calculated again to see if they are satisfied.

This is a common design process and seems reasonable. That is, since the original design is not satisfied based on building codes, increasing damping is a good concept. On the other hand, since nonlinear dampers often have more than two control parameters, this design procedure may make damper selection difficult for inexperienced designers.

More importantly, since the design object is a nonlinear system, the control parameters, i.e., the effective damping ratios, are essentially functions of the design parameters, i.e., structural displacements. In the literature, iterative methods are used to address this chicken-and-egg problem. However, many of the iterative algorithms used in the literature either do not converge, or do not converge to the correct values. In this chapter, it is attempted to provide certain formulas to simplify the iterations.

6.1.1.1.2.2 Forward Design Procedure The second design procedure may be referred to as the design parameter-based "direct" design, or more specifically, the displacement-based design, which contains the following steps:

1. Select proper dampers with their specifications parallel to the design of the structure.
2. Use information about the structures and dampers to determine the control parameters, the period, and the damping ratio.
3. Use the control parameters to compute the design parameters of the structure.

The forward design requires computation of the structural response directly, which can increase computation burdens and design costs, especially for initial estimations. This is because the computation deals with nonlinear integrations. In this chapter, several design principles are listed to help making certain decision in order to simplify the computations.

6.1.1.1.2.3 Modified Spectra-based Design Designers can use nonlinear design spectra to seek a reasonable selection of dampers and locate the responses at the same time. Specifically, this method utilizes the precalculated nonlinear response spectra. In the literature, debate on using the nonlinear spectra frequently appears. One of the disadvantages is that the general nonlinear spectra need more than two control parameters. Exhausting all the control parameters is very costly and inconvenient. However, once the nonlinear damping is classified into the bilinear and the sublinear cases, only four parameters remain for each individual situation. This makes using the nonlinear spectra more feasible. In Chapter 8, two kinds of nonlinear design spectra for the sublinear and bilinear damping respectively will be suggested.

The major disadvantage of using these tables is that the approach is for SDOF systems only, since it is impossible to list all of the response data for multi-degree-of-freedom (MDOF) systems. In order to build a bridge between the SDOF and the MDOF systems, empirical formulas are used. Note that once again, these empirical formulas do not provide accurate solutions, and thus are for initial estimation purposes only. Experienced designers may choose this second design path, which provides more design freedom, especially in choosing suitable combinations for the structures and dampers.

6.1.1.2 Nonlinear Damper Classification

In both the common and the forward design procedures, designers need to account for the interpretation from the damper specifications to the structural control parameters. It is therefore helpful to continue the discussion on the characteristics of various dampers.

6.1.1.2.1 Basic Categories

By examining the control parameters of an "equivalent linear" system reduced from a nonlinear system, the classification of bilinear damping and sublinear damping can be convenient for practical damper design.

First, this classification is based on the fact that bilinear damping will notably affect the control parameter of period, whereas the influence of sublinear damping is limited. Secondly, the peak input–peak output relationship of a bilinear system is rather proportional, whereas sublinear damping is nonproportional.

Bilinear damping, as mentioned before, is also called hysteretic damping or structural damping. Meanwhile, sublinear damping is mainly contributed by viscous devices, which are primarily hydraulic dampers.

6.1.1.2.1.1 Bilinear Damping Bilinear damping is mainly contributed by using bilinear dampers. Installation of these devices will increase both the capability of nonlinear energy dissipation, primarily through the effective damping coefficient, and some of the elastic restoring force, in terms of the effective stiffness. In this chapter, the general concept of bilinear dampers is discussed and in Chapter 8 conventional design principles for bilinear dampers are discussed. When the damping effect, or effective damping ratio, is assumed to be small so that the approximation of Timoshenko damping is appropriate. The bilinear damping can be large. In this case, an alternative approach to account for the damping effect based on the force ratio is suggested.

Bilinear damping can also be contributed by the effect of actual dry friction between the members of a structure. In many cases, this friction effect can drastically increase damping. However, the stiffness of the structure is often not significantly affected. Quantifying these damping effects is still a research topic for various types of member connections and structures. This friction usually takes place during initial movement of the structure, but may be restricted due to limitations where large relative displacements among the structural elements are not possible. In other words, the bilinear damping occurs when the vibration level is small and is limited when the vibration becomes large. It is often worth applying a large damping at the beginning of the vibration, which can suppress further growth of the motion. Secondly, the friction forces of structural connections are often not controllable, and can be varied due to many factors such as the age of the connections and structures, the pair of friction materials, and certain environmental effects.

Although bilinear damping due to structural connections is not controllable, it should be considered for damping design after the initial damping is estimated. Certain structures contain a large amount of friction and others do not. In a structure with large initial friction, the damping ratio will be higher. Recall that higher initial damping also induces the effectiveness of damping devices.

Bilinear damping can also contribute to the design of structures, which are deliberately designed with friction pairs or viscoelastic walls that may be modeled as bilinear energy dissipators. Bilinear damping can also occur when the structure enters the inelastic range. In this case, the damping effect will become notably larger, but the stiffness will be reduced. Today, bilinear damping approaches to account for inelastic actions for structures with added damping devices have been written into building codes (e.g., NEHRP 2003).

Although using bilinear damping is less effective than using sublinear damping in earthquake response control owing to the uncertainty of the random excitations, it provides the most effective displacement reduction in base isolations. The results are significant when compared with other types of damping.

Bilinear dampers are not the most effective devices. They can be used for earthquake protection and have been applied in actual structures. However, for frequent large structural motion, such as wind- and traffic-induced excitations, the durability of bilinear devices is not as good as viscous dampers. Another drawback is that bilinear dampers often result in permanent displacements of the structure, which is not desirable in many applications. The major advantage of bilinear dampers is their low cost. For structures with a limited budget for earthquake protection, bilinear dampers are often a good choice. In Chapter 8, the advantages and disadvantages of bilinear damping is further discussed.

6.1.1.2.1.2 Sublinear Damping Sublinear damping is mainly contributed by hydraulic devices. Although a hydraulic damper can provide both a superlinear and sublinear effect, sublinear damping is the most commonly used form. Dampers with slight sublinear effects do not increase the cost of manufacturing. Higher degrees of sublinear damping, i.e., with a damping exponent smaller than 0.5, will increase the cost, but in many cases remain cost-effective.

In Chapter 8, Section 8.1, the advantages and disadvantages of sublinear damping are further discussed; here, the major advantages they offer, such as high-level effectiveness and self-centering, are briefly pointed out. Because of these advantages, sublinear damping should be one of the main considerations for use in controlling earthquake-induced vibration. Compared to bilinear and linear dampings, these advantages may be significant.

6.1.1.2.2 Practical Terminology

In the above discussion, the types of dampers are classified according to the nature of the corresponding damping forces and damper displacement (or velocity). Namely, when the relationship between the damping force and the damper displacement is linear, it is referred to as linear damping. When the relationship between the damping force and the damper displacement is close to a bilinear parallelogram, it is referred to as bilinear damping. Finally, when the relationship between the damping force and the damper velocity is sublinear, it is referred to as sublinear damping.

However, in practice, a pure linear damper is rarely encountered. In this case, the most commonly used bilinear dampers can be called *dry dampers*, since these devices often are not of a hydraulic type. On the other hand, most sublinear dampers are hydraulic devices, and may therefore be called *wet dampers*.

6.1.2 Conventional Preliminary Estimation

When a structure can be modeled by an SDOF system, or its response can be represented by the first effective mode of an MDOF system, the following conventional treatment based on simplified linearizations can be used. To better understand this phenomenon, nonlinear design approaches are briefly reviewed.

6.1.2.1 Nonlinear Dynamics

6.1.2.1.1 Response Spectra-based Design

Response spectra-based design is more generic phrase for the approach described in commonly used building codes. Generally speaking, the modified spectra-based design using the linearized model with the two basic control parameters can be considered as the response spectra-based design. On the other hand, if time history analysis is used, the essence of the design procedure is different.

As mentioned above, two control parameters are needed to use the design spectrum. In current building codes, to deal with the structural nonlinearity, the major change is the computation of the effective fundamental period, T_{eff}. This is obtained by multiplying the square root of the effective ductility demand due to the design earthquake, μ_D, by the period of the elastic structure T_1. This equation was implied before and is clearly written as

$$T_{eff} = \sqrt{\mu_D}\, T_1 \tag{6.1}$$

Equation 6.1 is further generalized as

$$T_{effi} = R T_i \tag{6.2}$$

where T_{effi} and T_i are the ith effective and original period, respectively. Here, R is the response modification factor for the SDOF system. As an example, for elasto-perfectly-plastic (EPP) systems,

$$R = \mu_D = \frac{k_u}{k_{eff}} \tag{6.3}$$

Here, k_u and k_{eff} are, respectively, the *unloading and the effective stiffness* of the system, instead of for individual dampers, μ_D is the *structural ductility.*

In Chapter 7, it will be shown that the numerical damping coefficient B can be written as

$$B \approx 3\xi + 0.9 \tag{6.4}$$

Therefore, the effective numerical damping coefficient can be calculated using the concept of Timoshenko damping, that is,

$$B_i = 3\xi_{effi} + 0.9 = (0.24)\frac{E_{di}}{E_{pi}} + 3\xi_{0i} + 0.9 \tag{6.5}$$

Here, E_{di} and E_{pi} are the dissipative and potential energies, respectively, of the ith mode, and ξ_{0i} is the original damping ratio of the ith mode.

Or, by using the force-based effective damping ratio,

$$B_i = 3\xi_{effi} + 0.9 = 1.5\frac{f_{di}}{f_{ri}} + 3\xi_{0i} + 0.9 \tag{6.6}$$

Here, f_{di} and f_{ri} are the damping and restoring forces of the ith mode, respectively.

Once the effective period and damping are found, the spectral value of displacement, d_{iD}, and pseudo acceleration, a_{iD}, can be determined. For example, according to NEHRP 2003/2009,

$$d_{iD} = \left(\frac{g}{4\pi^2}\right)\Gamma_i \frac{S_{D1} T_{effi}}{B_i} \le \left(\frac{g}{4\pi^2}\right)\Gamma_i \frac{S_{DS} T_{effi}^2}{B_i} \quad (m), \quad i = 1, ..., S \tag{6.7}$$

and

$$a_{iD} = \frac{4\,\pi^2 d_{iD}}{T_{effi}^2} \quad (m/s^2), \quad i = 1, ..., S \tag{6.8}$$

S_{D1} and S_{DS} are defined in Chapter 2. Furthermore, Γ_i is the modal participation factor defined in Chapter 3, which is repeated in a later section, and S is the total number of interesting modes. Determining this number by using certain modal contribution indicators was also discussed in Chapters 3 and 4.

Note that the displacement and acceleration are those of the roof of the structure, if the system is an MDOF structure. To compute the structural displacement, called the floor deflection of the first "mode," the following equations can be used so that the displacement vector is

$$\mathbf{d}_{max} = \{d_j\} = d_{1D}\boldsymbol{p}_1 \quad (m), \quad j = 1, ..., n \tag{6.9}$$

The story pseudo-acceleration vector is

$$\mathbf{a}_{s_1} = \{a_{s_{j1}}\} = a_{1D}\boldsymbol{p}_1 \quad (m/s^2), \quad j = 1, ..., n \tag{6.10}$$

and so on for higher "modes" with

$$\mathbf{d}_{iD} = \{d_{ij}\} = d_{iD}\boldsymbol{p}_i \quad (m), \quad j = 1, ..., n; \quad i = 2, ..., S \tag{6.11}$$

for displacements vector and

$$\mathbf{a}_{s_i} = \{a_{s_{ji}}\} = a_{iD}\boldsymbol{p}_i \quad (m/s^2), \quad j = 1, ..., n; \quad i = 2, ..., S \tag{6.12}$$

for story acceleration vector. Here, d_{ij} and a_{Sij} are, respectively, the ith relative displacement and pseudo acceleration at the jth floor; and S is again the number of modes of interest. Equations 6.9 through 6.12 indicate that the modal displacement and acceleration will be distributed from the roof movement to the total stories.

Note that in Equations 6.9 through 6.12, the mode shape $\boldsymbol{p}_i$ belongs to the elastic structure, which is rewritten as follows:

$$\boldsymbol{p}_i = \frac{1}{p_{ni}}\{p_{ji}\}, \quad j = 1, ..., n; \quad i = 1, ..., S \tag{6.13}$$

In other words, $\boldsymbol{p}_i$ belongs to a linear system. It is understandable that when the structure exceeds its yielding point and becomes inelastic, or when nonlinear dampers are installed, the entire system is nonlinear, so that, theoretically, there are no modes or mode shapes.

Thus, for a more accurate computation of a nonlinear structure, in addition to modifying the period as described in Equations 6.1 and 6.2, the damping ratio and the "mode" shape should also be managed. In this chapter, these two basic issues are discussed in detail for the spectra-based design.

Once the displacement and the acceleration vectors are calculated, modal combination methods can be used to obtain the total responses, such as the square-root-of-the-sum-of-squares (SRSS) or the complete quadratic combination (CQC).

Example 6.1

Suppose a structure can be treated as an SDOF system with a fundamental period of T = 1.5 (s) and an original damping ratio of $\xi_0 = 5\%$. When the structure yields under earthquake loadings, it has a maximum ductility of $\mu_D = 5$. In this case, the energy dissipated by the nonlinear response in the half cycle of the maximum deformation is $E_d = 402$ (kN-m) and the damping force is $f_d = 3000$ (kN). At the maximum deformation, the remaining potential energy is $E_p = 2.399$ (kN-m) and the restoring force is $f_r = 56{,}250$ (kN).

The effective period and numerical damping coefficient need to be determined.

According to Equation 6.1, the effective period is calculated to be 3.35 (s). According to Equations 6.2 and 6.3, the effective period is calculated to be 7.5 (s). According to Equation 6.4, the effective damping coefficient is calculated to be 17.0%. According to Equation 6.5, the effective damping coefficient is calculated to be 13.0%.

6.1.2.1.2 Time History Analysis

As mentioned previously, the spectra-based design for nonlinear system is not quite accurate, which thus is useful for initial estimations, but may not be suitable for final designs. In fact, many codes suggest using time history analysis as a complementary measure. In Chapter 2, several important issues for time history analysis were discussed.

6.1.2.2 Nonlinear Statics

In addition to nonlinear dynamic analysis, the static approach is still used in practice. The major design procedure of the nonlinear static approach is briefly reviewed as follows:

6.1.2.2.1 FEMA Method 1

The Federal Emergency Management Agency (FEMA) method 1 uses the data of the linear SDOF system to determine the target displacement of a nonlinear MDOF system. Baseline data used to determine the target displacements are obtained through statistical studies on bilinear and trilinear non-strength-degrading SDOF systems with 5% damping. The general nonlinear force–displacement relationship is represented with a bilinear model.

By simply employing a coefficient, C_N, the target displacement, vector, $\mathbf{d}_{1D}$, of the yielding structure is described by the following equation:

$$\mathbf{d}_{1D} = C_N S_A\left(\xi_{eff}, T_{eff}\right) T_{eff}^2 \boldsymbol{p}_1 \quad (m) \tag{6.14}$$

Comparing Equation 6.14 with the combination of Equations 6.7 and 6.9, the key issue is to find the coefficient C_N. Here, C_N contains a series of coefficients, especially the function of a ratio of maximum inelastic displacement and the peak elastic spectral displacement. Different researchers have suggested different formulas for C_N. Each equation formula has its own working condition and it is difficult to precisely specify or quantify these conditions. Interested readers can consult NEHRP 2003/2009 for details on how to calculate this coefficient.

6.1.2.2.2 FEMA Method 2

Unlike method 1, which uses the initial effective stiffness, method 2 determines maximum response based on the displacement corresponding to the intersection of the capacity curve of the structure and the spectral demand curve characterizing the design seismic load. The target displacement is iteratively calculated through the effective stiffness based on nonlinear deformations. This method is then more suitable for larger damping.

To carry out this analysis, a pushover curve is needed, i.e., a static load vs. deflection curve. The philosophy of the pushover curve is based on the softening spring caused by the gradual procession of yielding from an initial proportional relationship. Thus, a rather constant stiffness following Hooke's law is observed at the initial stage. After the load passes the yield point of the load-deformation curve, the instantaneous stiffness begins to drop. As the load is further increased, the increase of displacement becomes nonproportionally larger. The pushover curve is the plot of the load vs. the corresponding deformation relationship in Cartesian coordinates that represent the property change of the structural model, accompanied by the corresponding changes to its dynamic properties.

An MDOF structure will have multiple input loads because it has multiple lumped masses. To achieve the pushover curve of the entire structure, the load configuration or load pattern must be determined. That is, the pattern of pushover degradation depends on the pattern of lateral load. In the literature, many researchers have expressed concern regarding the topic of distributed seismic load on yielding structures (e.g., Ramirez et al. 2000). One of the simplest approaches is the design procedure suggested by FEMA (1997), which requires the use of at least two different patterns for the lateral loads. The idea is to produce the bounds of the structural responses. The first pattern, called the uniform pattern, is based on lateral loads proportional to the total weight at each floor level. The second pattern, called the modal pattern, is closely proportional to the fundamental mode shape. FEMA (1997) also suggested calculating the load pattern by combining the modal responses according to the response spectrum. The total number of modes used should capture 90% of the total modal mass ratios.

Chopra and Goel (2002) suggest a modal pushover analysis that combines the fundamental and several higher effective modes. This method can be more accurate, especially when the higher modes contribute more significantly to the total deformation. However, the above-mentioned load patterns are still used.

Figure 6.1 illustrates this procedure to find the pushover curve, an example of which is shown in Figure 6.2a. Note that in Figure 6.2, k_u and k_d are, respectively, the unloading and the yield stiffness of the system instead of the individual dampers, while f_m and f_y are, respectively, the maximum and yield force of the system. In addition, the effective stiffness is taken to be the secant stiffness, that is,

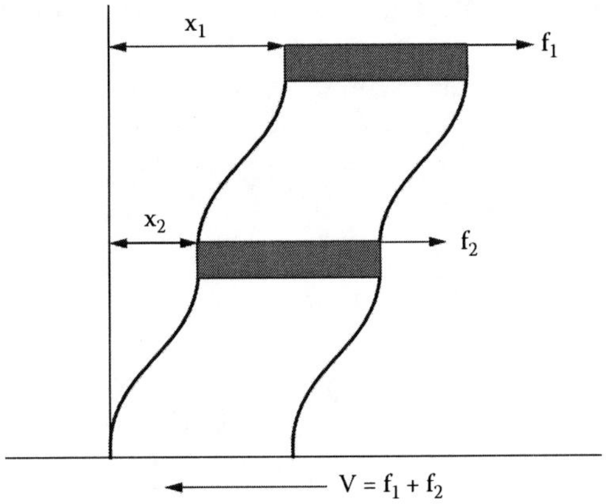

FIGURE 6.1 Obtaining the pushover curve.

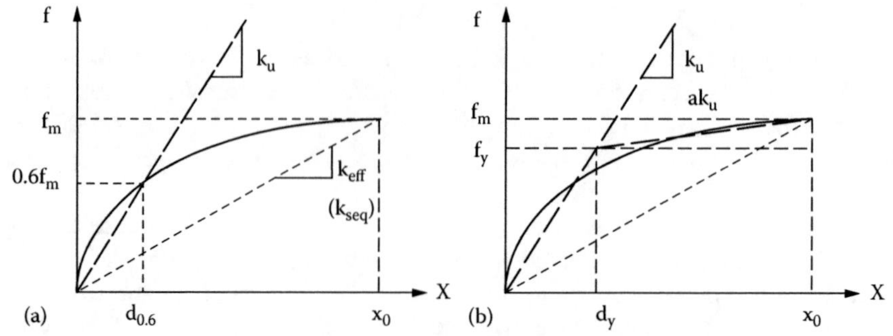

FIGURE 6.2 (a, b) Pushover curve and bilinear approximation.

$$k_{eff} = k_{sec} \quad (kN/m) \tag{6.15}$$

Suppose the pushover curve is obtained, which is the summation of the lateral force or the base shear vs. the roof displacement. The pushover curve is intersected with the capacity curve, which provides the basis for the following procedure:

1. Estimate the target roof displacement x_0, based on either an initial assumption or information obtained from previous iterations of this procedure. Once the displacement is obtained, either estimated initially or through iterative calculation, the base shear V can be determined through the pushover curve.

   Furthermore, the acceleration is computed by using the fundamental modal parameters. Thus,

$$a_{1D} = \frac{V}{m_{eff1}} \quad (m/s^2) \tag{6.16}$$

   Here m_{eff1} is the effective mass for the first mode, which is repeated as follows:

$$m_{eff1} = \frac{\left[\boldsymbol{p}_1^T \mathbf{MJ}\right]^2}{\boldsymbol{p}_1^T \mathbf{M} \boldsymbol{p}_1} \tag{t}$$

   In the above equation, $\boldsymbol{p}_1$ is the fundamental mode shape of the first effective mode that was defined previously.

2. The effective stiffness and viscous damping ratio can be determined as a function of the ductility and the expected shape of the hysteresis relationship for the response at the ductility level using either an explicit calculation or tabulated data for different seismic framing systems (Applied Technology Council 1996).

   For example, since in this step the displacement x_0 is given, the corresponding stiffness of the system can be calculated. The unloading stiffness k_u (initial stiffness when the system is still in elastic range) is

$$k_u = \frac{0.6f_m}{d_{0.6}} \quad (kN/m) \tag{6.17}$$

   The effective stiffness and damping ratio can also be calculated. The secant stiffness is assumed to be the effective stiffness k_{eff}. Thus,

$$k_{eff} = f_m/x_0 \quad (kN/m) \tag{6.18}$$

The secant stiffness, k_{eff}, is the slope of the line from the origin to the point at the target displacement x_0, which is shown in Figure 6.2a. The corresponding structural ductility, μ_D, is then written as

$$\mu_D = \frac{k_u}{k_{eff}} = \frac{k_u}{k_{sec}} \tag{6.19}$$

From Equation 6.1, it is seen that,

$$T_{eff} = \sqrt{\frac{k_u}{k_{eff}}} T_1 = 2\pi\sqrt{\frac{f_m m}{x_0}} \quad (s) \tag{6.20}$$

3. Generally speaking, a standard design spectral value determined by T_1 as the period of the structure in linear range with k_u and $\xi = 0.05$ can be compared. In this case, it is known that the structure is in the inelastic range or has a large amount of bilinear dampers installed, so that the total system is nonlinear. In other words, the "yielding point" has been passed. By using effective period T_{eff} and the damping ratio ξ_{eff} obtained above, values through the design response spectra of both the acceleration $S_A(T_{eff}, \xi_{eff})$ and the displacement $S_D(T_{eff}, \xi_{eff})$ are constructed.

   Note that FEMA suggests using Equation 6.20 to estimate T_{eff}, and using Timoshenko damping to estimate ξ_{eff}. Yet, the relationship between S_D and S_A is

$$S_D = \frac{T^2}{4\pi^2} S_A\, g \quad (m) \tag{6.21}$$

Suppose S_A and S_D are now available, which are the demand values calculated in this step. Further, calculate a_{1D} and d_{1D} as follows and compare them with these demand values. The acceleration and the displacement of the effective SDOF are converted, to further determine the responses of the nonlinear MDOF system.

The roof displacement d_{1D} for the fundamental modal representation of the structure is

$$d_{1D} = \frac{x_{11}}{\Gamma_1 p_{11}} \quad (m) \tag{6.22}$$

where x_{11} is the fundamental modal displacement at location 1, i.e., the roof. Meanwhile, Γ_1 is the modal participation factor of the fundamental mode and p_{11} is the first element of the fundamental mode shape. Note that in this case, x_{11} is taken to be x_0, which is obtained through step 1 (see Figure 6.3).

By using the computed quantities a_{1D} and d_{1D} from Equations 6.22 and 6.16, the spectral capacity curve can be plotted, as shown in Figure 6.3. The intersection point of the spectral capacity curve and the design demand curve can be found to determine the design displacement. Note that the design spectrum was discussed in Chapter 2. In Figure 6.3, the Y-axis of the capacity spectrum in Figure 6.3b is represented by S_A and shown by the broken curve. This indicates the statistically determined spectral value. The design spectrum represents the relationship between the spectral acceleration and the spectral displacement. The Y-axis for this pushover-induced response is displayed as the acceleration a_{1D}. To find the response

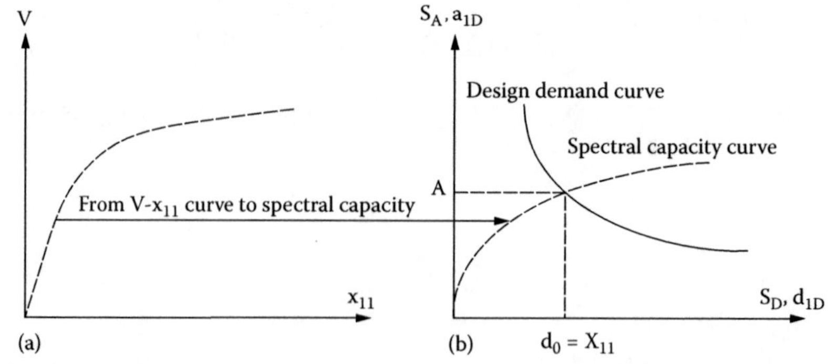

FIGURE 6.3 Pushover analysis: (a) force vs. displacement, (b) demand and capacity curves.

acceleration, S_A and a_{1D} have the same meaning (note that the difference is the constant g). Therefore, both S_A and a_{1D} are used in this figure. Similarly, the X-axis is marked by both S_D and d_{1D}. Note that, either a_{1D} nor d_{1D} contains a modal participation factor.

4. Compare the displacement amplitude calculated from the assumed secant stiffness and the damping ratio with the displacement assumed in step 1. If the values differ by more than 10%, then k_u and k_{eff} from step 1 need to be iteratively recalculated until the result converges.
5. For higher modes, this procedure is followed and the ith damping ratio is calculated by using the Timoshenko damping as

$$\xi_i = \frac{E_{d_i}}{4\pi E_{p_i}} \tag{6.23}$$

Here, subscript i denotes the ith mode and, for convenience, subscript "eff" is omitted.

The ith relationship between the spectral acceleration and the displacement, a_{iD} and d_{iD}, respectively, is repeated as

$$d_{iD} = \frac{T_i^2}{4\pi^2} a_{iD} \quad (m) \tag{6.24}$$

In Equation 6.24, T_i is the ith period. Meanwhile, the spectral acceleration is computed by using the ith modal parameters, which are given by

$$a_{iD} = \frac{V_i}{m_{effi}} \quad (m/s^2) \tag{6.25}$$

Here V_i is the base shear of the ith mode and m_{effi} is the effective mass for the ith mode, which is repeated as

$$m_{effi} = \frac{\left[\boldsymbol{p}_i^T \mathbf{M}\, \mathbf{J}\right]^2}{\boldsymbol{p}_i^T \mathbf{M}\, \boldsymbol{p}_i} \quad (t) \tag{6.26}$$

where $\boldsymbol{p}_i$ is the ith mode shape defined previously.

The spectral displacement d_{iD} for the fundamental modal representation of the structure is

$$d_{iD} = \frac{x_{1i}}{\Gamma_i p_{1i}} \quad (m) \tag{6.27}$$

where x_{1i} is the ith modal displacement at location 1, the roof, while Γ_i is the modal participation factor of the ith mode and p_{1i} is the first element of the ith mode shape.

Once the displacement response amplitude is calculated, it must be compared with that obtained from the assumed secant stiffness and the damping ratio with the displacement assumed in step 1. If the values differ by more than 10%, then k_u and k_{eff} from step 1 need to be iteratively recalculated until the result converges.

6. The total response is finally calculated through proper combinations, such as the SRSS method mentioned earlier.

   There are alternative ways to determine the effective stiffness. This method can be seen from Figure 6.2b, where the piecewise lines with slopes k_u and ak_u are used to represent the pushover curve and $a < 1$ is a proportionality coefficient. The intersection of the two lines is the yielding point. From Figure 6.2b, it is seen that, in this case,

$$k_{eff} = \frac{1+a(\mu-1)}{\mu} k_u \quad (kN/m) \tag{6.28}$$

where μ is the *displacement ductility* and

$$\mu = \frac{x_0}{d_y} \tag{6.29}$$

In Equation 6.28, the following relationship between k_d and k_u is used:

$$k_d = ak_u \quad (kN/m) \tag{6.30}$$

where a is a proportionality coefficient called the *stiffness yielding ratio* and, usually, $a \ll 1$.

The corresponding structural ductility μ_D is then written as

$$\mu_D = \frac{k_u}{k_{eff}} = \frac{\mu}{1+a(\mu-1)} \tag{6.31}$$

Furthermore, from Equation 6.1, it is seen that,

$$T_{eff} = \sqrt{\frac{\mu}{1+a(\mu-1)}} T_1 = 2\pi \sqrt{\frac{\mu m}{[1+a(\mu-1)]\, k_u}} \quad (s) \tag{6.32}$$

6.1.2.3 Engineering Issues

In the literature, it is shown that pushover analysis can provide more accurate results than other static nonlinear analysis. Furthermore, this method is claimed to be suitable for structures with large damping. However, pushover analysis is based on the static pushover curve and the design demand curve. Therefore, it has at least three inherent issues that will affect the design accuracy, especially when large damping is present.

6.1.2.3.1 Issues of Estimation of Effective Stiffness

First, pushover analysis uses the secant stiffness k_{eff} (see Figure 6.2). This is equivalent to using the diagonal line of a parallelogram. Thus, the stiffness can be overestimated and the damping is often underestimated.

6.1.2.3.1.1 Conservative Energy Consider the definition of stiffness in a linear system. It is well known that the stiffness of a linear system defines a unique relationship between the force and the displacement. Suppose under a force f_m, the system has a deformation x_0, and the rate defines the stiffness. That is, for static cases,

$$k = f_m/x_0 \quad (kN/m)$$

For a given displacement x_0, the system will have a potential energy $E_p = kx_0^2/2$ (see Equation 1.184). Therefore, another expression for the stiffness is

$$k = \frac{2E_p}{x_0^2} \quad (kN/m)$$

Apparently, in a linear system, the above two expressions are identical. This is because the potential energy can be written as

$$E_p = \frac{f_m x_0}{2} \quad (kN\text{-}m)$$

However, in a nonlinear system, this equation will no longer hold, because the maximum force, f_m, can contain two components: the dissipative force and the conservative force: $f_m = f_c + f_d$ (see Equation 1.228). Only the conservative force contributes to the potential energy, E_p. That is,

$$E_p = \frac{f_c x_0}{2} < \frac{f_m x_0}{2} \quad (kN\text{-}m) \tag{6.33}$$

6.1.2.3.1.2 Estimation of Effective Stiffness When we use an effective linear system to represent a nonlinear system, the effective stiffness should satisfy the following:

$$k_{eff} = \frac{2E_p}{x_0^2} \quad (kN/m) \tag{6.34}$$

$$k_{eff} = f_c/x_0 \quad (kN/m) \tag{6.35}$$

Therefore, the effective stiffness, k_{eff}, will be smaller than the secant stiffness, k_{sec}.

In Chapter 1, it was discussed that a vibration is caused by the energy exchange between potential and kinetic energies. It is seen that the natural frequency of a linear system can be obtained by letting the maximum potential energy equal the maximum kinetic energy, that is, through the relationship

$$\frac{kx_0^2}{2} = \frac{mv_0^2}{2} = \frac{m\omega_n^2 x_0^2}{2}$$

In nonlinear systems, the above equation is modified as

$$\frac{k_{eff}x_0^2}{2} = \frac{m\omega_{eff}^2 x_0^2}{2} \tag{6.36}$$

or

$$\omega_{eff} = \sqrt{\frac{k_{eff}}{m}} = \sqrt{\frac{f_c/x_0}{m}} < \sqrt{\frac{f_m/x_0}{m}} \quad (\text{rad/s}) \tag{6.37}$$

In other words, considering the dynamic property of a nonlinear system, the secant stiffness generally estimates the accuracy and the effective stiffness of a nonlinear system as defined in Equations 6.28 and 6.29.

In the bilinear case (see the shaded areas in Figure 6.4), when the system moves from 0 to x_0, the potential energy is

$$E_p = \frac{k_u d_y^2 + k_d\left(x_0 - d_y\right)^2}{2} \quad (\text{kN-m}) \tag{6.38}$$

Therefore,

$$k_{eff} = \frac{2E_p}{x_0^2} = \frac{k_u d_y^2 + k_d\left(x_0 - d_y\right)^2}{x_0^2} \quad (\text{kN/m}) \tag{6.39}$$

By using the displacement ductility μ,

$$k_{eff} = \frac{k_u + k_d\left(\mu - 1\right)^2}{\mu^2} \quad (\text{kN/m}) \tag{6.40}$$

Using the notation introduced in Equation 6.30,

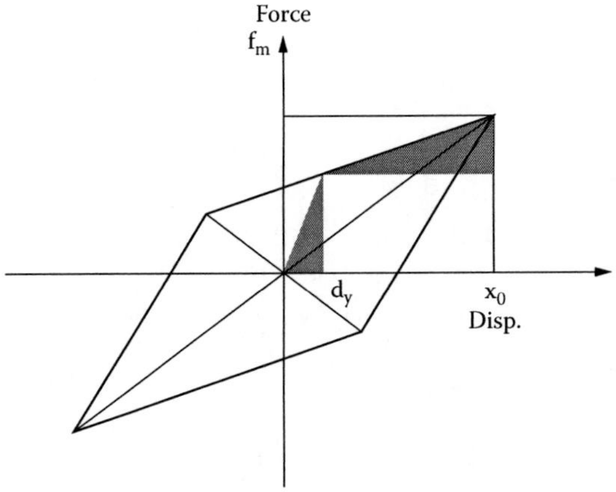

FIGURE 6.4 Maximum potential energy of a bilinear system.

$$k_{eff} = \frac{1+a(\mu-1)^2}{\mu^2} k_u \quad (kN/m) \tag{6.41}$$

Therefore, the corresponding effective period is

$$T_{eff} = \sqrt{\frac{\mu^2}{1+a(\mu-1)^2}} T_1 = 2\pi\mu \sqrt{\frac{m}{\left[1+a(\mu-1)^2\right] k_u}} \quad (s) \tag{6.42}$$

Comparing Equation 6.41 with Equation 6.28, it is seen that the effective stiffness estimated by the secant method overestimates by the factor

$$\frac{\dfrac{1+a(\mu-1)}{\mu}}{\dfrac{1+a(\mu-1)^2}{\mu^2}} = \frac{1+a(\mu-1)}{1+a(\mu-1)^2}\mu \tag{6.43}$$

6.1.2.3.1.3 Comparisons of Damping Ratios

Timoshenko Damping

In the following discussion, the effective damping ratio of the entire bilinear system is derived by using several different approximations of the effective stiffness through the Timoshenko damping approach.

First, by using the approach based on the quantities k_u and ak_u for Timoshenko damping, it can be written that

$$\xi_{eff} = \frac{E_d}{4\pi E_p} = \frac{2(\mu-1)(1-a)}{\pi\mu(1+a\mu-a)} \tag{6.44}$$

In Chapter 5, the formula for potential energy given in Equation 6.38 was used to derive the damping ratio as

$$\xi_{eff} = \frac{2q_d(x_0-d_y)}{\pi\left[k_u d_y^2 + k_d(x_0-d_y)^2 + kx_0^2\right]}$$

From the viewpoint of the entire system, this can be rewritten as

$$\xi_{eff} = \frac{2q_d(x_0-d_y)}{\pi\left[k_u d_y^2 + k_d(x_0-d_y)^2\right]} \tag{6.45}$$

Since the characteristic strength, q_d, for the bilinear system can be written as

$$q_d = (k_u - k_d)d_y \tag{6.46}$$

and further

$$\xi_{eff} = \frac{E_d}{4\pi E_p} = \frac{2(\mu-1)(1-a)}{\pi\left[1+a(\mu-1)^2\right]} \tag{6.47}$$

Force-based Effective Damping

Another way to calculate the effective damping ratio is through force-based effective damping, that is,

$$\xi_{eff} = \frac{f_d}{2f_m} \tag{6.48}$$

where f_d and f_m are the amplitudes of the damping and maximum forces that were defined previously.

Clearly defining the dissipative and restoring forces in a bilinear system is not a simple task. The following cases may exist: Case 1: friction dampers installed in a linear structure; Case 2: bilinear dampers installed in a linear structure; Case 3: friction dampers installed in a bilinear structure; and Case 4: bilinear dampers installed in a bilinear structure. In each case, the dissipative force, f_d, can be different due to the nonlinearity of the total system.

To simplify the study, assume that the damping force is $f_d = q_d = f_y - ak_u d_y$ and the equivalent restoring force is

$$f_m = (k_u - k_d)dy + k_d x_0 \quad (kN/m) \tag{6.49}$$

Therefore,

$$\xi_{eff} = \frac{f_d}{2f_m} = \frac{(1-a)}{2[1+a(\mu-1)]} \tag{6.50}$$

With different equations for the dissipative force, the calculated damping ratio will be slightly different.

6.1.2.3.2 Value of Design Spectrum

The second approach is an approximation of the design spectrum. The relationship between S_D and S_A is assumed to be proportional to the coefficient $T_{eff}^2/4\pi^2 g$, such as expressed by Equations 6.12 and 6.15. That is, the design demand curve is derived under the condition that the damping force is negligible. When damping becomes large, Equations 6.12 and 6.15 will no longer be valid. In this case, FEMA method 2 introduces another inaccuracy. Thus, when damping is large, the modification factor $\sqrt{1+4\xi_{eff}^2}$ should be considered. This was discussed in Chapter 5.

Example 6.2

A bilinear system with m = 1000 (t), k_u = 100,000 (kN/m), and a = 0.1 is used as an example. Also, suppose that the displacement ductility can be varied from 1 to 10. Equations 6.28, 6.32, and 6.44 are used to estimate the effective stiffness, the effective period, and the effective damping ratio, respectively, when the secant stiffness is used. The corresponding plots are shown, respectively, in Figures 6.5a through c. Furthermore, based on Equation 6.38, the effective stiffness, the effective period, and the effective damping ratio are also estimated through Equations 6.40, 6.42, and

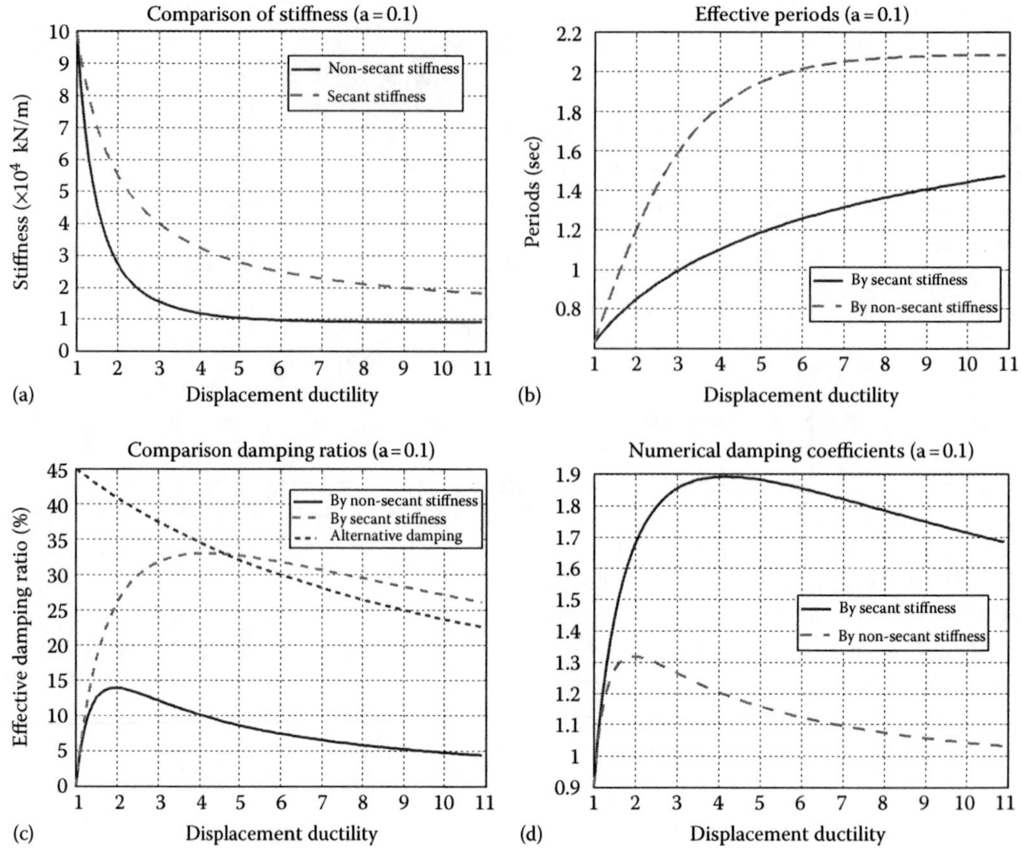

FIGURE 6.5 (a–d) Comparisons obtained using different effective stiffness.

6.47, respectively. The plots are shown in Figures 6.5a through c, respectively, as comparisons. In addition, in Figure 6.5b, the damping ratio obtained by using the approach of force-based effective damping is plotted.

In Figure 6.5d, the numerical damping coefficients are shown. From these plots, it is seen that the difference caused by using the secant and nonsecant stiffness is very significant. In order to observe the effects on the response estimations in Figure 6.6a and c, the seismic spectral coefficients and spectral displacements are plotted, based on the aforementioned different approaches. Here, the estimations are conducted without considering the factor $\sqrt{1+4\xi_{eff}^2}$.

For comparison purposes, in Figure 6.6b and d, the seismic coefficient factors and spectral displacement, respectively, are plotted along with the modification factor $\sqrt{1+4\xi_{eff}^2}$. From Figure 6.6, the differences under the various approaches can be seen.

In this example, the responses are estimated through simplified formulas described in Equations 6.7 and 6.8. In fact, Equations 2.336 and 2.338 are used with $A = S_i = 1.0$.

6.1.2.3.3 *Iterations*

The third approach to approximate the effective stiffness of a bilinear system is based on the fact that the effective period and the effective damping ratio are actually functions of the displacement x_0. However, these parameters are also needed to determine the displacement x_0. Therefore, FEMA method 2 must use iteration to determine the effective stiffness. Publications reporting nonconvergent cases (e.g., Nagarajaiah, Mao, and Saharabudhe 2006) can be found in the literature.

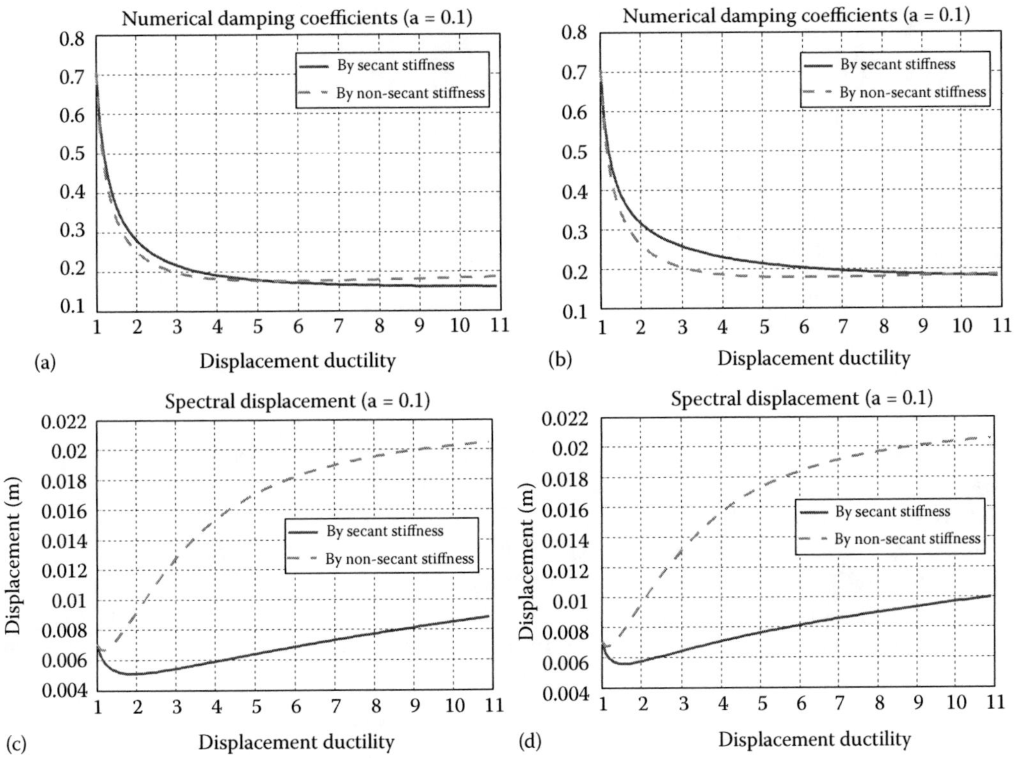

FIGURE 6.6 Response estimations affected by using different effective stiffness: (a) without modification, (b) with modification, (c) without modification, and (d) with modification.

Using pushover analysis for structures with large damping (i.e., the combined damping ratio ξ_Σ is greater than 30%) is not recommended. Therefore, the FEMA method should not be used for final design. A possible criterion is

$$\xi_\Sigma \leq 30\% \tag{6.51}$$

Note that FEMA methods 1 and 2 were not originally established for damper designs. It is particularly difficult to use the FEMA methods for sublinear damper design. Furthermore, in FEMA method 2, the pushover curve is only used for initial estimation. After several iterations, the deformation curve is considerably altered. In this case, it is not necessary to have an "accurate" initial guess, because the computational effort for developing the pushover curve is extensive.

6.1.2.4 Equivalent Linear SDOF Systems

6.1.2.4.1 Equal Energy Approach

The theory of *equal energy* assumes that the maximum deformation energy of an inelastic structure, E_{in}, is equal to the maximum potential energy of an elastic structure, E_{lin}, whose stiffness is identical to the inelastic structure before yielding.

$$E_{nonl} = E_{lin} \tag{6.52}$$

Figure 6.7a shows the relationship between an idealized SDOF EPP system and its corresponding equivalent SDOF linear system. Figure 6.7b shows a general bilinear system and its corresponding linear system. In these figures, f_N and d_N are the maximum force and displacement of the nonlinear

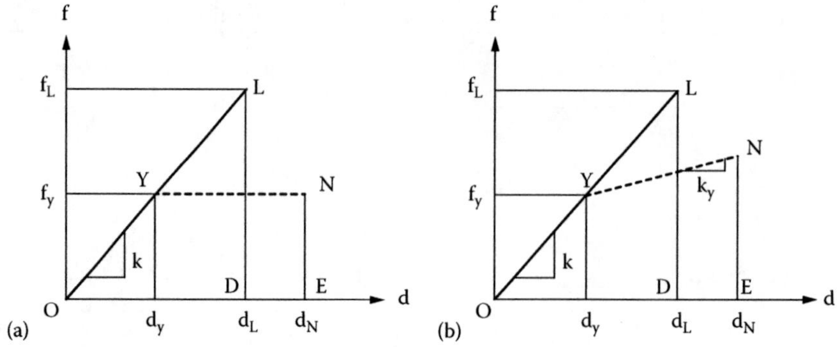

FIGURE 6.7 (a, b) Treatment of nonlinear stiffness.

system, respectively, while f_L and d_L are the maximum force and displacement of the corresponding linear system, respectively. In both cases, the stiffness of the linear system is identical to the unload stiffness (elastic stiffness) of the inelastic system.

According to the theory of equal energy, in Figure 6.7a, the maximum deformation energy is given by the area O-Y-N-E, whereas the maximum potential energy of the corresponding system is given by the area O-L-D. That is, denoting the area O-Y-N-E to be A_{OYNE} and the area O-L-D to be A_{OLD} to represent the amount of energy of the elastoplastic system and the corresponding linear system results in

$$A_{OYNE} = A_{OLD}$$

Equation 6.52 is used to estimate the maximum displacement of the inelastic system by using the corresponding linear system. Without a loss of generality, assume that the mass of these SDOF systems is unity, that is, these systems are all represented in monic form. Furthermore, let the linear systems have a typical damping ratio of 5%. In this case, since the stiffness k is known and using the design spectrum mentioned in Chapter 1, the displacement, d_L, and the force, f_L, are obtained.

Note that in both curves of Figures 6.7a and 6.7b at point Y, the structures start to yield. Thus, f_y and d_y are, respectively, the yielding strength and the corresponding deformation for both cases. Since f_y and d_y are fixed properties of the nonlinear system, f_y and d_y are also known to the designers.

For an EPP system, based on Equation 6.52,

$$d_N = \frac{d_L^2 + d_y^2}{2d_y} \tag{6.53}$$

This equation provides a preliminary estimate of the nonlinear response.

For a general bilinear system, the relationship becomes comparatively more complex. In this case,

$$d_N = d_y + \frac{-d_y + \sqrt{d_y^2 + a\left(d_L^2 - d_y^2\right)}}{a} \tag{6.54}$$

where a is the ratio of the yielding and unloading stiffness, defined previously.

If both the inelastic and equivalent systems have identical input energy under the same seismic excitations, then with equal energy dissipation the remaining vibration energy would be equal. In this case, it is acceptable to have a base using a linear system to represent the nonlinear system. However, it can be seen that the case of deterministic input energy is very rare. In fact, many simulated results can be listed to show that the estimated displacement, d_N, is problematic.

6.1.2.4.2 Equal Displacement and Response Modification Coefficient

The theory of *equal displacement* (Priestley 2000) assumes that the maximum deformation or displacement of an inelastic structure is equal to the maximum displacement of an elastic structure, whose stiffness is identical to the inelastic structure before yielding.

Figure 6.8a shows the relationship between an idealized SDOF EPP system and its corresponding SDOF linear system. Figure 6.8b shows a general bilinear structure and its corresponding linear system. In these figures, f_N and d_N are the maximum force and displacement of the nonlinear system, respectively, and f_L and d_L are the maximum force and displacement of the corresponding linear system, respectively.

According to the theory of equal displacement, in Figure 6.8,

$$d_N = d_L \tag{6.55}$$

In Figure 6.9, it is seen that when the maximum displacement d_L (or d_N) is reached, the linear system will have the maximum force f_L and the nonlinear system will have f_N. Clearly,

$$f_N < f_L \tag{6.56}$$

Therefore, if the structure is allowed to undergo nonlinear deformation, the maximum force will be considerably smaller than the corresponding linear system. The ratio of f_L and f_N is defined as the *response modification coefficient* denoted as *R* (also see Equation 6.2),

$$R = \frac{f_L}{f_N} \tag{6.57}$$

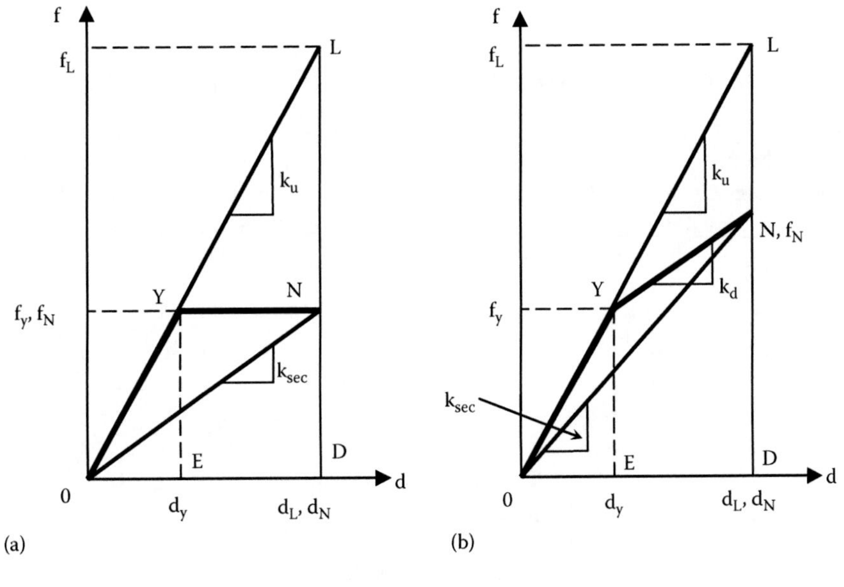

FIGURE 6.8 (a, b) Models of equivalent displacement.

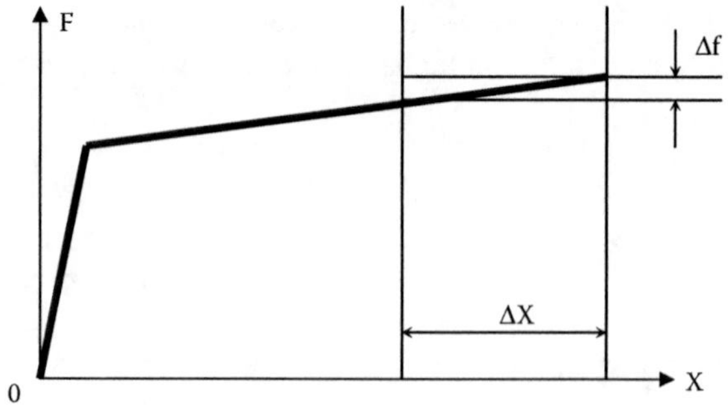

FIGURE 6.9 Variation of seismic force and displacement.

In NEHRP 2009, the *response modification coefficient* is used to reduce the level of seismic load. The idea is that structures designed to take large displacement ductility can have a large value of *R* and the corresponding seismic load will be *R* times smaller.

6.1.2.4.3 Equal Displacement and Force: Direct Displacement-based Design

Recently, in conjunction with the development of the performance-based design approaches, many researchers have pursued additional displacement requirements to ensure ductility. During the past few years, direct displacement-based design (DDBD) has been a subject of interest for the seismic design of buildings.

Figure 6.9 shows a conceptual nonlinear system with large inelastic displacement. It is seen that, when the displacement has a large variation Δx, the corresponding variation of force is relatively much smaller. Thus, when the displacement is fixed as a design target, it is workable in practice, then the design of damping force can be easier.

As noted earlier in this book, the traditional approach in earthquake resistant design of structures is forced-based and typically assumes small damping. Consideration of inelastic deformation capacity is an important additional criterion in performance-based design.

From the perspective of damping design, especially for a large value of damping to be delivered by added devices, the method to best handle the design of these nonlinear systems is not yet well established. There are many interesting and challenging questions that remain to be addressed.

6.1.3 Consideration of Nonlinear Damping for MDOF Systems

When a structure should not be modeled by an SDOF system, or its response cannot be completely represented by the first effective mode of an MDOF system, the methods described in Section 6.1.2 may introduce large errors. In these cases, the nonlinear damping and the nonlinear structural responses should be handled by first identifying the nature of the nonlinearity and then using specific methods to deal with nonlinearities in question.

6.1.3.1 Generic Nonlinear Damping

Although there are many commercially available dampers with subtle differences and specifications, dampers can be classified into dry and wet dampers for designs purposes.

Since both dry and wet dampers are nonlinear, when a structure is built with these dampers, the entire system becomes nonlinear, despite an apparent linearity of the structure itself. However, it is helpful to analyze the following four cases in order to provide an initial estimation of the effectiveness of the corresponding damper design;

1. Bilinear dampers installed in a linear structure
2. Sublinear dampers installed in a linear structure

3. Bilinear dampers installed in a nonlinear structure
4. Sublinear dampers installed in a nonlinear structure

In addition, there can also be cases of combined bilinear and sublinear dampers. However, this book focuses on the four cases mentioned above. More detailed discussions are provided for cases 1 and 2 in the next two subsections.

6.1.3.2 General Idea of Nonlinear Damping Design

Bilinear damping and sublinear damping are considered separately in the following based on their distinct natures.

6.1.3.2.1 Bilinear Damping

Bilinear damping and its design approach are more thoroughly discussed in Chapter 8. The basic idea is briefly described here as part of an overview of nonlinear damping design.

When a single bilinear damper is installed in a structure, the damping force vs. the related response of the structure can be modeled as a parallelogram. Based on the parallelogram, the linearization procedure can be estimated to determine the effective stiffness and damping ratio. The effective period and the numerical damping coefficient can then be determined, which is further used as a link with the design spectrum described in Chapter 2.

When a structure is an MDOF system and multiple bilinear dampers are installed, these dampers will reach their yielding point at different times so that the hysteresis loop will no longer be a parallelogram. In fact, although each individual damper will still have the parallelogram-shaped force–displacement relationship, the "effective modal" force and the corresponding "modal" displacement will no longer behave in accordance with a parallelogram. In many cases, the force–displacement relationship can become closer to the sublinear energy dissipation model. This creates difficulties in following the conventional treatment described above. Furthermore, when a large amount of bilinear damping is used, the issues of nonproportional damping and overdamped system response will occur. These situations are discussed in detail in Chapter 7.

On the other hand, when the dampers reach their yielding points at nearly the same time, the entire system is still treated as bilinear. Thus, in Chapter 8, based on this assumption, a corresponding design procedure is introduced, which is essentially different from these conventional treatments. This is because with the nonlinearity of the system, the effective stiffness and damping will depend on the response amplitude, which is, in turn, a function of the stiffness and damping.

Therefore, the approach of nonlinear response spectra is used. Although the nonlinear response can have considerable variation due to many parameters, focusing on the bilinear SDOF system will reduce the number of these parameters. Four parameters are sufficient to define the hysteresis loop of this SDOF system. For example, the amplitude of peak ground acceleration (PGA), the loading and unloading stiffness, and the characteristic strength can be used to generate the bilinear spectra.

An MDOF bilinear system, where it is assumed that the dampers will yield at the same time, can be linearized to have several effective "modes," each of which is assumed to be a bilinear SDOF system. Once the parameter of the SDOF system is determined, based on the bilinear spectra, the "modal" responses can be calculated. Then, the SRSS method can be used to summarize these responses.

In this process, the "mode" shape function or deformation shape function is needed. In Chapter 3, the shape functions are taken from an equivalent proportionally damped system. In Chapter 8, the generally damped system will be considered.

6.1.3.2.2 Sublinear Damping

Similar to the case of bilinear damping, the basic idea of sublinear damping is provided in the context of an overview for nonlinear damping design. The specific design procedure for sublinear damping is discussed in Chapter 8.

Unlike bilinear systems, where the basic force–displacement relationship can still be treated as a single parallelogram, it is difficult to find the equivalent damping coefficient for these 'effective modes" of sublinear MDOF systems. Therefore, finding the first several "modal" responses is not attempted. Instead, only the fundamental effective mode is considered for the determination of the equivalent damping coefficient. Once this quantity is calculated, the equivalent linear system and the responses of the first several modes can be constructed. Since the equivalent system is linear, the system can be decoupled and the modes can be accurately determined.

If the degree of nonlinearity is high, similar to the case of bilinear damping design, the nonlinear response spectra will be used. Again, focusing on the sublinear damping, only four parameters are needed. These parameters are the level of PGA, the modal stiffness of the original system without the sublinear damper, the equivalent damping coefficient, and the damping exponent.

6.1.4 Yield Structure with Supplemental Damping: Simplified Approach

In the above discussion, the state-of-the-practice for simplified nonlinear analysis of a structure that can enter a yielding state, which has been used in many codes, was briefly introduced. However, the method for adding damping devices to a structure has not yet been addressed. In the following discussion, a simplified method for adding dampers to nonlinear structures is provided.

The basic idea of the simplified methods is to use an equivalent linear SDOF system to approximate the MDOF nonlinear structure with dampers. Among the aforementioned methods, the pushover analysis is often deemed to be the best choice among the simplified analyses. To carry out the analysis, both the effective damping ratio and the effective period of the corresponding SDOF system need to be determined.

6.1.4.1 Damping Ratio Summability

When dampers are installed into a structure, there are typically three types of energy dissipations. To count all of these energy dissipations, the concept of combined effective damping ratio, ξ_Σ, is used, where it is assumed that

$$\xi_\Sigma \approx \xi_i + \xi_d + \xi_s \tag{6.58}$$

with ξ_Σ as the combined effective damping ratio. This can consist of the damping ξ_i initially possessed by the structure, also called the *inherent damping ratio*; the damping ξ_d contributed by the added devices, called the *device damping*; as well as the damping ξ_s caused by the nonlinear deformation of the total systems, which is sometimes referred to as the *structural damping*. In fact, all of the "effective damping ratios," ξ_{eff}, described in the previous section are actually the values of ξ_Σ in Equation 6.58. That is,

$$\xi_s = \xi_{eff} \tag{6.59a}$$

Liang and Lee (1991) have shown that when damping is not very large, i.e., the criterion in Equation 6.51 is satisfied, then Equation 6.58 will approximately hold.

Thus, Equation 6.58 is the basis for the design of the dampers for inelastic structures with simplified methods. Based on this assumption, the damping contributions by the energy dissipation capability of the elastic structure itself, the added damper, and the inelastic deformation of the yield components of the structures can be accounted for separately. After finishing all the individual computations, they are combined to obtain the equivalent damping ratio ξ_Σ.

Note that the effective damping ratios used in Equation 6.59a are all based on the assumption that the bilinear deformation can form a full parallelogram. However, practically speaking, when the deformation of a structure enters the inelastic range, the structure may not be fully plastic, nor can it be fully elastic. Therefore, there will always be a certain amount of energy dissipation, but it may not reach a full parallelogram. In this case, ATC-40 (Applied Technology Council 1996) suggests improving Equation 6.59a as

$$\xi_s = \kappa\xi_{eff} \tag{6.59b}$$

Here κ is a modification coefficient, which is a function of the effective damping ratio contributed by the structural yielding. ACT-40 suggests three methods to determine κ.

Method 1: When the energy dissipation loop is close to a full parallelogram, method 1 is used. Namely, when $\xi_{eff} \le 16.25\%$, then $\kappa = 1$. When $\xi_{eff} \ge 45\%$, then $\kappa = 0.77$. When $16.25\% < \xi_{eff} < 45\%$, then κ is determined by linear interpolation.

Method 2: When the structural deformation is not a full parallelogram, but a reasonable amount of energy is still dissipated, method 2 is used. Namely, when $\xi_{eff} \le 25\%$, then $\kappa = 0.67$. When $\xi_{eff} \ge 45\%$, then $\kappa = 0.54$, and when $25\% < \xi_{eff} < 45\%$, then κ is determined by linear interpolation.

Method 3: When the structural deformation is close to elastic and the energy dissipation is low, method 3 is used. In this case, κ is a constant with the value 0.34.

Note that when the effective damping ratio of the yielding structure is calculated, it is not suggested to use those based on the secant stiffness. Namely, Equation 6.38 for the Timoshenko damping and Equation 6.40 for the force-based damping are considered more suitable than the other equations.

6.1.4.2 Period Determination

In the previous section, the modification factor, *R*, was introduced to determine the effective period of a yielding structure from its original period, T_1. In this case, no dampers are added.

When bilinear dampers are used, they can provide certain stiffness. In order to simplify the analysis procedure, assume that a conceptual bilinear damper is installed into an SDOF bilinear structure. The bilinear damper has an effective stiffness denoted by k_{effd}; and the structure has an effective stiffness denoted by k_{effs}. In Chapter 8, a more detailed method to combine these two ports of stiffness is given. Generally speaking, these two quantities cannot be simply added to estimate the total effective stiffness, because the system is nonlinear.

However, in simplified design, the yielding of the damper should take place well before the yielding of the structure. In this case, the stiffness contribution of the damper for the total system can be considered to be a constant value. This value will be greater than k_{dd}, the yielding stiffness of the damper. However, in many cases, it is acceptable to let

$$k_{effd} \approx k_{dd} \quad (\text{kN/m}) \tag{6.60}$$

and let the total effective stiffness of the structure with the damper be

$$k_{eff} \approx k_{dd} + k_{effs} \quad (\text{kN/m}) \tag{6.61}$$

Here k_{eff} is the effective stiffness of the structure with supplemental dampers. It should not be confused with the effective stiffness used above for the bilinear structure only.

Therefore, the effective period is written as

$$T_{eff} \approx 2\pi\sqrt{\frac{m}{k_{dd} + k_{effs}}} \quad (s) \tag{6.62a}$$

Note that when compared with the structural stiffness k_{effd}, the effective stiffness of the damper, k_{dd}, is often very small, that is,

$$k_{effd} >> k_{dd}$$

Therefore, it is often acceptable to have

$$T_{eff} \approx 2\pi\sqrt{\frac{m}{k_{effs}}} \quad (s) \tag{6.62b}$$

Using sublinear dampers, Equation 6.62b also holds.

In practice, designers may attempt to use a bilinear damper to increase the lateral stiffness of a structure. However, the authors do not suggest this idea for aseismic design, because the stiffness of the yielding structure can significantly overestimate the value when cases i and ii, as described in ATC-40, are considered.

6.2 IRREGULAR MDOF SYSTEM

Most building codes address the issue of vertical and plan structural irregularities, including unevenly distributed weight. Buildings with irregularities are very common, and can often be interpreted as *mass and stiffness irregularities* from the structural dynamics' perspective. When dampers are installed in a structure, the distribution of the damping is not likely to be consistent with the distribution of the stiffness. If the added damping is large, the dynamic responses of the structure may be affected by the uneven energy dissipation, which is caused by damping as well as mass and stiffness irregularities. In practice, *irregular damping* distribution can be more difficult to account for than irregular mass and stiffness distributions.

The issues of irregular damping distribution and generally damped systems are closely related topics. The former is the main cause of generally damped systems. The latter is the theoretical modeling and formulation to account for irregular damping (Liang and Lee, 1991).

6.2.1 IRREGULAR STRUCTURES

6.2.1.1 Effect of Structural Configuration

In the previous chapter, the basic procedure for simplified damper design was discussed, which can be used for regular structures with proportional damping. In the design of dampers, it is very difficult to configure the devices to achieve "regular" damping in most cases. Therefore, these irregular cases must be addressed. Structures with irregular stiffness, irregular mass, and/or irregular damping are referred to as *irregular structures* in this book.

The effect of the irregularity in aseismic designs is considered in this section. Experience dictates that buildings that have irregular configurations can suffer greater damage than those with regular configurations. In NEHRP 2003/2009, the reasons for such phenomena are addressed as follows:

First, in a regular structure, inelastic demands produced by strong ground motion tend to be well distributed throughout the structure, resulting in an even distribution of energy dissipation and damage. In an irregular structure, however, inelastic behavior can concentrate in the zone of irregularity, resulting in rapid failure of structural elements in these areas.

Secondly, some irregularities introduce unanticipated stress into the structure that designers frequently overlook when detailing the structural system.

Thirdly, the elastic analysis methods typically employed in the design of a structure often cannot adequately predict the distribution of earthquake demands in an irregular structure, leading to an unsuitable design.

In addition to the above reasons, irregularities in a structure may introduce significant torsion that can considerably increase the resulting stress in structural elements. An irregular structure will also have nonproportional damping, which may noticeably reduce the effectiveness of vibration control, whereas the damping ratio can be still be very large.

Therefore, for a more comprehensive damper design, the first step is to check the structural irregularity. To do so, the conventional definitions of plane and vertical irregularities are considered and converted into the stiffness, mass, and damping irregularities for better understanding and modeling purposes. In order to involve the mass, damping, and stiffness matrices, a quantitative criterion, based on the concept of mode shapes, is considered.

The most "regular" structure can be described as a simple uniformed cantilever beam. Its mode shapes are pure sinusoidal functions. For example, the fundamental mode shape is a quarter of the sine wave. If the cantilever beam has irregularly distributed mass, it will still have these normal mode shapes. However, for each single mode, the mode shape will no longer be a pure sinusoidal function. Next, suppose the mass is still evenly distributed but the cross section of the cantilever beam changes, so that its stiffness becomes unevenly distributed. The resulting mode shape will not be a pure sine function either. In these two cases, a single mode shape can be expressed as a combination of a series of sinusoidal functions. It is apparent that the more the mode shape departs from the pure sine wave, the more irregularity the cantilever beam can have. Therefore, the mode shape may be used to quantify the structural irregularity.

It was seen that the mode shapes of either a cantilever beam or a structure can be expressed by using the equations introduced in Chapter 2. Here, for convenience, this equation is repeated, that is, the jth element of the ith eigenvector, p_{ji}, should be described as

$$p_{ji} = (-1)^{i+1} \sin\left[\frac{(2i-1)(n+1-j)\pi}{2n}\right], \quad i = 1,\ldots, S; \quad j = 1,\ldots, n \tag{6.63}$$

where S is the total number of modes of interest, n is the number of stories, and i denotes the ith modes, whereas j denotes the jth story. Note that when $j = 1$, p_{1i} denotes the model displacement of the top story of the ith mode, which is always unity due to normalization. The entire i^{th} mode shape can be denoted as

$$\boldsymbol{p}_i = \{p_{ji}\}, \quad i = 1,\ldots, S; \quad j = 1,\ldots, n \tag{6.64a}$$

When the structure has a certain stiffness irregularity, the corresponding mode shapes are disturbed from the pure sine functions. In other words, the mode shape for a single mode will no longer be sinusoidal, but instead is represented by several sinusoidal functions. In this case, the mode shape is denoted as

$$\bar{\boldsymbol{p}}_i = \{\bar{p}_{ji}\}, \quad i = 1,\ldots, L; \quad j = 1,\ldots, n \tag{6.64b}$$

Denote r_p as the maximum percentage difference between p_{j1} and $\bar{p}_{j1}$, $j = 2,\ldots, n$, to be

$$r_P = \max_j \left(\left| \frac{\bar{p}_{j1} - p_{j1}}{p_{j1}} \right| \right) \geq G_p \quad (6.65)$$

where r_P can be used as an indicator to determine if the structure has a severe mass-stiffness irregularity. Namely, a preset value, G_p, can be used to determine if the structure has a strong mass-stiffness irregularity. Usually, G_p is chosen to be about 20%.

Note that the indicator r_p is a nonlinear function of the mass-stiffness irregularity, which can be expressed as the change in mass and/or stiffness at specific locations, as mentioned in NEHRP 2003/2009. As the mass and stiffness start to vary, the indicator will grow very quickly. Normally, when a 150% change in mass or stiffness at a location occurs, r_P will become close to 20%. However, after this point, the growth of r_P will slow down. Fortunately, to determine if a structure is mass-stiffness irregular, only the change in r_P from zero to about 20% is needed.

In Equation 6.65, only the fundamental mode is considered. It is possible to involve higher modes with a different set of the criteria, G_p. For damper design, however, using the fundamental mode is sufficient.

Note that in addition to an unevenly distributed mass and stiffness, nonproportional damping will also make the mode shape different from the pure sinusoidal function. To clearly understand these phenomena, the effect of the mass and stiffness is considered first.

6.2.1.2 Conventional Definitions

In most building codes, the concepts of irregular stiffness and mass are described as the vertical irregularity and the plane irregularity. For example, as given in Chapter 2, NEHRP 2003/2009 defines them as follows:

1. The vertical irregularity is classified as stiffness irregularity, soft story, and extreme soft story; weight irregularity, vertical geometric irregularity, in-plane discontinuity in vertical lateral force resisting elements, and discontinuity in capacity. These were explained in detail in Table 2.2A.
2. Plane irregularity is classified as torsional irregularity, extreme torsional irregularity, re-entrant corners, diaphragm discontinuity, out-of-plane offsets, and nonparallel system. These were explained in detail in Table 2.2B.

In the above classification, the structural irregularities are explained in design terms that can be easily understood. However, they are not expressed in terms of basic dynamic parameters, namely, the mass, stiffness, and damping. In the NEHRP classification, the stiffness and mass irregularities are mixed in both the plane and the vertical descriptions. The damping irregularity is not mentioned. In order to better model the structures, the mass M, the stiffness K, and the damping C should be considered for both vertical and plane irregularities as described in the following sections.

6.2.2 Vertical Irregularity

The vertical irregularity of a two-dimensional plan, namely, the Y-Z plan defined in Figure 6.10, will be considered first.

For the mass coefficient matrix, there can be two types. The first is the commonly used *diagonal mass or lumped mass*, which presses the mass at each story as a lumped parameter. That is,

$$\mathbf{M} = \text{diag}(m_i) \quad (t) \quad (6.66a)$$

In a some circumstances, the mass is evenly distributed, so that

$$\mathbf{M} = a\mathbf{I} \quad (t) \tag{6.66b}$$

where **I** is an identity matrix and a is the proportional scalar.

The second case of the mass matrix is the *consistent mass*, namely,

$$\mathbf{M} = \mathbf{V}_K \operatorname{diag}(m_i)\, \mathbf{V}_K^T = \mathbf{V}_K\, \mathbf{\Lambda}_M\, \mathbf{V}_K^T \quad (t) \tag{6.67a}$$

where $\mathbf{V}_K$ is the eigenvector of the stiffness matrix **K**, and $\mathbf{\Lambda}_M = \operatorname{diag}(m_i)$ is the eigenvalue matrix of **M**, that is,

$$\mathbf{K} = \mathbf{V}_K \operatorname{diag}(k_i)\, \mathbf{V}_K^T \quad (\text{kN/m}) \tag{6.67b}$$

In Equation 6.77, m_i and k_i are, respectively, the eigenvalues of the mass and the stiffness matrices. If the mass matrix is consistent, namely, the mass **M** and the stiffness **K** share the same eigenvector matrix, then **M** and **K** can commute:

$$\mathbf{MK} = \mathbf{KM} \tag{6.68}$$

For a consistent mass of an n-DOF, if

$$m_i = a k_i, \quad i = 1, \ldots, n \tag{6.69a}$$

or

$$\mathbf{M} = a\mathbf{K} \quad (t) \tag{6.69b}$$

and if the eigenvector matrix satisfies Equation 6.68, the regular mass-stiffness is obtained for the case of a consistent mass. Again, the mass irregularity in this case is not an independent concept, which also relies on the stiffness matrix. In a general case, the stiffness in Equation 6.67b is not regulated. Therefore, even if the mass is consistent with the stiffness, the stiffness matrix can still be irregular and the entire structure can also be irregular.

Practically speaking, however, to determine if the mass is regular, the diagonal mass in Equation 6.66a is often used. In NEHRP 2003/2009, if the lumped mass at any story is 150% of the adjacent story, the mass is considered to be irregular.

6.2.3 Plane Irregularity

6.2.3.1 Principal Axes of Structures

To consider the structural plane irregularity, the total three-dimensional model in the X-Y-Z coordinate, as shown in Figure 6.10, is used. Generally speaking, for a linear MDOF system, the governing equation of motion (see Chapter 3) is rewritten as follows:

$$\begin{bmatrix} \mathbf{M}_X & \\ & \mathbf{M}_Y \end{bmatrix} \begin{Bmatrix} \ddot{\mathbf{x}} \\ \ddot{\mathbf{y}} \end{Bmatrix} + \begin{bmatrix} \mathbf{C}_{XX} & \mathbf{C}_{XY} \\ \mathbf{C}_{YX} & \mathbf{C}_{YY} \end{bmatrix} \begin{Bmatrix} \dot{\mathbf{x}} \\ \dot{\mathbf{y}} \end{Bmatrix} + \begin{bmatrix} \mathbf{K}_{XX} & \mathbf{K}_{XY} \\ \mathbf{K}_{YX} & \mathbf{K}_{YY} \end{bmatrix} \begin{Bmatrix} \mathbf{x} \\ \mathbf{y} \end{Bmatrix} = -\begin{bmatrix} \mathbf{M}_X & \\ & \mathbf{M}_Y \end{bmatrix} \begin{Bmatrix} \mathbf{J}\ddot{x}_g \\ \mathbf{J}\ddot{y}_g \end{Bmatrix} \tag{6.70}$$

where $\ddot{\mathbf{x}}$, $\dot{\mathbf{x}}$, and $\mathbf{x}$; and $\ddot{\mathbf{y}}$, $\dot{\mathbf{y}}$, and $\mathbf{y}$ are acceleration, velocity, and displacement in the X- and Y-directions, respectively; **J** is the input vector, where $\mathbf{J} = \{1\}$ with proper dimensions; and $\ddot{x}_g$ and

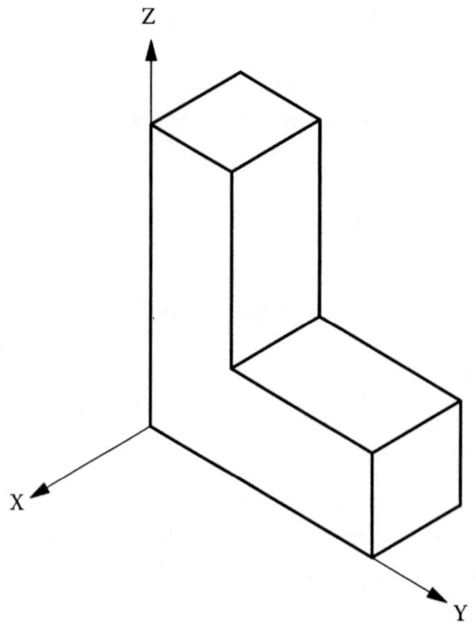

FIGURE 6.10 Coordinate system of structures.

$\ddot{y}_g$ are the ground accelerations in the X- and Y-directions. Note that for a general MDOF structure, if it has more than three degrees of freedom, its stiffness is likely to have no principal axes. Without principal axes, any pair of perpendicular directions can theoretically be selected as the X- and Y-directions. $\mathbf{K}_{XX}$ and $\mathbf{K}_{YY}$ are the stiffness coefficients in the X- and Y-directions, respectively. $\mathbf{K}_{XY}$ and $\mathbf{K}_{YX}$ are, respectively, the cross-stiffness coefficients in the X- and Y-directions and in the Y- and X-directions; $\mathbf{K}_{XY} = \mathbf{K}_{YX}^T$, $\mathbf{C}_{XX}$, and $\mathbf{C}_{YY}$ are the damping coefficients in the X- and Y-directions, respectively; $\mathbf{C}_{XY}$ and $\mathbf{C}_{YX}$ are, respectively, the damping coefficients across the X- and Y-directions and across the Y- and X-directions; and $\mathbf{C}_{XY} = \mathbf{C}_{YX}^T$. The natural directions such as east-west and north-south directions are often chosen as the X- and Y-axes. If in one direction there is n-DOF, the 2n × 2ns stiffness matrix **K** can be partitioned as

$$\mathbf{K} = \begin{bmatrix} \mathbf{K}_{XX} & \mathbf{K}_{XY} \\ \mathbf{K}_{YX} & \mathbf{K}_{YY} \end{bmatrix}_{2n\times 2n} \tag{6.71}$$

Thus, $\mathbf{K}_{XX}$ stands for the n × n submatrix of stiffness corresponding to responses and excitation in the X-direction. K_{XY} stands for the n × n submatrix of stiffness corresponding to responses and excitations in the X-direction and Y-direction, respectively, and so on. Similarly, the 2n × 2n damping matrix **C** can be partitioned as

$$\mathbf{C} = \begin{bmatrix} \mathbf{C}_{XX} & \mathbf{C}_{XY} \\ \mathbf{C}_{YX} & \mathbf{C}_{YY} \end{bmatrix}_{2n\times 2n} \tag{6.72}$$

In general, there can be two equal submatrices of the mass because there should be the same mass distributions in both the X- and Y-directions. For simplicity, the lumped mass matrix, which is diagonal, is taken as

$$\mathbf{M}_X = \mathbf{M}_Y = \text{diag}(m_j),\ j = 1, \ldots, n \tag{6.73}$$

where $\mathbf{M}_{(\cdot)}$ stands for the mass matrix in the (·) direction; and the total mass matrix $\mathbf{M}$ is

$$\mathbf{M} = \begin{bmatrix} \text{diag}(m_j) & \\ & \text{diag}(m_j) \end{bmatrix} \tag{6.74}$$

Note that in the first phase of the damper design, both the cross terms and the directional effects are ignored. That is, the MDOF system is described by using the normal mode approach and letting

$$\mathbf{K}_{XY} = \mathbf{K}_{YX}^{T} = 0 \tag{6.75}$$

and

$$\mathbf{C}_{XY} = \mathbf{C}_{YX}^{T} = 0 \tag{6.76}$$

In this case, the 2n-MDOF system can at least be decoupled into two sets of n-DOF equations. Namely,

$$\begin{bmatrix} \mathbf{M}_X & \\ & \mathbf{M}_Y \end{bmatrix} \begin{Bmatrix} \ddot{\tilde{\mathbf{x}}} \\ \ddot{\tilde{\mathbf{y}}} \end{Bmatrix} + \begin{bmatrix} \mathbf{C}_{XX} & \\ & \mathbf{C}_{YY} \end{bmatrix} \begin{Bmatrix} \dot{\tilde{\mathbf{x}}} \\ \dot{\tilde{\mathbf{y}}} \end{Bmatrix} + \begin{bmatrix} \mathbf{K}_{XX} & \\ & \mathbf{K}_{YY} \end{bmatrix} \begin{Bmatrix} \tilde{\mathbf{x}} \\ \tilde{\mathbf{y}} \end{Bmatrix} = -\begin{bmatrix} \mathbf{M}_X & \\ & \mathbf{M}_Y \end{bmatrix} \begin{Bmatrix} \mathbf{J}\ddot{\tilde{x}}_g \\ \mathbf{J}\ddot{\tilde{y}}_g \end{Bmatrix} \tag{6.77}$$

or

$$\mathbf{M}_X\ddot{\tilde{\mathbf{x}}} + \mathbf{C}_{XX}\dot{\tilde{\mathbf{x}}} + \mathbf{K}_{XX}\tilde{\mathbf{x}} = -\mathbf{M}_X\ \mathbf{J}\ \ddot{\tilde{x}}_g \tag{6.78}$$

and

$$\mathbf{M}_Y\ddot{\tilde{\mathbf{y}}} + \mathbf{C}_{YY}\dot{\tilde{\mathbf{y}}} + \mathbf{K}_{YY}\tilde{\mathbf{y}} = -\mathbf{M}_Y\ \mathbf{J}\ \ddot{\tilde{y}}_g \tag{6.79}$$

where $\ddot{\tilde{\mathbf{x}}}$, $\dot{\tilde{\mathbf{x}}}$, $\tilde{\mathbf{x}}$, and $\ddot{\tilde{x}}_g$, and $\ddot{\tilde{\mathbf{y}}}$, $\dot{\tilde{\mathbf{y}}}$, $\tilde{\mathbf{y}}$, and $\ddot{\tilde{y}}_g$ are respectively newly formed acceleration, velocity, displacement, and ground accelerations in the $\tilde{X}$- and $\tilde{Y}$-directions, in which Equations 6.75 and 6.76 hold.

In this circumstance, the earthquake input in the $\tilde{X}$-direction, $\ddot{\tilde{x}}_g$ will not affect the response in the $\tilde{Y}$-direction, and the earthquake input in the $\tilde{Y}$-direction, $\ddot{\tilde{y}}_g$ will not affect the response in the $\tilde{X}$-direction. The $\tilde{X}$-$\tilde{Y}$ axes are said to be the *dynamic principal axes* of the structure. In other words, if and only if a pair of axes $\tilde{X}$-$\tilde{Y}$ can be found so that the conditions of Equations 6.75 and 6.76 hold, the principal axes can exist. In certain cases, the condition in Equation 6.76 is violated; however, the condition in Equation 6.75 still holds. In this special case, a static input in the $\tilde{X}$-direction will not cause any response in the $\tilde{Y}$-direction, and a static input in the $\tilde{Y}$-direction will not affect the response in the $\tilde{X}$-direction. The $\tilde{X}$-$\tilde{Y}$ axes are the *static principal axes* or simply the *principal axes* of the structure.

6.2.3.2 Cross Effect

The conditions in Equation 6.76 are very difficult to realize in real-world structures. That is, there is rarely a structure that can have dynamic principal axes. In fact, the condition in Equation 6.75 under static loading can also be rather difficult to find. That is, statistically, structures with principal axes are rare. The response in the Y-direction caused by the input in the X-direction is referred to as the *simple cross effect* (Liang and Lee, 2002, 2003); for convenience, it is referred to as the *cross effect.*

Statically speaking, even if the structure does not have principal axes, the response in the Y-direction caused by input from the X-direction, namely, the cross effect, is usually quite small. In most cases, the cross effect can be ignored. However, under dynamic excitation, the cross effect can be cumulative and will significantly enlarge the response, due to the fact that a certain amount of vibration energy will transfer from one direction to its perpendicular direction. Such an energy transfer may sometimes reduce the vibration. However, this vibration reduction will largely depend on the nature of the ground excitation. The energy can also cause an increase in the vibration level. Based on numerical simulation, in many cases, this increase will be more than 30%.

Furthermore, even if the conditions in Equations 6.75 and 6.76 are satisfied, due to the uneven distribution of the mass, the structure can still have a rotation around the Z-axis. The rotation will further cause the response in the Y-direction due to the input in the X-direction. Lin and Chopra (2003) have shown examples of asymmetric structures to explain this phenomenon and modeled them with the following equation of motion:

$$\begin{bmatrix} \mathbf{M} & & \\ & \mathbf{M} & \\ & & \mathbf{I}_{XY} \end{bmatrix} \begin{Bmatrix} \ddot{\mathbf{x}} \\ \ddot{\mathbf{y}} \\ \ddot{\boldsymbol{\theta}} \end{Bmatrix} + \begin{bmatrix} \mathbf{K}_{XX} & 0 & \mathbf{K}_{X\Theta} \\ 0 & \mathbf{K}_{YY} & \mathbf{K}_{Y\Theta} \\ \mathbf{K}_{\Theta X} & \mathbf{K}_{\Theta Y} & \mathbf{K}_{\Theta\Theta} \end{bmatrix} \begin{Bmatrix} \mathbf{x} \\ \mathbf{y} \\ \boldsymbol{\theta} \end{Bmatrix} = -\begin{bmatrix} \mathbf{M} & & \\ & \mathbf{M} & \\ & & \mathbf{I}_{XY} \end{bmatrix} \begin{Bmatrix} \mathbf{J}\ddot{\mathrm{x}}_g \\ \mathbf{J}\ddot{\mathrm{y}}_g \\ \mathbf{J}\ddot{\theta}_g \end{Bmatrix} \tag{6.80}$$

Here, $\mathbf{I}_{XY}$ is the moment of inertia of the structure in the X-Y plan; $\mathbf{K}_{XX}$ and $\mathbf{K}_{YY}$ are as defined previously; $\mathbf{K}_{\Theta\Theta}$ is the submatrix of stiffness corresponding to the X-Y plan rotation; $\mathbf{K}_{\Theta X} = \mathbf{K}_{X\Theta}^T$ is the submatrix of stiffness corresponding to the displacement in the X-direction and the plan rotation; and $\mathbf{K}_{\Theta Y} = \mathbf{K}_{Y\Theta}^T$ is the submatrix of stiffness corresponding to the displacement in the Y-direction and the plan rotation. Additionally, $\boldsymbol{\theta}_g = \boldsymbol{\theta}_g$ (t) is the rotation vector, while $\ddot{\theta}_g = \ddot{\theta}_g(t)$ is the ground rotation acceleration, which is often very small and can be ignored.

From Equation 6.80, it is realized that even if $\mathbf{K}_{XY} = \mathbf{K}_{YX} = 0$, the input $\ddot{x}_g$ will cause a response in the Y-direction and $\ddot{y}_g$ can still cause a response in the X-direction through the rotation.

6.2.4 Irregular Damping

6.2.4.1 Regular Mass-Stiffness, Irregular Damping

In general, there are two types of irregular damping. The first is with irregular mass and stiffness, which has been discussed in the previous subsections. The second is when the mass and stiffness are regular, which will be discussed in the next subsection. In both cases, there is, as yet, no strictly theoretical treatment. Thus, examples are given to illustrate the issues and the process of dealing with the design of irregular building with enhanced damping.

To explain the nature of structures with regular mass and stiffness but irregular damping, an example of a 10-story building is considered. From this example, a pattern can be found and a method to deal with these systems is found.

Example 6.3

Suppose a structure with mass matrix to be $\mathbf{M} = 100\ \mathbf{I}$ (t) and the stiffness to be

$$\mathbf{K} = k_0 \begin{bmatrix} 2 & -1 & 0 \ldots & 0 \\ -1 & 2 & -1 \ldots & 0 \\ & \ldots & & \\ 0 & \ldots & 0 \ -1 & 1 \end{bmatrix}_{10\times10} \quad (\mathrm{kN/m})$$

$$\mathbf{K} = k_0 \begin{bmatrix} 2 & -1 & 0 & \cdots & 0 \\ -1 & 2 & -1 & \cdots & 0 \\ \vdots & \vdots & \vdots & \vdots & \vdots \\ 0 & \cdots & 0 & -1 & 1 \end{bmatrix}_{10\times 10} \quad (\text{kN/m})$$

Here, $k_0 = 1.11 \times 10^6$ (kN/m). It is seen that this example has regular mass and stiffness. It will be seen that even in the case of a structure with regular mass-stiffness, if the damping is irregular, care must be taken to prevent it from leading to nonproportional damping.

6.2.4.1.1 Regular Damping

First, suppose the inherent damping of the structure without added dampers is 1% for all the modes. Then, suppose dampers are added to this structure so that each story is installed with dampers and the damping coefficient is proportional to the stiffness of the story. With this configuration, the added damping will always be proportional. That is, the damping matrix for the added damping is proportional to the stiffness matrix, which is

$$\mathbf{C} = \alpha c_0 \begin{bmatrix} 2 & -1 & \dots & 0 \\ -1 & 2 & \dots & 0 \\ & \dots & & \\ 0 & \dots & -1 & 1 \end{bmatrix}_{10\times 10} \quad (\text{kN-s/m})$$

Here, α is a proportional factor. Initially, let $\alpha = 1$ and $c_0 = 3145$ (kN-s/m), so that the initial damping ratio for the first mode becomes 2% (see Table 6.1).

The proportional factor α is changed and the corresponding damping ratios for the first three modes are plotted in Figure 6.11a, where the X-axis represents the variation of α.

The responses of this structure are plotted in Figure 6.11b for the absolute acceleration and Figure 6.11c for the relative displacement, respectively. In Figure 6.11b, the unit of the Y-axes is (g); and in Figure 6.11c, the unit of the Y-axes is (in). These units are used to compare the reduction of responses vs. an increase in the damping coefficient to the cases in the above example. The responses are also the mean value plus one standard deviation under the 99 earthquake records used in this book. It can be seen that when the damping coefficient increases, the damping ratios of the first mode are always increased. The damping ratios for the second and third modes are roughly always increased as well.

Note that the modal mass ratios of this structure are close to those shown in Figure 3.4, so that only using [illegible] first mode is sufficient to represent the total responses. However, for the purpose of [illegible]e first three modes are listed anyway.

[illegible]s 6.11b and c, it is realized that adding damping always causes the responses to [illegible]gh the rate of decrease becomes smaller and smaller. With many other examples [illegible]h regular mass-stiffness and added proportional damping, when the values of the [illegible]e, the responses will always decrease. Such a structure is referred to as a *regular*

[illegible]ameters with Initially Added Damper

Mode	1st	2nd	3rd	4th	5th	6th	7th	8th	9th	10th
Natural freq. (Hz)	1.12	3.34	5.48	7.50	9.35	11.00	12.39	13.51	14.33	14.83
Damping ratio (%)	2.0	3.97	5.88	7.67	9.32	10.79	12.03	13.03	13.76	14.20

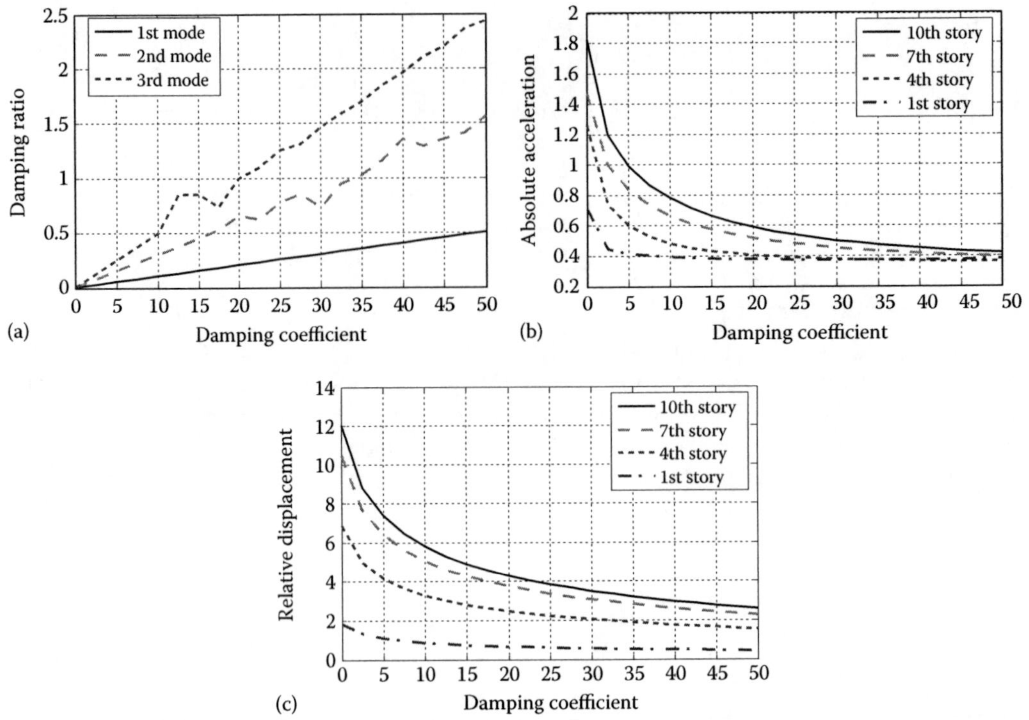

FIGURE 6.11 Damping ratio and responses of a regular structure: (a) damping ratio vs. damping coefficient, (b) absolute acceleration (g), and (c) relative displacement (in).

6.2.4.1.2 *Irregular Damping*

Next, with the same structure, nonproportional damping is added, and the damping matrix is

$$\mathbf{C} = \alpha c_0 \begin{bmatrix} 0 & 0 & \\ 0 & \cdots & \\ & & 1 \end{bmatrix} \quad (\text{kN-s/m})$$

where α is defined as before.

For comparison purposes, the initial damping ratio is set to be 1% for every mode. Then, dampers are continuously added. Initially, $\alpha = 1$, $c_0 = 3600$ (kN-s/m), and α increases. The added damping will make the damping coefficients and the modal energy transfer ratios vary, which are plotted in Figure 6.12a and b, respectively. From these two plots, it is seen that in this case, increasing the damping coefficient will not always increase the damping ratio, as is the case for regular structures.

In Figure 6.12c and d, the absolute acceleration and the relative displacement responses of this structure are plotted, respectively. In Figure 6.12c, the unit of the Y-axes is (g); and in Figure 6.12d, the unit of the Y-axes is (in) for comparison purposes. The responses are also the mean value plus one standard deviation under the aforementioned 99 earthquake records. From Figure 6.12c and d, it is realized that adding damping does not always cause the responses to decrease for all stories in irregular structures.

Since the mass and stiffness, as well as the damping configuration can greatly affect the reduction of the structural responses, it is necessary to study their behavior and suggest methods to deal with these phenomena, especially the effect of nonproportional damping.

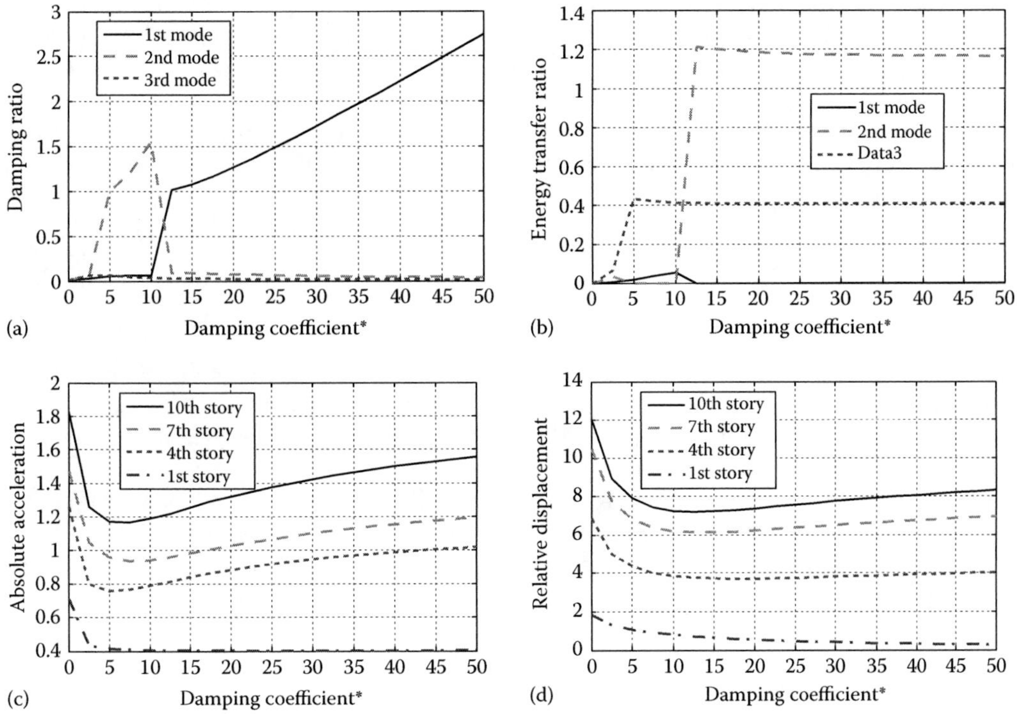

FIGURE 6.12 Damping ratio and responses of an irregular structure: (a) damping ratios, (b) modal energy transfer ratios, (c) absolute acceleration (g), and (d) relative displacement (in).

From the above description, the cause of damping irregularity is classified as follows:

1. Mass and stiffness are regular, but the damper installation is not proportional
2. Either the mass or stiffness or both are irregular and the damper installation is nonproportional

In other words, the damping irregularity is caused by nonproportional damping. That is, when a structure has irregular mass and stiffness, it is difficult to realize a proportionally damped system.

For irregular structures, if the normal mode approach is still used, then large errors may occur. To overcome this problem, the stiffness matrix must be examined. Note that, many commercial finite element programs, such as SAP 2000, do not allow users to directly handle the stiffness matrix.

One of the benefits of using the stiffness matrix is that the natural periods and mode shapes can be calculated for all modes interested. Various approximations do not need to be used. However, the traditional approach of using the stiffness matrix is to obtain the normal mode shapes. In this book, the complex mode approach is used to obtain considerably more accurate results.

6.2.5 Design Considerations for Nonproportionality Damping

From the above example, it is seen that for an irregular structure, the damping ratios of certain modes can quickly reach and exceed unity, whereas the damping ratio of the first mode is still rather small. Two cases may exist. In the first case, if more damping is added, the overdamped subsystems will continue to dissipate energy. Furthermore, the overall dynamic stiffness will continue to increase. Therefore, continuously adding damping will further reduce the structural responses, if the damping ratio of the fundamental mode is not satisfied. In the second case, however, continuously adding damping will no longer help further reduce the structural responses. Instead, the structural response will be magnified. This is because when the damping is unevenly distributed, at certain locations, the corresponding response will become very small, but at other locations the responses will remain large.

In practical damping design, the period and the damping ratio of a structure with damping irregularity may need to be estimated. To approximately quantify the error of estimation in the period, when the damping ratio and the energy transfer ratio of the fundamental modes are not very large, i.e., <35% and <3%, respectively,

$$\varepsilon_{T1} = 1 - \frac{T_{c1}}{T_{r1}} \approx \zeta_1 \tag{6.81}$$

Here ζ_1 is the fundamental energy transfer ratio; ε_{T1} is the percentage error of the fundamental period estimation; T_{r1} is the fundamental period estimated by the real mode approach; and $T_{c1} = T_1$ is the fundamental period estimated by the complex mode approach, which equals the true value of the fundamental period without errors.

The errors in estimating the damping ratio are more difficult to predict as the modal energy transfer ratio becomes larger. However, if the energy transfer ratio is notably smaller than 1% and no overdamped pseudo modes exists, using Rayleigh damping as an approximation,

$$\varepsilon_{d1} = \left| 1 - \frac{\xi_{r1}}{\xi_{c1}} \right| \approx \kappa \left| \zeta_1 \right| \tag{6.82}$$

Here, ε_{d1} is the percentage error of the fundamental damping ratio estimation; ξ_{r1} is the fundamental damping ratio estimated by the real mode approach; and $\xi_{c1} = \xi_1$ is the fundamental damping ratio estimated by the complex mode approach, which equals the true value of the fundamental damping ratio without errors; κ is in the range of 0.3~3.

From Equations 6.81 and 6.82, it is realized that the smaller the energy transfer ratio, the smaller the errors of the estimated period and damping ratio. In engineering practice, it is not necessary to minimize the energy transfer ratio to the lowest possible value, as such a minimization will be quite costly. In fact, if normal mode is to be used to approximate the generally damped MDOF systems, the following criterion (Liang, Tong and Lee 1991) may also be used:

$$\zeta_1 < 1\% \tag{6.83}$$

It can be realized that the higher the mode, the larger the error of the period and damping ratio estimation. However, these higher modes often contribute to the total response being much less than the fundamental mode. In practical damper design, the fundamental mode is of greatest interest.

Unlike mass and stiffness irregularities, which can be avoided by choosing the regular vertical and plan stiffness and weight distributions in many cases, the irregular damping is very difficult to prevent. This is why quantitative indices are used to describe the irregular damping. A qualitative explanation is also provided after the discussion on damping matrices.

6.2.6 Response Estimation Using Response Spectra

In the first section of this chapter, the key issues and the main challenges in coping with nonlinear systems were discussed. In practice, equivalent linear models are used to approximate nonlinear systems. Using response spectra to estimate the nonlinear responses is one possible means of measurement.

Using time history analysis to calculate the nonlinear response is a well-developed process. This direct way of coping with a nonlinear system is based on step-by-step temporal integration, which does not need any linearization and can provide very accurate results. However, the time history computation is extremely time consuming and is only for one set of input. Therefore, spectral analysis not only possesses clear physical meaning that is easier to understand and is familiar to designers, but it also results in simple and efficient designs.

To summarize these approaches for working with a nonlinear system with irregular damping, there are three basic approaches. The first is a direct linear model, which considers a predetermined load demand and a linear stiffness (load-deformation constitution), such as FEMA nonlinear method 1. Due to the broad possibility of the level and duration of random loading, this method may not be suitable for predicting the structural response, but rather may be used as a reference for comparison.

The second measure is to adopt an iterative procedure for response computation. At each iteration step, the piecewise linearization can be used to approximate the nonlinear constitution. Two uncertainties regarding the selection of the initial guess and the iteration model will affect the accuracy and computation burden. According to a previous study, the initial guess is less important since an "accurate guess" such as a pushover curve is not necessary. However, establishing the iteration model is vital, as publications on the nonconvergence of currently used iteration can be found in the recent literature. A method that is proven to be convergent and to converge to correct points calibrated through time history analysis has been developed. However, this method is only for SDOF systems. Thus, one of the future research objectives is to develop this method for MDOF structures.

The third approach is to use the nonlinear response spectra. Due to the wide variation of earthquakes, as well as many other types of dynamic loads, many parameters can make universal nonlinear response spectra extremely difficult to generate and use. However, if the role of damping is clearly quantified, it has been found that only two different types of damping can be employed to approximate the most commonly used damping; namely, bilinear damping and sublinear damping.

When a structure enters its inelastic range, due the formation of plastic hinges, the entire system usually behaves like a piecewise sublinear system. This book favors the use of nonlinear response spectra based on the piecewise sublinear system model.

6.3 MINIMIZING DAMPING NONPROPORTIONALITY

As pointed out earlier, for SDOF systems, increasing the damping coefficient instead of the damping ratio will usually result in vibration reduction, and it is not always the case for MDOF systems. Actual structures are often nonproportionally damped MDOF systems. When structures have large irregularities in mass, damping, and/or stiffness, nonproportional damping should be addressed in design. In general, the conservative energy of the vibrating structure should be minimized by maximizing the dynamic stiffness of the system.

In the previous section, the presence of irregular damping and nonproportional damping was discussed. This section focuses on minimizing the nonproportional damping. Strictly speaking, properly designed nonproportional damping can be used to effectively reduce the seismic responses of a structure. However, this type of design needs more sophisticated and careful computation, which is beyond the scope of this book. Therefore, without rigorous theoretical background and careful structural analysis, minimizing the nonproportional damping as a design principle is suggested herein.

6.3.1 Further Discussion on Conventional Energy Equation

6.3.1.1 Minimization of Conservative Energy

The input energy to the structural system from earthquake ground motion is not a constant value. The idea that the more energy dissipated by damping, the lower the remaining energy and the greater the reduction in the vibration level is not always true.

Liang et al. (1999) suggested an energy relationship, which is given in Equation 6.84 without proof. The energy equilibrium for the ith mode is

$$E_{mc_i}(t)+E_{mr_i}(t)+E_{dp_i}(t)+E_{dr_i}(t)+E_{kc_i}(t)+E_{kr_i}(t)=W_{e_i}(t)+W_{m_i}(t) \tag{6.84}$$

where superscript i denotes the ith mode. Here, $E_{mc_i}(t)$ is the conservative part and $E_{mr_i}(t)$ is the variable part of the vibratory kinetic energy; $E_{dp_i}(t)$ is the fixed portion and $E_{dr_i}(t)$ is the variable part of the energy dissipated by damping; $E_{kc_i}(t)$ is the conservative part and $E_{kp_i}(t)$ is the variable part of the vibratory potential energy; $W_{e_i}(t)$ is the work done by the external force and $W_{m_i}(t)$ is the energy transferred from (to) other modes. In the circumstance of pure passive control with fixed dampers,

$$E_{mr_i}(t)=E_{kr_i}(t)=0 \tag{6.85}$$

$E_{dr_i}(t)$ can be treated as the energy dissipated by the structure itself and $E_{kc_i}(t)$ is the energy dissipated by the supplemental dampers.

Note that, the work done by seismic force, denoted as $W_S(t)$, is

$$W_s(t)=\sum_i\left[E_{mc_i}(t)-W_{e_i}(t)\right] \tag{6.86}$$

In order to control the vibration induced by earthquake ground excitations or by other dynamic forces, the conservative portion of the energy, $E_c(t)$, stored in the structure needs to be minimized. Namely, let

$$\sum_i\left[E_{mc_i}(t)+E_{kc_i}(t)\right]=\left[E_c(t)\right]=\min\Big|_{\text{at all t}} \tag{6.87a}$$

Equation 6.87a is referred to as the *design principle of minimum conservative energy* for passive damping control. Note that with real-time structural parameter modification technologies (Liang and Lee 1999), $E_{mr_i}(t)$ and $E_{kr_i}(t)$ may become nonzero valued. However, $E_c(t) = \min$ still holds. From Equations 6.84 and 6.86a,

$$\min\left(E_c(t)\right)=\min\sum_i\left\{\left[W_{e_i}(t)+W_{m_i}(t)\right]-E_{dp_i}(t)-\left[E_{mr_i}(t)+E_{dr_i}(t)+E_{kr_i}(t)\right]\right\} \tag{6.87b}$$

In Equation 6.87b, the two terms in the first bracket on the right side are energy input. The third term represents the energy dissipated by the damping force. The remaining three terms on the right side of Equation 6.87b are the energy quantities, which can be removed by adjusting the mass, damping, and stiffness.

In Equation 6.87b, for proportionally damped system, the minimal amount of energy transfer possible is zero. Thus,

$$W_{m_i}(t)=0 \tag{6.88}$$

and

$$\min\left(E_c(t)\right)=\max\left[-W_{e_i}(t)+E_{dp_i}(t)+E_{dr_i}(t)\right] \tag{6.89a}$$

otherwise,

$$\min\left(E_c(t)\right)=\max\left[-W_{e_i}(t)-W_{m_i}(t)+E_{dp_i}(t)+E_{dr_i}(t)\right] \tag{6.89b}$$

In order to seek its maximum value, with the damping coefficient as a parameter, the derivative of the summation of the energy terms with respect to time is taken and the result is zero. That is,

$$C\dot{\mathbf{x}} - \mathbf{f} = 0$$

or

$$\dot{\mathbf{x}} = C^{-1}\mathbf{f}$$

Here **f** is the input force.

The above equation holds only if the response reaches the resonant steady state under sinusoidal excitation. And it is known that in this case, $E_c(t) = \max$. In other words, the summation of $[-W_{e_i}(t) + E_{dp_i}(t) + E_{dr_i}(t)]$ will not have a minimal value. That is, when displacement **x** changes, the amount of the variation of energy will not reach any particular point, so that a minimal value is expected, as seen in Equation 6.89. However, from the term $[-W_{e_i}(t) + E_{dp_i}(t) + E_{dr_i}(t)]$, it is also realized that the smaller the displacement **x**, the smaller the amount of energy, $E_c(t)$, will be. Note that for an MDOF system, **x** can be represented by

$$\mathbf{x} = \mathbf{K}_d^{-1}\mathbf{f} \tag{6.90}$$

where $\mathbf{K}_d$ is the dynamic stiffness. If the amplitude of the input force **f** remains the same and the dynamic stiffness $\mathbf{K}_d$ becomes larger, the displacement **x** will be smaller. Thus, for passive damping control, the maximum reduction of the displacement, the maximum amount of energy $\Sigma_i[-W_{e_i}(t) + E_{dp_i}(t) + E_{dr_i}(t)$, and the minimum amount of conservative energy are equivalent.

Equations 6.89 implies that the energy dissipated by the damping force, $E_{dp_i}(t)$, should be maximized. That is, it is desired that the damping effect is increased as much as possible, as long as it does not result in an increase in the input energy. In the following discussion, it is shown that for SDOF systems, with or without consideration of the supporting stiffness, increasing the damping coefficient will always reduce the vibration level. However, increasing the damping ratio will not always yield a good result for MDOF systems.

Next, all the other energy terms will be examined. First, the work done by the external force, $W_{e_i}(t)$, is considered. This energy can be affected by two terms. One term is the work done by the external static force only, which is a function of the static force and the corresponding static displacement. Most structures are designed based on static loads and the static stiffness will not be affected by adding dampers. In other words, the energy dissipation mechanism will not change the static configurations of force and displacement. This quantity of energy and the corresponding displacement is the lower limit of using any protective system. To evaluate a damper, the second term due to dynamic load must be considered and the deformation under dynamic loading can be compared with the static deformation.

In the discussion of Equation 6.89 above, the energy transfer term $W_{m_i}(t)$ was set to be zero. However, from a nonproportionally damped system, modal energy transfer cannot be ignored, because it can easily cause more than a 30% increase in vibration responses when the damping ratio of the structure is high.

The overall indication of whether a system is proportionally damped is given by the Caughey criterion previously discussed. However, this criterion does not work for individual modes. The direct indication of nonproportional damping for a mode is that the corresponding modal energy transfer ratio is not zero. From the above-mentioned theory of nonproportional damping, an overall nonproportionally damped system does not necessarily have all complex-valued modes. It is possible to isolate a particular normal mode, if this mode happens to be the dominant mode.

6.3.2 Minimization of Damping Nonproportionality

In actual structures, it is difficult to find a pure normal mode. However, some modes may be close to normal, or close to a proportionally damped mode. In order to minimize the effect of nonproportional damping, the structure and the dampers should be configured to have a vibrating mode close to a proportionally damped mode, which can be measured by letting

$$\zeta_i = \min, \quad \text{for } i = 1, \ldots, S \tag{6.91}$$

where S is the number of total modes of interest, as previously defined.

The quantity of complex mode indicators can also be used, that is,

$$r_{Ei} = \min, \quad \text{for } i = 1, \ldots, S \tag{6.92}$$

where the term r_{Ei} is defined in Equation 4.285.

It can be proven that minimizing nonproportional damping is not always beneficial. The reason for using nonproportional minimization is that properly utilizing nonproportional damping to optimize aseismic design is a very challenging task. For engineers who are not well versed in damping design, it is suggested that proportional damping be used, which is safer, but sometimes more conservative.

However, obtaining the modal energy transfer ratio is also difficult for inexperienced designers. Alternative measures are available (Liang et al. 1993; Warburton et al. 1984). Regardless of the approach used, a structure that contains large nonproportional damping in its dominant modes is often caused by improper damper design.

6.4 ROLE OF DAMPING IN NONLINEAR SYSTEMS

In Chapters 1 and 2, the effect of large damping on structural response and vibration suppression was discussed. The focus was on linear viscous damping. In actual structures, there are often nonlinear systems and various types of damping. Therefore, it is helpful to consider the details of two issues regarding the role of damping in supplemental damping design. First, for linear and viscous damping, using additional damping reduces vibration. The natural question to follow is: Can a similar vibration reduction occur in nonlinear systems? That is, what is the effectiveness of damping vibration control for nonlinear systems, especially when the structure enters the inelastic range? The second issue is, in linear systems, using larger damping to reduce the vibration response relies on an increase of the dynamic stiffness (rather than greater energy dissipation). The follow-up question is: What is the main reason for using supplemental damping to reduce the vibration of nonlinear systems? In this section, these two issues are discussed in detail by using the concept of energy.

6.4.1 Linear Systems

6.4.1.1 Response Spectra of Seismic Work and Energy

6.4.1.1.1 Maximum Seismic Work

In Chapter 2, it was observed that in the real spectrum of acceleration, the following equation can be used:

$$S_A = \sqrt{1 + 4\xi^2}\ \omega_n^2 S_D / g$$

It is also known that both S_A and S_D are inversely proportional to the numerical damping coefficient B, when the period is sufficiently long, that is,

$$S_A \propto \frac{1}{B}$$

$$S_D \propto \frac{1}{B}$$

In Chapter 7, it will be shown that the numerical damping coefficient B can be written as

$$B \approx 3\xi + 0.9 \tag{6.93}$$

Therefore, roughly speaking

$$S_A S_D \propto \frac{\sqrt{1+4\xi^2}}{B^2} = \frac{\sqrt{1+4\xi^2}}{(3\xi+0.9)^2} \propto \frac{1}{\xi} \tag{6.94}$$

From Equation 6.94, it seems that the product of the spectral values of the real acceleration and the displacement is approximately inversely proportional to the damping ratio. Thus, such a product, having the dimension of the work, can be a useful indicator of vibration reduction with supplemental damping. This point is further examined in the following paragraphs.

In the previous section, real and pseudo response spectra of acceleration, velocity, and displacement were used to show the effect of damping. In the literature, there are other approaches. For example, Housner (1956) was the first to use energy approach for aseismic design. This approach attempts to use the earthquake input energy to characterize the ground excitation regardless of the structural damping and period.

It is understandable that during an earthquake, the process of energy input is dynamic. That is, at a certain moment, there will be energy input to a structure. At other moments, there will be a certain amount of energy output from the structure, which is transmitted to the ground. There is no direct relationship between the maximum input energy and peak structure responses.

A different spectrum is considered, which is based on the work done by the seismic force. It is the spectrum of an SDOF system with linearly viscous damping, which can be established when the maximum acceleration and displacement are reached. The seismic force has been defined as the product of the mass and the absolute acceleration of an SDOF system, that is (also see Equation 1.204),

$$f = m\ddot{x}_A$$

Note that the ground force is different from the one defined in the above equation. For a monic system, the force becomes

$$f = \ddot{x}_A$$

Namely, the seismic force in the monic case equals the absolute acceleration. In earthquake engineering, it is well known that the damage to a structure or a member of the structure can be related to extra-large displacement or large force. At the same time, the damage to a structure can also be related to the energy or work done by forces. The advantage of using the maximum seismic work is that it contains the effects of both the force and the displacement.

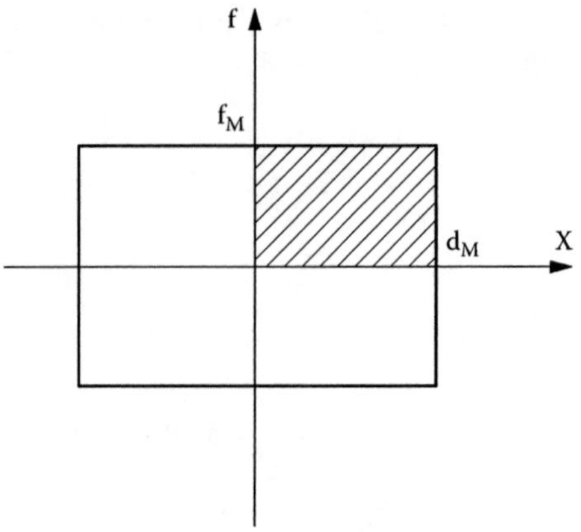

FIGURE 6.13 Rectangular energy dissipation.

In an idealized case, where the seismic force vs. the displacement has rectangular energy dissipation during an individual cycle, the amplitude of seismic force will remain constant, f_m, in this cycle. This case is shown in Figure 6.13.

The relationship of rectangular force–displacement can be significant in the evaluation of energy dissipation. In Chapter 5, it was shown that among all the damped structures, only the system with rectangular energy dissipation has the smallest displacement when the peak seismic force is specified. Therefore, it is helpful to study the spectral value of the rectangular energy dissipation.

To compare a generally damped system with a rectangular-damped system, an idealized *maximum work done by the seismic force* at the maximum displacement, denoted as E_{rec}, is defined as

$$E_{rec} = f_M d_{max} \tag{6.95}$$

where subscript "rec" represents rectangular work done or rectangular energy dissipation. That is, in Figure 6.13, the shaded area or the first quadrant of the rectangular energy dissipation represents this work.

In a linear SDOF system, the energy dissipation is no longer rectangular. Generally speaking, the relationship between maximum force and displacement of a linear SDOF system with viscous damping, which is excited by seismic ground motions, can approximately be expressed as a part of an elliptic curve. This part, i.e., in the first quadrant, can be expanded to the entire X-Y plan.

Example 6.4

A system with a period equal to 1 (s) and a damping ratio equal to 30% has a response history under the excitation of the El Centro earthquake. In Figure 6.14, the responses of the acceleration vs. displacement are plotted. In order to count the area covered by the displacement–acceleration plot, which is related to the energy dissipation, an envelope is also plotted, as shown in Figure 6.14. It is seen that the envelope is very close to an ellipse. The elliptic envelope shown in Figure 6.14 is not just a coincidence. More generally, the responses of a given system under many earthquake excitations can be plotted. In this case, we can also obtain the elliptic envelopes. To better understand this point, another example is given.

Example 6.5

In Figure 6.15, the responses of a system with a period equal to 2 (s) and a damping ratio equal to 50%, under the same 99 records are plotted. The envelope of these responses can also be represented by an elliptic curve, which is shown in Figure 6.15.

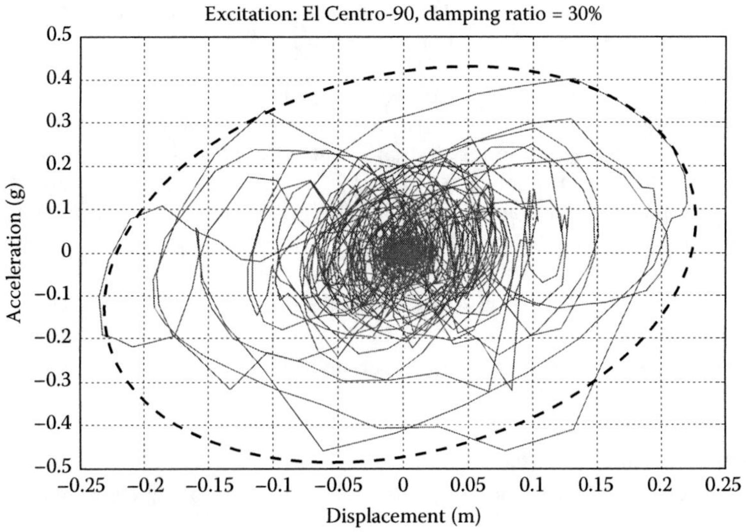

FIGURE 6.14 Earthquake responses.

Using an ellipse to approximate the envelope of the response curves has a special significance. When the group of earthquake records and the method of scaling the input amplitude are given, for a specific linear SDOF system defined by its period and damping ratio, these responses are determined. Therefore, the envelope of the ellipse is fixed. Suppose an ellipse is centered at the origin, which is the cross point of its short and long axes. Then, this ellipse can be determined by two parameters, the maximum values of the coordinates along the X-Y axes. In other words, the maximum seismic force and the maximum displacement of a given linear system, in the sense of statistical averaging under a group of earthquake excitations, define a unique ellipse.

Note that during an earthquake, the maximum force and displacement are likely to be reached only once. The remaining peak forces and displacements will be smaller than the maximum values. Therefore, using the envelope method shown in Figures 6.14 and 6.15 to estimate the corresponding parameters of the SDOF system, i.e., the damping, will not be exactly appropriate.

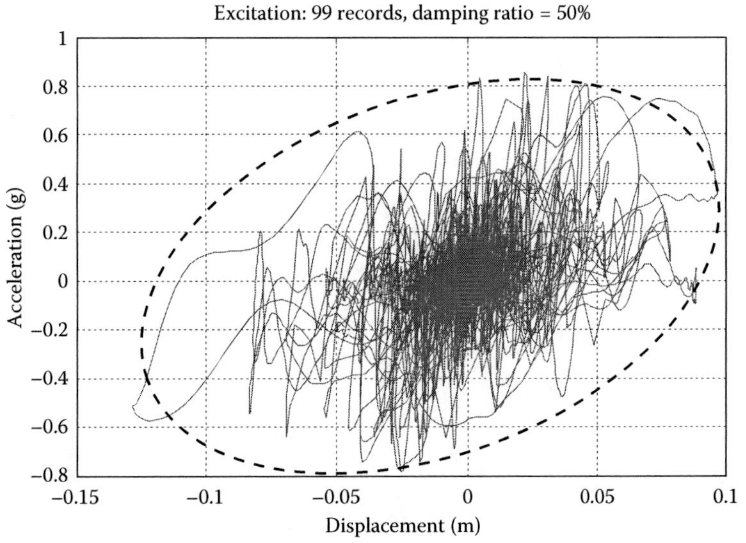

FIGURE 6.15 Envelope of linear responses (T = 2.0 sec).

In order to use the favorable characteristics represented by the ellipse envelope, modifications are needed. In the next subsection, a more detailed discussion using the "Penzien constant" is explained to account for this problem.

In any case, the area of the ellipse is the averaged *maximum energy* dissipated by the system in a specific first quadrant of a cycle where the maximum force and displacement are reached. This energy dissipation in the first quadrant is the *maximum work* done, denoted as $W_S(\Sigma_i W_{S_i} = \Sigma_i(E_{mc_i} - W_{e_i}))$, by the seismic force along the structural displacement. Thus, W_s can be determined by the maximum possible seismic force, denoted as f_M, and the maximum possible structural displacement, denoted as d_M.

In Chapter 1, the energy dissipation was discussed using an elliptic plot shown in Figure 1.19. This concept can be used to calculate the maximum seismic work through the given earthquake records, for the elliptic plots shown in Figures 6.14 and 6.15.

6.4.1.1.2 *Spectra of Seismic Work*

Since the maximum seismic work has significant meaning as discussed above, the issue of seismic work spectra requires additional study. That is, the relationship among the period, the damping, and the maximum seismic work is examined.

Using Equation 6.86, the maximum seismic work can be found through statistical analysis using the same 99 earthquake records, by considering the envelope method described in Figure 6.14, where the input acceleration levels of all the records are scaled to be 0.4 (g).

In Figure 6.16, the normalized maximum seismic work W_S, with periods equal to 0.4, 1.0, 2.0, and 3.0 (s) is plotted vs. the damping ratio. From this figure, it is seen that when the damping is increased, the maximum seismic work is increased. However, with different natural period, the rates of decrease are different.

If damping continues to be increased to a certain point, the maximum seismic work starts to decrease. This is because with the damping increased, the dynamic stiffness of the system is increased, which makes the input energy decrease. From Figure 6.16, it is seen that the shorter the period of the system, the lower the damping ratio at which the turning point will be reached. And after the turning point, the reduction of the maximum seismic work becomes more pronounced.

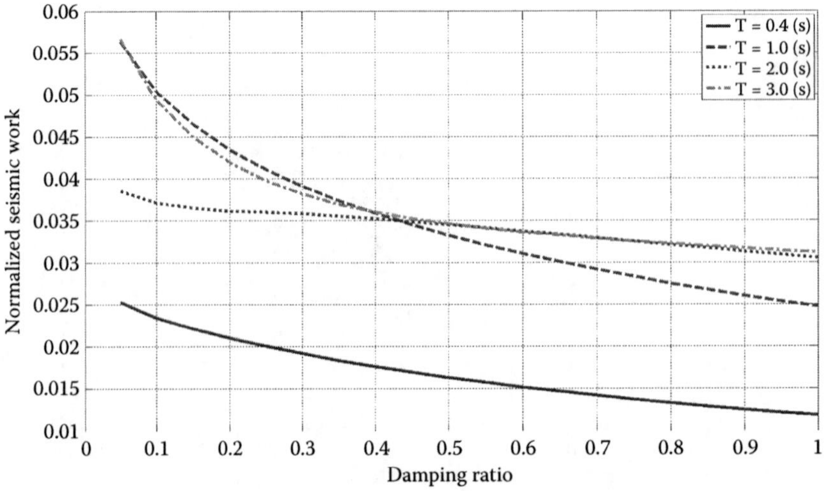

FIGURE 6.16 Maximum normalized seismic work vs. damping ratio.

The phenomenon exhibited in Figure 6.16 is important. It indicates that increasing the damping ratio may not always be helpful in reducing the work done by the seismic force, although Equation 6.95 shows that the product of the acceleration and the displacement is always decreased when the damping ratio becomes larger.

As mentioned earlier, in certain cases, the damage to the structure is more related to the maximum seismic work. Here, it is further seen that for a structure with its fundamental period, the effect of increased damping will be different from that of the shorter periods associated with higher modes.

Example 6.6

To further explain the above statements, consider the example shown in Figure 6.17, where a system has the normalized maximum seismic work, W_S, with damping ratios equal to 5%, 10%, 30%, and 50% are plotted vs. the period. From this figure, it is seen that when the period is increased, but below a certain value, the maximum seismic work will be significantly increased. For periods longer than that level, however, the maximum seismic work is generally reduced.

In the previous section, it was seen that the product of $a_{max}d_{max}$ might be independent of the period T. However, from Figure 6.18, it can be realized that this assumption is not true. That is, the maximum seismic work is a function of both the damping ratio and the period; it is not constant.

In Figure 6.18, the maximum seismic work is plotted vs. the absolute acceleration (mean plus one standard deviation). These curves are obtained by using the same 99 records with 0.4 (g) acceleration. The periods are taken to be 0.1 (s) to 5 (s). From these plots, it is seen that as the maximum acceleration increases, the maximum work is increased at the beginning and then gradually reduced. This is because the displacement or deformation of the structure is more significantly reduced.

In Figure 6.19, the maximum seismic work vs. the structural displacement (mean plus one standard deviation) is plotted. These curves are also obtained by using the same 99 records. Again, the periods are taken to be 0.1 (s) to 5 (s). From these plots, it is seen that when the displacement increases at the beginning but roughly below 0.07 (m), the maximum work significantly increases. Then, the maximum seismic work becomes fluctuated, depending on the selection of the damping ratio. In any case, when the displacement becomes sufficiently large, the seismic work starts to decrease.

From Figures 6.18 and 6.19, it is realized that the maximum seismic work can be expressed as a function of both the acceleration and the displacement. When these variables are relatively

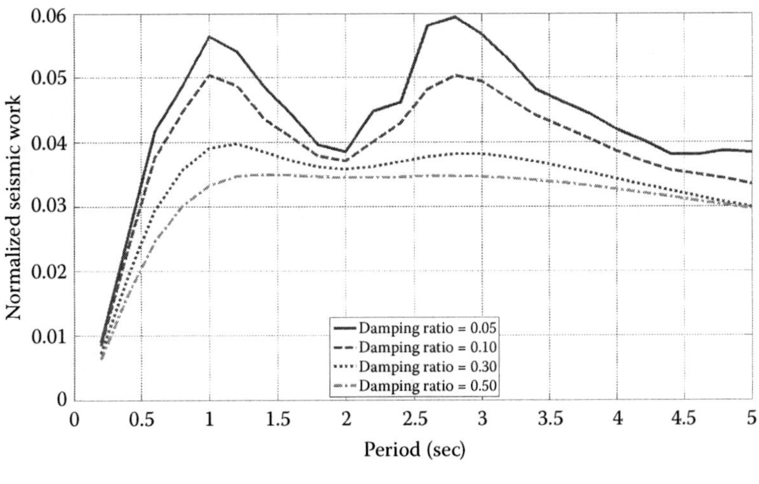

FIGURE 6.17 Maximum seismic work vs. periods.

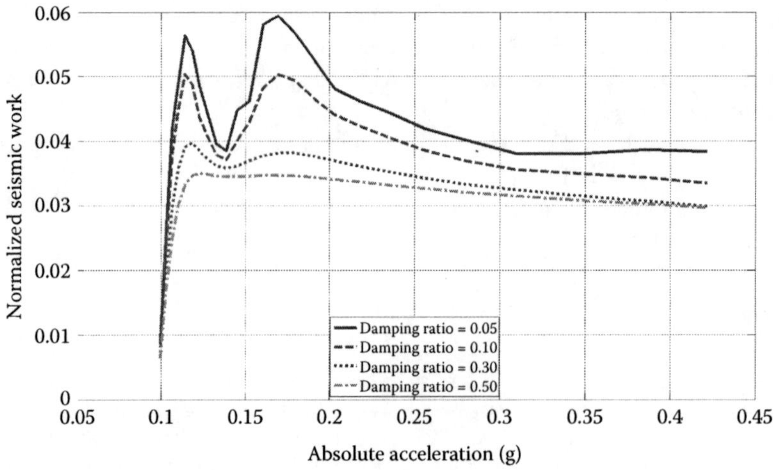

FIGURE 6.18 Maximum seismic work vs. absolute acceleration.

small or sufficiently large, the amount of work is also small. When the variables of acceleration and/or displacement reach an intermediate level, the maximum value of the seismic work is larger.

6.4.1.2 Brief Summary of Damping Effect for Linear Design Spectra

In order to further explore the role of damping in nonlinear systems, a brief review of the effect of damping on the seismic-induced vibration of linear SDOF structures is useful. The damping effect can be viewed in many ways. Here, the focus is on the scenario of design spectrum-based response estimations. Since a proportionally damped MDOF system can be accurately decoupled as several SDOF systems, the discussion also includes the corresponding case, and subscript i of the ith mode of an MDOF system is ignored.

1. For linear SDOF systems, both stiffness k and damping c are not functions of the displacement, and given unchanged mass and stiffness, the damping coefficient c is proportional to the damping ratio ξ. Note that this conclusion is not always true for nonlinear or generally damped systems.
2. In design spectra, the control parameter related to damping is the damping ratio.

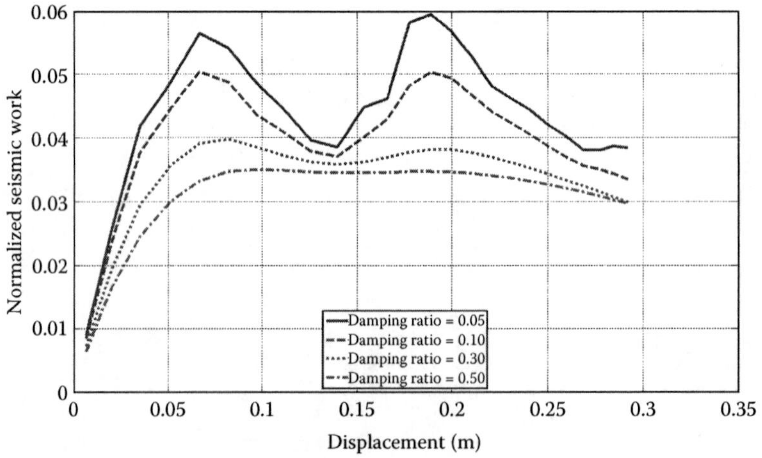

FIGURE 6.19 Seismic work vs. displacement.

3. The spectral value of displacement, S_D, will always be reduced when damping is increased. The reduction is inversely proportional to the numerical damping coefficient B.
4. The pseudo spectral value of acceleration, S_{PA}, will also always be reduced when damping is increased. The reduction is also inversely proportional to B. The pseudo spectral value of acceleration is inversely proportional to the square of period, T^2, and proportional to the spectral value of the displacement, S_D. The product of the pseudo acceleration and the displacement is independent of the period.
5. The actual spectral value, S_A, will not have the relationship described in item 4, especially when damping is large. The real spectral value of acceleration is not exactly proportional to the inverse square of the period. The products of the real acceleration and the displacement, especially the maximum seismic works, are both functions of the damping and the period. Generally,

$$S_A = \sqrt{1+4\xi^2}\,\omega_n^2 S_D/g = \frac{4\pi^2\sqrt{1+4\xi^2}}{T^2} S_D/g$$

6. The product of the mass and the pseudo acceleration, S_{PA}, is conventionally used as the base shear. The product of the mass and the real acceleration is used as the base shear V in this book. Thus, when damping is increased, the conventionally defined base shear will also always be reduced by the factor B. However, the newly defined total base shear will be more complex to use than the damping ratio ξ. When the period becomes large, the base shear reduction due to an increased damping ratio will be much less effective. For example, with the simplified expression based on AASHTO's formula,

$$V = mgS_A \quad (\text{kN})$$

so that

$$V \neq mgS_{PA} \quad (\text{kN}) \tag{6.96}$$

In design code, the spectral value of the absolute acceleration is often written as the elastic seismic coefficient, C_s. Therefore, different from the equation used in Chapter 2, there should be

$$C_s = \frac{\sqrt{1+4\xi^2}}{(3\xi_{eff}+0.9)R}\left(0.6\frac{S_{DS}}{T_0}T_{eff} + 0.4S_{DS}\right)I \tag{6.97a}$$

when $T_{eff} < T_0$,

$$C_s = \frac{S_{DS} I\sqrt{1+4\xi^2}}{(3\xi_{eff}+0.9)R} \tag{6.97b}$$

when $T_0 \leq T_{eff} \leq T_s$,

$$C_s = \frac{S_{D1} I\sqrt{1+4\xi^2}}{T_{eff}(3\xi_{eff}+0.9)R} \tag{6.97c}$$

when $T_s \le T_1 \le T_L$, and

$$C_s = \frac{S_{D1} I T_L \sqrt{1+4\xi^2}}{T_{eff}^2 (3\xi_{eff} + 0.9) R} \tag{6.97d}$$

when $T_1 > T_L$.

Here, T_1, T_0, T_s T_L, R, I, S_{DS}, and S_{D1} are defined in Chapter 2.

The spectral displacement d_{1D} is

$$d_{1D} = C_s \frac{T^2}{4\pi^2 \sqrt{1+4\xi^2}} g \quad (m) \tag{6.98a}$$

For example, when T_{eff} is greater than T_s, AASHTO's equation for displacement is

$$d = \frac{0.1 A S_i T_{eff}}{B_{eff}} \quad (mm) \tag{6.98b}$$

The base shear can be written as

$$V = C_s w \quad (kN) \tag{6.99}$$

7. Experience gained from recent earthquakes shows that the conventional equations of C_s and d may underestimate the structural acceleration and displacement. The calculated values are closer to the mean responses, instead of the values of the mean-plus-standard-deviation.
8. The spectrum of maximum seismic work is a newly suggested tool, which can be used to examine the effect of damping. Instead of structural energy accumulation and ground input energy, the maximum seismic work indicates the instantaneous energy dissipated only during the half cycle when the maximum displacement is reached. While the spectra of acceleration and displacement indicate that increasing the damping from a very small original value always reduces the spectral value, for maximum seismic work, the results are different.

6.4.1.3 Energy Equation in Linear Systems

The design spectrum approach has long been used in seismic design practice. In this book, as mentioned before, the spectrum-based approach is also used in the design of structures with supplemental damping.

As mentioned in previous discussion, there are at least two challenges in the design of supplemental dampers. The first is the fact that the design spectrum is obtained through linear systems, while structures with added dampers are most likely nonlinear systems. One way to address this issue is to use nonlinear response spectra, whose availability has been argued in the literature (e.g., Chopra 2002). This approach is discussed in more detail in the following paragraphs.

The second challenge is the treatment for nonproportional damping. Mathematically speaking, nonproportional damping (irregular damping) can be explained as energy transfer among normal modes.

Note that due to the nonlinearity of the systems, the "effective" modes can be more closely intertwined than linear cases, which results in greater irregular damping.

Both these challenges imply that with an increased energy dissipation capacity, there will be an increase in design errors in spectrum-based design for structures with supplemental damping.

Example 6.7

This example is examine the case of energy dissipation under earthquake excitation. Consider an SDOF system with a period equal to 1.0 (s). By varying the damping ratios from 1% to 50%, the energy dissipation vs. the damping ratio can be plotted as shown in Figure 6.20a. Suppose this system is subjected to the El Centro (1940) earthquake. It is seen that if the damping is small, i.e., from 1.0% to 3.0%, when the damping ratio is increased, the energy dissipation is increased accordingly. This is understandable because the appearance of damping dissipates energy. When the damping ratio is extremely small, i.e., near zero, the energy dissipation will also be near zero. As the damping increases, the energy dissipation grows. However, the maximum energy dissipation will soon reach its maximum value. In Figure 6.20a, the maximum value is used to normalize the entire course of energy dissipation. It is also seen that after the peak energy dissipation, as the damping ratio is increased, the energy dissipation starts to decrease. In Figure 6.20a, this point is at about 13%.

To determine if the energy dissipation pattern of increase-peak-decrease is relatively common, the Northridge earthquake (1994), which has the same type of plot, is used, and is shown in Figure 6.20b. It is seen that the pattern of energy dissipation is similar to the El Centro earthquake, except that the peak energy dissipation appears before 1%.

To examine the effect of increasing the damping ratio, the peak value of the accelerations and displacement are also plotted in Figure 6.20. It is seen that in these two cases, increasing the damping ratio always results in a reduction of the displacements. This phenomenon seems to contradict the above conclusion. However, by considering the case of an SDOF system rather

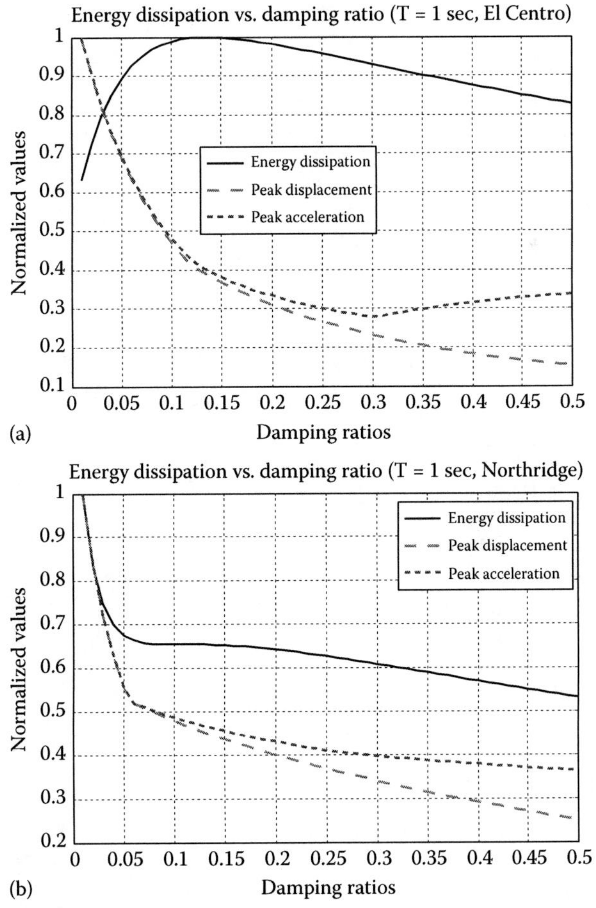

FIGURE 6.20 (a, b) Energy dissipation vs. damping ratio.

than an MDOF system, it is seen that the cross effect of the modal energy transfer and directional energy transfer does not exist. Note that the reason for using an SDOF system is to show clearly the energy dissipation, rather than the response reduction. From a different perspective, it can also be concluded that the energy dissipation does not directly correlate to the response reduction.

Furthermore, it is also seen that in the El Centro earthquake, when the damping ratio is greater than 30%, continuously adding damping also results in an increase in peak acceleration.

6.4.1.4 Limitation of Damping Control

In Chapters 2 and 5, it has been established that when the damping ratio is small, adding damping to a system can significantly reduce the level of vibration. However, when the original damping ratio is already large, using supplemental damping will be less effective. This implies one of the limitations of damping control.

In addition, NERRP 2009 specifies that when using damping control, the base shear should not be smaller than 0.75 times the design base shear V of systems without supplemental damping. In the following paragraph, this limitation is discussed in terms of design damping ratio.

According to NEHRP 2009, the total base shear with supplemental damping, denoted as V_d, can be written as

$$V_d = \frac{V}{B} \quad (\mathrm{kN})$$

Note that V is specified when the original damping ratio is 5%. This covers the case when the system becomes inelastic, which significantly increases the effective damping ratio. In the elastic range, the damping ratio of a structure is often considerably smaller than 5%. According to NEHRP 2009,

$$\frac{V/B_d}{V/B_0} \geq 0.75 \tag{6.100}$$

where B_d and B_0 denote the numerical damping coefficients with and without added damping.

From Equation 6.100 it is seen that,

$$B_d \leq 1.33$$

and with the help of Equation 6.93,

$$\xi_{\mathrm{design}} \leq 14.44\% \tag{6.101}$$

Considering various design safety factors, the total design damping ratio usually does not exceed 15%,

$$\xi_{\mathrm{design}} < 15\% \tag{6.102}$$

for the simplified design approach. Here, the simplified damping design means the approach of using the linear damping model, which is discussed in Chapter 7.

Liang and Lee (1991) suggest that when the original damping ratio is sufficiently small, adding damping to a system can boost the design damping ratio approximately as

$$\xi_{design} = \xi_0 + \xi_a \quad (6.103)$$

If the original damping is assumed to be 5%, for simplified approaches, the damping ratio contributed by supplemental dampers is recommended to be

$$\xi_a < 10\% \quad (6.104)$$

6.4.2 Nonlinear Damping and Nonlinear Systems

6.4.2.1 Energy Dissipation in Nonlinear Systems

In the above discussion, it was shown that in linear systems, increasing the damping coefficient may result in a decrease of the energy dissipation, instead of always increasing the energy dissipation. In other words, the correlation between response reduction and energy dissipation does not exist, so energy dissipation should not be used as a reference to determine if the damping design is correct.

For nonlinear systems, the problem of energy dissipation vs. increasing damping can also be seen. In these cases with a small damping ratio, at the beginning within a very short range, by increasing the damping, the energy dissipation is increased. However, in the remaining range, the energy dissipation will be reduced.

Chopra (2006) has shown that under white noise excitation, an increase in damping will always increase energy dissipation, which can be analytically proven. However, earthquakes have at least two characteristics that are essentially different from white noise. First, the duration of an earthquake is limited. Generally, an earthquake only lasts for a few dozen seconds or so. However, white noise can last forever. Second, the spectrum of an earthquake is also limited. Generally, the upper band is considered to be below 33 (Hz). On the other hand, white noise contains an infinite band of frequencies.

To examine the issues of energy dissipation vs. the increase of the effective damping ratio of nonlinear systems, the following examples are given.

Example 6.8

This example deals with an SDOF system with linear stiffness and nonlinear damping. The SDOF system has a mass equal to 2,000 (kg) and a stiffness chosen to have the following natural frequencies: 0.25, 0.5, 1, 2, 3, 4, and 5 (Hz). A sublinear viscous damper with the possible damping exponents of 0, 0.01, 0.025, 0.05, 0.1, 0.2, 0.3, 0.4, 0.5, 0.6, 0.7, 0.8, 0.9, and 1.0, is installed so that the system can have effective damping ratios from 1% up to 50%. The same 99 earthquake records are used as input to calculate the averaged energy dissipation and input energy. The averaged results are mean values plus one standard deviation. Note that in this numerical simulation, when the damping exponent is small, there are more significant response reductions and so for the corresponding energies. Thus, in the averages, when the damping exponent is close to 1, that is, when the system is close to linear, the largest responses will occur.

The resulting energy dissipation vs. the effective damping ratio (calculated through Timoshenko damping) is plotted in Figure 6.21a, where the nonlinear responses are computed through nonlinear integrations. For comparison, in Figure 6.21a, the energy dissipation of equivalent linear systems (which have identical damping ratios) is also plotted. From Figure 6.21a, it is seen that both the nonlinear and the equivalent linear systems with effective damping ratio and period have a similar pattern of energy dissipation vs. damping ratio. However, the nonlinear systems have steeper reduction curves than the linear systems.

Comparing the curves of nonlinear energy dissipations shown in Figure 6.21a with those of exact linear systems shown in Figure 6.20a, a difference is also apparent. That is, a system with

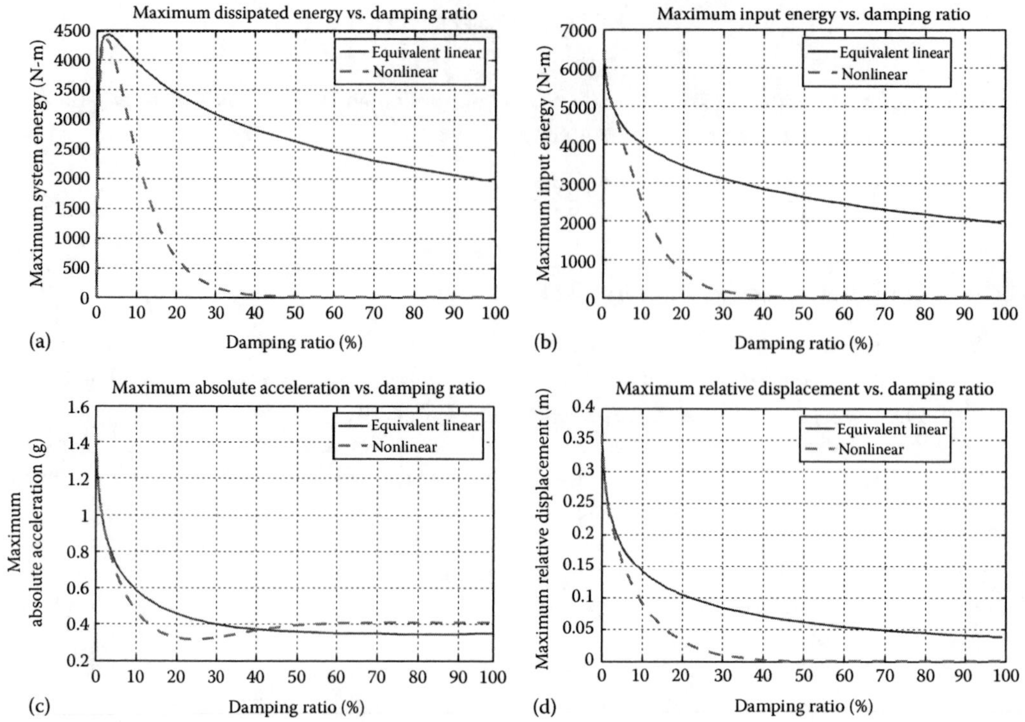

FIGURE 6.21 (a–d) Averaged responses of systems with sublinear damping.

nonlinear damping, or more precisely sublinear damping, will show a steeper drop of energy dissipation than systems with linear damping only. Furthermore, comparing the energy input shown in Figure 6.21 (b), the similar pattern can also be seen.

To review the effect of increasing the damping ratio, in Figures 6.21c and d, the absolute accelerations and the relative displacements vs. the effective damping ratio are also plotted. It is seen that as the damping ratio increases, the acceleration of the nonlinear system decreases quickly when the damping is relatively small but becomes slightly larger when large damping is added. On the other hand, when the damping ratio increases, the displacement is always reduced.

Again, for comparison, the responses of the equivalent linear systems are plotted and shown in Figures 6.21c and d, respectively. It is seen that using the equivalent linear system for this SDOF model, the responses are notably different.

From this example, it is further realized that using larger damping may dissipate less rather than more energy. And, for the sublinear damping, a more significant reduction of energy dissipation is seen than for linear systems. However, the responses of acceleration and displacement are reduced with different pattern as shown in Figure 6.21c and d. Thus, there is more evidence for the noncorrelation of energy dissipation vs. the damping ratio. From Figure 6.21b, it can be realized that the reason why increased damping results in response reduction while the energy dissipation continues to decrease is due to the significant reduction of input energies.

Thus, for a nonlinear system, although there is no analytical equation for the dynamic stiffness, it is still seen that increasing damping will cause an increase in the "effective" dynamic stiffness, which results in the response reduction. The point is that the response reduction of the nonlinear system is not caused by increased energy dissipation.

Example 6.9

In this example, a bilinear SDOF system is considered with nonlinear stiffness and linear supplemental damping. Suppose a bilinear SDOF system with mass = 300 (t). This system is installed with linear viscous dampers so that the designed supplemental damping ratio can be chosen as 5%, 10%, or 15%. Also, suppose that before the system enters the inelastic range, the original damping ratio is 1%. The yielding and unloading stiffness are, respectively, k_d = 48,000 (kN/m) and k_u = 480,000 (kN/m). By choosing its characteristic strength from 500 (kN) to 5,000 (kN), the effective damping ratio is increased from 12% up to 62%.

Now, if the system is subjected to El Centro earthquake excitations whose amplitude is modified to be 0.4 (g), the results of the maximum displacement and acceleration vs. the effective damping ratio are plotted in Figures 6.22a and b. In each figure, three curves are plotted according to the above-mentioned design supplemental damping ratio, which are marked as solid, broken, and dotted lines. From these two figures, it is found that when the effective damping is increased, the responses are reduced. Furthermore, when the designed damping ratio is taken to be a larger value, the corresponding response becomes smaller, which implies that even when a structure behaves in an elastic–plastic manner, using supplemental damping can still further reduce the seismic response. Apparently, the closer the structure is to a linear system, the greater the effectiveness of using supplemental damping, which is consistent with common sense.

However, the more important issue to realize is the trend of energy dissipations by both supplemental dampers and the elastic–plastic bilinear behavior of the structure vs. the increase of the effective damping ratio. By using the same earthquake record, the energy dissipation is calculated and the corresponding curves are plotted in Figure 6.22c.

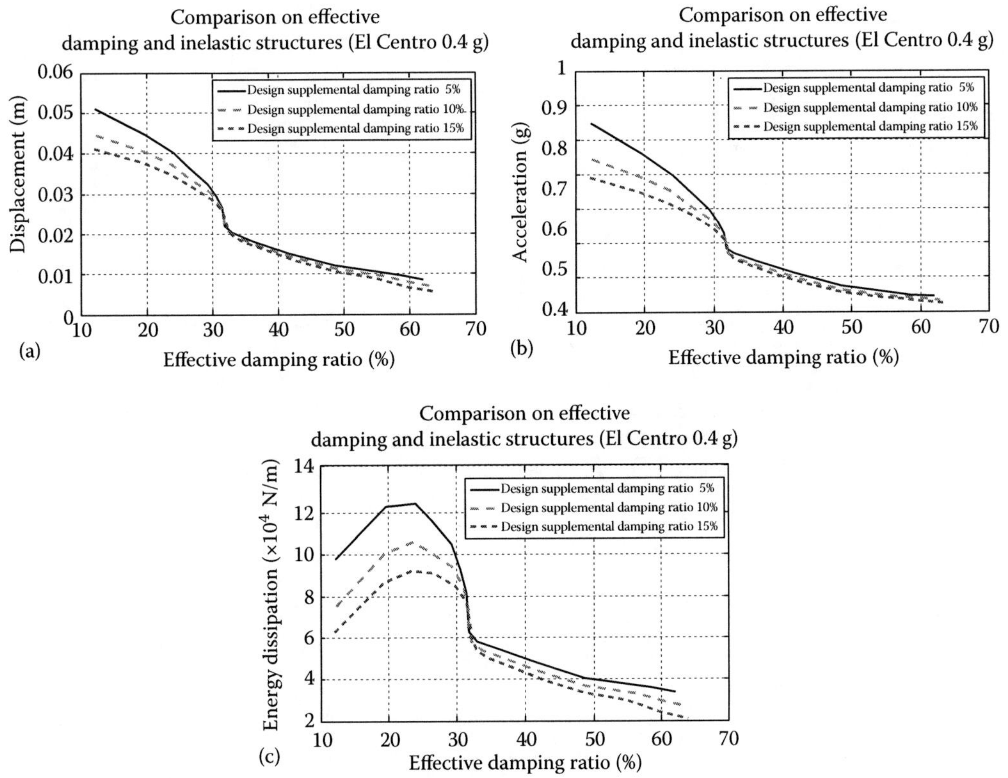

FIGURE 6.22 (a–c) Nonlinear response due to El Centro excitation.

From Figure 6.22c, it is seen that as the effective damping ratio is increased, the amount of energy dissipation is first increased before reaching a maximum value. After the peaks when the effective damping ratios increase, the amount of energy dissipation begins to drop. Comparing with Figure 6.21a, it is realized that both nonlinear systems have similar tendencies for energy dissipation, which also agrees with the case of the linear systems mentioned above. In other words, the energy dissipation responses of both linear and nonlinear systems do not maintain a monotonic relationship as the damping ratio increases.

Secondly, by comparing curves corresponding to 5%, 10%, and 15% design damping ratios, the same conclusion is reached; that is, increasing the damping ratio may result in a decrease, instead of an increase, in energy dissipation.

To determine with greater certainty whether such phenomena can be found in most earthquakes, the same 99 records are used again. Using the statistical values of the summation of mean plus one standard deviation, the energy dissipation, the input energy, as well as the peak displacement and acceleration are calculated.

To examine the relationship with larger effective damping ratios, the yielding and unloading stiffness are, respectively, chosen to be $k_d = 24{,}000$ (kN/m) and $k_u = 480{,}000$ (kN/m). By changing its characteristic strength to vary from 250 (kN) to 5,000 (kN), the effective damping ratio is increased from 10% up to 150%. With the same mass equal to 300 (t), this system is also installed with linear viscous dampers so that the designed supplemental damping ratios are chosen as 5%, 10%, and 15%. Also, suppose that before the system enters the inelastic range, the original damping ratio is 1%. The results are given in Figure 6.23.

In Figure 6.23a, it is again realized that as the effective damping ratio increases, the amount of energy dissipation first increases and then decreases. The statistical results agree with the case of the single earthquake excitation shown in Figure 6.21c. Secondly, by comparing the curves

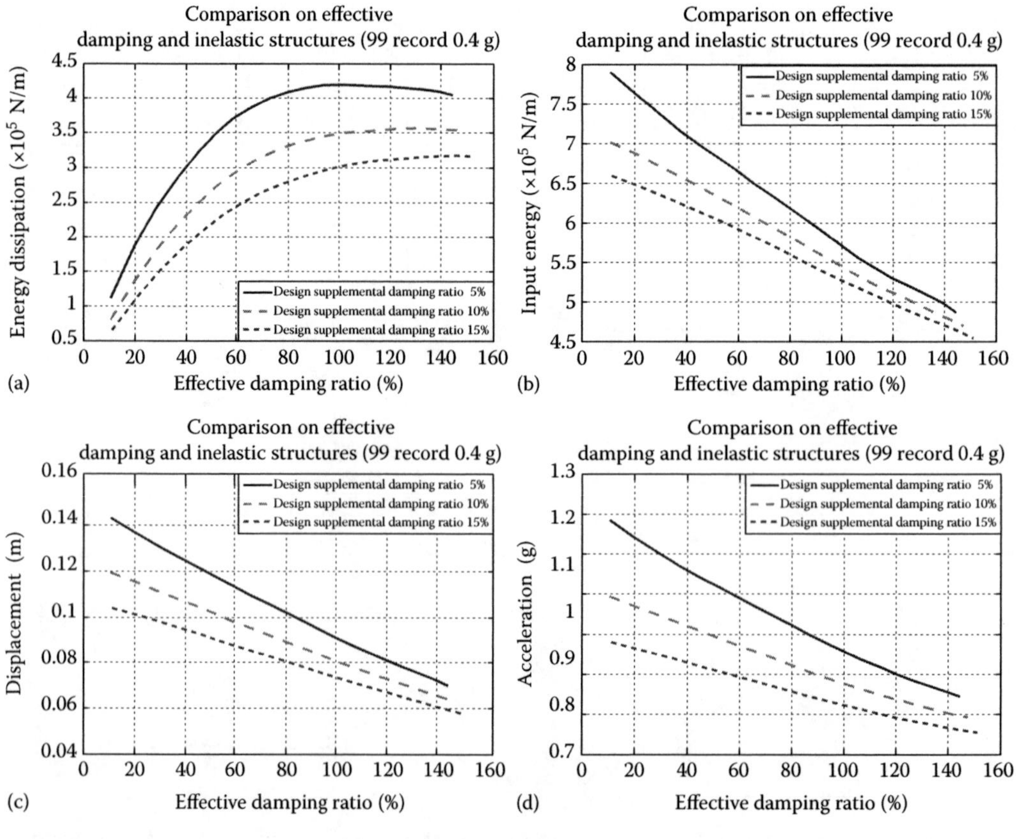

FIGURE 6.23 (a–d) Statistical nonlinear responses.

corresponding to 5%, 10%, and 15% of the design damping ratios, the same conclusion is reached again; that is, increasing the damping ratio may result in a decrease, rather than an increase, in energy dissipation.

To find the reason for the drop in energy dissipation, the energy input from the ground excitations is plotted in Figure 6.23b. It is seen that as the effective damping ratio increases, the input energy decreases. As the designed damping ratio increases, the input energy decreases as well.

In order to check if the increase of the effective damping can lead to a reduction of the responses, in Figure 6.23c and d, the maximum values of the displacement and the acceleration are also plotted. It is seen that as the effective damping ratio increases, the corresponding displacement and acceleration both decrease. And, as the designed damping ratio increases, the responses decrease as well.

Plots in Figure 6.23c and d do not exhibit the effectiveness of using larger damping or effective damping to control the seismically induced structural vibration, but clearly show that the reduction of the responses has no correlation with the energy dissipation.

Therefore, from these two cases, the limitations of using Equation 6.86 to explain the vibration reduction for nonlinear systems are illustrated again.

6.4.2.2 Notes on Nonlinear Response Spectra

In Example 6.9, the design spectrum is discussed, which is based on linear SDOF systems. However, in most cases, structures with added damping systems are nonlinear systems. Two approaches can be used to account for nonlinear systems. The first is to linearize a nonlinear system. A nonlinear system often has considerably more control parameters than linear systems. Using a two-parameter model to cover a multiple-parameter model may cause uncertainties and design errors. Generally speaking, the above-mentioned design spectrum can be used if the damping and the degree of nonlinearity are very small. Otherwise, nonlinear response spectra should be considered.

6.4.2.3 Rule 0.65 and the Penzien Constant

One of the fundamental differences between nonlinear and linear systems is that the damping effect of the nonlinear system varies according to the amplitude of the structural responses, whereas in a linear system, it is a constant value. Therefore, to evaluate the damping effect, either the level of the amplitude must be specified, which is awkward because the amplitude of the ground excitation is difficult to determine; or a rule needs to be devised for the standard of structural response estimation.

The meaning of the standard of the response estimation is examined first and the rule of this setup is subsequently discussed.

6.4.2.3.1 Penzien Constant

To realize the response estimation, two types of damper testing are considered as described in Figures 6.24a and b. Namely, a damper can be moved by using an actuator, as illustrated in Figure 6.24a, or deformed by seismic force, as depicted in Figure 6.24b.

In Figure 6.24a, k stands for the spring constant and c represents the damping coefficient of the damper. In Figure 6.24b, the SDOF system of a one-story structure is included for comparison. The dotted frames in Figure 6.24a and b emphasize the damper-spring system. It is seen that in (a), the force applied on the damper-spring system is applied by the actuator. In (b), the applied force is due to the product of the mass and the acceleration, namely, the inertial force caused by the ground motion. Now, it is seen that the difference between case (a) and (b) is that the former can have a controlled displacement.

That is, the displacement of the damper-spring system can be forced to have an exact position guaranteed by the accuracy of the actuator. When the driving frequency varies, the displacement can be controlled. In the second case, the inertia force is a function of not only the ground excitation,

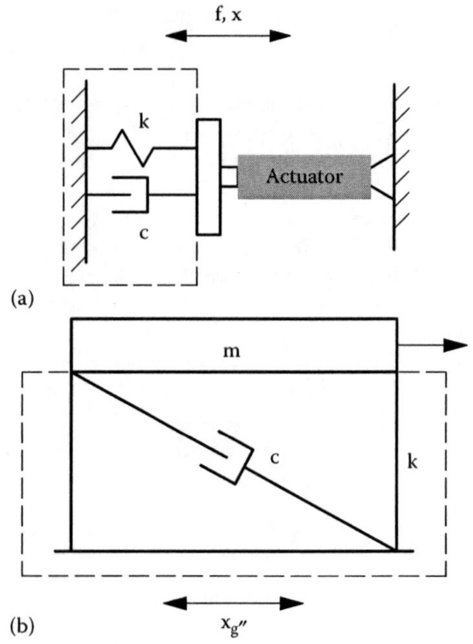

FIGURE 6.24 (a, b) Damping-spring system.

but also the dynamic stiffness of the system. When the driving frequency varies, the resulting displacement will likely vary accordingly. In this case, the energy dissipation loop obtained in case (a) can be very different from case (b). Practically speaking, case (a) is often used as the test setup to measure the properties of a damper. Case (b) is often used to simulate the situation that a structure experiences under earthquake ground excitations. In other words, the energy dissipation loop obtained during a damper test can be very different from a real earthquake vibration. Thus, care must be taken to distinguish between these two situations.

While both figures can be seen as experimental test setups, the difference between them is distinct. In Figure 6.24a, the displacement is preset. It is the actuator that decides the force–displacement parallelogram. In this way, near-perfect bilinear or sublinear model responses can be achieved. That is, the test repeatability can be quite good. However, in Figure 6.24b, the force, instead of the displacement, takes the actions. Since the force can be a seismic load, it is random. In this circumstance, we will not have perfect responses such as the parallelogram. In Figure 6.25, the bilinear parallelogram is used to conceptually explain the uncertainty of

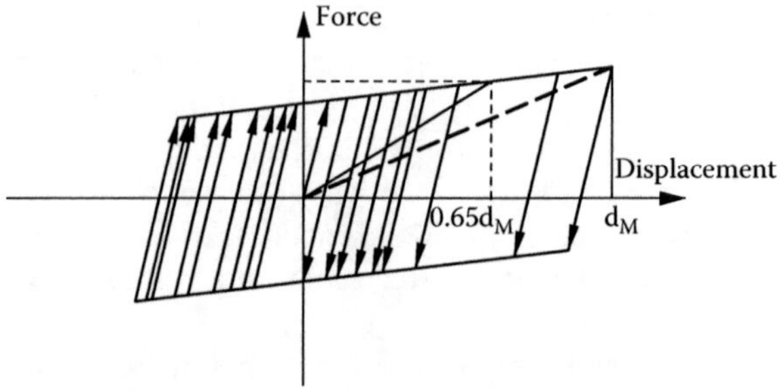

FIGURE 6.25 Bilinear model under random excitation.

the force–displacement relationship. The sublinear damping model can be realized to have the same problem.

Now, suppose the bilinear model is precisely correct. Namely, any force–displacement point will follow the parallelogram. However, since the amplitude of the force varies, the displacement will vary accordingly. This will form the plot shown in Figure 6.25, from which it is seen that the maximum displacement may only be reached once during the entire course of an earthquake excitation. In most cases, the displacements will be smaller than d_{max}.

Now, when both the stiffness and the damping are calculated using the parameters of maximum force and displacement, the stiffness calculated at the maximum displacement is smaller than that calculated anywhere else, as shown by the thick dotted and solid lines. Also, the damping ratio calculated at the maximum displacement is smaller than that calculated anywhere else.

Penzien et al. studied these phenomena and suggested that the stiffness and the damping should be computed at 65% of the maximum displacement, rather than exactly at the maximum displacement. This suggestion is adopted herein and 0.65 is referred to as the *Penzien constant*. That is, when the effective stiffness and damping are calculated, the displacement x_0 is used.

$$x_0 = 0.65d_{max} \tag{6.105}$$

Now, the displacement x_0 at p_c times the maximum value can be obtained:

$$x_0 = p_c d_{max} \tag{6.106}$$

Note that x_0 calculated by Equation 6.106 is frequently used as the effective design displacement in damper design. In most engineering practice, the Penzien constant p_c is taken as 0.65, so that it is also referred to as *Rule 0.65*.

6.4.2.3.2 Variation of Stiffness

There are basically two types of structures. The first type has virtually unchanged maximum value of stiffness under large inelastic deformations subjected to strong earthquakes. Many steel structures exhibit this constant maximum stiffness. The second type has decreased stiffness, when the inelastic range is reached and repeated. Most reinforced concrete (RC) structures have this type of stiffness. Rzhevsky and Lee (1998) studied the decreasing stiffness and found that it can be described as a function of the accumulation of inelastic deformation. Empirically, this can be expressed as follows:

$$k_n = k_0 e^{-x_0 sh^{-1}(\gamma_n)} \tag{6.107}$$

Here, k_0 and k_n are, respectively, the original stiffness and the stiffness after n-semicycle inelastic deformations; x_0 is the peak value of the first inelastic deformation. Subscript n denotes the total cycles of inelastic deformations, the term γ_n is defined as

$$\gamma_n = \left| \sum_{i=1}^{n} a_i \right| \tag{6.108}$$

where γ_n is called the *damage factor*, which is the summation of the absolute value of all the peak values of the inelastic deformations.

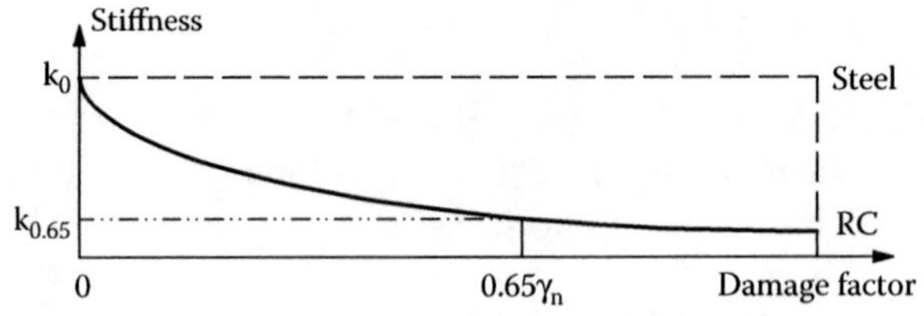

FIGURE 6.26 Stiffness vs. accumulation of inelastic deformation.

The initial stiffness, k_0, and the values of the inelastic deformation, a_i, can be specified in an experimental study of certain types of RC components and/or structures. Figure 6.26 conceptually shows the relationship between the decreased stiffness and the damage factor.

In Figure 6.26, for comparison, both the constant and the decreasing stiffness are plotted and, conceptually, both are initially given the same value. It is seen that as the inelastic formations accumulate, the constant stiffness keeps its value until its final failure stage. However, the decreased stiffness begins to diminish from the first semicycle, although it will eventually have a total failure, which is also conceptually plotted at the point when the constant stiffness fails. The decreasing curve is assumed to be exponential, as expressed in Equation 6.107. Note that for steel structures, the decrease is not as significant as in RC structures.

Suppose a predetermined value of the inelastic deformation is given, e.g., the deformation when the structure collapses. The corresponding cycle is denoted as n. At 65% of the accumulation of the total inelastic deformation, a special stiffness, denoted as $k_{0.65}$, is defined. This value will be used to approximate the effective stiffness of structures with decaying stiffness in a later section. That is,

$$k_{0.65} = k_0 e^{-a_0 sh^{-1}(\gamma_{0.65})} \tag{6.109}$$

where the term $\gamma_{0.65}$ is defined as

$$\gamma_{0.65} = \left| \sum_{i=1}^{0.65\,n} a_i \right| \tag{6.110}$$

By using the Penzien constant and corresponding estimation of the effective stiffness as well as damping, the linearized model can be further used to predict the structural responses of nonlinear systems.

6.4.2.4 Nonlinear Spectra

In addition to the above-mentioned approach, which utilizes the linear spectrum, another way to account for the nonlinear system, yet still based on the concept of statistical response spectra, is to generate nonlinear response spectra. In Chapter 8, the rationale, the generating procedure, and the utilization of nonlinear response spectra are discussed in detail.

Note that for linear response spectra, the spectral values of the absolute acceleration, S_A, and the relative displacement, S_D, have the relationship described in Chapter 2. However, for nonlinear response spectra, the spectral values of the absolute acceleration and the relative displacement may no longer have that relationship. Rather,

$$S_A = S_{\xi T}\sqrt{1+4\xi^2}\ \omega_n^2 S_D/g \qquad (6.111)$$

Here, the safety factor $\eta_{\xi T}$ is often not equal to 1 and it can be a function of the period, the damping ratio, and the number of the degrees-of-freedom. This is discussed in Chapter 7.

6.4.3 Inelastic Structure with Large Ductility

Under strong earthquake excitations, many types of structures will enter their inelastic range. Namely, their displacement ductility will be larger than unity. With the aseismic principle of noncollapse under strong earthquakes, many structures are purposely designed to allow large ductility. The role of supplemental damping in structures with large ductility is considered in the following.

6.4.3.1 Inelastic Structure with Supplemental Damping

Some believe in a theory that once a structure enters its inelastic range, it will dissipate a large amount of energy so that supplemental damping becomes comparatively insignificant; therefore, adding dampers to inelastic structures is not suggested; or, alternatively, a structure is designed with added dampers to prevent entrance into the large inelastic range.

Once a structure enters the inelastic range, that is, the system becomes nonlinear, a more complex situation occurs. Whether supplemental damping is less effective will depend on several factors, such as the degree of ductility, the type of supplemental damping, and the amount of damping adaptability. Another important issue is, when the structure enters its inelastic range, does the resulting damping remain in an underdamped mode, or does it become critical and overdamped? This issue will directly relate to the effectiveness of supplemental damping, as well as damping design for inelastic structures.

To understand these issues, a simple example is considered first.

Example 6.10

An SDOF EPP structure with characteristic strength q, yielding displacement d_y, maximum displacement d_m, and unloading stiffness k_u, is considered. The mass of the system is 1, that is, the monic system is chosen for simplicity.

Its damping force is $f_{EPP} = q = k_u d_y$, and the energy dissipated in a complete steady-state cycle under sinusoidal excitation is E_{EPP}. Suppose a linear viscous damper, with damping coefficient c and damping force f_D, has energy dissipated in a complete steady-state cycle under sinusoidal excitation E_D. Also, the maximum potential energy is $E_P = \frac{1}{2}k_u d_y^2$.

If the EPP system displacement is just d_y, then it remains linear with natural frequency $\omega_n = (k_u)^{1/2}$. When the driving frequency is equal to ω_n, it follows that $f_D = c\omega_n d_y = 2\xi\omega_n^2 d_y = 2\xi k_u d_y$ and $E_D = 2\pi\xi k_u d_y^2$.

When the EPP structure yields and reaches its maximum displacement μd_y, where μ is the displacement ductility, the energy dissipation is

$$E_{EPP} = 4qd_y(\mu - 1) = 4(\mu - 1)k_u d_y^2$$

The damping force contributed by the damper becomes $f_D = c\omega_n\mu d_y = 2\xi\mu k_u d_y$ and the energy dissipation is $E_D = 2\pi\xi\mu^2 k_u d_y^2$.

Therefore, for a viscously damped EPP system, the ratio of the energy dissipation of the damper and the maximum potential energy of the EPP system is

$$R_{D/P} = \frac{E_D}{E_P} = \frac{2\pi\xi\mu^2 k_u d_y^2}{1/2 k_u d_y^2} = 4\pi\xi\mu^2 \qquad (6.112a)$$

6.4.3.2 Criterion of Nonlinear Overdamping

When the system remains linear, $\mu = 1$, and is critically damped ($\xi = 1$), the corresponding energy dissipation ratio is

$$R_{D/P} = 4\pi \tag{6.112b}$$

That is, the energy dissipation is 4π times the maximum potential energy. It is known that for linear viscously damped systems, the criterion of critical damping is the unity damping ratio. However, this criterion is difficult to apply for nonlinear systems. To date, there are no commonly acceptable criteria to determine if general nonlinear systems become overdamped. For one of the possible judgments, Equation 6.112 can be used as an energy criterion. Other types of criteria can be the occurrence of free-decay vibration.

Using the energy criterion described in Equation 6.112, consider a pure EPP system without dampers. It is seen,

$$\frac{E_{EPP}}{E_P} = \frac{4(\mu - 1)k_u d_y^2}{1/2\,k_u d_y^2} = 8(\mu - 1) = R_{D/P} = 4\pi$$

Further when

$$\mu = \frac{\pi}{2} + 1 \approx 2.6$$

the EPP system has critical damping. It is understandable that when additional dampers are installed in an EPP system, less ductility is required to reach critical damping. Furthermore, when the structure is not an EPP system, namely, the yielding stiffness is not zero, large ductility is required to reach critical damping.

Note that the dynamic behaviors of underdamped and overdamped systems are very different. In Chapter 7, special treatment for an overdamped system and corresponding damping design is discussed.

6.4.3.3 Energy Dissipations by Viscous and Bilinear Damping

Next, consider the energy dissipation by viscous damping and by the yielding structure. A bilinear inelastic structure can be seen as a combination of a linear stiffness and an EPP system, and only the EPP system dissipates energy. Therefore, the energy dissipation described by Equation 6.112b can be seen as the maximum possible values for various inelastic models.

If a monic system is installed with sublinear viscous damping having a damping coefficient of c_{eq} and exponents α and β (see Equation 5.13), then when the system is under sinusoidal excitation with steady-state displacement x_0, the energy dissipation is $E_D = c_{eq}\omega_n^{\alpha} x_0^{\beta+1} A_\beta$ in general (see Example 5.3). When the displacement reaches the maximum value μd_y, and for convenience, let $\alpha = \beta$,

$$E_D = c_{eq}\omega_n^{\beta}\mu^{\beta+1}d_y^{\beta+1}A_\beta = c_{eq}k_u^{\beta/2}\mu^{\beta+1}d_y^{\beta+1}A_\beta \tag{6.113}$$

then the ratio of energy dissipation is

$$R_{D/P} = \frac{E_{EPP}}{E_D} = \frac{4(\mu - 1)k_u d_y^2}{c_{eq}k_u^{\beta/2}\mu^{\beta+1}d_y^{\beta+1}A_\beta} = \left(\frac{4}{c_{eq}A_\beta}k_u^{\frac{2-\beta}{2}}d_y^{1-\beta}\right)\left(\frac{\mu - 1}{\mu^{1+\beta}}\right) \tag{6.114}$$

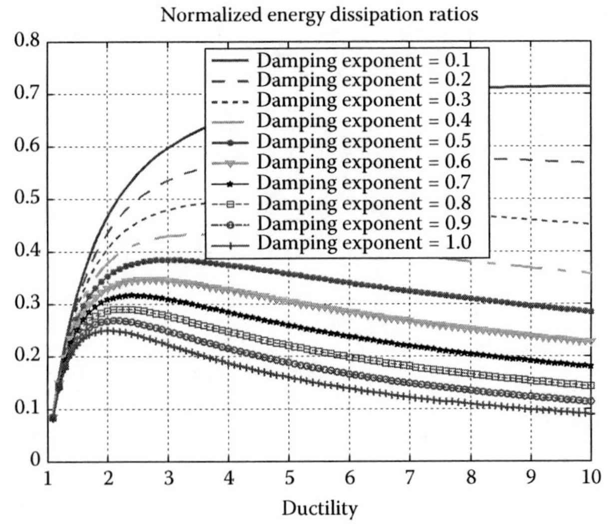

FIGURE 6.27 Normalized energy dissipation ratios.

On the right side of Equation 6.114, the terms in the first parentheses can be determined through the properties of the damper and the system. The value of the second parentheses is a function of ductility as well as the damping exponent. In Figure 6.27, the value in the first parentheses is equal to 1 and the normalized energy dissipation ratio is plotted to realize the influence of the damping exponent β and the ductility μ. It is seen that when β is small, i.e., close to 0.1, the damping adaptability of the damper is small; when ductility μ becomes larger, the ratio also becomes larger. This implies that the EPP system will dissipate more energy than the damper under the normalized condition. When β is larger than 0.1, an increase in the ductility will first increase the ratio and then it will decrease.

To further see the energy dissipations contributed by the damper and the EPP system, let us consider linear viscous damping, that is $\beta = 1$. It is seen that the ratio is

$$R_{D/P} = \frac{2(\mu - 1)}{\pi \xi \mu^2} \tag{6.115}$$

In Figure 6.28, the energy dissipation ratios, $R_{D/P}$, are plotted without normalization. Note that when the ratio is below unity, the damper will dissipate more energy than the EPP structure. For example, when the damping ratio is chosen to be 0.15, which is common in damping design, the damper will dissipate more energy as the ductility exceeds 2.6. Note that $\mu = 2.6$ is the critical point for a pure EPP system having critical damping. However, with additional viscous damping, the EPP system will enter the overdamped mode much sooner.

6.4.3.4 Seismic Responses of Inelastic Structures with Viscous Damping

Table 6.2 lists the seismic responses, including displacement ductility and acceleration, of an EPP system with linear viscous damping. This simulation is also carried out using the 99 ground motion records. The mass of the system is 200 (t), the unloading stiffness is 17.77 (MN/m), and the yielding displacement is 0.05 (m). For simplicity, the "damping ratio" is computed when the system remains elastic.

From Table 6.2, it is seen that when the "damping ratio" (in fact the corresponding damping coefficient of the damper) increases, the mean plus one standard deviation of displacement ductility always reduces. When the input level is small, i.e., 0.2 (g), increasing damping will cause the acceleration to decrease. When the input value is larger, however, increasing damping makes the

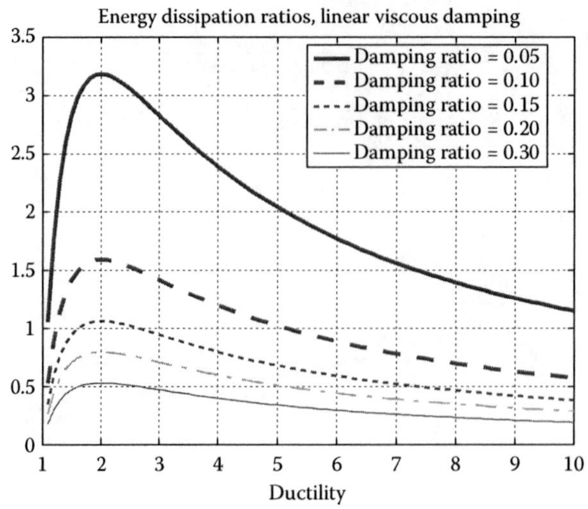

FIGURE 6.28 Energy dissipation ratio.

acceleration increase, instead of decrease (see the data expressed in italic in Table 6.2). This phenomenon is caused by overdamping. As mentioned above, with the presence of viscous damping, the system will enter the overdamped mode much easier.

6.4.3.5 Biased Deformation of Inelastic Structures

When a structure enters the inelastic range with a sufficiently large ductility under earthquake excitations, its dynamic equilibrium position may vary, instead of remaining in position. The biased deformations due to nonlinear response can also be seen in experimental studies. This biased deformation can enlarge the amplitude of the dynamic displacement as well as render a permanent structural deformation. That is, an inelastic structure with large ductility may introduce an unsafe situation, which should be prevented. To minimize the biased deformation, the design ductility can be reduced, or the amount of supplemental damping can be increased.

Figure 6.29 shows examples of an EPP system under Northridge earthquake excitations. Plot (a) is the displacement time history of the EPP system with a equal to 0.1 and the damping ratio equal to 0.01. Here the damping ratio is calculated when the system is linear. Plot (b) is the same system with a damping ratio equal to 0.2. It is seen that although the biased deformation still exists, the magnitude of the bias is significantly reduced.

TABLE 6.2
Examples of Seismic Response of Inelastic Systems

	Displacement Ductility					Acceleration (g)				
	Input Level (g)					Input Level (g)				
ξ \ PGA	0.20	0.40	0.60	0.80	1.00	0.20	0.40	0.60	0.80	1.00
0.01	*1.2648*	2.3954	3.8868	5.9542	8.4272	*0.4874*	0.5099	0.5071	0.5066	0.5099
0.025	*1.1360*	2.2089	3.5992	5.4867	7.7981	*0.4658*	0.5266	0.5310	0.5369	0.5445
0.05	*0.9836*	2.0071	3.2218	4.8781	6.9095	*0.4265*	0.5456	0.5651	0.5809	0.5990
0.10	*0.7886*	1.6012	2.7034	3.9752	5.6241	*0.3598*	0.5594	0.6177	0.6548	0.6918
0.15	*0.6638*	*1.3251*	2.2937	3.3621	4.7232	*0.3129*	*0.5536*	0.6550	0.7155	0.7699
0.20	*0.5786*	*1.1458*	1.9585	2.9039	4.0580	*0.2808*	*0.5344*	0.6739	0.7613	0.8337

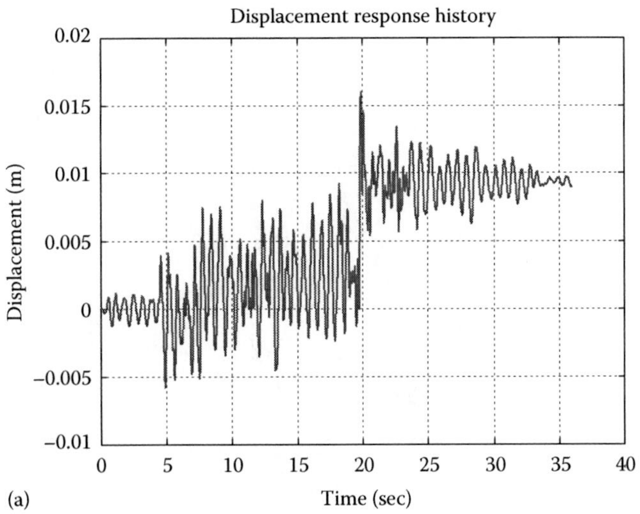

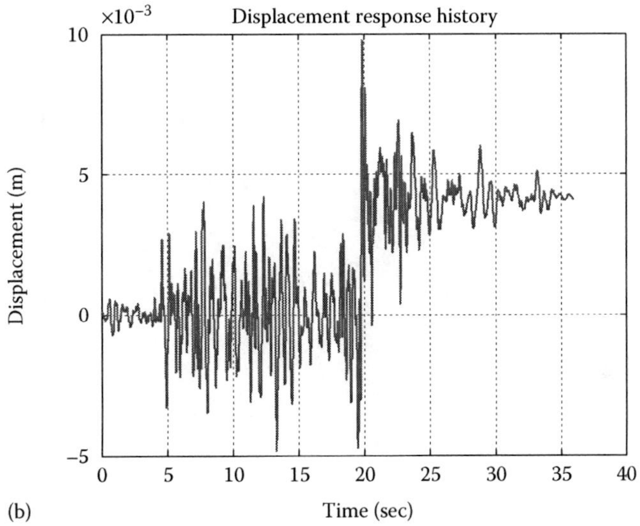

FIGURE 6.29 Biased deformation: (a) $\xi = 0.01$, a = 0.1 and (b) $\xi = 0.2$, a = 0.1.

6.4.3.6 Force–Displacement Loop of Inelastic Deformation

In the previous section, it was shown that the effective period and damping can be found by examining the force–displacement loop. It is known that a bilinear structure without viscous damping will have a parallelogram-shaped force–displacement loop. Now, consider a bilinear structure with supplemental damping. If the damping adaptability is not zero, then the damping force will grow when the displacement becomes larger. That is, when the structure has large ductility, the corresponding damping force can be very large. And, the damping force may be larger than the seismic force generated by the structure itself. In this case, the force–displacement loop will no longer be parallelogram shaped. In addition, an MDOF system in the yielding process due to large lateral seismic forces will behave differently from an SDOF system, since it may start to yield only at a certain story, whereas the rest of the stories remain elastic. In this case, the total structure behaves like a viscoelastic system, but will no longer be an idealized bilinear. In these situations, using the bilinear model may introduce errors.

Figure 6.30 shows examples of a bilinear system under Northridge earthquake excitations. Plot (a) is the force–displacement response of the bilinear system with a ratio of yield and original

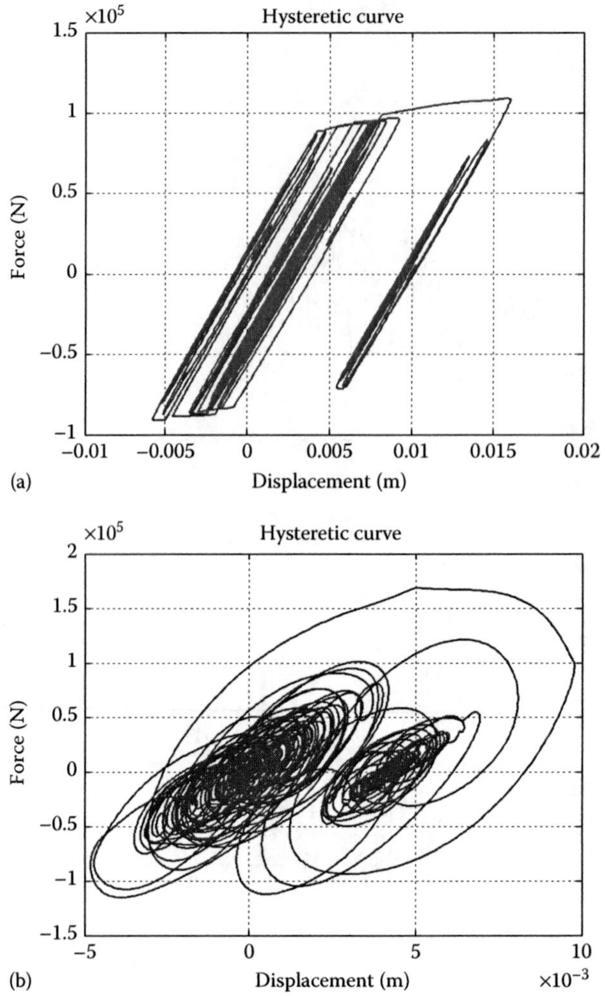

FIGURE 6.30 Examples of force–displacement loop: (a) $\xi = 0.01$, $a = 0.1$ and (b) $\xi = 0.2$, $a = 0.1$.

stiffness, denoted as $a = k_y/k = 0.01$. Here, the damping ratio is calculated when the system is linear. Plot (b) is the same system with a damping ratio equal to 0.2. It is seen that when the damping ratio is small, the force–displacement curve is very close to a parallelogram. However, when the damping ratio becomes larger, the shape of the loop is significantly altered.

6.5 SUMMARY

In this chapter, the basic principles of damping design are applied to nonlinear and irregular structures with the following emphases:

1. In terms of seismic response reduction as well as energy dissipation, using supplemental damping is beneficial, especially for displacement reduction.
2. Inelastic structures with supplemental damping can be easily overdamped. Using underdamped design spectra, the period and damping ratio need to be altered or redefined (see specific design considerations described in Chapter 8, Section 8.3).

3. Inelastic structures with large ductility may have significant biased deformation, which magnifies both the dynamic and the permanent displacements. Increasing damping can be one of the methods to control the inelastic deformations.
4. When a large amount of damping is added to an inelastic structure, the force–displacement (hysteretic) loop is likely to be altered.

The fundamental theories of vibration systems and important design principles for damper designs have been discussed in Parts I and II of this book. In the next two chapters, Part III, practical design procedures for linear and nonlinear damping systems are discussed with selected examples.

REFERENCES

Applied Technology Council. 1996. *ATC-40-Seismic Evaluation and Retrofit of Concrete Buildings*, SSC Report 96-01, Seismic Safety Commission, State of California, Sacramento, CA.

Building Seismic Safety Council (BSSC). 2003. *NEHRP Recommended Provisions for Seismic Regulation for New Buildings and Other Structures, 2003 Edition*, Report No. FEMA 450, Federal Emergency Management Agency, Washington, DC.

Building Seismic Safety Council (BSSC). 2009. *NEHRP Recommended Seismic Provisions for New Buildings and Other Structures, 2009 Edition*, Report Nos. FEMA P-750, Federal Emergency Management Agency, Washington, DC.

Chopra, A.K. 2006. *Dynamics of Structures: Theory and Applications to Earthquake Engineering*. 3rd ed. Upper Saddle River, NJ: Prentice Hall.

Chopra, A.K. and Goel, R. 2002. A modal pushover analysis procedure for estimating seismic demands for buildings. *Earthquake Engineering and Structure Dynamics* 31:561–82.

Constantinou, M.C., Soong, T.T., and Dargush, G.F. 1998. Passive energy dissipation systems for structural design and retrofit. Monograph No. 1, Multidisciplinary Center for Earthquake Engineering Research, University at Buffalo, NY.

Earthquake Engineering Research Institute. 1995. Northridge Earthquake of January 17, 1994 Reconnaissance Report. Supplement C to Vol. 11, Oakland, CA.

Federal Emergency Management Agency. 1997. *NCHRP Guidelines for the Seismic Rehabilitation of Buildings*, Report No. FEMA-273 and *NECHRP Commentary on the Guidelines for the Seismic Rehabilitation of Buildings*, Report No. FEMA-274, Washington, DC.

Housner, G.W. 1956. Limit design of structure to resist earthquakes, Proceeding of 1st World Conference on Earthquake Engineering, 5-1 to 5-13, Berkeley, CA.

Lazan, B.J. 1968. *Damping of Materials and Members in Structural Mechanics.* NY: Pergamon Press.

Liang, Z. and Lee, G.C. 1991. *Damping of Structures: Part 1 – Theory of Complex Damping*, NCEER Report 91-0014 NCEER, University at Buffalo, Buffalo, NY.

———. 2002. On principal axes of M-DOF structures: Static loading. *Journal of Earthquake Engineering and Engineering Vibration* 1 (2): 293–302.

Liang, Z., Tong, M., and Lee, G.C. 1991. An application of complex damping coefficients. Proceedings of Damping 91 Conference, Sponsored by Air Force Wright Laboratory and Flight Dynamics Directorate, Feb 13–15, 1991, San Diego, California.

Liang, Z., Tong, M., and Lee, G.C. 1993. An application of complex damping coefficients. *Proceedings of Damping 91 Conference*, Kissimmee, FL, Feb. 1–4, 1993.

Liang, Z. and Lee, G.C. 2002. On principal axes of M-DOF structures: Static loading. *Journal of Earthquake Engineering and Engineering Vibration* 1 (2).

Liang, Z. and Lee, G.C. 2003. On principal axes of M-DOF structures: Dynamic loading. *Journal of Earthquake Engineering and Engineering Vibration*, June 2003, 2 (1).

Liang, Z., Tong, M., and Lee, G.C. 1999. A real-time structural parameter modification approach for random vibration reduction, Part I: Principle. *Journal of Probabilistic Mechanics* 14:349–62.

Lin, W.H. and Chopra, A.K. 2003. *Earthquake Response of Symmetric and Asymmetric One-Story Elastic Systems with Nonlinear Fluid Viscous Dampers or Nonlinear Viscoelastic Dampers*, EERC Report No. EERC-2003-02, Berkeley, CA.

Nagarajaiah, S., Mao, Y., and Saharabudhe, S. 2006. Nonlinear, seismic response spectra of smart sliding isolated structures with independently variable MR dampers and variable stiffness SAIVS system. *Structural Engineering & Mechanics* 24 (3): 375–93.

Priestley, M.J.N. 2000. Performance based seismic design. Keynote address, Proceedings of 12th World Conference on Earthquake Engineering, Auckland, New Zealand (CD-ROM).

Ramirez, O.M., Constantinou, M.C., Kircher, C.A., Whittaker, A.S., Johnson, M.W., Gomez, J.D., and Chrysostomou, C.Z. 2000. *Development and Evaluation of Simplified Procedures for Analysis and Design of Buildings with Passive Energy Dissipation Systems*, MCEER Report 00-0010 Revision 1, MCEER, University at Buffalo, Buffalo, NY.

Rzhevsky, V. and Lee, G.C. 1998. Quantification of damage accumulation of moment resisting frames under earthquake ground motions. (Unpublished manuscript), MCEER, University at Buffalo, Buffalo, NY.

Tong, M., Liang, Z., and Lee, G.C. 1992. Correction criteria of finite-element modeling in structural damping. *Journal of Engineering Mechanics* 118 (4): 663–82.

Warburton, G.B. and Edney, S.L. 1984. Vibrations of rectangular plates with elastically restrained edges. *Journal of Sound and Vibration* 95 (4): 537–52.

Part III

Design of Supplemental Damping

Part III of this book presents selected design procedures and examples for supplemental damping. Guidelines for the design of structures with added damping devices have been included in current aseismic building codes (NEHRP 2000, NEHRP 2003, NEHRP 2009). In general, supplemental damping design approaches can be classified into five types: A: linear SDOF and MSSP (multiple-story-single-period) models; B: linear proportionally damped MDOF systems; C: linear generally damped MDOF structures; D: nonlinear damped elastic structures; and E: nonlinear damped inelastic structures. In this book, a selected set of design approaches are addressed in Chapters 7 and 8. These approaches are given in the current NEHRP Provisions. A number of departures from the NEHRP Guidelines are included to take care of large damping and nonlinear issues.

These additional issues included in Chapters 7 and 8 are noted below:

1. A more rigorous criterion to determine if a linear model can be used is based on system behavior instead of on the level of force in individual devices.
2. Dampers are classified based on rate-dependency (rather than on displacement- and velocity-dependency).
3. Estimation of the period utilizes the conservative force only. Estimation of the damping ratios includes both Timoshenko and force-based effective damping.
4. Relative displacements and absolute accelerations are no longer only proportional to the square of the natural period. For simplified design, an additional parameter is used to account for the absolute acceleration.
5. Mode shapes of acceleration and displacement are treated differently.
6. The Penzien constant is used to account for randomness of the structural response.
7. Irregular damping is considered by using a generally damped model, including nonproportional damping and overdamped subsystems.
8. Nonlinear design response spectra are used.
9. Supporting stiffness of damping devices is included.
10. Numerical damping parameters are modified.

7 Linear Damping Design

In the first part of this chapter, an overview of design procedure using three linear models is presented in Section 7.1. The design philosophy, criteria, and primary logic are presented and briefly explained. The design procedures for multi-story-single-period (MSSP) systems, and proportionally and nonproportionally damped systems are also discussed step-by-step in Section 7.2, 7.3, and 7.4. The second part of this chapter is concerned with damper installation issues within the context of structural dynamics. Furthermore, issues related to the complementary explanation for the basic idea of damping control used by NEHRP 2009 (BSSC 2009) are also discussed at the end of the chapter.

7.1 OVERVIEW OF DESIGN APPROACHES

When the damping ratio is small that can be modeled by Timoshenko damping (see Equation 1.227) and when the damping can be approximated as linear viscous damping, a linear approach can be used to simplify the damping design when the base structure is designed to be primarily within the elastic range. In this section, three design phases are described and step-by-step procedures are given for the MSSP model, the proportionally and generally damped systems, respectively. The design is based on spectral analysis, rather than time history computations.

7.1.1 Design Philosophy

The basic three-phase design procedure for linear structural systems is as follows: (1) evaluating the base structure to determine if supplemental damping can be effectively used; (2) calculating the structural responses, which is the primary effort of damping design (response computation); and (3) determining the specifications of the selected dampers and the damping matrix based on the required amount of supplemental damping when the damper configuration is predetermined.

Iterative evaluation of the effectiveness of a damping design may be carried out once the damping matrix is computed. Dividing the design procedure into three phases is mainly to promote a better understanding of the design process. Interactive consideration often exists among the specific design steps.

The seismic response computation is a key phase in damping design. The computation can be carried out through spectral analysis and/or through time history analysis. The proposed design in this book focuses on the former, which is compatible with NEHRP 2009.

Determining if a structure is suitable to use supplemental damping is based on the amount of the fundamental modal damping ratio of the base structure ξ_0, i.e., $\xi_0 < 2\%–5\%$. The total damping ratio ξ_{design} of the structure with supplemental damping devices is approximated by (Liang and Lee 1991)

$$\xi_{design} = \xi_0 + \xi_a \tag{7.1}$$

where ξ_a is the damping ratio contributed by the added dampers.

The criteria of using specific models, namely, the MSSP model, and the proportionally and nonproportionally damped models, are discussed in the following subsections. The spectral response estimation is based on modal analysis and superposition. That is, the seismic responses of the first several modes are calculated and combined to obtain the total responses. Design based on the MSSP model involves the fundamental mode only.

Generally speaking, the modal responses include peak floor displacements, absolute accelerations, lateral forces, and total base shear. The peak displacement and the acceleration are two basic quantities, from which the remaining responses can all be determined. Suggested by NEHRP 2009, the total base shear is taken to be the criterion to determine if the design is credible. The peak model displacement of the ith mode that covers each floor is a vector and is generally written as

$$\mathbf{d}_{\max_i} = \Gamma_i d_{iD} \boldsymbol{p}_i \quad (\mathrm{m}) \tag{7.2}$$

and the ith peak pseudo acceleration, which is used to approximate absolute acceleration, can be written as

$$\mathbf{a}_{s_i} = \Gamma_i C_{si} g \boldsymbol{p}_i \quad (\mathrm{m/s^2}) \tag{7.3a}$$

When damping is small, the ith peak absolute acceleration can be written as

$$\mathbf{a}_{a_i} \approx \Gamma_i C_{si} g \boldsymbol{p}_i \quad (\mathrm{m/s^2})$$

The modal responses are then combined to obtain the total relative displacement and absolute acceleration together with safety (modification) factors. The computation of these two key quantities essentially follows the same logic: find the products of the modal displacement d_{iD} and/or seismic response factor C_{si}, the modal participation factor Γ_i and mode shape $\boldsymbol{p}_i$.

Both the terms d_{iD} and C_{si} are functions of the period T_i, damping ratio ξ_i, and corresponding values taken from design spectra as well as information from seismic zoning, site specification, and type of structures. Γ_i is a function of mode shape $\boldsymbol{p}_i$ and mass matrix **M**. To calculate the modal parameters T_i, ξ_i, and $\boldsymbol{p}_i$, as basic control parameters, partial or total information about the stiffness **K** and the damping matrix **C** may be needed, depending on which model is used.

7.1.2 Design Methods for Linear Systems

Tables 7.1 through 7.3 provide a summary view of the design methods and corresponding design parameters for linear systems.

TABLE 7.1
Damping Designs for Linear Systems

System		Notes and Suggestions
MSSP (1st mode only)	*MC*	Mass matrix only. Inaccurate. Sufficient for initial response estimation and design
	MA	Mass and stiffness matrices needed. Most accurate in MSSP design
	MB	Mass and flexibility matrices needed, easier than M1
	MD	Mass matrix only. Inaccurate, for initial design only
PM (Need higher modes)	M1	Real mode only. Good for regular building and proportionally or lightly damped
	M2	Real mode only. Inaccurate
NM (Need higher modes)		Complex mode. Most accurate for linear system, not necessary for lightly damped

TABLE 7.2
Required Basic Parameters for System Modeling

System		Mass Matrix	Damp. Matrix	Stiffness Matrix	Mode Shape	Period	Damp. Ratio	Mdl Contribution
MSSP	M*C*	Lumped mass 7.6			$\boldsymbol{p}_1$ 7.12*C*	T_1 7.12*C*	ξ_1 7.35	γ_{ml} 7.9
	M*A*	Lumped mass 7.6		Lumped or FEM	$\boldsymbol{p}_1$ 7.12*A*	T_1 7.12*A*	ξ_1 7.35	γ_{ml} 7.9
	M*B*	Lumped mass 7.6		S-matrix	$\boldsymbol{p}_1$ 7.12*B*	T_1 7.12*B*	ξ_1 7.35	γ_{ml} 7.9
	M*D*	Lumped mass 7.6			$\boldsymbol{p}_1$ 7.12*D*	T_1 7.12*D*	ξ_1 7.35	
PM	M1	Lumped mass 7.6	7.34[a]	Lumped or FEM	$\boldsymbol{p}_i$ 7.53	T_i 7.53	ξ_i 7.76	γ_{mS} 7.49
	M2	Lumped mass 7.6		Lumped or FEM	$\boldsymbol{p}_i$ 7.53 p_{Rj} 7.54	T_i 7.53, T_R 7.54	ξ_i 7.76	
NM		FEM lumped	7.34	Lumped or FEM	$\boldsymbol{p}_i$ $\boldsymbol{q}_i$ 7.79, 7.84	T_i 7.81, 7.82	ξ_i 7.83	γ_{mS} 7.49

[a] Not necessary.

TABLE 7.3
Parameters of Response Estimation

System		Seismic Respon Coeff	Spectral Disp	Floor Disp	Mdl Participat. Factor	Effctv Mass	Base Shear	Floor Acc.	Lateral Force	Damper Parameter
MSSP	*MC*	C_s 7.10c	d_{1D} 7.10c	d_{max} 7.14 *ABC*	Γ_1 7.13		V 7.23 *ABC*	a_{aj} 7.19 *ABC*	f_{Lj} 7.22 *ABC*	c_{est} 7.30
	MA	C_s 7.10a	d_{1D} 7.10a	d_{max} 7.14 *ABC*	Γ_1 7.13		V 7.23 *ABC*	a_{aj} 7.19 *ABC*	f_{Lj} 7.22 *ABC*	C_{con} 7.31
	MB	C_s 7.10b	d_{1D} 7.10b	d_{max} 7.14 *ABC*	Γ_1 7.13		V 7.23 *ABC*	a_{aj} 7.19 *ABC*	f_{Lj} 7.22 *ABC*	$C^{(\cdot)}$ 7.34
	MD	C_s 7.10d	d_{1D} 7.10d	d_{max} 7.14 *D*	Γ_1 7.13	m_{eff} 7.24	V 7.23 *D*	a_{aj} 7.19 *D*	f_{Lj} 7.22 *D*	c_{design} 7.36 Φ 7.41 d_r 7.48
PM	M1	C_{si} 7.51a~d	d_{iD} 7.51a~d	d_{maxi} 7.57, 7.58	Γ_i 7.55		V 7.70	a_{si} 7.61 a_a 7.62*A*	f_{Lj} 7.66	Same as MSSP
	M2	C_{si} 7.51a~d C_{sR} 7.51e	d_{iD} 7.51a~d 7.51e	d_{maxi} 7.57, 7.58	Γ_i 7.55, Γ_R 7.56	m_{effi} 7.74	V_i 7.71, V_R 7.73 V 7.72	a_{si} 7.61 a_a 7.62*D*	f_{Lji} 7.67, 7.68, 7.69	
NM		C_{si} 7.51a~d C^R_{sk} 7.94 S_{V0} 7.93	d_{iD} 7.51a~d d^R_{iD} 7.92 S_{D0}, 7.91	d_{maxi} 7.57, 7.95, 7.97, 7.98, 7.99	Γ_i 7.55, 7.88 Γ^R_i 7.90		V 7.70	a_{si} 7.61 a_{si} 7.100 a^R_{si} 7.102 a_a 7.103	f_{Lji} 7.66	

In Table 7.1, several possible damping design methods are listed along with the required computation of parameters. The first column lists the models and corresponding methods. The first model is the MSSP for which four methods, denoted by MC, MA, MB, and MD, are available. The second model is a proportionally damped multi-degree-of-freedom (MDOF) system, denoted by PM for two methods, M1 and M2. The third model is the nonproportionally damped MDOF system, which is denoted by NM.

It is seen that the MSSP model requires the computation of the first (fundamental) mode only, which is the simplest method. The condition to accurately use this method is the high modal mass ratio (modal contribution factor) of the first mode. Unfortunately, such a condition is rarely satisfied in practical design. However, the MSSP model can provide an initial estimate of the effectiveness and cost of damping control. In this case, method C is often sufficient for the response estimation. Note that since higher modes will affect the acceleration (force) more than displacement, for displacement-based design, the MSSP model can be more appropriate; the requirement of the modal mass ratio can be relatively lower. Also, note that after the damper design is completed, the damping matrix for the supplemental dampers will be available, which can be further used in PM and NM designs.

Design based on a proportionally damped MDOF system (PM) includes higher modes and yields more accurate response estimation. This method needs both mass and stiffness matrices as well as the calculation of eigen-decompositions. Method 1 (M1) can provide a more accurate estimation of accelerations, but requires proper safety factors to calculate both acceleration and displacement. Method 2 (M2) utilizes the concept of residue mode, which is comparatively simpler than method 1. Discussion of PM design follows the one for MSSP system; and useful material presented in the section on MSSP will not be repeated. The design logic will become evident in the discussions regarding Table 7.2.

Design based on a nonproportionally damped MDOF system (NM) is suggested for systems with sufficiently high damping only. This method needs eigen-decompositions of a $2n \times 2n$ state matrix and the corresponding complex-valued mode shape as well as the accompanist matrices. The computation burden is large and the improvement is often not quite phenomenal, except in structures with extremely high levels of mass, damping, and stiffness irregularities. A discussion about NM design is introduced at the end of this section. The criterion for using this method is based on a sufficiently high modal energy transfer ratio.

Table 7.2 lists the required design parameters for the damping design methods. The first column is identical to Table 7.1. The second through the seventh columns are, respectively, the mass matrix, the damping matrix, the stiffness matrix, the real-valued and the complex-valued mode shapes, the periods, the damping ratios, the modal mass ratios, and the modal energy transfer ratios. The parameters listed in Table 7.2 will help a designer determine whether specific parameters are needed, and if so, the corresponding equations are given. For mode shapes, natural periods, and damping ratios, the requirements for the first modes are denoted by subscript 1, and higher modes are denoted by subscript i.

Table 7.3 lists computations of terms needed for response estimations, including the numbers of the corresponding equations. Tables 7.1 through 7.3 show the design logic and required computations.

7.2 MSSP SYSTEMS SIMPLIFIED APPROACH

7.2.1 General Description

Simplified design discussed here refers to structures that behave linearly with comparatively small damping so that they can be modeled as linear systems (a single-degree-of-freedom [SDOF] or MSSP system). The design logic is based on the philosophy of NEHRP 2009, which is applicable to structural vibration reduction design when the structures are more or less regularly shaped with evenly distributed mass. Under these conditions, the stiffness matrix and the exact mode shape often do not have to be used. However, the mass matrix is needed to estimate the fundamental mode shape. The simplified design procedure is described in a form very close to the approaches established in the current building codes, except for estimating absolute acceleration and lateral forces.

Spectra-based design for SDOF and MSSP systems relies on the estimation of modal parameters of the first mode, namely, the natural period, the damping ratio, and the mode shape. Simplified design includes four specific steps:

1. The basic structural parameters such as the natural period (or frequency), the fundamental mode shape, and the original damping ratio of the structures without added dampers, as the primary control parameters, are estimated. Using these parameters and the design spectrum, it can be determined whether or not dampers will be needed.
2. It is necessary to determine whether the simplified method can be adequately carried out with sufficient accuracy. A method of using the percentage modal mass ratio is proposed for this purpose, which is also a primary control parameter.
3. The design spectrum is used to determine the proper damping ratios. In NEHRP 2009, the design parameters, such as the base shear, the lateral forces, and/or the story deformations, are directly related to the magnitude of the damping ratio through the numerical damping coefficient B. Therefore, the damping ratio is considered to be the control parameter.
4. The appropriate dampers are chosen to realize the theoretical damping ratio. Criteria for selecting different types of dampers were provided in Chapters 5 and 6. The equations to calculate the parameters of individual linear dampers are given in the simplified design approach.

If the total damping ratio given by Equation 7.1 exceeds 15%,

$$\xi_{\text{design}} > 15\% \tag{7.4}$$

then the simplified design approach that assumes a linear system may not result in a good design. More appropriate design approaches are recommended in the next section and in Chapter 8.

In this book, only the complete design steps and procedure for an MSSP model are discussed. For proportionally and generally damped MDOF models, only the differences in the design procedures from that of the MSSP systems are addressed.

7.2.2 Feasibility of Damping Control

To determine whether or not using supplemental damping is beneficial, the following steps may be taken:

1. The first step is to understand the key control and design parameters designated by local and national building codes, such as T_0, T_s, T_L, S_{DS}, S_{D1}, and I (see Chapter 2, Section 2.4 for definitions). In Chapters 7 and 8, several simplified examples are illustrated by assuming that $T_0 = 0.16$ (s), $T_s = 0.8$ (s), $T_L = 2.5$ (s), $S_{DS} = 0.5$, $S_{D1} = 0.4$, and I = 1.
   The values listed above are, of course, variables; in the actual design process, their values will depend on the location, the type of structure, etc.
2. The second step is to have an initial estimation of the targeted structure, especially the fundamental natural frequency T_1, the mass matrix **M**, the possible damping ratio ξ_0 of the base structure and the *R* factor, and the approximate fundamental mode shape $\boldsymbol{p}_1$. In the following paragraphs, for the sake of simplicity, it is assumed that

$$\xi_0 < 3\% \text{ and } R = 1. \tag{7.5}$$

Generally, the diagonal mass matrix is sufficient for damping design, that is,

$$\mathbf{M} = \text{diag}\left(\mathrm{m_j}\right) \quad (\mathrm{t}) \tag{7.6}$$

where m_j is equal to the jth vertical load w_i divided by g.

3. To determine if damping control is required, two independent criteria can be used. First, if the natural period of the targeted structure is longer than T_s, increasing the stiffness will further shorten the period. This is not beneficial in reducing the seismic response. Therefore, the first criterion for damping design is

$$T_1 > T_s \quad (\mathrm{s}) \tag{7.7}$$

The second criterion is whether the original damping ratio is sufficiently small, or

$$\xi_0 < 2\% - 5\% \tag{7.8}$$

If Equations 7.7 and 7.8 are satisfied, damping control can offer a cost-effective method to reduce the seismic responses of the structure.

A further step is to consider which design strategy should be used. If the structure is designed to remain in the elastic range (R = 1) and the levels of its mass, damping, and stiffness irregularities are low, linear approaches can be used. In the following subsection, criteria are provided to determine if the damping is irregular. The simplified design method is discussed in terms of SDOF and/or MSSP models.

7.2.3 SDOF and MSSP Systems

In many practical situations, the responses of a multiple-story structure can be approximated by its first mode only. Therefore, to estimate the responses, a single period of this mode is all that is needed.

Mathematically speaking, an MSSP system is the first mode of an MDOF structure. Therefore, the criterion for using an MSSP model can be established by evaluating the modal contribution of the first mode (modal mass ratio) (see Equation 3.184):

$$\gamma_{m1} = \frac{\left[\boldsymbol{p}_1^T \mathbf{M J}\right]^2}{\boldsymbol{p}_1^T \mathbf{M}\, \boldsymbol{p}_1 \sum_{j=1}^{n} \mathrm{m_j}} \geq 85\% - 95\% \tag{7.9}$$

Note that in NEHRP 2009, the value of γ_{m1} is required to be at least 90%. This criterion is often overlooked so that MSSP systems are used. Practically speaking, however, not many structures can be modeled as linear MSSP systems, despite the fact that many designers use the first mode only to calculate the seismic responses. Note that when a structure yields and has a large displacement the MSSP model will be more accurate than linear systems, because the first "effective" mode contains the dominant deformation energy.

Both SDOF and MSSP systems have a single period and a single damping ratio. Therefore, the design spectrum (see Chapter 2) can be used directly, based on which, the acceleration (as well as the lateral force) and the lateral displacement can be calculated. The estimation of the values of the period and damping ratio as well as the mode shape can be carried out as outlined in Chapters 2 and 3. The computations of the force and displacement can be performed based on the equations given in Chapter 6. The details of derivations are not repeated in this chapter. For instance, Equations 2.334

and 2.337, discussed in Chapter 2, can be used to calculate the acceleration, C_s, and displacement, S_D, respectively.

7.2.4 Basic Design Procedure

Basic procedures for the simplified damper design are summarized with applicable equations for MSSP models. The design through an SDOF model is rarely used because it often does not satisfy the criterion in Equation 7.9. However, if an SDOF model is used, the same procedure can be followed by letting $\boldsymbol{p}_1 = 1$.

7.2.4.1 Estimation of Seismic Response of Original Structure

If the criterion described in Equation 7.9 is satisfied, then the simplified damping design is carried out using the following steps.

7.2.4.1.1 Seismic Response Coefficient and Spectral Displacement

The seismic response coefficient, C_s, and the spectral displacement for the first mode, d_{1D}, are calculated by

$$C_s = \frac{\sqrt{1+4\xi^2}}{BR}\left(0.6\frac{S_{DS}}{T_0}T_1 + 0.4S_{DS}\right)I \quad \text{and} \quad d_{1D} = \frac{C_s T_1^2}{4\pi^2\sqrt{1+4\xi^2}}g \ \ (\text{m}) \tag{7.10a}$$

when $T_1 < T_0$,

$$C_s = \frac{S_{DS}I\sqrt{1+4\xi^2}}{BR} \quad \text{and} \quad d_{1D} = \frac{C_s T_1^2}{4\pi^2\sqrt{1+4\xi^2}}g \ \ (\text{m}) \tag{7.10b}$$

when $T_0 \le T_1 \le T_s$,

$$C_s = \frac{S_{D1}I\sqrt{1+4\xi^2}}{T_1BR} \quad \text{and} \quad d_{1D} = \frac{C_s T_1^2}{4\pi^2\sqrt{1+4\xi^2}}g \ \ (\text{m}) \tag{7.10c}$$

when $T_s \le T_1 \le T_L$, and

$$C_s = \frac{S_{D1}IT_L\sqrt{1+4\xi^2}}{T_1^2BR} \quad \text{and} \quad d_{1D} = \frac{C_s T_1^2}{4\pi^2\sqrt{1+4\xi^2}}g \ \ (\text{m}) \tag{7.10d}$$

when $T_1 > T_L$.

In Equations 7.10, C_s is dimensionless and the unit of d_{1D} is (m). Compare Equations 7.10 and 6.7. In Equations 7.10, the design parameters are based on real spectrum and the modal participation factors are not included. These will also be true in Equation 7.51 for higher modes in the next subsection as well as in the next chapter.

In Equations 7.10, T_1 is the first (fundamental) period; ξ_{design} is denoted as ξ for simplicity; T_0, T_s, T_L, R, I, S_{DS}, and S_{D1} are defined in Chapter 2, Sections 2.4.2 and 2.4.3; and for elastic (linear) structures, $R = 1$. The numerical damping coefficient is

$$B = 3\xi + 0.9 \tag{7.11}$$

In Section 7.5, Equation 7.11 is explained in more detail. Note that, for the initial design, ξ is often assumed as 2%–5%, depending on the type of structure. For steel frame structures, the damping ratio is smaller and for concrete structures, there can be a larger value.

7.2.4.1.2 Mode Shape and Natural Period of Fundamental Mode

To calculate the mode shape, $\boldsymbol{p}_1$, and the period, T_1, the method of eigen-decomposition can be used if both the **M** and **K** matrices are available, which is referred to as *Method A* and denoted as M*A* in Tables 7.1 through 7.3, that is,

$$\boldsymbol{p}_1 = \mathbf{P}(:,1) = \begin{Bmatrix} p_1 \\ p_2 \\ \cdots \\ 1 \end{Bmatrix} \quad \text{and} \quad T_1 = \frac{2\pi}{\sqrt{\lambda_1\left(\mathbf{M}^{-1}\mathbf{K}\right)}} = 2\pi\sqrt{\frac{\boldsymbol{p}_1^{\mathrm{T}}\mathbf{M}\,\boldsymbol{p}_1}{\boldsymbol{p}_1^{\mathrm{T}}\mathbf{K}\,\boldsymbol{p}_1}} \quad (\mathrm{s}) \tag{7.12A}$$

or through the flexibility matrix **S**, which is referred to as *Method B* and denoted as M*B* in Tables 7.1 through 7.3:

$$\boldsymbol{p}_1 = \mathrm{g}\mathbf{SMJ} = \begin{Bmatrix} p_1 \\ p_2 \\ \cdots \\ p_n \end{Bmatrix} \quad \text{and} \quad T_1 = 2\pi\sqrt{\frac{\boldsymbol{p}_1^{\mathrm{T}}\mathbf{M}\,\boldsymbol{p}_1}{\boldsymbol{p}_1^{\mathrm{T}}\mathbf{MJ}\mathrm{g}}} \quad (\mathrm{s}) \tag{7.12B}$$

or through mode shape approximation, which is referred to as *Method C* and denoted as M*C* in Tables 7.1 through 7.3:

$$\boldsymbol{p}_1 = \left\{\sin\frac{\mathrm{j}\pi}{2\mathrm{n}}\right\} = \begin{Bmatrix} p_1 \\ p_2 \\ \cdots \\ 1 \end{Bmatrix} \quad \text{and} \quad T_1 = 0.1\mathrm{n} \quad (\mathrm{s}) \tag{7.12C}$$

Note that the estimation $T_1 = 0.1n$ (s) is used by NEHRP 2009, where the conditions are $n \leq 12$ and story height ≥ 3 (m). However, for initial response estimation in simplified design, this formula can be used for other types of buildings.

The triangle approximation used in NERHP 2009, referred to as *Method D* and denoted as M*D* in Tables 7.1 through 7.3, is given by

$$\boldsymbol{p}_1 = \{\mathrm{p}_\mathrm{j}\} = \left\{\frac{\mathrm{h}_\mathrm{j}}{\mathrm{h}_\mathrm{r}}\right\} = \begin{Bmatrix} p_1 \\ p_2 \\ \cdots \\ 1 \end{Bmatrix} \quad \text{and} \quad T_1 = 2\pi\sqrt{\frac{\sum_{\mathrm{j}=1}^{\mathrm{n}} \mathrm{w}_\mathrm{j}\delta_\mathrm{j}^2}{\mathrm{g}\sum_{\mathrm{j}=1}^{\mathrm{n}} \mathrm{f}_{\mathrm{Lj}}\delta_\mathrm{j}}} \quad (\mathrm{s}) \tag{7.12D}$$

Here, h_j is the height of the jth level above the ground and h_r is the height of the roof; w_j, f_{Lj}, and δ_j are, respectively, the part of the total gravity load W, the lateral load, and the elastic deflection, located at the jth story.

Note that in Equations 7.12*A*, 7.12*C*, and 7.12*D*, the mode shape is normalized so that the roof displacement is unity, which is not necessary. The significance of such a normalization is to make sure that the modal displacement of the roof has an identical sign.

7.2.4.1.3 Modal Participation Factor

In order to calculate the modal responses, the modal participation factor, Γ_1, is needed, which is given by

$$\Gamma_1 = \frac{\boldsymbol{p}_1^{\mathrm{T}} \mathbf{M}\, \mathbf{J}}{\boldsymbol{p}_1^{\mathrm{T}} \mathbf{M}\, \boldsymbol{p}_1} \tag{7.13}$$

7.2.4.1.4 Lateral Displacement

The maximum values of floor lateral displacement for methods *A*, *B*, and *C* are given by

$$\mathrm{d}_{max} = S_M S_{Dn} S_{D\xi} \Gamma_1 \mathrm{d}_{1D} \boldsymbol{p}_1 = \begin{Bmatrix} \mathrm{d}_1 \\ \mathrm{d}_2 \\ \ldots \\ \mathrm{d}_n \end{Bmatrix} \quad (\mathrm{m}) \tag{7.14ABC}$$

The jth maximum value of floor lateral displacement for method *D* is given by

$$\mathrm{d}_{max} = \Gamma_1 \mathrm{d}_{1D} \boldsymbol{p}_1 = \begin{Bmatrix} \mathrm{d}_1 \\ \mathrm{d}_2 \\ \ldots \\ \mathrm{d}_n \end{Bmatrix} \quad (\mathrm{m}) \tag{7.14D}$$

Note that in Equation 7.14*D*, the deflection δ_j can be calculated by

$$\delta_j = \mathrm{d}_j - \mathrm{d}_{j-1} \quad (\mathrm{m}), \quad j > 2 \tag{7.15a}$$

with the deflection of the first story

$$\delta_1 = \mathrm{d}_1 \quad (\mathrm{m}) \tag{7.15b}$$

In Equation 7.14*ABC*, S_M is a modification (or safety) factor regarding incomplete modes, and

$$S_M = (0.95 \sim 1)/\gamma_{m1} \tag{7.16}$$

In Equation 7.14*ABC*, S_{Dn} is a modification (or safety) factor for displacement estimation with regard to the number of stories, n. This modification considers possible vertical and plan irregularities of buildings. Buildings with different numbers of stories may have different degrees of irregularities. Many numerical simulations have been carried out by the authors, which suggest that Equation 7.17 is a reasonable estimation of the modification factor:

$$S_{Dn} = \frac{1}{a_4 n^4 + a_3 n^3 + a_2 n^2 + a_1 n + a_0} \tag{7.17}$$

where a_0 through a_4 are coefficients, which should be determined through local design spectrum and computer simulation. In Equation 7.16, the values 0.95 ~ 1 stand for the preset criterion for choosing the number of modes (see Equation 7.50 in the next subsection). Once this criterion is determined, the coefficients a_0 through a_4 can be determined. For example, if the same 99 earthquake records are used with $S_M = 0.95/\gamma_{m1}$,

$$S_{Dn} = \frac{1}{6.643\times10^{-8}n^4 - 1.1689\times10^{-5}n^3 + 7.405\times10^{-4}n^2 - 0.0214n + 0.7427}$$

In Equation 7.14*ABC*, $S_{D\xi}$ is a modification (or safety) factor for displacement estimation regarding the design damping ratio, ξ_{design}. This modification considers possible nonproportional damping as well as damping-induced cross effect. Again, based on limited numerical simulations, the following modification factor appears to be reasonable:

$$S_{D\xi} = \frac{1}{b_4\xi_{design}^4 + b_3\xi_{design}^3 + b_2\xi_{design}^2 + b_1\xi_{design} + b_0} \tag{7.18}$$

where b_0 through b_4 are coefficients, which should be determined through local design spectrum and computer simulation. For example, if the 99 earthquake records are used,

$$S_{D\xi} = \frac{1}{36.08\xi_{design}^3 - 23.30\xi_{design}^2 + 4.962\xi_{design} + 0.8057}$$

7.2.4.1.5 Lateral Absolute Acceleration

The lateral absolute acceleration vector is

$$\mathbf{a}_{a1} = S_M S_{An} S_{A\xi}\left[\left(\Gamma_1 C_s g \boldsymbol{p}_1\right) + \left(S_{D1}\mathbf{J}\right)\right] = \begin{Bmatrix} a_{a11} \\ a_{a21} \\ \vdots \\ a_{an1} \end{Bmatrix} \quad \left(\text{m/s}^2\right) \tag{7.19ABC}$$

and

$$\mathbf{a}_{a1} = \Gamma_1 C_s g \boldsymbol{p}_1 = \begin{Bmatrix} a_{a11} \\ a_{a21} \\ \vdots \\ a_{an1} \end{Bmatrix} \quad \left(\text{m/s}^2\right) \tag{7.19D}$$

In Equation 7.19*ABC*, S_M is defined in Equation 7.16, and S_{An} is a modification (or safety) factor for acceleration estimation regarding the number of stories, n. Based on limited numerical simulations, the following appears appropriate:

$$S_{An} = \frac{1}{a_4n^4 + a_3n^3 + a_2n^2 + a_1n + a_0} \tag{7.20}$$

where a_0 through a_4 are a set of coefficients different from those given in Equation 7.17, which should be determined through local design spectrum and computer simulation. For example, again using the 99 earthquake records,

$$S_{An} = \frac{1}{4.343\times10^{-6}n^3 - 7.793\times10^{-4}n^2 + 0.0433n + 0.8547}$$

In Equation 7.19*ABC*, $S_{A\xi}$ is a modification (or safety) factor for acceleration estimation with regard to the design damping ratio, ξ_{design}. Based on limited numerical simulations,

$$S_{A\xi} = \frac{1}{b_4\xi_{design}^4 + b_3\xi_{design}^3 + b_2\xi_{design}^2 + b_1\xi_{design} + b_0} \tag{7.21}$$

where, b_0 through b_4 are a group coefficients different from those given in Equation 7.18, which should be determined through local design spectrum and computer simulation. For example, using the 99 earthquake records, $S_{A\xi}$ is given by

$$S_{A\xi} = \frac{1}{53.95\xi_{design}^3 - 34.865\xi_{design}^2 + 7.987\xi_{design} + 0.6811}$$

7.2.4.1.6 Lateral Force

The lateral force of the jth floor can be calculated through one of the following equations:

$$f_{Lj} = m_j g a_{aj} \quad (kN) \tag{7.22ABC}$$

or

$$f_{Lj} = \frac{m_j h_j}{\sum_i^n m_i h_i} V \quad (kN) \tag{7.22D}$$

Here m_j is the mass of the jth level, a_{aj} is the peak acceleration at the jth level of a structure (see Equation 7.19), and h_j is the height of the jth level that was defined previously.

7.2.4.1.7 Base Shear

The total base shear can be calculated through the following equations. For methods *A*, *B*, and *C*,

$$V = \mathbf{f}_L^T \mathbf{J} \quad (kN) \tag{7.23ABC}$$

Here, $\mathbf{f}_L = \{f_{Lj}\}$ is the vector of lateral load (see Equation 7.22*ABC*). For method *D*,

$$V = m_{eff} g C_s \quad (kN) \tag{7.23D}$$

The effective mass of the first mode is

$$m_{eff} = \frac{\left(\boldsymbol{p}_1^T \mathbf{M} \mathbf{J}\right)^2}{\boldsymbol{p}_1^T \mathbf{M} \boldsymbol{p}_1} \quad (t) \tag{7.24}$$

Note that mathematically, Equations 7.23*ABC* and 7.23*D* are equivalent only if the fundamental mode is used. However, they are very different if more modes are involved. From the discussion in Chapter 4, Equation 7.23*ABC* can be shown to be more accurate and simpler.

7.2.4.2 Determination of Damping Ratio and Damping Coefficient

7.2.4.2.1 Damping Proportional Coefficient and Design Damping Ratio

If the base shear V, or the floor displacement d_j, or the lateral force f_{Lj} exceeds the allowed level, [V], [d_j], and [f_{Lj}], and if the original damping ratio is less than 5%, then, supplemental damping can be used. The criteria are

$$[V] \geq V, \quad [d_j] \geq d_j, \quad [f_{Lj}] \geq f_{Lj} \tag{7.25}$$

The allowed base shear [V] is denoted as

$$[V] = \alpha_V V \quad (kN) \tag{7.26}$$

the allowed displacement as

$$[d_j] = \alpha_{d_j} d_j \quad (m) \tag{7.27}$$

and the allowed lateral force as

$$[f_{Lj}] = \alpha_{f_{Lj}} f_{Lj} \quad (kN) \tag{7.28}$$

Here, the proportional coefficients α_V, α_{dj}, and α_{d_j}, which are now denoted as $\alpha_{(\cdot)}$ for convenience, are design parameters, which can be used to find the proper damping ratio ξ_a as

$$\xi_a = \frac{(1-\alpha_{(\cdot)})(3\xi_0 + 0.9)}{3\alpha_{(\cdot)}} \tag{7.29}$$

Note that once ξ_a is determined, the design damping ratio ξ_{design} can be calculated through Equation 7.1.

7.2.4.2.2 Damping Coefficient

For simplified design, a linear viscous damper is used first. The damping coefficient of an individual damper can be estimated as

$$c_{est} = \frac{4\pi\xi_a m_{eff1}}{NT_1} \quad (kN\text{-}s/m) \tag{7.30}$$

Here, N stands for the total N stories that are equipped with dampers. Note that the initial estimation does not have to be accurate.

7.2.4.2.3 Damper Configuration

Suppose the designer has a preconceived damper configuration in mind, which can be described mathematically by a damper configuration matrix, $\mathbf{C}_{con}$, and determined as follows:

$$\mathbf{C}_{con} = [c_{ij}], \quad i, j = 1, \ldots, n \tag{7.31}$$

Suppose there are p dampers that are connected to the ith mass from other p locations, the ii^{th} entry of $\mathbf{C}_{con}$ is given by

$$c_{ii} = p, \quad i=1,\ldots, p \tag{7.32}$$

The ij^{th} entry of $\mathbf{C}_{con}$ is

$$c_{ij} = -p, \quad i, j=1,\ldots, n, \quad i \neq j \tag{7.33}$$

7.2.4.2.4 Initial Damper Matrix

The estimated damping matrix is

$$\mathbf{C}^{(1)} = c_{est}\mathbf{C}_{con} \quad (\text{kN-s/m}) \tag{7.34}$$

7.2.4.2.5 Design Damping Ratio

The design damping ratio can be calculated as

$$\xi_a^{(1)} = \frac{T_1}{4\pi} \frac{\boldsymbol{p}_1^T \mathbf{C}^{(1)} \boldsymbol{p}_1}{\boldsymbol{p}_1^T \mathbf{M} \boldsymbol{p}_1} \tag{7.35}$$

For a general MDOF system, once the initially designed damping matrix is obtained, the modal damping ratios should be calculated and the modal responses should be reevaluated. This iterative procedure may yield the consequence of $\mathbf{C}^{(1)}$, $\mathbf{C}^{(2)}$,..., and $\xi_a^{(1)}$, $\xi_a^{(2)}$, etc. The iterative procedure is discussed in the next two subsections. Note that this iteration is not needed for MSSP systems.

7.2.4.3 Specifications of Dampers

The final step is to choose a proper damping device, determine its specifications and the corresponding configuration for installation. The major design parameters are the damping coefficient and stroke. Related issues include size, weight, installation issues such as connectors and supporting stiffness, working conditions such as temperature, monitoring, and maintenance, as well as design safety factors. The major design parameters are addressed in this section.

7.2.4.3.1 Nominal Damping Coefficient

The damping coefficient for an individual nominal damper can be calculated based on

$$c^{(1)} = G_{mj}^{-2} \frac{\xi_a}{\xi_a^{(1)}} c_{est} \quad (\text{kN-s/m}) \tag{7.36}$$

where G_{mj} is the geometrical magnification factor that relates the jth mass to the $(j-1)^{th}$ mass (see Equation 5.145). Note that damping devices must be installed. Otherwise, $G_{mj} = 0$.

7.2.4.3.2 Number of Dampers

Once the damping configuration is determined, the control parameters of the nominal dampers can be initially estimated. The initially determined damping coefficient can often be realized by more than one type of damper.

In order to avoid torsion, the damping forces must be balanced at each floor of a building. That is, symmetrically configured dampers are desirable. With multiple stories, there is an opportunity to install dampers in several stories to avoid the phenomenon of "man-made viscoelastic damping" due to the limited supporting stiffness, which is explained in Chapter 8. The strong degree of nonproportional damping due to unevenly distributed dampers should also be avoided. Dampers that provide a large amount of damping coefficients are often very large in size, which may not be suitable for certain structures; and, dampers with a large damping coefficient will apply a large damping force, which may not be appropriate in some specific applications. In the above, for each story as well as location there is only one damper theoretically determined, which may be called a *nominal* damper. Due to these reasons, the nominal damper must be divided into several feasible devices. In this subsection, the case when the damper configuration has already been determined is discussed. To achieve the nominal damper using commercially available dampers, consider the following strategy:

The specifications of the commercially available dampers need to be checked. Suppose the nominal damper has been replaced by several actual dampers. Although for a single mode the phase of damping force is almost 90° ahead of the restoring force, the peak values of both the damping and the restoring forces can occur almost simultaneously in the case of random excitation. Therefore, to compute the maximum allowed value, damping force can be treated as the regular internal force of the considered members of this structure.

Suppose an available linear viscous damper has a damping coefficient $c^{(1)}$. To determine if the selected damper is workable, its damping force is checked first. For linear viscous damping, the damping force of the jth damper, denoted as f_{dj}, can be calculated by

$$f_{dj} = c^{(1)} G_{mij}^{-2} \left| v_{i1} - v_{j1} \right| \quad (kN) \tag{7.37}$$

Here, v_{j1} and $v_{j-1,1}$ are the floor velocities relative to the ground and contributed by the fundamental mode, which can be approximated by

$$v_{j1} = 2\pi \frac{d_j}{T_1} \quad (m/s) \tag{7.38}$$

where d_j is the maximum floor displacement contributed by the fundamental mode defined in Equation 7.15.

In a more general case, a damper can be installed in between the ith and j^{th} locations, and the relative velocity is then calculated as $v_{i1}-v_{j1}$. This damping should be called the damping-installed-in-between-the-i^{th}-and-the-j^{th}-locations. For convenience, however, this damper can still be called the ith damper with the risk of confusion. In the following paragraphs, the ith or j^{th} damper is often used for many types of damping devices, which hopefully will not cause confusion. Thus,

$$f_{dj} = \frac{2\pi c^{(1)} G_{mij}^{-2} \left| \delta_{ij1} \right|}{T_1} \quad (kN) \tag{7.39}$$

where δ_{ij1} is the drift between ij^{th} locations, $\delta_{ij1} = d_{j1} - d_{i1}$ (m).

If the calculated damping force exceeds the allowed value $[f_{dj}]$, the damping coefficient should be reduced by

$$c_F^{(1)} \le c^{(1)} \frac{\left[f_{dj} \right]}{f_{dj}} \quad (kN\text{-}s/m) \tag{7.40a}$$

Otherwise, let

$$c_F^{(1)} = c^{(1)} \quad (\text{kN-s/m}) \tag{7.40b}$$

Here, subscript F denotes the case of the consideration of the damping forces. After carrying out the above-mentioned step, the working damping force that is equal to or less than the allowed level is obtained.

In addition, the size of the damper needs to be determined. Generally speaking, the length of the damper is often not a problem, unless specially specified. However, the diameter and the area of the cross section of the damper are usually limited. Roughly, once the damping coefficient of the linear viscous damping is determined, so will its diameter, which can be obtained from vendor specifications. As an estimate for initial design, the following equation can be used:

$$\Phi = 2\sqrt{\frac{f_{di}}{\pi p} + r^2} + B + U = 2\sqrt{\frac{2c^{(1)}\delta_{ij1}}{G_{mij}^2 T_1 p} + r^2} + B + U \quad (\text{cm}) \tag{7.41}$$

Here, Φ is the inner diameter of the damper and p is the working pressure of the hydraulic fluid inside the damper, usually $p \approx 21$ (MPa) (3000 (psi)); a specially designed high pressure, p, can be as high as 100 (MPa) (15000 (psi)). B is twice the thickness plus the outside fixtures of the hydraulic cylinder, usually $B = 1 \sim 6$ (cm); r is the diameter of the piston rod of the damper and usually $r = 1 \sim 5$ (cm). The larger the damping force f_{dj}, the larger the corresponding values of B and r. Furthermore, U is a tolerance; usually $U = 1 \sim 2$ (cm).

If the diameter is too large, several smaller dampers are used instead. Then, the damping coefficient of the damper is recalculated as follows:

$$c_\Phi^{(1)} \le c^{(1)} \frac{[\Phi] - B - U}{\Phi - B - U} \quad (\text{kN-s/m}) \tag{7.42a}$$

where $[\Phi]$ is the allowed diameter and Φ is the one calculated from Equation 7.41.

If the diameter is within the limits of the allowed level, the damper can be used. Thus,

$$c_\Phi^{(1)} = c^{(1)} \quad (\text{kN-s/m}) \tag{7.42b}$$

where subscript Φ stands for the consideration of the diameter of the damper. The corresponding damping coefficient is then chosen from the smaller value of the damping coefficient between c_F and C_Φ, and the initial estimate of the number of dampers, denoted by the symbol $N^{(1)}$, is

$$N^{(1)} \ge \frac{c^{(1)}}{\min\left[c_F^{(1)}, c_\Phi^{(1)}\right]} \tag{7.43}$$

The actual number must be an integer. In many cases, the number must also be even to prevent damping force-induced torsion.

The resulting design damping coefficient $c^{(2)}$ now becomes

$$c^{(2)} = \frac{S_\upsilon c^{(1)}}{N^{(1)}} \quad (\text{kN-s/m}) \tag{7.44}$$

Here S_υ is a safety factor to account for the uneven distribution of damping force divided by $N^{(1)}$. Based on a general treatment of unevenly distributed forces from design handbooks of mechanical elements (i.e., Avallone and Baumeister 1996), S_υ taken to be 1.05–1.2, should be considered.

The corresponding damping matrix is

$$\mathbf{C}^{(2)} = c^{(2)}\mathbf{C}_{con} \quad (\text{kN-s/m}) \tag{7.45}$$

Note that the idea described in Equation 7.41 can also be used to estimate the size of the dampers when the number of actual dampers is predetermined. In this case,

$$\Phi = 2\sqrt{\frac{2c^{(2)}\delta_{ij1}}{G_{mij}^2 T_1 p} + r^2} + B + U \quad (\text{cm}) \tag{7.46}$$

7.2.4.3.3 Damper Stroke

Suppose a damper is installed in between levels i and j. The floor drift is $(d_i - d_j)$. The stroke (relative displacement) between the two ends of the damper j, d_{rj}, is given by

$$d_{rj} = 2(d_i - d_j)G_{mj} \quad (\text{m}) \tag{7.47}$$

If all the dampers are identical, the required stroke is

$$d_r = S_\sigma \max(d_{rj}) \quad (\text{m}) \tag{7.48}$$

Here, S_σ is a safety factor, which may be taken to be 2.0 ~ 3.0 for structures designed to vibrate within the elastic range. The value should be considerably larger for structures that are allowed to develop large ductile deformation. The design displacement ductility is denoted as [μ]. Then, S_σ is taken to be 1.5 ~ 2.5 [μ], for inelastic structures.

Example 7.1

A simple example is presented to summarize the simplified design procedure. The same ten-story-seven-bay shear model is considered with the stiffness as given in Example 3.5, the roof displacement is x_{10}, the displacement of the first floor is x_1, etc. The mass matrix is assumed to be twice the value as that of Example 3.6. Further, the original damping ratio is assumed to be 2.5%. One nominal damper is installed in each level (note that to prevent torsion due to unbalanced damping force, more than one damper is needed per level; so that in this example, at least eight dampers are needed in each level). The damper configuration matrix is given by

$$\mathbf{C}_{con} = \begin{bmatrix} 2 & -1 & \dots & 0 \\ -1 & 2 & \dots & 0 \\ & \dots & & \\ 0 & \dots & -1 & 1 \end{bmatrix}_{10\times 10}$$

Since the story height is H = 3 (m) and the bay width is L = 10 (m), G_{mj} is taken to be 3/10.

Note that the modal mass ratios calculated by these methods are all below 90%. Based on NEHRP 2009, the MSSP model cannot be used due to small modal mass ratios. However, by letting the requirement relax to allow 85% modal mass ratio, the design computations can be carried out to explain the importance of the criterion of MSSP modal defined by Equation 7.9. Here, for comparison purposes, the calculation based on method *D* is included. The results given in Tables 7.5 and 7.6 can be carried out.

As a comparison, numerical simulations with the same 99 earthquake records are used again. Note that each record is scaled to have a peak ground acceleration (PGA) equal to 0.4 (g). The base shear calculated through the simulation is 64.63 (MN).

Table 7.4 lists the modal mass ratio γ_{m1}, natural period T_1, modal participation factor Γ_1, seismic response coefficient C_s, peak spectral displacement d_{1D}, effective mass m_{eff}, and base shear V calculated through the above-mentioned methods *A*, *B*, *C*, and *D*. It is seen that since only one mode is involved, the calculated base shears are rather small.

Here, the input level is assumed to be 0.4 (g). The seismic response coefficient C_s is calculated through a simplified equation described in Equation 7.51f, by letting $AS_i = 1$ and $B = 0.97$. In Example 7.2, the calculation of C_s through Equation 7.51 is shown.

Method *A* provides a comparatively larger base shear. This is the method preferred by the authors, though it requires information of **K**. Method *B* can also provide a good estimation and only needs the flexibility matrix **S**, which may be easier to obtain than **K**. Methods *C* and *D* only need the mass matrix. In this particular example, it seems that both *C* and *D* can also provide good estimations. However, in many cases, the corresponding estimation is poor. For example, if the mass matrix used in this example is exactly the same as used in Example 3.6, the natural period must be 0.75 (s), instead of 1.0 (s). The resulting error can be quite large, if $T_1 = 0.1n$ (s) is used.

To use method *D* as suggested by NEHRP 2009, since the lateral load, f_{Lj}, and the elastic deflection, δ_j, of the jth story in Equation 7.10*D* are not available before the computation of the period T_1, iterative computations are needed. However, in some cases, the iteration may not be convergent. Therefore, $T_1 = 1.0$ (s) is used in this example. The fundamental mode shapes calculated through methods *A*, *B*, *C*, and *D* are listed in Table 7.5. Table 7.6 lists the absolute accelerations, lateral forces, and floor displacements.

The quantities calculated by methods *A*, *B*, *C*, and *D* are denoted, respectively, by the corresponding superscripts. Here, the safety modification S_M of methods *A*, *B*, and *C* is taken to be $1/\gamma_{m1} = 1/0.85 = 1.18$. In addition, the safety factors $S_{Dn} = 1.684$ and $S_{D\xi} = 1.120$, respectively. As a comparison, the quantities obtained through numerical simulations based on the 99 earthquake records are also listed and denoted by superscript "sim."

From Table 7.6, it is seen that the computations of all the methods yield comparatively smaller quantities than the simulated results, because only the first modal responses are used and the modal mass ratios are rather small. It is also seen that methods *A*, *B*, and *C* provide closer results, whereas the data of method *D* are quite small. Generally speaking, when the model mass ratio is below the required value, the MDOF model should be used instead. In this case, using the MSSP model with methods *A* through *C* can provide a relatively good estimation of the displacement, but not the acceleration, lateral force, or base shear.

Next, the base shear is used as the design criterion to carry out the damper design. Since the base shear changes according to the methods used to calculate it, the data obtained from method *A* are

TABLE 7.4
Basic Parameters (Numerically Simulated V = 64.63 [MN])

	γ_{ml}	T_1 (s)	Γ_1	C_s	d_{1D} (cm)	m_{eff} (kt)	V (MN)
A	0.85	1.085	1.259	0.381	11.14	8.31	59.00
B	0.87	1.079	1.237	0.381	11.07	8.53	55.09
C	0.85	1.000	1.235	0.411	10.26	8.39	58.46
D	0.79	1.000	1.413	0.411	10.26	7.73	31.29

TABLE 7.5
Mode Shapes

	Method *A*	Method *B*	Method *C*	Method *D*
1st floor	0.1475	0.1564	0.1783	0.1000
2nd floor	0.2920	0.3090	0.3405	0.2000
3rd floor	0.4301	0.4540	0.4845	0.3000
4th floor	0.5587	0.5878	0.6105	0.4000
5th floor	0.6768	0.7071	0.7212	0.5000
6th floor	0.7794	0.8090	0.8131	0.6000
7th floor	0.8664	0.8910	0.8887	0.7000
8thfloor	0.9314	0.9511	0.9434	0.8000
9th floor	0.9772	0.9877	0.9813	0.9000
Roof	1.0000	1.0000	1.0000	1.0000

used and the equation is to be 60×10^3 (kN). Suppose the base shear must be the design parameter and the allowed value is $[V] = 48.65 \times 10^3$ (kN). The proportional coefficients α_v is

$$\alpha_V = \frac{[V]}{V} = 0.8108.$$

Therefore, the design damping ratio is 10.08% and $\xi_a = 7.58\%$. Namely,

$$\xi_a = \frac{\left(1-\alpha_{(\cdot)}\right)\left(3\xi_0+0.9\right)}{3\alpha_{(\cdot)}} = \frac{\left(1-0.8188\right)\left(3\times 0.025+0.9\right)}{3\times 0.8188} = 0.0758$$

Using Equation 7.30, c_{est} is about 799 (kN-s/m).
Suppose one nominal damper is installed in every story. Then, matrix $\mathbf{C} = c_{est}\, \mathbf{C}_{con}$ is

$$C = 799 \times \begin{bmatrix} 2 & -1 & 0 & \dots & 0 \\ -1 & 2 & -1 & \dots & 0 \\ & & \dots & & \\ 0 & & \dots & 2 & -1 \\ 0 & & \dots & -1 & 1 \end{bmatrix}_{10\times 10} \quad (\text{kN-s/m})$$

Using Equation 7.29, the initially designed damping ratio is $\xi_a^{(1)} = 0.15\%$. Thus, the nominal damping coefficient $c^{(1)}$ is 39.28 (MN-s/m).

Since this structure has 7 bays, suppose 16 realistic dampers are used (2 dampers are used in each of the 8 walls) and the safety factor S_υ is taken to be 1.1. Then, $c^{(2)}$ is 2.7 (MN-s/m). The diameter of the damper can be calculated using Equation 7.41 with the diameter of the piston rod r, the thickness of cylinder wall of the damper B/2, and the tolerance U assumed to be 2.0, 2.25, and 2.5 (cm), respectively (which are commonly used for dampers); and the pressure is assumed to be 100 (MPa). Using method *A*, it is seen that the maximum floor drift δ_j is 3.81 (cm). That is, $\Phi = 0.375$ (m) (if Φ is too large for a special design, then $N^{(1)}$ must be larger than 16, i.e., 18 or more) and the maximum nominal damper stroke is $2 \times 3.81 \times 0.3 = 2.29$ (cm). With a safety factor $S_\sigma = 4.0$, the designed damper stroke is 9.14 (cm).

There are issues remaining for damper installation, such as supporting stiffness and safety factors. These issues are considered after the discussion on linear damping design in Section 7.5.

TABLE 7.6
Calculated Responses

	Absolute Acceleration (m/s²)					Lateral Force (MN)					Lateral Displacement (m)				
	a_a^A	a_a^B	a_a^C	a_a^D	a_a^{sim}	f_L^A	f_L^B	f_L^C	f_L^D	f_L^{sim}	d_{max}^A	d_{max}^B	d_{max}^C	d_{max}^D	d_{max}^{sim}
1st floor	2.407	2.116	2.131	0.569	2.651	2.118	1.861	1.875	0.501	2.333	0.046	0.051	0.041	0.015	0.048
2nd floor	3.167	2.467	2.488	1.138	4.406	3.161	2.462	2.483	1.135	4.397	0.091	0.097	0.082	0.029	0.095
3rd floor	4.085	2.932	2.918	1.707	5.633	4.022	2.885	2.871	1.679	5.543	0.134	0.138	0.120	0.044	0.139
4th floor	5.017	3.429	3.348	2.275	6.485	4.212	2.879	2.811	1.910	5.445	0.174	0.174	0.155	0.058	0.180
5th floor	5.908	3.907	3.754	2.844	7.035	6.061	4.008	3.851	2.918	7.217	0.211	0.206	0.187	0.073	0.218
6th floor	6.699	4.333	4.105	3.413	7.405	5.993	3.875	3.672	3.053	6.623	0.243	0.232	0.214	0.087	0.250
7th floor	7.378	4.684	4.401	3.982	7.689	8.423	5.347	5.024	4.546	8.778	0.270	0.254	0.235	0.102	0.277
8thfloor	7.888	4.945	4.618	4.551	7.932	7.253	4.546	4.246	4.184	7.293	0.290	0.269	0.251	0.116	0.298
9th floor	8.251	5.105	4.770	5.120	8.262	8.686	5.374	5.022	5.390	8.698	0.304	0.280	0.261	0.130	0.312
Roof	8.431	5.159	4.846	5.688	8.571	8.616	5.273	4.952	5.814	8.759	0.311	0.286	0.264	0.145	0.319

7.3 PROPORTIONALLY DAMPED MDOF SYSTEMS APPROACH

7.3.1 General Description

If the criterion provided by Equation 7.9 is not satisfied, more modes need to be included in the response computations. When the damping irregularity of the targeted structure is sufficiently low, proportionally damped models can be used. Mathematically speaking, each modal response can be treated as a single MSSP system. Therefore, the difference between an MSSP system and a proportionally damped MDOF system in general is that in the latter case, more than one mode is considered. In this case, for these individual modes, the aforementioned equations can be used by simply replacing subscript 1, which stands for the first mode, by i, which stands for the ith mode.

After the first S modal responses are obtained, they are combined together to estimate the total responses. For a proportionally damped system, a modified SRSS method is used for simplicity, and is often of sufficient accuracy.

Note that the damping design for an MDOF system needs both the mass and the stiffness matrices. Therefore, the simplified approach that does not use the exact period and mode shape is abandoned. In this subsection, the corresponding design procedure is discussed, which is virtually the same as that for MSSP, except that more modes are considered.

7.3.2 Criterion for Modal Selection

The first step in the design process is the determination of the number of modes. Practically speaking, for all MDOF systems, modal truncation can be used and only the first S modes are considered. The criterion is based on the value of the accumulated modal mass ratio, γ_{mS}, as

$$\gamma_{mS} = \sum_{i=1}^{S} \gamma_{mi} \geq 90\% \sim 95\% \tag{7.49}$$

where the ith modal mass ratio γ_{mi} is given by

$$\gamma_{mi} = \frac{\left[\boldsymbol{p}_i^T \mathbf{MJ}\right]^2}{\boldsymbol{p}_i^T \mathbf{M}\, \boldsymbol{p}_i \sum_{j=1}^{n} m_j} \tag{7.50}$$

7.3.3 Basic Design Procedure

Once it is decided to use the proportionally damped model with the first S modes, the design procedure is as follows.

7.3.3.1 Estimation of Seismic Response of Original Structure

7.3.3.1.1 Seismic Response Coefficients and Spectral Displacements

The seismic response coefficient, C_{si}, and the spectral displacement, d_{iD} of the ith mode are given by

$$C_{si} = \frac{\sqrt{1+4\xi_i^2}}{B_i R}\left(0.6\frac{S_{DS}}{T_0}T_i + 0.4S_{DS}\right)I \quad \text{and} \quad d_{iD} = \frac{C_{si}T_i^2}{4\pi^2\sqrt{1+4\xi_i^2}}g \ (\mathrm{m}) \tag{7.51a}$$

when $T_i \leq T_0$,

$$C_{si} = \frac{S_{DS} I \sqrt{1+4\xi_i^2}}{B_i R} \quad \text{and} \quad d_{iD} = \frac{C_{si} T_i^2}{4\pi^2 \sqrt{1+4\xi_i^2}} g \ (m) \tag{7.51b}$$

when $T_0 < T_i \le T_s$,

$$C_{si} = \frac{S_{D1} I \sqrt{1+4\xi_i^2}}{T_i B_i R} \quad \text{and} \quad d_{iD} = \frac{C_{si} T_i^2}{4\pi^2 \sqrt{1+4\xi_i^2}} g \ (m) \tag{7.51c}$$

when $T_s < T_i \le T_L$, and

$$C_{si} = \frac{S_{D1} I T_L \sqrt{1+4\xi_i^2}}{T_i^2 B_i R} \quad \text{and} \quad d_{iD} = \frac{C_{si} T_i^2}{4\pi^2 \sqrt{1+4\xi_i^2}} g \ (m) \tag{7.51d}$$

when $T_i > T_L$.

In addition, NEHRP 2009 introduces a concept called *residual mode*. The seismic response coefficient, C_{sR}, and the spectral displacement, d_{RD}, are given by

$$C_{sR} = \frac{R S_{DS}}{C_d \Omega_0 B_R} \quad \text{and} \quad d_{RD} = \frac{\Gamma_R S_{D1} T_R}{4\pi^2 B_R} g \ (m) \tag{7.51e}$$

Furthermore, for long period structure, the simplified formula described in Equation 2.334 is used and is repeated as follows:

$$C_s = \sqrt{1+4\xi_i^2} \, \frac{0.4 A S_i}{B T_{eff}} \quad \text{and} \quad d_{iD} = \frac{C_{si} T_i^2}{4\pi^2 \sqrt{1+4\xi_i^2}} g \ (m) \tag{7.51f}$$

Here, C_{si} is dimensionless and the unit of d_{iD} is (m).

In Equation 7.51, T_i is the ith natural period; ξ_i stands for the ith design damping ratio; and T_0, T_s, T_L, R, I, S_{DS}, and S_{D1} are as defined previously. The numerical damping coefficient is

$$B_i = 3\xi_i + 0.9 \tag{7.52}$$

In Equation 7.51e, C_d and Ω_0 are the deflection amplification factor and overstrength factor, respectively, which are provided in NEHRP 2009.

Note that ξ_i is assumed to be the original damping ratio of the base structures. Without detailed design, the initial damping ratio of the ith mode can be chosen equally for each mode as the one estimated for the first mode, say 2%–5%. The resulting response computation will not yield a large error because the initial damping force is small.

After proper damper design, the damping matrix $\mathbf{C}^{(1)}$ and the modal damping ratios $\xi_i^{(1)}$ and $\xi_i^{(2)}$ will be determined. For general MDOF systems, often further evaluation of the design is needed until the design criteria, such as previously described in Equations 7.25 through 7.28, are satisfied. Once the initially designed damping matrix is obtained, the modal damping ratios are calculated and the modal responses are evaluated. This iterative procedure may yield the consequence of $\mathbf{C}^{(1)}$, $\mathbf{C}^{(2)}$,..., and $\xi_i^{(1)}$, $\xi_i^{(2)}$,..., and so on. The equations to find the damping ratios for the proportionally damped systems are given in the following subsections.

7.3.3.1.2 Mode Shapes and Natural Periods

To calculate the mode shape, $\boldsymbol{p}_i$, and the period, T_i, the eigen-decomposition method is used, that is,

$$\boldsymbol{p}_i = \mathbf{P}(:,i) = \begin{Bmatrix} 1 \\ p_{i2} \\ \dots \\ p_{in} \end{Bmatrix} \quad \text{and} \quad T_i = \frac{2\pi}{\sqrt{\lambda_i\left(\mathbf{M}^{-1}\mathbf{K}\right)}} = 2\pi\sqrt{\frac{\boldsymbol{p}_i^T\mathbf{M}\,\boldsymbol{p}_i}{\boldsymbol{p}_i^T\mathbf{K}\,\boldsymbol{p}_i}} \quad (s) \tag{7.53}$$

In Equation 7.53, **P** is first calculated through the eigen-decomposition of $\mathbf{M}^{-1}\mathbf{K}$. Note that in the ith mode shape vector $\boldsymbol{p}_i$, the jth element is denoted as p_{ji}, whereas in the simplified design for MSSP systems, the jth element is denoted as p_j. After the damping matrix $\mathbf{C}^{(p)}$ is determined, **P** can be recalculated through the $\mathbf{M}$-$\mathbf{C}^{(p)}$-$\mathbf{K}$ system, which is discussed in the next subsection.

In addition, the jth element of the residual mode shape, p_{Rj}, and the period of the residual mode, T_R, respectively, are given by

$$p_{Rj} = \frac{1-\Gamma_1 p_j}{1-\Gamma_1} \quad \text{and } T_R = 0.4T_1 \quad (s) \tag{7.54}$$

Here, p_j is the jth element of the fundamental mode shape given by using Equation 7.10*D*.

7.3.3.1.3 Modal Participation Factors

The modal participation factor Γ_i is given by

$$\Gamma_i = \frac{\boldsymbol{p}_i^T\mathbf{M}\,\mathbf{J}}{\boldsymbol{p}_i^T\mathbf{M}\,\boldsymbol{p}_i} \tag{7.55}$$

The residual modal participation factor in NEHRP 2009 is

$$\Gamma_R = 1-\Gamma_1 \tag{7.56}$$

7.3.3.1.4 Lateral Displacements

The maximum values of lateral floor displacement of the ith mode are given by

$$\mathbf{d}_{max_i} = \Gamma_i d_{iD}\boldsymbol{p}_i = \begin{Bmatrix} d_{1i} \\ d_{2i} \\ \dots \\ d_{ni} \end{Bmatrix} \quad (m) \tag{7.57}$$

The total maximum lateral displacement can be calculated using SRSS as

$$\mathbf{d}_{max} = S_M S_{Dn} S_{D\xi}\left(\sum_{i=1}^{S}\mathbf{d}_{max_i}^{\bullet 2}\right)^{\bullet 1/2} \quad (m) \tag{7.58}$$

In Equation 7.58, $S_M = 1/\,\gamma_{mS}$ (see the discussion of MSSP systems in the previous subsection) and S_{Dn} is given as follows:

$$S_{Dn} = \frac{1}{a_4 n^4 + a_3 n^3 + a_2 n^2 + a_1 n + a_0} \tag{7.59}$$

where a_0 through a_4 are coefficients, which should be determined through local design spectrum and computer simulation. For example, again using the 99 earthquake records, results in

$$S_{Dn} = \frac{1}{-6.806\times10^{-8}n^4 + 1.257\times10^{-5}n^3 - 7.763\times10^{-4}n^2 + 0.0173n + 0.4288}$$

$S_{D\xi}$ is recommend to be

$$S_{D\xi} = \frac{1}{b_4\xi^4_{design} + b_3\xi^3_{design} + b_2\xi^2_{design} + b_1\xi_{design} + b_0} \quad (7.60)$$

where b_0 through b_4 are coefficients, which should be determined through local design spectrum and computer simulation. For example, using the 99 earthquake records,

$$S_{D\xi} = \frac{1}{29.28\xi^3_{design} - 20.05\xi^2_{design} + 4.559\xi_{design} + 0.8185}$$

7.3.3.1.5 Lateral Absolute Accelerations

The pseudo lateral acceleration vector of the ith mode is approximately given by

$$\mathbf{a}_{s_i} = \Gamma_i C_{si} g \boldsymbol{p}_i = \begin{Bmatrix} a_{s_{1i}} \\ a_{s_{2i}} \\ \cdots \\ a_{s_{ni}} \end{Bmatrix} \quad (m/s^2) \quad (7.61)$$

The total maximum lateral absolute acceleration can be calculated through modified SRSS as

$$\mathbf{a}_a = S_M S_{An} S_{A\xi} \left[\sum_{i=1}^{S} \mathbf{a}_{s_i}^{\bullet 2} + (S_{Ai}\mathbf{J}) \right]^{\bullet 1/2} = \begin{Bmatrix} a_{a_1} \\ a_{a_2} \\ \cdots \\ a_{a_n} \end{Bmatrix} \quad (m/s^2) \quad (7.62A)$$

Or the method used by NERHP 2009

$$\mathbf{a}_a = \left[\sum_{i=1}^{S} \mathbf{a}_{s_i}^{\bullet 2} \right]^{\bullet 1/2} = \begin{Bmatrix} a_{a_1} \\ a_{a_2} \\ \cdots \\ a_{a_n} \end{Bmatrix} \quad (m/s^2) \quad (7.62D)$$

Note that the jth element in the acceleration vector $\mathbf{a}_a$ is denoted by a_{a_j} whereas in Equation 7.61 for the ith modal acceleration, the jth element is denoted by a_{aji}.

In Equation 7.62*A*, S_M is the same as defined in Equation 7.56, and S_{Ai} can be determined as

$$S_{Ai} = \Gamma_i \left(0.6 \frac{S_{DS}}{T_0} T_i + 0.4 S_{DS} \right), T_i \le T_0 \quad (7.63a)$$

$$S_{Ai} = \Gamma_i S_{DS}, T_0 < T_i \le T_s \tag{7.63b}$$

$$S_{Ai} = \Gamma_i \frac{S_{D1}}{T_1}, T_s < T_i \le T_L \tag{7.63c}$$

$$S_{Ai} = \Gamma_i \frac{S_{D1} T_L}{T_i^2}, T_i > T_L \tag{7.63d}$$

In Equation 7.61, S_{An} is given as follows:

$$S_{An} = \frac{1}{a_4 n^4 + a_3 n^3 + a_2 n^2 + a_1 n + a_0} \tag{7.64}$$

where a_0 through a_4 are a set of coefficients different from those given in Equation 7.16, which should be determined through local design spectrum and computer simulation. For example, using the 99 earthquake records,

$$S_{An} = \frac{1}{4.5738 \times 10^{-6} n^3 - 7.0650 \times 10^{-4} n^2 + 0.0328n + 0.3631}$$

and

$$S_{A\xi} = \frac{1}{b_4 \xi_{design}^4 + b_3 \xi_{design}^3 + b_2 \xi_{design}^2 + b_1 \xi_{design} + b_0} \tag{7.65}$$

where b_0 through b_4 are a set of coefficients different from those in Equation 7.17, which should be determined through local design spectrum and computer simulation. For example, using the 99 earthquake records,

$$S_{A\xi} = \frac{1}{-8.51 \xi_{design}^2 + 4.114 \xi_{design} + 0.815}$$

7.3.3.1.6 Lateral Forces

The lateral force at the jth floor can be calculated using the following equation:

$$f_{Lj} = m_j g a_{aj} \quad (kN) \tag{7.66}$$

Or it can be calculated using the equation provided by NEHRP 2009:

$$f_{Lji} = m_j g C_{si} p_{ji} \quad (kN) \tag{7.67}$$

In Equation 7.67, p_{ji} is the jth element of the ith displacement mode shape (see Equation 7.53). In addition, if the residual mode is used, NEHRP 2009 gives

$$f_{Lji} = m_j g C_{sR} \left(1 - \Gamma_1 p_{ji}\right) \quad (kN) \tag{7.68}$$

The total lateral force calculated by the modal force described in Equation 7.67 in NEHRP 2009 is given by

$$f_{Lj} = \sqrt{\sum_{i=1}^{S} f_{Lji}^2} \quad (kN) \tag{7.69}$$

7.3.3.1.7 Base Shear

The base shear of the ith mode can be calculated by the following equations:

$$V = \mathbf{f}_L^T \mathbf{J} \quad (kN) \tag{7.70}$$

$$V_i = m_{effi} g C_{si} \quad (kN) \tag{7.71}$$

$$V = \sqrt{\sum_{i=1}^{S} V_i^2} \quad (kN) \tag{7.72}$$

Using the residual modal response, NEHRP 2009 provides:

$$V_R = \left(m_\Sigma - m_{eff}\right) g C_{sR} \quad (kN) \tag{7.73}$$

where m_Σ is the total mass of the structure.

The ith effective mass of the first mode is given by

$$m_{effi} = \frac{\left(\boldsymbol{p}_i^T \mathbf{M}\,\mathbf{J}\right)^2}{\boldsymbol{p}_i^T \mathbf{M}\,\boldsymbol{p}_i} \quad (t) \tag{7.74}$$

7.3.3.2 Determination of Damping Ratio and Damping Coefficient

7.3.3.2.1 Damping Proportionality Coefficient and Design Damping Ratio

Similar to MSSP systems, using supplemental damping can be determined by Equations 7.25 through 7.28. The only difference is that the first modal damping ratio ξ_1, contributed by the supplemental dampers to replace ξ_a in Equation 7.29, is used. That is,

$$\xi_1 = \frac{\left(1-\alpha_{(\cdot)}\right)\left(3\xi_0 + 0.9\right)}{3\alpha_{(\cdot)}} \tag{7.75}$$

7.3.3.2.2 Damping Coefficient, Configuration, Matrix, and Initial Damping Ratio

The determination of the design damping ratio for a proportionally damped MDOF system can employ all the equations explained for MSSP systems (Equations 7.25 through 7.29, except ξ_1 is used to replace ξ_a in Equation 7.29). Similar to Equation 7.29, the modal damping ratio can be calculated as

$$\xi_i^{(1)} = \frac{T_i}{4\pi} \frac{\boldsymbol{p}_i^T \mathbf{C}^{(1)} \boldsymbol{p}_i}{\boldsymbol{p}_i^T \mathbf{M}\,\boldsymbol{p}_i}, \quad i = 1,\ldots, S \tag{7.76}$$

Once the modal damping ratios are obtained, there is often a need to go back to the first step, starting from Equation 7.51, and check if the criteria described in Equations 7.25 through 7.28 are satisfied. This iterative procedure may yield a sequences of $\mathbf{C}^{(1)}, \mathbf{C}^{(2)},\ldots, \mathbf{C}^{(p)}$ and $\xi_i^{(1)}, \xi_i^{(2)},\ldots, \xi_i^{(p)}$, as well as damping coefficient $c^{(1)}, c^{(2)},\ldots, c^{(p)}$.

Such a computation will likely yield complex-valued mode shapes. An initial estimation of a damping matrix is often needed. The method to change the complex-valued mode shape into a real-valued one and the way to estimate the initial damping matrix are discussed in the following subsection.

7.3.3.3 Selection of Damper

Suppose $\mathbf{C}^{(p)}$ satisfies the design criteria. The damping coefficient $c^{(p)}$ is thus used for damper selection. The procedure described in Equations 7.36 through 7.48 is for this purpose.

Example 7.2

A simple example is presented to summarize the suggested design procedure as well as the method through residue mode, for proportionally damped system. Consider the same ten-story-seven-bay shear model with identical damping and stiffness as given in Example 7.1. The roof displacement and the displacement of the first floor are denoted as x_{10}, and x_1, respectively, and so on. Similarly, the geometrical magnification factor, G_{mj}, is 3/10, the occupancy importance factor I is taken to be 1.12, and the response modification factor *R* is chosen to be 1.0 for simplicity.

Note that in Example 7.1, only the first mode of the same system was used to show the design based on the MSSP model, although it was known that the modal mass ratios of the first mode calculated with several methods of estimation were all below 90%. Now, in Table 7.7, basic parameters up to the fourth mode are given. It is seen that up to the second and third modes, the cumulated modal mass ratios are, respectively, 0.94 and 0.97. Therefore, if in Equation 7.49 the criterion is chosen to be 0.95, the first, second, and third modes for the response estimation of the damper design should be chosen for the proportionally damped MDOF system. In the following discussion, the damper design based on the proportionally damped MDOF system is compared with the MSSP model. The natural periods and mode shapes of the first three modes are calculated according to Equation 7.53. The periods as well as the period of the residue mode are listed in Table 7.7, respectively. The mode shapes of the first three modes are listed in Table 7.8. The mode shape of the residue mode, calculated by Equation 7.54, is also listed in Table 7.8.

As a comparison, the numerical simulations are repeated using the 99 earthquake records. However, this time, each record is scaled by letting the spectral value of acceleration at period = 1 (s) (see S_{D1} calculated below) be 0.36 (g).

Using Equation 7.51, the seismic response factors and the spectral displacements are calculated by assuming that the original damping ratios of these three modes are all 2.5%. Note that since T_1 = 1.085 (s), the parameter T_L is not needed in this example.

To determine T_0 and T_s, information about the spectral response acceleration at short periods (S_s) and at 1 (s) (S_1) is needed. Suppose S_s = 0.75 and S_1 = 0.3 are given. From Tables 2.1a and b, the spectral response acceleration parameters can be calculated at short periods (S_{MS}) and at 1 (s)

TABLE 7.7
Basic Parameters

Mode	1st	2nd	3rd	Residue
Mode mass ratio γ_{mi}	0.85	0.091	0.029	—
Cumulated mode mass ratio γ_{mS}	0.85	0.94	0.97	—
Period T_i (s)	1.085	0.359	0.222	0.434
Seismic response coefficient C_{si}	0.382	0.691	0.691	0.263
Special displacement d_{iD} (m)	0.125	0.025	0.011	0.01
Modal participation factor Γ_i	1.259	−0.398	0.217	−0.259
Effective mass m_{effi} (10^6 kg)	8.311	0.884	0.286	1.447

TABLE 7.8
Mode Shapes

	Suggested Method			
	1st Mode	**2nd Mode**	**3rd Mode**	**Residue Mode**
1st floor	0.1475	0.4534	0.7075	−3.1442
2nd floor	0.2920	−0.8253	1.0809	−2.4415
3rd floor	0.4301	−1.0290	0.8755	−1.7704
4th floor	0.5587	−1.0258	0.2077	−1.1453
5th floor	0.6768	−0.8467	−0.5537	−0.5712
6th floor	0.7794	−0.4902	−1.0102	−0.0724
7th floor	0.8664	−0.0441	−0.9819	0.3507
8th floor	0.9314	0.4123	−0.3520	0.6663
9th floor	0.9772	0.7912	0.4515	0.8890
Roof	1.0000	1.0000	1.0000	1.0000

(S_{M1}) for a risk-targeted maximum considered earthquake (MCE_R) based on site coefficients F_a corresponding to S_s and F_v corresponding to S_1. Suppose Site Class D is considered. From Tables 2.1a and b, $F_a = 1.2$ and $F_v = 1.8$. Therefore,

$$S_{MS} = F_a S_s = 1.2 \times 0.75 = 0.9$$

and

$$S_{M1} = F_v S_1 = 1.8 \times 0.3 = 0.54$$

Furthermore, the design earthquake spectral response acceleration parameters at short period (S_{DS}) and at 1 sec (S_{D1}) can be calculated as

$$S_{DS} = \frac{2}{3} S_{MS} = 0.6$$

and

$$S_{D1} = \frac{2}{3} S_{M1} = 0.36$$

Note that the value I in Equation 7.51 is chosen to be 1.95. According to NERHP 2009, T_0 and T_s are calculated as

$$T_0 = 0.2 \frac{S_{D1}}{S_{DS}} = 0.12 \quad (s)$$

and

$$T_s = \frac{S_{D1}}{S_{DS}} = 0.6 \quad (s)$$

Note that in this example, $T_s < T_1 < T_L$ and $T_0 < T_3 < T_2 < T_R < T_s$. Here, T_R is the period of the residue mode, which is 0.4 T_1.

In NEHRP 2009, the damping ratio is specified as 5%. In this example, however, the value of 2.5% is used. In this case, $B_1 = B_2 = B_3 = 0.975$. Based on the above parameters, as well as the required natural periods of the structure, also listed in Table 7.7, the seismic response factors and the spectral displacements are calculated as follows, which are also listed in Table 7.7.

$$C_{s1} = \frac{S_{D1} I \sqrt{1+4\xi_i^2}}{T_1 B_1 R} = 0.382$$

$$C_{s2} = C_{s3} = \frac{S_{DS} I \sqrt{1+4\xi_i^2}}{B_i R} = 0.691$$

In order to use Equation 7.51e, the values of R, C_d, and Ω_0 need to be specified, specifically for the residue mode. Suppose the building in this example is "steel eccentrically braced frames, non-moment resisting, connections at columns away from links" (BSSC/NEHRP 2009), then $R = 7$, $C_d = 8$, and $\Omega_0 = 2$. In this case, the frame is considered to yield, so the damping ratio must be significantly higher. Let ξ_R and $B_R = 1.0$. Thus,

$$C_{sR} = \frac{R S_{DS}}{C_d \Omega_0 B_R} = \frac{7 \times 0.6}{8 \times 2 \times 1.0} = 0.2625$$

(Note that the seismic response factors, C_{si}, are dimensionless. However, it is necessary to consider using $g = 9.8$ m/s^2 to calculate the accelerations and lateral forces.)

$$d_{1D} = \frac{C_{s1} T_1^2}{4\pi^2 \sqrt{1+4\xi_1^2}} g = 0.125 \quad (m)$$

$$d_{2D} = \frac{C_{s2} T_2^2}{4\pi^2 \sqrt{1+4\xi_2^2}} g = 0.025 \quad (m)$$

$$d_{3D} = \frac{C_{s3} T_3^2}{4\pi^2 \sqrt{1+4\xi_3^2}} g = 0.011 \quad (m)$$

and

$$d_{RD} = \frac{\Gamma_R S_{D1} T_R}{4\pi^2 B_R} g = 0.01 \quad (m)$$

The modal participation factors are calculated according to Equations 7.55 and 7.56, which are also listed in Table 7.7.

Using Equations 7.57 through 7.60, the lateral displacement is calculated. It is seen that $S_M = 1.0309$, $S_{Dn} = 1.8655$, and $S_{D\xi} = 1.0865$. The estimated displacements based on the suggested method and the method through the residue mode are denoted by $\mathbf{d}^S_{max}$ and $\mathbf{d}^R_{max}$ and are listed in Table 7.9, columns 2 and 3, respectively.

Using Equations 7.61 through 7.65, the lateral absolute acceleration is calculated.

It is seen that $S_{A1} = 0.42$, $S_{A2} = 0.24$, and $S_{A3} = 0.13$. Also $S_M = 1.0309$, $S_{An} = 1.6001$, and $S_{A\xi} = 1.0959$. The estimated accelerations based on the suggested method and the method through the residue mode are denoted as $\mathbf{a}^S_a$ and $\mathbf{a}^R_a$ and are listed in Table 7.9, columns 8 and 9, respectively.

TABLE 7.9
Calculated Responses

	Lateral Displacement (m)			Absolute Acceleration (m/s²)			Lateral Force (MN)		
	$\mathbf{d}^{S}_{max}$	$\mathbf{d}^{R}_{max}$	$\mathbf{d}^{Sim}_{max}$	$\mathbf{a}^{S}_{a}$	$\mathbf{a}^{R}_{a}$	$\mathbf{a}^{Sim}_{a}$	$\mathbf{f}^{S}_{L}$	$\mathbf{f}^{R}_{L}$	$\mathbf{f}^{Sim}_{L}$
1st floor	0.027	0.025	0.029	9.378	2.211	10.287	8.251	1.945	9.050
2nd floor	0.053	0.046	0.057	10.419	2.132	11.834	10.398	2.128	11.811
3rd floor	0.078	0.068	0.080	11.042	2.344	11.267	10.870	2.307	11.091
4th floor	0.100	0.088	0.101	11.220	2.738	10.957	9.420	2.299	9.199
5th floor	0.121	0.106	0.121	11.419	3.208	11.502	11.713	3.291	11.798
6th floor	0.139	0.123	0.137	11.617	3.669	10.791	10.390	3.281	9.651
7th floor	0.154	0.136	0.151	11.802	4.085	11.701	13.474	4.663	13.357
8thfloor	0.166	0.146	0.163	12.076	4.406	9.214	11.103	4.051	8.471
9th floor	0.174	0.154	0.172	12.790	4.637	10.338	13.465	4.882	10.884
Roof	0.178	0.157	0.177	13.466	4.754	14.487	13.763	4.858	14.806

Similar to Example 7.1, as a comparison, the displacement, acceleration, and force obtained through numerical simulations based on the 99 earthquake records are also listed in Table 7.9, columns 4, 7, and 10, and denoted as $\mathbf{d}^{Sim}_{max}$, $\mathbf{a}^{Sim}_{a}$, and $\mathbf{f}^{Sim}_{L}$, respectively.

The base shear is calculated using Equations 7.70 through 7.73. Here, it is necessary to find the effective mass m_{effi}, which is calculated using Equation 7.74 and listed in Table 7.7. The base shear calculated by the numerical simulation is listed in Table 7.10, column 1. The base shear calculated by the suggested method based on Equation 7.70 is listed in column 2. The base shear calculated by the method based on Equation 7.71 for modal base shear and Equation 7.72, the SRSS method, is listed in column 3. The base shear calculated based on the concept of residue mode Equation 7.73 is listed in column 4.

The calculated results in Examples 7.1 and 7.2 are compared based on the suggested methods (method A for MSSP and method 1 for proportionally damped MDOF system) with the other approaches. It is seen that the response estimations based on the suggested methods can be comparatively larger than by using other approaches. Although the response estimations based on the suggested methods can be conservative, the authors prefer to recommend using them to create safer designs.

Secondly, the responses calculated based on MSSP and MDOF systems are compared. Note that the modal mass ratio γ_{ml} of this particular structure is smaller than 90%. A small γ_{ml} does not necessarily mean that using MSSP will always yield smaller values of responses, if the safety factor $SM = 1/\gamma_{ml}$ is used. However, using the MDOF model always results in more accurate estimations.

Similar to the case shown in Example 7.1, base shears can be used as the design criterion to carry out the damper design (see Equation 7.75).

Once the damping matrix **C** is determined in the first round, the corresponding damping ratios for the first three modes can be calculated through eigen-decompositions and the newly obtained numerical damping coefficient B_i can be used to reestimate the accelerations, forces, displacements, and base shears, as mentioned previously, until the design criterion is satisfied.

TABLE 7.10
Total Base Shear (MN)

Simulated	(7.70)	(7.71) (7.72)	(7.73) (7.72)
110.12	112.85	31.70	31.30

7.4 DESIGN OF GENERALLY DAMPED SYSTEMS

7.4.1 Criteria for Generally Damped Systems

If the damping irregularity is high, the system may need to be treated as generally damped. In this case, using the following procedure may improve the design accuracy.

Based on NEHRP 2009, the proportional coefficient, $\alpha_{(.)}$, is often less than 0.75. In this case, if ξ_0 in Equation 7.8 is 5%, the largest design damping ratio of the first mode is about 16.67%. With certain safety considerations, the design damping ratio is chosen to be less than 20% based on linear approaches. Part of the reason to keep the damping ratio less than 20% is that, practically speaking, most commercially available dampers are nonlinear, which is discussed in Chapter 8.

Similar to proportionally damped MDOF systems, before checking if the system is generally damped, the first task is to check the criterion described in Equation 7.49 and find the number of terms, S, for those modes that need to be considered.

In generally damped systems, complex modes and overdamped subsystems can exist to determine if the fundamental mode of the system is complex, the criterion is used.

$$\zeta_1 = 1 - \frac{T_1^{(p)}}{T_1^{(n)}} < 1\% \tag{7.77}$$

In addition, the ratios of r_{E1} and/or r_{V1} can also be used (see Equations 4.288 and 4.302).

Here, $T_1^{(p)}$ and $T_1^{(n)}$ are, respectively, the fundamental periods of system **M-K** and **M-C$^{(\cdot)}$-K**. That is, the first term is calculated by using Equation 7.51 and the second term by using the state matrix $\mathcal{A}$ (see Equations 7.81 and 7.82).

To check the existence of overdamped subsystems in the first S modes, it is necessary to see if the ith eigenvalue is real, that is,

$$\text{Im}(\lambda_i) \neq 0, \quad i = 1, \ldots, S \tag{7.78}$$

7.4.2 Basic Design Procedure

If one or both of the above criteria are not satisfied, the system is generally damped. The design procedure of this system is similar to the proportionally damped system, except for several steps involving modal response estimation and modal combinations.

The first few steps in response estimation (to determine if damping control is needed) are basically identical to the case of proportionally damped systems, because the original damping is small and the system can be treated as proportionally damped. Equations 7.49 through 7.75 can be used, except the equations for residual modes (Equations 7.54, 7.60, and 7.68).

7.4.2.1 Estimation of Seismic Response of Structure with Dampers

Once the design damping ratio is estimated, the steps to determine the damping matrix are identical to those for MSSP systems (see Equations 7.36 through 7.45). The only difference is reestimation of the response once the damping matrix $C^{(p)}$ is determined. In the following subsections, these steps are discussed. Note that for convenience, the order of the required calculations is different from those mentioned for MSSP and proportionally damped systems.

7.4.2.1.1 Eigenvalues, Periods, and Damping Ratios

Different from proportionally damped systems, a nonproportionally damped model needs more information about the eigen-parameters of the systems. The eigenvalues λ_i and eigenvectors $\mathcal{P}$ can be obtained by using the eigen-decomposition of the state matrix $\mathcal{A}$ (see Equation 4.38):

$$\mathcal{A} = \mathcal{P}\Lambda\mathcal{P}^{-1} \tag{7.79}$$

and $\mathcal{a}$ is given by (see Equation 4.5)

$$\mathcal{a} = \begin{bmatrix} -\mathbf{M}^{-1}\mathbf{C}_{n\times n} & -\mathbf{M}^{-1}\mathbf{K}_{n\times n} \\ \mathbf{I}_{n\times n} & \mathbf{0}_{n\times n} \end{bmatrix}_{2n\times 2n} \tag{7.80}$$

Practically speaking, the eigenvalues and eigenvectors of the first few modes can be obtained using finite element programs. Here, to specifically denote the shape vector $\boldsymbol{p}_i^R$ for overdamped subsystems, superscript R is used.

The ith period $T_i^{(n)}$ can be calculated from

$$T_i^{(n)} = \frac{2\pi}{|\lambda_i|} \quad (s) \tag{7.81}$$

In the following discussion, superscript (n) of $T_i^{(n)}$ is omitted for convenience. If overdamped subsystems exist in the first S modes, the eigenvalue becomes real-valued λ_i^R and the pseudo period (see Equation 4.224):

$$T_i^R = -\frac{2\pi}{\lambda_i^R} \quad (s) \tag{7.82}$$

Furthermore, the ith damping ratio ξ_i can be calculated from (see Equation 4.16)

$$\xi_i = -\frac{\mathrm{Re}(\lambda_i)}{|\lambda_i|} \tag{7.83}$$

In addition to the periods, damping ratios, and mode shapes of a generally damped system, information about the corresponding normal mode system, especially for the period $T_i^{(p)}$ (see Equation 7.77) and the mode shape $\boldsymbol{p}_{ri}$, is needed. The several terms have been discussed in the previous section on proportionally damped models, where $\boldsymbol{p}_{ri}$ was used instead of $\boldsymbol{p}_i$ to specially denote the real-valued mode shapes.

7.4.2.1.2 Mode Shapes and Accompanist Vectors

Once the mode shapes $\mathbf{P} = [\boldsymbol{p}_1, \boldsymbol{p}_2, \ldots]$ are obtained through the eigenvector matrix $\mathcal{P}$ (see Equation 4.34), the corresponding accompanist vector for the ith mode is given by

$$\boldsymbol{q}_i = \frac{\boldsymbol{p}_i^T\mathbf{M}}{\boldsymbol{p}_i^T(2\lambda_i\mathbf{M}+\mathbf{C})\boldsymbol{p}_i} = \frac{\lambda_i\boldsymbol{p}_i^T\mathbf{M}}{\boldsymbol{p}_i^T(\lambda_i^2\mathbf{M}-\mathbf{K})\boldsymbol{p}_i} \tag{7.84}$$

Note that Equation 7.84 is also valid for overdamped subsystems; where λ_i is replaced by real-valued eigenvalue λ_i^R, and $\boldsymbol{q}_j^R$ is used to denote the corresponding accompanist vector.

Three cases can occur for generally damped systems: normal models, complex modes, and pseudo modes. Therefore, there are specific types of mode shapes for each case, which are discussed below.

The next step is to compute the real-valued vectors $\boldsymbol{\Phi}_i$ and $\boldsymbol{\Psi}_i$ (see Equation 4.174) given by

$$\boldsymbol{\Phi}_i + j\boldsymbol{\Psi}_i = \boldsymbol{p}_i\boldsymbol{q}_i\mathbf{J} \tag{7.85}$$

where

$$\boldsymbol{\Phi}_i = \{\phi_{ij}\} \tag{7.86}$$

and

$$\boldsymbol{\Psi}_i = \{\psi_{ij}\} \tag{7.87}$$

7.4.2.1.3 Modal Participation Factors

As mentioned in Chapter 4, for a generally damped system, the modal participation factor of the ith mode is no longer a scalar but a vector. For complex modes, the jth element of the vector is given by (see Equation 4.306)

$$\Gamma_{ij} = \frac{\psi_{ij}}{p_{ij}} = \begin{cases} 2\omega_i\left\{\left|-\xi_i\varphi_{ij} + \sqrt{1-\xi_i^2}\,\psi_{ji}\right| + 0.4\left|\varphi_{ij}\right|\right\}\Big/p_{ij}, \text{when } \left|-\xi_i\varphi_{ij} + \sqrt{1-\xi_i^2}\,\psi_{ij}\right| > \left|\varphi_{ij}\right| \\ 2\omega_i\left\{0.4\left|-\xi_i\varphi_{ij} + \sqrt{1-\xi_i^2}\,\psi_{ij}\right| + \left|\varphi_{ij}\right|\right\}\Big/p_{ij}, \text{when } \left|-\xi_i\varphi_i + \sqrt{1-\xi_i^2}\,\psi_{ij}\right| \le \left|\varphi_{ij}\right| \end{cases} \tag{7.88}$$

The total modal participation vector of the ith mode is

$$\boldsymbol{\Gamma}_i = \{\Gamma_{ji}\} = \begin{Bmatrix} \Gamma_{1i} \\ \Gamma_{2i} \\ \ldots \\ \Gamma_{ni} \end{Bmatrix} \tag{7.89}$$

For an overdamped subsystem, the modal participation factor is still a scalar, given by

$$\Gamma_i^R = \boldsymbol{q}_j^R \mathbf{J} \tag{7.90}$$

Note that the accompanist vector $\boldsymbol{q}_i^R$ is real valued this time. If the mode is normal, the modal participation factor is still a scalar, given by Equation 7.59.

7.4.2.1.4 Seismic Response Factors and Modal Displacements

The seismic response factors C_{si} and spectral displacement d_{iD} for vibration modes (real and complex) can be calculated using Equations 7.51. The required numerical damping coefficient B_i is given by Equation 7.52.

If in the first S "modes," there are one or more overdamped pseudo modes, the seismic response factors C_{sk}^R and spectral displacement d_{kD}^R for the corresponding overdamped subsystems need to be specified.

Overdamped displacement spectral values are needed to compute the term d_{kD}^R, which can be obtained by using Equation 4.233. A simplified approximation can be obtained through numerical simulations. For example, using the 99 earthquake records, the special displacement S_{D0} is established as

$$S_{D0} = 0.0748\left(T^R\right)^{1/2} + 0.0145\ln\left(T^R\right) - 0.0128T^R + 0.0252 \tag{7.91}$$

Thus, the spectral displacement d_{iD}^R can be written as

$$d_{iD}^R = S_{D1} S_{D0} \quad (m) \tag{7.92}$$

Figure 7.1a plots the normalize pseudo displacements ($S_{D1} = 1$) of the numerically simulated value and approximations by using Equation 7.91.

Overdamped velocity spectral values are needed to compute the term C_{sk}^R, which can also be obtained by using numerical simulations. For example, using the 99 earthquake records, the special displacement S_{v0} can be established as

$$S_{V0} = -0.857\left(T^R\right)^{1/2} + 0.447\ln\left(T^R\right) + 1.928, \quad T^R \le 1.35 \quad (s) \tag{7.93a}$$

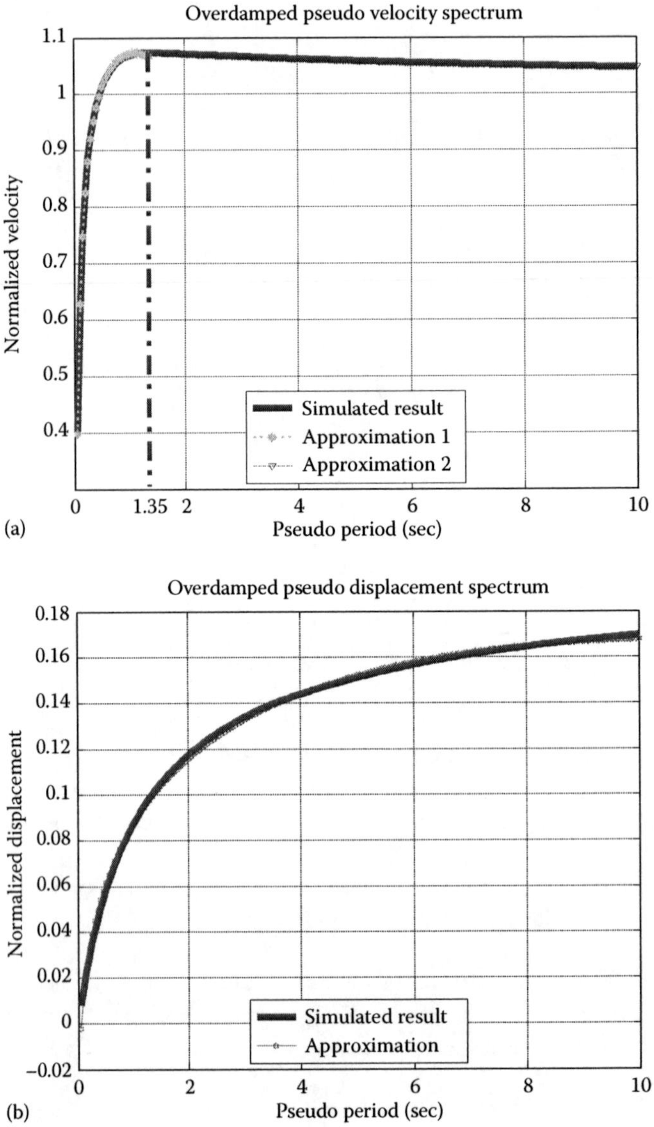

FIGURE 7.1 (a, b) Overdamped spectral values.

$$S_{V0} = 0.0002\left(T^R\right)^2 - 0.0054T^R + 1.0814, \quad T^R > 1.35 \quad (s) \tag{7.93b}$$

Figure 7.1b plots the normalized pseudo velocities ($S_{D1} = 1$) of the numerically simulated value and approximations by using Equations 7.93a and 7.93b.

Based on the discussion of overdamped pseudo modes in Section 4.3.1 (i.e., see Equation 4.233, etc.), the spectral acceleration Cs_i^R can be written as

$$C_{sk}^R = S_{D1}\frac{2\pi}{T_k^R}S_{V0} \tag{7.94}$$

Figure 7.1b plots the spectral acceleration when $S_{D1} = 0.4$.

7.4.2.1.5 Lateral Displacements

The lateral displacement of the ith complex mode is given by (see Equation 4.269)

$$\mathbf{d}_{\max_i} = d_{iD}\mathbf{\Gamma}_i \bullet \bar{\boldsymbol{p}}_i \quad (m) \tag{7.95}$$

where the shape function $\bar{\boldsymbol{p}}_i$ is given by (see Equation 4.305)

$$\bar{\boldsymbol{p}}_i = \left\{\frac{|\boldsymbol{p}_i|}{p_{ni}}\right\} \bullet \operatorname{sgn}\left(\boldsymbol{p}_{n_i}\right) = \left\{\bar{p}_{ij}\right\}, \quad j = 1, \ldots, n \tag{7.96}$$

Note that, when $-90° < \angle(p_{ij}) < 90°$, $\operatorname{sgn}(p_{ij}) = 1$. Otherwise, $\operatorname{sgn}(p_{ij}) = -1$.

The lateral displacement of the jth normal mode given in Equation 7.55 is repeated as

$$\mathbf{d}_{\max_j} = \Gamma_j d_{jD}\boldsymbol{p}_j \quad (m) \tag{7.97}$$

Here, the mode shape vector $\boldsymbol{p}_j$ is taken from the normal mode and is real valued.

The lateral displacement of the kth overdamped subsystem is given by

$$\mathbf{d}_{\max_k}^R = \Gamma_i^R d_{kD}\boldsymbol{p}_k^R \quad (m) \tag{7.98}$$

Similar to the proportionally damped system, the total maximum lateral displacement can be calculated by using SRSS as

$$\mathbf{d}_{\max} = S_M S_{Dn} S_{D\xi}\left(\sum_{i=1}^{S}\mathbf{d}_{\max_i}^{\bullet 2}\right)^{\bullet 1/2} \quad (m) \tag{7.99}$$

Note that, in Equation 7.99, the term $\mathbf{d}_{\max_i}$ includes the overdamped displacement.

Here, the safety factors S_M, S_{Dn}, and $S_{D\xi}$ are defined in Equations 7.56 through 7.58, respectively. Here,

$$S_{Dn} = \frac{1}{6.108\times10^{-7}n^3 - 5.719\times10^{-5}n^2 + 0.0015n + 0.525}$$

$$S_{D\xi} = \frac{1}{33.972\xi_{design}^3 - 20.578\xi_{design}^2 + 4.379\xi_{design} + 0.828}.$$

7.4.2.1.6 Lateral Absolute Accelerations

The pseudo acceleration of the ith complex mode is given by

$$\mathbf{a}_{s_i} = C_{si} g \boldsymbol{\Gamma}_i \bullet \bar{\boldsymbol{p}}_i \quad \left(m/s^2\right) \tag{7.100}$$

where the shape function $\bar{\boldsymbol{p}}_i$ is given by Equation 7.96.

The pseudo acceleration of the jth normal mode given in Equation 7.61 is

$$\mathbf{a}_{s_j} = \Gamma_j C_{s_j} g\, \boldsymbol{p}_j = \begin{Bmatrix} a_{s_{1j}} \\ a_{s_{2j}} \\ \cdots \\ a_{s_{nj}} \end{Bmatrix} \quad \left(m/s^2\right) \tag{7.101}$$

The pseudo acceleration of the kth overdamped pseudo mode is given by

$$\mathbf{a}_{s_k}^{R} = \Gamma_k^R C_{sk}^R\, g\, \boldsymbol{p}_k^R \quad \left(m/s^2\right) \tag{7.102}$$

Note that once an overdamped pseudo mode is considered, for more accurate computation of the acceleration described in Equation 7.102, it is better to include its companion subsystem. As mentioned in Chapter 4, two overdamped pseudo modes, which are companions, can be seen as a development from a single vibration mode due to the addition of sufficiently large damping. Therefore, the overdamped sub-subsystems must exist in pairs, one with a longer pseudo period and the other with a shorter period. The one with a longer period will contribute more to displacement and the other will contribute more to acceleration, so that excluding the second one may cause errors. Since the second one has a shorter period, it is easier to ignore. However, it is often difficult to identify the second one. Therefore, practically speaking, this situation can be compensated for by choosing more modes and/or overdamped subsystems.

Similar to the proportionally damped system, the total maximum absolute acceleration can be calculated by using the modified SRSS:

$$\mathbf{a}_a = S_M S_{An} S_{A\xi} \left[\sum_{i=1}^{S} \mathbf{a}_{s_i}^{\bullet 2} + \left(S_{Ai}\mathbf{J}\right)^{\bullet 2} \right]^{\bullet 1/2} = \begin{Bmatrix} a_{a_1} \\ a_{a_2} \\ \cdots \\ a_{a_n} \end{Bmatrix} \quad \left(m/s^2\right) \tag{7.103}$$

In Equation 7.103, the term a_{s_i} includes the acceleration $a_{s_i}^R$. Also in Equation 7.103, S_M is the same as defined in Equation 7.58 and S_{Ai} is determined by Equation 7.63. And, the safety factors S_{An} and $S_{A\xi}$ are defined in Equations 7.64 and 7.65, respectively. Here,

$$S_{An} = \frac{1}{5.9868\times10^{-6} n^3 - 9.0063\times10^{-4} n^2 + 0.0395n + 0.4092};$$

$$S_{A\xi} = \frac{1}{43.6574\xi_{design}^3 - 27.3608\xi_{design}^2 + 6.1865\xi_{design} + 0.7536}.$$

7.4.2.1.7 Lateral Forces and Base Shear

For nonproportionally damped systems, the lateral force can also be calculated through Equation 7.66 and the base shear is calculated through Equation 7.70.

7.4.2.4 Redesign of Damping Devices

On completion of the above procedure, the design criteria described by Equation 7.25 should be rechecked. If they are not satisfied, the damping ratio and/or damper configuration need to be reconsidered and a new damping matrix should be determined. The response estimation is revisited until the results are satisfactory.

7.4.2.5 Selection of Dampers

If the above-mentioned criteria are satisfied, similar to the design for the MSSP system, the dampers can be selected as discussed above.

The discussion on response estimation of nonproportionally damped systems has been completed in the preceding sections. In the following discussion, a simplified example is given to illustrate the design procedure and to compare methods in dealing with proportionally and nonproportionally damped MDOF systems.

Example 7.3

Assume that the design base shear in Example 7.1 is [V] = 48.65 (kN). The base shear calculated through earthquake simulations is 64.64 (kN) with $\alpha_v = 0.7527 > 0.75$.

Therefore, damping control can be used to reduce the seismic responses. Suppose the damping matrix of the base structure is

$$C_0 = 12.95 \times \begin{bmatrix} 2 & -1 & 0 & \dots & 0 \\ -1 & 2 & -1 & \dots & 0 \\ & & \dots & & \\ 0 & & \dots & 2 & -1 \\ 0 & & \dots & -1 & 1 \end{bmatrix}_{10\ 10} \quad (\text{MN-s/m})$$

which provides 0.025% damping ratio for the first mode.

Now, suppose the dampers are only installed in the first three stories. The corresponding damper configuration matrix is given by

$$\mathbf{C}_{con} = \begin{bmatrix} 2 & -1 & 0 & \dots & 0 \\ -1 & 2 & -1 & \dots & 0 \\ 0 & -1 & 1 & \dots & 0 \\ & & & \dots & \\ 0 & & \dots & & 0 \end{bmatrix}_{10\times 10}$$

Similar to Example 7.1, $c^{(2)} = 106.9$ (MN-s/m) can be calculated. The supplemental damping matrix is $c^{(2)}\,\mathbf{C}_{con}$. The total damping matrix is $\mathbf{C}_0 + \mathbf{C}^{(2)}$. Now, consider response estimations based on both proportionally and nonproportionally damped models. For the estimations based on the proportionally damped models, two cases are possible: the design using method *A* and the design using method *D*.

The basic parameters of the accumulated modal mass ratio, the periods, and damping ratios of the proportionally and nonproportionally damped systems are given in Table 7.11. Note that the nonproportionally damped system is obtained after the supplemental dampers are installed.

TABLE 7.11
Basic Parameters

				Period T_i (s)		Damping Ratio ξ_i	
Mode	Cumulated Mdl Mass Ratio	ETR ζ_{ι}	Mdl Participation Factor	Proportionally Damped	NonProportionally Damped	Proportionally Damped	NonProportionally Damped
1st	0.8517	0.024	1.2589	1.0854	1.0598	0.1013	0.131
2nd	0.9422	—	–0.3981	0.3589	0.4858	0.3065	>1
3rd	0.9715	0.411	0.2170	0.2215	0.3366	0.4967	0.325
4th	0.9839	0.337	–0.1242	0.1604	0.2261	0.6857	0.288
5th	0.9912	0.207	0.0933	0.1255	0.1549	0.8767	0.235

In the second column of Table 7.11, the cumulated modal mass ratio of the corresponding proportionally damped system is listed. It is seen that up to the third mode, $\gamma_{mS} = 0.9715 > 0.95$. Note that to choose the exact value of S, the number of modes needed for a nonproportionally damped system, the quantity γ_{mS} given in Equation 7.49 cannot be used. However, for estimation of S, which includes several more modes, γ_{mS} is often sufficient. In this example, S = 5 is chosen. The safety modification factor S_M is taken to be 1.0.

The third column of Table 7.11 is the modal energy transfer ratio. It is seen that $\zeta_1 = 2.4\%$, which is larger than 1%. Furthermore, it is realized that the second "mode" is actually an overdamped subsystem. Thus, the model of a generally damped system can be considered. The fourth column lists the periods calculated through the proportionally damped and nonproportionally damped model, and the fifth column lists the damping ratios of the proportionally damped and nonproportionally damped mode. From these columns, the differences between the proportionally and nonproportionally damped systems are evident.

The mode shapes of the proportionally and nonproportionally damped system are listed in Table 7.12. For the generally damped system, the accompanist vectors are also listed. Due to space limitations, only data from the first four modes are presented. For the proportionally damped systems, the first three mode shapes are listed in Table 7.5. In the last column of Table 7.8, the shape function of the residual mode is also used. Note that the modal participation factor of the residual mode is – 0.413.

The estimated modal responses are listed in Tables 7.13 and 7.14. Comparing the responses among these methods and the numerically simulated results, it is seen that those obtained from the nonproportionally model have the closest results. Those obtained from the proportionally damped model with method A are also good estimations. On the other hand, the data obtained from the concept of residual mode seem to be not compatible.

According to Example 7.1, supplemental damping is used to reduce the base shear from 60 (MN) (64.63 (MN) according to simulated result) to less than 48.65 (MN). The simulated result is based on a statistical time history analysis and is considered to be more accurate. Seen from Table 7.14, all the methods indicate that the design is satisfactory. However, it seems that the proportionally damped models, especially those with inaccurate mode shapes, do not provide accurate results.

7.5 DAMPER DESIGN ISSUES

Many practical issues in damper installation cannot be covered using only the principles of structural dynamics, such as supporting stiffness, non-Timoshenko damping, and safety and maintenance issues. Here, certain important details are discussed. Most of these concepts also apply to nonlinear damping design.

7.5.1 Supporting Stiffness

7.5.1.1 General Requirement

In Chapter 5, the effect of supporting stiffness on damper design was discussed. Practically speaking, once a specific type of damper is decided, the damper supporting system needs to be considered. As noted earlier, the calculation of the influence of the stiffness of the supporting member connecting the dampers to the structure is not included in the simplified design. Equations 5.141 and 5.144 can be used for this purpose. The concept of these two equations can be described by

$$k_s > \gamma_\xi \xi k \tag{7.104}$$

where k_s is the supporting stiffness and γ_ξ is a proportional coefficient.

In NEHRP 2009, it is assumed that structures to be installed with dampers are single-bay structures, as shown in Figure 7.2a. However, structures may have multiple bays, and dampers may not necessarily be installed in all bays, such as shown in Figure 7.2b.

TABLE 7.12
Modal Shapes and Accompanist Vectors

	Complex Mode Shape p_i				Accompanist Vector q_i				Normal p_4	Residue Shape
	1st Mode	2nd Mode	3rd Mode	4th Mode	1st Mode	2nd Mode	3rd Mode	4th Mode	4th Mode	p_r
1st floor	0.146∠16°	76.779	0.540∠−156°	0.563∠−23°	0.0022 ∠101°	0	0.0023∠−73°	0.0016∠94°	−0.990	−2.079
2nd floor	0.290∠15°	66.748	1.020∠−158°	1.052∠−27°	0.0050∠101°	0	0.0049∠−75°	0.0033∠89°	−1.089	−1.737
3rd floor	0.427∠15°	6.982	1.377∠−162°	1.395∠−37°	0.0073∠101°	0	0.0065∠−79°	0.0044∠80°	−0.077	−1.395
4th floor	0.552∠ 9°	5.001	1.329∠179°	0.952∠82°	0.0080∠95°	0	0.0053∠−98°	0.0025 ∠35°	1.012	−1.053
5th floor	0.670∠6°	3.549	1.152∠164°	1.004∠−141°	0.0119∠91°	0	0.0056∠−113°	0.0033∠−25°	1.233	−0.711
6th floor	0.774∠3°	2.554	0.804∠146°	1.188∠−175°	0.0120∠89°	0	0.0034∠−131°	0.0034∠−58°	0.160	−0.369
7th floor	0.863∠2°	1.848	0.444∠109°	1.065∠165°	0.0170∠87°	0	0.0024∠−168°	0.0039∠−79°	−1.059	−0.026
8th floor	0.929∠1°	1.407	0.457∠37°	0.507∠135°	0.0148∠86°	0	0.0020∠119°	0.0013 ∠−108°	−1.042	0.316
9th floor	0.976∠0°	1.127	0.785∠8°	0.507∠19°	0.0178∠86°	0	0.0039∠91°	0.0017∠136°	−0.045	0.658
Roof	1.000∠0°	1.000	1.000∠0°	1.000∠0°	0.0177∠85°	0	0.0049∠83°	0.0032∠117°	1.000	1.000

TABLE 7.13
Seismic Responses

	Absolute Acceleration (m/sec²)					Lateral Force (MN)					Lateral Displacement (m)				
	a^{Pro}	a_a^B	a_a^{res}	a_a^{Gn}	a_a^{sim}	f_L^{Pro}	f_L^B	f_L^{Res}	f_L^{Gen}	f_L^{sim}	d_m^{Pro}	d_m^B	d_m^{res}	d_m^{Gn}	d_m^{sim}
1st floor	2.775	0.796	8.479	2.904	2.466	2.483	0.700	7.460	2.555	2.170	0.025	0.015	0.035	0.030	0.026
2nd floor	3.254	1.356	7.109	4.035	3.072	3.259	1.354	7.095	4.027	3.066	0.049	0.030	0.033	0.058	0.052
3rd floor	3.393	1.714	5.768	5.047	3.562	3.560	1.687	5.676	4.966	3.505	0.073	0.044	0.034	0.086	0.076
4th floor	3.925	2.113	4.484	4.960	4.220	3.314	1.774	3.765	4.165	3.543	0.094	0.057	0.037	0.107	0.100
5th floor	4.274	2.536	3.323	4.922	4.779	4.388	2.601	3.409	5.049	4.902	0.114	0.070	0.043	0.126	0.122
6th floor	4.555	2.884	2.466	4.952	5.235	4.096	2.579	2.206	4.429	4.683	0.131	0.080	0.050	0.143	0.140
7th floor	4.864	3.200	2.285	5.080	5.658	5.554	3.653	2.608	5.799	6.459	0.146	0.089	0.058	0.157	0.157
8thfloor	5.090	3.391	2.908	5.128	6.039	4.700	3.117	2.674	4.715	5.553	0.157	0.096	0.066	0.168	0.169
9th floor	5.339	3.555	3.975	5.407	6.380	5.629	3.744	4.184	5.692	6.717	0.165	0.100	0.075	0.176	0.177
Roof	5.576	3.705	5.219	5.820	6.586	5.713	3.786	5.334	5.948	6.731	0.168	0.103	0.084	0.181	0.181

TABLE 7.14
Base Shear

	Proportionally Damped Model	Method B	Method of Residual Mode	Generally Damped Model	Simulated Results
Base shear (MN)	42.70	24.99	44.410	47.345	47.327

In this case, the idea of limited supporting stiffness can be extended, as conceptually shown in Figure 7.2c. That is, for a simplified design, a structure with dampers, as shown in Figure 7.2b, can be represented by an equivalent single-bay model. Using this model, the mass and stiffness condensed from multiple bays to a single bay must be addressed. In addition, the effectiveness of damping for this equivalent single bay must be addressed. In NEHRP 2009, the energy absorption by an individual damper is thoroughly described. Here, vibration energy in bays without dampers is absorbed by the structural components in these bays and by dampers installed in adjacent bays. These energy quantities can be calculated through MDOF systems.

In multiple-bay structures, the values of γ_ξ and k will vary at different locations. Thus, the supporting stiffness may be written as

$$k_{sj} \geq \gamma_{kj}\, \xi\, k_j \quad (kN/m) \tag{7.105}$$

where subscript j stands for the jth story where dampers are installed.

With multiple-bay frames, the apparent lateral stiffness is considerably stronger than that of the single-bay frames and the resulting damping ratio is much lower. Thus, the product of ξk_j may not change significantly, if the "additional supporting" stiffness is assumed to be infinite. In order to determine the required supporting stiffness, it can be assumed that the lateral stiffness is k_j. To have the same value of damping ratio ξ, means a larger supporting stiffness, g_i, is required. Based on limited numerical simulation, the coefficient γ_{kj} may be modified as follows:

$$\gamma_{kj} = \left(1 + \sqrt{\frac{\Delta k_j}{k_j}}\right)\gamma_{j0} \tag{7.106}$$

Here, k_j is the lateral stiffness of the jth story with unit (kN/m); Δk_j is the additional lateral stiffness of "empty" bays with unit (kN/m). γ_{j0} is the coefficient for single-bay structures. For example, with the help of Equations 5.141 and 7.106,

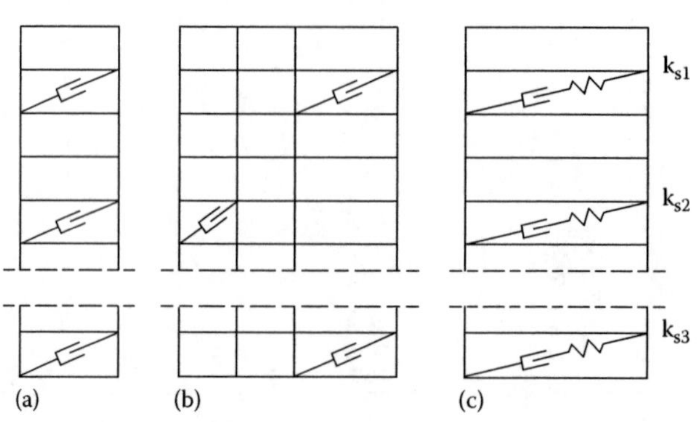

FIGURE 7.2 (a–c) Single-bay and multiple-bay structures.

$$\gamma_{kj} = 10\left(1+\sqrt{\frac{\Delta k_j}{k_j}}\right)$$

Example 7.4

Suppose that in the jth story of an eight-bay building, dampers are installed in only two bays. The lateral stiffness contributed by each bay is about 100,000 (kN/m).

Approximately, Δk_j is taken as the stiffness of "empty" bays,

$$\Delta k_j (8-2)\times 100{,}000 = 600{,}000 \quad (\text{kN/m}).$$

Furthermore, since two bays are installed with dampers,

$$k_j = 2\times 100{,}000 = 200{,}000 \quad (\text{kN/m}).$$

Therefore, when Equation 7.29 is used,

$$\gamma_{kj} = \left(1+\sqrt{\frac{\Delta k_j}{k_j}}\right)\gamma_{j0} = 2.7\gamma_0$$

This result implies that with the "empty" bays, the supporting stiffness needs to be considerably larger. In practice, the resulting supporting stiffness can be too large to realize. In this case, installing the dampers in more bays should be considered.

7.5.2 Modification of Non-Timoshenko Damping

Computation based on Timoshenko damping is used more often in this chapter. For the case of using the force-based effective damping, modification of the damping ratio computation is needed as follows:

$$\xi_n = \sqrt{p_c}\,\xi_T \tag{7.107}$$

in which ξ_n is the non-Timoshenko damping and ξ_T is the Timoshenko damping. Both can be the damping ratios of an SDOF system; or they can also be the ith damping ratio of the ith mode of an MDOF system, respectively. Furthermore, p_c is the Penzien constant, which is explained in Chapter 6.

In practical applications, the vendors of dampers cannot provide information on the nature of the damping associated with their devices, namely, Timoshenko or non-Timoshenko damping. Generally, when the damping ratio is small, i.e., $\xi < 0.1$, all the damping ratios can be treated as Timoshenko damping. Otherwise, if the damping adaptability is high (see Equation 5.59), it is better to modify the Timoshenko damping by using Equation 7.107.

Example 7.5

Suppose a system has nonlinear damping. Using the equation $\xi_T = E_d/4\pi E_p$, the damping ratio can be estimated to be 30%, which is the Timoshenko damping.

As a rough estimation, the non-Timoshenko damping can be calculated by letting $p_c = 0.65$, in this case,

$$\xi_n = \sqrt{0.65} \times 0.3 = 0.24$$

Thus, the force-based effective damping ratio is estimated to be a smaller value, which is expected in most cases.

7.5.3 Safety, Reliability, and Maintenance Issues

Damping devices, which are in fact mechanical elements, are added to structures for the purpose of reducing the seismic response of the structures at the same time. They may also create complications in structural responses, primarily due to the damper-induced cross effects and other nonlinear responses of the structures. These issues are not fully quantified at present. Therefore, safety factors commonly used for mechanical elements may be considered. Based on a common treatment of mechanical elements, some ranges of these safety factors are discussed and suggested only to qualitatively illustrate safety issues.

7.5.3.1 Fail-Safe Concept

Earthquake protective systems are meant to reduce the vibration level of the structure induced by seismic ground motions. Therefore, the protective system itself should, in principle, be designed with a larger safety factor. On the other hand, a fail-safe mode of design may be established for stiffness when the protective system fails. In any case, several aspects of dampers must be considered in damper design. These are briefly discussed in the following paragraphs.

7.5.3.2 Maximum Force in Dampers

The maximum force to be experienced by a damper is briefly examined. Figure 7.3 shows examples of typical damper installation details. Figure 7.4 shows the additional force applied to a damper at

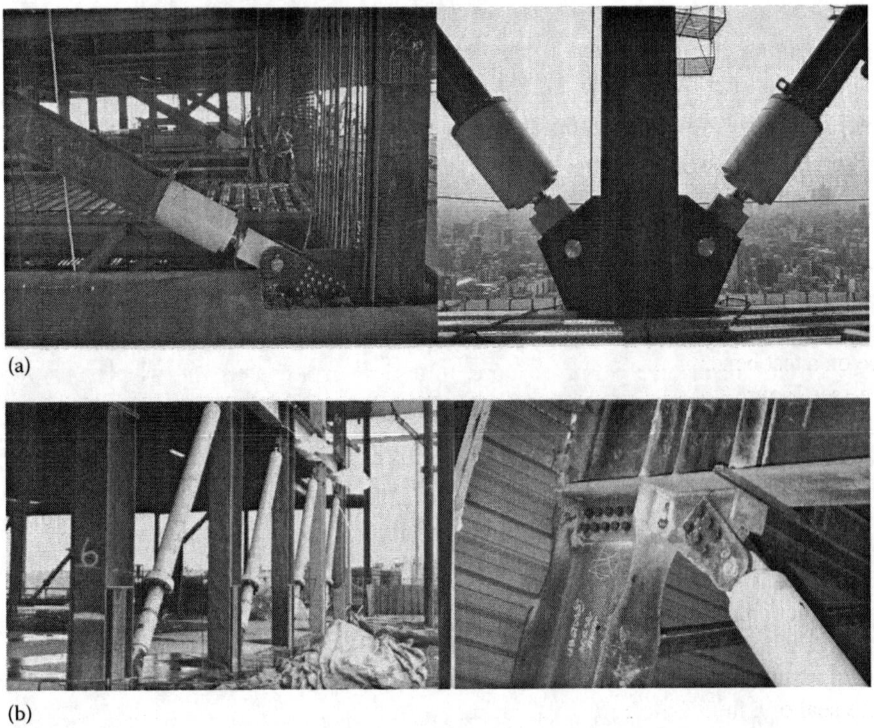

(a)

(b)

FIGURE 7.3 (a, b) Damper connection details. ((a) Courtesy of Taylor Devices, Inc. North Tonawanda, NY. and (b) Courtesy of ITT Enidine, Inc. Orchard Park, NY.)

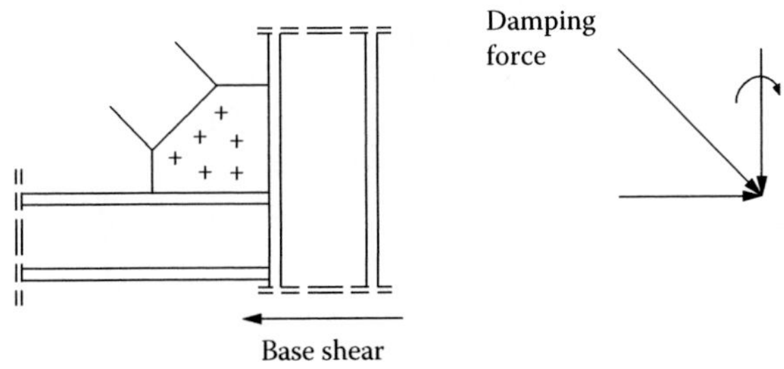

FIGURE 7.4 Damping force on joints.

the conceptual connection of the original beam and column. This additional force does not statically exist. The amplitudes of the additional damping force and the storage forces and torque of the beam and column are time variables. In most cases, they will not reach their peak value at the same time.

If the base shear is calculated by primarily considering the restoring force, the influence of the damping force should be considered. That is,

$$\sigma_{\Sigma} = \sigma_s + \kappa\sigma_d \tag{7.108}$$

$$\tau_{\Sigma} = \tau_s + \kappa\tau_d \tag{7.109}$$

where $\sigma_{(\cdot)}$ and $\tau_{(\cdot)}$ are the maximum normal and shear stress, respectively with proper units, the subscripts Σ, s, and d stand for the total stress, including the stress caused by storage force and the stress caused by damping force, respectively; and κ is a coefficient, depending on different situations. When the stress caused by the storage force reaches its peak values, those caused by the damping force should be close to zero. In Chapter 4, it was shown that this is not always true for MDOF systems. Based on limited simulations, the authors suggest that κ be given a value of

$$\kappa = 20\% - 50\%. \tag{7.110}$$

rather than zero.

Larger values are taken when the levels of weight, plan, damping irregularity, and/or the damping adaptability f_{adp} (see Equation 5.59) are high.

7.5.3.3 Stability of Damper System

When a damper is installed in a structure, it can be subjected to forces more complex in nature than the force on a test bench in the laboratory. Some of these forces may cause unstable working conditions (i.e., see Figure 7.6). On the other hand, when a damper is installed in a particular location of a structure, and damping forces are applied to the structure at this location, they may help the overall system to be more stable due to the constraints of the damper locations, as discussed below.

7.5.3.3.1 Side Load of Dampers

When a damper is tested in the laboratory, most likely it is only subjected to pure axial load. In actual installations, side load perpendicular to the axial load may exist. Figure 7.5 conceptually shows this situation.

The side load may increase the friction force of the fluid damper seals. In the extreme case, side force may lock the piston head of a fluid damper, or it may damage or accelerate the rate of damage of the seal of a fluid damper. The side load may be caused by a bending moment due to multi-directional input, or by an inertia force of the damper, or simply by misalignment, etc. In order to reduce the side load, universal connectors have been used. The disadvantage is discussed in the next subsection.

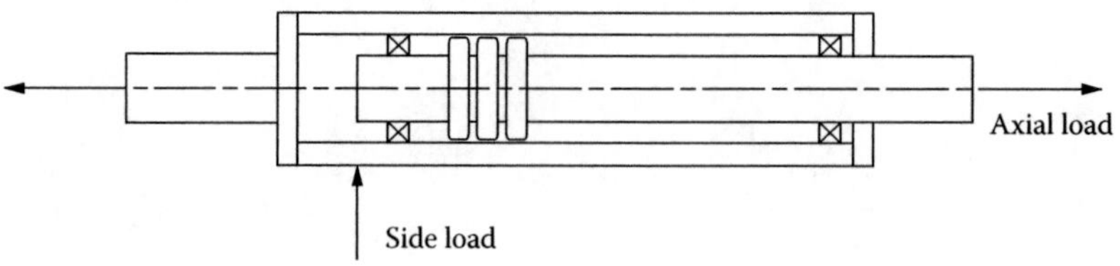

FIGURE 7.5 Side load on a damper.

The side load may change the friction coefficient of a friction damper by a factor of between –10% and –30%, based on limited numerical simulations, depending on the amplitude and frequency of the side load.

If the side load is simply caused by the gravity force, creep may occur, causing an unstable condition of the damper system, because VE material has near zero stiffness under static force. Figure 7.6 shows this problem.

In addition, since the stiffness of a damper perpendicular to its axial direction can be considerably less than in the axial direction, even though the mass of the damper and its support is small, the corresponding vibration frequency can fall into the major part of an earthquake spectrum. As a result, a local vibration mode may develop (see Figure 7.6b).

7.5.3.3.2 Buckling Condition of Damper Systems

A damper is typically installed in a diagonal configuration as shown in Figure 7.6. When the connection of the damper support to the frame of the structure is a hinge type, an additional degree-of-freedom is introduced. This weakens the buckling condition of the damper system.

7.5.3.3.3 Buckling Condition of the Overall System

The introduction of dampers may change the buckling condition of the overall frame. When a column is subjected to vertical loading, it may fail by in-plane buckling. The device configuration may provide more lateral deformation constraints to the frame, but at the same time decreases the buckling load of the column. Therefore, to determine whether the damper system contributes to the frame buckling condition, a detailed analysis needs to be conducted. Furthermore, as previously discussed, the combination of damping and spring forces provided by the damper and the damper-supporting member should be considered. It is safe to assume that both forces reach their peak value at the same time during an earthquake.

To date, sufficient data do not exist on the effect of damper systems on the buckling strength of structures. In fact, design parameters are not defined.

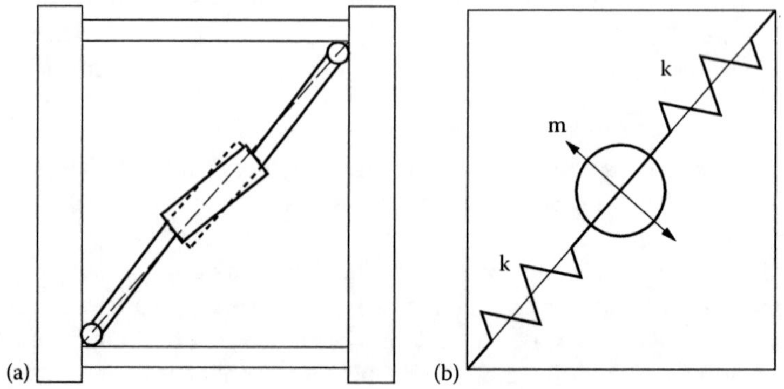

FIGURE 7.6 (a, b) Side load of a VE damper and local vibration mode.

7.5.3.4 Combinations of Different Devices

Different types of devices can be used together for multiple objectives. When the loading or deformation exceeds the level permitted by the normal design of one damper (either fails or behaves like a different type of device), the remaining devices may still be functional. The way to analyze this situation is briefly discussed herein.

For linear and proportionally damped systems, this problem can be simply solved by adding the effectiveness of each damper. In most cases, damper design involves nonlinear and nonproportional damping properties for which accurate calculation is more complex. Generally, numerical calculations for nonlinear systems have to be carried out and 2n state-space needs to be used to calculate the results of modal superposition.

However, if the damping ratio is not very large, i.e., less than 15%, the equivalent damping method can yield fairly good accuracy. In this case,

$$\frac{\xi_1}{\xi_2} \approx \frac{\|\mathbf{C}_1\|}{\|\mathbf{C}_2\|} \tag{7.111}$$

and

$$\xi_1 + \xi_2 \approx \|\mathbf{C}_1\| + \|\mathbf{C}_2\| \tag{7.112}$$

where ξ_1 and ξ_2 are damping ratios in different cases corresponding to different damping matrices $\mathbf{C}_1$ and $\mathbf{C}_2$. The symbol $\|(\bullet)\|$ stands for a particular norm of the matrix $(\bullet)$.

For example, suppose a damping matrix $\mathbf{C}_1$ provides the damping ratio for the first mode to be ξ_1. Later, the damping coefficient matrix is increased α times so that

$$\mathbf{C}_2 = \alpha \mathbf{C}_1 \quad (\text{kN-s/m}) \tag{7.113}$$

Then,

$$\xi_2 = \alpha \xi_1 \tag{7.114}$$

Next, suppose damper 1 gives a damping coefficient $\mathbf{C}_1$ and damper 2 gives a damping coefficient $\mathbf{C}_2$. Individually, damper 1 yields a damping ratio ξ_1, and damper 2 yields a damping ratio ξ_2. Then, when the two types of dampers are used together, with υ representing the number of the first type of damper and ν the second damper, the approximate total damping ratio, ξ_{total}, is given by

$$\xi_{total} \approx \upsilon \xi_1 + \nu \xi_2 \tag{7.115}$$

7.5.3.5 Safety Factors

Since using a damping system to control the dynamic responses of structures is more complex than the design of structures to withstand static load without dampers, the considerations of design safety and reliability for these cases are different. This is briefly discussed in the following subsection.

7.5.3.5.1 Types of Safety Factors

There are two types of safety factors for the design of dampers: damping and structural response reduction.

If a structure with supplemental damping is proportionally damped, adding more damping for each important mode will further reduce the corresponding modal responses and therefore the total structural response. However, for complex modes, the opposite may result. In this case, the *damping safety factors* can be used.

The advantages of damping safety factors are obvious. The concept of using a higher value of damping to guarantee vibration control is easier to understand. Using the damping safety factors can simplify the design procedure. This is discussed in the following paragraphs.

When structures with added dampers have high damping nonlinearity, the concept of a special safety measure, the *response safety factors*, are to be used.

When the response safety factors are employed, the initially calculated structural responses, such as the total base shear, floor drift, story force, overturning moment, foundation uplifting force, etc., are magnified, resulting in a more conservative design.

7.5.3.5.2 Damping Safety Factors

With the damping safety factors, the control parameter ξ will be modified to require a higher value.

7.5.3.5.2.1 Damper and Installation Factor S_ϖ The quality control of the damper installation as well as the quality of the damper itself should also be considered by using the damper and installation factor, S_ϖ, which consists of several components.

The first safety consideration in damper installation is the supporting stiffness, which is usually not infinitely large. For example, a fluid damper can be compressed more than the designed value if there is too much air in the fluid; a viscoelastic damper can have a smaller storage modulus due to higher temperature; etc. Thus, a safety factor, $S_{\varpi 1}$, should be used.

The second component in safety consideration is the space tolerance for the installation of bolts or pivots. Although this space tolerance is small, it can alter the damping force-damper stroke constitution, which is equivalent to a reduction in damping.

When a certain tolerance, denoted by *g*, exists between the damper and the structure, the designer may use a second safety factor $S_{\varpi 2}$.

The third safety consideration by using a safety factor $S_{\varpi 3}$ accounts for the errors introduced in decoupling a nonproportionally damped MDOF system into the SDOF models when nonproportional damping is insignificant.

The fourth safety consideration is the uncertainties introduced by using idealized assumptions about the dampers. Due to various imperfections, dampers may not deliver the desired damping force. For example, a hydraulic damper may have unwanted leakage or the fluid viscosity may become thinner; a friction damper may be overly worn to maintain correct normal force and the friction coefficient may become smaller; a viscoelastic damper may have a larger thickness or its loss factor may become smaller due to higher working temperature; etc. Due to these imperfections of the dampers, the total energy dissipation, which is the area enclosed by the hysteric loop, can be smaller than the idealized theoretical value. A quality control factor must then be considered to account for those uncertainties (Ramirez et al. 2000).

$$S_{\varpi 4} = 1/q_H \tag{7.116}$$

Here, q_H is the quality factor or the hysteresis loop adjustment factor, which is defined in Equation 5.44 and repeated as follows:

$$q_H = \frac{E_r}{E_d} \tag{7.117}$$

Thus, the total safety factor for structure-damper system with high nonlinearity is given by

$$S_\varpi = S_{\varpi 1} S_{\varpi 2} S_{\varpi 3} S_{\varpi 4} \tag{7.118}$$

7.5.3.5.2.2 Uneven Modal Distribution Factor S_ξ In simplified design, it is often assumed that the higher modes can have the same value of the damping ratio as the fundamental mode. However, it is possible that the higher modes may have a smaller damping ratio. For this situation, a safety factor S_ξ should be to account for the effects of uneven modal combinations.

7.5.3.5.2.3 Damper Realization Factor S_υ Some uncertainties exist between the design (of idealized damping devices) and installation of dampers. Safety factors, called the *damper realization factor* and denoted as $S_{\upsilon 1}$, are established by considering several components.

Second, when the fundamental mode only is considered, an error may be introduced by ignoring the higher modes. A safety factor $S_{\upsilon 2}$ may be used.

Third, earthquake excitations do not necessarily occur in one principal direction of the structure. Orthogonal effects by an input from the perpendicular direction should be considered. This is accounted for by yet another safety factor, $S_{\upsilon 3}$.

Fourth, when a larger damping force is needed but large size dampers are not available (or cannot be used due to space limitations), several smaller-sized dampers are used in parallel. Since these smaller dampers cannot have exactly the same specifications or the same installation conditions, the damping force can be unevenly distributed to these dampers. It is thus necessary to introduce the uneven distribution factor, $S_{\upsilon 4}$.

The total safety factor for the above additional considerations is

$$S_\upsilon = S_{\upsilon 1} S_{\upsilon 2} S_{\upsilon 3} \tag{7.119}$$

7.5.3.5.3 Response Safety Factors

In addition to the standard safety factors used in seismic design (BSSC/NEHRP 2009), there are other response safety factors related to damper design, with which the design parameters of responses will be modified to have higher values.

7.5.3.5.3.1 Response Modification (Safety) Factors The response modification factors S_{Dn}, $S_{D\xi}$, S_{An}, and $S_{A\xi}$ are also important factors, which were mentioned in Section 7.1.

7.5.3.5.3.2 Stroke Safety Factor S_σ The stroke safety factor, denoted as S_σ, is introduced in Equation 7.48 to ensure that dampers do not suffer damage due to extremely strong earthquakes. It also covers the design uncertainty as well as damper performance uncertainty. To ensure that failure of dampers does not occur, a large safety factor is used; especially if the structure is designed to be highly ductile.

7.5.3.6 Reliability and Maintenance

The issue of reliability and maintenance of the damping devices must be carefully considered in design. This is important, but beyond the scope of this book.

7.5.4 Numerical Damping Coefficient

Numerical damping coefficient B is a function of the damping ratios, which are used to modify the value of the design spectrum. Therefore, the accuracy of B will directly affect the design value of the seismic responses. Several issues can affect the estimation of B, which are discussed below.

7.5.4.1 Concept of Numerical Damping Coefficient B

The numerical damping coefficient B is inversely proportional to the system base shear, expressed by Equation 2.319. The basic idea suggested by NEHRP 2009 is that when the numerical damping coefficient increases, the system base shear will be reduced proportionally. The same idea is also applicable to the lateral forces, f_L, and floor deflections, δ, that is,

$$V \propto \frac{1}{B}, \tag{7.120}$$

$$f_L \propto \frac{1}{B}, \tag{7.121}$$

$$\delta \propto \frac{1}{B}. \tag{7.122}$$

In these three relationships, the numerical damping coefficient B can be defined as a ratio between the spectral value with 5% damping ratio and a damping ratio ξ at period T. That is,

$$B = \frac{S_A(T, 0.05)}{S_A(T, \xi)} \tag{7.123}$$

From Equation 7.123, the factor B is a function of the damping ratio ξ. And, the larger the damping ratio, the larger the factor B. In Table 7.15, the value of B is listed according to NEHRP 2009.

Using the numerical damping coefficient, the spectral acceleration of a system whose damping ratio is different from 5% should be modified. That is, from Equation 7.123,

$$S_A(T, \xi) = \frac{S_A(T, 0.05)}{B} \tag{7.124}$$

Similarly, for spectral displacement,

$$S_D(T, \xi) = \frac{S_D(T, 0.05)}{B} \tag{7.125}$$

Table 7.15 implies that the numerical damping coefficient is only a function of the damping ratio, that is,

$$B = f(\xi) \tag{7.126}$$

For convenience of analysis and design, some empirical equations can be used to approximate NEHRP's table and/or Equation 7.126. As can be seen in Figure 7.7, the relationship between the numerical damping coefficient B and the damping ratio ξ is very close to a straight line. Therefore, a simplified linear relationship between B and ξ can be approximated. That is,

$$B = f(\xi) = a\xi + b \tag{7.127}$$

TABLE 7.15
Numerical Damping Ratio

ξ	0.02	0.05	0.10	0.20	0.30	0.40	0.50	0.60	0.70	0.80	0.90	1.00
B	0.8	1.0	1.2	1.5	1.8	2.1	2.4	2.7	3.0	3.3	3.6	4.0

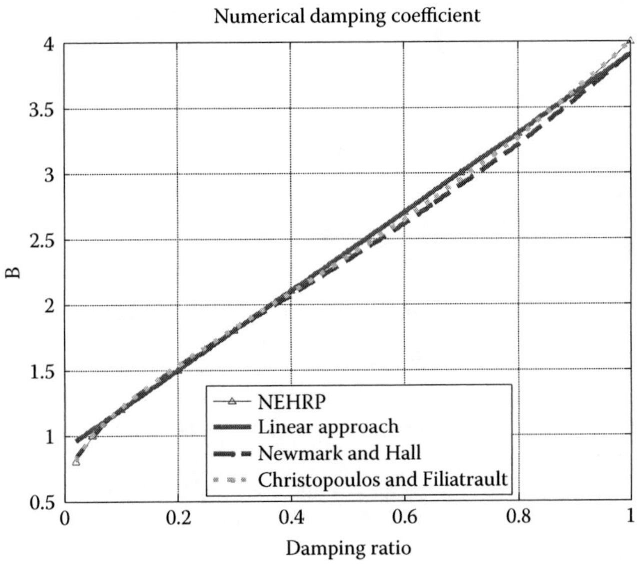

FIGURE 7.7 Numerical damping coefficients.

Based on the least square method, the coefficients a and b can be calculated as a = 3.0 and b = 0.9. Therefore,

$$B \approx 0.3\xi + 0.9 \tag{7.128}$$

In the literature, there are other expressions for the numerical damping coefficient. Newmark and Hall (1982) suggested that for the constant velocity region of the design spectrum, the amplitude factor based on the damping ratio, denoted as A_ξ, can be written as

$$A_\xi = \frac{2.31 - 0.41\ln(100\xi)}{1.65} \tag{7.129}$$

Furthermore, in the constant acceleration region of the design spectrum, the amplitude factor based on the damping ratio can be written as

$$A_\xi = \frac{3.21 - 0.68\ln(100\xi)}{2.12} \tag{7.130}$$

Because the numerical damping coefficient can be written as

$$B = \frac{S_A(T, 0.05)}{A_\xi} \tag{7.131}$$

in this case, described by Equation 7.129,

$$B = \frac{1.65}{2.31 - 0.41\ln(100\xi)} \tag{7.132}$$

Christopoulos and Filiatrault (2006) propose another formula for the numerical damping coefficient, called the *damping reduction factor*, which is written as

$$B = \frac{4}{1 - \ln(\xi)} \tag{7.133}$$

The curves of B vs. ξ described by Equations 7.132 and 7.133 are also plotted in Figure 7.7 for comparison.

7.5.4.2 Modification of Design Spectrum Based on Period Range

7.5.4.2.1 Modification with Respect to Period

The relationship described by Equation 7.126 has some limitations. First, this relationship, developed for the design spectrum of an SDOF system, does not consider the existence of multiple modes. Thus, it is not a function of period T_i.

However, from Chapters 1 and 2, it is known that when the major driving frequency is considerably lower than the natural frequency of a system, the most effective way to reduce the vibration level is to increase the stiffness of the system. The corresponding frequency range is referred to as the range of stiffness control. When the major driving frequency is considerably higher than the natural frequency of a system, to reduce the vibration level, the mass of the system is increased. The corresponding frequency range is referred to as the range of mass control. When the value of the major driving frequency is in the vicinity of the natural frequency of a system (the system is close to resonant), the damping of the system can be increased to reduce the vibration level. The corresponding frequency range is referred to as the range of damping control. These facts indicate that at different frequencies or periods, the effect of the damping will be different. Therefore, the numerical damping coefficient B should also be a function of the period. NEHRP 2009 (BSSC 2009) does not include these detailed considerations in favor of simplicity.

Ramirez et al. (2000) suggested a method to improve the value of the numerical damping coefficient by considering the plot shown in Figure 7.8, where the value of B is a function of the period. In fact, when the system is very stiff, that is, the period is shorter then $T_s/5$, B will have a small value; this means that damping will play a less important role. In the range between $T_s/5$ and T_s, there will be larger values of B, which continuously increase until they reach the value B_l. After T_s, the damping coefficient becomes constant and equals B_l.

Thus, according to Ramirez et al., the value of the numerical damping coefficient B should be a function of both the damping ratio and the period. By using the 99 earthquake records, the mean response spectra can be computed. Then, the spectrum of 5% damping is used as a reference to find the value of B. The resulting curves are plotted in Figure 7.9 so that the relationship between B and the period can be examined.

If B is not a function of the period, then, in Figure 7.9, horizontal flat lines with different values of the damping ratio should be seen. However, there are a group of curves. These curves, when

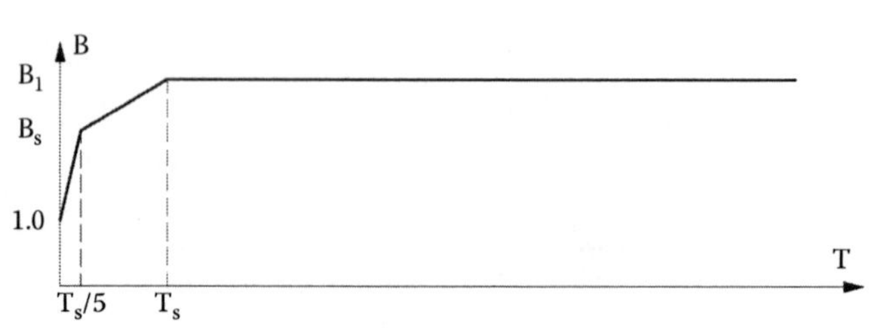

FIGURE 7.8 Damping coefficient spectrum.

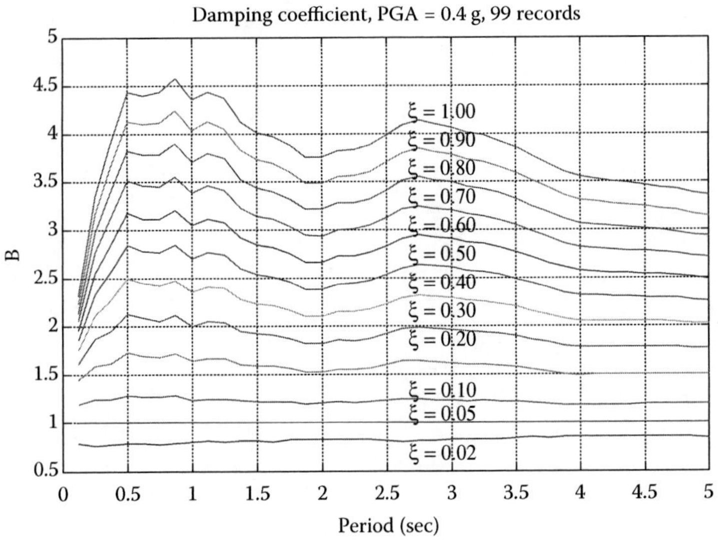

FIGURE 7.9 Computed numerical damping coefficient vs. period.

damping is small, are close to Ramirez's plot, shown in Figure 7.8. When damping becomes larger, the differences are obvious. In order to examine this issue further, a group of curves of B vs. the damping ratio are plotted and shown in Figure 7.10. For comparison, the curve based on Table 7.6, which is marked as the break line, is also plotted.

To further illustrate that the values of B are affected by the variation of the period, the 99 earthquake records were used to obtain the mean response spectra with different damping ratios, as shown in Table 7.16. In this table, selected values of the damping coefficient B are listed for different periods and damping ratios. As a comparison, the values given by NEHRP 2009 are also given.

Based on the principle that larger damping will reduce more vibration, the value of NEHRP 2009 is safe to apply. Thus, for a simplified damping design, the NEHRP data can be used directly without modification.

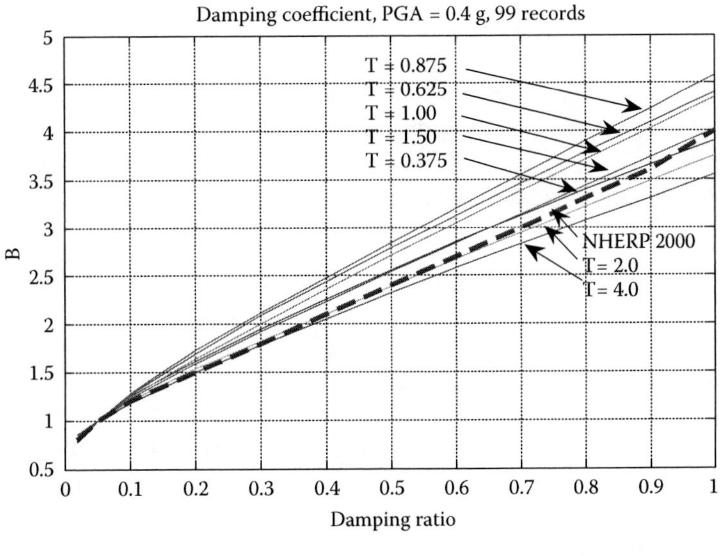

FIGURE 7.10 Computed numerical damping coefficient vs. damping ratio.

TABLE 7.16
Numerical Damping Coefficients

	T = 0.375 (s)	T = 0.625 (s)	T = 0.875 (s)	T = 1.00 (s)	T = 1.50 (s)	T = 2.00 (s)	T = 4.00 (s)	NHERP
$\xi = 2\%$	*0.7731*	*0.7918*	*0.7915*	0.8049	0.8089	0.8267	0.8533	0.8
$\xi = 10\%$	1.2482	1.2711	1.2803	1.2383	1.2233	1.2051	*1.1794*	1.2
$\xi = 20\%$	1.6205	1.6990	1.7259	1.6378	1.5938	1.5316	*1.4962*	1.5
$\xi = 30\%$	1.9544	2.0867	2.1128	2.0020	1.9245	1.8224	*1.7806*	1.8
$\xi = 40\%$	2.2658	2.4441	2.4804	2.3580	2.2356	2.1074	*2.0571*	2.1
$\xi = 50\%$	2.5619	2.7842	2.8427	2.7072	2.5397	*2.3905*	*2.3247*	2.4
$\xi = 60\%$	2.8503	3.1206	3.1990	3.0488	2.8447	*2.6685*	*2.5774*	2.7
$\xi = 70\%$	3.1297	3.4537	3.5512	3.3791	3.1426	*2.9415*	*2.8255*	3.0
$\xi = 80\%$	3.3973	3.7810	3.8958	3.7081	3.4330	*3.2110*	*3.0692*	3.3
$\xi = 90\%$	3.6545	4.0939	4.2388	4.0326	3.7248	*3.4812*	*3.3104*	3.6
$\xi = 100\%$	*3.8984*	4.3951	4.5760	4.3504	4.0152	*3.7482*	*3.5488*	4.0

In Table 7.16, the values that are smaller than the corresponding data of NHERP 2009 are denoted by italic letters. In such cases, however, using the value of the NEHRP 2009 numerical damping coefficient will overestimate the effect of damping and result in a design that may be unsafe.

It is noted that Table 7.16 is obtained by using only 99 earthquake records, but it still shows that the damping coefficients are ground motion dependent. In seismic design, the local earthquake histories should be carefully examined.

As mentioned in Section 7.1, the period of a building is also a function of the number of stories of the structure. That is,

$$T = f(n) \quad (s) \tag{7.134}$$

Therefore, in order to reflect the effect of period in practical design, modification (safety) factors S_{Dn} and S_{An}, described in Equations 7.16 and 7.22, are recommended.

7.5.4.2.2 Modification with Respect to Damping Ratio

In the previous chapters, structural dynamics principles of acceleration and displacement controls were discussed. Now, their corresponding design issues are discussed.

First, since the relationship between absolute acceleration and relative displacement needs to be modified by the factor $\sqrt{1+4\xi^2}$, the damping factors for acceleration and displacement reductions, denoted as B_{acc} and B_{disp}, should be different, namely,

$$B_{acc} = \sqrt{1+4\xi^2}\, B_{disp} \tag{7.135}$$

In Figure 7.11, curves of the inverse of the numerical damping factor are plotted. The line marked with triangles is based on Table 7.15 and the solid line is based on Equation 7.128, the linear approach; it is seen that the difference is small, especially when the damping ratio becomes larger.

When the reduction effect described by Equation 7.126 based on limited numerical simulations of displacement and acceleration with the factor $\sqrt{1+4\xi^2}$ is plotted and marked by dotted and break lines, respectively, the difference is also small. This validates Equation 7.128. To write the design equations compatible with NEHRP 2009, the number of damping factors for acceleration and displacement reductions are not distinguished but only one parameter B is used. Therefore, the

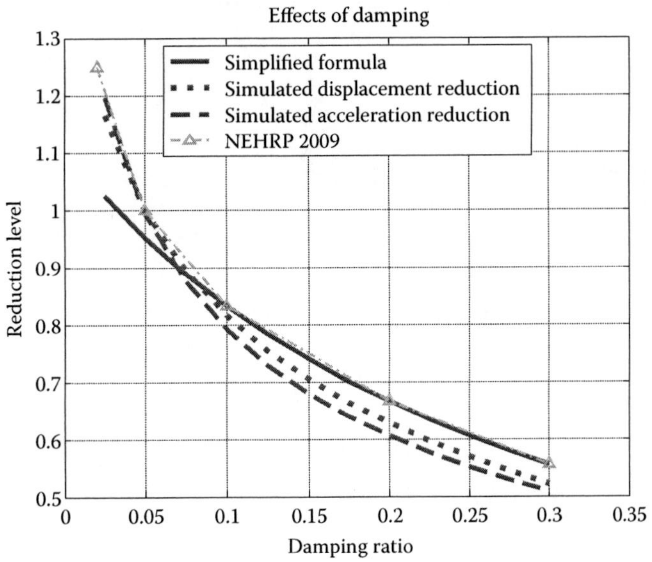

FIGURE 7.11 Effects on response reductions by damping ratios.

factor $\sqrt{1+4\xi^2}$ is directly used in the equations of the seismic response factors C_{Si} (see Equations 7.10 and 7.51).

However, from Figure 7.11, a difference in the reductions is found when compared with the reduction based on Table 7.15 and numerical simulations. It is understood that the concept of using a numerical damping factor is to realize the corresponding reduction, which is referenced at $\xi = 5\%$. As previously mentioned, a 5% damping ratio covers certain inelastic deformations; whereas when a structure remains in the elastic range, the damping ratio can be considerably smaller than 5%. To minimize the differences, the modification (or safety) factors $S_{A\xi}$ and $S_{D\xi}$ are introduced (see Equations 7.23 and 7.17). These two modifications are only valid in the range of the damping ratio between 2% and 30%.

7.5.5 Modified SRSS

In the above discussion of NEHRP 2009 on the computation of SRSS, it was mentioned that it is better to avoid using the computation with SRSS as well as CQC, as this may introduce errors.

If the peak values at different stories can be counted at an identical time, then using SRSS can be avoided. If direct modal superposition can be used, then such approximations can also be avoided.

However, in many cases, the moment when a peak response occurs cannot be identified; e.g., using the values of design spectra. In this case, to increase the accuracy, an improved CQC that considers nonproportional damping and overdamping can be used. The interested reader may refer to Song et al. (2007) for a more detailed explanation.

7.5.5.1 Absolute Acceleration

To increase the computational accuracy of the absolute acceleration, repeat the M-C-K equation once as described in Equation 3.99:

$$\mathbf{M}\ddot{\mathbf{x}}_r(t) + \mathbf{C}\dot{\mathbf{x}}_r(t) + \mathbf{K}\mathbf{x}_r(t) = -\mathbf{MJ}\ddot{x}_g(t) \tag{7.136}$$

where, to emphasize that the variables are relative, subscript r is used. Equation 7.136 can also be written as

$$\mathbf{M}\ddot{\mathbf{x}}_a(t) + \mathbf{C}\dot{\mathbf{x}}_r(t) + \mathbf{K}\mathbf{x}_r(t) = 0 \tag{7.137}$$

where the absolute acceleration $\ddot{\mathbf{x}}_a(t)$ is

$$\ddot{\mathbf{x}}_a(t) = \ddot{\mathbf{x}}_r(t) + \mathbf{J}\ddot{\mathbf{x}}_g(t) \tag{7.138}$$

When the damping force is small, that is,

$$\mathbf{C}\dot{\mathbf{x}}_r(t) = 0 \tag{7.139}$$

then

$$\mathbf{M}\ddot{\mathbf{x}}_a(t) \approx -\mathbf{K}\mathbf{x}_r(t) \tag{7.140}$$

Note that Equation 7.140 does not necessarily mean that the absolute acceleration $\ddot{\mathbf{x}}_a(t)$ and the relative displacement $\mathbf{x}_r(t)$ share common shape vectors, even if Equation 7.139 holds, unless the response contains only one mode (MSSP system).

Similarly, using spectral values, denoted as $\mathbf{a}_a$ and $\mathbf{d}_{max}$, respectively, for the responses in vector form, Equation 7.140 can be rewritten as

$$\mathbf{M}\mathbf{a}_a \approx \mathbf{K}\mathbf{d}_{max} \tag{7.141}$$

Equation 7.141 does not necessarily mean that the absolute acceleration $\mathbf{a}_a$ and the relative displacement $\mathbf{d}_{max}$ share common shape vectors. The reason $\mathbf{a}_a$ and $\mathbf{d}_{max}$ in general do not share common shape vectors is obvious, because both $\mathbf{a}_a$ and $\mathbf{d}_{max}$ result from different values of modal combinations. In this case, the equations suggested by NEHRP 2009, see Equations 7.21*D* and 7.62*D*, do not have a solid mathematical base and should be improved.

However, for each mode, the pseudo acceleration $\ddot{\mathbf{x}}_{s_i}(t)$ and the relative displacement $\mathbf{x}_{r_i}(t)$ share the same shape function, and approximately

$$\ddot{\mathbf{x}}_{s_i}(t) \approx \omega_i^2 \mathbf{x}_{r_i}(t) \tag{7.142}$$

or the amplitude of pseudo acceleration can be written as

$$\mathbf{a}_{s_i} \approx \omega_i^2 \mathbf{d}_{max_i} \tag{7.143}$$

where $\mathbf{a}_{s_i}$ and $\mathbf{d}_{max_i}$ are defined in Section 7.1.

Generally speaking, the shape function of $\mathbf{d}_{max}$ is closer to triangular function, as suggested by NERP 2009, whereas the shape function of $\mathbf{a}_a$ is closer to rectangular. Therefore, to improve the accuracy of acceleration estimation, Equations 7.62*A* and 7.103 are suggested.

7.5.5.2 Incomplete Modes

In practice, it is often impossible to include all the modes in response estimations; and furthermore, it is not necessary to do so. Therefore, the first S modes, such as the one used in Equation 7.103, etc., are often used. In so doing, a small amount of response energy will be excluded. To improve this situation, a modification factor, S_M, is needed, especially for design using MSSP models. Generally speaking, S_M can be seen as a function of the cumulated modal contribution factors, such as the modal mass ratio, $\gamma_{m\Sigma}$, that is,

$$S_M = f(\gamma_{m\Sigma}) \tag{7.144}$$

Based on limited simulations, for the MSSP model,

$$1.0 \le S_M = (0.95 \sim 1)/\gamma_{m1} \tag{7.145}$$

and for the MDOF system,

$$1.0 \le S_M = (0.95 \sim 1)/\gamma_{m\Sigma} \tag{7.146}$$

In Equations 7.145 and 7.146, more modes are taken, a smaller coefficient is used, and the smallest values equal unity.

7.6 DAMPER DESIGN CODES

In Section 7.1, the criteria to determine the need to use damping control for structural aseismic design ware discussed, mainly based on the principles of structural dynamics. In actual design, the requirements of design codes must be followed.

Damping design is not the design of dampers, but the design of structures with added damping devices. Basically, the structure itself must have the capacity to resist gravity loads and other types of normal loads. In addition, the structure should withstand lateral forces due to earthquake ground accelerations. The design requirements are provided in design codes, such as NEHRP 2009. In the following discussion, some basic requirements of NEHRP 2009 for the design of structures with supplemental damping are briefly summarized. Major differences between the equations in NEHRP 2009 and those given in Section 7.1 and/or additional provisions are also briefly discussed. Note that the language used in NEHRP 2009 is not exactly repeated or followed closely in this book.

In NEHRP 2009, for linear systems, the damper design is based on the response spectrum. The fundamental control parameter is the base shear. The design procedure mentioned in the previous sections of this chapter exactly follows the logic of NEHRP 2009. It is known that the expected spectral value is determined by both the period T and the numerical damping ratio B. Since the system is linear, not much can be done on the period. In order to achieve the desired base shear, the damping is increased, which defines the corresponding design parameters of the dampers.

As a complete aseismic design, NEHRP 2009 starts with the Seismic Use Group, followed by the Seismic Design Category. In the previous sections, however, due to space limitations in this book, the focus was on specific damper design only. Thus, the completeness of the design procedure is one of the major differences between this book and NEHRP 2009. For practical aseismic design, engineers should follow NEHRP 2009.

Calculating the period, especially for MDOF systems, however, is another major difference. In NEHRP 2009, the periods are functions of floor displacement. These displacements, on the other hand, are also functions of the periods. Therefore, to obtain the period in order to use the design spectrum, some iterative computations are needed when exactly following the NEHRP 2009 design logic. Since accurate convergence of such iterations cannot be guaranteed, approaches are suggested in the previous section of this book.

In addition, NEHRP 2009 uses a linear shift of the fundamental mode shape to calculate the residual mode shape and approximate the influences of higher modes. In the previous section, an alternative method to find higher mode shapes was suggested, which is another major difference.

7.7 BRIEF SUMMARY OF DAMPING DESIGN OF LINEAR SYSTEMS

7.7.1 Major Step (1) Decision Making

The first major step is to decide on whether or not to use supplemental damping. This is graphically shown in Figure 7.12. This step is design phase I.

First, the use of damping control depends on two conditions: (A) the original damping ratio of the base structure is sufficiently small, which is the main reason to add supplemental damping; and (B) the fundamental period of the base structure is larger than T_s, in which case, increasing stiffness will magnify the seismic load so that damping control is more effective. Condition B is not always necessary; detailed modeling, such as through finite element analysis, may be needed to make a final decision.

Second, it is necessary to determine if the structure does not allow large ductile deformation and if the damping devices can be modeled as linear viscous; otherwise, nonlinear damping design must be considered if supplemental damping is to be used, which is discussed in Chapter 8.

Third, if the total damping ratio is sufficiently small and the fundamental mode contains the main part of the vibration energy, or if only a quick initial estimation is needed without detailed calculation; the MSSP model can be marked as type (1) in Figure 7.12. Note that design based on an SDOF model is very similar to that through the MSSP model. In this chapter, SDOF systems are not discussed due to space limitations.

If more detailed design is needed, more modes must be involved. However, the supplemental damping is proportional to stiffness and/or mass; in other words, if the mass, damping, and stiffness irregularities are sufficiently small, the model of a proportionally damped system can be used, marked as type (2) in Figure 7.12. Note that if both the original and the supplemental dampings are not very high, this method is often acceptable for linear damping design.

In certain cases, more sophisticated design is necessary, e.g., careful reevaluation of the initial design or larger damping requirements and/or both damping and mass/stiffness irregularities are large. The model of generally damped MDOF systems should then be considered, marked as type (3) in Figure 7.12.

7.7.2 Major Step (2) Modal Analysis

The second major step is design phase II, which consists of steps (2) through (4), which are for structural response estimations. Figure 7.13 shows all three design phases and six design steps.

MDOF systems are the most typical structures in practice. Therefore, in addition to the period and damping ratio of the fundamental mode of structures, the mode shape as well as other model parameters of high modes are needed, which are obtained in this major step. In Figure 7.13, the major design steps are shown graphically as flowcharts.

For MSSP systems, the fundamental period T_1 and mode shape $\boldsymbol{p}_1$, as well as modal parameter Γ_1 need to be calculated. The original damping ratio of the base structure ξ_0 is also needed, which usually cannot be accurately calculated through initial modeling, such as finite element modeling. Thus, the damping ratio of the base structure, ξ_0, is often roughly estimated by experience. For a more accurate response estimation, more sophisticated methods are needed, except in the case of eigen-decompositions for simplified MSSP design, where method *C* mentioned in Section 7.1 can be acceptable. Therefore, only the mass matrix **M** may be needed.

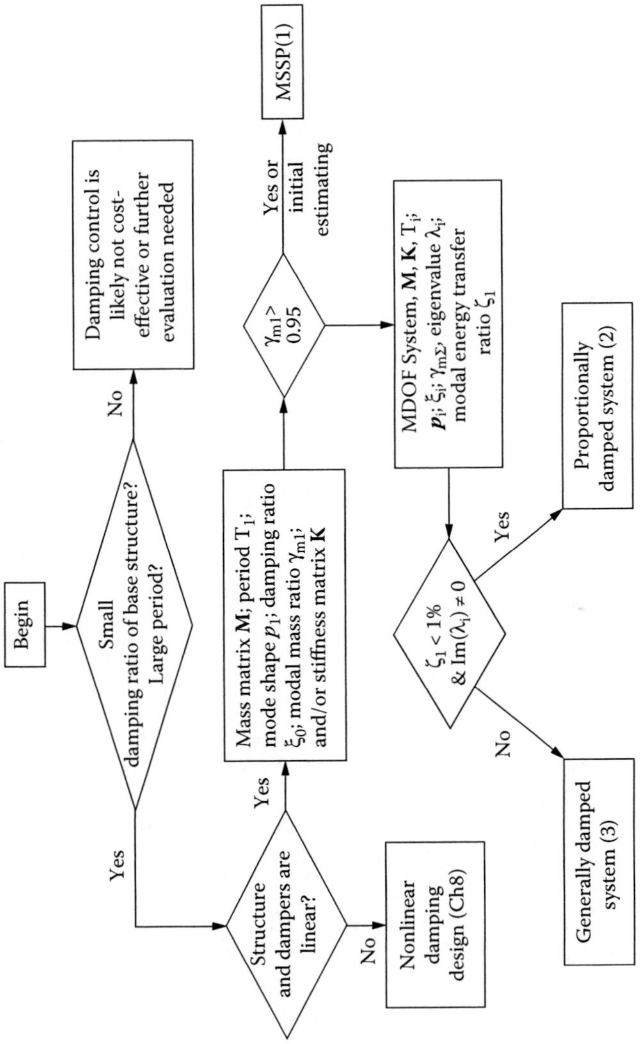

FIGURE 7.12 Phase I ((step (1)): decision making.

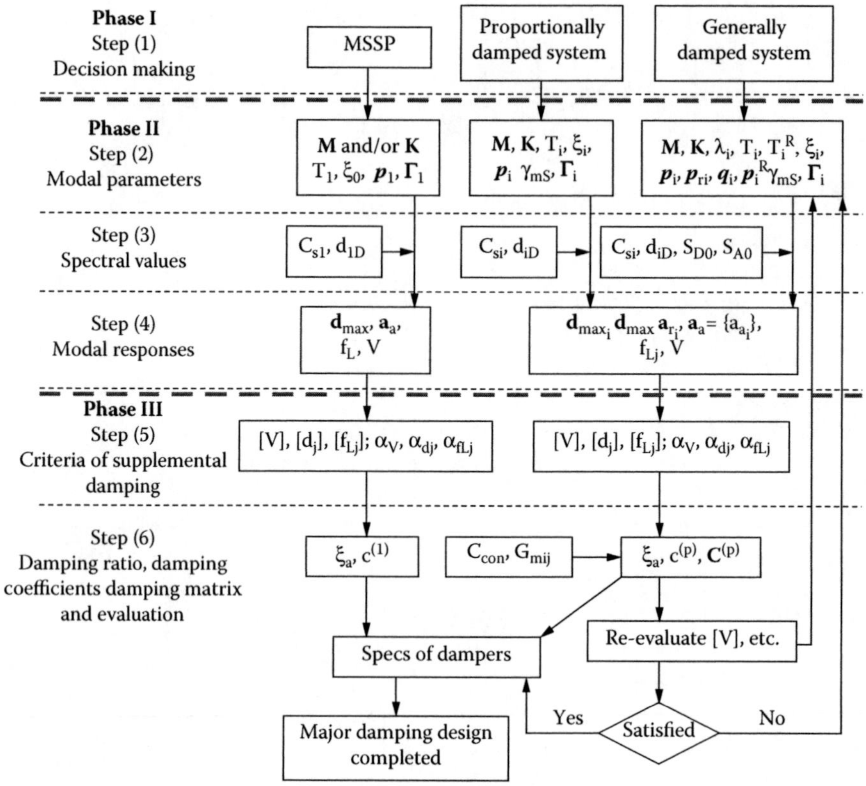

FIGURE 7.13 Design phases and major steps.

For proportionally damped MDOF systems, the modal parameters must be further calculated more accurately than for the MSSP model and more modes must be accounted for, such as the period, damping ratio, and mode shape of the ith mode, T_i, ξ_i, and $\boldsymbol{p}_i$, as well as modal participation factor Γ_i; and, in addition to the period (frequency), damping ratio, and mode shape for each mode, the modal contribution indicator, such as modal mass ratio γ_{mS}, is also needed. Thus, not only the mass matrix **M**, but also the stiffness matrix **K** is needed. Note that to use the methods suggested by NEHRP 2009, several extra equations, such as the modal effective mass m_{eff1} and m_{effi} and/or residual modal period T_R, modal shape p_{Rj}, and modal participation factor Γ_R, are needed.

For a generally damped system, which further involves damping matrix **C**, the mode parameters of the corresponding proportionally damped system, whose period and mode shape are now denoted by $T_i^{(p)}$ and $\mathbf{p}_{ri}$, need to be computed, as well as those from the generally damped system. Therefore, different sets of period, damping ratio, and mode shape $T_i^{(n)}$ (or T_i), ξ_i, and $\boldsymbol{p}_i$, as well as ζ_l and γ_{mS} are calculated. In addition, the eigenvalue λ_i, accompanist vector $\boldsymbol{q}_{i.}$, and the modal energy transfer ratio ζ_l must be obtained. Note that this time, the period may include the pseudo period T_j^R; and to find Γ_i, the parameters $\boldsymbol{\Phi}_i$ and $\boldsymbol{\Psi}_i$ need to be calculated as well.

7.7.3 Major Step (3) Spectral Values

The third major step is to calculate the spectral values, the modal seismic response factor C_{si}, and modal displacement d_{iD}. This step is familiar to earthquake engineers and will not be explained in detail. However, if overdamped subsystems are involved, the response spectra S_{D0}, S_{A0} is needed, which must be determined according to local seismic zones.

7.7.4 Major Step (4) Model Responses

In the fourth major step, the modal responses, such as modal displacement, modal acceleration, and modal lateral forces, are calculated. Other quantities such as floor drift, base shear, etc., may also be needed. The modal displacement and acceleration are the basic design parameters from which other responses can be calculated. Both the modal displacement $\mathbf{d}_{\max_i}$ and the acceleration $\mathbf{a}_{a_i}$ or $\mathbf{a}_{s_i}$ are productions of A: spectral values such as d_{iD} and C_{si}; B: modal participation factor Γ_i; C: mode shape $\boldsymbol{p}_i$; and the recommended response modification (safety) factors S_M, S_{Dn}, $S_{D\xi}$, S_{An}, and $S_{A\xi}$. Note that NEHRP 2009 does not require these factors in method *D*.

It is not necessary to accurately decide the exact numbers of modes. In general, they can be roughly estimated and several more modal responses are then added.

7.7.5 Major Step (5) Criteria of Supplemental Damping

The fifth major step is design phase III, which consists of steps (5) and (6). This is the last but often not the final design phase because iteration is often needed to determine if the structural responses are successfully controlled. Figure 7.14 shows the detailed steps for design phase III.

In step (5) two tasks are carried out. First the total seismic responses, such base shear V, floor displacement d_{ji}, lateral force f_{Lj}, etc., are calculated through a combination of modal responses obtained in step (4). Although the SRSS method is used, care must be taken to deal with the randomness introduced by seismic excitations. Generally speaking, SRSS is avoided whenever possible; and for absolute acceleration, modifications are used.

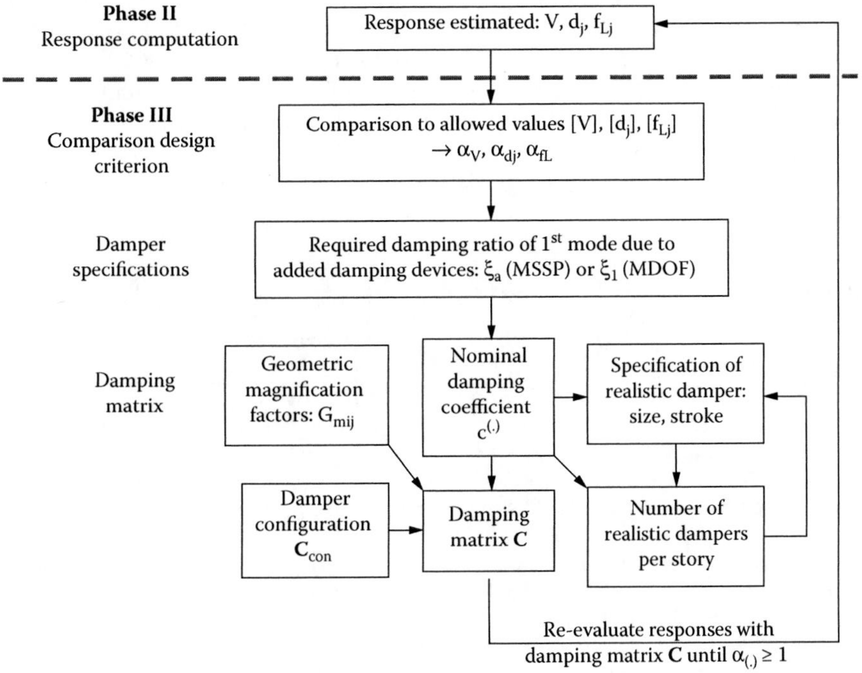

FIGURE 7.14 Design phase III, damping design criteria, damper specifications, and damping matrix.

Secondly, the estimated seismic responses are compiled with the allowed values, such as allowed base shear [V], allowed floor displacement [d_{ji}], allowed lateral force [f_{Lj}], etc. The calculated differences $\alpha_{(\cdot)}$ are the criteria used to determine the proper amount of supplemental damping.

7.7.6 Major Step (6) Design Damping Ratio, Specifications of Damping Devices, Reevaluation of Total Design

In this major step, based on the factors $\alpha_{(\bullet)}$, the design damping ratio ξ_a is first calculated. By using the damping ratio of the systems, ξ_a, the nominal damping coefficient $c^{(\bullet)}$ is further calculated, from which the specifications of the damping devices can be calculated, such as damping coefficient, size, and stroke of dampers, etc. For MSSP models, the design is now complete. However, for MDOF systems, more work may be needed.

That is, parallel to a system's damping ratio ξ_a as well as ξ_{design}, the damper configuration needs to be considered, which will further provide the geometrical magnification factor G_{mij}. With information from these two rounds of computation, the damping matrix $C^{(\cdot)}$ can be calculated, which includes the coefficient matrix that contributes to the original damping ratio ξ_0 of the base structure and the one that describes the added dampers. The reason to have the damping matrix is to reevaluate the total damping design by reestimating the seismic responses of the newly formed $\mathbf{M}$-$\mathbf{C}^{(\cdot)}$-$\mathbf{K}$ system. This task can be carried out through proportionally damped MDOF systems. However, using a proportionally damped model can be somewhat inaccurate. Therefore, the evaluation is more appropriate when the generally damped model is used.

In addition to the above six major design steps, several miscellaneous yet important issues are also considered in this chapter. First, the supporting stiffness of a damper is often not infinite, so it needs to be designed, or the effectiveness of the supplemental damping will be undermined.

The design safety and safety factors, in many cases, need to be carefully considered due to the randomness of seismic excitations and structure-device uncertainty. One of the concerns comes from the nonlinearity of dampers as well as structures. The linear approach used in this chapter is simple but sometimes not accurate. In fact, most dampers are nonlinear and the assumption of linear viscous damping is based on certain linearization computations. Linearization may introduce unsafe errors, such as directly using the approach of Timoshenko damping. Thus, a corresponding safety factor may be used to improve the situation. Another commonly accepted safety treatment is to account for the quality of commercially available damping devices. In addition, several miscellaneous safety issues were introduced, and most of them need to be further studied and determined.

REFERENCES

Avallone, E.A. and Baumeister III, H. 1996. *Mark's Standard Handbook for Mechanical Engineering*. 10th ed. NY: McGraw-Hill.

Building Seismic Safety Council (BSSC). 2000. *NEHRP Recommended Provisions for Seismic Regulation for New Buildings and Other Structures, 2000 Edition*, Report Nos. FEMA 368, Federal Emergency Management Agency, Washington, DC.

———. 2003. *NEHRP Recommended Provisions for Seismic Regulation for New Buildings and Other Structures, 2003 Edition*, Report Nos. FEMA 450, Federal Emergency Management Agency, Washington, DC.

———. 2009. *NEHRP Recommended Seismic Provisions for New Buildings and Other Structures, 2009 Edition*, Report Nos. FEMA P-750, Federal Emergency Management Agency, Washington, DC.

Christopoulos, C. and Fillatrault, A. 2006. *Principles and Passive Supplemental Damping and Seismic Isolation*. Pavia: IUASS Press.

Liang, Z. and Lee, G.C. 1991. Representation of damping matrix. *Journal of Engineering Mechanics* 117 (5): 1005–20.

Newmark, N.M. and Hall, W.J. 1982. *Earthquake Spectra and Design*. Oakland, CA: Earthquake Engineering Research Institute.

Ramirez, O.M., Constantinou, M.C., Kircher, C.A., Whittaker, A.S., Johnson, M.W., Gomez, J.D., and Chrysostomou, C.Z. 2000. Development and evaluation of simplified procedures for analysis and design of buildings with passive energy dissipation systems, MCEER Report 00-0010 Revision 1, MCEER, University at Buffalo, Buffalo, NY.

Song, J., Chu, Y.-L., Liang, Z., and Lee, G.C. 2007. Estimation of peak relative velocity and peak absolute acceleration of linear SDOF systems. *Earthquake Engineering and Engineering Vibration* 6 (1): 1–10.

8 Nonlinear Damping

Similar to linear damping devices, major decisions for nonlinear damping are based on the anticipated dynamic behavior of a given system, associated damping ratios and periods. In addition, specific types of damping are chosen to satisfy the basic requirements of structural response, as well as types of construction.

In Chapter 7, the design procedures for linear structures with linear damping were discussed. Although the linear assumptions cannot easily be satisfied, when both the amount of damping and the design ductility of the structures are small, the linear approach can provide simple yet acceptable designs. Otherwise, nonlinear design must be considered, which is the focus of this chapter.

8.1 OVERVIEW OF DESIGN APPROACHES

As with the linear system approaches, there are three major phases for nonlinear damping design: the choice of damping type and the corresponding model, the response estimation, and the damping specification.

8.1.1 General Description

For spectra-based deisgn of nonlinear systems, the total structural response is also treated as a product of the response of a single-degree-of-freedom (SDOF) system and a shape function. The SDOF system, often referred to as a "substitute" system, is one of the modes of a linear system. The approach evolves from the first mode of the linear system due to the addition of nonlinear damping, and/or due to inelastic deformation of the system. If only nonlinear damping is used and the base structure remains linear, higher "effective" modes can be considered. In this case, more than one SDOF system needs to be considered, along with responses and the corresponding shape functions.

To determine the response of the substitute SDOF system, the first approach is to linearize the nonlinear damping through the concept of effective damping and stiffness. By using Timoshenko damping and/or force-based effective damping, the effective damping ratio, ξ_{effi}, and period, T_{effi}, for the ith mode of a multi-degree-of-freedom (MDOF) system can be obtained. In Chapter 6, the corresponding formulas were discussed. After the effective modal damping ratio and period are calculated, the linear damping design can be achieved exactly as discussed in Chapter 7. To simplify the design procedure, the nonlinear damping is classified into bilinear and sublinear dampings, which are discussed in Sections 8.2 and 8.3, respectively.

A second approach to determine the response of the substitute SDOF system is based on nonlinear design spectra, which covers the effects on an elastic structure when nonlinear damping is installed. In this case, while the structure remains mainly elastic, its damping and stiffness are altered. These cases should be addressed when the amount of nonlinear damping added to a structure becomes sufficiently high. This second approach is considered for sublinear damping within a linear structure in Section 8.4.

The third approach deals with supplemental damping installed in an inelastic structure. This also employs the nonlinear design spectra and is described in Section 8.5.

In each case, after determining the response(s) of SDOF system(s), the deformation shape functions need to be determined to distribute the response to each story of the building. As mentioned in Chapter 5, when bilinear damping is used, its damping adaptability is zero, so the damping force is close to constant. The deformation shape functions will not be significantly affected by the damping force if the supplemental damping is comparatively small. Thus, the same shape functions are used to estimate the seismic response with different amounts of damping.

Sublinear damping has nonzero damping adaptability. The damping force will increase when the displacement increases. This makes the shape function to have a larger variation than bilinear damping. This is the main reason why the damping design is classified into bilinear and sublinear.

Note that the damping adaptability of linear viscoelastic damping is unity, but its energy dissipation loop is close to bilinear. The corresponding shape function will change because its damping adaptability is unity. However, the bilinear model can be used to design this type of damping. On the other hand, most viscoelastic damping devices are not suitable for aseismic building because they often do not allow sufficient damper displacement. Therefore, only bilinear and sublinear dampings will be discussed.

8.1.1.1 Condition of Using Supplemental Damping

Similar to the situation for a linear system, the first step of design is to evaluate the base structure to determine whether it has a small original damping ratio and whether the fundamental period indicates that using damping will be beneficial.

8.1.1.2 Amount of Damping

Once the need to use supplemental damping is established, the seismic response can be estimated and the criteria can be checked in a similar way as in the evaluation of linear systems. Detailed steps are described by Equations 7.25 through 7.29.

8.1.1.3 Type of Damping Devices

The third major task is the selection of the type of damping devices.

1. The advantages and disadvantages of linear damping as well as its limitations have been discussed in Chapter 7.
2. To decide if bilinear damping should be used, two criteria are considered: damping force and cost. If the structure is designed to have large ductility, dampers with larger damping adaptability will eventually provide a great amount of damping force, whereas bilinear damping will not; thus, it can be a better choice. In addition, many types of bilinear dampers are comparatively inexpensive. Thus, for construction projects with limited budgets, bilinear damping may be a better selection. A more detailed discussion on the pros and cons of bilinear damping is presented in Section 8.2.1.
3. The decision to use sublinear damping can primarily be based on the required quality control of the devices. Sublinear damping is often provided by hydraulic devices, whose design parameters, such as working forces and damping coefficient, are usually more accurately guaranteed by experienced vendors. A more detailed discussion on the pros and cons of sublinear dampers is presented in Section 8.3.1.

8.1.1.4 Models for Reevaluation of Structural Responses

1. If the design damping ratio is sufficiently small, the linear proportionally damped model can be used. This was discussed in Chapter 7.
2. If the design damping ratio is greater than 10%, but more and less smaller than 15%, the linear nonproportionally damped model can be used, which was also discussed in Chapter 7.
3. If the damping ratio is large, especially greater than 15%, and a bilinear type of damper is used, a bilinear damping model with a linear base structure should be used. This is discussed in Section 8.2, where the structure is designed to remain in the elastic range.
4. If the damping ratio is large, especially greater than 15%, and a sublinear type of damper is used, the sublinear damping model with a linear base structure should be considered. This is discussed in Sections 8.3 and 8.4, where again the structure is designed to remain in the elastic range.

5. If the damping ratio is large, especially greater than 15%, and the structure is designed to allow large inelastic deformations, a nonlinear model should be used. This is discussed in Section 8.5 for the case of bilinear damping.

As mentioned earlier, the above are only the cases discussed in this book as examples to deal with nonlinear damping design. Many other possible cases are not addressed for lack of sufficient knowledge at present.

8.1.1.5 Structural Ductility

In building seismic design, the principle of life safety or noncollapse under strong earthquakes is typically followed. Thus, in strong earthquake zones, large ductility is a standard requirement.

While design with large ductility in general is beyond the scope of this book, it is fair to say that the inelastic dynamic response of structures due to large earthquake ground motions is still a current research frontier (Aref et al. 2001, Whittaker et al. 2003, Filiatrault et al. 2004, Grant et al. 2005, Deierlein et al. 2010, and Bruneau 2011). Here, the only concern is for the design of a ductile structures with added damping devices.

8.1.2 Types of Damping

8.1.2.1 Bilinear Damping

The energy dissipation loop of a bilinear damper is close to a rectangular shape, which is often the most effective damping, as discussed in Chapter 5. However, its damping adaptability is close to zero and it only works well if the dynamic range of the vibration is small. Therefore, bilinear damping is more suitable for applications when the vibration amplitude is predictable and the dynamic range is not very large. In addition, bilinear damping devices require more precise computation, especially for the selection of the strength characteristic.

The use of bilinear dampers is often cost-effective, because these damping devices are relatively inexpensive. Normally, the price per bilinear damper can be one-third to one-fifth that of hydraulic dampers. However, the cost of installation is virtually the same for all types of dampers, as are business interruption costs.

Some bilinear dampers can deliver large displacement, such as most friction dampers. A metallic damper, which is also bilinear, cannot satisfy very large damper displacement. Viscoelastic devices, which can provide somewhat good bilinear damping, have a more limited capability for large damper displacement.

Some bilinear dampers can be relatively small. For example, a special friction damper utilizes a V-shaped friction surface to magnify the equivalent friction coefficient to 5–10 times and its friction force remains stable (Lee et al. 2006). However, other types of friction dampers with a planar surface do not offer this advantage. Metallic dampers, on the other hand, typically occupy more space.

Bilinear dampers typically have an excellent capability to resist forces in the lateral direction. This is because these are often made of materials with good strength, such as steel.

Another notable advantage of bilinear dampers is the capability to suppress the first extra-large peak response. Often, strong earthquake ground excitations reach the structure with an increased peak after a few seconds, which introduces a large response of the structure. Generally speaking, using pure damping does not provide much help in limiting the response due to a large peak excitation that occurs during the initial records of the ground motions; the stiffness of the structure will need to be increased to limit such responses. Pure passive sublinear dampers, which typically use hydraulic fluid, cannot contribute to the effective stiffness of the structure. Bilinear dampers, on the other hand, may provide a certain amount of stiffness, depending on the strength characteristic design. Therefore, in earthquake zones where sudden ground motion pulses may occur, bilinear damping is a better choice.

Bilinear damping can provide additional stiffness before yielding; that is, it contributes to the stiffness to help reduce structural displacement during small levels of earthquakes, winds, and traffic-induced vibration. However, during strong earthquakes, the contribution to the total yielding stiffness by bilinear damping is often negligible.

Another main advantage of bilinear damping is that its damping force will not increase significantly when the structure becomes inelastic with large ductility. This is discussed in Section 8.5.

The major disadvantage, besides the near-zero damping adaptability, is that most bilinear dampers do not have the capability of self-centering. After a strong earthquake with a large permanent displacement, the dampers must be readjusted.

The other major disadvantage of bilinear dampers is that they are not as stable as hydraulic devices. The friction surface can wear out and the friction force will drop accordingly, or they can develop adhesion after many years in a stationary state. For example, the moving surface of a friction damper can be locked due to corrosion. The yielding metallic dampers can have significant low-cycle fatigue problems, so that they would need to be replaced after a strong earthquake.

Briefly, bilinear dampers, especially friction dampers, are more suitable for structures that are located near strong seismic zones with a history of long period ground motion and where the primary budget for the structures is limited. Bilinear dampers are not as suitable for structures subjected to frequent dynamic loading, such as heavy wind loads, or regions with frequent seismic activity.

8.1.2.2 Sublinear Damping

The major advantage of using sublinear dampers is their effectiveness in vibration reduction when compared to other types of dampers. Due to the uncertainty of the excitation amplitude of the seismic ground motion, the dynamic range of the earthquake-induced vibration should be considered as large. Sublinear damping can reduce vibration more effectively than bilinear damping, though its damping adaptability is still less than that of linear damping. Therefore, if the predicted dynamic range is not extremely high, which is often true in earthquake-induced vibration, it is better to use sublinear damping to effectively reduce the structural responses. The second advantage of sublinear damping is that it has a self-centering capability after the earthquake. The third advantage is its durability. Sublinear dampers can be used for multiple hazards, such as earthquakes, wind, and traffic-induced vibration, as well as shock absorption.

The main disadvantage of sublinear dampers is the cost, which can be significantly more expensive than bilinear dampers. However, most sublinear dampers are reusable after strong earthquakes, whereas many bilinear dampers have a limited lifespan. Therefore, the cost of sublinear dampers should be recalculated considering the replacement cost of bilinear dampers. Furthermore, as mentioned previously, installation and business interruption costs are comparable for all damper types and often dominate the overall costs.

8.1.2.3 Basic Differences between Bilinear and Sublinear Damping

In terms of damping design, the major differences between bilinear and sublinear dampings may be summarized by the following three items.

First, bilinear damping will apply additional lateral stiffness to the structure, whereas sublinear damping does not contribute extra stiffness. Therefore, in bilinear damping design, the total stiffness matrix needs to be modified. The periods of the system will also be altered. These effects are not significant for sublinear systems.

Secondly, if the amount of supplemental damping is not sufficiently large, i.e., the resulting design damping ratio is less than 20%, i.e.,

$$\xi_{\text{design}} \approx \xi_0 + \xi_a < 20\% \tag{8.1}$$

then for bilinear damping, the mode shape of the base structure can be used. Here, ξ_0 is the damping ratio of the original base structure and ξ_a is the additional damping ratio contributed by the supplemental dampers. Thus, the mode shapes of systems with sublinear damping will not significantly

change. Since the base structure can usually be modeled as a proportionally damped system, the mode shapes are real valued and much easier to obtain and handle.
On the other hand, with sublinear damping, especially when the damping adaptability is

$$f_{\text{adp}} > 0.4 \sim 0.5 \tag{8.2}$$

it is better not to use the normal mode shape as the shape functions. Consequently, this will increase the computational complexity for sublinear systems.

Thirdly, to achieve more accurate response estimation, more effective modes need to be employed for sublinear systems, whereas for structures with bilinear damping, fewer "modes" are used. In many cases, it is possible to only use the first effective mode in bilinear damping design.

8.1.2.4 Other Types of Dampers

In addition to linear, bilinear, and sublinear dampers, there are other types of damping devices, such as the viscoelastic (VE) dampers. Since most aseismic designs for structures need to consider relatively large damper displacement, VE dampers are not suitable due to their limitations in providing large displacement.

8.1.3 Design Procedures

Compared to the linear systems discussed in Chapter 7, nonlinear damping design requires more careful selection of the dampers and thus has a more complicated design procedure. In Table 8.1, the basic design approaches and corresponding sections where they are discussed are listed. In the following paragraph, the criteria for choosing the approaches listed in Table 8.1 are considered.

8.1.3.1 Equivalent Linear Systems

If the design damping ratio is not sufficiently high, i.e.,

$$\xi_{\text{design}} < 15\% \sim 30\% \tag{8.3}$$

then the method of equivalent linear systems can be chosen. Namely, an equivalent linear SDOF system or systems are found first and the corresponding effective period and effective damping ratio are calculated. Then, the effective mode shape functions are found and the "modal" displacement and acceleration are distributed to each story using shape functions. This method is discussed in Sections 8.2 and 8.3.

8.1.3.2 Nonlinear Response Spectra

When the damping ratio is large, the equivalent linear system approach may introduce significant errors due to nonproportional damping. Alternatively, a nonlinear response spectra approach can be adopted.

TABLE 8.1
Design Approaches for Nonlinear Systems

	Nonlinear Damping Type	
Design Approach	**Bilinear**	**Sublinear**
Equivalent linear system	Linear structure (8-2)	Linear structure (8-3)
Nonlinear response spectra	Nonlinear structure (8-5)	Linear structure (8-4)

On the other hand, for a significant amount of bilinear damping, the entire system becomes inelastic while the base structure remains elastic. For example, suppose bilinear dampers with a yielding ratio of a = 5, yielding ductility of $\mu = 4$, and assume that the additional stiffness contributing to the base structure is twice the yielding stiffness of the damper. When a 35% damping ratio is designed, the supplemental bilinear damper can contribute about 10% more stiffness, which affects the elastic behavior of the entire system. In this case, to estimate the response more accurately, the nonlinear response spectra are used, which is discussed in Section 8.5.

8.2 EQUIVALENT LINEAR SYSTEMS APPROACH WITH BILINEAR DAMPERS

When the effective damping ratio is relatively small, the design procedure for bilinear damper is identical to that for linear systems. The conceptual decision making, the response estimation, and the damper specification are the basic design phases. For response estimation, a linearized damping ratio, as well as an effective period, is considered so that the design spectra can be adopted. Bilinear devices have zero damping adaptability, which simplifies the design procedure so that for MDOF systems, the shape function obtained through the base structure can be used. Since a bilinear damping system is essentially nonlinear, special considerations must be given when a large amount of damping exists.

8.2.1 General Description

8.2.1.1 Selection of Design Models

In Section 8.1, the main approach for selecting the type of dampers, such as bilinear or sublinear dampers, was discussed. Suppose a bilinear damper is chosen to be used. The model and method for response estimations must now be specified. In addition, more detailed consideration must be given to some practical issues as well as the damper specifications.

8.2.1.2 Response Estimation and First Round of Damper Design

8.2.1.2.1 Estimation of Seismic Response of Original Structure

Estimation of the seismic response of the original base structure is carried out first to determine the amount of damping, which is the same approach used for linear systems. In Chapter 7, the procedure for response estimation was discussed.

On completion of this task, the period T_i and mode shape $\boldsymbol{p}_i$ of the base structure are obtained. However, unlike for a linear system, in bilinear damping design, the period often needs to be recalculated, which is discussed after the damper specifications are selected.

8.2.1.2.2 Determination of Damping Ratio and Damping Configuration

After the initial response estimation, the design damping ratio is determined as shown in Equations 7.29 and 7.75.

The damping configurations also need to be determined so that the configuration matrix $\mathbf{C}_{con}$ and geometric magnification factors G_{mj} are available.

8.2.1.2.3 Damper Specifications

Different from linear damping, which is defined by a damping coefficient, a bilinear damper is identified by characteristic strength q_d, unloading stiffness k_u, yielding stiffness k_d, and the yielding ratio a. Among these five parameters, only three are independent. For example, yielding displacement, yielding ratio, and unloading stiffness are often used to characterize a bilinear damper. These parameters should be provided in the specification sheets from the vendor.

8.2.1.2.3.1 Required Damping Ratio When the supplemental damping is not very large, the Timoskenko damping approach can be used. That is, the effective damping of the first mode of the equivalent linear system can be used as the damping ratio for a structure with bilinear dampers. This was introduced in Chapter 5 and is repeated as follows:

$$\xi_{\mathrm{eff}_i} = \frac{2q_d \sum_{j=1}^{L} \left(x_{ij} - d_y\right)}{\pi \mathbf{d}_{\max_i}^T \mathbf{K} \mathbf{d}_{\max_i}} \qquad i = 1,2,\ldots,S \tag{8.4a}$$

where S is the number of modes considered. If only the multi-story-single-period (MSSP) system is considered, then in Equation 8.4a, i = S = 1.

Note that for nonlinear damping design, the Penzien constant, p_c, often needs to be considered for the first effective mode. Then, Equation 8.4a can be rewritten as

$$\xi_{\mathrm{eff}_1} = \frac{2q_d \sum_{j=1}^{L} \left(p_c\, x_{1j} - d_y\right)}{\pi p_c^2 \mathbf{d}_{\max_1}^T \mathbf{K} \mathbf{d}_{\max_1}} \tag{8.4b}$$

where x_{ij} and $\mathbf{d}_{\max_i}$ are the maximum relative displacement of the ith effective modes between the jth and $(j-1)^{th}$ ends of the corresponding damper and the floor displacement, respectively. Meanwhile, d_y is the yielding displacement of the bilinear damping; and L is the number of nominal dampers.

Furthermore,

$$\mathbf{d}_{\max_i} = \Gamma_i\, d_{iD}\, \boldsymbol{p}_i = \begin{Bmatrix} d_{1i} \\ d_{2i} \\ \ldots \\ d_{ni} \end{Bmatrix} \quad (m) \tag{8.5}$$

That is, the maximum relative displacement is now computed without safety factor *S* (see Equation 7.14*D*). However, when only the first effective modal response is included, the safety factor should be considered.

Note that in Equation 8.5, $\boldsymbol{p}_i$ is the ith mode shape of the original structure, which is assumed to be linear. As previously defined,

$$p_{ni} = 1$$

For the sake of design simplicity, for most bilinear damping design, the first effective mode is the focus and the MSSP approach is often used. In this special case, i = 1, and in Equation 8.4b, the number of modes, S, is equal to 1. In a later discussion, the feasibility of using S = 1 is explained.

To calculate the effective damping ratio $\xi_{\mathrm{eff}i}$, the maximum displacements x_{ij} at each of the jth stories are needed. However, without knowing the damping ratio, x_{ij} cannot be determined. Here, an iterative approach may be used.

To avoid iteration, the method similar to that for the SDOF systems is used. In order to carry out this approach, Equation 8.4b is rewritten as

$$\xi_{\text{eff}_i} = \frac{2q_d\left\{\mathbf{J}^T\left[(\boldsymbol{C}_{\text{con}}\boldsymbol{q}_c)\bullet(\mathbf{R}_s\,\boldsymbol{p}_i)\right]\,\Gamma_i d_{iD} - \mathbf{J}^T\boldsymbol{q}_c d_y/p_c\right\}}{\pi p_c\,\omega_{\text{effi}}^2\,m_{\text{effi}}\,d_{iD}^2}$$

$$= \frac{q_d/p_c\left(c_{\text{con}_i}\Gamma_i\,d_{iD} - g_y d_y\right)T_{\text{effi}}^2}{2\pi^3\,m_{\text{effi}}\,d_{iD}^2} \tag{8.6}$$

where $\boldsymbol{C}_{\text{con}}$ is a damper configuration matrix, in which the geometric magnification factors are considered. (Note that $\boldsymbol{C}_{\text{con}}$ is different from the configuration matrix $\mathbf{C}_{\text{con}}$ mentioned in Chapter 7.) If all the dampers are installed in only the adjacent levels,

$$\boldsymbol{C}_{\text{con}} = \text{diag}\left(\left[G_{mj}^2\,h_j\right]\right) \tag{8.7}$$

where the Heaviside function, h_j, used as a locator and is defined as

$$h_j = \begin{cases} 1, & \text{damper installed in jth story} \\ 0, & \text{no damper installed} \end{cases} \tag{8.8}$$

For example, suppose all the geometric magnification factors are identical and denoted as G_m, then $\boldsymbol{C}_{\text{con}}$ can be seen as shown in Example 8.1. In Equation 8.6, $\boldsymbol{q}_c$ is the damper configuration vector, whose j[th] element is equal to h_j. Also in Equation 8.6, $\mathbf{R}_s$ is a square matrix, which transfers the relative displacement vector $\boldsymbol{p}_i$ into floor drift. That is,

$$\mathbf{R}_s = \begin{bmatrix} 1 & 0 & 0 & \cdots & 0 \\ -1 & 1 & 0 & \cdots & 0 \\ 0 & -1 & 1 & \cdots & 0 \\ & & & \cdots & \\ & & & -1 & 1 \end{bmatrix} \tag{8.9}$$

Furthermore, in Equation 8.6,

$$c_{\text{con}_i} = \mathbf{J}^T\boldsymbol{C}_{\text{con}}\,\mathbf{R}_s\boldsymbol{p}_i \tag{8.10}$$

$$g_y = \mathbf{J}^T\boldsymbol{q}_c/p_c \tag{8.11}$$

and, in Equation 8.6, Γ_i is the modal participation factor, repeated as

$$\Gamma_i = \frac{\boldsymbol{p}_i^T\mathbf{MJ}}{\boldsymbol{p}_i^T\mathbf{M}\boldsymbol{p}_i}$$

and the effective modal mass is repeated as

$$m_{\text{effi}} = \frac{\left(\boldsymbol{p}_i^T\mathbf{MJ}\right)^2}{\boldsymbol{p}_i^T\mathbf{M}\boldsymbol{p}_i} \tag{8.12}$$

Example 8.1

A four-story building structure, with one nominal damper installed in the first and second stories only (see Figure 8.1), is used as an example. The damper configuration matrix and vector are, respectively, given as

$$\boldsymbol{C}_{\text{con}} = \begin{bmatrix} G_m^2 & 0 & 0 & 0 \\ 0 & G_m^2 & 0 & 0 \\ 0 & 0 & 0 & 0 \\ 0 & 0 & 0 & 0 \end{bmatrix} \quad \text{and} \quad \boldsymbol{q}_c = \begin{Bmatrix} 1 \\ 1 \\ 0 \\ 0 \end{Bmatrix}$$

8.2.1.2.3.2 Direct Computation of Displacement Note that Equation 8.6 can be rewritten as a quadratic equation in ξ_{effi}. Thus, by using these notations, the effective damping ratio of the ith mode can be calculated as

$$\xi_{\text{effi}} = \frac{-B_i \pm \sqrt{B_i^2 - 4A_iC_i}}{2A_i} \tag{8.13}$$

To evaluate the effective damping ratio, the parameters A_i, B_i, and C_i need to be calculated. In the following discussion, some examples are used to demonstrate this idea. For simplicity, the geometric magnification is $G_m = 1$ and the safety factor is $S = 1$. When the effective period T_{effi} is greater than T_S and less than T_L, the parameters can be written as

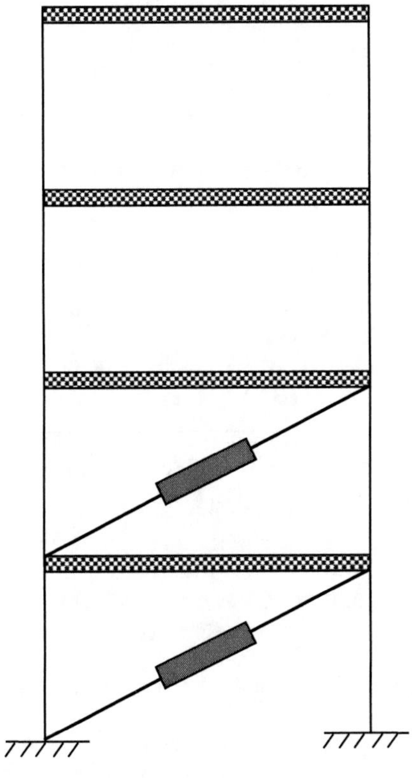

FIGURE 8.1 Four-story structure with two dampers.

$$A_i = -36\pi^2 g_y d_y \quad (8.14a)$$

$$B_i = 3c_{con_i} T_{effi} \Gamma_i I_g - 24\pi^2 g_y d_y (3\xi_0 + 0.9) - \pi m_{effi} I_g^2 p_c / 2q_d \quad (8.14b)$$

$$C_i = (3\xi_0 + 0.9)\left[T_{effi} c_{con_i} \Gamma_i I_g - 4\pi^2 g_y d_y (3\xi_0 + 0.9)\right] \quad (8.14c)$$

In Equations 8.14b and 8.14c, I_g is the input level,

$$I_g = S_{D1} Ig / R \quad (m/s^2) \quad (8.15)$$

The parameters S_{D1}, I, and R are defined in Chapter 7. For an elastic structure, $R = 1$.

When the effective period, T_{effi}, is equal to or less than T_s,

$$A_i = -36\pi^2 g_y d_y \quad (8.16a)$$

$$B_i = 3c_{con_i} T_{effi}^2 \Gamma_i I_g - 24\pi^2 g_y d_y (3\xi_0 + 0.9) - \pi m_{effi} T_{effi}^2 I_g^2 p_c / 2q_d \quad (8.16b)$$

$$C_i = (3\xi_0 + 0.9)\left[T_{effi}^2 c_{con_i} \Gamma_i I_g - 4\pi^2 g_y d_y (3\xi_0 + 0.9)\right] \quad (8.16c)$$

Note that in this case,

$$I_g = S_{Ds} Ig / R \quad (m/s^2) \quad (8.17)$$

where S_{Ds} is defined in Chapter 7.

In real applications, $G_m \neq 1$, T_{effi} can be less than T_0 or greater than T_L. In these cases, the direct computation of the effective damping ratio can also be obtained following the same procedure.

Equation 8.13 has two solutions, but only one solution makes engineering sense.

Example 8.2

Suppose in Example 8.1, **M** = diag([100.8089, 99.51468, 96.4605, 92.0809]) (t)

$$K = \begin{bmatrix} 2.7844 & -1.3921 & 0.0004 & -0.0002 \\ -1.3921 & 2.7840 & -1.3917 & 0.0001 \\ 0.0004 & -1.3917 & 2.7841 & -1.3919 \\ -0.0002 & 0.0001 & -1.3919 & 1.3923 \end{bmatrix} \times 10^4 \quad (kN/m)$$

Originally, the system has proportional damping and the damping ratio of the first mode is 0.02. It can be calculated that $\Gamma_1 = 1.2478$ and $\boldsymbol{p}_1^T = [0.3520,\ 0.6593,\ 0.8835,\ 1.0000]$. The effective mass, m_{eff1}, is 347.37 (t). The input factor $I_g = 2.9542$ (m/s²). Suppose the response is too high so that an additional damping ratio of approximately 15% is used for the first mode. The bilinear damper is designed to have $d_y = 0.02$ (m); $q_d = 700$ (kN); $a = 0$; $G_m = 1$; and $p_c = 0.65$. Since only two nominal dampers are used and installed in the first and second stories, $g_y = 2/0.65 = 3.077$ and $c_{con_i} = 0.6593$.

Suppose the period of the base structure is not affected by the two supplemental dampers, and its fundamental period is still around $T_{eff1} = 1.50$ (s). For simplicity, let $S = 1$, thus Equations 8.14 through 8.16 can be used to calculate $A_1 = -21.8650$, $B_1 = -7.4792$, and $C_1 = 1.2603$. Equation 8.13

has two roots: −0.4658 and 0.1237. The design damping ratio is 0.02 + 0.124 = 14.4%. Note that if $p_c = 1$, the design damping ratio will become 26.3%, which overestimates the damping ratio.

With the effective damping ratio calculated, the numerical damping coefficient and the effective modal responses can also be calculated. That is,

$$\xi_{designi} \approx \xi_{effi} + \xi_{oi} \tag{8.18}$$

where $\xi_{designi}$, ξ_{effi}, and ξ_{oi} are, respectively, the ith design damping ratio, the effective damping ratio contributed by the bilinear damper, and the damping ratio contributed by the base structure. Also,

$$B_i = 3\xi_{designi} + 0.9 \tag{8.19}$$

Now, the parameters needed to realize the design damping ratio have been calculated. As seen from Equation 8.6, the characteristic strength, q_d, is proportional to the required effective damping ratio, that is,

$$q_d \propto \xi_{effi} \tag{8.20}$$

From the above notation, it is seen that if the required damping ratio is not satisfied in the initial estimation, the needed quantity, q_d, can be obtained. Note that the yielding displacement, d_y, is often predetermined according to the specific type of damping materials. These parameters are nominal terms and are not yet included in the specifications of individual dampers.

8.2.1.2.3.3 Equivalent Stiffness and Characteristic Strength of Bilinear Dampers Typically, the elastic or unloading stiffness, k_u, the yielding or damper stiffness, k_d, and the characteristic strength, q_d, of a bilinear damper are provided by vendors. These are marked with italic letters. After installation, the values of these parameters change in most applications. Therefore, the *equivalent unloading stiffness*, $\mathrm{k_u}$, *damper stiffness*, $\mathrm{k_d}$, *and characteristic strength*, $\mathrm{q_d}$, which are the system parameters, need to be clearly distinguished for bilinear damper design.

To establish the relationship between the structural parameters $\mathrm{k_u}$, $\mathrm{k_d}$, and $\mathrm{q_d}$, and the damper parameters k_u, k_d, and q_d, recall the concept of the geometrical magnification factor, G_m, introduced in Chapter 5.

Therefore, the initial damper stiffness is affected from the viewpoint of the total structural system. When a bilinear damper is installed with the geometrical magnification factor, the equivalent damping stiffness for the structure should be written as

$$\mathrm{k_d} = \mathrm{G_m^2} k_d \quad (\mathrm{kN/m}) \tag{8.21}$$

The equivalent characteristic strength can be written as

$$\mathrm{q_d} = \mathrm{G_m} q_d \quad (\mathrm{kN}) \tag{8.22}$$

In certain cases, the *equivalent yielding displacement*, $\mathrm{d_y}$, should also be used. This can be written as

$$\mathrm{d_y} = d_y/\mathrm{G_m} \quad (\mathrm{m}) \tag{8.23}$$

where d_y is the yielding displacement of the bilinear damper.

8.2.1.2.3.4 Structural Parameters Suppose the original structure is an SDOF system with stiffness k. The yielding and unloading stiffness of the structure, denoted as k_D and k_U, respectively, will be

$$k_D = k_d + k \quad (kN/m) \tag{8.24}$$

$$k_U = k_u + k \quad (kN/m) \tag{8.25}$$

The characteristic strength, q_D, and yielding displacement, d_Y, remain unchanged, that is,

$$q_D = q_d \quad (kN) \tag{8.26}$$

and

$$d_Y = d_y \quad (m) \tag{8.27}$$

8.2.1.2.3.5 Equivalent System for Bilinear Damper with Finite Supporting Stiffness When a bilinear damper is installed, the supporting stiffness is likely a finite value. Practically speaking, a finite supporting stiffness can significantly reduce the effectiveness of the bilinear damper. Therefore, at the design stage, it is necessary to quantify the effect of finite supporting stiffness.

If the damping is purely bilinear, then for an SDOF structure with finite supporting stiffness, the entire system can also be treated as a bilinear system. Two typical cases of finite supporting stiffness are shown in Figures 8.2a and b. In these figures, k_s stands for the value of the supporting stiffness of the system. In Figure 8.2b, the stiffness k_{11} and k_{12} represent the structural stiffnesses. The system shown in Figure 8.2a can be replaced by the equivalent bilinear system shown in Figure 8.2c with the equivalent structural stiffness k_{E1} and the equivalent damper stiffness k_{E2}, which are expressed as

$$k_{E1} = \frac{k_1 k_s}{k_1 + k_s} \quad (kN/m) \tag{8.28}$$

and

$$k_{E2} = \frac{k_2 k_s^2}{(k_1 + k_2 + k_s)(k_1 + k_s)} \quad (kN/m) \tag{8.29}$$

Furthermore, the equivalent yielding displacement, d_{Ye}, is given by

$$d_{Ye} = \frac{k_1 + k_2 + k_s}{k_s} d_y \quad (m) \tag{8.30}$$

where d_y is the yielding displacement as previously defined. The equivalent characteristic strength q_{De} can be written as

$$q_{De} = d_{Ye} k_{E2} \quad (kN) \tag{8.31}$$

Using Equations 8.28 through 8.31, a new bilinear system with the control parameters k_{E1}, k_{E2}, q_{De}, and/or d_{Ye} is defined. And, it is relatively simple to use Equations 8.28 through 8.31 to account for the case described in Figure 8.2b.

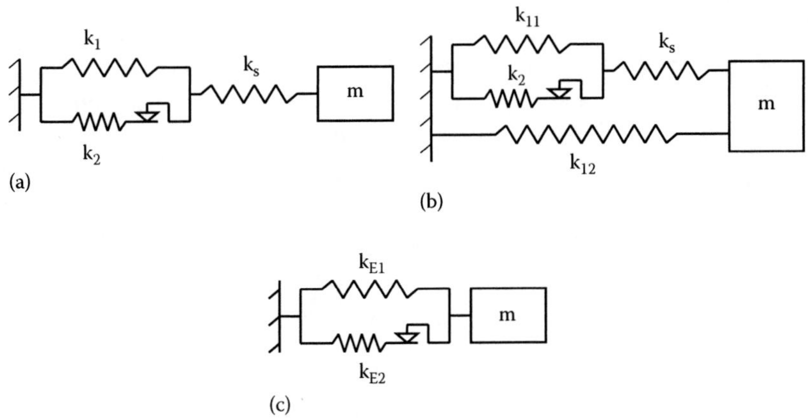

FIGURE 8.2 (a–c) Finite supporting stiffness and equivalent system.

Damper design for an SDOF system is a systematic procedure based on predetermined control parameters. However, design of nonlinear dampers for an MDOF structure is not a systematic process. The differences between SDOF and MDOF systems with added bilinear dampers are discussed in the following paragraphs.

First, the control parameters of an SDOF system, namely, the unloading stiffness k_u and the characteristic strength q_d, are fixed values and directly depend on the properties of the bilinear damper. Even the effective period T_{eff}, which must be determined by considering the structural stiffness and deformation, can be obtained from the structural parameters and by assuming the maximum deformation to be tolerated.

For MDOF systems, these parameters will largely depend on the dynamic process of the entire system. Suppose several dampers are installed at different locations in a structure. The yielding points of these dampers are probably not at the same level, which must not only depend on the amplitude of the ground excitations, but on the vibration prior to given time points. In other words, the yield point of identical bilinear dampers installed at different locations cannot be predicted. Furthermore, different ground excitations will dictate the occurrence of the yield point of these dampers.

Therefore, the unloading stiffness k_U, the characteristic strength q_D, as well as the effective period T_{eff}, obtained through the approach described above, will no longer have fixed values, but are time variables:

$$k_U = k_U(t) \quad (kN/m) \tag{8.32}$$

$$q_D = q_D(t) \quad (kN) \tag{8.33}$$

and

$$T_{eff} = T_{eff}(t) \quad (s) \tag{8.34}$$

However, using an SDOF system to design the dampers for an MDOF system is the essence of the method based on the response spectrum. In the next subsection, determining these time-variable design parameters is discussed in detail. Included in the discussion is a method to determine the uncertain deformation shape function in bilinear damper design.

8.2.2 Response Estimation

8.2.2.1 Equation of Motion

After the number configurations and specifications of the dampers are determined, the stiffness matrix $\mathbf{K}_\delta$ contributed by the bilinear dampers can be found,

$$\mathbf{K}_\delta = \left[k_{Dij} \right] \quad (\text{kN/m}), \quad i, j = 1, 2, \ldots, n, \tag{8.35}$$

In Equation 8.35,

$$k_{Dij} = \begin{cases} (-1)^{\delta ij} \bar{k}_{dij} G_{mij}^2, & \text{damper installed between the i}^{\text{th}} \text{ and j}^{\text{th}} \text{ floor} \\ 0, & \text{no damper installed between the i}^{\text{th}} \text{ and j}^{\text{th}} \text{ floor} \end{cases} \tag{8.36}$$

where G_{mij} is the geometric magnification factor in between the ith and j[th] floors as previously defined, and $\bar{k}_{dij}$ is the effective damper stiffness of the bilinear damper installed between the ith and j[th] floors. Note that $\bar{k}_{dij}$ is neither the unloading stiffness k_u nor yielding k_d of the specific damper. Furthermore, $\bar{k}_{dij}$ is not the secant stiffness of the damper. The effective stiffness should be determined from Equation 6.41. That is,

$$\bar{k}_{dij} = \frac{1 + a\left(\mu_{ij} - 1\right)^2}{\mu_{ij}^2} k_u \quad (\text{kN/m}) \tag{8.37}$$

where a is the stiffness yielding ratio and μ_{ij} is the corresponding ductility of the damper. Since μ_{ij} is a variable, the stiffness k_{Dij} will also be a variable. A rough estimation of the ductility can be given by using the Penzien constant p_c, that is,

$$\mu_{ij} = \frac{p_c G_{mij} \left| x_i - x_j \right|}{d_y} = 0.65 G_{mij} \frac{\left| x_i - x_j \right|}{d_y} \tag{8.38}$$

Note that in Equation 8.36, δ_{ij} is a Kronecker delta function defined as

$$\delta_{ij} = \begin{cases} 0 & i = j \\ 1 & i \neq j \end{cases} \tag{8.39}$$

For a multiple-story structure with bilinear dampers installed, the following simplified governing equation can be obtained:

$$\mathbf{M}\ddot{\mathbf{x}} + \mathbf{C}\dot{\mathbf{x}} + \left(\mathbf{K} + \mathbf{K}_\delta\right)\mathbf{x} = -\mathbf{MJ}\ddot{\mathbf{x}}_g - \mathbf{f}_\delta \tag{8.40}$$

where the coefficient matrices **M**, **C**, and **K** are as previously defined for linear systems. In particular, **C** can be treated as a proportional damping matrix that provides a small amount of original damping, i.e., the original damping ratio is less than 5%. The stiffness $\mathbf{K}_\delta$ is defined in Equation 8.35. The term $\mathbf{f}_\delta$ is the damping force, which is a function related to the sign of the relative velocity and related to whether the relative displacement is greater than or equal to the yielding displacement, d_y.

Next, the transpose of the deformation shape function $\boldsymbol{p}^T$ is premultiplied on both sides of Equation 8.40,

$$\boldsymbol{p}^T\mathbf{M}\ddot{\mathbf{x}}+\boldsymbol{p}^T\mathbf{C}\dot{\mathbf{x}}+\boldsymbol{p}^T\left(\mathbf{K}+\mathbf{K}_\delta\right)\mathbf{x}=-\boldsymbol{p}^T\mathbf{MJ}\ddot{x}_g-\boldsymbol{p}^T\mathbf{f}_\delta \tag{8.41}$$

where $\boldsymbol{p}$ is in fact a mode shape of the **M**-**C**-(**K** + $\mathbf{K}_\delta$) system. Note that the nth element of $\boldsymbol{p}$, which is the roof displacement, is normalized to be unity and $\boldsymbol{p}$ is a dimensionless vector.

Again, let

$$\mathbf{x}(t)=\boldsymbol{p}\mathbf{y}(t) \tag{8.42}$$

$$\left(\boldsymbol{p}^T\mathbf{M}\boldsymbol{p}\right)\ddot{\mathbf{y}}+\left(\boldsymbol{p}^T\mathbf{C}\boldsymbol{p}\right)\dot{\mathbf{y}}+\left[\left(\boldsymbol{p}^T\mathbf{K}\boldsymbol{p}\right)+\left(\boldsymbol{p}^T\mathbf{K}_\delta\boldsymbol{p}\right)\right]\mathbf{y}=-\boldsymbol{p}^T\mathbf{MJ}\ddot{x}_g-\boldsymbol{p}^T\mathbf{f}_\delta \tag{8.43}$$

Dividing the scalar quantity $\boldsymbol{p}^T\mathbf{M}\boldsymbol{p}$ on both sides of Equation 8.43 results in

$$\ddot{y}+2\xi_0\omega_{eff}\dot{y}+\omega_{eff}^2 y=-\Gamma_d\ddot{x}_g-\frac{\boldsymbol{p}^T\mathbf{f}_\delta}{\boldsymbol{p}^T\mathbf{M}\boldsymbol{p}} \tag{8.44}$$

In Equation 8.44, the effective frequency ω_{eff}^2 is defined as

$$\omega_{eff}=\left[\frac{\boldsymbol{p}^T\left(\mathbf{K}+\mathbf{K}_\delta\right)\boldsymbol{p}}{\boldsymbol{p}^T\mathbf{M}\boldsymbol{p}}\right]^{1/2}=\left[\frac{\boldsymbol{p}^T\left(\mathbf{K}+\mathbf{K}_\delta\right)\boldsymbol{p}}{m_{e1}}\right]^{1/2}\ (\text{rad/s}) \tag{8.45}$$

Also, in Equation 8.44, ξ_0 is the original damping ratio and Γ_d is a scalar, which can be seen as an effective mode participation factor.

In Equation 8.45, for convenience,

$$m_{e1}=\boldsymbol{p}^T\mathbf{M}\boldsymbol{p}\quad(t) \tag{8.46}$$

Note that both **M** and **K** are positive definite. Thus, $\omega_{eff} > 0$. The effective period is

$$T_{eff}=\frac{2\pi}{\omega_{eff}}\quad(s) \tag{8.47}$$

On introducing the linear MDOF system, it is seen that the natural period T and mode shape $\boldsymbol{p}$ can be further written as T_i and $\boldsymbol{p}_i$ specifically for the ith modal parameters. Furthermore, the ith modal participation factor, denoted as Γ_{di}, is also obtained.

Now, consider the term $\boldsymbol{p}_i^T\mathbf{f}_\delta$, which is only related to the ith mode and not intertwined with other modes. Therefore, Equation 8.44 actually only describes the ith mode. Since the system is nonlinear, the concept of "mode" is actually the "effective mode." In practice, only the fundamental effective mode is considered.

Example 8.3

The four-story system described by Equation 8.40 is used as an example, where the geometric magnification factors are identical in each story, denoted as G_m (see Figure 8.1). Consider the first mode only. Thus,

$$\boldsymbol{p}_1^{\mathrm{T}} \boldsymbol{F}_\delta = \left[p_{11}, p_{21}, p_{31}, p_{41} \right] G_m^2 q_d \begin{bmatrix} 1 & -1 & 0 & 0 \\ 0 & 1 & 0 & 0 \\ 0 & 0 & 0 & 0 \\ 0 & 0 & 0 & 0 \end{bmatrix} \begin{pmatrix} \operatorname{sgn}(\dot{x}_1)\, h(|x_1| \ge d_y) \\ \operatorname{sgn}(\dot{x}_2 - \dot{x}_1)\, h(|x_2 - x_1| \ge d_y) \\ \operatorname{sgn}(\dot{x}_3 - \dot{x}_2)\, h(|x_3 - x_2| \ge d_y) \\ \operatorname{sgn}(\dot{x}_4 - \dot{x}_3)\, h(|x_4 - x_3| \ge d_y) \end{pmatrix}$$

Here, the term $h(|(\cdot)| \ge d_y)$ is a Heaviside function. When $|(\cdot)| \ge d_y$, it equals unity; otherwise, it equals zero.

In most cases, for the first mode,

$$p_{11} < p_{21} < p_{31} < p_{41} = 1$$

Therefore, in a single quarter cycle, i.e., the structure moves toward a positive direction, assuming that all the Heaviside functions are $h(|(\cdot)| \ge d_y) = 1$,

$$\boldsymbol{p}_1^{\mathrm{T}} \mathbf{f}_\delta = G_m^2 q_d \left[p_{11} - (p_{21} - p_{11}) + (p_{21} - p_{11}) \right] = G_m^2 q_d p_{11} \quad (\mathrm{kN})$$

Consider the entry cycle. When all the Heaviside functions are $h(|(\cdot)| \ge d_y) = 1$,

$$\boldsymbol{p}_1^{\mathrm{T}} \mathbf{f}_\delta = G_m^2 q_d p_{11} \operatorname{sgn}(\dot{x}_1) \approx G_m^2 q_d p_{11} \operatorname{sgn}(\dot{y}_1) \quad (\mathrm{kN})$$

It is seen that the term $G_m^2 q_d p_{11} \operatorname{sgn}(\dot{x}_1)$ is a time variable and a function of velocity $\dot{x}_1$, which makes the following equation nonlinear:

$$\ddot{y} + 2\xi_0 \omega_{eff}\, \dot{y} + \omega_{eff}^2 y = -\Gamma_d \ddot{x}_g - \frac{G_m^2 q_d p_{11} \operatorname{sgn}(\dot{x}_1)}{m_{e1}} \approx -\Gamma_d \ddot{x}_g - \tilde{q}_d \operatorname{sgn}(\dot{y}_1)$$

Thus, linear modes do not exist. However, when the bilinear damping $G_m^2 q_d p_{11} \operatorname{sgn}(\dot{x}_1)$ is small, a near constant shape function $\boldsymbol{p}_1$ and effective natural period T_1 can be approximated. Similarly, a near constant shape function $\boldsymbol{p}_i$ and effective natural period T_i are obtained. In the above equation, the term $\tilde{q}_d$ is a normalized *modal characteristic strength*:

$$\tilde{q}_d = \frac{G_m^2 p_{11} q_d}{m_{e1}} \quad (g)$$

Note that the value p_{11} here is the first element of the dimensionless shape function $\boldsymbol{p}_1$. Therefore, $\tilde{q}_d$ should be a deterministic value with units of (m/s^2) or (g).

From this example, it is also realized that for higher modes, the effect of $\boldsymbol{p}_i^{\mathrm{T}} \mathbf{f}_\delta$, $i > 1$ will be comparatively smaller, because the mode displacement has different signs, whereas the modal displacement of the fundamental mode has the identical sign. Furthermore, consider the value of the modal displacement of a damper, $G_m (p_{ji} - p_{j-1,i})$. Compared to the yielding displacement d_y, it can be understood that the difference is small; and in many cases, $d_y > G_m (p_{ji} - p_{j-1,i})$. In this case, the bilinear damper does not dissipate a significant amount of energy in the higher modes. The first effective mode is often chosen and the MSSP approach is used for bilinear design.

8.2.2.2 Response Reevaluation

8.2.2.2.1 Mode Shape and Natural Period of Fundamental Mode

In the first steps in damper design, after the response of the base system is calculated and the required damping ratio is found, the nominal damper specifications q_d, k_u, k_d, d_y, and a for individual dampers and damping configuration of the total system are determined. Next, the additional stiffness matrix $\mathbf{K}_\delta$ is calculated. Then, the corresponding effective period T_{effi}, damping ratio ξ_{effi}, participation factor Γ_{di}, and shape function $\boldsymbol{p}_i$ are determined. The accumulated modal mass ratio, γ_{mS}, is also available.

From these parameters, the responses of effective SDOF systems are calculated. That is, from Equation 7.11,

$$B_i = 3(\xi_{effi} + \xi_{oi}) + 0.9 \quad (8.48)$$

8.2.2.2.2 Lateral Displacement

The maximum values of lateral floor displacement of the ith effective mode are given by

$$\mathbf{d}_{max_i} = \Gamma_{di} d_{iD} \boldsymbol{p}_i = \begin{Bmatrix} d_{1i} \\ d_{2i} \\ \dots \\ d_{ni} \end{Bmatrix} \ (m) \quad (8.49)$$

Here, the parameters Γ_{di} and $\boldsymbol{p}_i$ are calculated in the previous steps, while the spectral displacement, d_{iD}, is determined with the help of Equation 7.51 in Chapter 7.

The total maximum lateral displacement is calculated using the square-root-of-the-sum-of-squares (SRSS) as

$$\mathbf{d}_{max} = S_M S_{Dn} S_{D\xi} \left(\sum_{i=1}^{S} \mathbf{d}_{max_i}^{\bullet 2} \right)^{\bullet 1/2} \ (m) \quad (8.50)$$

In Equation 8.50, similar to the linear systems discussed in Chapter 7,

$$S_M = (0.9 \sim 1)/\gamma_{mS} \quad (8.51)$$

where S_{Dn} is given as in Equation 7.56. For example, again using the 99 earthquake records scaled by PGA,

$$S_{Dn} = 1/\left(-6.806 \times 10^{-8} n^4 + 1.257 \times 10^{-5} n^2 - 7.763 \times 10^{-4} n + 0.0173 n + 0.4288\right)$$

Meanwhile, $S_{D\xi}$ is recommended as in Equation 7.58. For example, using the 99 earthquake records,

$$S_{D\xi} = 1/\left(0.9697 \xi_{design} + 0.6987\right)$$

8.2.2.2.3 Absolute Lateral Acceleration

The pseudo lateral acceleration vector of the ith effective mode is as approximately given by

$$\mathbf{a}_{s_i} = \Gamma_i C_{s_i} g \boldsymbol{p}_i = \begin{Bmatrix} a_{s_{1i}} \\ a_{s_{2i}} \\ \dots \\ a_{s_{ni}} \end{Bmatrix} \ (m/s^2) \quad (8.52)$$

The total maximum absolute lateral acceleration can be calculated through modified SRSS as

$$\mathbf{a}_a = S_M S_{An} S_{A\xi}\left[\sum_{i=1}^{S} \mathbf{a}_{s_i}^{\bullet 2} + (S_{Ai}\mathbf{J})\right]^{\bullet 1/2} = \begin{Bmatrix} a_{a1} \\ a_{a2} \\ \dots \\ a_{an} \end{Bmatrix} \ (m/s^2) \tag{8.53}$$

In Equation 8.53, S_M is the same as defined in Equation 8.51 and S_{Ai} can be determined using Equation 7.63.

Similar to linear systems, the jth element in the acceleration vector $\mathbf{a}_a$ is denoted as $\mathbf{a}_{aj}$, whereas in the equation for the ith effective modal acceleration, the jth element is denoted as $\mathbf{a}_{aji}$.

In Equation 8.54, S_{An} is given by Equation 7.64. For example, using the 99 earthquake records results in

$$S_{An} = 1/\left(4.825\times10^{-6}n^3 - 7.453\times10^{-4}n^2 + 0.0346\,n + 0.383\right)$$

and $S_{A\xi}$ is given in Equation 7.65. Again, using the 99 earthquake records,

$$S_{A\xi} = 1/\left(0.8754\,\xi_{design} + 0.9596\right)$$

8.2.2.2.4 Lateral Force and Base Shear

The lateral force at the jth floor can be calculated using Equation 7.66, while the base shear of the ith effective mode can be evaluated from Equation 7.70.

8.2.3 Design Issues

8.2.3.1 SDOF Systems

In Chapters 1 and 2, linear damping for SDOF systems was discussed. When the original damping of a structure is small, it is always beneficial to use linear dampers. In Chapter 5, the basic design concepts from currently used codes, such as NEHRP 2009 (BSSC 2009), were explained. Essentially, if the structural responses are larger than the desired value, it is safe to add damping, regardless of whether the damping is linear or nonlinear. (All nonlinear damping is linearized).

However, the above statement is not always true, especially for bilinear dampers. This can be shown by using numerical simulations, where a linear structure with very little damping is used as a reference. Consider bilinear dampers with 5% linear damping, making the corresponding effective damping equal to or greater than 5%. The effective damping is calculated using both Timoshenko damping and force-based effective damping. Numerical simulation results from the responses of the linear structure and the structure with bilinear damping are compared in Figure 8.3. The calculations are made by using the following procedure.

Suppose a structure has 5% viscous damping, under the aforementioned 99 earthquake record (see Example 4.7 in Chapter 4). The absolute acceleration or the structural forces are calculated by using the mean-plus-one-standard-deviation. For comparison, consider the same structure installed with some bilinear dampers with an initial viscous damping ratio assumed to be zero. Such an assumption is practical for many structures whose damping ratios are rather small, for example, 2%–3%. By changing the characteristic strength and the yielding stiffness of the bilinear dampers, and by varying the period of the base structure, the responses of the corresponding structural forces and displacement can be calculated.

Three parameters are needed to describe the properties of a structure with bilinear damping. For convenience, the characteristic strength q_d, the unloading stiffness of the damper k_u, as well as the period of the base structure, T, are choosen to calculate the lateral stiffness of the structure k. In this process, a proportional parameter of k_u/k is actually used. For simplicity, the geometric magnification factor is assumed to be unity. In order to compare the responses of structures with linear viscous damping and with bilinear damping, the ratios of the corresponding structural forces of the two cases are calculated and the ratios of the corresponding displacements are also evaluated. To visualize the corresponding effective damping ratios, the Timoshenko damping ratio and the force-based effective damping ratio are also computed. These ratios are then plotted vs. the characteristic strength. In these computations, the structure is assumed to have unit mass. Thus, the characteristic strength can be represented with the unit (g).

In each of the comparisons given in Figure 8.3, four curves are plotted. The solid line is the ratio of the displacements. To obtain this ratio, the mean-plus-one-standard-deviation values of the displacement of the structure with bilinear dampers, d_B, and that of a structure with 5% linear damping, d_L, are first calculated.

Note that for linearized SDOF systems, the ratio of d_B/d_L is exactly the numerical damping coefficient B. However, in order to improve the process of the linearization procedures, the concept of a numerical damping coefficient is not initially used. Instead, the reciprocal ratio of d_L/d_B is used. The ratio d_L/d_B vs. the characteristic strength q_d are plotted under different conditions with a given value of the bilinear damping stiffness, k_d. It is understandable that the inequality,

$$d_L/d_B > 1 \tag{8.54}$$

means the displacement of the structure when the bilinear damping is smaller than for the corresponding linear system. Hence, using the bilinear damper is beneficial.

Similarly, the mean-plus-one-standard-deviation values of the structural force (the absolute acceleration) of the structure with bilinear dampers, a_B, and that of the structure with 5% linear damping, a_L, are also calculated. Then, the ratio a_L/a_B vs. the characteristic strength q_d are also plotted and shown with broken lines, which are also under different conditions with a given value of the bilinear damping stiffness, k_d. When

$$a_L/a_B > 1 \tag{8.55}$$

the structural force of the structure with bilinear damping is smaller than for the corresponding linear system. Hence, using bilinear dampers is effective. Also, note that using bilinear dampers enlarges the responses.

For reference, the calculated effective damping ratios are also plotted in Figure 8.3. The dotted lines represent Timoshenko damping, while the dot-dash lines indicate force-based effective damping.

In Figures 8.3a through d, the period of the structure is fixed to be 0.4 (s). The corresponding damper stiffness is, respectively, 0.32, 0.72, 1.28, and 2.0 times the stiffness of the structure. It is seen that when the damper stiffness is small, i.e., 0.32 k, both the force ratio and the displacement ratio are smaller than 1. This means that both the force (or absolute acceleration) and the displacement of the structure are actually greater than the 1 with 5% linear damping. Note that in these cases, the effective damping calculated by Timoshenko and force-based effective damping are almost always greater than 5%.

When the damper stiffness of the bilinear damper becomes larger, i.e., 0.72 k, the displacements are reduced; however, the force is still magnified. As the damper stiffness becomes larger, i.e., greater than 1.28 k, both the force and the displacement are reduced.

From Figures 8.3e through h, a similar tendency is observed; except in this case, the period is chosen to be 1.0 sec and thus the required damper stiffness should be higher.

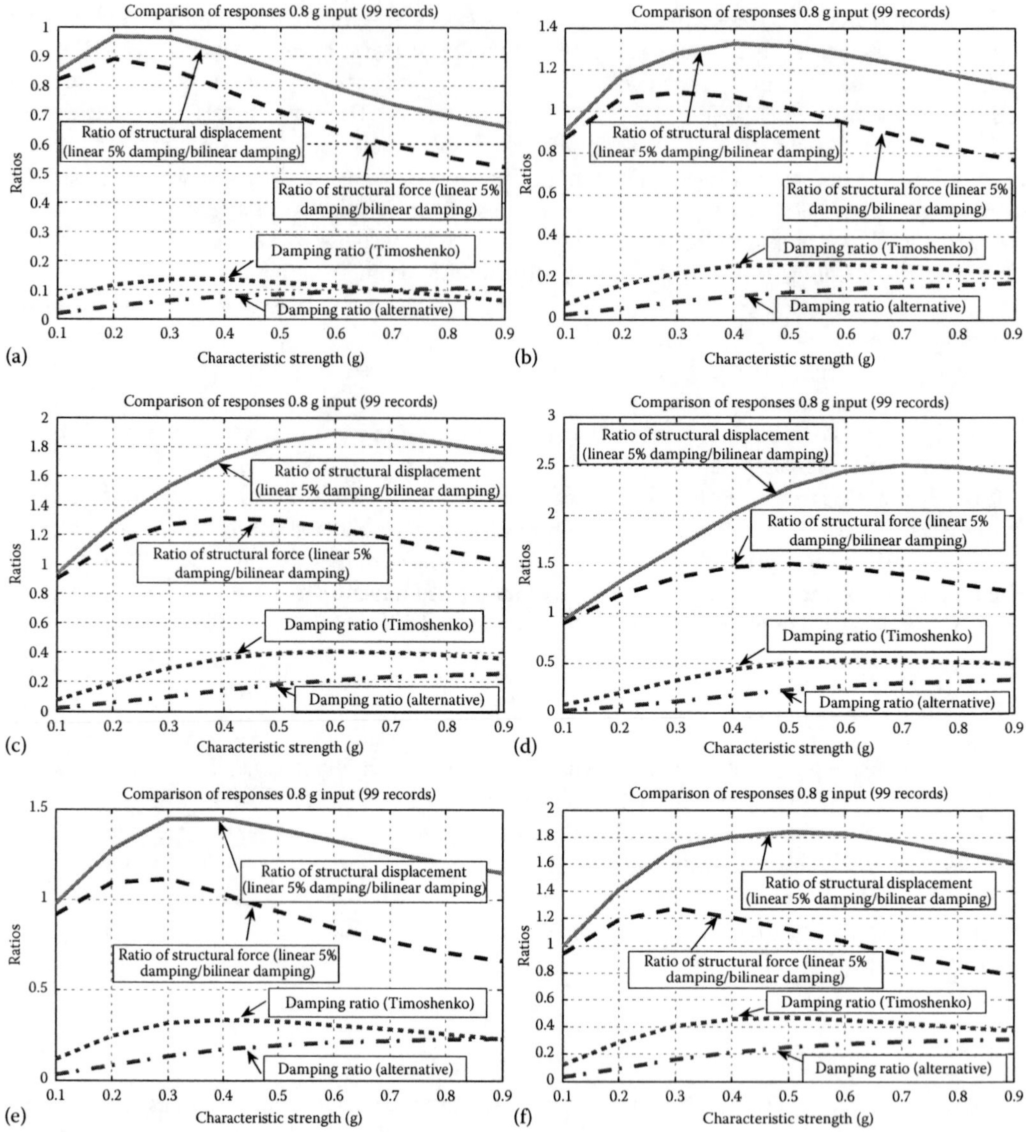

FIGURE 8.3 Comparisons of responses, 5% viscous vs. bilinear damping: (a) $k_u = 0.32$ k $T = 0.4$ (s); (b) $k_u = 0.72$ k $T = 0.4$ (s); (c) $k_u = 1.28$ k $T = 0.4$ (s); (d) $k_u = 2.0$ k $T = 0.4$ (s); (e) $k_u = 0.98$ k $T = 1.0$ (s); (f) $k_u = 1.62$ k $T = 1.0$ (s); (g) $k_u = 3.38$ k $T = 1.0$ (s); (h) $k_u = 5.12$ k $T = 1.0$ (s); (i) $k_u = 1.28$ k $T = 2.0$ (s); (j) $k_u = 5.12$ k $T = 2.0$ (s).

From Figures 8.3i and j, it is further noted that the tendency of requiring a higher damper stiffness to reduce both the force and the displacement is also true. In addition, it is seen that when the period is increased, the requirement becomes more demanding.

Furthermore, the characteristic strength, q_d, of the bilinear damping can also greatly influence the effectiveness of the reduction of both the force and the displacement. From Figure 8.3, it is seen that when q_d is very small, increasing its value will help further reduce the structure force and the displacement. However, when q_d is sufficiently large, continuously increasing its value becomes less effective.

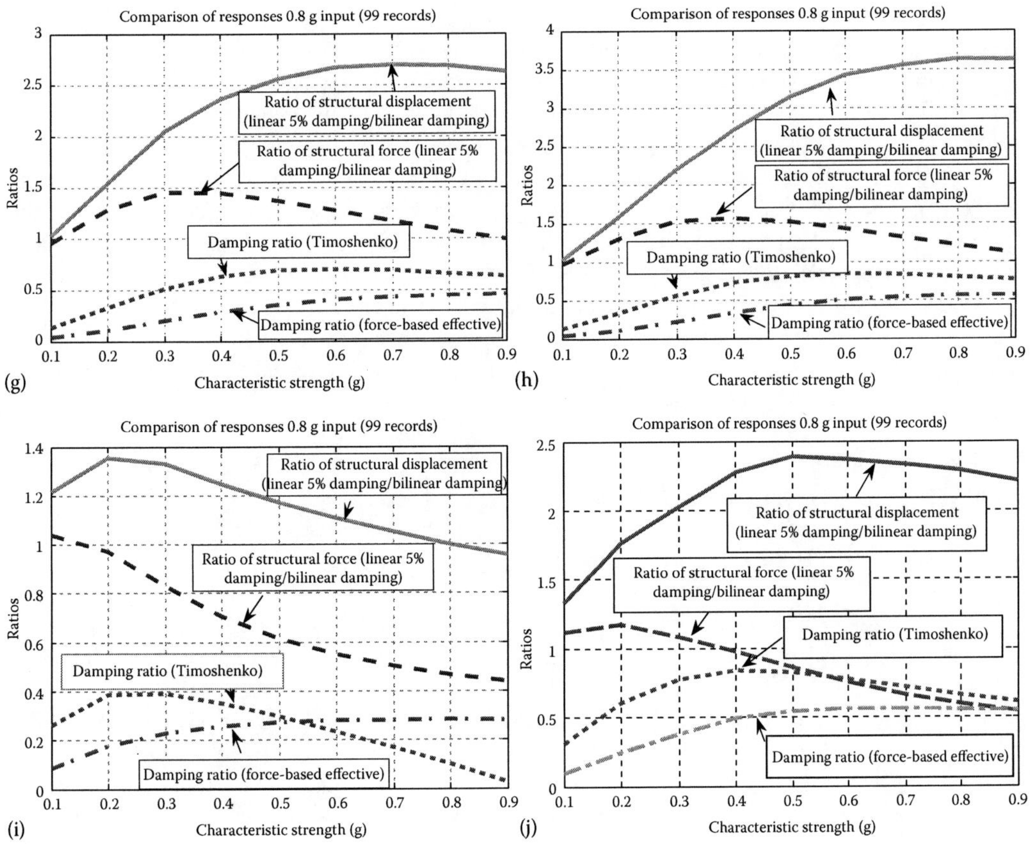

FIGURE 8.3 (*Continued*).

Increasing the elastic (unloading) stiffness of the structure k_u will first lower the effectiveness of the force reduction, followed by lowering the effectiveness of the displacement reduction.

Some further observations can be made from Figures 8.3a through d. When the period is 0.4 (s), the characteristic strength can be selected to be higher than 0.4 (g). Thus, practically speaking, if a structure has mass m, or weight W,

$$q_d = 0.4\ mg = 0.4\ W \quad (kN) \tag{8.56}$$

When the period is 1.0 (s), the best characteristic strength that can be chosen is 0.3 (g). Thus, $q_d = 0.3$ mg = 0.3 W. Therefore, in design, the following formula may be used to make the initial choice of the characteristic strength for a simplified SDOF model:

$$q_d = I_g\, m_{eff1}\, g \quad (kN) \tag{8.57}$$

Although the phenomena observed from Figure 8.3 are the results from a particular example, they are somewhat general. The authors have used other groups of earthquake records, and different combinations of the period, damper stiffness, and characteristics strength have been tried. In most cases, using a statistical approach, similar phenomena have been obtained, but are not reported herein.

8.2.3.2 MDOF Systems

Bilinear damper design for the seismic response reduction of MDOF structures involves a complex relationship between the seismic forces and the deformations of the structures. An SDOF system has only a single relationship to define energy dissipation. MDOF systems with multiple bilinear dampers have many energy dissipations paths, because dampers can have different yielding points. Therefore, using a unified loop to represent these relationships is only an approximation. Furthermore, both the deformation shape function and the acceleration shape function differ from the mode shape function obtained from their corresponding linear MDOF systems. These two nonlinear shape functions are also distinct from each other. Therefore, using the linearized approach, although very simple, is difficult to estimate the errors involved in the approximation. In Section 8.5, a number of issues and approaches to bilinear damper design for the response reduction of MDOF structures are discussed, using a multi-story frame as the base structure. Whenever possible, the design spectrum approach is used.

8.2.3.3 Brief Summary

Several statements can now be made based on the above discussions with respect to using bilinear dampers to control the earthquake-induced vibrations for structures whose original damping is small. General statements to aid in developing nonlinear design approaches include:

1. For bilinear damping, if the design damping ratio is larger than 5%, the structural responses may not necessarily be reduced when compared to a structure with 5% linear damping.
2. Elastic SDOF structures with linear damping can be characterized by two control parameters, whereas those with bilinear damping require three parameters, e.g., the period of the original structure T, the yielding stiffness k_d, and the characteristic strength, q_d, of the bilinear damper. The effects of these parameters on damping reduction require further study with respect to damping design of structures with supplemental damping.
3. Structural force and displacement are typically used to analyze the effects of these control parameters. In general, the reduction of the displacement is easier to accomplish than the reduction of structural force.
4. A threshold value of the unloading damper stiffness, k_u, exists. If k_u is smaller than a given threshold, the vibration cannot be reduced. Once the threshold is reached, increasing the damper stiffness, k_u, seems to always be beneficial in reducing both the displacement and the force. Note that varying k_u, though beneficial, is not significant. In addition, k_u is often specified by vendors, so the designer does not have many choices. Therefore, in practical design, less attention is given to the value of the elastic damper stiffness.
5. The yielding displacement d_y, which is closely related to the characteristic strength q_d and the stiffness k_u, plays a more significant role in response reduction than k_u. Generally, the smaller the value of d_y, the better the results that can be achieved. However, d_y is related to the material properties of the damper and thus it is also often specified by vendors.
6. The effect of the period T is more complex to visualize directly. In general, to reduce the force, if T is increased, stronger unloading stiffness k_u is needed. (These are has some exceptions, especially when the characteristic q_d is comparatively small.) This statement particularly applies to the case of displacement reduction. Generally speaking, when q_d is small, increasing T will make the displacement reduction more effective. However, with larger q_d, increasing T will make the displacement reduction less effective. In the next subsection, these issues are discussed from a quantitative perspective.
7. For the value of the damper unloading stiffness k_u, larger is better for response reduction. For the characteristic strength of the bilinear damper, the value should be neither too large nor too small for vibration reduction. Thus, q_d has optimal value when the damping

stiffness k_d and the period T are given. This is directly related to the quality of the damper design and should be the first consideration in bilinear design. In the next subsection, these issues are also quantitatively described.

8. When the effective damping ratio is large, using an equivalent linear approach may introduce large errors.

8.2.4 Damper Specification

Several practical issues concerning the choice of damper specifications are discussed in the following paragraphs.

8.2.4.1 Basic Parameters

8.2.4.1.1 Characteristic Strength

In SDOF systems, characteristic strength can have an optimal value. In MDOF systems, this is also true. With a full damper configuration, the following formula can be used to estimate the optimal value of the characteristic strength:

$$\bar{q}_D \geq \frac{a}{n} + b \quad (g) \tag{8.58}$$

where n is the number of stories of the structure, and a and b are generic parameters. Equation 8.58 is obtained based on empirical regression from limited computer simulations. As an example of using Equation 8.58, a 10-story building with bilinear dampers installed in every story is given in the following. In this example, a is chosen to be 0.72 and b is chosen to be 0.1 ± 0.05, and the normalized characteristic strength is chosen to be between 0.122 and 0.222 (g).

Similar to the case of SDOF systems, increasing the value of the unloading stiffness k_u always reduces the displacement, although the acceleration can be increased.

8.2.4.1.2 Influences of the Structural Stories

When a bilinear damper is installed in an SDOF system, the total damping of the system is likely to be bilinear. On the other hand, bilinear dampers installed in MDOF systems do not necessarily yield bilinear damping. To understand this point, Figure 8.4 shows several plots of the acceleration (seismic force) vs. the displacement. These curves are obtained by using different numbers of stories, marked above each energy dissipation plot. When the number of stories is small, a near perfect parallelogram can result. However, as the number of stories becomes larger and larger, the force–displacement relationship progressively deviates from the typical bilinear plot and becomes increasingly similar to sublinear plots, which are described in Section 8.3.

These phenomena are understandable, because the yielding point of the bilinear dampers installed at different locations will, in general, not be reached at the same time. As the number of stories increases, each with a damper fully installed, the point of change of the stiffness becomes less obvious.

In general, this effect can also be determined by considering different damper configurations. In any case, when the total damping of the structure is bilinear, it is best to use the bilinear model to calculate the responses. When the total damping differs from the bilinear damping, such as the last two plots shown in Figure 8.4, using the bilinear damping model may introduce errors.

A simple rule of thumb that may be used to determine the suitability of using the bilinear model for multi-story structures is given by

$$D_C < n \tag{8.59}$$

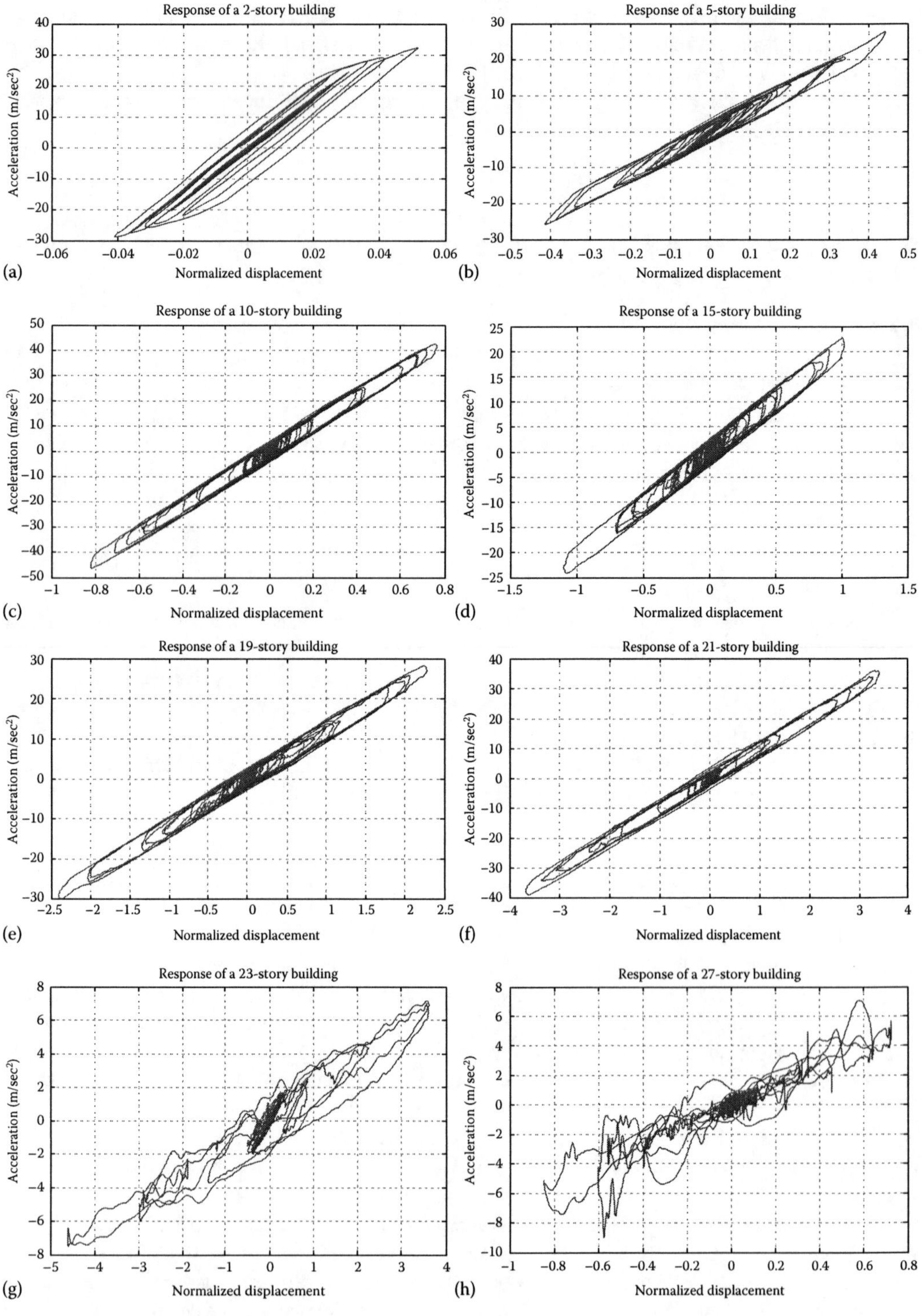

FIGURE 8.4 Influence of number of stories on force–displacement relationships: (a) 2-story; (b) 5-story; (c) 10-story; (d) 15-story; (e) 19-story; (f) 21-story; (g) 23-story; (h) 27-story.

where D_C is a design control parameter and n is the preset number of stories of a structure. If the inequality is satisfied, then the bilinear model is used for the damper design. Otherwise, an alternative formula of sublinear damping should be used. Based on limited computational simulations, D_C is suggested to be 10 for the estimation of the acceleration and 20 for the estimation of the displacement. However, this is based on a limited number of numerical simulations.

8.2.4.1.3 Selection of Damper Stiffness

According to the rectangular law expressed in Chapter 5, the minimal possible peak displacement, denoted as d_{min}, is defined by:

$$E_d = 4\bar{q}_d g d_{min} \tag{8.60}$$

From Equation 8.60, a minimal peak displacement cannot be obtained unless the unloading stiffness of the bilinear damper is infinitely large or the yielding displacement, d_y, is zero. Otherwise,

$$E_d = 4\bar{q}_d g\left(d_{min} - d_y\right) \tag{8.61}$$

Suppose the seismic work is kept constant. To obtain the smallest peak displacement, the smallest possible yielding displacement should be,

$$d_y \rightarrow 0 \tag{8.62}$$

and/or the largest possible unloading stiffness. The interpretation of the best possible bilinear damper design also relates to the largest possible unloading stiffness,

$$k_u \rightarrow \infty \tag{8.63}$$

Equations 8.62 and 8.63 define the initial criteria for selecting the stiffness of a bilinear damper.

8.2.4.1.4 Selection of Characteristic Strength

From the nonlinear spectra, it is seen that in order to lower the acceleration, the characteristic strength, $\bar{q}_d$, should be as small as possible. However, from Equation 8.60, it can also be seen that the demand increases the characteristic strength. Actually, for given period and damper stiffness, the optimal characteristic strength can be obtained. Figure 8.5 shows the curves of the product of ($a_a d_{max}$) vs. the normalized characteristic strength $\bar{q}_d$, when the period is chosen to be 0.4 (s) and the input acceleration level is 0.4 (g).

Figure 8.5 shows that it is better to limit the characteristic strength to a given range, especially when the damper yielding stiffness is small compared to the structural stiffness.

8.2.4.1.5 Compromise between the Acceleration and the Displacement

In typical seismic response reduction design, compromise needs to be made between the design parameters of acceleration and displacement. A smaller acceleration can imply a larger displacement. This characteristic is described by the following notation for linear damping:

$$W_s \propto a_a \times d_{max} \tag{8.64}$$

Equation 8.64 is also true for bilinear systems. Thus, the design must be a compromise solution between the acceleration and the displacement.

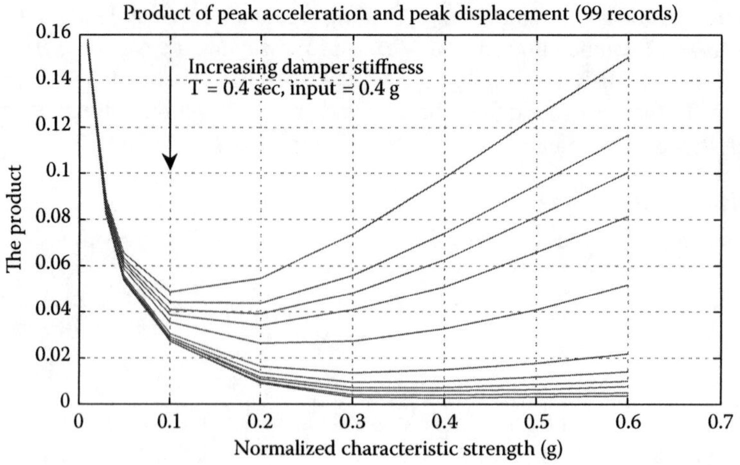

FIGURE 8.5 Optimal characteristic strength.

8.2.4.2 Selection of Bilinear Dampers

Once the basic parameters are determined, individual dampers are selected to realize these parameters. This procedure is the same in principle as the selection of linear viscous devices. During the selection, safety issues must also be considered.

8.3 EQUIVALENT LINEAR SYSTEMS APPROACH WITH SUBLINEAR DAMPERS

Sublinear damping can provide a significant amount of energy dissipation and the quality of the dampers is easier to control. In most applications, the dampers are reusable after earthquakes. Similar to linear damper design, the conceptual decision making, response estimation, and damper specification are also the basic design phases. For response estimation, a linearized damping ratio is used so that linear design spectra can be adopted. Sublinear devices have nonzero damping adaptability, making the shape functions comparatively more difficult to obtain. A more precise approach for sublinear damping design can be carried out using a nonlinear spectra approach, which is illustrated in Section 8.4.

8.3.1 General Description

In this section, the design of MDOF structures with added sublinear dampers is discussed. A more complex procedure than currently used for nonlinear damper design is required. Nonlinear design is based on the assumption of proportional damping and Timoshenko damping ratios, and uses direct iteration to account for the structural nonlinearity. However, in practical applications of hydraulic damper design, the equivalent linear system as an intermediate product is often nonproportionally damped and often contains overdamped pseudo modes. In addition, the direct iterations rarely converge to accurate values. To overcome these difficulties, several different approaches are proposed in this section. First, the effective damping ratio, the fundamental period, and the damping exponent are specified as control parameters to establish the corresponding equivalent damping matrix. The approximation of a linearized model based on Timoshenko and force-based effective damping is used to find the corresponding shape functions, so that the parameters of the SDOF system can be used for the MDOF system.

The task of conceptual decison making for linear and bilinear dampings was discussed in Sections 8.1 and 8.2. Consequently, only the basic formulas are presented in this section.

To use simplified sublinear design, namely, to use the concept of the effective damping ratio and equivalent linear approach, the corresponding damping ratio should be comparatively small, that is,

$$\xi_{design} = \xi_0 + \xi_a < 20\% \sim 25\% \tag{8.65}$$

Otherwise, a nonlinear response spectra-based approach may be tried, as illustrated in Section 8.4.

8.3.1.1 Response Estimation and First Round of Damper Design

Similar to bilinear damping design, the following tasks are carried out.

8.3.1.1.1 Estimation of Seismic Response of Original Structure

This procedure is identical to the linear damping design approach.

8.3.1.1.2 Determination of Damping Ratio and Damping Configuration

This procedure is identical to the bilinear damping design approach. After this step, the required damping ratio for the fundamental effective mode, ξ_{eff1}, is obtained as well as the damper configuration matrix, $\mathbf{C}_{con}$.

Information on the base structure, namely, the mass, damping and stiffness matrices, **M**, **C** and **K** is also needed. The damping matrix does not need to be accurate. A small amount of damping is assumed, so that **C** can be generated to provide small and proportional damping, i.e., the corresponding damping ratio is about 2% ~ 5%.

From the proportionally damped system, **M**, **C** and **K**, the modal parameters, including natural period T_i and mode shape $\boldsymbol{p}_i$, as well as modal participation factor Γ_i and accumulated modal mass ratio γ_{mS} are obtained.

8.3.1.1.3 Damper Specifications

Based on these parameters, the initial damper decision, ξ_{eff1} and $\mathbf{C}_{con}$, can be selected. Generally speaking, the equivalent damping coefficient, c_{eq}, the damping exponent, β, and the number of dampers need to be considered.

In most cases, the exact damping exponent β cannot be arbitrarily selected, for it is provided by vendors. Therefore, based on the required effective damping ratio, the equivalent damping coefficient for individual dampers as well as the total number of dampers need to be determined as design parameters.

Based on the formula of the effective damping ratio, ξ_{eff_1}, through Timoshenko damping as shown in Equation 5.15, the required equivalent damping coefficient for an MSSP system is

$$c_{eq} = \frac{2\pi\xi_{eff_1}\mathbf{d}_{max_1}^{T}\mathbf{K}\mathbf{d}_{max_1}}{\omega_1^{\alpha_1}A_\beta\sum_{j=1}^{S}x_{j1}^{\beta_1+1}} \quad \left(\text{kN-s}^\beta/\text{m}^\beta\right) \tag{8.66}$$

Here it is supposed that a total of S dampers are installed and $\mathbf{d}_{max_1}$ is the displacement vector of the first effective mode. See Equation 8.7 for a general description of the displacement vector of the ith effective mode. In Equation 8.66, ω_1 and x_{j1} are the first natural frequency or effective natural frequency and modal amplitude at the jth location, respectively, and

$$x_{j1} = G_{mj}\left(d_{j1} - d_{j-1,1}\right) \quad (\text{m}) \tag{8.67}$$

Here d_{j1} and $d_{j-1,1}$ are the floor displacements of the first effective mode, and G_{mj} is the geometrical magnification factor.

The exponents α_i and β_i are taken as constant for each mode, which is found from the vendor specification sheet. For simplicity, however, in Equation 8.66 as well as in the equations for the ith effective damping ratio (discussed later),

$$\alpha_1 = \alpha_i = \beta_1 = \beta_i = \beta \tag{8.68}$$

and approximately (see Equation 5.19),

$$A_\beta = A_{\beta_1} = 0.298\,\beta^2 - 1.147\beta + 4 \tag{8.69}$$

Similar to the design of linear viscous dampers described earlier, once the nominal parameter c_{eq} is calculated through Equation 8.66, the number of corresponding bays and the size of the dampers can be examined to determine the number of dampers.

8.3.2 Response Estimation

8.3.2.1 Mode Shape Computations

Different from bilinear damping, the design for sublinear damping may require recalculation of the mode shapes, which are not likely to be normal.

To construct an equivalent linear damping matrix, $\mathbf{C}_{eq}$, the initial estimated damping matrix is denoted as

$$\mathbf{C}_{eq}^{(1)} = c_{eq}\mathbf{C}_{con} \quad \left(\text{kN-s}^\beta/\text{m}^\beta\right) \tag{8.70}$$

Using $\mathbf{C}_{eq}^{(1)}$ as well as the mass $\mathbf{M}$ and stiffness $\mathbf{K}$ matrices, the corresponding damping ratio $\xi_1^{(1)}$ of the first mode can be calculated. Then, the equivalent linear damping matrix, $\mathbf{C}_{eq}$, is determined as

$$\mathbf{C}_{eq}^{(2)} = \frac{\xi_{eff_1}}{\xi_1^{(1)}}\mathbf{C}_{eq}^{(1)} \quad \left(\text{kN-s}^\beta/\text{m}^\beta\right) \tag{8.71}$$

Now, based on the equivalent linear system, $\mathbf{M}\text{-}\mathbf{C}_{eq}^{(2)}\text{-}\mathbf{K}$, the mode shape $\boldsymbol{p}_i$ can be calculated. Note that in the initial response estimation discussed above, the mode shape $\boldsymbol{p}_i$ is taken through the linear system $\mathbf{M}\text{-}\mathbf{C}\text{-}\mathbf{K}$ and is real valued. However, in this case, it is likely to be complex valued.

Therefore, if the design damping ratio is relatively large, i.e.,

$$\xi_{eff_1} > 10\% \sim 15\% \tag{8.72}$$

the eigen-decomposition through the $\mathbf{M}\text{-}\mathbf{C}_{eq}^{(2)}\text{-}\mathbf{K}$ system should be used.

In addition, different from the bilinear damping design, more than one effective mode may have to be chosen. In this case, overdamped effective modes are likely, and using the $\mathbf{M}\text{-}\mathbf{C}_{eq}^{(2)}\text{-}\mathbf{K}$ system is suggested.

Similar to the case of generally damped linear MDOF systems, once the modal parameters ω_i, ξ_i, and $\boldsymbol{p}_i$ are calculated, the accompanist vector, $\boldsymbol{q}_i$, as well as vectors $\boldsymbol{\Phi}_i$ and $\boldsymbol{\Psi}_i$ can be obtained (see Equation 7.85).

8.3.2.2 Response Reevaluation

8.3.2.2.1 Effective Damping Ratios

In the first round of damper design, the response of the base system is calculated and the required damping ratio is determined. Then, the nominal damper specifications, c_{eq}, as well as the damping exponent β for individual dampers and the damping configuration of the total system, are obtained. The damping configuration matrix $\mathbf{C}_{con}$ and equivalent damping matrix $\mathbf{C}_{eq}^{(2)}$ are determined by the corresponding effective period T_i, participation factor Γ_i, and shape function $\boldsymbol{p}_i$, as well as the vectors Φ_i and Ψ_i. The accumulated modal mass ratio, γ_{mS}, is also available.

From the complex-valued mode shape $\boldsymbol{p}_i$ (if it is indeed complex valued), it is possible to work with the corresponding real-valued shape function $\bar{\boldsymbol{p}}_i$, given by (see Equation 7.96)

$$\bar{\boldsymbol{p}}_i = \left\{\frac{|\boldsymbol{p}_i|}{p_{ni}}\right\} \bullet \text{sgn}(\boldsymbol{p}_i) = \{\bar{p}_{ij}\}, \quad j = 1,2,\ldots,n \tag{8.73}$$

where the shape function $\boldsymbol{p}_i$ is obtained through the proportionally damped **M**-**C**-**K** system. Note that, when $-90^\circ < \angle(p_{ij}) < 90$, $\text{sgn}(p_{ij}) = 1$. Otherwise, $\text{sgn}(p_{ij}) = -1$.

If the displacement vector $\mathbf{d}_{max_i}$ is known, the ith effective damping ratio can be calculated by

$$\xi_{eff_i}^{(.)} = \frac{c_{eq}\omega_i^{\beta} A_{\beta} \sum_{j=1}^{S} x_{ji}^{\beta+1}}{2\pi \mathbf{d}_{max_i}^{T} \mathbf{K} \mathbf{d}_{max_i}} \tag{8.74a}$$

Here superscript (.) of the damping ratio $\xi_{eff_i}^{(.)}$ means that an iterative approach may be needed to achieve the proper effective damping ratios. For simplicity, in the following discussion, superscript (.) is omitted. The design engineer should be careful to identify whether the $(.)^{th}$ iteration is needed.

However, $\mathbf{d}_{max_i}$ is often unknown before the structural responses are calculated. Unlike bilinear design where the quadratic Equation 8.6 and its solution Equation 8.13 can be used, $\mathbf{d}_{max_i}$ and the other responses in Equation 8.74a have to be found by iterative approaches. This is discussed in the next subsection (see, e.g., Equation 8.96b) and the iterative approach can provide the value of the effective damping ratios.

Note that not using the Penzien constant often affects the first effective mode, so that if the Penzien constant, p_c, is considered, Equation 8.74a can be rewritten as

$$\xi_{eff_1}^{(.)} = \frac{(p_c)^{\beta-1} c_{eq}\omega_1^{\beta} A_{\beta} \sum_{j=1}^{S} x_{j1}^{\beta+1}}{2\pi \mathbf{d}_{max_1}^{T} \mathbf{K} \mathbf{d}_{max_1}} \tag{8.74b}$$

In Equations 8.74, it is supposed that a total of S dampers are installed. Here ω_i and x_{ji} are the ith effective natural frequency and modal amplitude at the jth location, respectively, and

$$x_{ji} = G_{mj}\left(d_{ji} - d_{j-1,i}\right) \quad (m) \tag{8.75a}$$

When the Penzien constant is considered,

$$x_{j1} = p_c G_{mj}\left(d_{j1} - d_{j-1,1}\right) \quad (m) \tag{8.75b}$$

8.3.2.2.2 Numerical Damping Coefficient

Equations 8.74 and 8.75, the responses of the effective SDOF systems, are calculated by finding the ith numerical damping coefficient, etc. That is,

$$B_i = 3(\xi_{effi} + \xi_{oi}) + 0.9 \tag{8.76}$$

8.3.2.2.3 Modal Participation Factor

As mentioned in Chapter 4, for generally damped systems, the modal participation factor of the ith mode is no longer a scalar, but a vector. For the case of complex modes, the jth element of the vector is given by Equation 7.88. That is,

$$\Gamma_{ij} = \frac{\psi_{ij}}{\bar{p}_{ij}} = \begin{cases} 2\omega_i \left\{ \left| -\xi_i \varphi_{ij} + \sqrt{1-\xi_i^2}\, \psi_{ji} \right| + 0.4 \left| \varphi_{ij} \right| \right\} \Big/ p_{ij}, & \text{when } \left| -\xi_i \varphi_{ij} + \sqrt{1-\xi_i^2}\, \psi_{ij} \right| > \left| \varphi_{ij} \right| \\ 2\omega_i \left\{ 0.4 \left| -\xi_i \varphi_{ij} + \sqrt{1-\xi_i^2}\, \psi_{ij} \right| + \left| \varphi_{ij} \right| \right\} \Big/ p_{ij}, & \text{when } \left| -\xi_i \varphi_i + \sqrt{1-\xi_i^2}\, \psi_{ij} \right| \le \left| \varphi_{ij} \right| \end{cases} \tag{8.77}$$

The total modal participation vector of the ith mode is

$$\mathbf{\Gamma}_i = \{\Gamma_{ji}\} = \begin{Bmatrix} \Gamma_{1i} \\ \Gamma_{2i} \\ \dots \\ \Gamma_{ni} \end{Bmatrix} \tag{8.78}$$

Note that if the effective damping ratio to be designed is not very large, a normalized shape function can be used to calculate the modal participation factor, which reduces to a scalar. In an example given in Section 8.3.3, this simplified approach is discussed in more detail.

For an overdamped subsystem, the modal participation factor is still a scalar, given by

$$\Gamma_i^R = \boldsymbol{q}_j^R \, \mathbf{J} \tag{8.79}$$

8.3.2.2.4 Seismic Response Factors and Modal Displacement

The seismic response factors, C_{si}, and spectral displacement, d_{iD}, for vibration modes (real and complex) can be calculated using Equation 7.51. For an elastic structure, $R = 1$.

Similar to linear systems described in Chapter 7, if the first S "modes" contain one or more overdamped subsystems, the seismic response factors, C_{sk}^R, and spectral displacement, d_{kD}^R, for the corresponding overdamped subsystems need to be specified.

And overdamped displacement spectral values are needed to compute the term d_{kD}^R, which can be obtained using Equation 4.233. A simplified approximation can be obtained through numerical simulations. For example, using the 99 earthquake records, the special displacement, S_{Do}, can be established as

$$S_{Do} = 0.0748\left(T^R\right)^{1/2} + 0.0145\ln\left(T^R\right) - 0.0128T^R + 0.0252 \tag{8.80}$$

Thus, the spectral displacement, d_{iD}^R, can be written as

$$d_{iD}^R = S_{D1}\, S_{Do} \quad (m) \tag{8.81}$$

Overdamped velocity spectral values are needed to compute the term C^R_{sk}, which can also be obtained by using numerical simulations. For example, using again the 99 earthquake records, the special displacement, S_{Vo}, can be established as

$$S_{Vo} = -0.857\left(T^R\right)^{1/2} + 0.447\ln\left(T^R\right) + 1.928, \quad T^R \le 1.35 \quad (s) \tag{8.82a}$$

$$S_{Vo} = 0.0002\left(T^R\right)^2 - 0.0054\,T^R + 1.0814, \quad T^R > 1.35 \quad (s) \tag{8.82b}$$

The spectral acceleration, C^R_{si}, for the overdamped case can be written as

$$C^R_{sk} = S_{D1}\frac{2\pi}{T^R_k}S_{Vo} \tag{8.83}$$

8.3.2.2.5 *Lateral Displacement*

The lateral displacement of the ith complex mode is given by Equation 4.269. That is,

$$\mathbf{d}_{max_i} = d_{iD}\boldsymbol{\Gamma}_i \bullet \bar{\boldsymbol{p}}_i \quad (m) \tag{8.84}$$

where the shape function $\bar{p}_i$ is given by Equation 8.73. Meanwhile, the lateral displacement of the jth normal mode is repeated as

$$\mathbf{d}_{max_j} = \boldsymbol{\Gamma}_j d_{jD} \bullet \boldsymbol{p}_j \quad (m) \tag{8.85}$$

Here, the mode shape vector $\boldsymbol{p}_j$ is taken from the normal mode and is real valued.

The lateral displacement of the kth overdamped subsystem is given by

$$\mathbf{d}^R_{max_k} = d^R_{kD}\boldsymbol{\Gamma}^R_i \bullet \boldsymbol{p}^R_k \quad (m) \tag{8.86}$$

Similar to the proportionally damped system case, the total maximum lateral displacement can be calculated by using SRSS as

$$\mathbf{d}_{max} = S_M S_{Dn} S_{D\xi}\left(\sum_{i=1}^{S}\mathbf{d}^{\bullet 2}_{max_i}\right)^{\bullet 1/2} \quad (m) \tag{8.87}$$

in which the safety factors S_M, S_{Dn}, and $S_{D\xi}$ are defined in Equations 7.16, 7.59, and 7.60, respectively and the overdamped cases are included as mentioned in Chapter 7. Here,

$$S_{Dn} = 1\Big/\left(6.108\times10^{-7}n^3 - 5.719\times10^{-5}n^2 + 0.0015n + 0.525\right)$$

$$S_{D\xi} = 1\Big/\left(-3\,\xi_{design} + 3.12\right)$$

8.3.2.2.6 *Absolute Acceleration*

The calculated pseudo acceleration of the ith complex mode is given by

$$\mathbf{a}_{a_i} = C_{si}\,\boldsymbol{\Gamma}_i g \bullet \bar{\boldsymbol{p}}_i \quad \left(m/s^2\right) \tag{8.88}$$

where the shape function $\bar{\boldsymbol{p}}_i$ is given by Equation 8.73.

The pseudo acceleration of the ith normal mode is given as

$$\mathbf{a}_{s_i} = C_{si}\Gamma_i g \boldsymbol{p}_i = \begin{Bmatrix} a_{s1i} \\ a_{s2i} \\ \cdots \\ a_{sni} \end{Bmatrix} \quad (m/s^2) \tag{8.89}$$

The pseudo acceleration of the kth overdamped subsystem is given by

$$\mathbf{a}^R_{s_k} = C^R_{sk}\Gamma^R_k g \boldsymbol{p}^R_k \quad (m/s^2) \tag{8.90}$$

The procedure for an overdamped subsystem for sublinear damping can be carried out as described for linear generally damped systems.

Similar to the case of a proportionally damped system, the total maximum absolute acceleration can be calculated by using modified SRSS in the form

$$\mathbf{a}_a = S_M S_{An} S_{A\xi}\left[\sum_{i=1}^{S} \mathbf{a}^{\bullet 2}_{s_i} + (S_{Ai}\mathbf{J})\right]^{\bullet 1/2} = \begin{Bmatrix} a_{a1} \\ a_{a2} \\ \cdots \\ a_{an} \end{Bmatrix} \quad (m/s^2) \tag{8.91}$$

In Equation 8.91, the term $\mathbf{a}_{s_i}$ includes the acceleration $\mathbf{a}^R_{s_i}$. Also, in Equation 8.91, S_M is the same as defined in Equation 7.16 and S_{Ai} is determined by Equation 7.63. The safety factors S_{An} and $S_{A\xi}$ are defined in Equations 7.64 and 7.65, respectively. Again,

$$S_{An} = 1/\left(5.9868\times10^{-6}n^3 - 9.0063\times10^{-4}n^2 + 0.0395\,n + 0.4092\right);$$

$$S_{A\xi} = 1/\left(-193.7\,\xi^2_{design} - 25.2\,\xi_{design} + 2.76\right)$$

8.3.2.2.7 Lateral Force and Base Shear

For nonproportionally damped systems, the lateral force can also be calculated through Equation 7.66 and the base shear is calculated through Equation 7.70.

8.3.2.3 Summary of Simplified Design

The design procedure may be stated in the following three steps.

First, suppose the mass, stiffness, and the damping exponent β of the structure have been determined. Then, from the mass **M** and stiffness **K**, the fundamental period T_1, the corresponding mode shape $\boldsymbol{p}_1$ can be determined. In previous chapters, the method of how to find T_1 and $\boldsymbol{p}_1$ was discussed. With given parameters T_1 and β, the corresponding displacement d_1 and acceleration a_1 can be determined, as well as the effective damping ratio ξ_{eff1}. Consequently, the equivalent damping coefficient, c_{eq}, can be obtained.

Second, through the individual damping coefficient c_{eq}, the damping coefficient matrix, $\mathbf{C}_{eq}$, for the corresponding linear system can be constructed. The system has the identical damper configuration and identical damping ratio of the first mode (or effective mode for the sublinear system). Similar to the system with bilinear damping, the corresponding linear system provides more information about the higher modes, when the first effective mode is not sufficient to represent the total responses. By using mass **M** and stiffness **K**, the first several modal parameters can be obtained. Practically speaking, the newly formed $\mathbf{M}$-$\mathbf{C}_{eq}$-$\mathbf{K}$ system is generally damped.

Third, by using response estimation, such as SRSS, the responses of the higher modes of the newly generated system can be included. An alternative approach is to employ all the modal contributions of the corresponding linear system, instead of counting the second and higher modes, whereas the fundamental "modal" contribution is taken from the nonlinear system. This approach is simpler. When the nonlinearity of the system is not too high, say the damping exponent is larger than 0.75 and the supplemental damping is not very heavy, this approach often yields acceptable results.

If the computed responses are not satisfactory, trial-and-error iteration must be carried out until good design parameters are determined. A good estimation of the response is needed in order to reduce the computational burden.

8.3.3 Design Issues

In the above discussion, the procedure for simplified design when the required effective damping ratio is comparatively small, was presented. One of the issues is the expected errors if the effective damping ratio is not negligibly small. In the following paragraph, these possible errors are summarized.

8.3.3.1 SDOF Systems: Effective Mode

8.3.3.1.1 Iterations

Unlike bilinear damping, an iterative method to calculate the response of a structure for sublinear damping is necessary. Associated with iteration is the issue of converging to incorrect values or even nonconvergence. The methods of convergence are briefly discussed in the following paragraphs.

The computation of the effective damping ratio ξ_{eff} usually employs the Timoshenko equation, as shown in Equation 5.15. For convenience, Equation 8.74a is rewritten using Equation 8.84 as follows:

$$\xi_{eff_i} = \frac{c_{eq}\omega^{\beta}A_{\beta}\sum_{j=1}^{S} x_j^{\beta+1}}{2\pi \mathbf{d}_{max_i}^T \mathbf{K} \mathbf{d}_{max_i}} = \frac{c_{eq}\omega_i^{\beta-2}A_{\beta}\sum_{j=1}^{S} x_{ji}^{\beta+1}}{2\pi m_{eff_i} d_{iD}^2} \tag{8.92}$$

where $T_s < T_{eff_i} < T_L$

$$\begin{aligned}\xi_{eff_i} &= \frac{c_{eq}T_{eff_i}\,A_{\beta}\left(\Gamma_i G_{m_{ij}}\right)^{\beta+1}\sum_{j=1}^{S}\Delta\bar{p}_{ji}^{\beta+1}}{(2\pi)^{\beta+1}\,m_{eff_i}}\;\frac{\left[3\left(\xi_{eff_i}+\xi_{0_i}\right)+0.9\right]^{1-\beta}}{I_g^{1-\beta}}\\ &= \left\{\frac{c_{eq}T_{eff_i}\,I_g^{\beta-1}A_{\beta}\left(\Gamma_i G_{m_{ij}}\right)^{\beta+1}\sum_{j=1}^{S}\Delta\bar{p}_{ji}^{\beta+1}}{(2\pi)^{\beta+1}\,m_{eff_i}}\right\}\left[3\left(\xi_{eff_i}+\xi_{0_i}\right)+0.9\right]^{1-\beta}\end{aligned} \tag{8.93}$$

where

$$\Delta\bar{p}_{ji} = \bar{p}_{ji} - \bar{p}_{j-1,i} \quad (m) \tag{8.94}$$

Denote

$$a_i = \frac{c_{eq} T_{eff_i} \, I_g^{\beta-1} A_\beta \left(\Gamma_i G_{m_{ij}}\right)^{\beta+1} \sum_{j=1}^{S} \Delta\bar{p}_{ji}^{\beta+1}}{(2\pi)^{\beta+1} m_{eff_i}} \tag{8.95}$$

Thus,

$$\xi_{effi} = a_i \left[3\left(\xi_{eff_i} + \xi_{0_i}\right) + 0.9\right]^{1-\beta} \tag{8.96a}$$

This is a single variable nonlinear equation, which can be solved by iterative computations. That is, an assumed value $\xi_{effi} = \xi_{effi}^{(1)}$ is substituted into the right side of Equation 9.86a and the second value of $\xi_{effi}^{(2)}$ is calculated. The kth iteration is written as

$$\xi_{effi}^{(k+1)} = a_i \left[3\left(\xi_{eff_i}^{(k)} + \xi_{0_i}\right) + 0.9\right]^{1-\beta} \tag{8.96b}$$

The iteration will be stopped if a particular preset criterion of the iteration is satisfied. For example,

$$\left|\xi_{effi}^{(i+1)} - \xi_{effi}^{(i)}\right| \Big/ \xi_{effi}^{(i)} \le \varepsilon \tag{8.97}$$

as the criterion, where ε is a small number, i.e.,

$$\varepsilon = 0.1\% \tag{8.98}$$

In normal cases, the convergence is fast and usually, Equation 8.96b can be satisfied after 3 ~ 5 iterations.

Example 8.4

A base structure with identical mass and stiffness matrices as given in Example 8.2 is used in this example. Its original damping ratio is 2%. Two nominal dampers are installed in the first and second floors for a design damping ratio equal to 15% (see Figure 8.1). With supplemental dampers, the effective mode shape will be altered.

The shape function is approximated by using an equivalent linearly damped system with two linear viscous dampers installed in the first and second stories. The corresponding first mode shape and the shape function calculated using Equation 8.73 are, respectively,

$$\boldsymbol{p}_1 = \begin{Bmatrix} 0.3544 + 0.0476j \\ 0.6639 + 0.0858j \\ 0.8855 + 0.0305j \\ 1.0000 \end{Bmatrix} \quad \text{and} \quad \bar{\boldsymbol{p}}_1 = \begin{Bmatrix} 0.3576 \\ 0.6694 \\ 0.8861 \\ 1.0000 \end{Bmatrix}$$

Using this shape function, the modal participation factor, effective mass, and effective period of the first effective mode can be found as

$$\Gamma_1 = \frac{\bar{p}_i^T \mathbf{MJ}}{\bar{p}_i^T \mathbf{M}\bar{p}_i}\Big|_{i=1} = 1.2438; \quad m_{eff1} = \frac{\left(\bar{p}_i^T \mathbf{MJ}\right)^2}{\bar{p}_i^T \mathbf{M}\bar{p}_i}\Big|_{i=1} = 348.5253 \ \ (t),$$

$$\text{and } T_{eff1} = \frac{\bar{p}_i^T \mathbf{K}\bar{p}_i}{\bar{p}_i^T \mathbf{M}\bar{p}_i}\Big|_{i=1} = 1.496 \ \ (s).$$

Suppose the ground input level, I_g=2.634 (m/s^2), and the nominal sublinear damper has $\beta = 0.3$; and $c_{eq} = 500$ (kN/(m/s)$^{0.3}$). The iteration in Equation 8.95b is used to calculate the desired effective damping ratio.

First, according to Equation 8.95, $a_1 = 0.1074$. Assume that $\xi_{effi}^{(1)} = 0.13$. Then, $\xi_{effi}^{(2)} = 0.1325$, $\xi_{effi}^{(3)} = 0.1330$, and the iteration converges to $\xi_{effi}^{(4)} = 0.1332$. For comparison, assume that $\xi_{effi}^{(1)} = 0.05$. Then, $\xi_{effi}^{(2)} = 0.1295$; $\xi_{effi}^{(3)} = 0.1324$; $\xi_{effi}^{(4)} = 0.1331$; and the iteration converges to $\xi_{effi}^{(5)} = 0.1332$. As a first comparison, assume that $\xi_{effi}^{(1)} = 1.0$. Then, $\xi_{effi}^{(2)} = 0.2815$; $\xi_{effi}^{(3)} = 0.1624$; $\xi_{effi}^{(4)} = 0.1391$; $\xi_{effi}^{(5)} = 0.1344$, and eventually the iteration converges to $\xi_{effi}^{(7)} = 0.1332$.

Equations 8.92 and 8.95 only work when $T_s < T_{effi} < T_L$. The iterative formulas for the cases of $T_{effi} < T_0$, $T_0 < T_{effi} < T_s$ as well as $T_L > T_{effi}$ can also be found. Once the proper value of the effective damping ratio is calculated, Equation 7.51 can be used to calculate the seismic response factors, C_{si}, and the spectral displacement, d_{iD}. Then, the procedure discussed in the above subsection can be followed to estimate the structural responses. Since this method does not directly calculate the displacement, it is referred to as *indirect iteration*.

In the literature, many authors suggest that the displacement can be found iteratively by first assuming an initial displacement $\mathbf{d}_{max_i}^{(1)}$ and using Equation 8.74 to calculate the effective damping ratio $\xi_{maxi}^{(1)}$. Then, the numerical damping coefficient $B^{(1)}$ is found, which is used to provide the second round of iterative value $\mathbf{d}_{max_i}^{(2)}$. However, such a *direct iteration* may converge to incorrect points.

8.3.3.1.2 Displacement and Acceleration

It has been shown that when damping becomes larger in linear SDOF systems, the relationship between the relative displacement, d_{max}, and the absolute acceleration, a_a, expressed in Equation 8.99 can yield large errors:

$$a_a = \omega_n^2 d_{max} \tag{8.99}$$

Thus, an alternative approach is recommended as

$$a_a = \sqrt{1+4\xi^2}\ \omega_n^2 d_{max} \tag{8.100}$$

to improve the estimation. When the damping exponent is close to unity, Equation 8.100 can be used as a simplified approach. However, when the damping exponent is closer to zero, Equation 8.100 will be insufficient. A practical way to improve this situation is to introduce a factor that is a function of the damping exponent, effective damping, and the number of the degrees-of-freedom, which is discussed in the next section.

8.3.3.1.3 Effective Damping Ratio of SDOF System

Problems in bilinear damping design on the modeling or formulation of effective damping also exist in sublinear damping. In general, Timoshenko damping often overestimates the damping effect.

Thus, if the nonlinearity introduced by the sublinear dampers is high, the currently used design approach can be improved by introducing a twofold effort. The first is using the force-based effective damping ratio to reduce overestimation. The second is to manipulate the MDOF system through specific sublinear response spectra. When nonlinearity is not sufficiently high, however, other methods can be used, as discussed in Section 8.2.

The formula of force-based effective damping is used for the sublinear damping as the secondary choice, which is repeated here as

$$\xi_{eff} = \frac{c_\beta x_0^{\beta-1}}{2\,k} \tag{8.101}$$

8.3.3.1.4 Justification for Computation of Displacement due to Damping Exponent

In the above discussion, the effective damping ratio is formulated by equating the linear and nonlinear damping forces. From the concept of Timoshenko damping, if the damping force of a sublinear system is equal to a linear system, the displacement of the sublinear system should be smaller than in the linear system. This is conceptually seen in Figure 8.6, where the areas under the curves of the damping force vs. displacement represent energy dissipated by damping and the two areas are equal.

In Figure 8.6, the sublinear system is generated with a damping exponent of 0.2. In order to simplify the comparison, the damping force is normalized to be unity and the displacement of the linear system is also normalized to be unity. Note that Figure 8.6 only shows the energy dissipation in the first quadrant.

With respect to the issue of equal energy dissipation, when the damping exponent β equals unity, the displacement is normalized to be unity. When the damping exponent β is equal to zero, the displacements will be smaller. The corresponding factor is $4/\pi \approx 0.79$. If the damping exponent β is between 0 and 1, the two displacements of the linear and sublinear system will have approximately the following ratio, given by

$$\frac{d_l}{d_\beta} = \frac{4}{\pi}\int_0^{\frac{\pi}{2\omega}} \cos^{\beta+1}(\omega t)\, d\omega t = \frac{2}{\sqrt{\pi}} \frac{\Gamma\left(\frac{\beta}{2}+1\right)}{\Gamma\left(\frac{\beta}{2}+1.5\right)} \tag{8.102}$$

where d_β and d_l denote the displacements of the sublinear and linear systems, respectively.

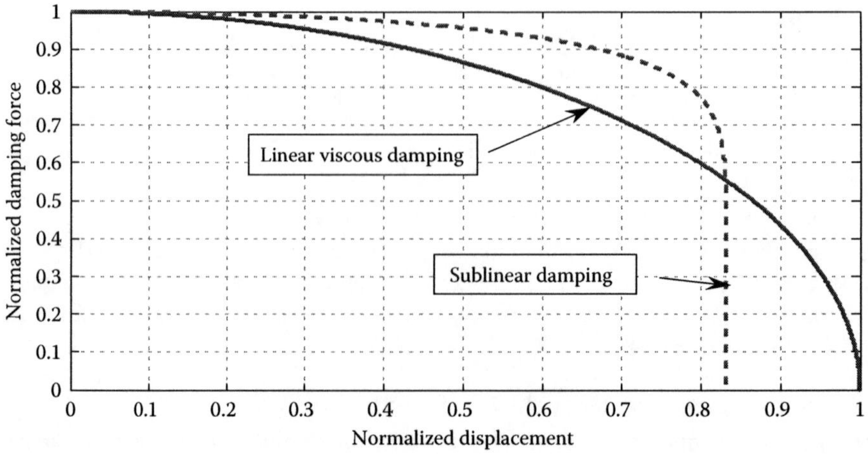

FIGURE 8.6 Comparison of linear and nonlinear displacements.

Equation 8.102 is further approximated by using a linear function,

$$\frac{d_\beta}{d_l} \approx a\beta + b \tag{8.103}$$

where a and b are constant coefficients.

In conclusion, by considering the effect of the damping exponent on the displacement computation, the calculated response based on the effective damping mentioned in the previous section needs to be justified.

8.3.3.1.5 *Justification for Computation of Displacement due to Effective Damping Ratio*

In the above discussion, the necessary modification of the displacement caused by the nonlinear effect on the damping exponent was explained. A second necessary modification due to the linearization procedure on the effective damping ratio is now considered.

In Chapter 7, it was mentioned that when the damping ratio is not 5%, the spectral value needs to be modified by the numerical damping coefficient B, approximately defined as

$$B \approx 0.3\xi + 0.9$$

This damping ratio is taken from a linear system with viscous damping. When the damping ratio is taken from a sublinear system, it should also be modified, and the modification factor can be written as a function of the effective damping ratio.

Similar to the modification expressed in Equation 8.103, it appears reasonable that

$$\frac{d_\xi}{d_l} \approx a\xi_{eff} + b \tag{8.104}$$

where a and b are coefficients. In Equation 8.104, a and b are also generic coefficients, unrelated to those used in Equation 8.103. Furthermore, in Equation 8.104, d_ξ and d_l denote the displacements of the sublinear and linear systems, respectively. Subscript ξ denotes the consideration of the effective damping exponent ξ_{eff}.

Practically speaking, the effect of both the effective damping ratio and the damping exponent on the displacement estimation for a sublinear system can be combined to simplify the design procedure. For this purpose, the following expression may be used:

$$\frac{d_{\xi\beta}}{d_l} \approx a\,\xi_{eff}^2 + b\,\xi_{eff}\,\beta + c\,\beta^2 + d\,\xi_{eff} + e\beta + f \tag{8.105}$$

where $d_{\xi b}$ stands for the estimated displacement affected by the damping ratio ξ and the damping exponent β, which is seen as the estimated displacement of the sublinear system through the linearization approaches. Here, a, b, c, d, e, and f are coefficients.

8.3.3.1.6 *Further Justification for Computation of Displacement due to Effective Frequency*

The above justifications are based on the logic that effective damping is obtained by equating the maximum sublinear and linear viscous damping forces by assuming that the responses are steady-state sinusoidal displacements. In addition, Timoshenko damping is determined by equating the energy dissipations of the sublinear and linear systems with the same assumption of sinusoidal displacements.

However, neither the force-equating approach nor the energy-equating approach can precisely represent equal displacement, because the system is nonlinear, for which the pure sinusoidal approach is only an approximation. In addition, the responses of the structure are not steady-state responses. In Timoshenko damping, the influence of the natural frequency is not considered. However, under earthquake excitations, the influence of the effective frequency is a concern. It can be expressed as the third modification factor, which may be approximately written as a linear function of the effective natural frequency as

$$\frac{d_\omega}{d_l} \approx a\,\omega_{eff} + b \tag{8.106}$$

where a and b are also generic coefficients. Also in Equation 8.106, d_ω and d_l denote the displacements of the sublinear and linear systems, respectively. Subscript ω denotes the consideration of the effective frequency ω_{eff}.

If the influence of the frequency is considered independently, the following approximation can be used:

$$\frac{d_\omega}{d_l} = \eta_\omega = a\,f_{eff} + b \approx 0.49\,f_{eff} + 2.12$$

where f_{eff} is the effective natural frequency.

In the previous chapters, it was shown that the natural frequency or period is affected by the number of stories n. That is,

$$T = 0.1\,n \quad (s)$$

In sublinear systems, since the value of the first effective period is taken from the original structure, the above equation can still be used. In fact, the proportional relationship between the period and the number of stories often provides a valid approximation. Replacing T by T_{eff}, the above equation is rewritten as

$$T_{eff} = 1/f_{eff} = a\,n \quad (s) \tag{8.107}$$

where a is again a generic coefficient.

8.3.3.1.7 Damping Nonproportionality and Overdamping

Since the dampers are sublinear, the total system must be nonlinear and the conventional approach to determine if the system is generally damped cannot be used. However, similar effects of nonproportional damping and overdamping still exist for nonlinear MDOF systems. These effects include some of the energy transferred between the effective modes and contained in the overdamped pseudo modes.

As mentioned in Chapter 4, it is the energy transfer among vibration "modes" that results in the complex modal parameters. This energy transfer results from damping nonproportionality. Thus, it is understandable that in nonlinear damping, for the most part, energy transfers are magnified. In addition, it is rare that the linearized damping matrix, $\mathbf{C}_{eff}$, which was obtained in the previous section, is purely proportional.

In addition, certain modes of the MDOF system with sublinear damping are likely to be overdamped. It is quite common that with the linearized damping matrix, $\mathbf{C}_{eff}$, obtained in the previous section, the equivalent linear system contains overdamped subsystems.

To determine if the system is overdamped, it is neccessary to see if real-valued eigenvalues exist. To check if the system has complex modes, several indicators can be used also. These indicators are discussed previously.

Since the effective natural periods are not significantly affected by using the normal mode approach, the initial estimation of the periods can be carried out by assuming proportional damping. To use this approach, the complex-valued mode shape must be rewritten with pure real numbers.

8.3.3.1.8 Cross Effect

In Chapter 6, cross effect in linear systems, especially those in between two horizontal directions, was only briefly introduced. A structure with sublinear dampers cannot be decoupled as a linear system, because energy transfer always exists. In this case, cross effect may be magnified. A practical way to consider cross effect for the design of sublinear dampers is to use larger safety factors.

8.3.3.1.9 Supporting Stiffness

Similar to linear and bilinear systems, the supporting stiffness is also an important issue for sublinear systems. Compared with linear viscous damping, the damping force of a sublinear damper will not vary as much as a linear damper. Therefore, a sublinear damper can enter the range of insufficient supporting stiffness considerably earlier than linear viscous dampers. This, therefore, demands a stronger supporting stiffness, as is the case for bilinear dampers.

8.3.3.2 MDOF Systems

8.3.3.2.1 Mode Shapes

From SDOF to MDOF systems, mode shapes are used to distribute the effective modal responses. Similar to a linear MDOF system, choosing an incorrect shape function will be a major source of inaccuracies in sublinear damping design. To date, there is no effective method to reduce the design error. Therefore, for precise design, time history analysis should be used together with spectra-based design.

8.3.3.2.2 Damping Exponent

It is assumed that all sublinear dampers for an MDOF system have the same damping exponent β. When the entire system is represented by several effective modes, i.e., S effective modes of interest, they are all affected by an identical damping exponent β. That is,

$$\beta_i = \beta \equiv \text{cont. } i = 1,2,\ldots,S \tag{8.108}$$

Practically speaking, this assumption is not exactly correct. The damping exponents of individual dampers are rarely identical. The authors have investigated the possible errors by using Equation 8.92 and have concluded that the errors are comparatively small. Figure 8.7 shows an example of simulated earthquake response of damping force vs. displacement of a 55-story building installed with sublinear dampers, whose damping exponent is equal to 0.15 and the effective damping ratio of the first effective mode is 15%. From this example, it is realized that the plot aggregates well with the damping force, $f_d = c_{eq}|\dot{x}|^{0.15}\text{sgn}(\dot{x})$.

8.3.3.2.3 Period

When the mass and stiffness matrices of a structure are given, the undamped period, T_{oi}, is determined. It is assumed that the effective period, T_{effi}, for the first several effective modes is fixed to equal T_{0i}. That is, they are not affected by damping, as expressed by Equation 8.109:

$$T_{effi} \equiv T_{oi} \quad (s) \quad i = 1,2,\ldots,S \tag{8.109}$$

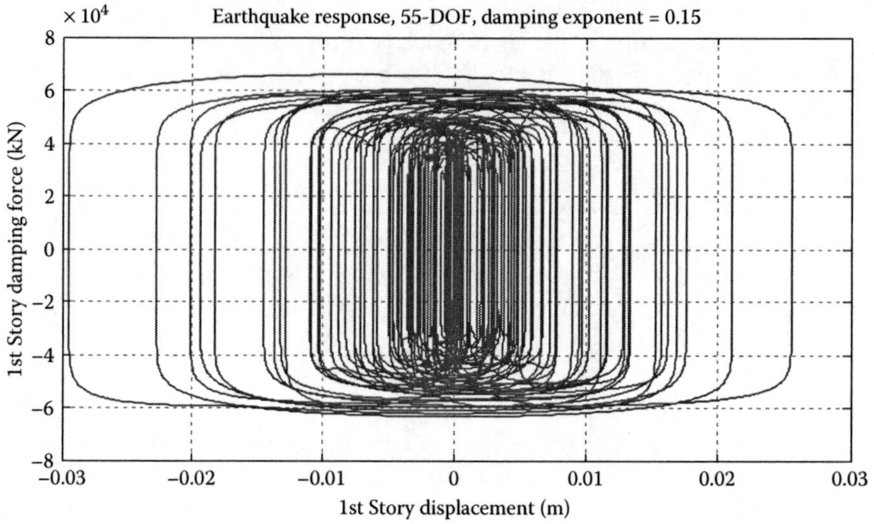

FIGURE 8.7 Damping force vs. displacement of a sublinear system.

This assumption is also not exactly correct. The effective periods will vary for these effective modes when the displacements vary. Therefore, T_{oi} cannot be defined exactly. In addition, the supplemental damping will also change the value of T_{oi}. The authors have investigated the errors introduced by Equation 8.109. In general, these errors are also of negligible magnitude. In the following discussion, for simplicity, T_{effi} is not distinguished from T_{oi} and T_i unless specifically addressed.

8.3.3.2.4 Modal Participation Factor

The above discussion actually implies an important assumption; that is, the total responses of a sublinear system can be represented by its first several effective modes and those effective modes do exist. Thus, to identify these modes, the modal participation factors obtained by assuming proportional damping are not supposed to change; that is,

$$\Gamma_i \equiv \text{cont. } i = 1,2,\ldots,S \tag{8.110}$$

Again, this assumption is an approximation with acceptable errors.

With the above assumptions, the basic approach of the sublinear damper design can be performed based on the response spectra approach. Since the system can be highly nonlinear, similar to a bilinear system, nonlinear response spectra may need to be generated. Once the nonlinear, specifically sublinear, response spectra are generated, the remaining design procedure is similar to the spectra design approach commonly used in the design profession.

The authors did not exhaust all possible combinations of earthquake records in this formulation. Because sublinear response spectra are highly dependent on the selection of input earthquakes, the numbers used in this book should be treated as illustrations of a generic methodology, for which the step-by-step procedure of the sublinear spectrum generation is given so that users can generate their own sublinear response spectra. In the next section, these spectra are given. Similar to bilinear damping design, the nonlinear response spectra should be generated by using code-compatible ground excitations specified for local applications.

8.3.4 Damper Specifications

Similar to linear viscous dampers and bilinear dampers, once the response estimation is satisfactory and the basic parameters of a nominal damper are determined, the next step is to choose the devices. Since sublinear dampers are likely to be hydraulic devices, the damper specifications and selections are similar to those used for linear viscous dampers, which are considered in Section 7.2.

8.4 NONLINEAR RESPONSE SPECTRA APPROACH WITH SUBLINEAR DAMPERS

8.4.1 General Description

In the above section, the simplified design procedure was presented and the possible errors were discussed. It is seen that one of the most serious sources of errors comes from the iterative approach, when the effective modal response is calculated. When the required damping ratio is large, this type of error can be significant. To improve the design, nonlinear spectra can be used to avoid the iteration process.

A major step in damper design is to determine the proper damping coefficient, c_β, and to calculate the corresponding relative displacement vector, $\mathbf{d}_{max}$, and absolute acceleration, $\mathbf{a}_a$, of the structure. From $\mathbf{d}_{max}$ and $\mathbf{a}_a$, other control parameters, such as base shear, floor drift, and overturning moment, can be determined. In addition, the amount of supporting stiffness may also be needed. Thus, determining the effective modal response is one of the key steps.

The sublinear response spectra can be used to calculate the effective modal response, which can reduce the errors generated through iterative approaches.

8.4.1.1 Control Parameters and Response Spectra

Suppose the parameters of the ith effective mode have already been determined, namely, the effective period T_i, the effective damping ratio ξ_{effi}, and the damping exponent β_i. The next step is to estimate the modal response. This task can be carried out using the sublinear response spectra.

Alternatively, proper design parameters of the ith effective mode based on an estimated response can be obtained, followed by using the sublinear spectra.

8.4.2 Response Estimation

For an SDOF sublinear system, the governing equation can be written as

$$m\ddot{x} + c\dot{x} + c_\beta |\dot{x}|^\beta \operatorname{sgn}(\dot{x}) + kx = -mx_{g''} \tag{8.111}$$

where c_β is the damping coefficient of the sublinear damper, which can be expressed as

$$c_\beta = 2\xi_{eq}\,\omega_{eff}\,m \quad \left(\text{kN-s}^\beta/\text{m}^\beta\right) \tag{8.112}$$

where ω_{eff} is the effective "natural" frequency with

$$\omega_{eff} \approx \sqrt{\frac{k}{m}} \quad (\text{rad/s})$$

Substituting Equation 8.112 into Equation 8.111 and dividing mass m on both sides of the equation yields

$$\ddot{x} + 2\xi\,\omega_{eff}\,\dot{x} + 2\,\xi_{eq}\,\omega_{eff}\,|\dot{x}|^{\beta}\,\mathrm{sgn}(\dot{x}) + \omega_{eff}^{2}x = -x_{g''} \tag{8.113}$$

in which the effective "natural" frequency can be replaced by the effective period as

$$\omega_{eff} = \frac{2\pi}{T_{eff}} \quad (\mathrm{rad/s})$$

Equation 8.113 contains three control parameters: the effective period T_{eff}, the equivalent damping ratio ξ_{eq}, and the damping exponent β. In other words, to generate the sublinear response spectra, these three parameters are sufficient. The effective damping ratio, ξ_{eff}, is used under the consideration described in Equation 8.112. Now, by using a group of prespecified earthquake records, the relative displacement, absolute acceleration, and damping force of the mean value plus one standard deviation can be obtained. Table 8.2 shows an example of these factors.

In Table 8.2, the damping exponent is $\beta = 0.1$. Similar tables are developed for different values of β. Table 8.2A lists the displacement with unit (cm) and Table 8.2b gives the acceleration with unit (g). Note that when the responses are calculated, the mass of the SDOF system is assumed to be m = 100 (t). Thus, the damping force can also be provided as a ratio of 100 (kN)/100 (t).

In both tables, the columns are marked by different periods from 0.4 to 3.0 (s), and the rows provide different equivalent damping ratios from 0.01 to 1.00.

The sublinear responses and the linear responses are examined first. As an example, consider a system with a damping ratio of 0.05, a period of 1 sec, and 0.4 (g) input. In this case, the displacement is 15.8489 (cm), the acceleration is 0.6862, which implies 0.6862 (g), and the damping force is 586.4 (kN) with the mass equal to 100 (t). If the system is linear with the viscous damping, it is seen that according to the equations suggested by NEHRP 2003 (BSSC, 2003), d = 0.1 T = 10 (cm), which is considerably smaller than 15.8 (cm).

Additionally, the acceleration is a = 0.4/T = 0.4, also much smaller than 0.69 (g). On the other hand, the amplitude of the damping force is $f_d = 2 \times 0.05 \times (2\pi/T)\ m = 6.28$ (kN), which is obviously larger than 586 (kN).

Furthermore, if the system is linear, the relationship between the displacement and the acceleration is given approximately by

$$a = \sqrt{1+4\xi^2}\,\frac{4\pi^2}{9.8\,T^2}\,d \quad (g) \tag{8.114}$$

TABLE 8.2A
Displacement (cm), β = 0.1, Input Level = 0.4 (g)

ξ_{eq} \ T (s)	0.4	0.6	0.8	1.0	1.5	2.0	3.0
0.01	5.7088	12.9384	19.6412	26.9801	37.0336	45.1123	83.3525
0.05	2.5065	6.2830	9.9336	15.8489	22.4376	30.3342	56.0481
0.10	0.9362	2.9292	5.5945	9.2355	15.6599	22.1647	40.0841
0.20	0.0436	0.5341	1.6672	3.2021	7.8998	12.7648	22.9006
0.40	0.0128	0.0184	0.0476	0.2549	1.6871	3.9294	10.2584
0.60	0.0088	0.0128	0.0166	0.0204	0.2714	1.1991	4.3906
0.80	0.0068	0.0098	0.0128	0.0157	0.0303	0.2780	1.9457
1.00	0.0056	0.0080	0.0104	0.0128	0.0185	0.0492	0.7984

TABLE 8.2B
Acceleration (g), β = 0.1, Input Level = 0.4 (g)

ξ_{eq} \ T (s)	0.4	0.6	0.8	1.0	1.5	2.0	3.0
0.01	1.4590	1.4626	1.2475	1.0958	0.6690	0.4588	0.3761
0.05	0.7441	0.7808	0.6842	0.6862	0.4337	0.3295	0.2672
0.10	0.4608	0.4845	0.4715	0.4692	0.3461	0.2727	0.2131
0.20	0.3940	0.3631	0.3439	0.3246	0.2755	0.2301	0.1714
0.40	0.4003	0.4000	0.3924	0.3657	0.2953	0.2449	0.1868
0.60	0.4004	0.4002	0.4001	0.3994	0.3636	0.3093	0.2303
0.80	0.4006	0.4003	0.4002	0.4001	0.3962	0.3629	0.2821
1.00	0.4007	0.4004	0.4004	0.4002	0.4001	0.3919	0.3276

With $\xi_{eq} = 0.05$ and $T_{eff} = 1$ (s), the calculated value is $(1 + 0.05^2)^{1/2}(4\pi^2)0.1585/9.8 = 0.6549$ (g). Compared to the value of 0.6862 (g), the calculated acceleration is 4.6% smaller.

In fact, all the other pairs of displacements, d, and accelerations, a, can be examined, but none will exactly follow the description shown in Equation 8.114. Rather, for a sublinear system, a factor of $\eta > 1$ is chosen, such that

$$a = \eta\sqrt{1+4\xi_{eff}^2}\,\frac{4\pi^2}{9.8\,T_{eff}^2}\,d \quad (g) \tag{8.115}$$

Note that each pair of d and a are actually the "modal" displacement and acceleration. For simplicity, these notations are used for now and the main design procedures will be revisited in Section 8.5.

8.4.2.1 Linear Interpolations

In the above tables, only eight equivalent damping ratios and seven damping exponents were provided. When values are not given, linear interpolations can be used.

As stated earlier, there is a basic difference between a sublinear system and a bilinear system. In an MDOF bilinear system, although several dampers may enter yielding status at different times, it can still be assumed that the entire system will have an equivalent parallelogram force–displacement relationship without introducing significant errors. Therefore, the control parameters of the equivalent parallelogram can be directly determined based on the parameters of the stiffness of the structure and the parameters of these dampers. Sublinear dampers, on the other hand, do not allow us to directly link the control parameters of the SDOF system to the real-world MDOF system.

The reason Tables 8.2A and 8.2B are generated is to find the so-called "equivalent damping ratio" ξ_{eq}, instead of using the control parameter ξ_{eff}. However, the effective damping ratio ξ_{eff}, (instead of the equivalent damping ratio ξ_{eq}) should be used as the control parameter. To account for this issue, several steps are required. The first is to assume that the four input levels of 0.4, 0.6, 0.8, and 1.0 (g) are sufficient to represent all the practical cases. If the four levels are insufficient, linear interpolation can be used as described by the following example. Assume the input level is, for example, specified as 0.4 (g).

The second step is to use the specific parameters of the effective period and damping exponent to generate two columns of vectors. The first column is the displacement at specific periods and damping exponent (including the input level) through linear interpolations. The second column is the corresponding effective damping ratios using the formula linking the effective and equivalent dampings.

The third step is to use interpolation again to finally determine the corresponding displacement, followed by the acceleration and the damping force.

This simplified approach can be summarized quantitatively as follows:

1. Specify the input level, for example, 0.4 (g).
2. With the specific period T and damping exponent b, find four columns of vectors, namely, $D(T_1,\beta_1)$, $D(T_2,\beta_1)$, $D(T_1,\beta_2)$, and $D(T_2, \beta_2)$. Here

$$T_1 < T < T_2 \quad (s) \tag{8.116a}$$

and

$$\beta_1 < \beta < \beta_2 \tag{8.116b}$$

In the above paragraph, $\boldsymbol{d}(T_1, \beta_1)$ is the column of displacement, when the period and the damping exponent are T_1 and β_1, respectively, and so on.

3. Using linear interpolation,

$$\boldsymbol{d}_1 = \frac{\boldsymbol{d}(T_2,\beta_1) - \boldsymbol{d}(T_1,\beta_1)}{T_2 - T_1}(T - T_1) + \boldsymbol{d}(T_1,\beta_1) \quad (\text{cm}) \tag{8.117a}$$

and

$$\boldsymbol{d}_2 = \frac{\boldsymbol{d}(T_2,\beta_2) - \boldsymbol{d}(T_1,\beta_2)}{T_2 - T_1}(T - T_1) + \boldsymbol{d}(T_1,\beta_2) \quad (\text{cm}) \tag{8.117b}$$

Furthermore, the displacement vector is determined by

$$\boldsymbol{d} = \frac{\boldsymbol{d}_2 - \boldsymbol{d}_1}{\beta_2 - \beta_1}(\beta - \beta_1) + \boldsymbol{d}_1 \quad (\text{cm}) \tag{8.117c}$$

Note that the vector $\boldsymbol{d}$ has eight elements in this case, that is,

$$\boldsymbol{d} = \{d_i\} \quad (\text{cm}), \quad i = 1, 2, \ldots, 8 \tag{8.118}$$

Each element d_i is associated with the ith equivalent damping ratio ξ_{eqi}.

4. The column of equivalent damping ratios ξ_{eqi} is now used to find the corresponding column vector with elements ξ_{effi}. In more general cases,

$$\frac{c_\beta}{c_{eq}} = \beta_1^\beta \, p_{11}^{\beta-1} \tag{8.119}$$

where p_{11} is the first element of the mode shape of the M-K system; p_{11} is not necessarily equal to the first story modal displacement under the triangle assumption such that $p_{11} = 1/n$. The shape function should be normalized to let the top level (the roof) modal displacement equal unity. That is,

$$p_{n1} = 1$$

Now, since

$$c_\beta = 2\,\xi_{eff}\,\omega_{eff} \quad \left(\text{kN-s}^\beta/\text{m}^\beta\right)$$

and

$$c_{eq} = 2\,\xi_{eq}\,\omega_{eff} \quad \left(\text{kN-s}^\beta/\text{m}^\beta\right)$$

results in

$$\xi_{eff} \approx \xi_{eq}\omega_1^\beta\, p_{11}^{\beta-1} \tag{8.120}$$

Therefore, for the eight elements,

$$\xi_{effi} \approx \xi_{eqi}\omega_1^\beta\, p_{11}^{\beta-1}\ i = 1, 2, \ldots, 8 \tag{8.121}$$

5. Using the given design value of ξ_{eff}, the interpolation method is used again to finally determine the required displacement $\boldsymbol{d}$, as follows:

$$\boldsymbol{d} = \frac{\boldsymbol{d}\left(\xi_{eff1}\right) - \boldsymbol{d}\left(\xi_{eff2}\right)}{\xi_{eff2} - \xi_{eff1}}\left(\xi_{eff2} - \xi_{eff}\right) + \boldsymbol{d}\left(\xi_{eff2}\right) \quad (\text{cm}) \tag{8.122}$$

Here, $\boldsymbol{d}(\xi_{eff(.)})$ is the displacement, when the effective damping $\xi_{eff(.)}$ is used, and

$$\xi_{eff1} < \xi_{eff} < \xi_{eff2} \tag{8.123}$$

6. The corresponding acceleration can be calculated as follows:

$$\boldsymbol{a}_1 = \frac{\boldsymbol{a}\left(T_1,\beta_1\right) - \boldsymbol{a}\left(T_2,\beta_1\right)}{T_2 - T_1}\left(T - T_2\right) + \boldsymbol{a}\left(T_2,\beta_1\right) \tag{8.124a}$$

$$\boldsymbol{a}_2 = \frac{\boldsymbol{a}\left(T_1,\beta_2\right) - \boldsymbol{a}\left(T_2,\beta_2\right)}{T_2 - T_1}\left(T - T_2\right) + \boldsymbol{a}\left(T_2,\beta_1\right) \tag{8.124b}$$

and

$$\boldsymbol{a} = \frac{\boldsymbol{a}_2 - \boldsymbol{a}_1}{\beta_2 - \beta_1}\left(\beta - \beta_1\right) + \boldsymbol{a}_1 \tag{8.124c}$$

where $\boldsymbol{a}(T_1,\beta_1)$ is the column of acceleration, when the period and the damping exponent are T_1 and β_1, respectively. Vector $\boldsymbol{a}$ has eight elements in this case. That is,

$$\boldsymbol{a} = \{a_i\} \quad (g), \quad i = 1, 2, \ldots, 8 \tag{8.125}$$

Each element a_i is determined by using the ith effective damping ratio ξ_{effi}. The required acceleration of the fundamental effective mode is given by

$$\boldsymbol{d} = \frac{\boldsymbol{a}(\xi_{eff1}) - \boldsymbol{a}(\xi_{eff2})}{\xi_{eff2} - \xi_{eff1}}(\xi_{eff2} - \xi_{eff}) + \boldsymbol{a}(\xi_{eff2}) \quad (g) \tag{8.126}$$

where $\boldsymbol{a}(\xi_{eff(.)})$ is the displacement, when the effective damping $\xi_{eff(.)}$ is used.

7. The damping force can be similarly calculated.

8.4.2.2 MDOF Systems

Returning to the main design procedures, the nonlinear spectra approach outlined above actually obtains the "modal" displacement, d_{iD}, and acceleration, C_{si}. That is,

$$d_{iD} = d \quad (m) \tag{8.127}$$

and

$$C_{si} = a/g \tag{8.128}$$

Once the effective modal responses are determined, Equations 8.84 through 8.91 can be used to determine the corresponding displacement, acceleration, and lateral force vectors, as well as base shear.

8.4.3 Design Issues

In the above procedure, the use of nonlinear spectra to determine the seismic responses can be seen as effective modal responses. These responses belong to some SDOF systems. To obtain these SDOF systems from an MDOF structure, certain conditions should first be satisfied. Without loss of generality, consider a simple example shown in Figure 8.8.

In Figure 8.8a, three sublinear dampers are installed, one in each story. For the sake of simplicity, suppose the geometric magnification factors in each floor are identical and equal to unity. Also, suppose all the dampers have the same specifications with an identical damping coefficient c_{eq} and exponent β. The mass, velocity, and the ith mode displacement (mode shape function) are marked in Figure 8.8a, and are also used in Figures 8.8b and c. Also, for simplicity, assume that the base structure is proportionally damped. The equation of motion can be written as

$$\mathbf{M}\ddot{\mathbf{x}} + \mathbf{C}\dot{\mathbf{x}} + \mathbf{K}\mathbf{x} = -\mathbf{M}\mathbf{J}\ddot{x}_g - \begin{Bmatrix} c_{eq}|\dot{x}_1|^{\beta}\operatorname{sgn}(\dot{x}_1) + c_{eq}|\dot{x}_1 - \dot{x}_2|^{\beta}\operatorname{sgn}(\dot{x}_1 - \dot{x}_2) \\ c_{eq}|\dot{x}_1 - \dot{x}_2|^{\beta}\operatorname{sgn}(\dot{x}_2 - \dot{x}_1) + c_{eq}|\dot{x}_2 - \dot{x}_3|^{\beta}\operatorname{sgn}(\dot{x}_2 - \dot{x}_3) \\ c_{eq}|\dot{x}_3 - \dot{x}_2|^{\beta}\operatorname{sgn}(\dot{x}_3 - \dot{x}_2) \end{Bmatrix} \tag{8.129}$$

Based on the method of modal decoupling discussed in Chapter 4, letting $x = \boldsymbol{p}_i\, y$, and premultiplying $\boldsymbol{p}_i^T$ on both sides of Equation 8.129 results in

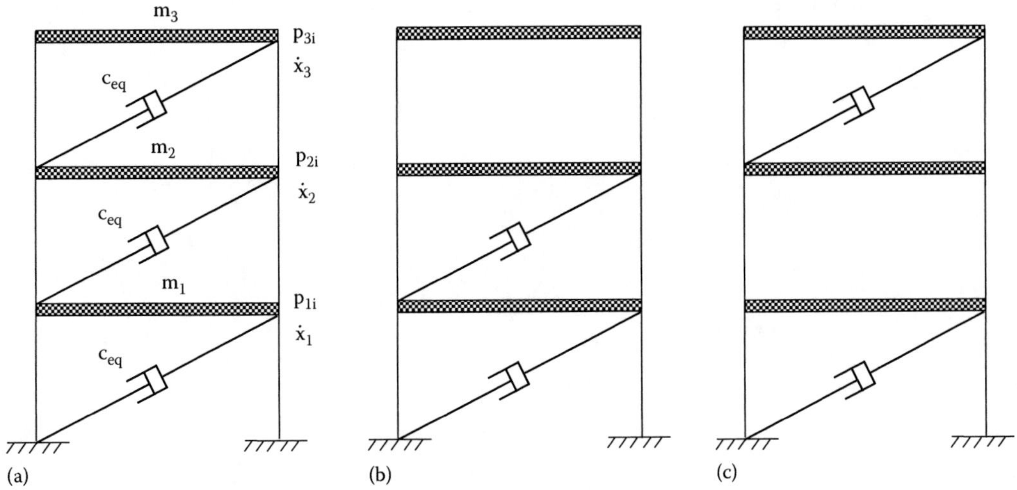

FIGURE 8.8 (a–c) Examples of a structure with dampers.

$$
\begin{aligned}
&\boldsymbol{p}_i^T\mathbf{M}\boldsymbol{p}_i\ddot{y}_i + \boldsymbol{p}_i^T\mathbf{C}\boldsymbol{p}_i\dot{y}_i + \boldsymbol{p}_i^T\mathbf{K}\boldsymbol{p}_i y_i = -\boldsymbol{p}_i^T\mathbf{MJ}\,\ddot{x}_g \\
&-p_{1i}\left[c_{eq}\left|\dot{x}_1\right|^{\beta}\operatorname{sgn}(\dot{x}_1) + c_{eq}\left|\dot{x}_1-\dot{x}_2\right|^{\beta}\operatorname{sgn}(\dot{x}_1-\dot{x}_2)\right] \\
&-p_{2i}\left[c_{eq}\left|\dot{x}_1-\dot{x}_2\right|^{\beta}\operatorname{sgn}(\dot{x}_2-\dot{x}_1) + c_{eq}\left|\dot{x}_2-\dot{x}_3\right|^{\beta}\operatorname{sgn}(\dot{x}_2-\dot{x}_3)\right] \\
&-p_{3i}\left[c_{eq}\left|\dot{x}_3-\dot{x}_2\right|^{\beta}\operatorname{sgn}(\dot{x}_3-\dot{x}_2)\right]
\end{aligned}
\tag{8.130}
$$

It is seen that although the base system is decoupleable, with sublinear dampers, the system cannot be decoupled. In other words, the corresponding SDOF systems cannot be obtained as pure modes (effective modes) and the above-mentioned spectra-based design cannot be carried out.

Next, assume that the fundamental mode shape can be written as a triangular-shaped modal displacement (BSSC/NEHRP 2009):

$$
\boldsymbol{p}_1 = \begin{Bmatrix} p_{11} \\ p_{21} \\ p_{31} \end{Bmatrix} = \frac{1}{3}\begin{Bmatrix} 1 \\ 2 \\ 3 \end{Bmatrix}
\tag{8.131a}
$$

and the velocity vector can be written as a triangular-shaped velocity vector:

$$
\dot{\mathbf{x}} = \begin{Bmatrix} \dot{x}_1 \\ \dot{x}_2 \\ \dot{x}_3 \end{Bmatrix} = \begin{Bmatrix} \dot{y}_1 \\ 2\dot{y}_1 \\ 3\dot{y}_1 \end{Bmatrix}
\tag{8.131b}
$$

Substituting Equations 8.131a and 8.131b into Equation 8.130 with i = 1 results in

$$\ddot{y}_1 + 2\xi\ \omega_{eff}\dot{y}_1 + \omega_{eff}^2 y_1 = -\Gamma_1\ \ddot{x}_g - \frac{1}{3\boldsymbol{p}_1^T\mathbf{M}\boldsymbol{p}_1}\left\{\left[c_{eq}\left|\dot{y}_1\right|^\beta \operatorname{sgn}(\dot{y}_1) + c_{eq}\left|\dot{y}_1\right|^\beta \operatorname{sgn}(-\dot{y}_1)\right]\right.$$

$$\left. -2\left[c_{eq}\left|\dot{y}_1\right|^\beta \operatorname{sgn}(\dot{y}_1) + c_{eq}\left|\dot{x}_1\right|^\beta \operatorname{sgn}(-\dot{y}_1)\right] - 3\left[c_{eq}\left|\dot{y}_1\right|^\beta \operatorname{sgn}(\dot{y}_1)\right]\right\}$$

$$= -\Gamma_1\ \ddot{x}_g - \frac{c_{eq}\left|\dot{y}_1\right|^\beta \operatorname{sgn}(\dot{y}_1)}{\boldsymbol{p}_1^T\mathbf{M}\boldsymbol{p}_1}$$

Here,

$$\frac{\boldsymbol{p}_1^T\mathbf{C}\boldsymbol{p}_1}{\boldsymbol{p}_1^T\mathbf{M}\boldsymbol{p}_1} = 2\xi\,\omega_{eff}$$

$$\frac{\boldsymbol{p}_1^T\mathbf{K}\boldsymbol{p}_1}{\boldsymbol{p}_1^T\mathbf{M}\boldsymbol{p}_1} = \omega_{eff}^2$$

$$\frac{\boldsymbol{p}_1^T\mathbf{MJ}}{\boldsymbol{p}_1^T\mathbf{M}\boldsymbol{p}_1} = \Gamma_1$$

Furthermore, letting

$$\frac{c_{eq}}{\boldsymbol{p}_1^T\mathbf{M}\boldsymbol{p}_1} = 2\xi_{eq}\omega_{eff}$$

results in

$$\ddot{y}_1 + 2\xi\ \omega_{eff}\dot{y}_1 + 2\xi_{eq}\omega_{eff}\left|\dot{y}_1\right|^\beta \operatorname{sgn}(\dot{y}_1) + \omega_{eff}^2 y_1 = -\Gamma_1\ \ddot{x}_g \tag{8.132}$$

Comparing Equations 8.132 and 8.113, the only difference is the proportionality factor Γ_1. Therefore, under the "triangular" assumption described by Equation 8.131, the desired SDOF system can be obtained. Therefore, in the case of two dampers, shown in Figure 8.8b or c, Equation 8.132 becomes

$$\ddot{y}_1 + 2\xi\ \omega_{eff}\dot{y}_1 + \frac{2}{3}\left[2\xi_{eq}\omega_{eff}\left|\dot{y}_1\right|^\beta \operatorname{sgn}(\dot{y}_1)\right] + \omega_{eff}^2 y_1 = -\Gamma_1\ \ddot{x}_g$$

and if only one damper is used,

$$\ddot{y}_1 + 2\xi\ \omega_{eff}\dot{y}_1 + \frac{1}{3}\left[2\xi_{eq}\omega_{eff}\left|\dot{y}_1\right|^\beta \operatorname{sgn}(\dot{y}_1)\right] + \omega_{eff}^2 y_1 = -\Gamma_1\ \ddot{x}_g$$

Generally speaking, in an n-story building with N nominal dampers,

$$\ddot{y}_1 + 2\xi\ \omega_{eff}\dot{y}_1 + \frac{N}{n}\left[2\xi_{eq}\omega_{eff}\left|\dot{y}_1\right|^{\beta}\text{sgn}\left(\dot{y}_1\right)\right] + \omega_{eff}^2 y_1 = -\Gamma_1\ \ddot{x}_g$$

or

$$\ddot{y}_1 + 2\xi\omega_{eff}\dot{y}_1 + 2\tilde{\xi}_{eq}\omega_{eff}\left|\dot{y}_1\right|^{\beta}\text{sgn}\left(\dot{y}_1\right) + \omega_{eff}^2 y_1 = -\Gamma_1\ \ddot{x}_g \tag{8.133}$$

where

$$\tilde{\xi}_{eq} = \frac{N}{n}\xi_{eq} \tag{8.134}$$

In this case, it is assumed that the shape function is written in a triangular form as

$$\boldsymbol{p}_1 = \begin{Bmatrix} p_{11} \\ p_{21} \\ \cdots \\ p_{n1} \end{Bmatrix} = \frac{1}{n}\begin{Bmatrix} 1 \\ 2 \\ \cdots \\ n \end{Bmatrix} \tag{8.135a}$$

and the velocity vector is written as

$$\dot{\mathbf{x}} = \begin{Bmatrix} \dot{x}_1 \\ \dot{x}_2 \\ \cdots \\ \dot{x}_n \end{Bmatrix} \approx \begin{Bmatrix} \dot{y}_1 \\ 2\dot{y}_1 \\ \cdots \\ n\dot{y}_1 \end{Bmatrix} \tag{8.135b}$$

The requirements described in Equations 8.135a and 8.135b represent a restrictive condition. A slightly relaxed condition can be given as

$$\boldsymbol{p}_1 = \frac{1}{p_{n1}}\begin{Bmatrix} s_1 \\ s_2 \\ \cdots \\ s_n \end{Bmatrix} \tag{8.136a}$$

and the velocity vector written as

$$\dot{\mathbf{x}} = \begin{Bmatrix} \dot{x}_1 \\ \dot{x}_2 \\ \cdots \\ \dot{x}_n \end{Bmatrix} \approx \begin{Bmatrix} s_1\dot{y}_1 \\ s_2\dot{y}_1 \\ \cdots \\ s_n\dot{y}_1 \end{Bmatrix}, \quad s_1 \leq s_2 \cdots \leq s_n \tag{8.136b}$$

Here, the proportionality coefficients $s_{(.)}$ are constant.

In this case, the virtual work done by the sublinear damping force and the virtual displacement $\boldsymbol{p}_i^T$ is also related only to the fundamental effective mode. For example, consider premultiplying the last term in Equation 8.129 by $\boldsymbol{p}_i^T$:

$$\boldsymbol{p}_i^T \begin{Bmatrix} c_{eq}|\dot{x}_1|^\beta \operatorname{sgn}(\dot{x}_1) + c_{eq}|\dot{x}_1 - \dot{x}_2|^\beta \operatorname{sgn}(\dot{x}_1 - \dot{x}_2) \\ c_{eq}|\dot{x}_1 - \dot{x}_2|^\beta \operatorname{sgn}(\dot{x}_2 - \dot{x}_1) + c_{eq}|\dot{x}_2 - \dot{x}_3|^\beta \operatorname{sgn}(\dot{x}_2 - \dot{x}_3) \\ c_{eq}|\dot{x}_3 - \dot{x}_2|^\beta \operatorname{sgn}(\dot{x}_3 - \dot{x}_2) \end{Bmatrix}$$

$$= c_{eq}|\dot{x}_1|^\beta \operatorname{sgn}(\dot{x}_1) + c_{eq}\operatorname{sgn}(\dot{x}_1 - \dot{x}_2)$$

$$+ (p_{21} - p_{31}) c_{eq}|\dot{x}_2 - \dot{x}_3|^\beta \operatorname{sgn}(\dot{x}_2 - \dot{x}_3)$$

$$= c_{eq}|\dot{x}_1|^\beta \operatorname{sgn}(\dot{x}_1)\left[p_{11} + (p_{21} - p_{11})|s_1 - s_2|^\beta + (p_{31} - p_{21})|s_2 - s_3|^\beta\right]$$

$$= \tilde{c}_{eq}|\dot{x}_1|^\beta \operatorname{sgn}(\dot{x}_1)$$

Since all the elements in shape function $\boldsymbol{p}_i$ are constant and both $s_{(.)}$ and β are constant,

$$\tilde{c}_{eq} = c_{eq}\left[p_{11} + (p_{21} - p_{11})|s_1 - s_2|^\beta + (p_{31} - p_{21})|s_2 - s_3|^\beta\right] = \text{const} \quad \left(\text{kN-s}^\beta/\text{m}^\beta\right). \tag{8.137}$$

Although Equation 8.137 only describes the case shown in Figure 8.8, it can easily be extended to n-DOF sublinear systems, provided that Equation 8.136b holds. That is, as long as the conditions in Equations 8.136a and 8.136b are satisfied, Equation 8.133 can always be obtained, where the term $\tilde{\xi}_{eq}$ is redefined as

$$\tilde{\xi}_{eq} = \frac{\tilde{c}_{eq}}{2\omega_{eff}\boldsymbol{p}_1^T\mathbf{M}\boldsymbol{p}_1} \tag{8.138}$$

Example 8.5

In order to use Equation 8.138, the term $\tilde{c}_{eq}$ needs to be evaluated. Furthermore, from Equation 8.137, the equivalent damping coefficient is c_{eq}. The nonlinear spectra is used to determine c_{eq}.

A four-story system with an identical stiffness matrix as given in Example 8.1 is used to demonstrate this procedure. The system has $T_1 = 1.5$ (s), and the shape function $\boldsymbol{p}_1$ is identical to $\bar{\boldsymbol{p}}_1$ used in Example 8.4. The system is subjected to a ground excitation with a level 0.4 (g) as well as $I = R = 1$. Four sublinear dampers are chosen for each story with $\alpha = \beta = 0.1$ and $G_{m(.)} = 1$; and $\xi_{eq} = 0.10$. For convenience, it is assumed the modal participation factor $\Gamma_1 = 1.0$ and all the safety factors $S_{D(.)} = 1.0$.

If $\mathbf{d}_{max_1}$ as well as x_{ji}, $j = 1,\ldots,4$ can be found, then from Equation 8.66, this term can be calculated.

From Table 8.2a, it is seen that $d_{1D} = 15.67$ (cm). Note that $A_\beta = 3.89$. Using normal mode approach, from Equation 8.85,

$$\mathbf{d}_{max_1} = \Gamma_1 d_{1D}\boldsymbol{p}_1 = 15.67\begin{Bmatrix} 0.3576 \\ 0.6694 \\ 0.8861 \\ 1.0000 \end{Bmatrix} \approx \begin{Bmatrix} 5.60 \\ 10.49 \\ 13.89 \\ 15.67 \end{Bmatrix} \text{ (cm)}$$

Note that for the nonlinear systems, the symbol ξ_{eq} is used instead of ξ_{eff1} to denote the damping ratio in Equation 8.66 and

$$c_{eq} = \frac{2\pi\xi_{eq}\ \mathbf{d}_{max_1}^T\ \mathbf{K}\mathbf{d}_{max_1}}{\omega_1^{\alpha_1} A_{\beta_1} \sum_{j=1}^{S} x_{j1}^{\beta_1+1}} = 185.0 \quad \left(\text{kN-s}^{\beta}/\text{m}^{\beta}\right)$$

Practically speaking, it is very rare for the conditions in Equations 8.135 and 8.136 to be fully satisfied, so that an MDOF system cannot be exactly decoupled. In these cases, continuing to use the method based on the nonlinear spectra still will introduce errors. This is especially true for structures with significantly higher modes. Thus, if the method based on nonlinear spectra is used, only one effective mode is often used.

If using methods based on linearization of the damping force to achieve effective damping ratios, such as Timoshenko or force-based effective damping, more modes are needed. However, the linearization approach also introduces errors as mentioned above.

In the authors' experience, if a building is somewhat regular and the effective damping ratio is comparatively high, the method based on the nonlinear spectra provides better response estimation. If the structure enters the inelastic range, a nonlinear spectra approach is also recommended. This is discussed in the next section.

If a building has a large mass, stiffness, and damping irregularities (e.g., weight, plan irregularities), the effective modal combination method is suggested.

The above statements are based on a limited number of numerical simulations and the accuracy of the response estimation is highly input sensitive. To be more confident of its use in damping design, additional intensive theoretical and experimental research is needed.

8.5 NONLINEAR RESPONSE SPECTRA APPROACH WITH BILINEAR DAMPERS

When a structure enters the inelastic range, the effective period becomes much longer, and a significant amount of energy can be dissipated with virtually zero damping adaptability. If supplemental damping is used, it can absorb another large amount of energy, which helps protect the structure from further damage. In many cases, the entire system becomes overdamped. The basic idea of response estimation is still the production of the effective modal response and the shape function to distribute the response to each story. A highly ductile system can be treated as an MSSP model, because its fundamental effective mode is most dominant. This simplifies the design, if the fundamental shape function can be found. Since the system is highly nonlinear, care must be given to the use of spectra-based design approach, and to the response modification coefficient R that is greater than unity.

8.5.1 General Description

8.5.1.1 Supplemental Damping and Inelastic Structures

In Chapter 6, it was mentioned briefly that when a structure enters the inelastic range, supplemental damping can absorb a great deal of energy. Depending on the degree of damping adaptability, as the value of f_{adp} increases, the ratio of energy dissipated by supplemental damping and the energy dissipated by the base structure also increases.

As discussed in previous chapters, the energy dissipation capability of a system is directly related to its dynamic stiffness. The larger the dynamic stiffness is, the lower the seismic response that can be achieved. Therefore, using supplemental damping can also be beneficial for structures that can enter the inelastic range.

A simpler approach is to let the response modification coefficient $R > 1$ reduce the level of load and then use the same procedure described in Sections 8.2 and 8.3. Due to space limitations, these design details are not repeated, but note that in this case, the period can be estimated by

$$T_{eff} = T_1\sqrt{\mu} \quad (s) \tag{8.139}$$

where μ is the design ductility and T_1 is the fundamental period when the structure remains elastic.

Since the damping adaptability is not zero, when the ductility is large, the damping force can be quite large. Thus, for structures designed with large ductility, sublinear damping with $\beta > 0.3$ is not recommended.

8.5.1.2 Displacement-Based Design

Until now, all the damping design procedures in Chapters 7 and 8 have been based on force. In Chapter 6, the idea of equal displacement and equal energy was introduced, which may lead to displacement- and energy-based design.

8.5.1.2.1 Indirect Displacement-Based Design

One of the displacement-based design procedures for aseismic structures is the indirect approach. The method using pushover curves, mentioned in Chapter 6, is an indirect displacement-based design (IDBD) approach. Indirect methods are based on monotonically increasing the static lateral forces. The corresponding displacements are then accumulated until a predefined targeted level is achieved. In the procedure of increasing lateral forces, it is difficult to distinguish the conservative and damping forces. It is also difficult to distinguish the resistance contributed by stiffness and the desired damping. Therefore, this method, at least to date, is not popular in damping design.

8.5.1.2.2 Direct Displacement-Based Design

Direct displacement-based design (DDBD) uses a different approach (e.g., Kowalsky 1994; Priestley 1996; Grant 2005). The DDBD uses an equivalent force–displacement linear SDOF system (Chapter 6), which is called the "substitute structure." It is has been claimed that the DDBD can be used for supplemental damping design (Lin et al. 2003). The main approach of this method also consists of conceptual decision making, response estimation, and damper specification. The basic idea of response estimation is still based on calculation of a damping ratio. The response is a product of the effective modal response, in this case called the response of the substitute structure, and the deformation shape function.

In the following discussion, the response of the SDOF system is considered first. The corresponding shape function to distribute the response to each story is determined next.

8.5.1.2.3 Response of Substitute Structure

8.5.1.2.3.1 Substitute Structure As shown in Figure 8.9, an MDOF structure with damping is replaced by an SDOF system with mass m_{eq}, damping c_{eq}, and stiffness k_{eq}. Here, the equivalent mass is

$$m_{eq} = \frac{\sum_{i=1}^{n} m_i h_i}{h_n} \quad (t) \tag{8.140}$$

where m_i and h_i are, respectively, the mass and the height of the ith floor, as defined in Figure 8.9.

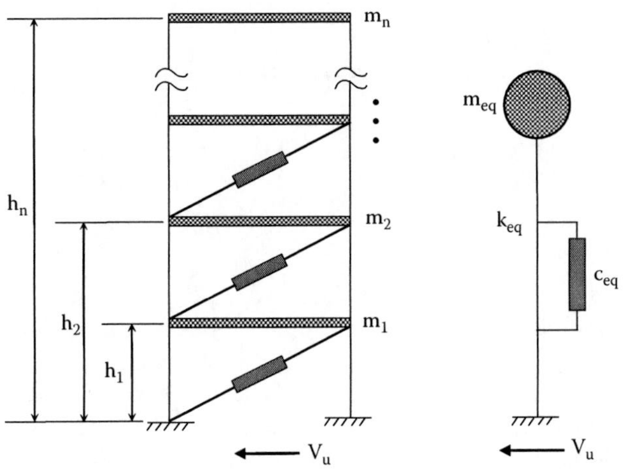

FIGURE 8.9 Substitute system.

The stiffness of the substitute system is given in Equation 6.57 and repeated as follows:

$$k_{eq} = m_{eq}\left(\frac{2\pi}{T_{eq}}\right)^2 \quad (\text{kN/m}) \tag{8.141}$$

Here, the term T_{eq} is the equivalent period, which is found through linear displacement spectra. In the initial design, the equivalent period can be estimated by Equation 8.139.

8.5.1.2.3.2 Design Damping Ratio The substitute structure has equivalent damping c_{eq}, which accounts for the capability of energy dissipation of the original linear structure, the supplemental dampers, and the yielding mechanism. That is, when the structure enters the inelastic range, the design damping ratio can be written as

$$\xi_{design} = \xi_0 + \xi_h + \xi_a \tag{8.142}$$

where ξ_0, ξ_h, and ξ_a are the damping ratios contributed by the elastic base structure, the yielding structure, and the supplement dampers, respectively.

While ξ_0 is assumed as a small value, say 2%–5%, ξ_a and ξ_h are calculated by a direct assignment, namely, the targeted maximum displacement x_0. Suppose the yielding displacements of each story in the structure are identical, denoted as d_y, and the corresponding story yielding occurs simultaneously. The displacement ductility μ is defined in Equation 6.29.

The damping ratio ξ_h contributed by the inelastic deformation is given by the following empirical formula:

$$\xi_h = a\left(1 - \frac{1}{\mu^b}\right)\left[1 + \frac{1}{\left(T_{eq} + c\right)^d}\right] \tag{8.143}$$

Here, the parameters a, b, c, and d can be found in Table 8.3 (Grant et al. 2005).

TABLE 8.3
Parameters in Equation 8.143

Model	a	b	c	d
Bilinear (EPP)	0.222	0.397	0.287	1.295
Bilinear (general)	0.161	0.952	0.945	2.684

The damping ratio ξ_a contributed by supplemental damping can be calculated through the concept of effective damping, such as Timoshenko damping. For example, if bilinear damping is specified, Equation 8.4 can be used. If sublinear damping is considered, Equation 8.74 with i = 1 is used.

The shape function is needed in both Equations 8.4 and 8.74. Note that when a structure yields, its first effective mode is dominant. Therefore, the MSSP approach is used so that one shape function is needed. In the literature about DDBD, the shape function is suggested to be triangular. However, the triangularly shaped deformation is not always accurate because when dampers are installed in a structure, they are likely to alter the deformation shapes. Thus, using the shape function mentioned in Section 8.3 is recommended. In addition, an equivalent damping matrix $\mathbf{C}_{eq}^{(2)}$ is defined as

$$\mathbf{C}_{eq}^{(2)} = c_{eq}\mathbf{C}_{con} \quad \left(\text{kN-s}^{\beta}/\text{m}^{\beta}\right) \tag{8.144}$$

where c_{eq} is obtained though the sublinear damping design approach mentioned in Section 8.3. Note that to actually determine c_{eq}, it is necessary to find out if the system is overdamped. As mentioned in Chapter 6, when a structure enters the inelastic range, it is quite easy to be overdamped. However, at this time, whether or not the first effective mode is an overdamped system needs to be known. Although precise judgment needs to be carried out through an iterative approach, the occurrence of overdamping can be roughly estimated by the value of the ductility. For example, if $\mu > 5$, the first effective mode is likely to be overdamped. In this case, the value of c_{eq} can continue to be increased until the resulting shape function of the fundamental mode becomes real. Otherwise, Equation 8.73 can be used to obtain real-valued shaped functions.

8.5.1.2.3.3 Displacement Iteration Once the damping ratios are calculated, ξ_{design} and x_0 will be obtained. Recall the displacement design spectrum, which is used to find the spectral displacement d_{1D}. Now this term is redenoted by symbol x_0, if the set of parameters T and ξ are given. This procedure indicates that given any pair of parameters among the triple $< T, \xi, X_0 >$, the third quantity can be determined. In this case, ξ_{design} and x_0 are known. Then, the period T_{eq} can be figured out. For convenience, in the iterative approach, the initial guess of the equivalent period is denoted as $T_{eq}^{(1)}$. The next one obtained in the displacement spectrum is denoted as $T_{eq}^{(2)}$. Substituting $T_{eq}^{(2)}$ into Equation 8.143 achieves a more accurate damping ratio, now denoted as $\xi_h^{(2)}$; and $\xi_{design}^{(2)}$ and x_0 can be used to further define $T_{eq}^{(3)}$, if necessary. Once the period is determined, the equivalent stiffness k_{eq} is finally established.

Although designing structures with k_{eq} is beyond the scope of this book, in the above design procedure, the parameters of bilinear and/or sublinear dampers can be adjusted to enable more accurate structural design.

8.5.2 Response Estimation

8.5.2.1 Overview

A different approach to design inelastic structures is to use nonlinear spectra. Compared to DDBD, this method can be more accurate and simpler. However, methods to generate the nonlinear spectra need to be thoroughly investigated. This includes selection and scaling of the ground excitations, and the selection of peak responses. In this section, the 99 earthquake records used in previous discussion in this book are used again used to generate the nonlinear spectra to illustrate the concept of nonlinear spectra-based design for bilinear dampers.

When a structure yields, if it can be modeled as a bilinear system, then with bilinear supplemental damping, the total system is still bilinear. Detailed equations of the specific characteristic strength, the unloading and yielding stiffness, etc., are not given herein due to space limitations.

All spectra, including the nonlinear spectra, are obtained through SDOF systems. In Section 8.4.2 on sublinear damping design, it was shown that this SDOF system can be obtained though modal decoupling under specific conditions. The same type of conditions can also be applied to bilinear systems. Following the same logic described by Equations 8.129 through 8.133, the virtual work done by the damping force F_δ and displacement $\boldsymbol{p}^T$ is briefly considered, namely, the term $\boldsymbol{p}^T F_\delta$ (see Equation 8.41). It is seen that when the conditions in Equations 8.135a and 8.135b hold,

$$\boldsymbol{p}^T\mathbf{f}_\delta = G_m^2 q_d \frac{N}{n}\operatorname{sgn}(\dot{y}_1)\delta\left(|y_1| \geq d_y\right) \quad (\text{kN}) \tag{8.145}$$

where y_1 is the temporal variable of the fundamental effective modal displacement.

An SDOF bilinear system has now been achieved. With the more relaxed conditions,

$$\dot{x}_1 \approx s\,\dot{y}_1 \quad (\text{m/s}) \tag{8.146a}$$

$$\dot{x}_j \geq \dot{x}_{j-1} \quad (\text{m/s}), \quad j = 2,\ldots,n \tag{8.146b}$$

results in

$$\boldsymbol{p}^T\mathbf{f}_\delta = \tilde{q}_d \operatorname{sgn}(\dot{y}_1)\, h\left(|y_1| \geq d_y\right) m_{e1} \quad (\text{kN}) \tag{8.147}$$

where

$$\tilde{q}_d = \frac{G_m^2 \left[h_1 p_{11} + \sum_{j=0}^{n-2} h_{n\text{-}j}\left(p_{n-j,1} - p_{n-j\text{-}1,1}\right)\right]}{m_{e1}} q_d \quad (g) \tag{8.148}$$

and here the Heaviside function $h_{(.)}$ is defined in Example 8.8.

Again, the conditions described by Equation 8.146 are not always satisfied, and thus using this method may introduce errors. Similar to the case of sublinear damping design, the irregularity of a structure needs to be considered and a decision on whether or not the method based on nonlinear spectra can be used needs to be made.

In the following discussion, several tables are given to list the bilinear spectra generated by the 99 records. From these tables, if the ground input level, the normalized characteristic strength, and the yielding ratio are known, the corresponding displacement, d_{1D}, and the acceleration, C_s, can be determined. The participation factor (see Equation 8.44) Γ_d can be written as

$$\Gamma_d = \frac{\boldsymbol{p}^T\mathbf{M}\mathbf{J}}{\boldsymbol{p}^T\mathbf{M}\boldsymbol{p}} \tag{8.149}$$

With the shape functions, the displacement and acceleration vectors are obtained, and the lateral force, base shear, etc. (see Equations 8.49 through 8.53), can be computed.

8.5.2.2 Response Estimation Based on Bilinear Spectra

The response estimation based on the bilinear spectra consists of first replacing the MDOF system with an SDOF system, which are both bilinear, to determine the required three basic parameters, such as yielding stiffness k_D, characteristic strength q_D, and yielding displacement d_Y. The effective modal participation factor, Γ_d, should also be calculated. The next task is to find the response d_{1D} and C_s. In the third phase, the shape function is used to estimate the response of each story.

The steps for using nonlinear spectra to find d_{1D} and C_s are as follows:

First, determine a specific group of earthquake records to generate the bilinear spectra.

Second, use the bilinear SDOF system that assumes small initial damping ratios and given period ranges to compute the responses of the displacements, the accelerations, and the damper forces. The statistical mean value plus one standard deviation is then calculated and listed in Tables 8.4 through 8.9. For example, the responses are listed in Tables 8.4 through 8.9.

Third, use the method of interpolation to determine the corresponding displacement, acceleration, and the damper force from the values given in these tables.

If the computation provides satisfactory responses, the corresponding damper design is workable. The next step is to realize these parameters in dampers, as discussed in Section 8.2.1. If the responses are still too large, a further round of trial-and-error must be carried out.

8.5.2.3 Bilinear Design Spectra

The tables are examined next. Using Table 8.4, the characteristic strength q_D can be calculated as 0.03 mg, where m is the mass of the SDOF system. Note that in Table 8.4, for convenience, the characteristic strength is actually taken to be q_D/m, denoted as $\bar{q}_D$ with unit (g). And, from Tables 8.4 through 8.9, q_D can be taken to be 0.04 mg up to 0.4 mg. Similarly, $\bar{q}_D$, which are 0.04 (g) to 0.4 (g), etc., are listed. In each group of tables, (A) lists the displacement with unit (cm) etc., (B) lists the acceleration with unit (g), and (C) lists the damper forces with unit (MN). In each table, the columns are arranged according to a different period. From a period of 0.4 ~ 3 (s), a total of seven columns are listed. Additionally, in each table, the rows are arranged according to the unloading stiffness of the damper, which is achieved by listing the ratio of the unloading stiffness and the structural stiffness. For example, the number 0.4 means that k_D = 0.4 (kN/m).

TABLE 8.4A
Displacement (cm) $\bar{q}_D$ = 0.03 (g), Input = 0.4 (g)

k_D/k \ T (s)	0.4	0.6	0.8	1.0	1.5	2.0	3.0
0.40	5.8848	11.7410	16.4819	22.9247	28.8176	35.0578	62.4114
0.50	5.8336	11.6386	16.3339	22.6634	28.2036	34.1123	59.7268
0.60	5.8040	11.5536	16.2249	22.4613	27.7814	33.4514	58.0277
0.70	5.7842	11.4844	16.1472	22.3079	27.4995	33.0146	56.7065
0.98	5.7534	11.3833	16.0237	22.0361	27.0710	32.3593	54.3306
2.00	5.7013	11.2988	15.8266	21.8415	26.4927	31.2979	51.3725
2.88	5.6871	11.2714	15.7778	21.7385	26.2156	31.0312	50.5114
3.92	5.6760	11.2599	15.7432	21.6823	26.0177	30.8374	49.8384
5.12	5.6685	11.2529	15.7279	21.6458	25.8786	30.7634	49.3354
8.00	5.6593	11.2408	15.715	21.5894	25.7499	30.6866	48.7470
12.5	5.6563	11.2354	15.7057	21.5523	25.6193	30.6249	48.4165

TABLE 8.4B
Acceleration (g) $\bar{q}_D = 0.03$ (g), Input = 0.4 (g)

k_D/k \ T(s)	0.4	0.6	0.8	1.0	1.5	2.0	3.0
0.40	1.5116	1.3438	1.0674	0.9536	0.5460	0.3834	0.3096
0.50	1.4988	1.3324	1.0581	0.9430	0.5350	0.3737	0.2975
0.60	1.4913	1.3228	1.0513	0.9348	0.5274	0.3670	0.2898
0.70	1.4863	1.3151	1.0464	0.9287	0.5224	0.3625	0.2839
0.98	1.4786	1.3038	1.0386	0.9177	0.5147	0.3560	0.2733
2.00	1.4655	1.2943	1.0262	0.9099	0.5043	0.3452	0.2599
2.88	1.4619	1.2913	1.0231	0.9057	0.4994	0.3425	0.2561
3.92	1.4591	1.2900	1.0209	0.9035	0.4958	0.3406	0.2531
5.12	1.4572	1.2892	1.0200	0.9020	0.4933	0.3398	0.2508
8.00	1.4549	1.2878	1.0192	0.8997	0.4910	0.3390	0.2482
12.5	1.4541	1.2872	1.0186	0.8982	0.4887	0.3384	0.2467

8.5.2.3.1 Period

Using these tables, if a group of design parameters can directly match the parameters listed, the corresponding responses can be obtained. For example, if in a design for a structure with a period of T = 0.4 (s), the characteristic strength, q_D, and the damping stiffness, k_D, of the bilinear damper are, respectively, 0.01mg and 0.5k. Then the displacement can be found from Table 8.4A to be 7.8925 (cm), the acceleration from Table 8.4B is 1.9971, which means 1.9971 (g) and the damper force is 26.1415 (MN).

When the control parameters are not exactly listed in these tables, the interpolation method can be used. To determine the displacement for a period not listed in Tables 8.4A through 8.9A, Equation 8.150 can be used. That is,

$$d(T) = a_T T^3 + b_T T^2 + c_T T + d_T \quad (cm) \tag{8.150}$$

where d(T) is the displacement at the specific value of the period T. The coefficients a_T, b_T, c_T, and d_T are determined by the following least square method as

TABLE 8.4C
Normalized force (MN) $\bar{q}_D = 0.03$ (g), Input = 0.4 (g)

k_D/k \ T(s)	0.4	0.6	0.8	1.0	1.5	2.0	3.0
0.40	20.3911	17.3191	13.6555	11.3139	6.8752	4.8926	3.5619
0.50	20.1781	17.1449	13.4651	11.1606	6.6801	4.7315	3.4041
0.60	20.0519	17.0217	13.3400	11.0119	6.5375	4.6159	3.3020
0.70	19.9723	16.8994	13.2458	10.8917	6.4408	4.5486	3.2208
0.98	19.8314	16.6938	13.1001	10.6973	6.2887	4.4037	3.0719
2.00	19.6330	16.4536	12.8546	10.4951	6.0748	4.1688	2.8780
2.88	19.5765	16.3902	12.8038	10.4089	6.0059	4.1058	2.8240
3.92	19.5319	16.3565	12.7639	10.3573	5.9491	4.0622	2.7788
5.12	19.5012	16.3322	12.7395	10.3195	5.9093	4.0422	2.7432
8.00	19.4691	16.2896	12.7146	10.2645	5.8742	4.0175	2.6965
12.5	19.4548	16.2640	12.6914	10.2225	5.8426	4.0002	2.6689

TABLE 8.5A
Displacement (cm) $\bar{q}_D = 0.05$ (g), Input = 0.4 (g)

k_D/k \ T(s)	0.4	0.6	0.8	1.0	1.5	2.0	3.0
0.40	4.9945	10.2026	14.5974	20.2292	27.3763	34.3455	61.1032
0.50	4.8948	10.0002	14.1762	19.6955	26.4092	32.6520	56.8997
0.60	4.8395	9.8465	13.9019	19.3221	25.7328	31.4758	53.7137
0.70	4.7998	9.7177	13.7277	19.0680	25.2157	30.6070	51.1950
0.98	4.7239	9.5514	13.4095	18.6024	24.2381	29.0557	47.0976
2.00	4.5998	9.3784	12.9588	18.0145	22.7472	26.9840	41.6483
2.88	4.5588	9.3119	12.8123	17.7942	22.2470	26.3575	39.9497
3.92	4.5387	9.2789	12.7271	17.6913	21.8361	25.9268	38.8616
5.12	4.5231	9.2423	12.6583	17.5898	21.5496	25.6995	38.0041
8.00	4.5117	9.1941	12.5658	17.4992	21.2358	25.4410	37.0315
12.5	4.5057	9.1564	12.5087	17.4681	21.0106	25.1607	36.5365

$$\begin{Bmatrix} a_T \\ b_T \\ c_T \\ d_T \end{Bmatrix} = \begin{bmatrix} T_1^3 & T_1^2 & T_1 & 1 \\ T_2^3 & T_2^2 & T_2 & 1 \\ & & \dots & \\ T_7^3 & T_7^2 & T_7 & 1 \end{bmatrix}^+ \begin{Bmatrix} d_1 \\ d_2 \\ \dots \\ d_7 \end{Bmatrix} \tag{8.151}$$

where

$$\left[T_1, T_2, T_3, T_4, T_5, T_6, T_7\right] = \left[0.4, 0.6, 0.8, 1.0, 1.5, 2.0, 3.0\right] \quad (s)$$

and $d_{(.)}$ are the corresponding displacements listed in the specific tables at the specific period $T_{(.)}$. For example, if under the condition $q_D = 0.01$ mg and $k_D = 0.4\,k$,

$$\left[d_1, d_2, d_3, d_4, d_5, d_6, d_7\right] = \left[7.9024,\ 15.6856,\ 21.9754,\ 28.2762,\ 36.3029,\ 41.6237,\ 73.7052\right] \quad (cm)$$

TABLE 8.5B
Acceleration (g) $\bar{q}_D = 0.05$ (g), Input = 0.4 (g)

k_D/k \ T(s)	0.4	0.6	0.8	1.0	1.5	2.0	3.0
0.40	1.3075	1.1917	0.9690	0.8652	0.5406	0.3976	0.3244
0.50	1.2824	1.1690	0.9423	0.8435	0.5231	0.3801	0.3058
0.60	1.2685	1.1518	0.9250	0.8284	0.5110	0.3679	0.2915
0.70	1.2585	1.1374	0.9141	0.8182	0.5018	0.3587	0.2800
0.98	1.2394	1.1188	0.8940	0.7994	0.4840	0.3430	0.2614
2.00	1.2081	1.0994	0.8657	0.7757	0.4573	0.3218	0.2365
2.88	1.1978	1.0920	0.8565	0.7668	0.4483	0.3154	0.2288
3.92	1.1927	1.0883	0.8511	0.7627	0.4410	0.3111	0.2240
5.12	1.1888	1.0842	0.8468	0.7586	0.4358	0.3088	0.2201
8.00	1.1859	1.0788	0.8409	0.7549	0.4302	0.3062	0.2158
12.5	1.1844	1.0746	0.8373	0.7537	0.4262	0.3034	0.2135

TABLE 8.5C
Normalized force (MN) $\bar{q}_D$=0.05 (g), Input = 0.4 (g)

k_D/k \ T (s)	0.4	0.6	0.8	1.0	1.5	2.0	3.0
0.40	18.0535	15.8532	12.7899	10.9206	7.1692	5.2767	3.9114
0.50	17.6387	15.4577	12.3753	10.6136	6.8944	5.0714	3.7280
0.60	17.4357	15.1698	12.1000	10.3559	6.6840	4.9216	3.5786
0.70	17.2959	14.9387	11.9049	10.1565	6.5068	4.8046	3.4342
0.98	16.9770	14.5500	11.5646	9.8048	6.2202	4.5359	3.2266
2.00	16.5399	14.1352	11.0887	9.3264	5.7730	4.1400	2.9817
2.88	16.4181	14.0045	10.9377	9.1470	5.6447	4.0137	2.8710
3.92	16.3340	13.9350	10.8406	9.0495	5.5290	3.9273	2.8024
5.12	16.2712	13.8557	10.7850	8.9656	5.4443	3.8878	2.7422
8.00	16.2314	13.7491	10.6967	8.8725	5.3420	3.8394	2.6591
12.5	16.2191	13.6729	10.6243	8.8223	5.2794	3.7800	2.6203

Then under this condition, as an example, according to Equation 8.60, the following can be obtained:

$$\left[a_T, b_T, c_T, d_T\right] = \left[8.0401,\ \ -39.5763,\ \ 76.7307,\ \ -17.4354\right]$$

Therefore, the following equation can be used to determine the displacement with the period not listed in the table:

$$d(T) = 8.0401\,T^3 - 39.5763\,T^2 + 76.7307\,T - 17.4354 \quad (\text{cm})$$

For example, if T=1.25 (s), d(1.25) can be computed to be 32.34 (cm). Note that this value is very close to the result obtained from linear interpolation:

$$(28.2762 + 36.3029)/2 = 32.29 \quad (\text{cm})$$

Thus, linear interpolation can also be used for simplicity.

TABLE 8.6A
Displacement (cm) $\bar{q}_D$ = 0.1 (g), Input = 0.4 (g)

k_D/k \ T(s)	0.4	0.6	0.8	1.0	1.5	2.0	3.0
0.40	4.1987	9.1297	12.7663	18.5619	28.2884	38.5718	68.6337
0.50	3.9799	8.6406	11.8782	17.3423	26.3839	35.4735	60.2889
0.60	3.8168	8.2667	11.3611	16.3433	24.6886	32.8917	54.4997
0.70	3.7224	8.0018	10.9495	15.6371	23.4167	30.8668	50.4883
0.98	3.5372	7.5996	10.2209	14.2105	21.3388	27.4345	44.1303
2.00	3.2691	6.9401	9.1019	12.3839	18.2601	22.9228	32.8726
2.88	3.1846	6.7422	8.7450	11.8841	17.2873	21.4385	30.1138
3.92	3.1450	6.5817	8.5096	11.6447	16.5663	20.2701	27.5374
5.12	3.1310	6.4808	8.3087	11.4985	15.9838	19.4066	26.2544
8.00	3.0985	6.3511	8.1418	11.2475	15.2179	18.4886	25.0700
12.5	3.0933	6.2536	7.9991	11.0930	14.6404	17.7083	23.6746

TABLE 8.6B
Acceleration (g) $\bar{q}_D = 0.1$ (g), Input = 0.4 (g)

k_D/k \ T (s)	0.4	0.6	0.8	1.0	1.5	2.0	3.0
0.40	1.1571	1.1219	0.9058	0.8503	0.6112	0.4939	0.4035
0.50	1.1020	1.0672	0.8486	0.7999	0.5762	0.4634	0.3677
0.60	1.0610	1.0250	0.8157	0.7593	0.5453	0.4368	0.3448
0.70	1.0372	0.9954	0.7895	0.7305	0.5223	0.4159	0.3284
0.98	0.9906	0.9504	0.7433	0.6731	0.4837	0.3798	0.3019
2.00	0.9231	0.8766	0.6729	0.5989	0.4278	0.3321	0.2499
2.88	0.9018	0.8545	0.6504	0.5787	0.4096	0.3163	0.2365
3.92	0.8918	0.8365	0.6356	0.5691	0.3967	0.3045	0.2245
5.12	0.8883	0.8252	0.6230	0.5632	0.3862	0.2955	0.2179
8.00	0.8801	0.8107	0.6125	0.5531	0.3725	0.2862	0.2129
12.5	0.8788	0.7998	0.6035	0.5469	0.3621	0.2783	0.2060

To determine the acceleration for a period not listed in Tables 8.4B through 9.8B, Equation 8.152 can be used. That is,

$$a(T) = a_T T^4 + b_T T^3 + c_T T^2 + d_T T + e_T \quad (g) \tag{8.152}$$

where a(T) is the acceleration at the specific value of the period T. The coefficients a_T, b_T, c_T, d_T, and e_T are determined by the following least square method as

$$\begin{Bmatrix} a_T \\ b_T \\ c_T \\ d_T \\ e_T \end{Bmatrix} = \begin{bmatrix} T_1^4 & T_1^3 & T_1^2 & T_1 & 1 \\ T_2^4 & T_2^3 & T_2^2 & T_2 & 1 \\ & & & \cdots & \\ T_7^4 & T_7^3 & T_7^2 & T_7 & 1 \end{bmatrix}^+ \begin{Bmatrix} a_1 \\ a_2 \\ \cdots \\ a_7 \end{Bmatrix} \tag{8.153}$$

TABLE 8.6C
Normalized force (MN) $\bar{q}_D = 0.1$ (g), Input = 0.4 (g)

k_D/k \ T (s)	0.4	0.6	0.8	1.0	1.5	2.0	3.0
0.40	17.0783	15.9117	12.9075	11.8822	8.5267	6.5506	4.8196
0.50	16.1968	14.9694	12.1092	11.1972	8.1426	6.2960	4.5138
0.60	15.5907	14.2525	11.6031	10.5998	7.7305	6.0426	4.3485
0.70	15.1722	13.7668	11.2441	10.1830	7.4461	5.8620	4.1993
0.98	14.4139	12.9348	10.5118	9.3152	6.8318	5.4341	4.0078
2.00	13.3597	11.8039	9.4287	8.2342	6.0222	4.8223	3.6438
2.88	13.0188	11.4985	9.0335	7.8395	5.7934	4.5821	3.5657
3.92	12.7893	11.2276	8.8066	7.6523	5.5561	4.3978	3.4082
5.12	12.6836	11.0530	8.6086	7.5009	5.4128	4.3001	3.3390
8.00	12.5568	10.8389	8.4196	7.2878	5.1977	4.1253	3.2168
12.5	12.5042	10.6232	8.2631	7.1414	5.0305	4.0157	3.1441

TABLE 8.7A
Displacement (cm) $\bar{q}_D = 0.2$ (g), input = 0.4 (g)

k_D/k \ T (s)	0.4	0.6	0.8	1.0	1.5	2.0	3.0
0.40	4.2586	9.4290	13.8618	20.4512	34.4728	47.4664	80.6420
0.50	3.7769	8.3606	12.4086	18.1947	31.4078	43.7763	71.6059
0.60	3.5606	7.6282	11.1748	16.0857	28.6111	40.1462	64.2992
0.70	3.2996	7.0104	10.3709	14.6659	26.2925	37.2988	61.0113
0.98	2.8427	5.9819	8.8084	12.5516	22.5591	31.3549	52.2255
2.00	2.1524	4.3445	6.6792	9.2502	16.3105	22.1821	36.7146
2.88	1.9289	3.9064	5.8659	8.1236	13.7997	18.8831	30.6252
3.92	1.7973	3.5629	5.2851	7.3202	11.8692	16.6110	26.1828
5.12	1.6910	3.2494	4.9664	6.7106	10.8639	14.9997	23.7877
8.00	1.5752	2.8857	4.5724	5.8196	9.3133	12.8058	19.0533
12.5	1.5219	2.6672	4.0722	5.3585	8.2674	10.8742	16.3006

Here, $T_{(.)}$ is defined in Equation 8.152:

$$[T_1, T_2, T_3, T_4, T_5, T_6, T_7] = [0.4, 0.6, 0.8, 1.0, 1.5, 2.0, 3.0] \quad (s)$$

and $a_{(.)}$ are the corresponding accelerations listed in the specific table at the specific period (.). For example, if under the condition $\bar{q}_D = 0.01(g)$ and $\bar{k}_D = 0.4\,k$,

$$[a_1, a_2, a_3, a_4, a_5, a_6, a_7] = [1.9996, 1.7652, 1.3932, 1.1491, 0.6600, 0.4292, 0.3399]$$

then the following can be calculated:

$$[a_T, b_T, c_T, d_T, e_T] = [-0.1206, 0.7204, -0.9949, -0.9698, 2.5195]$$

Therefore, the following equation can be used to determine the acceleration for a period that is not listed in the table:

TABLE 8.7B
Acceleration (g) $\bar{q}_D = 0.2$ (g), Input = 0.4 (g)

k_D/k \ T (s)	0.4	0.6	0.8	1.0	1.5	2.0	3.0
0.40	1.2724	1.2596	1.0856	1.0341	0.8190	0.6529	0.4986
0.50	1.1509	1.1393	0.9920	0.9440	0.7709	0.6319	0.4724
0.60	1.0965	1.0554	0.9120	0.8582	0.7242	0.6077	0.4482
0.70	1.0308	0.9847	0.8598	0.7996	0.6826	0.5838	0.4431
0.98	0.9157	0.8702	0.7562	0.7121	0.6147	0.5298	0.4260
2.00	0.7419	0.6861	0.6204	0.5727	0.4990	0.4353	0.3782
2.88	0.6856	0.6371	0.5692	0.5275	0.4491	0.3983	0.3507
3.92	0.6525	0.5987	0.5327	0.4949	0.4164	0.3750	0.3304
5.12	0.6258	0.5636	0.5126	0.4703	0.3956	0.3536	0.3157
8.00	0.5966	0.5229	0.4878	0.4344	0.3667	0.3304	0.2950
12.5	0.5832	0.4985	0.4563	0.4159	0.3480	0.3097	0.2763

TABLE 8.7C
Normalized force (MN) $\bar{q}_D = 0.2$ (g), Input = 0.4 (g)

k_D/k \ T(s)	0.4	0.6	0.8	1.0	1.5	2.0	3.0
0.40	20.3282	19.1746	16.0540	14.6693	10.9338	8.0348	5.5316
0.50	18.6638	17.3429	15.0257	13.8005	10.6657	8.0753	5.3758
0.60	17.6047	16.2153	14.0559	12.8560	10.1597	8.0032	5.2451
0.70	16.5160	15.3043	13.3502	12.1294	9.7343	7.8691	5.2384
0.98	14.5253	13.3958	11.9035	11.0018	8.9159	7.4724	5.3098
2.00	11.7124	10.9310	9.9302	9.1089	7.7807	6.7585	5.3217
2.88	10.7921	10.2023	9.0708	8.4159	7.2220	6.3322	5.2676
3.92	10.4257	9.5934	8.5316	7.8989	6.7382	6.0017	5.1148
5.12	10.0248	9.0780	8.2594	7.5230	6.5574	5.9255	5.1504
8.00	9.5139	8.3355	7.7451	7.0489	6.1537	5.6002	4.9154
12.5	9.2194	7.9769	7.2508	6.6864	5.8142	5.3644	4.8554

$$a(T) = -0.1206\,T^4 + 0.7204\,T^3 - 0.9949\,T^2 - 0.9698\,T + 2.5195 \quad (g)$$

For example, if T = 1.25 (s), a(1.25) can be computed to be 0.87 (g). Note that this value is very close to the result obtained from the linear interpolation:

$$(1.1491 + 0.6600)/2 = 0.9045 \quad (g)$$

Thus, for simplicity, linear interpolation can also be used to obtain rough estimates.

In the design of the damper force, since a sufficient safety factor must be used to guarantee the working condition of a bilinear damper, the linear interpolation method can be used.

8.5.2.3.2 *Characteristic Strength $\bar{q}_D$*

To determine the displacement for a characteristic strength not listed in Tables 8.4A through 8.9A, Equation 8.154 can be used, that is,

TABLE 8.8A
Displacement (cm) $\bar{q}_D = 0.2$ (g), Input = 0.4 (g)

k_D/k \ T(s)	0.4	0.6	0.8	1.0	1.5	2.0	3.0
0.40	4.8302	10.5588	15.9089	23.0275	39.6296	52.6675	84.3503
0.50	4.1464	8.8663	14.2601	20.8185	36.5847	50.4254	75.4187
0.60	3.8022	7.9141	12.5394	18.4238	33.1689	46.3210	67.6530
0.70	3.4629	7.1070	11.5038	16.5569	30.0974	42.5714	65.4271
0.98	2.7336	5.7327	9.3418	13.6825	25.3216	36.3202	58.7662
2.00	1.7734	3.7836	6.2423	8.8246	16.7531	24.7004	41.5828
2.88	1.4027	3.1480	5.0424	7.4112	13.1278	20.1380	34.6287
3.92	1.1778	2.5818	4.2437	6.2210	10.7895	16.7074	28.9539
5.12	1.0116	2.2356	3.7372	5.3086	9.7029	14.4339	24.5457
8.00	0.8072	1.7467	2.9622	4.2221	7.7366	11.0264	19.0756
12.5	0.6685	1.4250	2.3727	3.4927	6.1621	8.9224	14.7352

TABLE 8.8B
Acceleration (g) $\bar{q}_D = 0.3$ (g), input = 0.4 (g)

k_D/k \ T(s)	0.4	0.6	0.8	1.0	1.5	2.0	3.0
0.40	1.5225	1.4963	1.3170	1.2329	0.9706	0.7364	0.5284
0.50	1.3452	1.3076	1.2200	1.1537	0.9419	0.7476	0.5060
0.60	1.2575	1.1968	1.1114	1.0624	0.8918	0.7306	0.4832
0.70	1.1720	1.1008	1.0449	0.9887	0.8429	0.7040	0.4921
0.98	0.9883	0.9492	0.9011	0.8696	0.7685	0.6726	0.5030
2.00	0.7465	0.7234	0.6931	0.6610	0.6204	0.5714	0.4934
2.88	0.6532	0.6523	0.6200	0.6017	0.5476	0.5234	0.4771
3.92	0.5965	0.5889	0.5671	0.5507	0.5086	0.4921	0.4592
5.12	0.5547	0.5502	0.5352	0.5154	0.4825	0.4586	0.4353
8.00	0.5032	0.4955	0.4865	0.4701	0.4385	0.4225	0.4093
12.5	0.4683	0.4595	0.4493	0.4407	0.4111	0.3952	0.3812

$$d\left(\bar{q}_D\right) = a_q\,\bar{q}_D^{-3} + b_q\,\bar{q}_D^{-2} + c_q\,\bar{q}_D^{-1} + d_q\,\bar{q}_D + e_q \quad (\text{cm}) \tag{8.154}$$

where $d(\bar{q}_D)$ is the displacement at a specific value of the characteristic strength $\bar{q}_D$. The coefficients a_q, b_q, c_q, d_q, e_q are determined by the following least square method as

$$\begin{Bmatrix} a_q \\ b_q \\ c_q \\ d_q \\ e_q \end{Bmatrix} = \begin{bmatrix} \bar{q}_{D1}^{-3} & \bar{q}_{D1}^{-2} & \bar{q}_{D1}^{-1} & \bar{q}_{D1} & 1 \\ \bar{q}_{D2}^{-3} & \bar{q}_{D2}^{-2} & \bar{q}_{D2}^{-1} & \bar{q}_{D2} & 1 \\ & & & \cdots & \\ \bar{q}_{D9}^{-3} & \bar{q}_{D9}^{-2} & \bar{q}_{D9}^{-1} & \bar{q}_{D9} & 1 \end{bmatrix}^{+} \begin{Bmatrix} d_1 \\ d_2 \\ \cdots \\ d_9 \end{Bmatrix} \tag{8.155}$$

Here, $\bar{q}_{D(.)}$ is defined in Equation 8.154, for example,

TABLE 8.8C
Normalized force (MN) $\bar{q}_D = 0.3$ (g), Input = 0.4 (g)

k_D/k \ T(s)	0.4	0.6	0.8	1.0	1.5	2.0	3.0
0.40	24.3322	22.7880	18.8792	16.8065	12.2353	8.5982	5.6891
0.50	22.3537	20.5157	18.1786	16.2546	12.3064	8.9890	5.5729
0.60	21.0621	19.2141	17.0044	15.4689	11.8867	9.0747	5.4629
0.70	19.7303	18.0751	16.1756	14.7115	11.4621	8.9875	5.5732
0.98	16.6976	15.5941	14.4004	13.4530	10.7980	9.0192	5.9356
2.00	12.7897	12.3524	11.8211	10.9862	9.7444	8.5551	6.5764
2.88	11.3905	11.2572	10.5612	10.2683	9.0722	8.1453	6.8010
3.92	10.4687	10.3317	9.8915	9.5247	8.4909	7.7977	6.8155
5.12	9.7806	9.6332	9.3791	8.9432	8.2828	7.7864	6.8409
8.00	8.9099	8.8086	8.6307	8.3562	7.8705	7.3403	6.6696
12.5	8.4144	8.2089	8.0414	7.8570	7.3890	7.0997	6.6762

TABLE 8.9A
Displacement (cm) $\bar{q}_D = 0.4$ (g), Input = 0.4 (g)

k_D/k \ T(s)	0.4	0.6	0.8	1.0	1.5	2.0	3.0
0.40	5.4756	11.8611	17.8607	25.5747	41.8835	55.8070	84.3502
0.50	4.6668	9.7775	16.1698	23.4702	39.7146	53.9463	75.5723
0.60	4.2374	8.7505	14.1549	20.9629	36.7739	49.2080	67.8694
0.70	3.7621	7.8095	12.8276	18.7596	33.1123	46.1179	67.3518
0.98	2.8734	6.1984	10.3165	15.1668	28.2927	40.1913	64.0741
2.00	1.7424	3.7279	6.3723	9.2669	18.1414	27.3652	46.0316
2.88	1.3090	3.0604	4.9817	7.4855	13.9643	21.7038	37.7465
3.92	1.0629	2.3772	3.9652	5.9427	11.2308	17.8822	31.1057
5.12	0.8794	2.0191	3.4026	5.0112	9.7156	14.9490	27.2378
8.00	0.6662	1.4155	2.4763	3.7051	7.2597	11.1080	20.2075
12.5	0.4714	1.0452	1.8792	2.7242	5.4946	8.4424	15.2690

$$\left[\bar{q}_{D1}, \bar{q}_{D2}, \bar{q}_{D2}, \bar{q}_{D3}, \bar{q}_{D4}, \bar{q}_{D5}, \bar{q}_{D6}, \bar{q}_{D7}, \bar{q}_{D8}, \bar{q}_{D9}\right]$$
$$= \left[0.01, 0.03, 0.05, 0.1, 0.2, 0.3, 0.4, 0.5, 0.6\right] \quad (g)$$

and $d_{(.)}$ are corresponding displacements listed in the specific tables at the specific characteristic strength $\bar{q}_{D(.)}$. For example, if under the condition T = 0.4 (s) and $\bar{k}_D = 0.4$ k, then $d_1 = 7.9024$ (cm), $d_2 = 5.8848$ (cm), and so on.

Under this condition, for example, the following equation can be used to determine the displacement for a characteristic strength not listed in the table as

$$d\left(\bar{q}_D\right) = a_T\,\bar{q}_D^{-3} + b_T\,\bar{q}_D^{-2} + c_T\,\bar{q}_D^{-1} + d_T\,\bar{q}_D + e_T \quad (cm) \tag{8.156}$$

For example, if $\bar{q}_D = 0.015$ (g), then d(0.015) can be computed to be 7.260 (cm). Note that this calculated value 6.8935 (cm) is also close to the result of the linear interpolation. Thus, for rough estimation, linear interpolation can also be used for the sake of simplicity.

TABLE 8.9B
Acceleration (g) $\bar{q}_D = 0.4$ (g), input = 0.4 (g)

k_D/k \ T(s)	0.4	0.6	0.8	1.0	1.5	2.0	3.0
0.40	1.7899	1.7425	1.5224	1.4051	1.0424	0.7845	0.5286
0.50	1.5856	1.5206	1.4398	1.3522	1.0473	0.8091	0.5074
0.60	1.4740	1.4023	1.3177	1.2640	1.0198	0.7888	0.4861
0.70	1.3513	1.2929	1.2367	1.1815	0.9615	0.7760	0.5108
0.98	1.1242	1.1171	1.0769	1.0364	0.9068	0.7753	0.5586
2.00	0.8387	0.8172	0.8044	0.7927	0.7558	0.6991	0.5790
2.88	0.7296	0.7425	0.7243	0.7122	0.6784	0.6497	0.5831
3.92	0.6676	0.6660	0.6496	0.6432	0.6318	0.6182	0.5710
5.12	0.6214	0.6259	0.6142	0.6104	0.5971	0.5795	0.5590
8.00	0.5677	0.5584	0.5559	0.5493	0.5337	0.5383	0.5261
12.5	0.5187	0.5170	0.5183	0.5097	0.5065	0.5032	0.4987

TABLE 8.9C
Normalized force (MN) $\bar{q}_D = 0.4$ (g), Input = 0.4 (g)

k_D/k \ T(s)	0.4	0.6	0.8	1.0	1.5	2.0	3.0
0.40	27.4833	25.7009	20.9474	18.3748	12.7291	8.8560	5.6897
0.50	25.7338	23.4725	20.6831	18.2995	13.1518	9.3811	5.5800
0.60	24.8310	22.1334	19.5119	17.7352	13.0046	9.4969	5.4777
0.70	23.0659	20.8959	18.6165	16.9926	12.5210	9.5287	5.6715
0.98	19.6642	18.3068	16.7967	15.5888	12.2098	9.9617	6.3009
2.00	14.8820	14.4857	14.0817	13.0850	11.5267	10.0464	7.3784
2.88	13.1875	13.2527	12.5746	12.3722	10.9654	9.7506	7.9511
3.92	12.1205	12.0552	11.7341	11.4815	10.4179	9.5071	8.1586
5.12	11.3252	11.3257	11.1840	10.8363	10.1262	9.5881	8.4204
8.00	10.3478	10.3361	10.2600	10.1252	9.7715	9.1768	8.2161
12.5	9.5916	9.6291	9.6679	9.5294	9.2199	8.9054	8.4515

To determine the acceleration for a characteristic strength not listed in Tables 8.4B through 8.9B, Equation 8.157 can be used. That is,

$$a(\bar{q}_D) = a_q \bar{q}_D^{-3} + b_q \bar{q}_D^{-2} + c_q \bar{q}_D^{-1} + d_q \bar{q}_D + e_q \tag{8.157}$$

where $a(\bar{q}_D)$ is the acceleration at the specific value of the characteristic strength $\bar{q}_D$. The coefficients a_q, b_q, c_q, d_q, and e_q are determined by the following least square method as

$$\begin{Bmatrix} a_q \\ b_q \\ c_q \\ d_q \\ d_q \end{Bmatrix} = \begin{bmatrix} \bar{q}_{D1}^{-3} & \bar{q}_{D1}^{-2} & \bar{q}_{D1}^{-1} & \bar{q}_{D1} & 1 \\ \bar{q}_{D2}^{-3} & \bar{q}_{D2}^{-2} & \bar{q}_{D2}^{-1} & \bar{q}_{D2} & 1 \\ & & & \cdots & \\ \bar{q}_{D9}^{-3} & \bar{q}_{D9}^{-2} & \bar{q}_{D9}^{-1} & \bar{q}_{D9} & 1 \end{bmatrix}^{+} \begin{Bmatrix} a_1 \\ a_2 \\ \cdots \\ a_9 \end{Bmatrix} \tag{8.158}$$

Here, $\bar{q}_{D(.)}$ is defined in Equation 8.157 and $a_{(.)}$ are the corresponding accelerations listed in the specific tables at the specific characteristic strength $\bar{q}_{D(.)}$. For example, if under the condition T = 0.4 (s) and $k_d = 0.4$ k, $a_1 = 1.9996$ (g), $a_2 = 1.51116$ (g), then under this condition, for example, according to Equation 8.67, the following equation can be used to determine the acceleration with a characteristic strength not listed in the table:

$$a(\bar{q}_D) = 0.413\times10^{-7}\,\bar{q}_D^{-3} - 2.8830\times10^{-4}\,\bar{q}_D^{-2} + 0.03437\,\bar{q}_D^{-1} + 2.7603\bar{q}_D + 0.5767 \quad (g)$$

For example, if $\bar{q}_D = 0.015$, a(0.015) can be computed to be 1.8773 (g). Note that this value is somewhat close to the result of the linear interpolation 1.7556 (g).

8.5.2.3.3 Damper Stiffness k_D

To determine the displacement for a coefficient of damper stiffness, k_D/k, not listed in Tables 8.4A through 8.9A, Equation 8.159 can be used. That is,

$$d(\kappa_d) = a_K \kappa_D^{-3} + b_K \kappa_D^{-2} + c_K \kappa_D^{-1} + d_K \kappa_D + e_K \quad (\text{cm}) \tag{8.159}$$

where $d(\kappa_D)$ is the displacement at the specific value of the coefficient κ_D, with κ_D used to denote the ratio k_D/k. That is,

$$\kappa_D = k_D/k \tag{8.160}$$

The coefficients a_K, b_K, c_K, d_K, and e_K are determined by the following least square method as

$$\begin{Bmatrix} a_K \\ b_K \\ c_K \\ d_K \\ e_K \end{Bmatrix} = \begin{bmatrix} \kappa_{D1}^{-3} & \kappa_{D1}^{-2} & \kappa_{D1}^{-1} & \kappa_{D1} & 1 \\ \kappa_{D2}^{-3} & \kappa_{D2}^{-2} & \kappa_{D2}^{-1} & \kappa_{D2} & 1 \\ & & & \cdots & \\ \kappa_{D11}^{-3} & \kappa_{D11}^{-2} & \kappa_{D11}^{-1} & \kappa_{D11} & 1 \end{bmatrix}^{+} \begin{Bmatrix} d_1 \\ d_2 \\ \cdots \\ d_{11} \end{Bmatrix} \tag{8.161}$$

Here, $\kappa_{D(.)}$ is defined in Equation 8.159:

$$\left[\kappa_{D1}, \kappa_{D2}, \kappa_{D3}, \kappa_{D4}, \kappa_{D5}, \kappa_{D6}, \kappa_{D7}, \kappa_{D8}, \kappa_{D9}, \kappa_{D10}, \kappa_{D11}\right]$$
$$= \left[0.4, 0.5\, 0.6\, 0.7, 0.98, 2.0, 2.88, 3.92, 5.12, 8.0, 12.5\right] \tag{8.162}$$

and $d_{(.)}$ are the corresponding displacements listed in the specific tables at the stiffness coefficient $\kappa_{D(.)}$. For example, if under the condition $T = 0.4$ (s) and $q_D = 0.01$ mg = 0.01 (kN), then $d_1 = 7.9024$ (cm), $d_2 = 7.8925$ (cm).

Under this condition, for example, according to Equation 8.161, the following equation can be used to determine the displacement for a characteristic strength not listed in the table:

$$d(\kappa_D) = -0.0642\times10^{-2}\kappa_D^{-3} + 0.0224\times10^{-2}\kappa_D^{-2} + 0.0292\kappa_D^{-1} - 0.0304\times10^{-2}\kappa_D + 7.838 \quad (\text{cm})$$

For example, if $\kappa_D = 0.45$, d(0.45) can be computed to be 7.9 (cm). Note that this calculated value is very close to the result 7.9 (cm) from linear interpolation. Thus, linear interpolation can also be used for simplicity.

To determine the acceleration with a damper stiffness k_D not listed in Tables 8.4B through 8.9B, Equation 8.163 can be used. That is,

$$a(\kappa_D) = a_K\,\kappa_D^{-3} + b_K\,\kappa_D^{-2} + c_K\,\kappa_D^{-1} + d_K\,\kappa_D + e_K \quad (\text{g}) \tag{8.163}$$

where $a(\kappa_D)$ is the acceleration at the specific value of κ_D. The coefficients a_K, b_K, c_K, d_K, and e_K are determined by the following least square method as

$$\begin{Bmatrix} a_K \\ b_K \\ c_K \\ d_K \\ e_K \end{Bmatrix} = \begin{bmatrix} \kappa_{D1}^{-3} & \kappa_{D1}^{-2} & \kappa_{D1}^{-1} & \kappa_{D1} & 1 \\ \kappa_{D2}^{-3} & \kappa_{D2}^{-2} & \kappa_{D2}^{-1} & \kappa_{D2} & 1 \\ & & & \cdots & \\ \kappa_{D11}^{-3} & \kappa_{D11}^{-2} & \kappa_{D11}^{-1} & \kappa_{D11} & 1 \end{bmatrix}^{+} \begin{Bmatrix} a_1 \\ a_2 \\ \cdots \\ a_{11} \end{Bmatrix} \tag{8.164}$$

where, $\kappa_{D(.)}$ is defined in Equation 8.163 and $a_{(.)}$ are the corresponding accelerations listed in the specific tables at the specific value of $\kappa_{D(.)}$. For example, if under the condition $T = 0.4$ (s) and $k_D = 0.4$ k, then $a_1 = 1.9996$ (g), $a_2 = 1.5116$ (g).

Under this condition, for example, the following equation can be used to determine the acceleration for a value of κ_D not listed in the table as

$$a(\kappa_D) = -0.0162\times10^{-2}\kappa_D^{-3} + 0.056\times10^{-2}\kappa_D^{-2} + 0.7357\times10^{-2}\kappa_D^{-1} - 0.0077\times10^{-2}\kappa_D + 1.9835 \quad (g)$$

For example, if $\kappa_D = 0.45$, a(0.015) can be computed to be 1.9983 (g). Note that this value is very close to the result obtained by linear interpolation, 1.9984 (g).

8.5.2.3.5 *Input Level*

In the above discussion, using Tables 8.4 through 8.9, the input level is 0.4 (g). When the input level is not 0.4 (g), additional computations must be performed.

If the input level does not exactly equal 0.4, 0.6, 0.8, or 1.0 (g), then an interpolation method must be used again. In this case, both displacement and acceleration can share the same form as described in Equation 8.165, that is,

$$x(I) = a_I I^2 + b_I I + c_I \tag{8.165}$$

where x(I) can be either the displacement or the acceleration at the specific input level I_g. The coefficients a_I, b_I, and c_I can be calculated through the following equation:

$$\begin{Bmatrix} a_I \\ b_I \\ c_I \end{Bmatrix} = \begin{bmatrix} 0.16 & 0.4 & 1 \\ 0.36 & 0.6 & 1 \\ 0.64 & 0.8 & 1 \\ 1.0 & 1.0 & 1 \end{bmatrix}^{+} \begin{Bmatrix} x_{0.4} \\ x_{0.6} \\ x_{0.8} \\ x_{1.0} \end{Bmatrix} = \begin{bmatrix} 6.25 & -6.25 & -6.25 & 6.25 \\ -10.25 & 8.25 & 9.25 & -7.25 \\ 4.05 & -2.15 & -2.85 & 1.95 \end{bmatrix} \begin{Bmatrix} x_{0.4} \\ x_{0.6} \\ x_{0.8} \\ x_{1.0} \end{Bmatrix} \tag{8.166}$$

Here, x(.) is the displacement or acceleration at input level (.). For example, when $q_D = 0.01$ mg, T = 0.4 (s), and $k_D = 0.4$ k at levels of 0.4, 0.6, 0.8, and 1.0 (g), the displacements are 7.90, 12.73, 17.66, and 22.72 (cm), respectively. From these data, a rather good linearity is observed between the input levels and displacements. However, the reason the quadratic interpolation described in Equation 8.95 is used instead of linear interpolation is because when the period becomes longer, the relationship between the input and the response will become nonlinear. In order to unify the interpolation, Equation 8.165 is used.

With these data and by using Equation 8.166, the result for displacement can be

$$d(I) = 1.4183\, I^2 + 22.7047\, I - 1.4049 \quad (cm)$$

Another example is found when $q_D = 0.01$ mg, T = 0.4 (s), and $k_D = 0.4$ (kN) at levels of 0.4, 0.6, 0.8, and 1.0 (g), the accelerations are 2.00, 3.22, 4.46, and 5.73 (g), respectively. By the same token, the results for acceleration can be

$$a(I) = 0.3571\, I^2 + 5.7165\, I - 0.3437 \quad (g)$$

8.5.2.3.6 *Higher-Dimensional Search*

All the above-mentioned interpolations are focused on one variable. For example, when the period T is considered, other possible variables, such as the characteristic strength q_D, the damper stiffness k_D, as well as the input level I, are assumed to have the values listed in the corresponding tables. In this case, finding the response at a specific value of the variable is regarded as a one-dimensional search. However, in many cases, there can be more than one variable, which is not listed in the tables. For example, suppose not only the period T, but also the characteristic strength q_D, the damper stiffness k_D, and the input level I are not listed. In this case, a higher-dimensional or multidimensional search must be carried out.

From the discussion on one-dimensional searches, it is seen that over a relatively small range, linear interpolation can be used to calculate the response with acceptable accuracy. Therefore, for simplicity, linear interpolation can be used for multidimensional searches.

For example, suppose the parameter κ_D is taken between κ_{Di} and κ_{Di+1}, T is taken between T_j and T_{j+1}, and $\bar{q}_D$ is taken between $\bar{q}_{Dk}$and $\bar{q}_{Dk+1}$. Note that κ_{Di} and $\kappa_{Di+1,}$ T_j and T_{j+1}, and also $\bar{q}_{Dk}$ and $\bar{q}_{Dk+1}$ are listed in the table with fixed input I, i.e., I = 0.4 (g). Thus, the corresponding displacements d_{ijk}, $d_{i+1,j,k}$ $d_{i,j+1,k}$, $d_{i,j+1,k}$, $d_{i+1,\,j+1}$, k $d_{i,\,j+1,\,k+1}$, and $d_{i+1,\,j+1,\,k+1}$ can be found in the tables. Here, subscripts i, j, and k are used and correspond to κ_{di}, T_j, $\bar{q}_{Dk}$. In this case,

$$d\left(\kappa_D, T, \bar{q}_{Dk}\right) = \frac{D_2 - D_1}{\bar{q}_{Dk+1} - \bar{q}_{Dk}}\left(\bar{q}_D - \bar{q}_{Dk}\right) + D_1 \quad (\text{cm}) \tag{8.167}$$

where

$$D_1 = \frac{d_2 - d_1}{T_{j+1} - T_j}\left(T - T_j\right) + d_1 \quad (\text{cm}) \tag{8.168}$$

$$D_2 = \frac{d_4 - d_3}{T_{j+1} - T_j}\left(T - T_j\right) + d_3 \quad (\text{cm}) \tag{8.169}$$

In Equations 8.168 and 8.169,

$$d_1 = \frac{d_{i+1,j,k} - d_{i,j,k}}{\kappa_{Di+1} - \kappa_{Di}}\left(\kappa_D - \kappa_{Di}\right) + d_{i,j,k} \quad (\text{cm}) \tag{8.170a}$$

$$d_2 = \frac{d_{i+1,j+1,k} - d_{i,j+1,k}}{\kappa_{Di+1} - \kappa_{Di}}\left(\kappa_D - \kappa_{Di}\right) + d_{i,j+1,k} \quad (\text{cm}) \tag{8.170b}$$

$$d_3 = \frac{d_{i+1,j+1,k} - d_{i,j+1,k}}{\kappa_{Di+1} - \kappa_{Di}}\left(\kappa_D - \kappa_{Di}\right) + d_{i,j+1,k} \quad (\text{cm}) \tag{8.170c}$$

$$d_4 = \frac{d_{i+1,j+1,k} - d_{i,j+1,k}}{\kappa_{Di+1} - \kappa_{Di}}\left(\kappa_D - \kappa_{Di}\right) + d_{i,j+1,k} \quad (\text{cm}) \tag{8.170d}$$

Similarly, to find the acceleration $a\left(\kappa_D, T, \bar{q}_D\right)(g)$,

$$a\left(\kappa_D, T, \bar{q}_D\right) = \frac{A_2 - A_1}{\bar{q}_{Dk+1} - \bar{q}_{Dk}}\left(\bar{q}_D - \bar{q}_{Dk}\right) + A_1 \quad (g) \tag{8.171}$$

where

$$A_1 = \frac{a_2 - a_1}{T_{j+1} - T_j}\left(T - T_j\right) + a_1 \quad (g) \tag{8.172a}$$

$$A_2 = \frac{a_4 - a_3}{T_{j+1} - T_j}\left(T - T_j\right) + a_3 \quad (g) \tag{8.172b}$$

and, in Equations 8.172,

$$a_1 = \frac{a_{i+1,j,k} - a_{i,j,k}}{\kappa_{Di+1} - \kappa_{Di}}\left(\kappa_D - \kappa_{Di}\right) + a_{i,j,k} \quad (g) \tag{8.173a}$$

$$a_2 = \frac{a_{i+1,j+1,k} - a_{i,j+1,k}}{\kappa_{Di+1} - \kappa_{Di}}\left(\kappa_D - \kappa_{Di}\right) + a_{i,j+1,k} \quad (g) \tag{8.173b}$$

$$a_3 = \frac{a_{i+1,j+1,k} - a_{i,j+1,k}}{\kappa_{Di+1} - \kappa_{Di}}\left(\kappa_D - \kappa_{Di}\right) + a_{i,j+1,k} \quad (g) \tag{8.173c}$$

$$a_4 = \frac{a_{i+1,j+1,k} - a_{i,j+1,k}}{\kappa_{Di+1} - \kappa_{Di}}\left(\kappa_D - \kappa_{Di}\right) + a_{i,j+1,k} \quad (g) \tag{8.173d}$$

Here, a_{ijk}, $a_{i+1,j,k}$ $a_{i,j+1,k}$, $a_{i,j+1,k}$, $a_{i+1,j+1,k}$ $a_{i,j+1,k+1}$, and $a_{i+1,j+1,k+1}$ are accelerations according to κ_{Di}, T_j, and $\bar{q}_{Dk}$. These values are found in the corresponding tables.

8.5.3 Design Issues

When a structure deforms into the inelastic range, its dynamic stiffness for acceleration reduction becomes much larger. From the viewpoint of dynamics, the fundamental effective period of the system becomes drastically elongated, so that the seismic load is reduced. However, on the other hand, its dynamic stiffness for displacement decreases and a very large displacement can be observed.

For highly nonlinear systems, the experience gained from a static point of view is important but with limitations. Drifting of the equilibrium points may significantly enlarge the displacement. There are other problems associated with dynamic excitations, such as resonance, cross effects, three-dimensional stress, possible continuation of stiffness reduction, etc.

Using supplemental damping is an effective way to regulate the large structural displacement for inelastic structures. When a structure yields, its fundamental effective mode is dominant. In this case, the single period vibration becomes easier to realize. This is the main reason why the MSSP model and the DDBD method can be used to find the "substitute" SDOF responses first and then the shape function can be used to estimate the seismic responses in each story, to simplify the design procedure. However, a method to determine if the system becomes overdamped is also needed.

Example 8.6

The four-story structure shown in Figure 8.1 is used to illustrate how the response spectra can be used to estimate the responses. Suppose a structure yields when it is subjected to strong earthquake load. With bilinear dampers installed in the lower two stories, the entire system behaves like a bilinear structure. The modeling of this system can be described by Equation 8.40. However, in this case, the stiffness **K** and $\mathbf{K}_\delta$ should be the yielding stiffness of the base structure and the damper system.

Suppose the yielding stiffness, unloading stiffness, and characteristic strength of the damper are, respectively, $k_d = 1.317$ (MN/m), $k_u = 3.776$ (MN/m), and $q_d = 5.625$ (kN). The mass matrix is 100 **I** (t) and the yield stiffness matrices **K** and $\mathbf{K}_\delta$ are given as

$$\mathbf{K} = \begin{bmatrix} 6.766 & -3.947 & 0 & 0 \\ -3.947 & 9.587 & -5.640 & 0 \\ 0 & -5.640 & 13.159 & -7.520 \\ 0 & 0 & -7.520 & 9.400 \end{bmatrix} (\text{MN/m}), \ \mathbf{K}_\delta = \begin{bmatrix} 2.634 & 1.317 & 0 & 0 \\ -1.317 & 1.317 & 0 & 0 \\ 0 & 0 & 0 & 0 \\ 0 & 0 & 0 & 0 \end{bmatrix} (\text{MN/m})$$

Furthermore, the unloading stiffness matrices of the base structure and the bilinear damper systems are, respectively, denoted as K_U and K_u and given by

$$\mathbf{K}_U = \begin{bmatrix} 60.133 & -30.067 & 0 & 0 \\ -30.067 & 60.133 & -30.067 & 0 \\ 0 & -30.067 & 60.133 & -30.067 \\ 0 & 0 & -30.067 & 30.067 \end{bmatrix} (\text{MN/m}), \ \mathbf{K}_u = \begin{bmatrix} 7.553 & -3.776 & 0 & 0 \\ -3.776 & 3.776 & 0 & 0 \\ 0 & 0 & 0 & 0 \\ 0 & 0 & 0 & 0 \end{bmatrix} (\text{MN/m}),$$

Suppose the dimensionless shape function can be approximately described as $\boldsymbol{p}_1^T = [0.25\ 0.5\ 0.75\ 1]$. The equivalent mass m_{e1} is $\boldsymbol{p}_1^T \mathbf{M} \boldsymbol{p}_1 = 187.5$ (t). The equivalent period is $\boldsymbol{p}_1^T(\mathbf{K}+\mathbf{K}_\delta)\boldsymbol{p}_1/m_{e1} = 1.5$ (s). The ratio of yielding and unloading stiffness is $\boldsymbol{p}_1^T(\mathbf{K}+\mathbf{K}_\delta)\boldsymbol{p}_1/\boldsymbol{p}_1^T(\mathbf{K}_U+\mathbf{K}_u)\boldsymbol{p}_1 = 0.4$. The normalized modal characteristic strength $\bar{q}_D = q_d/m_{e1} = 0.03$ (g).

If this structure is subjected to input of 0.4 (g), then from Tables 8.4a through c, it is seen that the spectral displacement is $d_{1D} = 0.288$ (m), the seismic response factor is $C_s = 0.541$, and the damper force is $f_D = 6.88$ (MN). For simplicity, let all the safety factors, denoted as S, G_m, and p_c, be unity. Note that $\Gamma_1 = 1.333$. Therefore, the displacement, acceleration, and lateral force are:

$$\mathbf{d}_{max} = S\, d_{1D} \Gamma_1 \boldsymbol{p}_1 = \begin{Bmatrix} 0.096 \\ 0.192 \\ 0.288 \\ 0.384 \end{Bmatrix} (\text{m})$$

and

$$\mathbf{a}_a \approx \mathbf{a}_s = S C_s \Gamma_1 \boldsymbol{p}_1 = \begin{Bmatrix} 0.180 \\ 0.361 \\ 0.541 \\ 0.721 \end{Bmatrix} (\text{g})$$

8.6 SUMMARY

Several important issues in nonlinear damping design are discussed in this chapter. The focus is given to response estimations through response spectra analysis. A design procedure for structures with nonlinear damping is briefly summarized in the following paragraphs.

8.6.1 Preliminary Decision Making

As mentioned in Chapter 7 and earlier in this chapter, there are many issues to consider in preliminary decision making. Here, the focus is on identification of the types of damping, as well as computational approaches.

At the outset, designers should be aware of the type of damping to be used, including the type of supplemental damper and the response of the estimated structure with added dampers.

The simplest linearized type of realistic structure with supplemental damping is the proportionally damped linear viscous system. Although this case is rare, when the damping ratio is low, i.e., 5%, this model of damping can be used, regardless of the type of dampers that are actually installed. The 5% damping ratio is assumed by most building codes, which considers the possible inelastic deformation of the total structure. Actually, when a structure remains linear, the damping ratio of the fundamental mode is rarely as high as 5%. However, a 5% damping ratio for an assumed linear system is still a possible and acceptable design target.

Another simple linearized type of system is a generally damped system with linear viscous damping. This model improves the design accuracy and eliminates some underestimation of the resulting structural responses. From the viewpoint of structural design, nonproportional damping is the result of damping irregularity. This damping model has been discussed earlier. Note that the assumption of linear damping has limitations. Therefore, the condition for using this model is, again, the amount of damping should not be too high, i.e., a damping ratio of less than 7%~20%, which depends on damping irregularity of the structure.

Dampers installed in a structure can produce a parallelogram-shaped relationship between the damping force of the first effective "mode" and the displacement. These damping devices are referred to as bilinear dampers. To estimate the system responses, specific bilinear response spectra are used. To further obtain the response vectors of the nonlinear MDOF systems, separation of the displacement and the acceleration are carried out and the combination of the first several effective "modes" is obtained. One of the major criteria to use the bilinear damping model is the degree of structural inelastic deformation. When the structure yields with sufficient ductile deformation, it is better to consider the bilinear damping model, regardless of the actual type of damping device.

Another major type of damping is known as sublinear damping. Even if the selected damping devices are of the bilinear type, when the dampers are installed in a structure with different yielding time points, the total structure can behave as a sublinear system. In this chapter, the problem associated with direct iteration is analyzed, and the use of sublinear response spectra is encouraged. Different from bilinear damping, where the interested effective "modes" of interest can all be treated bilinearly, sublinear damping rarely contributes accurate information for higher "modes." Therefore, an alternative approach for an equivalent linear MDOF system, likely to be a generally damped system, is presented.

8.6.2 Initial Damping Design

A systematic procedure for damping design of linear systems was discussed in Chapter 7 for a base structure that needs supplemental damping. In general, there is no straightforward design process for nonlinear damping. A two-step design procedure is generally used.

The first step is to use a linearized damping model with either Timoshenko or alternative force-based approximations to estimate the effective damping ratio. The design procedure for selecting damper specifications, optimal damper configurations, general damping matrices, mode shape normalization, as well as response estimation is then carried out. This is referred to as the initial damping design.

8.6.3 Response Estimation with Proper Model of Damping

8.6.3.1 Parameters of an SDOF System

For either linear or nonlinear damping design, the key challenge is response estimation. Generally speaking, if supplemental dampers are already designed and selected, response estimation becomes more complex. Here, the estimations are based on design spectra. For linear or equivalent linear systems, only two parameters are needed. For nonlinear spectra, more parameters are required, so that to use the design spectra, these parameters need to be calculated ahead of time.

In linear damping design, the response estimation is comparatively simpler than for nonlinear dampers. The design parameters of nonlinear dampers cannot usually be directly determined in one step. In order to calculate the damping ratio, the structural responses are needed, which in turn are also functions of the damping ratio. For bilinear dampers, the relationship of the damping ratio and the response can be determined by quadratic equations, so that the results can be directly calculated. For sublinear dampers, on the other hand, iterative computations are involved.

In most cases, for equivalent linear systems, Timoshenko damping and a linear damping model are used to calculate the structural response. This can introduce errors. Therefore, after the initial design, the structural responses need to be reexamined, based on the nonlinear design spectra. This approach is established by using real-world earthquake records to check the seismic response via time history analyses and is recommended for the initial design. When the system becomes nonlinear, a simplified approach by letting the response modification coefficient $R > 1$ to reduce the level of load is recommended, before proceeding with further design steps as described in Sections 8.2 and 8.3. This approach ensures a good design by using the response spectra-based estimation, along with validation by time history analysis.

8.6.3.2 Shape Function and Modal Participation Factor

Once the effective modal responses for an SDOF system are obtained, they need to be distributed to each story via shape functions. Thus, the estimation of shape functions is equally important. In addition, the corresponding effective modal participation factor also needs to be determined.

8.6.3.3 Trial-and-Error Iteration

When the peak response is considered to be safe, by using both the design spectra-based analysis and the time history analysis, the initial damper design may be accepted.

If one set of peak responses, either forces or deformations, exceeds the preset safety level, then the iterative process must be carried out to redesign the dampers. The criteria are

$$\left[f_j\right] \geq \max\left\{f_j^{(S)},\ f_j^{(T)}\right\} \tag{8.174}$$

$$\left[d_j\right] \geq \max\left\{d_j^{(S)}, d_j^{(T)}\right\} \tag{8.175}$$

where $[f_j]$ and $[d_j]$ are, respectively, the allowed force (moment, shear, stress, etc.) and allowed deformation (lateral displacement, rotation, shear strain, etc.) at the jth story. Meanwhile $f_j^{(S)}$ and $f_j^{(T)}$ are, respectively, the estimated maximum response of force under spectra analysis and time history analysis at the jth story. Finally, $d_j^{(S)}$ and $d_j^{(T)}$ represent, respectively, the estimated maximum response of deformation under spectral analysis and time history analysis at the jth story.

If Equations 8.174 and 8.175 are not satisfied, the process should return to step 1 to reselect damping devices and the corresponding modeling approach, followed by step 2 (simplified damper design), and step 3 (response estimation), until these criteria are satisfied.

8.6.4 Selection of Damping Devices

The final task of damper design is to select proper devices that can fully realize the design parameters. In most cases, the dampers in response design are only nominal or theoretical. To realize the nominal dampers, see Section 7.2.4, and safety issues should also be considered.

REFERENCES

Aref, A.J. and Guo, Z. 2001. A Framework of a Finite Element-Based Large Increment Method for Nonlinear Structural Problems. *Journal of Engineering Mechanics, ASCE*, Vol. 127, No. 7, 739–746.

Bruneau, M., Uang, C.M., and Sabelli, R. July, 2011. Ductile Design of Steel Structures (2nd Edition), McGraw-Hill Professional.

Building Seismic Safety Council (BSSC). 2003. *NEHRP Recommended Provisions for Seismic Regulation for New Buildings and Other Structures, 2003 Edition*, Report Nos. FEMA 450, Federal Emergency Management Agency, Washington, DC.

———. 2009. *NEHRP Recommended Seismic Provisions for New Buildings and Other Structures, 2009 Edition*, Report Nos. FEMA P-750, Federal Emergency Management Agency, Washington, DC.

Deierlein, G.G., Reinhorn, A.M., and Willford, M.R. 2010. *Nonlinear Structural Analysis for Seismic Design-A Guide for Practicing Engineers,* NEHRP Seismic Design Technical Brief No. 4, Gaithersburg, MD, NIST GCR 10-917-5.

Filiatrault, A., Epperson, M., and Folz, B. 2004. On the Equivalent Elastic Modeling for the Direct–Displacement Based Seismic Design of Wood Structures. *Special Edition of the ISET Journal of Earthquake Technology on Performance Based Design*, Vol. 41, No. 1, 75–100.

Grant, D.N., Blandon, C.A., and Priestley, M.J.N. July, 2005. *Modeling Inelastic Response in Direct Displacement-Based Design.* Research Report No. ROSE-2005/03, ROSE School, Pavia, Italy, IUSS Press.

Grant, D.N., Fenves, G.L., and Auricchio, F. 2005. *Modeling and Analysis of High-Damping Rubber Bearings for the Seismic Protection of Bridge*, IUSS Report 88-7358-028-9, IUSS Press, Milan, Italy.

Lee, G.C. and Liang, Z. 2008. Friction dampers. US Patent 7,419,145.

Lin, W.H. and Chopra, A.K. 2003. *Earthquake Response of Symmetric and Asymmetric One-Story Elastic Systems with Nonlinear Fluid Viscous Dampers or Nonlinear Viscoelastic Dampers*, EERC Report No. EERC-2003-02, Berkeley, CA.

Kowalsky, M., Priestley, M.J.N., and MacRae, G.A. 1994. Displacement-based design of RC bridge columns. *Proceedings, 2nd International Workshop on the Seismic Design of Bridges*, Queenstown, New Zealand 1:138–63.

Prestley, M.J.N., Seible, F., and Calvi, G.M. 1996. *Seismic Design and Retrofit of Bridges*. NY: John Wiley.

Whittaker, A.S., Constantinou, M.C., Ramirez, O.M., Johnson, M.W., and Chrysostomou, C.Z. November 2003. *Equivalent lateral force and modal analysis procedures of the 2000 NEHRP Provisions for buildings with damping systems*. Vol. 19, No. 4, 959–980, *Earthquake Spectra*, Oakland, CA.

Index

C

D

M